Introduction to Texture Analysis

Reflecting emerging methods and the evolution of the field, *Introduction to Texture Analysis: Macrotexture, Microtexture, and Orientation Mapping* keeps mathematics to a minimum in covering both traditional macrotexture analysis and more advanced electron-microscopy-based microtexture analysis. The authors integrate the two techniques and address the subsequent need for a more detailed explanation of philosophy, practice, and analysis associated with texture analysis. The book illustrates approaches to orientation measurement and interpretation and elucidates the fundamental principles on which measurements are based. Thoroughly updated, this Third Edition of a bestseller is a rare introductory-level guide to texture analysis.

- Discusses terminology associated with orientations, texture, and their representation, as well as the diffraction of radiation, a phenomenon that is the basis for almost all texture analysis.
- Covers data acquisition, as well as representation and evaluation related to the well-established methods of macrotexture analysis.
- Updated to include experimental details of the latest transmission or scanning electron microscope-based techniques for microstructure analysis, including electron backscatter diffraction (EBSD).
- Describes how microtexture data are evaluated and represented and emphasizes the advances in orientation microscopy and mapping, and advanced issues concerning crystallographic aspects of interfaces and connectivity.
- Offers new and innovative grain boundary descriptions and examples.

This book is an ideal tool to help readers in the materials sciences develop a working understanding of the practice and applications of texture.

Introduction to Texture Analysis

Macrotexture, Microtexture, and Orientation Mapping

Third Edition

Olaf Engler, Stefan Zaefferer, and Valerie Randle

CRC Press
Taylor & Francis Group
Boca Raton New York London

CRC Press is an imprint of the
Taylor & Francis Group, an **informa** business

Third edition published 2024
by CRC Press
2385 Executive Center Drive, Suite 320, Boca Raton, FL 33431

and by CRC Press
4 Park Square, Milton Park, Abingdon, Oxon, OX14 4RN

CRC Press is an imprint of Taylor & Francis Group, LLC

Second edition published by CRC Press 2010

ISBN: 978-1-032-19248-2 (hbk)
ISBN: 978-1-032-18942-0 (pbk)
ISBN: 978-1-003-25833-9 (ebk)

DOI: 10.1201/9781003258339

Typeset in Times
by KnowledgeWorks Global Ltd.

Contents

PART III Microtexture Analysis

Preface

Most solid-state materials, including metals, ceramics, and minerals, have a polycrystalline structure in that they are composed of a multitude of individual crystallites or "grains." This book is concerned with a specific aspect of such materials—the crystallographic orientation of its components or the crystallographic texture, or simply *texture*, of the polycrystalline compound. The significance of texture lies in the anisotropy of many material properties; that is, the value of this property depends on the crystallographic direction in which it is measured. In most cases, grain orientations in polycrystals, whether naturally occurring or technologically fabricated, are not randomly distributed and the preference of certain orientations may indeed affect material properties by as much as 20%–50% of the property value. Therefore, the determination and interpretation of texture are of fundamental importance in materials technology. Furthermore, analysis of the texture changes during the thermomechanical treatment of materials yields valuable information about the underlying mechanisms, including deformation, recrystallization, or phase transformations. In geology, texture analysis can provide insight into the geological processes that led to rock formations over geological ages.

Nowadays there is a selection of techniques available to analyze the texture of materials. Besides the well-established techniques based on x-ray or neutron diffraction, known as *macrotexture* methods, there are a number of techniques that bridge the gap between microscopic mechanisms of formation of textures and microstructures and the resulting macroscopic properties. The latter are based on measurements of individual orientations in transmission or scanning electron microscopes and directly related to the microstructure, which has given rise to the term *microtexture*. Microtexture practice has grown principally through the application of electron backscatter diffraction (EBSD), and it is now possible to measure orientations automatically from predetermined coordinates in the microstructure, which is known generically as *orientation mapping*. From the full range of texture techniques now available, insights can be gained into material processing, corrosion, cracking, fatigue, grain boundary properties, and other phenomena with a crystallographic component.

Over the past 70 years, a large number of publications on texture analysis have appeared in the literature. However, there are only a few monographs on the subject, many of which are highly specialized with a strong focus on the mathematical aspects of texture. We have written the present book to provide comprehensive coverage of the range of concepts, practices, and applications of the techniques for determining and representing texture. The mathematics of the subject has been kept to the minimum necessary to understand the scientific principles. For a more complete treatment, a comprehensive bibliography directs the reader to more specialized texts. The text is inclined toward microtexture analysis, reflecting both the growing emphasis on this modern approach to texture analysis and the greater requirement for detailed explanation of the philosophy, practice, and analysis associated with microtexture. The book is intended for materials scientists, physicists, and geologists—both nonspecialists, including students, and those with more experience—who wish to learn about the approaches to orientation measurement and interpretation or to understand the fundamental principles on which measurements are based to gain a working understanding of the practice and applications of texture.

The sequence of the book is as follows. Part I, Fundamental Issues, addresses the descriptors and terminology associated with orientations and texture and their representation in general. This part concludes with an introduction to the diffraction of radiation, since this phenomenon forms the basis of almost all texture analysis. Part II, Macrotexture Analysis, covers both data acquisition and representation. Part III, Microtexture Analysis, provides experimental details of the transmission or scanning electron microscope-based techniques for microtexture analysis, followed by a description of how microtexture data are evaluated and represented. The innovative topics of orientation microscopy and orientation mapping are introduced, and more advanced issues concerning crystallographic aspects of interfaces and connectivity are treated.

The present third edition of *Introduction to Texture Analysis* is written about 15 years after the second edition. In these years EBSD has become the standard tool for microtexture measurements and microstructure characterization. The EBSD technique has developed a lot of different applications, including measurement on crystallographically more complex materials, stress and strain measurements, in-situ measurements of materials processes, and three-dimensional measurements. This is mainly because accuracy, crystallographic versatility, and acquisition speed of EBSD are steadily increasing. On the other hand, EBSD even begins to challenge the classical X-ray macrotexture measurements. Also in the field of transmission electron microscopy (TEM)-based orientation mapping significant progress has been made and one standard technique has developed for texture and microstructure mapping at highest resolution. The new edition of this book, therefore, now focuses more on the theoretical fundaments and on standard and new analysis techniques of the EBSD- and TEM-based methods. Among others, we have added sections on the theory of electron diffraction for spot-type and Kikuchi-type diffraction patterns. The description of the microtexture measurement techniques is terminated with an example chapter illustrating the wide field of applications of these techniques and discusses which technique should be employed for which kind of applications.

Closely related to the determination of microtextures is the measurement of microstructures by the orientation mapping techniques. Characterization of microstructures is indispensable for the correct interpretation of microtextures, firstly because microstructures and the related thermodynamic and kinetic conditions are the final reason for texture formation. Secondly, microstructure—maybe even more than texture—determines the final properties of a material and, thus, textures cannot not be considered independently of microstructures. We, therefore, address several microstructural features in some details, in particular deformation structures, grain boundaries and their crystallographic interpretation, and phase boundaries and the related orientation relationships. For all these topics the focus is laid on the description of these features by crystallographic orientations and their spatial arrangement.

We are indebted to a large number of colleagues from whom we have learned, with whom we have discussed and interacted, or who have provided thoughtful comments on parts of this book. In particular, we would like to acknowledge Rainer Abart, Thierry Baudin, Nathalie Bozzolo, Hans-Joachim Bunge, Michael Dahms, Austin Day, David Dingley, Tadashi Furuhara, Günter Gottstein, Gerlinde Habler, Jürgen Hirsch, Martin Hölscher, Bevis Hutchinson, Dorte Juul Jensen, Jerzy Jura, Fred Kocks, Kurt Lücke, Tadashi Maki, Krzysztof Pawlik, Richard Penelle, Jan Pospiech, Dierk Raabe, Tony Rollett, Robert Schwarzer, Hans-Jürgen Ullrich, Steve Vale, Hasso Weiland, and Rudy Wenk.

Olaf Engler
Stefan Zaefferer
Valerie Randle

About the Authors

Olaf Engler received his PhD and habilitation at the RWTH University of Technology in Aachen, Germany. After working for four years at the Materials Science and Technology Division of Los Alamos National Laboratory in the United States, he joined the Research and Development of Speira in Bonn, Germany. From 2004 until 2011, he has also been an adjunct professor for texture and crystal plasticity at the NTNU Trondheim, Norway, and since 2012 he is adjunct professor at RWTH Aachen University. Prof. Engler is also a member of the international committee of the International Conference on Texture of Materials (ICOTOM). Prof. Engler has more than 30 years of experience in analysis, interpretation, and modeling of the development of microstructure and texture during the thermomechanical processing of metallic materials and control of the resulting materials properties.

Stefan Zaefferer received his PhD at the Technical University of Clausthal in Germany. Then he spent three years as a researcher at the University Paris XII in Paris, France, and two years at Kyoto University, Japan. Since 2000 he heads the group "Microscopy and Diffraction" at the Max-Planck-Institute for Iron Research in Düsseldorf, Germany, and is lecturer for materials characterization at RWTH Aachen University. He also teaches regularly at Vienna University. At all his research positions Dr. Zaefferer was involved with the development and application of techniques for individual orientation determination and lattice defect characterization, first by transmission electron microscopy (TEM) techniques and then, most prominently, by scanning electron microscopy (SEM) techniques. Besides development of diffraction techniques his research interests are the mechanisms of microstructure formation in metals and other materials, including deformation, recrystallization, and phase transformation.

Valerie Randle received her PhD at Cardiff University in Wales, United Kingdom. Eventually she was awarded a Royal Society Research Fellowship which took her to Swansea University in Wales, where she became a professor in the Department of Metals Engineering and headed the Materials Research Centre until her early retirement in 2013. Prof. Randle has been eminent in the field of electron backscatter diffraction for more than 30 years. Her other main research interest is grain boundary engineering in metals and alloys. She has written more than 300 scientific publications on these research topics, including five textbooks. She is a past Welsh Woman of the Year and recipient of the Institute of Materials' Rosenhain Medal.

Part I

Fundamental Issues

1 Introduction

1.1 THE CLASSICAL APPROACH TO TEXTURE

This book is concerned with a specific aspect of the structure of natural and technological materials—the crystallographic orientation of its component units. The *crystallographic orientation*, or in this context simply *orientation*, refers to how the atomic planes in a volume of crystal are positioned relative to a fixed reference. This characteristic applies to all solids whose structure is crystalline, including minerals, ceramics, semiconductors, superconductors, and metals. Almost all of these materials are polycrystalline (rather than mono- or bicrystalline), and their component units are referred to as crystals or "grains."

Grain orientations in polycrystals, whether naturally occurring or fabricated, are rarely randomly distributed. One of the few examples where grain orientations are randomly distributed is a polycrystalline aggregate that has been made by compression of a powder. However, in most materials, there is a pattern in the orientations that are present and a propensity for the occurrence of certain orientations caused first during crystallization from a melt or amorphous solid state and subsequently by further thermomechanical processes. This tendency is known as *preferred orientation* or, more concisely, *texture*. The importance of texture to materials lies in the fact that *many material properties are texture specific*. Indeed, it has been found that the influence of texture on material properties is, in many cases, 20%–50% of the property values (Bunge, 1987). As an example, Figure 1.1 shows the elastic modulus of iron single crystals as a function of crystal direction. Clearly, the elastic properties strongly differ from the well-known bulk modulus of 210 GPa (dotted line), which is only obtained in an isotropic, that is, texture-free material. Some examples of properties that depend on the average texture of a material are as follows:

- Young's modulus
- Poisson's ratio
- Strength
- Ductility
- Toughness
- Magnetic permeability
- Electrical conductivity
- Thermal expansion (in noncubic materials)

The effect of texture on properties is exploited in materials technology to produce materials with specific characteristics or behavior. In general, the exact mechanisms by which particular textures evolve are incompletely understood, although the empirical validation is sufficient for many processes to become established commercial practice. Four examples of texture/property links follow: The first two examples

DOI: 10.1201/9781003258339-2

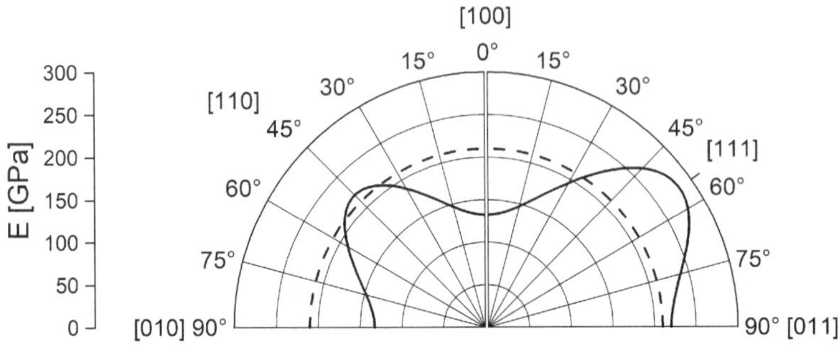

FIGURE 1.1 Elastic (Young's) modulus *E* of single-crystalline iron as a function of crystal direction. Clearly, the values strongly differ from the well-known bulk modulus of 210 GPa (dotted line), which is only obtained in isotropic, that is, texture-free material.

are well-established practices, whereas the third and fourth are examples of an application of texture control that is still evolving.

Grain-oriented silicon steel. It is well known that the optimal crystallographic direction of magnetization in Fe–3%Si steels for power transformers is ⟨001⟩. Figure 1.2a displays the magnetization curves of transformer steels for ⟨001⟩, ⟨011⟩, and ⟨111⟩, showing that the ⟨001⟩ direction is magnetically the softest; that is, it experiences minimum magnetization losses upon remagnetization. Hence, processing of these steels is directed toward making a product with a high proportion of grains having the so-called Goss orientation that has ⟨001⟩ coinciding with the direction in which the product

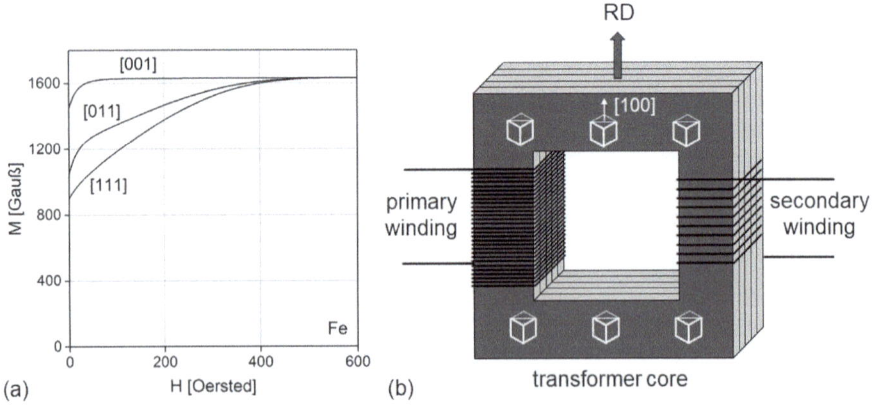

FIGURE 1.2 (a) Anisotropy of magnetization in different crystal directions of iron. The [001] crystal direction is magnetically the softest. (b) Sketch of a transformer made of sheets from grain oriented silicon steel. The magnetically soft [001] direction of the Goss-oriented grains is situated parallel to the rolling direction, RD. Hence, sheets with a pronounced Goss texture (110)[001] sustain minimum magnetization losses.

has been rolled to sheet (Figure 1.2b). Even though technologically this is a vitally important process, the exact mechanisms that control the texture formation are still not completely understood (Dorner et al., 2006).

Anisotropy in strip metal products. Very large tonnages of steel and aluminum alloys for deep-drawing applications, such as car bodies or beverage (drinks) cans, are produced every year. For steels, the deep-drawing process is greatly facilitated by maximizing the number of {111} planes that lie nearly parallel to the sheet surface before the drawing operation. For aluminum alloys, the required textures are rather more complex, and industrial processes concentrate on efficient use of material by balancing the texture effects so that a homogeneously shaped drawn product results (Knorr et al., 1994; Engler and Hirsch, 2007; Engler and Knarbakk, 2021). To illustrate the deleterious effect of pronounced texture on formability, Figure 1.3 shows a cylindrical cup that is deep-drawn from a circular blank of a heavily textured aluminum sheet, displaying the well-known *earing* behavior (Section 5.5).

Surface of ferritic stainless steels. Some grades of ferritic stainless steel (FSS) sheets are affected by an undesirable surface phenomenon known as ridging when the sheets are deformed in tension or in deep-drawing. The deformed sheets display corrugated surfaces, with ridges on one side of the sheet coinciding with troughs on the opposite side, which prevent use of such parts in exterior, visible applications. Ridging is generally assumed to

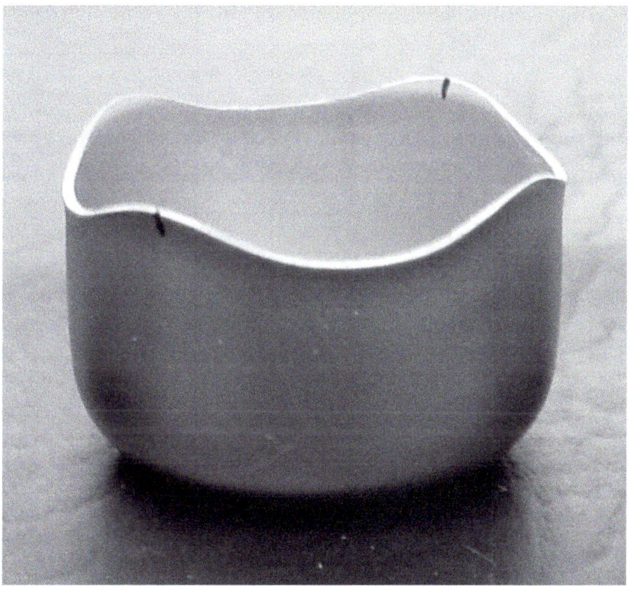

FIGURE 1.3 Typical appearance of a deep-drawn cup with ears under 0° and 90° to the RD (marked) from a heavily textured aluminum sheet (AA 1200-O, i.e., commercial purity aluminum, cold rolled, and soft annealed).

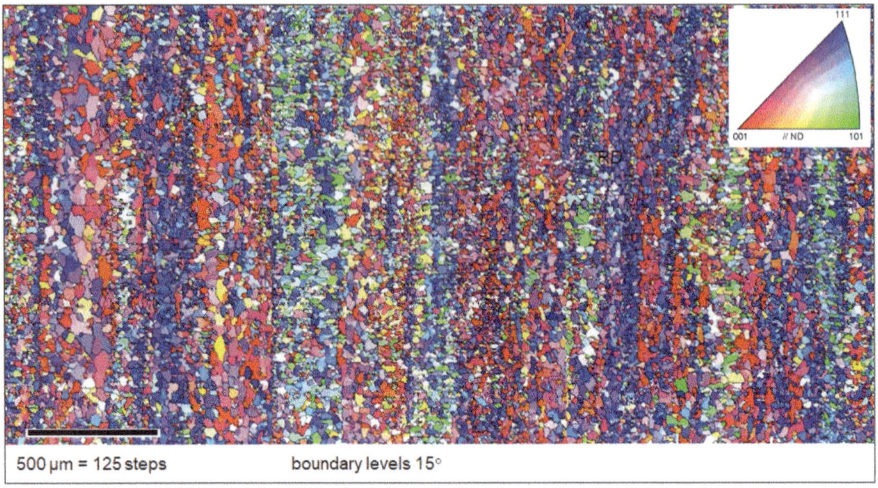

FIGURE 1.4 EBSD maps of a ferritic stainless steel sheet susceptible to ridging. The color scheme (stereographic triangle) is given as an inset. (Adapted from Engler et al., 2005b.)

be caused by the formation of grain colonies of similar crystal orientations (Brochu et al., 1997; Park et al., 2002; Huh et al., 2005). Figure 1.4 shows an electron backscatter diffraction (EBSD) orientation map of an affected FSS sheet, where the individual orientations are color-coded according to their crystallographic direction along the normal direction (ND), the color code (standard triangle) is given in the inset.

Superconductors. The practical application of bulk polycrystalline superconductors such as $YBa_2Cu_3O_{7-x}$, or short YBCO, depends on obtaining a high value of transport critical current density J_c. As illustrated in Figure 1.5, J_c has been found to depend critically on the degree of texturing in the material and, more precisely, on the type of grain boundaries that certain textures produce (Zhu et al., 1991). Manufacture of superconductors is tailored to produce textured materials (Obst et al., 2003; Koblischka-Veneva and Koblischka, 2008).

A natural or technological material will often exhibit a completely new texture after such processes as deformation, recrystallization, and phase transformations. Characterization of such changes can be used as an experimental probe to investigate the history of the material, especially where the changes involved are well defined. This widely used application of texture as a diagnostic tool constitutes the second reason why knowledge of texture is valuable. Hence, texture research has proceeded on two broad (and related) fronts:

- Technological control of material processing
- Further understanding of fundamental materials science and geological sciences

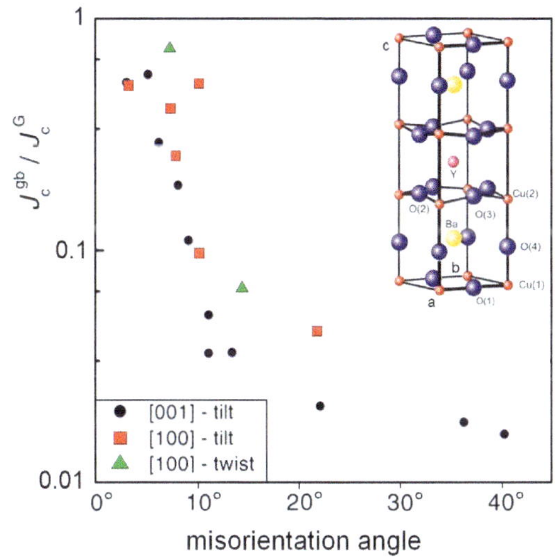

FIGURE 1.5 Current transmission, that is, the ratio of intergrain (Gb) and intragrain (G) critical current densities J_c across grain boundaries in YBCO superconductors as a function of misorientation angle (at 5 K). High superconducting current transmission requires maximum fraction of small-angle grain boundaries, that is, a very sharp texture. (Courtesy of B. Obst.)

The most established method for measuring textures is x-ray diffraction using a texture goniometer (Section 4.3), which gives a measure of the volume fraction of a particular family of planes, for example, {111}, {0001}, which are orientated for diffraction. Hence the texture that is obtained is an *average* value for the whole sampled volume, typically comprising thousands of grains. The x-ray technique for texture evaluation was first used several decades ago, and much effort has gone during the intervening years into refining it to produce a standard system to provide quantitative texture data in conventionally accepted formats, that is, *pole figures* (Sections 2.5, 4.3.2, and 5.2) or *orientation distribution functions* (ODFs) (Sections 5.3 and 5.4). Since the 1960s, neutron diffraction has also been employed to determine crystallographic textures, which offers several advantages especially in coarse-grained or multi-phase specimens (Sections 4.4 and 4.5).

This section has established that the importance of texture lies first in its direct relationship with material properties and second as a method of fingerprinting the history of a material. The dominant approach to texture research and practice before the early 1980s was the use of x-rays to probe the average texture of a specimen, and indeed this approach is still valid if knowledge of the average texture is all that is required.

1.2 THE MODERN APPROACH TO TEXTURE: MICROTEXTURE

The hallmark of the x-ray approach is that it provides efficiently an overview of the texture. In turn, this texture is compiled from numerous individual grains, each of which possesses a discrete orientation. The x-ray texture tells us what volume fraction of the specimen (obtained from the intensity of diffraction by particular planes) has a particular orientation. However, it does not tell us how these grains are distributed throughout the material (as e.g. in the example shown in Figure 1.4). An approach that deals with the orientation statistics of a population of individual grains, and usually encompasses the *spatial location* of these grains, that is, the *orientation topography*, has been termed *microtexture*—the conjoining of microstructure and texture (Randle, 2003). To avoid ambiguity with the traditional view of texture, a texture that reflects an average value obtained from many different grains is often termed "macrotexture" rather than just "texture."

Traditionally, texture and microstructure were considered to be two different entities which were studied by different techniques, e.g. optical or electron microscopy for the microstructure and x-ray diffraction for the texture. Correctly viewed, however, both are aspects of the same property, namely the crystalline nature of the material. While the microstructure is the collectivity of all crystalline defects (grains and different phases, together with their interfaces, dislocations and other deformation features, residual stresses), the texture is the distribution of the orientations resulting from the existence and evolution of microstructure. The microtexture measurement tools—foremost EBSD in the scanning electron microscopy (SEM), but also various tools in the transmission electron microscopy (TEM) if highest spatial resolution is required and, since a while, spatially resolved synchrotron diffraction—allow to measure both microstructure and texture simultaneously and have therefore become the most powerful tools for the research on crystalline materials. Several recent conferences, or parts of conferences, have been dedicated specifically to microtexture (see General Bibliography).

1.2.1 Applications of Microtexture

If the spatial location plus the orientation of individual grains in a sample population is known, access to a whole new stratum of knowledge is heralded, since the crystallographic and morphological aspects of structure evolution become fused. The following phenomena, which are all directly linked to microtexture (i.e., individual grain orientation), can be explored:

- Local property effects
- Interfacial parameters and properties
- Morphological/geometrical grain parameters, that is, size, shape, and location
- Orientation variations within individual grains
- Phase relationships
- Direct ODF measurement

The importance of the way in which grain orientations are distributed can be illustrated by the idealized sketch of a microstructure in Figure 1.6. Here a simplified case is considered where some grains have a strong *preferred orientation* (with a spread of, say, 15°) and the remaining grains have "random" (i.e., no discernable preferred orientation) textures. Two limiting spatial conditions are considered: In

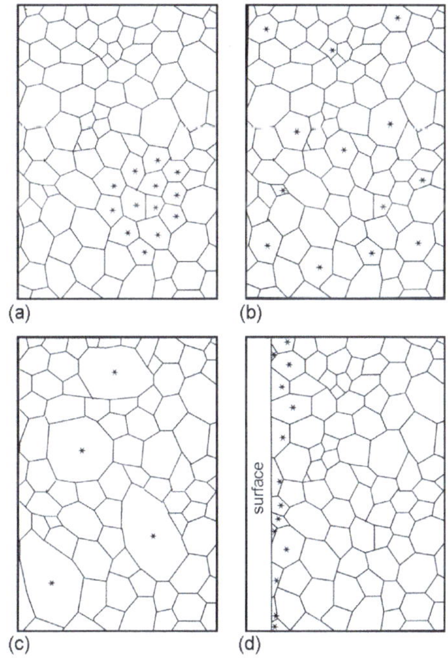

FIGURE 1.6 Idealized portrayal of microtexture or orientation topography. Grains having orientation X are (a) contiguous in the microstructure; (b) located singly in the microstructure; (c) in a different size class to the rest of the microtexture; (d) located in a special way with respect to the specimen, for example, near the surface.

Figure 1.6a, all the X grains are located in a contiguous patch, and in Figure 1.6b, the same number of X grains is spread singly throughout the microstructure.

Local property effects. The effect of property variations with orientation will be magnified where grains occur in clusters rather than singly. For example, slip transmission may be facilitated between grains of favorable orientations, such as the contiguous X grains in Figure 1.6a. Another example is that compressed plates of tantalum take up an "hourglass" shape due to the occurrence of two different microtextures, which in turn have different deformation characteristics, at the specimen surface and the center (Wright et al., 1994).

Interfacial parameters and properties. If individual orientations are known, it is possible to calculate the orientation difference or *misorientation* between neighboring grains and hence provide information about the distribution of grain boundary geometry, which is sometimes called *mesotexture*, that is, the texture between grains (Chapter 10). This is a very powerful aspect of microtexture since it allows access to grain boundary structure/property links, which is unavailable via a macrotexture approach (Chapter 11). In recent years, it has been realized that the properties of grain boundaries vary markedly with their structure, and moreover the structure of boundaries can be manipulated to produce boundaries that perform better.

With regard to the simple example in Figure 1.6, the misorientations between grains having orientations X will be small angle since the orientations are similar, and misorientations between X and other grains and also between non-X grains will be mainly large angle. Hence the small-angle boundaries—which have physical properties different from the boundaries that are large angle—are spatially connected. This grouping will have a major effect on inter-granular transport properties in the material, since some pathways in the grain boundary network will effectively be blocked (Rohrer et al., 2006).

Morphological/geometrical grain parameters. The morphological/geometrical characteristics of grains—grain size, shape, and macroscopic location—cannot be accessed by macrotexture measurements. A microtexture approach, however, allows concurrent access to both geometrical and orientational parameters, which in turn permits correlations to be made between the two. Referring again to Figure 1.6, the X grains may belong, for example, to a particular size class or be located in a specific way with respect to the specimen geometry, such as adjacent to the specimen surface, as shown in Figure 1.6c and d, respectively.

A grain or crystal is formally defined as a unit having a single orientation (but see the following text). A direct consequence of this definition is that the grains themselves can be depicted by knowing where in the microstructure orientation changes take place. This leads to "mapping" of microtexture (Section 1.2.3 and Chapter 9).

Orientation variations within individual grains. By classical definition, a grain has a single, distinct orientation. However, in practice, the orientation within a grain may vary in a continuous or discontinuous manner, which can be termed *orientation perturbation.* In particular, there may be perturbations close to interfaces and in deformed structures reflecting the distribution of strain. A microtexture technique is able to characterize these distributions on a local level, which is not a capability of a macrotexture technique.

Phase relationships. So far, we have considered that the hypothetical grains in Figure 1.6 are all the same phase. In exactly the same manner, multiphase materials can also be studied to produce concurrently the texture of each phase, the distribution of its orientations, and orientation relationships between individual crystals of each phase.

Direct ODF measurement. The traditional route for ODF measurement relies on calculation from several pole figures that have been determined by macrotexture methods (Chapter 5). These procedures admit inherent inaccuracies that are obviated when a microtexture technique is used to obtain the orientation distribution. For this case, full grain orientations are obtained *directly* rather than calculated from crystal plane distributions as in the x-ray case. Hence ODFs obtained by microtexture measurements give the *true* orientation distribution of the sampled grains.

Although the schematic illustration in Figure 1.6 might seem rather hypothetical, in practice such microstructures have been observed—for example, where clusters of similarly oriented grains aggregate into "supergrains" (Mason and Adams, 1994; Huh et al., 2005), in a superplastically deformed aluminum alloy (Randle, 1995a), and in various sheet metals and ceramics (Fionova et al., 1994)—and are a focus of research interest. More generally, microtextures may consist of several preferred orientations rather than only one as depicted in Figure 1.6. However, the preferred orientations tend to be spatially distributed nonrandomly, which gives rise to anisotropic distributions of grain boundary and grain properties as described earlier and demonstrates the cruciality of obtaining microtexture information to address these complexities.

In summary, this section has introduced the concept of microtexture, which is a recent approach to orientation-related analyses. Microtexture focuses texture analysis onto a local basis and allows it to be correlated with the microstructure. This, in turn, allows whole facets of the microstructure to be explored, which were inaccessible by macrotexture methods, and has indeed enlarged the view of microstructure to synthesize both the morphological and crystallographic aspects.

1.2.2 Electron Backscatter Diffraction

The ability to obtain microstructure-level information implies that the probe size formed by the exploring radiation (i.e., x-rays, neutrons, or electrons; Section 3.6 and Table 3.1) must be smaller than the size of the microstructural units themselves. This rules out x-ray diffraction as an experimental tool for microtexture measurement. Electrons are ideal for combined microstructural/crystallographic studies, and indeed until the 1980s, TEM was the major technique used for such work (Sections 6.2 and 8.3), with some input from Kossel diffraction (Section 6.3.5) and electron channeling pattern (ECP) (Section 6.3.4) in an SEM. Since that time a more convenient SEM-based technique for microtexture has been developed, known as electron backscatter diffraction (EBSD), which is now the backbone of most microtexture research. EBSD or, as it is equivalently known, backscatter Kikuchi diffraction (BKD), or electron backscatter patterns (EBSP), is an add-on package to an SEM. The most attractive feature of EBSD is its unique capability to perform concurrently rapid, automatic diffraction analysis to give crystallographic data and imaging with a spatial resolution in the nanometer range, combined with the regular capabilities of an SEM such as capacity for large specimens, concurrent chemical analysis, and the ability to image rough surfaces (Venables and Harland, 1973; Dingley, 1984; Randle, 2003).

EBSD is, without doubt, the most suitable and widely used experimental technique for the determination of microtexture. For this reason, and moreover because EBSD with all its latest ramifications is still a relatively new technique compared to x-ray methods, it is described in detail in Section 8.2. Briefly, the technique relies on positioning the specimen within the SEM sample chamber such that a small angle, typically 20°, is made between the incident electron beam and the specimen surface. This simple expedient enhances the proportion of backscattered electrons able to undergo diffraction and escape from the specimen surface. The resulting diffraction pattern is captured and interrogated in real time, and computer algorithms allow the orientation of each diffraction pattern to be obtained and stored. From this raw data a microtexture is constructed serially. A state-of-the-art EBSD system can automatically position a stationary probe on a suitable specimen, capture a diffraction pattern, index it, and store the result at rates well exceeding 1,000 patterns per second on suitable materials (Wright and Nowell, 2008; Schwarzer and Hjelen, 2010). In 2017, complementary metal oxide semiconductor (CMOS) imaging sensors were commercially introduced, providing acquisition speeds of up to 3,000 patterns per second (Nowell, 2022). Future-generation EBSD systems will achieve even faster data collection rates; at the time of writing, the fastest CMOS detectors from Gatan/EDAX and Oxford Instruments are said to achieve indexing rates of up to 6,700 patterns per second.

It is no exaggeration to say that since its inception, EBSD has revolutionized texture investigations in both research and industry. The technique has passed through various stages of development and, in its most refined form, provides a totally synthesized approach to orientation and microstructure. Needless to say, EBSD is not able to solve all problems related to microstructure and microtexture of materials. The most severe limitation is the limited spatial resolution, particularly on light materials. In this case transmission electron diffraction techniques, either in TEM or in SEM, become important which are nowadays also highly automatized and reach high measurement rates. Another important limitation of EBSD arises from the fact that it only observes the surface of the material; 3D analysis requires rather tedious destructive serial sectioning to look inside. This severely hampers the informational value of EBSD-based in-situ observations of crystallographic processes. Here spatially resolved synchrotron diffraction techniques with their high energy and brightness play a bigger and bigger role for microtexture and microstructure studies. We believe that this field will see a great increase in use with the advent of lab-sized radiation sources.

1.2.3 Orientation Microscopy and Orientation Mapping

Automated EBSD is exploited to select a grid of sampling points whose spacing is much finer than the grain size. If a representation of the orientation at each grid point is plotted, a crystallographic map of the microstructure is obtained. This is known variously as "orientation imaging microscopy" (OIM™) (Adams et al., 1993) or "automated crystal orientation mapping" (ACOM) (Schwarzer, 1997a, 1997b). The terms "crystal orientation mapping" (COM) (Davies and Randle, 1999) and "orientation scanning microscopy" (Gobernado et al., 2012) are other versions that have been adopted. In this book, the term *orientation microscopy* (shortly *ORM*) is adopted to describe the procedures involved in getting the data and *orientation mapping* is used to describe the output of data (Chapter 9).

Orientation microscopy/mapping is a landmark innovation in the field of EBSD and microtexture because it links orientation *directly* to the microstructure. In other words, the crystal orientation map portrays faithfully the orientational components, that is, those concerned with crystallography, of the microstructure in an analogous manner to the mapping of chemical elements in the microstructure via energy dispersive spectroscopy in an SEM. As the potential and capabilities of orientation mapping are considered in greater depth, it emerges that there are many subtle ramifications to its application, which introduce a degree of *quantification* into the definition of microstructure that was hitherto not possible. An orientation map can highlight different aspects of the microstructure, depending on the distribution/step size of the sampling schedule plus the scale and requirements of the inquiry, as specified by the user.

1.3 A GUIDE TO THE BOOK

This book is intended for the reader who is familiar with crystalline materials and wishes to gain a working understanding of the practice and applications of texture. The following prior knowledge, at an introductory level, is assumed:

- Basic crystallography (crystal planes and directions)
- Principles of electron microscopy (TEM and SEM)
- Familiarity with vectors and matrix algebra

Several standard textbooks cover different aspects of this information (see General Bibliography). In this book, the mathematics of the subject area has been kept to that required to understand the scientific principles. Many of the examples relate to metallic materials because this group is the largest to be affected by texture-associated phenomena. It is, however, recognized that semiconductors, superconductors, and geological materials are growing areas of interest, particularly where microtexture is concerned.

The sequence of the book is as follows. We begin in Chapter 2 by addressing the fundamental descriptors and terminology associated with orientations and their representation in general. Chapter 3 is an introduction to diffraction of radiation since this phenomenon forms the basis of almost all texture analysis. This completes Part I, which is on Fundamental Issues. Part II (Chapters 4 and 5) describes Macrotexture Analysis, both data acquisition (Chapter 4) and data representation (Chapter 5). This part is shorter than Part III on Microtexture Analysis, reflecting both the growing emphasis on the modern microtexture approach to texture analysis and the greater requirement for detailed explanation of the philosophy, practice, and analysis associated with microtexture. Chapters 6 and 7 describe the physical principles of electron diffraction and the evaluation procedures of electron diffraction patterns with the aim of orientation determination. These techniques form the basis of orientation mapping and orientation microscopy, which allows the reconstruction of microstructures and the determination of microtexture. Chapter 8 is dedicated to the practice of orientation measurements in SEM and TEM, followed in Chapter 9 by a description of orientation mapping and orientation microscopy in SEM and TEM, including a discussion when which technique should be applied. Orientation microscopy and mapping is an innovative topic, and Chapter 10 is devoted to a treatise on how the statistics of microtexture data are analyzed and represented. In Chapter 11 we will address crystallographic issues concerning grain boundaries, interfaces, surfaces, and connectivity. This chapter also covers the determination of three-dimensional microstructures and their textures by means of 3D EBSD serial sectioning techniques. Chapter 12 is another spin-off of the microtexture approach and describes orientation relationships between different crystallographic phases as well as mechanisms of texture formation involving phase transformations. Finally, Chapter 13 covers minority non-diffraction-based techniques for microtexture and microstructure measurements as well as the latest evolution in non-destructive three-dimensional or even four-dimensional orientation mapping using synchrotron radiation. Furthermore, Chapter 13 provides a detailed comparison between all the techniques for texture measurement.

2 Descriptors of Orientation

2.1 INTRODUCTION

In this chapter, after a short review of crystal structures and symmetries (Section 2.2), the main mathematical parameters that are used to describe an orientation are defined and explained. These are as follows:

- Orientation matrix (Section 2.3)
- "Ideal orientation," that is, Miller or Miller-Bravais indices (Section 2.4)
- Euler angles (Section 2.6)
- Angle/axis of rotation (Section 2.7)
- Rodrigues vector (Section 2.8)
- Quaternions (Section 2.9)

All these descriptors are employed to process and represent different aspects of macrotexture and microtexture measurements. This chapter will elucidate the theoretical aspects of these parameters and their interrelationships. For graphical representation, it is usual to plot orientations, represented as projected poles, Euler angles, angle/axis, or the Rodrigues vector, in an appropriate "space." These spaces—the pole figure or inverse pole figure "projected space," Euler space, cylindrical angle/axis space, and Rodrigues space—are explained in Sections 2.5 through 2.8. These explanations encompass only the basic theoretical aspects; in practice, their utilization is customized according to whether they are employed to display and analyze macrotexture or microtexture data. Therefore, these two divisions are discussed separately in Chapters 5 and 10.

The term *crystallographic orientation* can refer either to a crystal (grain), which is three-dimensional (3D) and therefore comprises three independent variables, or to a crystallographic plane, which comprises two independent variables. Where the term *orientation* is used throughout this book, it refers to the orientation of the full crystal, rather than of a plane, unless it is made clear otherwise.

2.2 CRYSTAL STRUCTURES AND CRYSTAL SYMMETRIES

Materials are called crystalline if the constituent atoms, molecules, or ions are packed in a regularly ordered, repeating pattern extending in all three spatial dimensions. Figure 2.1a shows an example of several millimeter-sized crystals of niobium, a lustrous gray transition metal with cubic crystal structure which is found in the vanadium group of the periodic table. The crystal structure can be idealized into having discrete translation symmetry at the atomic level. That is to say, the arrangement of

DOI: 10.1201/9781003258339-3

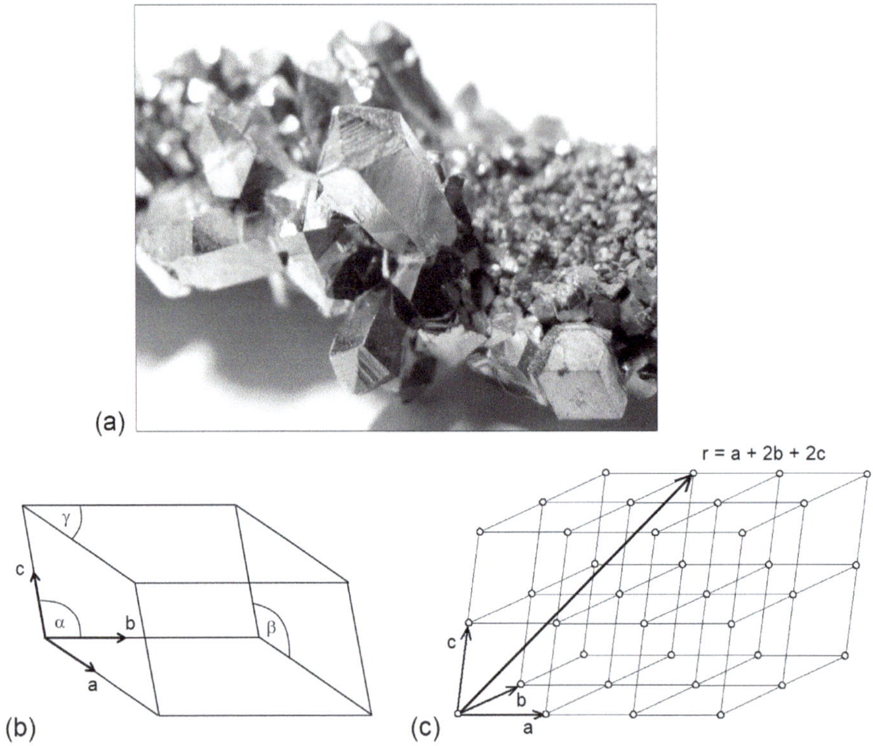

FIGURE 2.1 (a) Assemblage of several millimeter-sized niobium crystals. (From Wikimedia Commons, https://commons.wikimedia.org/wiki/File:Niobium_crystals_1.jpg, under CC BY-SA 3.0 license); (b) unit cell with the lattice parameters *a*, *b*, and *c* as well as the interaxial angles α, β, and γ; (c) construction of a 3D crystal lattice based on a set of three unit vectors *a*, *b*, and *c*.

the atoms is repeated periodically on a *lattice* in three dimensions. A lattice can be characterized by the geometry of its unit cell—a parallelogram or parallelepiped with the lattice parameters *a*, *b*, and *c* as well as the angles α, β, and γ between them (Figure 2.1b). As shown in Figure 2.1c, the unit cell generates the whole lattice using only translations by three noncoplanar unit vectors, *a*, *b*, and *c* and a set of integers *l*, *m*, and *n* such that each lattice point, identified by a vector *r*, can be obtained from

$$r = l \cdot a + m \cdot b + n \cdot c \tag{2.1}$$

Thus, a lattice is an infinite set of points generated by a set of discrete translation operations. A crystal is made up of one or more atoms that are repeated at each lattice point. The crystal then looks the same when viewed from any of these lattice points. In most of the common metals, the lattice points are occupied by single atoms. In more complex structures, the lattice points contain identical groups of atoms, called a *motif* (or *basis*) of atoms. That is to say, the crystal structure consists of the same atoms, or motifs, positioned around each and every lattice point. This group of atoms, therefore, repeats indefinitely in three dimensions according to the arrangement of the *crystal lattices*. In total, there are 14 lattices that fill the 3D space, which are the well-known 14 *Bravais lattices* presented in Figure 2.2. The characteristic rotation and mirror symmetries of the atoms or motifs are described by the crystallographic *point groups*, of which there are 32, and *space groups*, of which there are 230 (Hahn, 2005).

The defining property of a crystal is its inherent symmetry, which means that certain operations result in an atomic configuration identical to the original configuration. For example, rotating the crystal by 90° or 180° about a certain axis may leave the crystal unchanged. The crystal is then said to have, respectively, a fourfold or twofold rotational symmetry about this axis. In addition to rotational symmetries like this, a crystal may have symmetries in the form of mirror planes

Name	Conditions	Primitive	Base centered	Body centered	Face centered
Triclinic	$a \neq b \neq c$ $\alpha \neq \beta \neq \gamma$	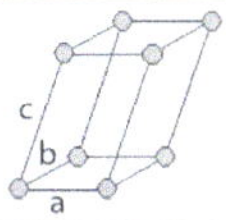			
Monoclinic	$a \neq b \neq c$ $\alpha = \gamma = 90° \neq \beta$	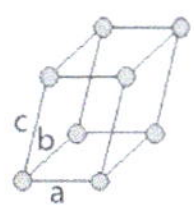	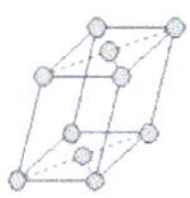		
Orthomombic	$a \neq b \neq c$ $\alpha = \beta = \gamma = 90°$	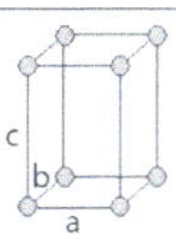	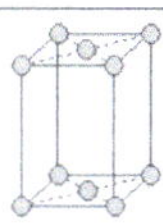	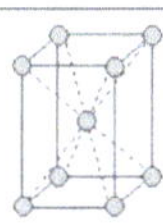	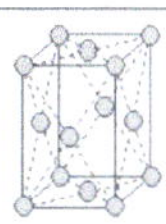
Tetragonal	$a = b \neq c$ $\alpha = \beta = \gamma = 90°$	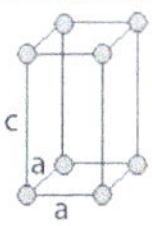		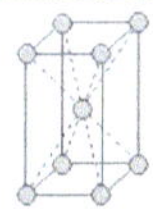	
Rhombohedral (trigonal)	$a = b = c$ $\alpha = \beta = \gamma \neq 90°$	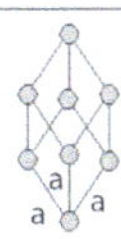			
Hexagonal	$a = b \neq c$ $\alpha = \beta = 90°, \gamma = 120°$	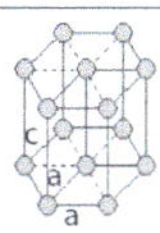			
Cubic	$a = b = c$ $\alpha = \beta = \gamma = 90°$	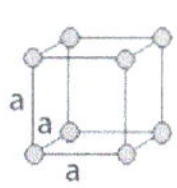		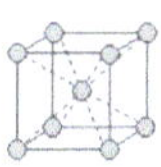	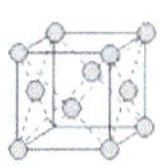

FIGURE 2.2 Unit cells of the 14 Bravais lattices, that is, possible combinations of the 7 lattice types with one of the lattice centerings.

and translational symmetries, and also the so-called compound symmetries that are a combination of translation and rotation/mirror symmetries. A full classification of a crystal is achieved when all of these inherent symmetries of the crystal are identified.

A crystal lattice is conveniently described by the unit cell, with the atoms (or motifs) situated on its corners (Figure 2.1b, c). There may be additional lattice points at the center of certain faces or at the center of volume within the unit cell. The positions of the atoms inside the unit cell are described by the set of atomic positions (x_i, y_i, z_i) measured from the origin, a lattice point at the corner of the unit cell. There are only seven possible crystal systems that atoms can pack together to produce an infinite 3D space lattice in such a way that each lattice point has an identical environment to that around every other lattice point. The crystal systems are a grouping of crystal structures according to the axial system used to describe their lattice. Each crystal system consists of a set of three axes in a particular geometrical arrangement:

- *Triclinic.* No symmetry other than translational symmetry
- *Monoclinic.* One twofold rotation axis or one mirror plane
- *Orthorhombic.* Three twofold rotation axes or one twofold rotation axis and two mirror planes
- *Tetragonal.* One fourfold rotation axis
- *Rhombohedral (also called trigonal).* One threefold rotation axis
- *Hexagonal.* One sixfold rotation axis
- *Cubic (also called isometric).* Four threefold rotation axes

The vast majority of metals crystallize in either cubic or hexagonal crystal structure. Examples for metals with tetragonal crystal structure include tin and indium, whereas gallium, uranium, and neptunium are orthorhombic. Plutonium has, among several others, a monoclinic allotrope, whereas (solid) mercury as well as several metalloids like antimony, bismuth, and arsenic are rhombohedral.

Combination of the seven crystal systems with the various lattice centerings yields the well-known 14 *Bravais lattices* (or *space lattices*) shown in Figure 2.2. They describe the geometric arrangement of the lattice points, and thereby the translational symmetry of the crystal. The possible lattice centerings are as follows:

- *Primitive (P).* Lattice points on the cell corners only, that is, at the coordinates $(x,y,z) = (0,0,0)$
- *Body centered (I).* One additional lattice point at the center of the cell so that for each atom at (0,0,0) there is another one at (1/2,1/2,1/2)
- *Face centered (F).* One additional lattice point at the center of each of the faces of the cell, that is, there are also atoms at (1/2,1/2,0), (1/2,0,1/2), and (0,1/2,1/2)
- *Base centered (C).* One additional lattice point at the center of the base of the cell (the so-called C-plane containing *a* and *b*), that is, at (1/2,1/2,0)

It is seen that not all combinations of the crystal systems and lattice centerings are needed to describe the possible space lattices. There are in total $7 \times 6 = 42$

combinations, but it can be shown that several of these are in fact equivalent to each other, reducing the number of combinations to the 14 Bravais lattices shown in Figure 2.2. With a view to texture analysis, the expansion of the seven crystal systems into the 14 Bravais lattices is of importance since the latter affect diffraction measurements (Sections 3.2 and 3.3) and evolution of textures through activation of slip systems (Section 5.5).

The existence of the translation lattice limits the number of symmetry elements that can be present. There are 32 possible combinations of rotations, mirror planes, and an inversion center, collectively called the 32 point groups or crystal classes. When these point groups are combined with the translation symmetries, we arrive at the 230 possible space groups, containing all symmetry elements that transform every crystal lattice into itself. However, since the space groups are generally not of interest in texture analysis, they will not be discussed here any further.

For texture analysis by diffraction techniques, it is more important whether or not the crystals are centrosymmetric, that is, contain an inversion center so that for every point (x,y,z) in the unit cell there is an indistinguishable point $(-x,-y,-z)$ (Weiss and Wenk, 1985; Matthies et al., 1987; Kocks, 1998a; Man, 2022). A diffraction (Laue) experiment produces a projection of the lattice planes (Chapter 3). Hence, the resulting reflection diagram must comprise the symmetry of the crystal structure. However, a diffraction diagram will always be centrosymmetric, even if the crystal is not. This is caused by the fact that the intensities reflected at two opposite planes (hkl) and $(\bar{h}\,\bar{k}\,\bar{l})$ are equal (Friedel's law; Friedel, 1913). Hence, diffraction experiments are only able to distinguish between the various centrosymmetric crystal classes. These 11 so-called *Laue classes* or *Laue groups* containing an inversion center are listed in Table 2.1, whereas Figure 2.3 shows their symmetry elements. As a matter of fact, all crystal lattices and the majority of crystals are centrosymmetric, that is, they do contain an inversion center. Crystal structures with a multiatom motif generally have a lower symmetry than that of the lattice, some of them lacking an inversion center. Anyhow, the diffraction patterns of these 21 non-centrosymmetric classes become centrosymmetric through addition of an inversion center to the corresponding crystal class, such that the apparent symmetry of the crystal is then one of the 11 Laue classes (Table 2.1). In texture analysis, symmetry of the Laue groups feeds through to the size and symmetry of orientation space necessary to represent texture information of the corresponding crystal structure (Section 2.6.2).

2.3 TRANSFORMATION BETWEEN COORDINATE SYSTEMS: THE ROTATION MATRIX

2.3.1 Coordinate Systems

To specify an orientation, it is necessary to set up terms of reference, each of which is known as a *coordinate system* (Bollmann, 1970; McKie and McKie, 1974; Hansen et al., 1978; Bunge, 1982; Kocks, 1998b). Two systems are required: one relating to the whole specimen and the other relating to the crystal. Both systems are Cartesian

TABLE 2.1
The Eleven Laue Groups and Their Non-centrosymmetric Subgroups in Hermann-Mauguin Notation

Crystal System	Space Group G_C (Schönflies Notation)	Laue Class	Noncentro-symmetric Classes	Number of Symmetry Elements
Triclinic	C_1	$\bar{1}$	1	1
Monoclinic	C_2	2/m	2, m	2
Orthorhombic	D_2	Mmm	222, mm2	4
Trigonal	C_3	$\bar{3}$	3	3
	D_3	$\bar{3}$m	3m, 32	6
Tetragonal	C_4	4/m	4, $\bar{4}$	4
	D_4	4/mmm	4mm, $\bar{4}$m2, 422	8
Hexagonal	C_6	6/m	6, $\bar{6}$	6
	D_6	6/mmm	6mm, $\bar{6}m2$, 622	12
Cubic	T	m$\bar{3}$	23	12
	O	m$\bar{3}$m	$\bar{4}$3m, 432	24

Source: Data from Kleber et al. (2002).

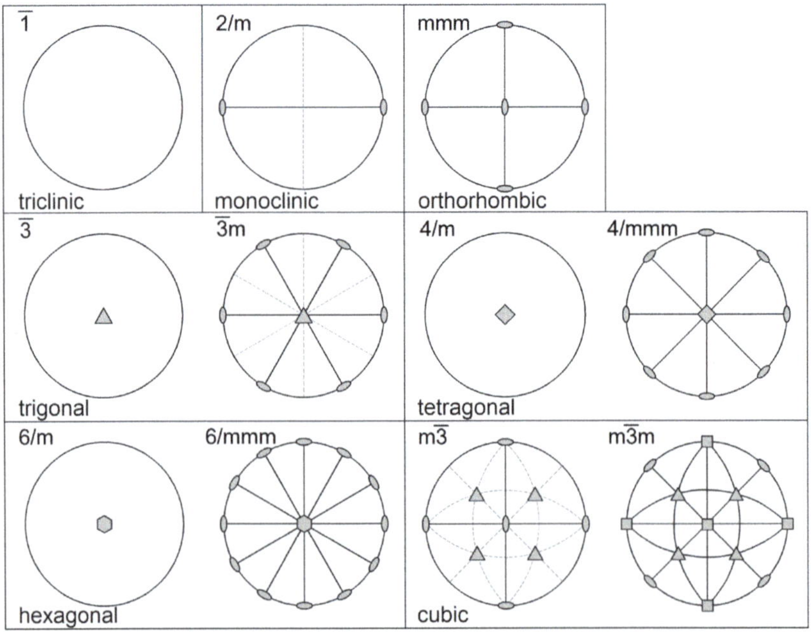

FIGURE 2.3 Symmetry elements in the 11 Laue groups.

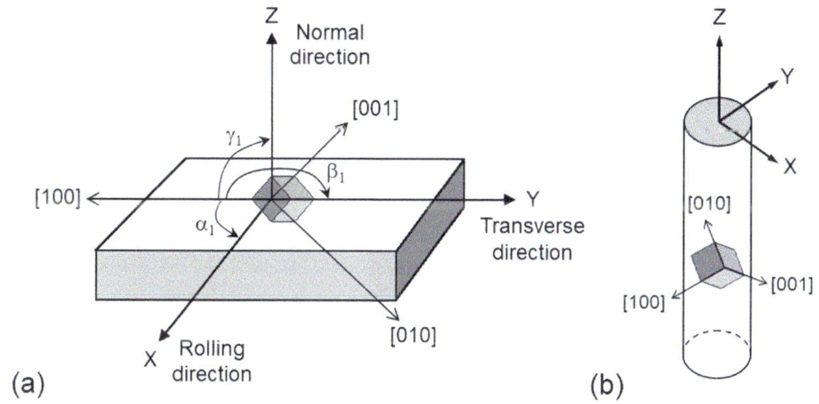

FIGURE 2.4 Relationship between the specimen coordinate system {X, Y, Z} and the crystal coordinate system {100, 010, 001} where the (cubic) unit cell of one crystal in the specimen is depicted. (a) Rolled product with RD, TD, and ND. The cosines of the angles α_1, β_1, γ_1 give the first row of the orientation matrix (see text). (b) Cylindrical specimen where Z is oriented parallel to the long axis while the choice of X and Y is arbitrary.

(i.e., having axes at right angles) and preferably right-handed. The axes of the *sample* or *specimen coordinate system* $S = \{s_1, s_2, s_3\}$ are chosen according to important surfaces or directions associated with the external form or shape of the specimen. For example, for a fabricated material, there are obvious choices defined by the processing geometry. One of the most common of these relates to a rolled product, and hence the directions associated with the external shape are the rolling direction (RD); the through-thickness direction, that is, the direction normal to the rolling plane (ND); and the transverse direction (TD). These directions are illustrated in Figure 2.4a. Other specimens, such as a tensile test piece, a rod, or a wire, have only uniaxial symmetry; hence it is only necessary to specify one axis in the specimen coordinate system and the other two axes can be chosen arbitrarily, as shown in Figure 2.4b. In natural rocks, the plane of foliation and a line of lineation within that plane often make a natural choice for specimen axes. Sometimes there are no clearly defined directions associated with the specimen. For these cases, the specimen coordinate system can be chosen arbitrarily and it is conventional in experimental work for the normal to the "principal" specimen surface, from which diffraction information is collected, to be labeled Z. The X and Y axes are both perpendicular to Z and should form a right-handed set.

The second coordinate system, the *crystal coordinate system* $C = \{c_1, c_2, c_3\}$, is specified by directions in the crystal. The choice of directions is in principle arbitrary, although it is convenient to adapt it to the crystal symmetry. For crystals of the orthogonal branch (cubic, tetragonal, orthorhombic), the axes [100], [010], [001] are mutually perpendicular and are hence readily adopted as the crystal coordinate system. Subsequently the crystal axes are made orthonormal, that is, normalized with the lattice parameters *a*, *b*, *c*, to be all the same length (Figure 2.5a). For non-orthogonal crystals a Cartesian frame needs to be associated with the crystal axes.

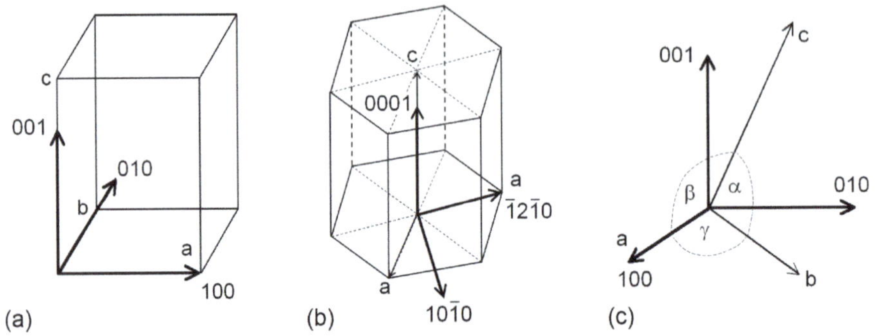

FIGURE 2.5 Orthonormalized crystal coordinate systems for (a) orthotropic, (b) hexagonal, and (c) general (triclinic) crystal symmetries.

As indicated in Figures 2.5b and I.1b, for hexagonal and trigonal symmetry, the two obvious choices are

$$X_1 = [10\bar{1}0], \quad Y_1 = [\bar{1}2\bar{1}0], \quad Z = [0001] \tag{2.2a}$$

or

$$X_2 = [2\bar{1}\bar{1}0], \quad Y_2 = [01\bar{1}0], \quad Z = [0001] \tag{2.2b}$$

Figure 2.5c shows a triclinic frame formed by the crystal axes a, b, and c and a normalized Cartesian frame {[100], [010], [001]} where both frames share the direction a = [100]. For indexing of Kikuchi patterns, etc., it is necessary to transform vectors—e.g. zones or zone axes—from the non-orthogonal crystal frame to the (normalized) Cartesian frame. This can be carried out by premultiplying a vector r^c in the crystal frame with the so-called crystal matrix L, i.e. $r^o = L \cdot r^c$, where the matrix L defines the base axes of the crystal (superscript c) columnwise with respect to the orthonormal base (superscript o). The crystal matrix L has the following elements (Young and Lytton, 1972):

$$L = \begin{pmatrix} a & b\cos\gamma & c\cos\beta \\ 0 & b\sin\gamma & \dfrac{c\left(\cos\alpha - \cos\beta\cos\gamma\right)}{\sin\gamma} \\ 0 & 0 & \dfrac{c\left(1 + 2\cos\alpha\cos\beta\cos\gamma - \cos^2\alpha - \cos^2\beta - \cos^2\gamma\right)^{1/2}}{\sin\gamma} \end{pmatrix} \tag{2.3a}$$

where a, b, c and α, β, γ denote the lattice parameters of the crystal system. In this way all crystal systems have the same form of crystal coordinate system, that is, orthonormal. The general matrix given in Equation 2.3a–c can be simplified in all, but the triclinic case. For example, for orthorhombic crystals where all angles are 90°, the crystal matrix simplifies to

$$L = \begin{pmatrix} a & 0 & 0 \\ 0 & b & 0 \\ 0 & 0 & c \end{pmatrix} \tag{2.3b}$$

For hexagonal crystals, the crystal matrix becomes

$$L = \begin{pmatrix} a & -a/2 & 0 \\ 0 & a\sqrt{3}/2 & 0 \\ 0 & 0 & c \end{pmatrix} \tag{2.3c}$$

Conversely, to transform from orthonormal coordinates back to the crystal reference system, it is necessary to premultiply by the inverse crystal matrix L^{-1}, i.e. $r^c = L^{-1} \cdot r^o$. Directions and planes in hexagonal or trigonal crystals can be described using either Miller or Miller-Bravais indices (Appendix I).

2.3.2 The Rotation (Orientation) Matrix

Having specified the specimen and crystal coordinate systems, an orientation is then defined as "the position of the crystal coordinate system with respect to the specimen coordinate system," that is,

$$C_C = g \cdot C_S \tag{2.4}$$

where C_C and C_S are the crystal and specimen coordinate systems, respectively, and g is the orientation. g can be expressed in several different ways, and these are listed in Section 2.1. The fundamental means for expressing g is the *rotation* or *orientation matrix*, which embodies the rotation of the specimen coordinates onto the crystal coordinates.

The orientation matrix is a square matrix of nine numbers and is obtained as follows. The first row of the matrix is given by the cosines of the angles between the first crystal axis, [100], and each of the three specimen axes, *X*, *Y*, *Z*, in turn. These three angles, α_1, β_1, γ_1, are labeled in Figure 2.4. The second row of the orientation matrix is given by the cosines of the angles α_2, β_2, γ_2 between [010] and *X*, *Y*, *Z*, in turn. Similarly, the third row of the matrix comprises the cosines of the angles α_3, β_3, γ_3 between [001] and *X*, *Y*, *Z*. Equation 2.5 shows the complete orientation matrix g in terms of the direction cosines, elements g_{ij}, and a numerical example:

$$g = \begin{pmatrix} \cos\alpha_1 & \cos\beta_1 & \cos\gamma_1 \\ \cos\alpha_2 & \cos\beta_2 & \cos\gamma_2 \\ \cos\alpha_3 & \cos\beta_3 & \cos\gamma_3 \end{pmatrix} = \begin{pmatrix} g_{11} & g_{11} & g_{13} \\ g_{21} & g_{22} & g_{23} \\ g_{31} & g_{32} & g_{33} \end{pmatrix} = \begin{pmatrix} 0.768 & -0.582 & 0.267 \\ 0.384 & 0.753 & 0.535 \\ -0.512 & -0.308 & 0.802 \end{pmatrix} \tag{2.5}$$

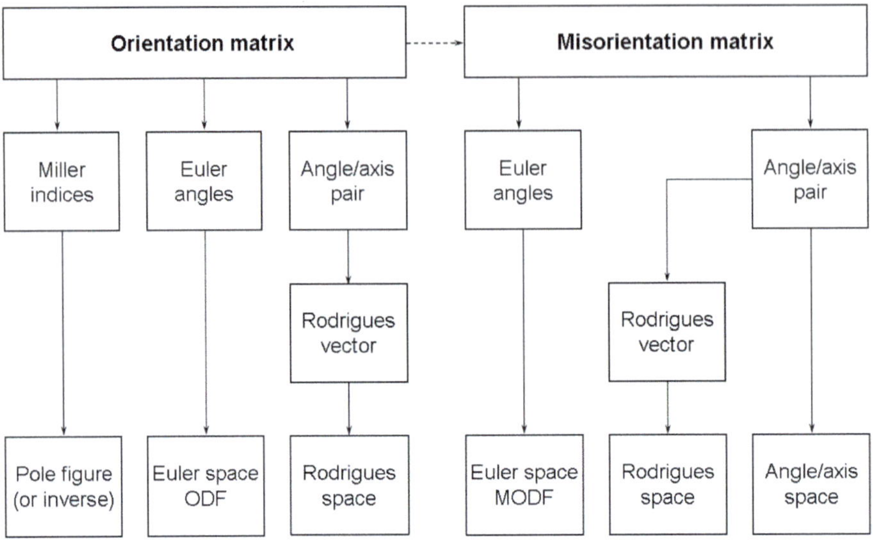

FIGURE 2.6 Relationship between the orientation matrix and the most commonly used orientation descriptors.

The orientation matrix allows a crystal direction to be expressed in terms of the specimen direction to which it is parallel and vice versa. Both the rows and columns of the matrix are unit vectors, that is, the matrix is orthonormal and the inverse of the matrix is equal to its transpose. Since a crystal orientation needs only three independent variables to specify it (Section 2.1), it is clear that the matrix, having nine numbers, contains non-independent elements. In fact, the cross product of any two rows, or columns, gives the third and for any row or column the sum of the squares of the three elements is equal to unity. These properties can be checked using the example in Equation 2.5. The pertinence of the orientation matrix is that it is the mathematical tool for calculation of all the other descriptors of orientation, as shown in Figure 2.6.

2.3.3 Crystallographically Related Solutions

We have stated that the crystal coordinate system and specimen coordinate system are related by the orientation (rotation) matrix. However, specification of both these coordinate systems is not usually unique, and a number of solutions can exist depending on the symmetry of both the crystal and the specimen. To consider the crystal symmetry, there are 24 different ways in which a crystal with cubic symmetry can be arranged. These 24 possibilities are illustrated in Figure 2.7a in the form of a set of 24 isometric dice. Consequently, there are 24 solutions for an orientation matrix of a material having cubic symmetry. The 24 *crystallographically related solutions* for the example matrix given in Equation 2.5 are listed in Table 2.2, and it can be seen that the rows of the matrix, and the signs of the elements, interchange. The full set of solutions is obtained by premultiplying the orientation matrix by each of 24 "symmetry matrices," which are listed in Appendix II, in turn. These matrices describe

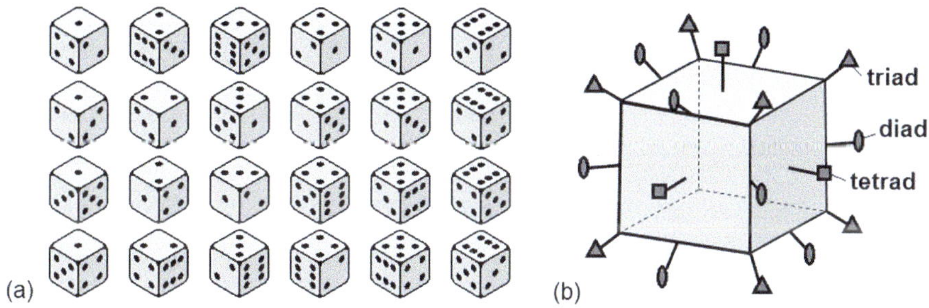

FIGURE 2.7 (a) Set of 24 isometric dice and (b) position of twofold (diad), threefold (triad), and fourfold (tetrad) symmetry elements in a cubic crystal.

TABLE 2.2
Twenty-Four Equivalent Descriptions for the Orientation (123)[634]

Ideal Orientation	Orientation Matrix	Angle/Axis Pair ($\theta/u\ v\ w$)	Rodrigues Vector ($R_1\ R_2\ R_3$)	Euler Angles ($\varphi_1\ \Phi\ \varphi_2$)
$(123)[63\bar{4}]$	0.768 –0.582 0.267 0.384 0.753 0.535 –0.512 –0.308 0.802	48.6° 0.562 –0.520 –0.644	0.254 –0.235 –0.291	307.0° 36.7° 26.6°
$(32\bar{1})[\bar{4}3\bar{6}]$	–0.512 –0.308 0.802 0.384 0.753 0.535 –0.768 0.582 –0.267	120.9° –0.028 –0.915 –0.403	–0.048 –1.613 –0.711	232.9° 105.5° 56.3°
$(\bar{1}2\bar{3})[\bar{6}34]$	–0.768 0.582 –0.267 0.384 0.753 0.535 0.512 0.308 –0.802	155.3° 0.271 0.933 0.237	1.239 4.263 1.081	121.0° 143.3° 333.4°
$(\bar{3}21)[436]$	0.512 0.308 –0.802 0.384 0.753 0.535 0.768 –0.582 0.267	74.6° 0.579 0.814 –0.039	0.441 0.620 –0.030	52.9° 74.5° 303.7°
$(13\bar{2})[6\bar{4}\bar{3}]$	0.768 –0.582 0.267 –0.512 –0.308 0.802 –0.384 –0.753 –0.535	122.5° 0.922 –0.386 –0.041	1.679 –0.704 –0.075	333.0° 122.3° 18.4°
$(1\bar{2}\bar{3})[6\bar{3}4]$	0.768 –0.582 0.267 –0.384 –0.753 –0.535 0.512 0.308 –0.802	153.3° –0.937 0.272 –0.220	–3.944 1.146 –0.925	121.0° 143.3° 153.4°
$(1\bar{3}2)[643]$	0.768 –0.582 0.267 0.512 0.308 –0.802 0.384 0.753 0.535	72.2° –0.816 0.061 –0.574	–0.595 0.045 –0.419	153.0° 57.7° 161.6°
$(2\bar{1}3)[3\bar{6}\bar{4}]$	0.384 0.753 0.535 –0.768 0.582 –0.267 –0.512 –0.308 0.802	67.4° 0.022 –0.567 0.824	0.015 –0.378 0.550	301.0° 36.7° 116.6°
$(\bar{1}\bar{2}3)[\bar{6}\bar{3}\bar{4}]$	–0.768 0.582 –0.267 –0.384 –0.753 –0.535 –0.512 –0.308 0.802	149.3° –0.222 –0.240 0.945	–0.807 –0.872 3.440	301.0° 36.7° 206.6°

(*Continued*)

TABLE 2.2 (*Continued*)
Twenty-Four Equivalent Descriptions for the Orientation (123)[634]

Ideal Orientation	Orientation Matrix	Angle/Axis Pair ($\theta/u\ v\ w$)	Rodrigues Vector ($R_1\ R_2\ R_3$)	Euler Angles ($\varphi_1\ \Phi\ \varphi_2$)
$(\bar{2}13)[\bar{3}6\bar{4}]$	–0.384 –0.753 –0.535 0.768 –0.582 0.267 –0.512 –0.308 0.802	125.6° 0.354 0.014 –0.935	0.688 0.027 –1.820	301.0° 36.7° 296.6°
$(231)[3\bar{4}6]$	0.384 0.753 0.535 –0.512 –0.308 0.802 0.768 –0.582 0.267	109.2° 0.732 0.124 0.670	1.030 0.174 0.942	52.9° 74.5° 33.7°
$(312)[\bar{4}63]$	–0.512 –0.308 0.802 0.768 –0.582 0.267 0.384 0.753 0.535	141.2° –0.388 –0.334 –0.859	–1.102 –0.948 –2.442	153.0° 57.7° 71.6°
$(\bar{2}3\bar{1})[\bar{3}\bar{4}\bar{6}]$	–0.384 –0.753 –0.535 –0.512 –0.308 0.802 –0.768 0.582 –0.267	168.4° 0.548 –0.582 –0.600	5.413 –5.748 –5.920	239.2° 105.5° 326.3°
$(\bar{3}\bar{1}2)[4\bar{6}3]$	0.512 0.308 –0.802 –0.768 0.582 –0.267 0.384 0.757 0.535	71.7° –0.537 0.625 0.567	–0.388 0.451 0.409	153.0° 57.7° 251.6°
$(\bar{3}1\bar{2})[46\bar{3}]$	0.512 0.308 –0.802 0.768 –0.582 0.267 –0.384 –0.753 –0.535	143.3° 0.854 0.350 –0.385	2.577 1.055 –1.163	333.0° 122.3° 288.4°
$(2\bar{3}\bar{1})[34\bar{6}]$	0.384 0.753 0.535 0.512 0.308 –0.802 –0.768 0.582 –0.267	106.7° –0.722 –0.680 0.126	–0.971 –0.914 0.169	232.9° 105.5° 146.3°
$(\bar{2}\bar{3}1)[\bar{3}46]$	–0.384 –0.753 –0.535 0.512 0.308 –0.802 0.768 –0.582 0.267	113.9° –0.120 0.712 –0.692	–0.185 1.094 –1.062	52.9° 74.5° 213.7°
$(3\bar{1}\bar{2})[\bar{4}\bar{6}\bar{3}]$	–0.512 –0.308 0.802 –0.768 0.582 –0.267 –0.384 –0.753 –0.535	137.1° 0.357 –0.871 0.338	0.907 –2.216 0.860	333.0° 122.3° 108.4°
$(21\bar{3})[364]$	0.384 0.753 0.535 0.768 –0.582 0.267 0.512 0.308 –0.802	178.6° –0.832 –0.457 –0.315	–67.983 –37.363 –25.710	121.0° 143.3° 63.4°
$(\bar{1}32)[\bar{6}\bar{4}3]$	–0.768 0.582 –0.267 –0.512 –0.308 0.802 0.384 0.753 0.535	140.4° 0.038 0.511 0.859	0.107 1.421 2.387	153.0° 57.7° 341.6°
$(3\bar{2}1)[\bar{4}\bar{3}6]$	–0.512 –0.308 0.802 –0.384 –0.753 –0.535 0.768 –0.582 0.267	177.3° 0.493 –0.351 0.796	20.621 –14.661 33.260	52.9° 74.5° 123.7°
$(\bar{2}\bar{1}\bar{3})[\bar{3}\bar{6}4]$	–0.384 –0.753 –0.535 –0.768 0.582 –0.267 0.512 0.308 –0.802	143.3° –0.482 0.876 0.013	–1.453 2.644 0.039	121.0° 143.3° 243.4°
$(\bar{1}\bar{3}\bar{2})[\bar{6}4\bar{3}]$	–0.768 0.582 –0.267 0.512 0.308 –0.802 –0.384 –0.753 –0.535	175.9° –0.339 –0.808 0.481	–9.362 –22.343 13.305	333.0° 122.3° 198.4°
$(\bar{3}\bar{2}\bar{1})[4\bar{3}\bar{6}]$	0.512 0.308 –0.802 –0.384 –0.753 –0.535 –0.768 0.582 –0.267	138.9° –0.850 0.026 0.527	–2.269 0.068 1.407	232.9° 105.5° 236.3°

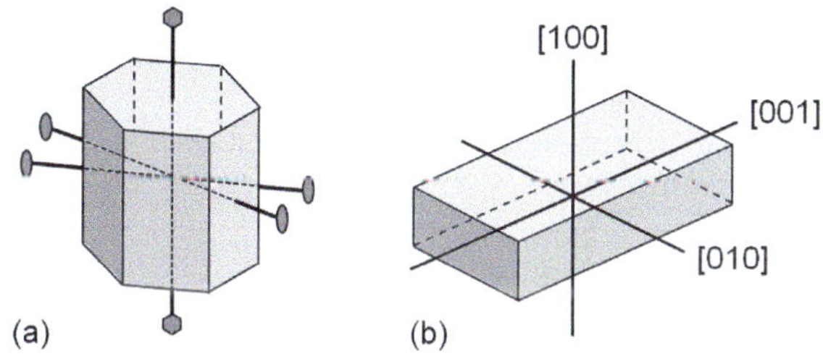

FIGURE 2.8 Symmetry elements of (a) hexagonal and (b) orthorhombic crystal structure.

the symmetry operations—two rotations of 120° about each of the four ⟨111⟩ crystal axes, three rotations of 90° about each of the three ⟨100⟩ axes and one rotation of 180° about each of the six ⟨110⟩ axes, plus the identity matrix (see Figure 2.7b). Crystal systems other than cubic have fewer crystallographically related solutions (Schmidt and Olesen, 1989). For example, the twelve and four variants for hexagonal and orthorhombic crystals are shown in Figure 2.8a and b, respectively, while the corresponding symmetry matrices are also given in Appendix II.

It is important to stress that each of the matrices in Table 2.2 describes the *same* orientation and so in general any one can be chosen to represent the orientation. However, depending on what method is chosen to represent an orientation or misorientation (i.e., Miller indices, angle/axis pair, Rodrigues vector, Euler angles, quaternions as described in the remainder of this chapter), the selection of a particular crystallographically related solution may both facilitate representation and allow insight into the physical meaning of the data. These points are expanded in Sections 2.7 and 10.6.

2.4 THE "IDEAL ORIENTATION" (MILLER OR MILLER–BRAVAIS INDICES) NOTATION

A practical way to denote an orientation is via the Miller indices. Formally, the direction cosines in both the first column (the specimen X direction expressed in crystal coordinates) and the last column (the specimen Z direction expressed in crystal coordinates) of the orientation matrix are multiplied by a suitable factor to bring them to whole numbers, then divided by their lowest common denominator and conventionally written as $(hkl)[uvw]$ or $\{hkl\}\langle uvw\rangle$ if nonspecific indices are quoted, for example, $(112)[11\bar{1}]$ and $\{112\}\langle 111\rangle$. These are the Miller indices, and for the example in Equation 2.5, the ideal orientation is $(123)[63\bar{4}]$. Thus, rotation matrix g and Miller indices $(hkl)[uvw]$ are related through

$$g = \begin{pmatrix} u/N_1 & q/N_2 & h/N_3 \\ v/N_1 & r/N_2 & k/N_3 \\ w/N_1 & s/N_2 & l/N_3 \end{pmatrix} \tag{2.6}$$

The center column $[qrs]$ corresponds to the specimen Y direction expressed in crystal coordinates (with $[qrs] = (hkl) \times [uvw]$). The constants N_i in Equation 2.6 are required to normalize the three columns of the matrix to unity. It should be remembered that it is generally only for cubic materials that the indices for a plane and a direction are identical and hence it is less straightforward for non-cubic symmetries to extract the ideal orientation from the rotation matrix (Partridge, 1969).

In practice, the direction cosines from the orientation matrix are "idealized" to the nearest low-index Miller indices. For example, the following orientation is expressed in both direction cosines and the nearest family of Miller indices and is 2° and 6° away from (*hkl*) and [*uvw*], respectively:

$$(0.662 \quad 0.684 \quad 0.306)[0.095 \quad -0.482 \quad 0.871] \rightarrow (221)[0\bar{1}2] \tag{2.7a}$$

Of course the idealized two vectors (*hkl*) and [*uvw*] should still be mutually perpendicular, which can readily be checked from

$$\cos\varphi = \frac{\begin{pmatrix} h \\ k \\ l \end{pmatrix} \cdot \begin{pmatrix} u \\ v \\ w \end{pmatrix}}{\sqrt{h^2+k^2+l^2} \cdot \sqrt{u^2+v^2+w^2}} = \frac{\begin{pmatrix} 2 \\ 2 \\ 1 \end{pmatrix} \cdot \begin{pmatrix} 0 \\ \bar{1} \\ 2 \end{pmatrix}}{\sqrt{9} \cdot \sqrt{5}} \overset{!}{=} 0 \tag{2.7b}$$

This shorthand method of representation is very popular in metallurgy. Most usually single digit Miller indices are used to formulate the ideal orientation, although the components of the direction cosines could also simply be multiplied by 100 or 1000. An ideal orientation is also commonly formulated for hexagonal metals, for example, $(1\bar{2}10)[10\bar{1}0]$.

The crystallographically related solutions for an orientation feed through to the ideal orientation notation, as shown in the 24 equivalent solutions for $(123)[63\bar{4}]$ listed in Table 2.2. However, there are also orientations that, when denoted in the ideal orientation notation, can appear confusingly similar. For example, the orientation $(123)[\bar{6}\bar{3}4]$ is *not* a crystallographically related solution of $(123)[63\bar{4}]$, and it does not therefore appear in the list in Table 2.2. Another way to express this is that the matrix associated with $(123)[\bar{6}\bar{3}4]$ is

$$g = \begin{pmatrix} -0.768 & 0.582 & 0.267 \\ -0.384 & -0.753 & 0.535 \\ 0.512 & 0.308 & 0.802 \end{pmatrix} \tag{2.8}$$

which cannot be transformed into the matrices in Table 2.2 by column swapping, that is, multiplying by the symmetry matrices in Appendix II. For the most general case, there are four different orientations that have the same Miller indices families, $\{hkl\}\langle uvw\rangle$. These are

$$\begin{aligned} &(hkl)[uv\bar{w}] \\ &(hkl)[\bar{u}\bar{v}w] \\ &(khl)[vu\bar{w}] \\ &(khl)[\bar{v}\bar{u}w] \end{aligned} \tag{2.9}$$

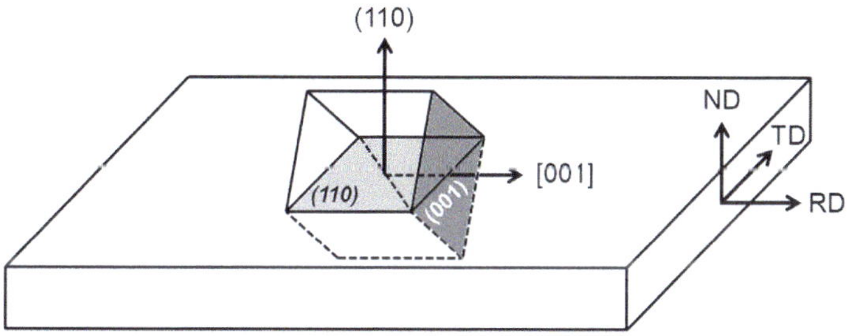

FIGURE 2.9 Schematic illustration of the relationship between the crystal and specimen axes for the (110) [001] Goss orientation, that is, the normal to (110) is parallel to the specimen ND, or Z axis and [001] is parallel to the specimen RD, or X axis.

Appendix III lists all 24 crystallographically related solutions for each of these, using (123)[63$\bar{4}$] as an example, illustrating that there is no duplication among the 24 × 4 solutions (Davies and Randle, 1996). These four different sets of solutions are related through symmetries of the specimen coordinate system (Sections 2.7, 5.4.2, and 10.4).

The main advantage of the ideal orientation notation is that it highlights important planes and directions that are parallel to principal directions in the specimen. These are often of great practical importance in texture analysis (Hatherly and Hutchinson, 1979). Furthermore, for cubic materials, it is usually possible to extract the ideal orientation "by eye" from the orientation matrix and it is easy to visualize the particular orientation. For example, Figure 2.9 shows a schematic diagram of the (110)[001] orientation that is called the Goss orientation and is very important for power transformer technology (Section 1.1). Here (110) lies in the plane of the sheet, that is, the direction normal to this plane is parallel to the specimen ND, or Z axis; [001] is parallel to the specimen RD or X axis. For hexagonal materials, the relationship between the ideal orientation expressed as Miller-Bravais indices and the orientation matrix is more complex (Section 2.3), for example

$$(\bar{5}7\bar{2}6)[2\bar{4}27] \Leftrightarrow (\bar{5}76)[0\bar{6}7] \Leftrightarrow \begin{pmatrix} 0.233 & 0.751 & -0.617 \\ -0.404 & 0.652 & 0.641 \\ 0.884 & 0.100 & 0.456 \end{pmatrix} \tag{2.10}$$

Here the first term is the orientation in Miller-Bravais indices, the second in Miller indices, and the third is the orientation matrix in Cartesian coordinates.

2.5 THE REFERENCE SPHERE, POLE FIGURE, AND INVERSE POLE FIGURE

The direction of any 3D vector in a crystal—a crystallographic direction or the normal to a crystal plane—can be described as a point on the *unit reference sphere*, that is, a sphere with radius 1 notionally residing around the crystal (Appendix IV). As an example, Figure 2.10 shows the (0001) plane in a hexagonal crystal. The point of intersection of the normal to this plane with the reference sphere, that is, its *pole*, is a measure for the arrangement of this plane in the crystal. Provided that the reference unit sphere is attached to an external frame, the position of the pole on the

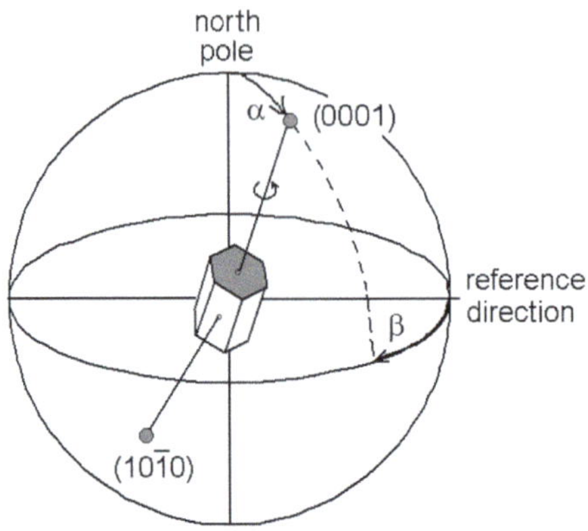

FIGURE 2.10 Orientation of the basal plane (0001) in a hexagonal crystal. The position of the (0001) pole on the unit sphere with regard to an external reference frame is described by the two angles α and β. However, since the crystal can still rotate about the (0001) pole, for an unequivocal definition of the orientation of the crystal more information, here the position of the (10$\bar{1}$0) pole is required.

sphere also provides information on the crystallographic orientation of the crystal with respect to this frame, although the crystal has still one degree of freedom by rotating around this particular axis.

Representation of the 3D orientation information on the unit sphere in a two-dimensional (2D) plane requires a projection of the sphere onto that plane. This problem is well known from the presentation of geographic features on the surface of the earth in a 2D map in earth science. In crystallography and metallurgy, most commonly the *stereographic projection* is used, the principle of which is shown in Figure 2.11 and explained in more detail in Appendix IV. In geology, the *equal area* (or *Lambert*) *projection*, which is also explained in Appendix IV, tends to be more popular. Derivatives of the stereographic or equal area projection (pole figures and inverse pole figures; see Sections 2.5.1 and 2.5.2) are discussed further with respect to macrotexture and microtexture in Sections 5.2 and 10.2, respectively.

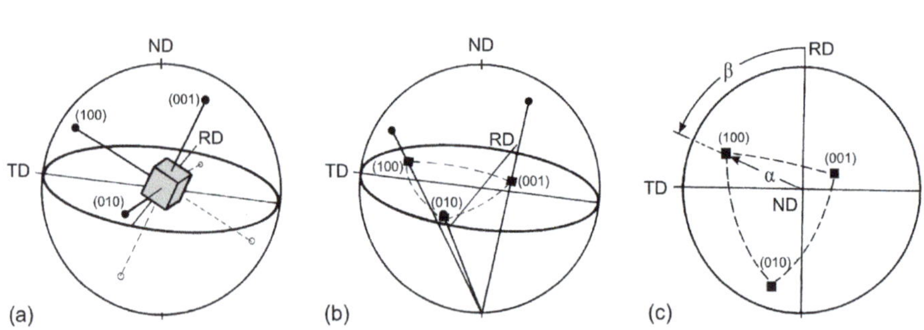

FIGURE 2.11 Presentation of the {100} poles of a cubic crystal in the stereographic projection. (a) Crystal in the unit sphere; (b) projection of the {100} poles onto the equator plane; (c) {100} pole figure and definition of the pole figure angles α and β for the (100) pole.

2.5.1 The Pole Figure

Poles are projected from the reference sphere onto a *pole figure* as follows. The position of a given pole on the sphere is commonly characterized in terms of two angles (Hansen et al., 1978): the angle α describes the azimuth of the pole, where $\alpha = 0°$ is the north pole of the unit sphere, and the angle β characterizes the rotation of the pole around the polar axis, starting from a specified reference direction (Figure 2.10). To characterize the crystallographic orientation of the crystal under investigation, the spatial arrangement of the corresponding poles in terms of the angles α and β has to be determined with respect to an external reference frame, that is, the sample or specimen coordinate system S (Section 2.3.1). For instance, for rolling symmetry, the sheet normal direction ND is typically chosen to be in the north pole of the sphere, so that $\alpha = 0°$ for ND, and the rotation angle β is 0° for the RD or, less frequently, the TD. For other deformation modes or specimen geometries an appropriate 3D, preferably right-handed, coordinate system has to be established (Section 2.3.1).

Thus, the reference system of the pole figure is the specimen coordinate system S given by the specimen axes $\{s_1 s_2 s_3\}$, and the crystal orientation given by the axes of the crystal coordinate system $C = \{c_1 c_2 c_3\}$ is projected into this frame. If R is a vector parallel to the pole of interest (XYZ), then it can be expressed in the two frames S and C according to

$$R = s_1 \sin\alpha\cos\beta + s_2 \sin\alpha\sin\beta + s_3 \cos\alpha \tag{2.11a}$$

and

$$R = \frac{1}{N}\left(c_1 X + c_2 Y + c_3 Z\right) \tag{2.11b}$$

(XYZ) are the coordinates of the pole in the crystal frame, for example, (111), and N is a constant with $N = \sqrt{X^2 + Y^2 + Z^2}$ to normalize R to unity. Scalar multiplication of Equations 2.11a and b in succession by the three vectors s_i under consideration of the definition of an orientation (Equation 2.4) finally yields

$$\begin{pmatrix} \sin\alpha\cos\beta \\ \sin\alpha\sin\beta \\ \cos\alpha \end{pmatrix} = \frac{1}{N}\begin{pmatrix} g_{11} & g_{21} & g_{31} \\ g_{12} & g_{22} & g_{32} \\ g_{13} & g_{23} & g_{33} \end{pmatrix} \cdot \begin{pmatrix} X \\ Y \\ Z \end{pmatrix} \tag{2.12}$$

(note that $g^{T} = g^{-1}$ appears here). Equation 2.12 yields nine equivalent expressions to derive the pole figure angles α and β for a given pole (XYZ) from the orientation matrix g (Hansen et al., 1978). The opposite way, that is, the indexing of the poles and determination of the orientation g from a given pole figure, is described in Appendix V.

So far, only individual poles have been considered. However, as already mentioned, one pole does not yield the entire orientation information, since the crystal can still rotate about this particular pole (Figure 2.10). For this reason, the distribution of the c axes in a hexagonal material, that is, its (0001) pole figure, does not provide an unambiguous representation of the texture of the sample; other poles have to be considered to represent unambiguously an orientation. In this example, the additional information about the position of, say, the $(10\bar{1}0)$ pole would determine the orientation in an unequivocal manner (Figure 2.10). Although in this example the orientation is fully characterized by two poles, in general—depending on the symmetry of the crystal or the poles—three poles are necessary to derive completely the orientation matrix g in Equation 2.12. Usually, the additional information is provided by other poles of the same family of planes (Appendix I), that is, different poles (hkl) of the family $\{hkl\}$. For instance, the hexagonal crystal has three equivalent {1120} planes, $(11\bar{2}0)$, $(1\bar{2}10)$, and $(\bar{2}110)$, so that each orientation in the corresponding {1120} pole figure is exactly characterized by the corresponding three poles. In pole figures like {1121} and {1101}, any orientation is defined by six equivalent poles. In cubic crystals, a given orientation is described by (not counting the same poles with negative signs) three {100}, four {111}, six {110}, twelve {012}, {112}, {113}, etc., and—in the most general case—twenty-four $\{hkl\}$ poles, which means that all pole figures of cubic crystals comprise enough poles to describe unambiguously an orientation.

2.5.2 The Inverse Pole Figure

Rather than representing the orientation of the crystal coordinate system in the specimen coordinate system, that is, in a pole figure, vice versa the orientation of the specimen coordinate system can be projected into the crystal coordinate system. This representation is called the *inverse pole figure.* Thus, the reference system of the inverse pole figure is the crystal coordinate system C, and the "orientation" is defined by the axes of the specimen coordinate system S, for example, RD, TD, and ND. In an analogy to Equation 2.11a, where the pole figure angles α and β of a unit vector parallel to the crystallographic axis (XYZ) have been considered in the frame S, now the angles γ and δ of a vector parallel to a specimen axis s_i in the coordinate system C must be introduced:

$$s_i = c_1 \sin\gamma_i \cos\delta_i + c_2 \sin\gamma_i \sin\delta_i + c_3 \cos\delta_i \tag{2.13}$$

Scalar multiplication now leads to (see Equation 2.12)

$$\begin{pmatrix} \sin\gamma\cos\delta \\ \sin\gamma\sin\delta \\ \cos\delta \end{pmatrix} = \begin{pmatrix} g_{11} & g_{12} & g_{13} \\ g_{21} & g_{22} & g_{23} \\ g_{31} & g_{32} & g_{33} \end{pmatrix} \cdot \begin{pmatrix} X_S \\ Y_S \\ Z_S \end{pmatrix} \tag{2.14}$$

These expressions describe the elements of the orientation matrix g in terms of the position of the specimen axes. Inverse pole figures are often used for axial symmetric specimens, where only one of the axes is prescribed. For instance, for compression or tensile specimens the orientation changes of the compression or tensile axis are plotted in the crystal coordinate system. According to the crystal symmetry it is not necessary to show the entire pole figure, but one unit triangle (Appendix IV) will suffice. In the case of cubic crystal symmetry, the well-known unit triangle ⟨100⟩, ⟨110⟩, and ⟨111⟩ is used.

2.6 THE EULER ANGLES AND EULER SPACE

Both the orientation matrix and the ideal orientation notation overdetermine the orientation, since only three variables are needed to specify an orientation. The most well-established method of expressing these three numbers is as the three *Euler angles*, which reside in a coordinate system known as *Euler space*.

2.6.1 The Euler Angles

The Euler angles refer to three rotations that, when performed in the correct sequence, transform the specimen coordinate system onto the crystal coordinate system—in other words, specify the orientation g (Euler, 1776). There are several different conventions for expressing the Euler angles. The most commonly used are those formulated by Bunge, as shown in Figure 2.12 (Bunge, 1965, 1982). The rotations are as follows:

1. φ_1 about the normal direction ND, transforming the TD into TD′ and the RD into RD′
2. Φ about the axis RD′ (in its new orientation)
3. φ_2 about ND″ (in its new orientation)

where φ_1, Φ, φ_2 are the Euler angles (Bunge definition). The effect of the operation sequence of these three rotations can be seen in Figure 2.12.

Analytically, the three rotations are expressed as

$$g_{\varphi_1} = \begin{pmatrix} \cos\varphi_1 & \sin\varphi_1 & 0 \\ -\sin\varphi_1 & \cos\varphi_1 & 0 \\ 0 & 0 & 1 \end{pmatrix} \tag{2.15a}$$

$$g_{\Phi} = \begin{pmatrix} 1 & 0 & 0 \\ 0 & \cos\Phi & \sin\Phi \\ 0 & -\sin\Phi & \cos\Phi \end{pmatrix} \tag{2.15b}$$

$$g_{\varphi_2} = \begin{pmatrix} \cos\varphi_2 & \sin\varphi_2 & 0 \\ -\sin\varphi_2 & \cos\varphi_2 & 0 \\ 0 & 0 & 1 \end{pmatrix} \tag{2.15c}$$

By multiplication of these three matrices in order, an expression is obtained, which links the rotation matrix g to the three Euler angles:

$$g = g_{\varphi_2} \cdot g_{\Phi} \cdot g_{\varphi_1} \tag{2.16a}$$

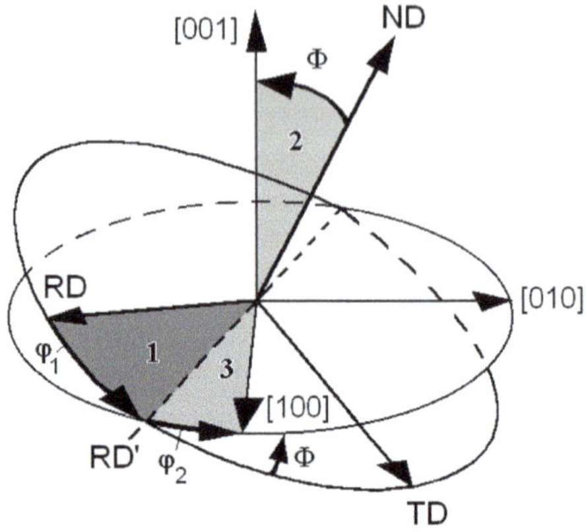

FIGURE 2.12 Diagram showing how rotation through the Euler angles φ_1, Φ, φ_2, in order 1, 2, 3 as shown describes the rotation between the specimen and crystal axes.

The elements of the matrix in terms of the Euler angles are, therefore, given by

$$\begin{aligned}
g_{11} &= \cos\varphi_1\cos\varphi_2 - \sin\varphi_1\sin\varphi_2\cos\Phi \\
g_{12} &= \sin\varphi_1\cos\varphi_2 + \cos\varphi_1\ \sin\varphi_2\cos\Phi \\
g_{13} &= \sin\varphi_2\sin\Phi \\
g_{21} &= -\cos\varphi_1\sin\varphi_2 - \sin\varphi_1\cos\varphi_2\cos\Phi \\
g_{22} &= -\sin\varphi_1\,\mathrm{sins}\,\varphi_2 + \cos\varphi_1\cos\varphi_2\cos\Phi \\
g_{23} &= \cos\varphi_2\sin\Phi \\
g_{31} &= \sin\varphi_1\sin\Phi \\
g_{32} &= -\cos\varphi_1\sin\Phi \\
g_{33} &= \cos\Phi
\end{aligned} \tag{2.16b}$$

The ideal orientation (123)[63$\bar{4}$], which is used as an example throughout this chapter, can equivalently be expressed as Euler angles φ_1 = 59.0°, Φ = 36.7°, φ_2 = 51.2°. Table 2.2 shows the effect of the symmetry-related solutions on these Euler angles.

From the definition of the Euler angles, it follows that they are periodic with period 2π. Thus, it holds:

$$g\{\varphi_1 + 2\pi, \Phi + 2\pi, \varphi_2 + 2\pi\} = g\{\varphi_1, \Phi, \varphi_2\} \tag{2.17}$$

Moreover, there is the identity

$$g\{\varphi_1 + \pi, 2\pi - \Phi, \varphi_2 + \pi\} = g\{\varphi_1, \Phi, \varphi_2\} \tag{2.18}$$

representing a reflection in the plane $\Phi = \pi$ with a simultaneous displacement through π in both φ_1 and φ_2. This means that there is a glide plane in the Euler angle space. Thus, in the most general case, the Euler angles are defined in the range $0° \le \varphi_1, \varphi_2 \le 360°$, and $0° \leq \Phi \leq 180°$. It is noted, however, that the sine and cosine functions in Equation 2.16b are only defined in the range $-90° \leq \varphi_1, \Phi, \varphi_2 \leq 90°$. Therefore, determination of the Euler angles from the rotation matrix (Equation 2.16b) is not straightforward, since the range of the angles has to be considered too. For example, let $\Phi = 35°$ and $g_{31} = 0.439$. In that case, $\varphi_1 = \arcsin(g_{31}/\sin\Phi)$ can be both 50° and 130°. The correct value has to be decided by taking the information in other matrix elements into account, for example, g_{32}.

Besides the convention of Euler angles according to Bunge, two other definitions are in use. Independently of Bunge, Roe (1965) proposed a set of Euler angles Ψ, Θ, Φ that differs from Bunge's definition by the second rotation being performed about the TD rather than the RD (note that the same definition is used by Matthies et al. [1987], but the angles are labeled α, β, γ). More recently, Kocks (1988) introduced a set of so-called symmetric Euler angles ψ Θ, φ. Although the rotations of Ψ and Θ are identical to those introduced by Roe, φ was defined as to be $\pi\text{-}\Phi_{\text{Roe}}$. In this set, the orientation matrix g becomes its transpose upon an interchange of Ψ and φ, which gives a symmetry between the "reference" and the "rotated" frame, which is not present in any of the other conventions.

The equivalence between these two sets and the Bunge angles is as follows:

Bunge–Roe:

$$\varphi_1 = \Psi + \pi/2 \quad \Phi = \Theta \quad \varphi_2 = \Phi - \pi/2 \tag{2.19a}$$

Bunge–Kocks:

$$\varphi_1 = \Psi + \pi/2 \quad \Phi = \Theta \quad \varphi_2 = \pi/2 - \phi \tag{2.19b}$$

A major disadvantage of the Euler angles is the fact that for $\Phi = 0°$ (and $\Phi = 180°$), φ_1 and φ_2 both describe an identical (or opposite) rotation about the normal direction, ND. Formally, for $\Phi = 0°$ Equation 2.15b becomes the identity matrix such that the rotation matrix g (Equation 2.16) simplifies to

$$g_{\Phi=0°} = \begin{pmatrix} \cos\varphi_1 \cdot \cos\varphi_2 - \sin\varphi_1 \cdot \sin\varphi_2 & \sin\varphi_1 \cdot \cos\varphi_2 + \cos\varphi_1 \cdot \sin\varphi_2 & 0 \\ -\cos\varphi_1 \cdot \sin\varphi_2 - \sin\varphi_1 \cdot \cos\varphi_2 & \cos\varphi_1 \cdot \cos\varphi_2 - \sin\varphi_1 \cdot \sin\varphi_2 & 0 \\ 0 & 0 & 1 \end{pmatrix} \tag{2.20a}$$

and, using the trigonometry formulae, finally to

$$g_{\Phi=0°} = \begin{pmatrix} \cos(\varphi_1 + \varphi_2) & \sin(\varphi_1 + \varphi_2) & 0 \\ -\sin(\varphi_1 + \varphi_2) & \cos(\varphi_1 + \varphi_2) & 0 \\ 0 & 0 & 1 \end{pmatrix} \tag{2.20b}$$

Equation 2.20b is a rotation matrix describing a rotation by the sum of $(\varphi_1 + \varphi_2)$ about the Z axis (ND), albeit independent of the particular values of φ_1 and φ_2. Thus, the rotation matrix loses one degree of freedom. In engineering applications, this problem is known as *gimbal lock* which endangered the flight of the lunar module of Apollo 10 (see Hanson, 2006). In computer programs this problem can be avoided by using quaternions instead of Euler angles (Section 2.9).

Despite this adverse effect, the Euler angles formed the backbone of texture research and practice for many years and have been particularly associated with x-ray analysis of textures. Details of their use in practice are given in Sections 5.4, 5.5, 10.3, and 10.6.1. Although the Euler angles are less intuitive than the ideal orientation nomenclature, they do present a fully quantitative description of the texture that is suitable for graphical representation.

2.6.2 The Euler Space

Any orientation expressed in terms of its Euler angles can be represented unequivocally as a point in a 3D coordinate system whose axes are spanned by the three Euler angles (Figure 2.13). The resulting space is referred to as *Euler space* (which is one example of an *orientation space* or *g-space*). Although the common Cartesian Euler space suffers from several disadvantages (see the following text), it is by far the most common orientation space in which to represent macrotexture data and hence we will encounter it throughout this book. In particular, the quantitative analysis of x-ray macrotextures is mainly based on the representation of textures in Euler space, and this issue will be discussed in detail in Sections 5.4 and 5.5. Accordingly, Euler space is also used to represent individual orientations and orientation distributions that are determined by techniques for single-grain orientation measurements (Section 10.3). For the representation of misorientations, other spaces are more advantageous; but nonetheless, misorientation distributions can be represented in Euler space, which warrants a short discussion (Section 10.6.1).

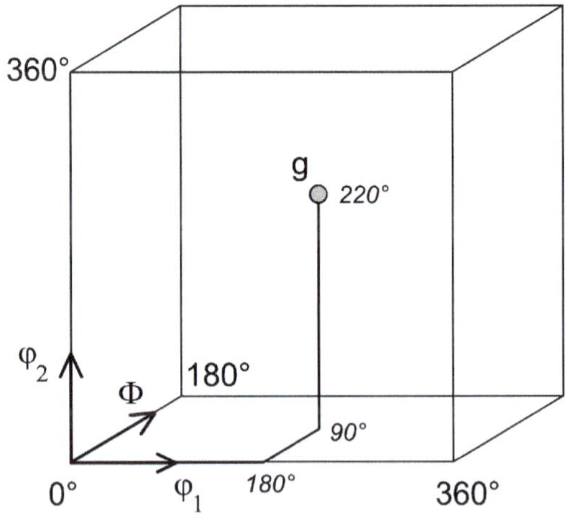

FIGURE 2.13 Representation of orientations in a three-dimensional orientation space defined by the Euler angles. Each orientation *g* corresponds to a point in the Euler space whose coordinates are given by the three Euler angles φ_x, Φ, φ_2 describing the orientation (Bunge's convention).

TABLE 2.3
Size of the Euler Space Necessary to Represent Unequivocally Orientations for Different Sample and Crystal Symmetries

Crystal System	Laue Class	Crystal Symmetry Φ	Crystal Symmetry φ_2	Sample Symmetry: Orthonormal φ_1	Sample Symmetry: Monoclinic φ_1	Sample Symmetry: None (Triclinic) φ_1
Triclinic	$\bar{1}$	180°	360°			
Monoclinic	2/m	180°	180°			
Orthorhombic	mmm	90°	180°			
Trigonal	$\bar{3}$	180°	120°			
	$\bar{3}$m	90°	120°	90°	180°	360°
Tetragonal	4/m	180°	90°			
	4/mmm	90°	90°			
Hexagonal	6/m	180°	60°			
	6/mmm	90°	60°			
Cubic	m$\bar{3}$	90°[a]	180°			
	m$\bar{3}$m	90°[a]	90°			

Note:
[a] Threefold symmetry due to 120°(111) symmetry.

As shown in the preceding section, in the most general case of triclinic crystal symmetry and no sample symmetry, the Euler angles are defined in the range $0° \leq \varphi_1, \varphi_2 \leq 360°$, and $0° \leq \Phi \leq 180°$, which in turn defines the maximum size of the Euler angle space, the so-called asymmetric unit. However, as already discussed (Sections 2.3 and 2.4), symmetries of the crystal or the specimen frame result in a number of different, but equivalent, descriptions of a given orientation. Such symmetries lead to a reduction in the size of the Euler space that is required to represent unequivocally all possible crystallographic orientations and, consequently, lead also to a number of symmetrically equivalent subspaces (Table 2.3). The necessary transformation laws for the various symmetry operations can be found in Hansen et al. (1978), Pospiech (1982), Bunge (1982), and Kocks (1998b).

Different *sample symmetries* affect the range of the angle φ_1. In the most general case, which is no sample symmetry (sometimes called triclinic symmetry), the full range $0° \leq \varphi_1 \leq 360°$ is required to represent all possible orientations, whereas any symmetry in the specimen coordinate system reduces the size of the Euler space. In specimens deformed by rolling—or more generally under a plane strain deformation state—it is usually assumed that two mirror planes exist, one in the RD and the other in the TD. In this case of *orthonormal sample symmetry*, the range of the Euler angle φ_1 is reduced to one quarter, that is, to $0° \leq \varphi_1 \leq 90°$. When only one symmetry plane is present, the Euler space is halved to $0° \leq \varphi_1 \leq 180°$. This *monoclinic sample symmetry* is given for specimens deformed by shear, for example, in torsion experiments, which comprise a twofold symmetry axis (diad) in the shear plane perpendicular to the cylinder axis. This symmetry also applies for the surface layers of a rolled sheet.

In contrast to the plane strain state at the sheet center, in the surface layers the deformation is no longer symmetrical with regard to the TD, but only with regard to the RD (Section 5.4.2).

Crystal symmetry further reduces the size of the Euler space by affecting the range of the angles Φ and φ_2. In general, an *n*-fold symmetry axis reduces the Euler space by a factor of *n*. This means that the Euler angles (φ_1, Φ, φ_2) and (φ_1, Φ, $\varphi_2 \pm 360°/n$) are equivalent. Thus, the threefold axis in trigonal crystal symmetry reduces the range by a factor of 3, which means that a range of $0° < \varphi_2 < 120°$ is sufficient to represent all possible orientations. Accordingly, for tetragonal (fourfold axis) and hexagonal crystals (sixfold axis), φ_2 can be reduced to 90° and 60°, respectively. An additional twofold symmetry or mirror planes affects Φ by reducing the range from 180° to 90°, which is given in all but triclinic crystal symmetry (Table 2.3).

For cubic crystal symmetry it has already been discussed in Section 2.3.3 that there are 24 possibilities to describe one and the same orientation of a cube in any reference frame, so that the Euler space can be subdivided into 24 equivalent subspaces. In accordance with these considerations, the three fourfold axes (i.e., 90° ⟨100⟩ rotations) and the six twofold axes (i.e., 180° ⟨110⟩ rotations) reduce the necessary Euler angle space to the range $0° \leq \varphi_1 \leq 360°$, $0° \leq \Phi, \varphi_2 \leq 90°$, which means that there are only eight equivalent subspaces (i.e., equivalent to tetragonal symmetry). However, this reduction does not consider the four threefold axes, that is, the 120° ⟨111⟩ rotations. Since a further reduction in the threefold symmetry would lead to a complex-shaped subspace, this operation is not usually carried out and, therefore, for cubic crystals each orientation appears three times in the reduced Euler angle space. Figure 2.14 shows the symmetry elements in the Euler space for cubic crystal symmetry and orthonormal sample symmetry (Hansen et al., 1978).

If we now consider *misorientations* between two crystals, as opposed to the orientation of a crystal with respect to an external reference frame, the Euler space is further reduced by applying the crystal symmetry of the second crystal rather than

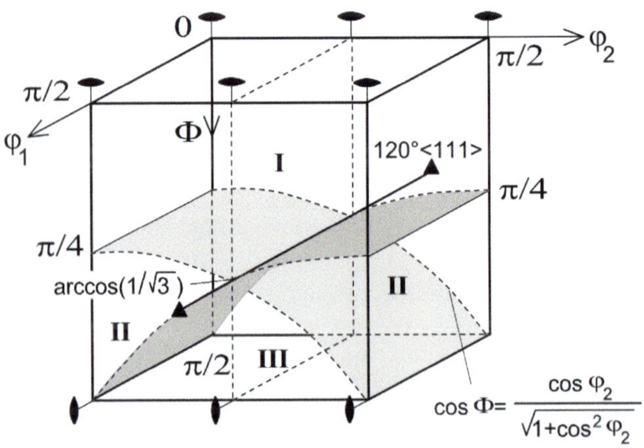

FIGURE 2.14 Symmetry elements in the Euler space for cubic crystal symmetry and orthonormal sample symmetry.

sample symmetry (Section 2.7.2). Furthermore, the Euler space is reduced by an additional factor of 2 due to the equivalence of the misorientation and the inverse misorientation, because with respect to the misorientation the two crystallites are indistinguishable. Thus, for cubic crystal symmetry, one has 24 × 24 × 2 equally valid descriptions of a given misorientation, which contain 24 generally different rotation angles θ. Section 10.6.1 discusses one of the resulting 1,152 equivalent subspaces of the Euler space.

2.7 THE ANGLE/AXIS OF ROTATION AND CYLINDRICAL ANGLE/AXIS SPACE

In Section 2.3 we showed how an orientation can be described by the concept of three rotations that transform the coordinate system of the crystal onto the reference, that is, a specimen, coordinate system. The same final transformation can be achieved if the crystal coordinate system is rotated through a *single* angle, provided that the rotation is performed about a *specific* axis. This angle and axis are known as the *angle of rotation* and *axis of rotation*, or more briefly the *angle/axis pair.*

2.7.1 Angle/Axis of Rotation

Figure 2.15a illustrates the angle/axis of rotation between two cubic coordinate systems. The left-hand cube (axes $X_1Y_1Z_1$) is chosen as the fixed, reference position and the right-hand cube (axes $X_2Y_2Z_2$) would need to be rotated through 50° about an axis that is vertical in the diagram and parallel to the common [001] to achieve the reference orientation. It is important to notice that the rotation axis [001] is *the same direction in both cubes.* Similarly, Figure 2.15b shows angle/axis of rotation between two hexagons.

If we now consider the orientation of crystal lattices, $X_1Y_1Z_1$ is replaced by the specimen coordinate system as a reference, for example, RD, TD, ND, and $X_2Y_2Z_2$ are replaced by the crystal axes 100,010,001. The three independent variables that embody an orientation comprise one variable for the rotation angle θ and two variables for the rotation axis *r*, since *r* is normalized such that $r_1^2 + r_2^2 + r_2^2 = 1$, and so the third index is redundant. The transformation from the angle/axis pair θ/*r* to the orientation matrix *g* is given by

$$
\begin{aligned}
g_{11} &= \left(1 - r_1^2\right) \cdot \cos\theta + r_1^2 \\
g_{12} &= r_1 r_2 \cdot (1 - \cos\theta) + r_3 \cdot \sin\theta \\
g_{13} &= r_1 r_3 \cdot (1 - \cos\theta) - r_2 \cdot \sin\theta \\
g_{21} &= r_1 r_2 \cdot (1 - \cos\theta) - r_3 \cdot \sin\theta \\
g_{22} &= \left(1 - r_2^2\right) \cdot \cos\theta + r_2^2 \\
g_{23} &= r_2 r_3 \cdot (1 - \cos\theta) + r_1 \cdot \sin\theta \\
g_{31} &= r_1 r_3 \cdot (1 - \cos\theta) + r_2 \cdot \sin\theta \\
g_{32} &= r_2 r_3 \cdot (1 - \cos\theta) - r_1 \cdot \sin\theta \\
g_{33} &= \left(1 - r_3^2\right) \cdot \cos\theta + r_3^2
\end{aligned}
\tag{2.21}
$$

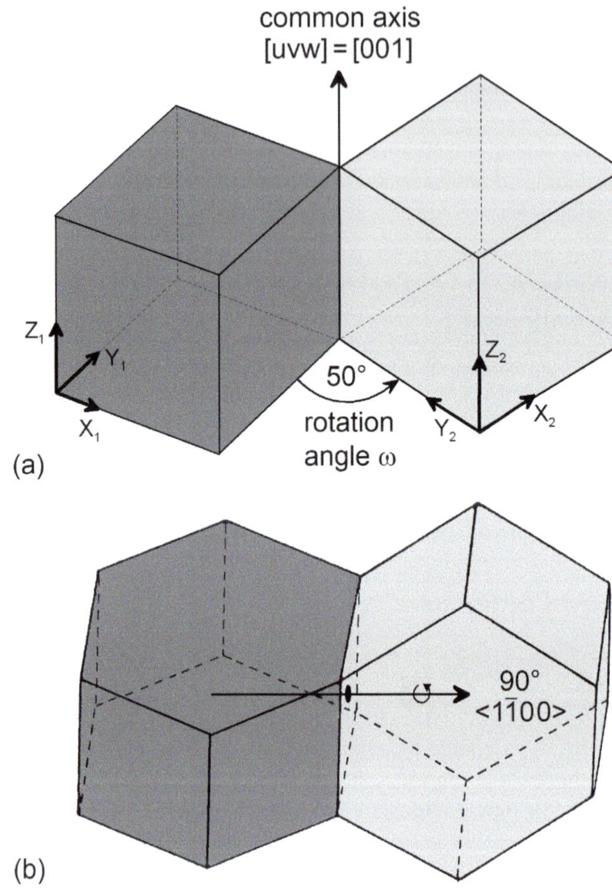

FIGURE 2.15 Diagram showing the angle/axis of rotation between (a) two cubes and (b) two hexagons.

Vice versa, the angle/axis of rotation is extracted from the orientation matrix as follows:

$$
\begin{aligned}
\cos\theta &= \frac{(g_{11} + g_{22} + g_{33} - 1)}{2} \\
r_1 &= \frac{(g_{23} - g_{32})}{(2\sin\theta)} \\
r_2 &= \frac{(g_{31} - g_{13})}{(2\sin\theta)} \\
r_3 &= \frac{(g_{12} - g_{21})}{(2\sin\theta)}
\end{aligned}
\tag{2.22a}
$$

If $\theta = 0°$ or $180°$, $\sin\theta$ becomes 0. In that case, r can be derived from

$$r_1 = \left(\frac{(g_{11}+1)}{2}\right)^{1/2}$$
$$r_2 = \left(\frac{(g_{22}+1)}{2}\right)^{1/2} \quad (2.22b)$$
$$r_3 = \left(\frac{(g_{33}+1)}{2}\right)^{1/2}$$

For example, the orientation matrix given in Equation 2.5 yields an angle/axis pair of 48.6° (0.562 –0.520 –0.644). The reference orientation, chosen to coincide with the specimen axes, is described by the identity matrix. It is not usually necessary to ascribe a handedness to the rotation, since conventionally the rotation is described as a right-handed screw operation and hence θ is always a positive angle (Mykura, 1980). A negative angle is equivalent to changing the sign of r. An example of where it is necessary to know the sense of a rotation is when measuring the orientation of adjacent subgrains (Sections 9.4.5 and 10.5.2). If the rotations are all in the same sense, then the total rotation is cumulative; otherwise, adjacent rotations of the same magnitude cancel each other out (Hansen and Juul Jensen, 1994; Randle et al., 1996).

Section 2.3.3 describes the crystallographically related solutions of a single orientation. It follows then that these solutions will give rise to the same number of axis/angle pairs, each of which is calculated from the crystallographically related solution of the matrix (Appendix II). Table 2.2 includes the 24 angle/axis pairs that accompany the ideal orientation (123)[$63\bar{4}$]. Inspection of Table 2.2 shows that, even though each of these angle/axis pairs describes exactly the same orientation, this could not have been known from the format of the angle/axis pairs themselves—numerically they appear to be quite unrelated. The relationship only becomes clear from concurrent inspection of the matrix or ideal orientation.

2.7.2 Angle/Axis Description of Misorientation

There is a second choice for designation of a reference orientation. Instead of choosing the specimen axes for the reference orientation, it could be the axes (orientation) of another grain in the specimen, usually the neighboring grain. The orientation of grain 2—a grain having crystal axes parallel to $X_2Y_2Z_2$—is then expressed relative to the orientation of grain 1, which is similarly a grain having crystal axes parallel to $X_1Y_1Z_1$. For the examples in Figure 2.15a and b, the angle/axis of misorientation is 50° ⟨001⟩ and 90° ⟨$1\bar{1}00$⟩, respectively, where the latter axis is given in Miller-Bravais indices. The purpose of such a descriptor of orientation is to characterize the orientation relationship (between different phases) or the misorientation (between two grains of the same phase) (Lange, 1967; Santoro and Mighell, 1973). For the latter case, specification of the misorientation gives access to the grain boundary crystallography, which is a very important subset of microtexture (Randle, 1993). Chapter 11 describes extraction and analysis of grain boundary parameters from the misorientation.

A misorientation is calculated from the orientations of grains 1 and 2 by

$$M_{12} = g_1^{-1} g_2 \tag{2.23}$$

where M_{12} is the matrix that embodies the misorientation between g_2 and g_1, where g_1 is arbitrarily chosen to be the reference orientation. The angle/axis of rotation (usually called the *angle/axis of misorientation* to distinguish it as a grain or phase boundary parameter) is calculated by using Equation 2.22 and substituting the elements of the misorientation matrix M_{12} for those of the orientation matrix g.

In summary, the essential difference between an angle/axis pair description of the orientation of a single grain and an angle/axis pair description of the misorientation between the crystal lattices of two grains is simply the choice of reference axes. This is illustrated graphically in the stereogram shown in Figure 2.16 (the stereographic projection is explained in Appendix IV). In Figure 2.16a, the {100} crystal axes of two grains 1 and 2 are represented relative to the specimen axes RD, TD, and ND; hence, the reference axes are the specimen axes. The location of r, common to both orientations, is plotted on the stereogram along with the angle θ through which $[100]_2$, $[010]_2$, $[001]_2$ must be rotated to make them parallel with the crystal axes $[100]_1$, $[010]_1$, $[001]_1$ of grain 1. The misorientation axis r can be constructed from the intersection of two great circles that bisect pairs of the same crystal axes in both grains, whereas the angle between the poles of two of these great circles gives the misorientation angle θ.

In Figure 2.16b, the reference axes are the crystal axes of grain 1. Accordingly, the crystal axes of grain 2 are expressed in the frame of grain 1, that is, $[221]_2$, $[\bar{1}22]_2$,

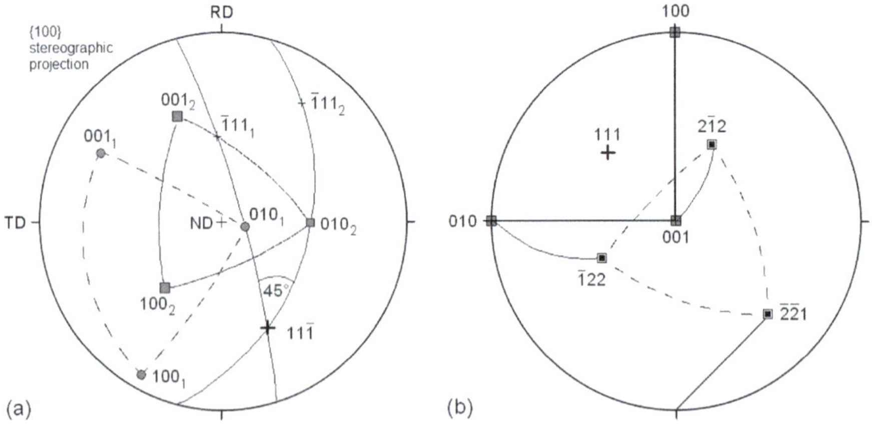

FIGURE 2.16 Stereographic representation of (a) an orientation and (b) a misorientation. In (a), the crystal axes of two grains 1 and 2 are represented relative to the specimen axes RD, TD, ND. The misorientation axis is given by the intersection of great circles that bisect pairs of the same crystal axes in both grains. The angle between the poles of two of these great circles gives the misorientation angle (Appendix IV). In (b) the misorientation axis (cross) and angle (arcs of small circles) are expressed relative to the crystal axes of one grain. This particular example is a twin. (Adapted from Randle and Ralph, 1988.)

$[2\bar{1}2]_2$. The misorientation axis (cross) and angle (arcs of small circles) between grains 2 and 1 are also expressed relative to the crystal axes of grain 1. In summary, Figure 2.16a depicts the orientation, sometimes called the *absolute orientation*, of grains 1 and 2 and Figure 2.16b depicts the misorientation between grains 1 and 2, both expressed as an angle/axis pair. Figure 2.6 includes the most common parameters that are derived from the misorientation matrix.

As it stands, the angle/axis pair is not a very useful representation of an orientation since it does not convey easily understood physical information about the relationship between the orientation and important axes in the specimen. However, the angle/axis pair is an extremely effective means for representation of a misorientation, since the description actually relates physically to the grain or phase boundary misorientation angle and axis, which in turn is used for categorization and subsequent analysis (Section 11.3). Frequently, the angle and axis of misorientation are represented separately, which is discussed in Section 10.5.1. Finally, it should be noted that the value of the angle/axis description of orientation is as a starting point for the Rodrigues vector, or *R*-vector formulation, which leads to a superior way of representing orientations and misorientations (Section 2.8).

2.7.3 The Cylindrical Angle/Axis Space

Orientations and misorientations can also be represented in a 3D space formed by the parameters of the angle/axis pair description. As the angle/axis representation is easy to visualize and convenient for analysis and interpretation of orientation correlations, the angle/axis space will mostly be used for the representation of misorientations, but the basic principles discussed in the following text also hold for the presentation of orientations.

As described in Section 2.7.1, both orientations and misorientations can be expressed in terms of rotational parameters, that is, the rotation axis r and the rotation angle θ. To represent the orientations or misorientations in a suitable angle/axis space, the rotation axis $r = (r_x, r_y, r_z)$ is expressed in terms of two angles ϑ and ψ:

$$r_x = \cos\psi \sin\vartheta, \quad r_y = \sin\psi \sin\vartheta, \quad r_z = \cos\vartheta \tag{2.24}$$

Thus, the angle/axis rotational parameters are now given by the three angles θ, ϑ, and ψ. The necessary analytical descriptions connected with this way of representing orientations and misorientations are given by Pospiech (1972) and Pospiech et al. (1986). In calculations, the relations between Euler angles (φ_1, Φ, φ_2) and the rotational parameters (θ, ϑ, ψ) might be useful:

$$\sin\frac{\theta}{2}\cdot\sin\vartheta = \sin\frac{\Phi}{2}, \quad \cos\frac{\theta}{2} = \cos\frac{\Phi}{2}\cdot\cos\frac{\varphi_1+\varphi_2}{2}, \quad \psi = \frac{1}{2}(\varphi_1 - \varphi_2) \tag{2.25}$$

All possible orientations can be presented in the interval ($0° \le \theta \le 360°$, $0° \le \vartheta \le 90°$, $0° \le \psi \le 360°$) into a space with the simple shape of a cylinder, that is, *cylindrical angle/axis space* (Figure 2.17). When the cylinder axis is chosen parallel to the

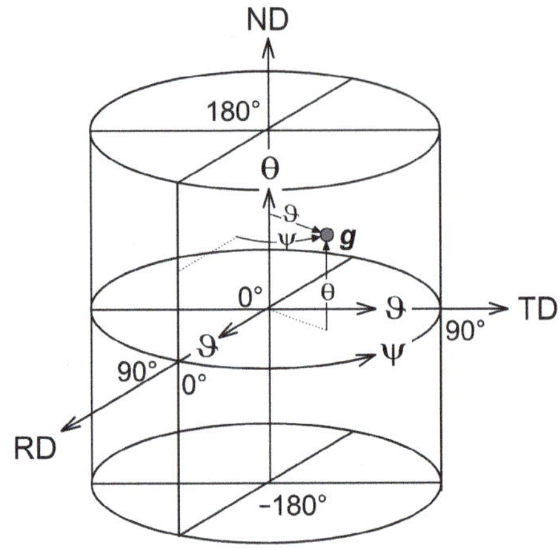

FIGURE 2.17 Representation of an orientation g given by the angle/axis description parameters r and θ in a cylindrical orientation space defined by the rotational parameters ϑ, ψ, and θ.

θ axis, the base plane of the cylinder corresponds to a stereographic projection of the rotation axis r. Thus for each misorientation angle θ, the corresponding rotation axis r can readily be derived from its position in the stereographic triangle, which is a very familiar way to evaluate misorientation data (Section 10.5.1). Examples of *misorientation distribution functions* (MODFs) in the cylindrical angle/axis space are shown in Sections 10.6.2 and 12.2.

2.8 THE RODRIGUES VECTOR AND RODRIGUES SPACE

In the previous section, we have seen that the angle/axis formulation is not very useful for representing macrotexture data since it does not convey easily understood physical information about the relationship between the crystal orientations and specimen axes. However, the angle/axis pair provides useful microtexture information, in that the description actually relates physically to the misorientation angles and axes, which in turn is used for categorization and subsequent analysis of grain or phase boundaries (Section 11.3). The values of the angle/axis description of orientation are also a starting point for the Rodrigues vector, or R-vector formulation (Rodrigues, 1840), which is used with great effect to convey and represent both orientations and, most notably, misorientations. In summary, the formulation of the angle/axis of misorientation is an end in itself and also used further to calculate R vectors, whereas the angle/axis of rotation for orientations is not a particularly effective parameter except to illustrate the principles of the R vector.

2.8.1 The Rodrigues Vector

An orientation can be expressed as a 3D vector, since such a vector requires three variables to specify its position in space and similarly an orientation requires three variables for its specification. Such an "orientation vector" is most easily described and visualized from the angle/axis descriptor: the rotation axis gives the *direction* of the vector and the rotation angle its *magnitude*. The *Rodrigues vector R* combines the angle and axis of rotation into one mathematical entity; thus,

$$R = \tan\left(\frac{\theta}{2}\right) \cdot r \tag{2.26}$$

For example, the first angle/axis pair entry in Table 2.2 gives a Rodrigues vector with elements $R_1R_2R_3$ given by

$$\begin{aligned} R_1 &= \tan\left(\frac{\theta}{2}\right) \cdot r_1 = 0.254 \\ R_2 &= \tan\left(\frac{\theta}{2}\right) \cdot r_2 = -0.235 \\ R_3 &= \tan\left(\frac{\theta}{2}\right) \cdot r_3 = -0.291 \end{aligned} \tag{2.27}$$

The attraction of this parameterization is that the three components of the *R* vector define a vector that lies in a Cartesian coordinate system whose axes can be chosen to correspond to either the specimen or the crystal axes. Hence, a population of *R* vectors, each of which represents an orientation or misorientation, can be represented in a 3D space known as *Rodrigues–Frank* (RF) *space* or *Rodrigues space*. Since the *R* vector is derived directly from the angle/axis pair, it can be described by a number of different crystallographically related solutions, with the precise number of solutions depending on the crystal system and symmetry (Section 2.3.3). The *R* vector is a function of θ, and so it follows that the smallest value of the vector is calculated from the smallest θ. Such *R* vectors will lie closest to the origin of RF space, and therefore the crystallographically related solution of an orientation or misorientation having the smallest angle (e.g., the first row in Table 2.2) is the most convenient one to select for representation in Rodrigues space.

Use of Rodrigues space is a relatively new field compared with the long-established Euler space approach for representation of rotations (Frank, 1988). The explanation for the late introduction of Rodrigues space is that its application is intimately connected with microtexture, that is, discrete orientation measurements, whereas representation of texture in Euler space relates to the representation of continuous orientation distributions as calculated from pole figures (Sections 2.6 and 5.3). Microtexture data can be, and frequently are, represented in either Euler space or Rodrigues space (Chapter 10), whereas macrotexture data tend to be represented only in Euler space because commercial systems are configured for this approach (Chapter 5).

The geometrical features of Rodrigues space offer considerable benefits for microtexture analysis. Here we will describe the essential features of Rodrigues space, and the methodology of its use in Sections 10.4 and 10.6.3. Mathematical formalisms are kept to a minimum since these are not essential to the understanding and use of Rodrigues space, and moreover details can be found elsewhere (Becker and Panchanadeeswaran, 1989; Neumann, 1991; Randle and Day, 1993; Field, 1995).

2.8.2 The Fundamental Zone

The entirety of Rodrigues space encompasses each one of the crystallographically related solutions of a rotation (Section 2.3.3). Since it is simpler for each orientation from a sample population to be represented only once, it is convenient to formulate the R vector from only one crystallographically related solution. That solution with the lowest rotation angle is chosen because it feeds through to give the smallest R vector. By use of the lowest-angle description, all orientations—represented by the endpoints of their R vectors—can be made to reside in a polyhedron known as the *fundamental zone of Rodrigues space*, which contains the origin of the space. Hence any orientations that lie outside the fundamental zone can be reexpressed by equivalent points lying within it by rechoosing the crystallographically related solution to be the lowest angle one. Exceptionally, an orientation may lie on the surface of the fundamental zone and will be matched by an equivalent point on the opposite face.

The shape of the fundamental zone is governed by the crystal symmetry of the material, and the geometry of these polygons has been derived for each crystal system (Morawiec and Field, 1996) and for combinations of lattices (Heinz and Neumann, 1991). For cubic symmetry, the shape of the zone is a cube with truncated corners such that the bounding surfaces are six octahedra and eight equilateral triangles as shown in Figure 2.18. The axes XYZ of the fundamental zone are also marked. For the representation of orientations, the XYZ axes are aligned with the specimen axes. Alternatively, XYZ can be aligned with the crystal axes, which is convenient for the representation of misorientations.

The distance of the point where the reference axes intercept the zone surface is represented, in terms of crystal coordinates, by the angle/axis pair 45° ⟨100⟩ that gives $R = (0.4142,0,0)$. The R vectors of the centers of the equilateral triangles that form the truncated corners of the zone are given by the angle/axis pair 60° ⟨111⟩ and

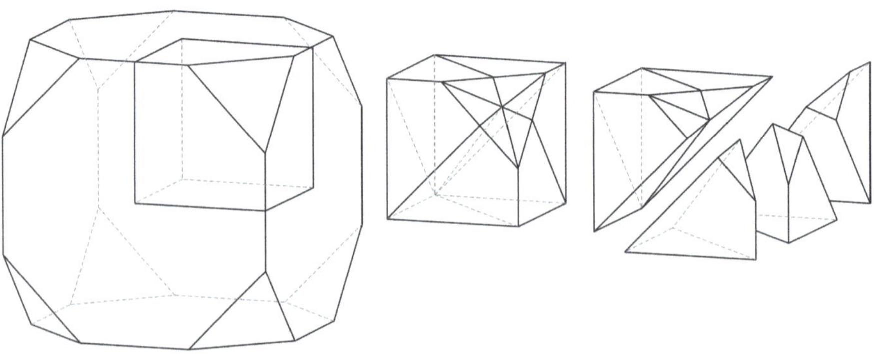

FIGURE 2.18 The fundamental zone of Rodrigues space for cubic symmetry, showing also the decomposition of the space into 48 subvolumes. (Adapted from Day, 1993.)

TABLE 2.4
Parameters of Rodrigues Space According to Crystal System

Crystal System	Distance of the Main Face from the Origin	Description	Subvolume Fraction
Cubic	tan π/8	Six octagonal faces normal to the fourfold axes; eight triangular faces normal to the threefold axes	1/48
Hexagonal	tan π/4	Two dodecagonal faces normal to the sixfold axis; 12 square faces normal to the diad axes	1/12
Trigonal	tan π/4	Two hexagonal faces normal to the threefold axis; six square faces normal to the diad axes	1/6
Tetragonal	tan π/6	Eight triangular faces perpendicular to the threefold axes (i.e., a regular octahedron)	1/24
Orthorhombic	tan π/4	Six faces normal to the diads (i.e., a square)	1/8
Icosahedral	tan π/10	Twelve pentagonal faces normal to the fivefold symmetry axes (i.e., a regular dodecahedron)	1/120

Source: Data from Morawiec and Field (1996).

a corresponding *R* vector (0.333,0.333,0.333). Another point that defines the shape of the fundamental zone for cubic crystals is an apex of the equilateral triangle, given by $62.8°/\langle 11(\sqrt{2}-1)\rangle$, which represents the greatest possible misorientation between two cubes, and yields an *R* vector of (0.4124,0.4124,0.1716). The shapes of fundamental zones for all crystal systems are described in Table 2.4. It can be seen that the shape of the zone is equivalent to the crystal symmetry. For monoclinic and triclinic symmetries, the fundamental zone corresponds to the entire Rodrigues space, and for this case the space has no boundaries, which is a disadvantage to the representation of orientations having this symmetry. However, most materials of interest from a texture point of view fall into other symmetry categories and can be represented readily in Rodrigues space, for example, trigonal Al_2O_3 (Morawiec and Field, 1996).

2.8.3 Properties of Rodrigues Space

The most powerful aspect of Rodrigues parameterization for the representation of orientations and misorientations is that *only rectilinear*, that is, *straight-line geometry* is involved, which renders the space easy to handle and visualize. The rectilinear geometry of Rodrigues space is manifest by the following:

- The axis of rotation gives the direction of the *R* vector. Hence rotations about the same axis of rotation lie on a straight line that passes through the origin.
- The angle of rotation gives the length of the *R* vector. Hence small-angle boundaries cluster close to the origin. These first two points are especially relevant to the analysis of misorientations (Section 10.6.3).

- Orientations that include a common direction, for example, a fiber texture, lie on a straight line that in general does not pass through the origin.
- The edges of zones in Rodrigues space are straight lines, and the faces of zones are planar.

In summary, every straight line in Rodrigues space represents rotations about a certain, fixed axis. These are called *geodesic lines* and remain straight even if the origin of the space is shifted (Frank, 1988; Neumann, 1991a,b). Another advantageous aspect of Rodrigues space is that it is nearly *homo-choric*, that is, a distribution of untextured orientations will be distributed almost uniformly throughout the space. The invariant orientation volume is actually within a factor of 2 (Neumann, 1991a,b), which is generally considered low enough for orientation distributions in the space to be evaluated in a straightforward manner.

The fundamental zone can be further divided into identical subvolumes, as shown in Figure 2.18 (Randle, 1990). There are 48 of these for the cubic system, and the subvolume fraction for other crystal systems is shown in Table 2.4. Subvolumes are used for the display of misorientation data.

2.9 QUATERNIONS

Among the various methods of tackling crystallographic orientations and rotations, the utilization of quaternion algebra, especially the use of unit quaternions, provides a numerically very efficient and convenient way (e.g. Altmann, 1986; Morawiec, 2004; Hanson, 2006). Based on earlier work by Gauss and Euler, quaternions were introduced by Rodrigues (1840) for the description of three-dimensional rotations and, independently, by Hamilton (1844) as an extension of complex numbers. Hamilton spent a large part of his scientific life on these numbers, which is why they are also called Hamilton numbers.

Quaternions are a non-commutative extension of the complex numbers, consisting of a real component and a vectorial component. In quaternion algebra, a rotation is described by a normalized four-component quaternion $Q = (q_0; q_1, q_2, q_3)$ which consists of a scalar q_0 ($q_0 \geq 0$) and a vector $\boldsymbol{q} = (q_1, q_2, q_3)$. For rotations, the quaternions are always of unit length, i.e. $|Q| = \left(q_0^2 + q_1^2 + q_2^2 + q_3^2\right)^{0.5} = 1$. These unit quaternions define the surface of a hypersphere in four-dimensional space (like unit vectors in three-dimensional space describe the surface of a sphere). Thus, at variance to the Euler angles, quaternion does not use axes but rotations are described by a set of four values that define a vector on a sphere and a single normalized angle around that vector. The multiplication of four-element quaternions corresponding to crystallographic rotations requires considerably smaller number of operations and the relations of quaternion elements with the rotation parameters are in general simpler and, therewith, faster than they are in the case of orthogonal matrices. However, quaternions have the disadvantage that with their four parameters a direct graphical representation of rotations or orientations in quaternion space is impossible. Accordingly, in many applications quaternions are used for internal computations before the final results are transferred into Euler space or Rodrigues space for visualization.

There is a direct relationship between the quaternion parameters $Q = (q_0; q_1, q_2, q_3)$ and the parameters of the angle/axis description, i.e. rotation angle ω and axis $\boldsymbol{n} = (n_1, n_2, n_3)$:

$$Q = \left\{ \cos\left(\frac{\omega}{2}\right); \quad n_1 \cdot \sin\left(\frac{\omega}{2}\right), \quad n_2 \cdot \sin\left(\frac{\omega}{2}\right), \quad n_3 \cdot \sin\left(\frac{\omega}{2}\right) \right\} \tag{2.28}$$

It is seen that, similarly as for the Rodrigues vector, unit quaternions are a unit vector representing the rotation axis relative to a reference frame that is common to all the orientations, where ω is the rotation angle. Accordingly, the relationship between quaternion parameters $(q_0; q_1, q_2, q_3)$ and the Rodrigues vector $R(\omega, r)$ are straightforward (Morawiec and Field, 1996):

$$R(\omega, r) = \tan\frac{\omega}{2} \cdot \frac{r}{|r|} = \frac{1}{q_0}(q_1, q_2, q_3) \tag{2.29}$$

The connection of quaternions to orthogonal orientation matrices and, especially, the Euler angles (φ_1, Φ, φ_2) was reviewed by Morawiec and Pospiech (1989):

$$Q = \left\{ \cos\left(\frac{\Phi}{2}\right) \cdot \cos\left(\frac{\varphi_1 + \varphi_2}{2}\right); \quad \sin\left(\frac{\Phi}{2}\right) \cdot \cos\left(\frac{\varphi_1 - \varphi_2}{2}\right), \quad \sin\left(\frac{\Phi}{2}\right) \cdot \sin\left(\frac{\varphi_1 - \varphi_2}{2}\right), \right.$$
$$\left. \cos\left(\frac{\Phi}{2}\right) \cdot \sin\left(\frac{\varphi_1 + \varphi_2}{2}\right) \right\} \tag{2.30}$$

As an example, our reference orientation $(123)[63\bar{4}]$ from Table 2.2 would read $Q = (0.924, 0.242, -0.202, -0.217)$. The reverse relations are obtained through the following transformations of Equation 2.30:

$$\begin{aligned} \tan\varphi_1 &= (q_1 q_3 + q_0 q_2)/(q_0 q_1 - q_2 q_3) \\ \cos\Phi &= q_0^2 - q_1^2 - q_2^2 + q_3^2 \\ \tan\varphi_2 &= (q_1 q_3 - q_0 q_2)/(q_0 q_1 + q_2 q_3) \end{aligned} \tag{2.31a}$$

These functions can readily be solved by applying the atan2 function in order to account for the full definition range 0° ··· 360° of the two Euler angles φ_1 and φ_2. If $q_1 = q_2 = 0$ the denominator in Equation 2.31a becomes zero. In that case the formulae read

$$\Phi = 0°, \quad \cos(\varphi_1 + \varphi_2) = q_0^2 - q_3^2, \quad \sin(\varphi_1 + \varphi_2) = 2q_0^2 q_3^2 \tag{2.31b}$$

If $q_0 = q_3 = 0$ we analogously get

$$\Phi = 180°, \quad \cos(\varphi_1 - \varphi_2) = q_1^2 - q_2^2, \quad \sin(\varphi_1 - \varphi_2) = 2q_1^2 q_2^2 \tag{2.31c}$$

Note that in these two cases, $\Phi = 0°$ and $\Phi = 180°$, φ_1 and φ_2 both describe a rotation about the normal direction, ND (Section 2.6.1). Accordingly, Equations 2.31b and c merely provide information about the sum (or difference) of the two angles $\varphi_1 \pm \varphi_2$.

A combination of two successive rotations g_1 and g_2 is obtained through a multiplication of the rotation matrices $g = g_2 \cdot g_1$. Using unit quaternions the resulting combined quaternion $Q \cdot P$ is obtained by multiplying the two quaternions Q and P according to

$$Q \cdot P = \left(q_0 p_0 - \boldsymbol{q} \circ \boldsymbol{p};\, q_0 p + p_0 q + \boldsymbol{q} \times \boldsymbol{p} \right) \tag{2.32}$$

Thus, computation of the combined quaternion $Q \cdot P$ involves both the dot product and the cross product of the vector part of the two quaternion vectors $\boldsymbol{q}$ and $\boldsymbol{p}$. Note that usually $Q \cdot P \neq P \cdot Q$. Likewise, the calculation of misorientations is straightforward, in that the misorientation quaternion Q^m between two orientations g_1 and g_2, expressed in form of the two quaternions Q and P, is

$$Q^m = Q^{-1} \cdot P \tag{2.33}$$

(see Equation 2.23). Here Q^{-1} is the inverse of the quaternion Q which is given as $Q^{-1} = \left(q_0; -q_1, -q_2, -q_3 \right)$. Thus, the quaternion describing the misorientation becomes

$$Q^m = Q^{-1} \cdot P = \left(q_0 p_0 + \boldsymbol{q} \circ \boldsymbol{p};\, q_0 p - p_0 q - \boldsymbol{q} \times \boldsymbol{p} \right) \tag{2.34}$$

If crystal symmetry is to be taken into consideration, the possible misorientations are derived as

$$M_i = Q^m \cdot S_i, \quad i = 1, \ldots, k \tag{2.35}$$

where k is the number of the crystal-symmetry dependent possibilities to express an orientation and S_i are the crystal-dependent symmetry operators, i.e. the symmetry matrices T_i given in Appendix II expressed in the form of quaternions (e.g. Cho et al., 2005). From the possible k misorientations, we conveniently chose the one characterized by the minimum rotation angle ω_i. Adams et al. (1990a) derived the following conditions for the fundamental zone, or asymmetric domain, upon the quaternion description of cubic-cubic misorientations:

$$q_1 \geq q_2 \geq q_3 \geq O, \quad q_0 \geq \left(\sqrt{2} + 1 \right) q_1, \quad q_0 \geq q_1 + q_2 + q_3 \tag{2.36}$$

Quaternions can readily be used to determine the average orientation, $\bar{g}$, e.g. for electron backscatter diffraction (EBSD) single grain orientation measurements of multiple subgrains in a single grain (Krieger Lassen et al., 1994b; Humbert et al., 1996). The quaternion $\bar{Q}$ associated with the mean orientation $\bar{g}$ is equal to the sum of the quaternions corresponding to the individual orientations g_i, divided by the norm of the sum:

$$\bar{Q} = \sum_{i=1}^{N} q_i \Bigg/ \left\| \sum_{i=1}^{N} q_i \right\| \tag{2.37}$$

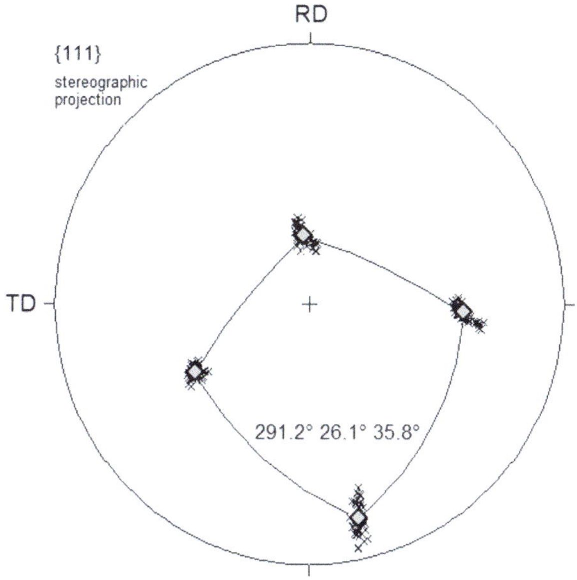

FIGURE 2.19 Set of 50 orientations obtained by EBSD in a {111} pole figure together with the mean orientation $\bar{g}$ obtained by averaging of the respective quaternions Q_i.

where N is the total number of measured data points. Figure 2.19 shows an example of 50 orientations which were determined by means of EBSD within a given deformed grain in a cold-rolled Al alloy. First, $\bar{Q}$ was computed by applying Equation 2.37 to average the quaternions Q_i associated with the individual orientations g_i. Then, the mean orientation $\bar{g}$ was obtained with Equation 2.31. Evidently, the resulting average orientation $\bar{g}$ fits the individual orientations very well. In the next step, the misorientation angles ω_i between the individual orientations g_i and the mean orientation $\bar{g}$ were computed (Equations 2.33 and 2.28) and evaluated statistically. In the present example, local misorientations up to 8.5° were observed; the average intra-grain misorientation was 4.3° with a standard deviation of 1.9°.

It shall be noted here that quaternions may also be used to produce data sets of random orientations, which are required e.g. for texture simulations, etc. A naïve attempt for the generation of random orientations might be to randomly sample each of the three Euler angles independently. However, this approach results in a distribution which is heavily biased toward polar regions (i.e. $\Phi \sim 0°$ and $\Phi \sim 90°$). Rather, the following algorithm due to Shoemake (1992) may be utilized, where random unit quaternions $Q = (r_2 \cdot \cos\theta_2,\ r_1 \cdot \sin\theta_1, r_1 \cdot \cos\theta_1, r_2 \cdot \sin\theta_2)$ are derived from three independent random variables $X_0, X_1, X_2 \in \mathrm{U}(0,1)$, where:

$$
\begin{aligned}
r_1 &= \sqrt{1 - X_o} \\
r_2 &= \sqrt{X_o} \\
\theta_1 &= 2\pi \cdot X_1 \\
\theta_2 &= 2\pi \cdot X_2
\end{aligned}
\tag{2.38}
$$

2.10 SUMMATION

This chapter has established the terminology associated with analysis, representation, and display of orientation and misorientation data. All the descriptors of orientation and their associated spaces that are commonly encountered in the field of texture—the orientation matrix, "ideal orientation"/Miller indices, Euler angles, angle/axis of rotation, Rodrigues vector, and quaternions plus the misorientation counterparts, where appropriate—have been explained. There is some variation in the popularity and patterns of usage for each descriptor. Briefly, the orientation matrix is the mathematical calculation instrument, and the ideal orientation notation and Euler angles have been traditional descriptors of texture and associated with x-ray diffraction, whereas the angles/axes of rotation, Rodrigues vectors, and quaternions have come to the fore much more recently and are used more for misorientations. However, a modern approach to texture can draw upon any of the texture descriptors, and so an introductory knowledge of each is desirable to choose the most convenient descriptor for a particular task.

3 Application of Diffraction to Texture Analysis

3.1 INTRODUCTION

The vast majority of techniques for texture analysis are founded on the diffraction of radiation by a crystal lattice, and so it is vital to understand this phenomenon to appreciate the principles on which the various techniques for experimental texture measurement are based. Radiation, which is diffracted by crystallographic lattice planes, is able to provide information on their arrangement and, consequently, on the orientation of the sampled volume of material with respect to some fixed reference axes (Section 2.3). To instigate diffraction of radiation at lattice planes, the wavelength of the incident radiation must be smaller than the lattice spacing, which for materials of interest is typically tenths of a nanometer. Table 3.1 shows the wavelengths, in addition to other characteristics, of various radiations that are commonly used for texture measurements. Data for light are included only for comparison, since x-rays, neutrons, and electrons are diffracted by lattice planes, whereas light is not.

The various kinds of radiation interact with matter in different ways. This manifests itself by substantial differences in the absorption of radiation by matter and, with regard to texture analysis, by the depth of penetration of radiation into the sample material. To illustrate this, the scale of penetration depths for the various radiations is also given in Table 3.1. Thus, the particular characteristics of radiation have strong impact on the application. This chapter addresses the fundamental aspects of diffraction, the characteristics of various radiations, and the way these relate to the experimental setups for texture measurement. It is pertinent to have some knowledge of subjects such as structure factor (Section 3.3) and characteristics of radiation (Section 3.6) to understand subsequently the principles of texture determination. (For further details, see Bacon, 1975; Cullity, 1978; Thomas and Goringe, 1979; Barrett and Massalski, 1980; Goodhew et al., 2001.)

3.2 DIFFRACTION OF RADIATION AND BRAGG'S LAW

Electromagnetic radiation, like light and x- or γ-rays, is diffracted by elastic scattering of the incident waves at the atoms of the sample material. Particle beams, like electrons and neutrons, can also be considered as waves of radiation, with their wavelength given by the de Broglie relation:

$$\lambda = \frac{h}{mv} \quad \text{or} \quad \lambda = \frac{h}{\sqrt{2mE_{\text{kin}}}} \tag{3.1}$$

DOI: 10.1201/9781003258339-4

TABLE 3.1
Average Diffraction Properties of Radiation Used for Texture Measurement by Diffraction

	Light	Neutrons	X-Rays	Electrons
Wavelength (nm)	400–700	0.05–0.3	0.05–0.3	0.001–0.01
Energy (eV)	1	10^{-2}	10^{4}	10^{5}
Charge (C)	0	0	0	-1.602×10^{-19}
Rest mass (g)	0	1.67×10^{-24}	0	9.11×10^{-28}
Penetration depth, absorption length (mm)	–	10–100	0.01–0.1	10^{-3}

Note: Light is included for comparison.

where λ = wavelength, h = Planck's constant, and m, v, and E_{kin} = mass, velocity, and kinetic energy of the particles, respectively.

When a plane wave of radiation hits an atom, this acts as a source for a spherical wave of the same wavelength (Figure 3.1). The efficiency of an atom in scattering radiation is usually described in terms of the atomic scattering amplitude (or scattering length) f. Since the intensity of a wave is the square of its amplitude, f^2 is a measure for the intensity of the scattered wave in dependence on the sort of the atoms, the scattering angle ϑ, the wavelength λ of the incident beam, and the kind of radiation (Doyle and Turner, 1968; Bacon, 1975; Cullity, 1978; Prince, 2004).

X-rays are scattered by the shell electrons of the atoms through an interaction between the charged electrons and the electromagnetic field of the x-rays (Section 3.6.1). Since the size of the atoms is of the same order as typical x-ray wavelengths, different waves that are scattered at the various electrons of the atom

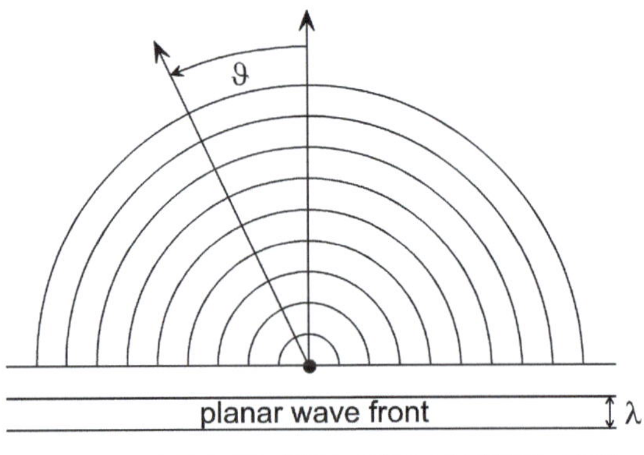

FIGURE 3.1 Scattering of a planar wave front at a point source giving rise to the formation of a Huygens spherical wave.

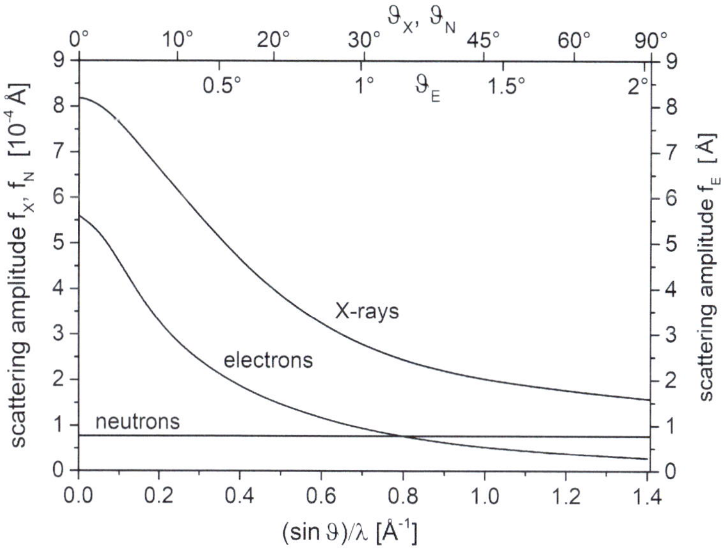

FIGURE 3.2 Angular variation of the atomic scattering amplitude f of copper for x-rays, electrons, and neutrons (note different scales for electrons). The reflection angles of the upper axis were calculated with $\lambda = 0.07107$ nm for x-rays and neutrons, and with $\lambda = 0.00251$ nm for electrons (see Table 3.1). Values for x-rays and electrons are taken from Doyle and Turner (1968); values for neutrons are taken from Bacon (1975).

interfere, which gives rise to a strong dependency of the scattering amplitude f_x on the atomic radius R and the scattering angle ϑ. Whereas near the incident beam, that is, for small angles ϑ, the individual scattered waves will all nearly be in phase and reinforce each other, for large angles ϑ, they are out of phase and hence reinforce each other much less. Accordingly, f_X strongly decreases with increasing scattering angle ϑ or, more precisely, with an increasing value of sin ϑ/λ, as shown in Figure 3.2. *Electrons* interact with both the shell electrons and the nucleus of the scattering atoms (Section 3.6.3). Their scattering amplitude f_E decreases with sinϑ/λ similarly as f_X does, though more rapidly, but the value of f_E is approximately four orders of magnitude greater than f_X (Figure 3.2). Thus, electrons scatter much more intensely than x-rays, and this is the reason why electrons can be used with such a high resolution in electron microscopy. *Neutrons*, in contrast, mainly interact with the nucleus of the atom (except for magnetic materials where magnetic scatter with the shell electrons takes place; Section 3.6.2). As the size of the atomic nucleus is negligible compared to the typical wavelengths in neutron diffraction experiments, the scattering amplitudes for neutrons f_N are virtually independent of the reflection angle ϑ (Figure 3.2).

When radiation interacts with matter, rather than with individual atoms, the individually scattered waves interfere to form a secondary wave (Huygens–Fresnel principle), as shown schematically in Figure 3.3. In most cases, the different waves will

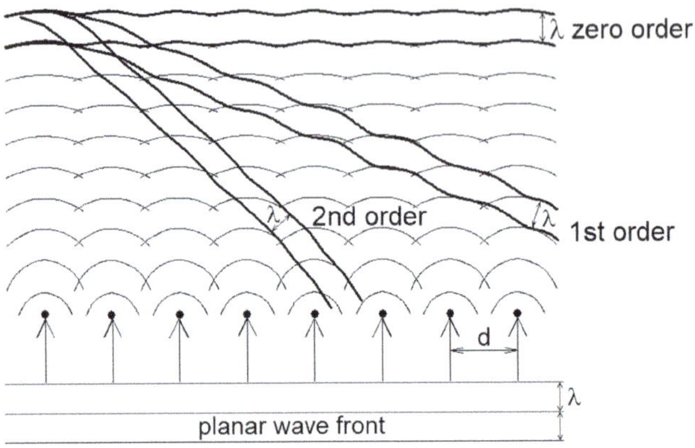

FIGURE 3.3 Reinforcement of scattered waves producing diffracted beams in the different orders. (Adapted from Barret and Massalski, 1980.)

be out of phase, which leads to an annihilation of the reflected intensity to zero. Only at specific angles will the wave fronts be in phase, which means that at those angles diffraction of the incoming radiation can be observed. For this to occur, three conditions must be fulfilled:

- The atomic arrangement must be ordered, that is, crystalline.
- The radiation must be monochromatic, that is, consist only of one wavelength λ.
- This wavelength must be of the same size order or smaller than the diffracting features.

Under these circumstances, independent of the kind of radiation, specific diffraction maxima are observed.

From Figure 3.3 it appears that the angles θ at which diffraction occurs depend on both the wavelength λ and the spacing of the scattering atoms d. To derive this dependency, *diffraction* of radiation at the individual atoms in a crystal can be considered to be *reflection* of radiation at a set of semitransparent "mirrors," at a distance d. Bragg and Bragg (1913) showed that these mirrors are formed by atomic planes, that is, the lattice planes $\{hkl\}$, which are considered to be geometrically smooth. Figure 3.4 shows a section through a lattice with three atom layers A, B, and C. Rays are incident on these planes in the direction LM at the angle θ. A small portion of the incident radiation is reflected at the angle 2θ at plane A, whereas the rest continues traveling into the lattice until it is reflected at layers further below. The line $L–L_2$ is drawn perpendicular to the incident beam to indicate one of the crests of the approaching inphase waves. To get a reinforced, reflected beam in the direction MN, the waves must again be in phase along the line $N_{2A}–N_{2C}$. To achieve this, the lengths of path for beams reflected at different, successive layers in the crystal must be equal to an integral number of wavelengths. In the example shown in Figure 3.4 this means that for instance the path difference between the two reflected beams

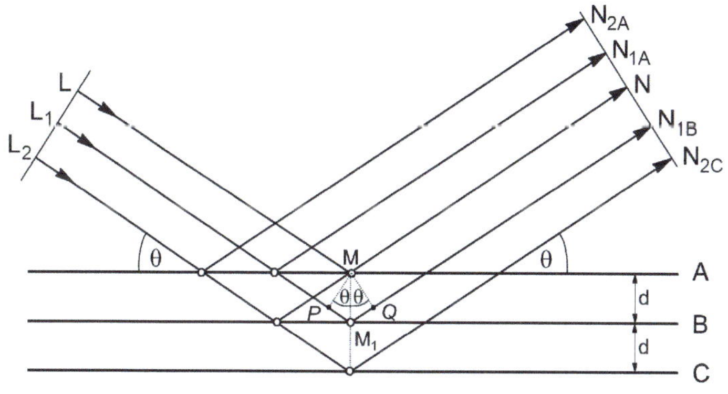

FIGURE 3.4 Diffraction from lattice planes, indicating the geometry that leads to the derivation of Bragg's law.

L–M–N and L_1–M_1–N_{1B}—that is, the distance P–M_1–Q—is either one wavelength λ or a multiple of it—$n\lambda$. It can be seen that

$$PM_1 = M_1Q = d\sin\theta \tag{3.2}$$

It is clear from Figure 3.4 that the other reflected beams bear the same geometrical relationships and, therefore, the condition for reinforcement of all reflected beams can be written as

$$n\lambda = 2d\sin\theta \tag{3.3a}$$

where n is the order of reflection and d is the interplanar spacing. This is *Bragg's law* and is usually written as

$$\lambda = 2d\sin\theta_B \tag{3.3b}$$

since often only first-order diffraction ($n = 1$) is considered. Accordingly, the specific angles at which reflection is observed are termed *Bragg angles* θ_B. Laue (1912) derived a similar relation from consideration of scattering at individual atoms.

Bragg's law is fundamental to texture work since, for radiation of known wavelength, lattice planes can be identified from measurement of the Bragg angles through which the waves are diffracted. Although the phenomenon that occurs is diffraction and not reflection, diffracted beams are frequently referred to as "reflected beams" and the lattice planes as "reflecting planes" or "reflectors." Note that it is convenient to distinguish between the reflecting planes {*hkl*}, written with braces, and the corresponding reflected beam *hkl*, written without braces. Because each set of lattice planes {*hkl*} has a different lattice spacing *d*, reflections from various sets of lattice planes can be distinguished by setting the detector to the corresponding angle 2θ with respect to the incident radiation. For cubic structures, the lattice spacing d of lattice planes $\{hkl\}$ is related to the material's lattice constant a through

$$d = \frac{a}{\sqrt{h^2 + k^2 + l^2}} \tag{3.4}$$

For noncubic structures, the relations are more complex. For orthorhombic structures, the relationship is

$$\frac{1}{d^2} = \frac{h^2}{a^2} + \frac{k^2}{b^2} + \frac{l^2}{c^2} \tag{3.5}$$

which contains Equation 3.4 as a special case.

Equation 3.4 entails that for cubic crystals symmetric reflections like 100, 010, and 001 cannot be differentiated since the related crystal planes (100), (010), and (001) are indistinguishable (Section 2.2). In crystal structures with a lower symmetry—e.g. tetragonal, orthorhombic, or lower—the respective lattice spacings d_{hkl} are different, in contrast (Equation 3.5), which may give rise to separate reflections with specific Bragg angles. In order to illustrate this, Figure 3.5a shows an idealized diffraction pattern from a cubic material with a 0.35-nm unit cell of a face-centered crystal, which reveals a phase transformation to a crystal structure with lower symmetry. Figure 3.5b shows the ensuing diffraction lines for a similar unit cell with a distortion from the cubic to a tetragonal crystal structure, which is achieved by a slight elongation of the c axis to 0.36 nm. The diffraction pattern displays a characteristic splitting of some of the diffraction lines from the original cubic pattern. The 200 cubic peak (which is due to reflections 200, 020, and 002) splits into two separate reflections 200/020 and 002 due to the fact that $a \neq c$. Likewise, the 220 cubic peak splits into the two reflections 220 and 202. The 111 cubic peak is slightly shifted but will *not* split under the cubic → tetragonal symmetry transformation, in contrast. Upon a phase transformation to an orthorhombic structure with $a = 0.35$ nm, $b = 0.34$ nm, and $c = 0.36$ nm, the 200 cubic peak splits

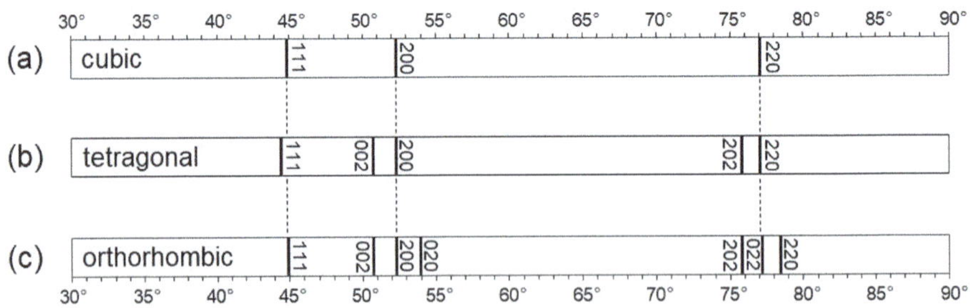

FIGURE 3.5 Idealized diffraction patterns showing the characteristic splitting of some of the diffraction lines during a phase transformation from a (a) face-centered cubic (fcc) structure with $a = b = c = 0.35$ nm to (b) a tetragonal structure with $a = b = 0.35$ nm and $c = 0.36$ nm and (c) an orthorhombic structure with $a = 0.35$ nm, $b = 0.34$ nm, and $c = 0.36$ nm (x-ray CuKα radiation, $\lambda = 0.1542$ nm).

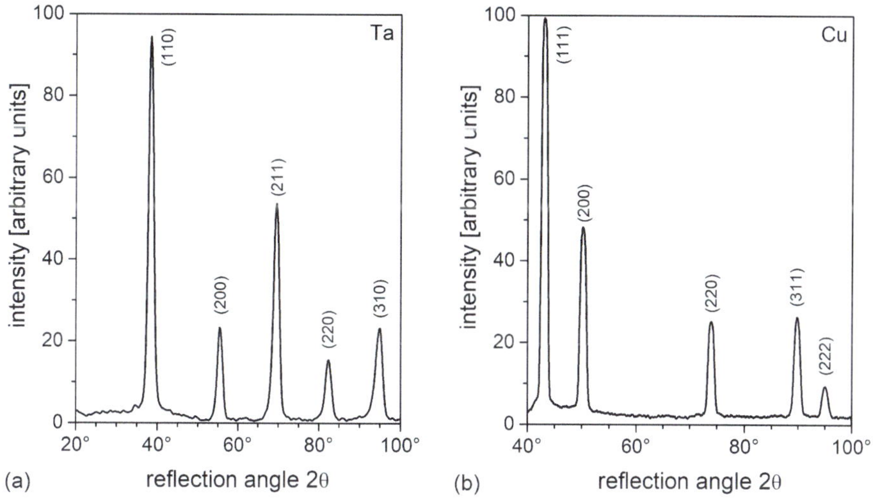

FIGURE 3.6 θ/2θ diffraction spectrum of (a) bcc tantalum and (b) fcc copper (powder samples; x-ray CuKα radiation, λ = 0.1542 nm).

into the three separate reflections 200, 020, and 002 (Figure 3.5c). Similarly, the 220 reflection splits into 220, 202, and 022, whereas the 111 cubic peak again remains solitary. Accordingly, materials with lower crystal symmetry are generally characterized by diffraction patterns with much more diffraction peaks.

Finally, for hexagonal close-packed (hcp) crystals, the relationship between the lattice spacing *d* and the crystal axes *a* and *c* is

$$\frac{1}{d^2} = \frac{4}{2}\left(\frac{h^2 + hk + k^2}{a^2}\right) + \frac{l^2}{c^2} \tag{3.6}$$

Figure 3.6a and b shows the diffraction spectra of tantalum with body-centered cubic (bcc) crystal structure and copper with face-centered cubic (fcc) crystal structure, respectively. Such spectra, which are well known from powder diffractometry, are routinely obtained by recording the reflected intensities as a function of the diffraction angle for gradually increasing angles θ and 2θ. Powder samples are used to eliminate possible effects of texture. Provided the crystal structure and its lattice parameters are known, the individual peaks can be indexed with the help of Bragg's law (Equations 3.3 and 3.4), for instance, by setting up a table containing all possible reflectors, as in Table 3.2 for cubic materials. However, comparison of the possible reflections in Table 3.2 with the actual diffraction spectra in Figure 3.6 shows that reflection through the respective Bragg angle is not observed for *all* possible sets of lattice planes *hkl*.

To illustrate this, we consider the reflection at the {100} planes in a body-centered crystal, as schematically shown in Figure 3.7. To obtain 100-reflection, the Bragg angle must be set so that the {100} planes are in reflection position, that is, for lattice spacing d_{100}. However, the {200} planes with spacing $d_{200} = d_{100}/2$ are situated exactly

TABLE 3.2
Bragg Angles θ for Diffraction of Radiation at Different Reflectors *hkl* in Tantalum (bcc, a = 0.331 nm) and Copper (fcc, a = 0.361 nm)

hkl	$h^2 + k^2 + l^2$	2θ, Ta	2θ, Cu
100	1	26.9	24.6
110	2	38.5	35.1
111	3	47.6	43.4
200	4	55.5	50.5
210	5	62.8	57.0
211	6	69.6	63.0
220	8	82.4	74.2
221/300	9	88.6	79.6
310	10	94.9	84.8
311	11	101.1	90.0
222	12	107.6	95.3

Note: CuKα radiation, λ = 0.1542 nm.

halfway between the {100} planes (in Figure 3.4, this would mean we consider reflection at the planes A and C with the plane B in-between). For these intermediate planes, the path difference P–M_1–Q in Figure 3.4 is exactly λ/2 (Equation 3.3), rather than an integer multiple of λ as required to reinforce the reflected beam. Hence, reflection at the {100} planes is rendered extinct by the antiphase reflection at the {200} planes in-between. This cancellation will not occur if the crystal is irradiated at the Bragg angle for second-order reflection (i.e., $n = 2$), as then the scattering from both the {100} and {200} planes will be in phase.

In general, extinction occurs when there is an equivalent plane halfway between the planes that are in the Bragg position for reflection. For body-centered crystal

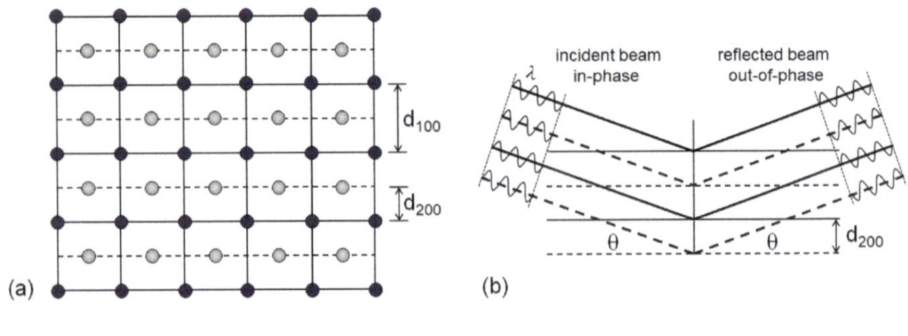

FIGURE 3.7 Extinction of the 100-reflection by antiphase reflection at the intermediate {200} plane in a body-centered crystal.

TABLE 3.3
Geometrical Rules for the Structure Factor *F*

Crystal Structure	Observed Reflection	Structure Factor *F*
Primitive	All h, k, l	f (1 atom per unit cell)
Body centered	$(h + k + l)$ even	$2f$ (2 atoms per unit cell)
Base centered	$(h + k)$ even	$2f$ (2 atoms per unit cell)
Face centered	h, k, l all odd or even	$4f$ (4 atoms per unit cell)
Hexagonal close packed	$h + 2k = 3n$, l even;	$2f$ (e.g., {0002}, {1120}, …);
	$h + 2k = 3n \pm 1$, l odd;	$\sqrt{3}f$ (e.g., {1011}, {1013}, …);
	$h + 2k = 3n \pm 1$, l even	f (e.g., {1010}, {1012}, …)

structures, this is always given when the sum over all Miller indices $h + k + l$ is odd. Vice versa, reflection is obtained from the planes {110}, {200}, {211}, {310}, etc. (Figure 3.6a and Table 3.3). For face-centered structures, reflection is observed when the individual Miller indices h, k, and l of the reflecting planes are either all odd or all even, that is, {111}, {200}, {220}, {311}, etc. (Figure 3.6b and Table 3.3). For the sake of completeness, the results for base-centered crystals are also included in Table 3.3. In general, the relative intensity of a given reflector and, consequently, the rules of extinction can be deduced from calculation of the *structure factor*, which is shown in Section 3.3.

3.3 STRUCTURE FACTOR

Diffraction of radiation at the various atoms in a crystalline structure is governed by the structure factor F, which is the unit cell equivalent of the atomic scattering amplitude f. Each atom j within a given unit cell scatters radiation with an amplitude proportional to f_j for that atom and with a phase difference with regard to the incident beam that is determined by the position (x_j,y_j,z_j) of that atom in the unit cell. The summation over the waves scattered by each of the atoms in the unit cell gives the amplitude of the wave being reflected at the plane $\{hkl\}$, which is called structure factor F:

$$F = \sum_j f_j \exp\left[2\pi i\left(hx_j + ky_j + lz_j\right)\right] \tag{3.7}$$

(i is the imaginary constant, i.e. $i = \sqrt{-1}$). Since the intensity of a wave is proportional to the square of its amplitude, the intensity of the reflected radiation is proportional to F^2 and, hence, reflection only occurs if the reflected waves are all in phase, that is, if $F \neq 0$.

The differences between the various radiations used for texture analysis, that is, x-rays, neutrons, and electrons, are represented by the atomic scattering amplitude f (Section 3.2), but the geometrical considerations associated with structure factor F hold for all kinds of radiation. In the following text, we will calculate the structure factors for several simple crystal structures and, therewith, derive the rules of extinction introduced in the preceding section. The results are summarized in Table 3.3.

If there is only one atom at the position $(x,y,z) = (0,0,0)$ in the unit cell, the structure factor F is independent of $\{hkl\}$ since for all combinations of h, k, and l, $F = f$ holds. Thus, in a crystal with primitive structure, all lattice planes would reflect radiation. For body-centered crystals, we have two atoms, one at $(x_1,y_1,z_1) = (0,0,0)$ and the other at $(x_2,y_2,z_2) = (½,½,½)$. Substituting these values for (x,y,z) in Equation 3.7, the structure factor F becomes

$$F = f \cdot \left[1 + \exp\left(2\pi i\left(\frac{h}{2} + \frac{k}{2} + \frac{l}{2}\right)\right)\right] \tag{3.8}$$

To derive the rules of extinction, the following relations with complex numbers are helpful:

$$\begin{aligned} \exp(\pi i) = \exp(3\pi i) = \exp((2n+1)\pi i) = -1 \\ \exp(2\pi i) = \exp(4\pi i) = \exp(2n\pi i) = +1 \end{aligned} \tag{3.9}$$

It is seen that if $(h + k + l)$ is odd, the exponential term in Equation 3.8 becomes –1, so that $F = 0$. However, if $(h + k + l)$ is even, then $F = 2f$, which means that in accordance with the qualitative considerations in Section 3.2, reflection in body-centered crystals only occurs at planes for which $(h + k + l)$ is even (Table 3.3).

For face-centered crystals, four atoms have to be considered, with positions (0,0,0), (½,½,0), (½,0,½), and (0,½,½). In this instance, the structure factor F is

$$F = f \cdot \left[1 + \exp(\pi i(h+k)) + \exp(\pi i(h+l)) + \exp(\pi i(k+l))\right] \tag{3.10}$$

Considering the possible combinations of h, k, and l, it turns out that if all three are either odd or even, then all of the exponential terms are $\exp(2n\pi i) = 1$. In those cases, the structure factor F equals $4f$ and reflection occurs (Table 3.3). In the other possible cases, however, two of the three phase factors will be odd multiples of π, giving two terms of –1 in Equation 3.10 and, therewith, $F = 0$. Note that these considerations are valid regardless of what structure the body- or face-centered crystal belongs to.

The hexagonal close-packed (hcp) cell contains two atoms with the coordinates (0,0,0) and (1/3,2/3,1/2). Inserting these values in Equation 3.7, we get

$$F = f \cdot \left[1 + \exp\left(2\pi i\left(\frac{h}{3} + \frac{2k}{3} + \frac{l}{2}\right)\right)\right] \tag{3.11}$$

Table 3.3 lists the results for F for the possible combinations of h, k, l. It turns out that extinction occurs if $h + 2k$ (or $h – k$) is a multiple of 3 and l is odd. Thus, {0002} and {1120} reflections occur, whereas {0001} and {1121} are absent. {101l} reflections are obtained for all values of l, albeit with different structure factors and, in consequence, intensities. As an example, Figure 3.8 shows the diffraction spectrum of the hexagonal metal titanium.

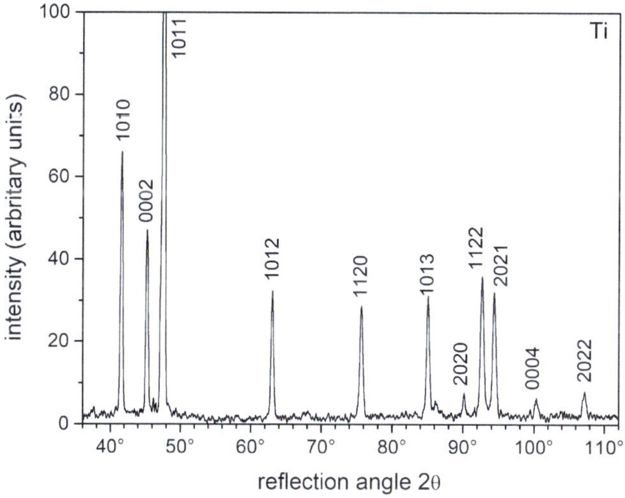

FIGURE 3.8 $\theta/2\theta$ diffraction spectrum of a hexagonal titanium powder sample (x-ray CoKα radiation).

So far, only reflection of radiation in pure metals has been considered, which means that the unit cell only contains atoms of one sort. For unit cells consisting of different species of atoms—alloys, intermetallic phases and ordered structures—the atomic scattering factors f_j in Equation 3.7 differ, so that the various phase factors do not necessarily cancel out. In random alloys all lattice points will be occupied stochastically with the same portion of alloy elements. For example, in a nickel alloy containing 25% aluminum the Ni and Al atoms will be randomly distributed in the corner positions and the centers of the faces of the unit cells. Hence, identical extinction rules as in pure face-centered cubic metals will hold. In ordered structures the different lattice points will systematically be occupied with different portions of the alloy elements, by contrast, and there will be only a weakening but no complete extinction of the forbidden reflections. Thus, in the ordered intermetallic compound Ni_3Al the 100-reflection—which is absent in random fcc alloys—is observed, though weak (the so-called superlattice peaks; Figure 3.9). The Ni_3Al unit cell has the Al atom on the (0,0,0) corner sites and the three Ni atoms on the face centers ($L1_2$-structure). Hence, the structure factor is given by

$$F = f_{Al} + f_{Ni} \cdot \left[\exp\left(\pi i (h+k)\right) + \exp\left(\pi i (h+l)\right) + \exp\left(\pi i (k+l)\right) \right] \qquad (3.12)$$

In analogy to the face-centered lattice, strong reflection occurs with $F = (f_{Al} + 3f_{Ni})$ if h, k, l are all either odd or even. However, for mixed h, k, l, the structure factor is not zero but $F = (f_{Al} - f_{Ni})$, so that the missing peaks occur as well, though with weaker intensity (Figure 3.9). As a matter of fact, the ratio of the intensity of the superlattice peaks to that of the fundamental peaks can be used as a measure for the degree of ordering (e.g. Ball et al., 1992; Kaneno et al., 2006).

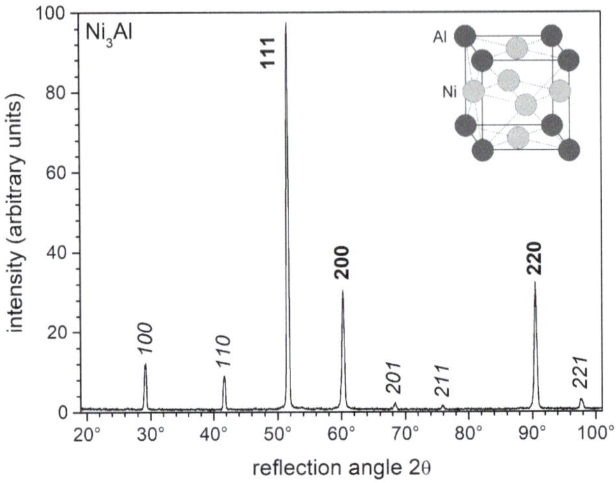

FIGURE 3.9 θ/2θ diffraction spectrum of ordered Ni_3Al ($L1_2$ structure) showing superstructure peaks (x-ray CoKα radiation).

3.4 LAUE AND DEBYE–SCHERRER METHODS

According to Bragg's equation (Equation 3.3), reflection of radiation at a given set of lattice planes {*hkl*} is observed under a certain angle θ for a given wavelength λ. In principle, crystallographic orientations can be studied by means of two different approaches:

- At constant θ, the wavelength λ is varied (Laue method).
- At constant λ (i.e., monochromatic radiation), θ is varied; either by rotation of a single crystal (rotating crystal method) or by analyzing a multitude of grains with random texture (powder method).

These two methods go back, respectively, to Laue (1912) and to Debye and Scherrer (1916, 1917), shortly after Röntgen (1896) discovered x-radiation. Accordingly, they are strongly linked to this kind of radiation, though their principles hold for other radiations as well. In fact, in Section 6.2.2 we will introduce an application of the Debye–Scherrer method in electron diffraction, namely, the determination of pole figures from small volumes in the transmission electron microscopy (TEM).

Until 1912, both the periodic lattice structure of crystals and the electromagnetic character of the x-radiation were only hypotheses, when Laue (1912) proved both presuppositions in one single experiment. They irradiated crystals of zinc blende and copper sulfate with a finely collimated, polychromatic (i.e., "white") x-ray beam and recorded the transmitted intensities on a film. Rather than finding homogeneously scattered intensities, they observed a large number of spots that were symmetrically arranged in a characteristic pattern (Figure 3.10), which looked as if the crystal contained an arrangement of "mirrors" at different angles. A few months later, Bragg and Bragg (1913) showed that these mirrors are formed by the various lattice planes in the crystal, which led to the formulation of Bragg's law, described in Section 3.2.

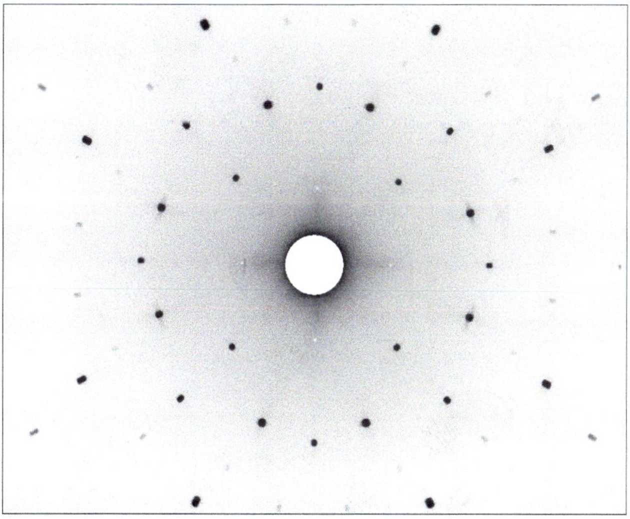

FIGURE 3.10 Laue diagram of a silicon single crystal (reflection technique).

To utilize the Laue method for orientation determination, a finely collimated x-ray beam is focused on the single crystal or the grain of interest in a polycrystalline sample. Because white, that is, polychromatic, x-radiation is used in this method, for each set of lattice planes {*hkl*}, there are x-rays that fulfill Bragg's condition for diffraction (Figure 3.11). The diffracted beams result in a point for each set of lattice planes in the Laue diagram (Figure 3.10). Provided the crystal structure is known, the spatial orientation of the corresponding planes and, thus, the crystallographic orientation can be evaluated from the position of the diffraction spots.

Probably the most momentous contribution to modern texture analysis was made by Debye and Scherrer (1916, 1917), who observed interference of x-rays in fine-grained polycrystalline powders that were irradiated with monochromatic x-radiation, that is, at constant wavelength λ. In a powder sample typically all possible orientations are present, so that always some crystals will have lattice planes that are oriented at the corresponding Bragg angle with respect to the incident beam (whereas in the Laue method, the lattice planes "select" the proper wavelength out

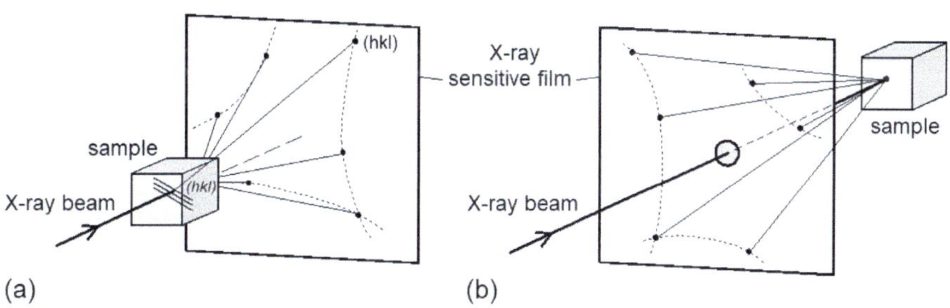

FIGURE 3.11 Measurement of individual orientations according to the Laue technique. (a) Transmission technique and (b) reflection technique.

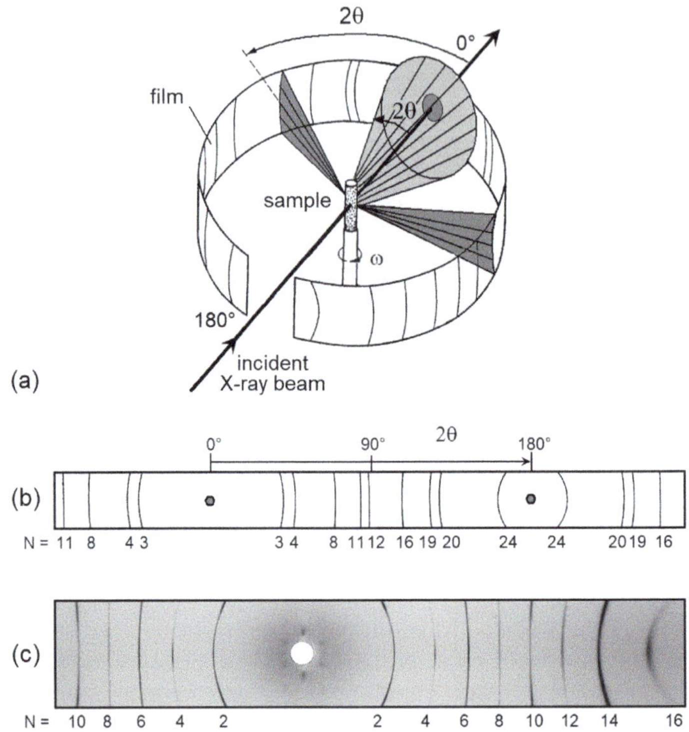

FIGURE 3.12 (a) Debye–Scherrer technique to determine the crystal structure and texture of a polycrystalline sample. Diffraction cones from two lattice planes are shown intersecting the film. (b) Schematic diffraction pattern of an fcc crystal on an unrolled film. (c) An example of a Debye–Scherrer diagram of bcc tungsten. (Courtesy of J. Ihringer.)

of the white x-radiation). The diffracted intensities lie on a cone with half apex angle 2θ about the direction of the incident beam. To produce a Debye–Scherrer pattern, the sample is placed in the center of a cylindrical camera and the diffraction cones are recorded on an x-ray sensitive film situated at the inner wall of the camera (Figure 3.12), where the cones form segments of circles with the radius 2θ that through Bragg's law is related to the lattice spacing d.

Powder samples having a random orientation distribution produce patterns with homogeneous Debye–Scherrer rings, as the number of reflecting planes is equal in all spatial directions. By contrast, in samples with a preferred orientation distribution, the number of reflecting planes varies in different directions. Hence, the intensity distribution along the Debye–Scherrer rings becomes inhomogeneous, and in heavily textured samples, some parts of the rings may even be completely missing (examples of such Debye–Scherrer diffraction patterns are shown in Figures 4.17b and 12.3b and, for electron diffraction, in Figure 8.4b and c). Wever (1924) utilized this fact to generate the first pole figures of heavily cold-rolled aluminum, before pole figure goniometers and Geiger counters were introduced in the 1940s to enable quantitative texture analysis. This is discussed in Chapter 4.

3.5 ABSORPTION AND DEPTH OF PENETRATION

In Section 3.2 we saw that the various kinds of radiation—x-rays, neutrons, and electrons—interact with atoms in different ways. With regard to texture analysis of large crystalline arrangements, these interactions manifest themselves as substantial differences in the absorption of radiation by matter (Table 3.1). On transmission of radiation through a sample with thickness t, the incident intensity I_0 is attenuated according to

$$I = I_0 \exp(-\mu t) \tag{3.13}$$

Here, μ is the linear absorption coefficient (commonly given in cm^{-1}) that depends on the radiation and the investigated material as well as the wavelength of the radiation. Often, the absorption coefficient is related to the specific weight ρ, which yields the mass absorption coefficient μ/ρ ($cm^2\ g^{-1}$). Table 3.4 gives data of μ/ρ for x-rays with different wavelengths as well as for neutrons in various pure metals. For alloys, μ/ρ can be calculated by a summation over the absorption coefficients $(\mu/\rho)_i$ of the individual components i with the weight fractions w_i:

$$\frac{\mu}{\rho} = \sum_i w_i \cdot \left(\frac{\mu}{\rho}\right)_i \tag{3.14}$$

As an example, the mass absorption coefficient of a ferritic stainless steel AISI 430 containing 17% Cr for Mo-Kα-radiation can be estimated as

$$\left(\frac{\mu}{\rho}\right)_{\text{AISI 430}} = w_{\text{Fe}} \cdot \left(\frac{\mu}{\rho}\right)_{\text{Fe}} + w_{\text{Cr}} \cdot \left(\frac{\mu}{\rho}\right)_{\text{Cr}} = 0.83 \cdot 38.5 + 0.17 \cdot 31.1 = 37.2 \frac{\text{cm}^2}{\text{g}} \tag{3.15}$$

The linear mass absorption coefficient μ is then obtained by multiplication of the mass absorption coefficient with the density of stainless steel grade 430, $\rho_{\text{AISI 430}} = 7.7$ g/cm³, i.e. $\mu = 286.4\ cm^{-1}$.

The mass absorption coefficients for x-rays are typically of the order of 100 $cm^2\ g^{-1}$ (Table 3.4), so that it readily follows from Equations 3.13 and 3.14 that x-rays are heavily absorbed by matter. For example, the intensity of x-radiation with wavelength 0.179 nm (CoKα) that passes through a 10-μm-thick zinc foil ($\rho = 7.13\ g\ cm^{-3}$) is reduced to approximately 50% of its initial intensity, whereas a 100-μm sheet would almost completely absorb the x-rays ($I/I_0 < 1\%$). Thus, the depth of penetration of x-ray in matter is limited to layers of 10–100 μm, which has a strong impact on the design of techniques for texture determination by means of x-rays (Chapter 4). Neutrons are much less absorbed by matter, and with a few exceptions (boron, cadmium), the mass absorption coefficients are of the order of 0.01 $cm^2\ g^{-1}$ (Table 3.4). Thus, to attenuate neutron radiation to 50% of its initial intensity as in the earlier example, a zinc block with a thickness of 17 cm is necessary, which means that the penetration depth of neutrons is larger by approximately four orders of magnitude than that of x-rays. Electrons, however, are much more strongly absorbed than

TABLE 3.4
Mass Absorption Coefficients (μ/ρ) for X-Rays and Neutrons of Different Wavelengths

Material (Atomic Number, Z)	X-Ray λ = 0.07107 nm (MoKα)	X-Ray λ = 0.15418 nm (CuKα)	X-Ray λ = 0.17902 nm (CoKα)	Neutrons λ = 0.108 nm
Be (4)	0.3	1.5	2.3	0.0003
B (5)	0.4	2.4	3.6	24
Mg (12)	4.1	38.6	59.5	0.001
Al (13)	5.16	48.6	74.8	0.003
Si (14)	6.44	60.6	93.3	0.002
Ti (22)	24.2	208	308	0.044
V (23)	27.5	233	343	0.033
Cr (24)	31.1	260	381	0.021
Mn (25)	34.7	285	414	0.083
Fe (26)	38.5	308	52.8	0.015
Co (27)	42.5	313	61.1	0.21
Ni (28)	46.6	45.7	70.5	0.028
Cu (29)	50.9	52.9	81.6	0.021
Zn (30)	55.4	60.3	93.0	0.0055
Sr (38)	95.0	125	190	0.005
Y (39)	100	134	203	0.006
Zr (40)	15.9	143	216	0.0006
Nb (41)	17.1	153	230	0.004
Mo (42)	18.4	162	243	0.009
Ag (47)	25.8	218	321	0.2
Cd (48)	27.5	231	338	14
W (74)	99.1	172	253	0.036
Au (79)	115	208	302	0.17
Pb (82)	120	232	334	0.0003

Source: Data for x-rays taken from Barrett and Massalski (1980); data for neutrons taken from Bacon (1975).

Note: Examples of absorption edges are underlined (Section 3.6.1).

x-rays. Here, mass absorption coefficients of 10^4 cm^2 g^{-1} are obtained, so that the penetration of electrons is limited to values as low as 1 μm (Table 3.1).

3.6 CHARACTERISTICS OF RADIATIONS USED FOR TEXTURE ANALYSIS

3.6.1 X-Rays

X-ray diffraction is the most established technique for texture measurement and reveals the integral macrotexture of a volume of material from a flat specimen by measuring the intensities of diffraction maxima. X-radiation suited for diffraction experiments is commonly produced with x-ray tubes (Section 4.3.1) or, for special applications, obtained from a synchrotron source (Section 13.2). As discussed in detail in Sections 4.2 and 4.3, application of x-ray diffraction to measure macrotextures in a material of known lattice parameter involves the use of a monochromatic beam of radiation of a given wavelength with the Bragg angle fixed for a chosen set of lattice planes {*hkl*}. If a polycrystal of this material is rotated in space, then when grains in the sampled volume become oriented such that their {*hkl*} planes coincide with the Bragg condition for their interplanar spacing, a diffracted intensity is measured. The direct output from an x-ray texture goniometer is a chart showing the diffracted intensity with respect to the rotation angles of the specimen, which can be represented in a pole figure (Sections 4.2 and 5.2) or used for calculation of the 3D ODF (Section 5.3).

In Section 3.5, we noted that the high absorption of x-rays in matter is a very important aspect of texture analysis using x-ray radiation, as it limits the penetration depth to approximately 100 μm and, as discussed in Chapter 4, it also influences the choice of an appropriate x-ray tube. The mass absorption coefficients of most metals are in the range 10–350 $cm^2\ g^{-1}$ (Table 3.4). Beryllium has a notably lower absorption of approximately 1 $cm^2\ g^{-1}$ and is therefore used as window in x-ray tubes and detectors. It is seen in Table 3.4 that the absorption in general increases with increasing atomic number Z of the analyzed sample material and the wavelength λ of the used radiation. For a given wavelength, however, there are steep drops in absorption by up to one order of magnitude between certain elements; some examples are underlined in Table 3.4. These *absorption edges* are caused by the efficiency of the x-rays in emitting photoelectrons from the electron shells of the sample material. For a given material, if the wavelength of the radiation is too long (i.e., the energy is too low) to eject photoelectrons, the radiation will pass through the material accompanied by quite low absorption. Waves with shorter wavelengths may have sufficient energy to eject an electron from one of the shells of the absorbing atom, however. The absorbed energy released by the recombination of the electron hole is emitted as *fluorescence*, which leads to a much higher absorption of such radiation. Accordingly, ejection of an electron from the K-shell results in K-excitation, and the discontinuity in absorption is called the K-absorption edge. As an example, the K- and L-absorption edges in zirconium are shown in Figure 3.13. This effect is utilized to produce (quasi) monochromatic radiation by eliminating Kβ-radiation from the x-ray spectrum by appropriate filters (Section 4.3.1). For texture analysis high absorption is undesirable, however, and hence the corresponding combinations of x-radiation and sample material should be avoided.

3.6.2 Neutrons

Neutron diffraction is used to obtain an average macrotexture in a manner very similar to that for x-rays (Section 4.4). However, the low availability of the specialized facility required (mostly a beam line on a nuclear reactor) means that neutrons are

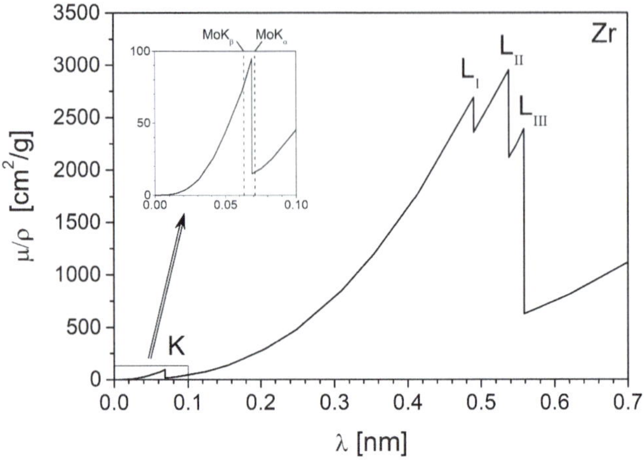

FIGURE 3.13 Variation of the mass absorption coefficient μ/ρ with wavelength λ in zirconium, showing the K- and L-absorption edges. The inlay shows the K_{Zr} absorption edge used to produce (quasi) monochromatic MoKα radiation for x-ray diffraction experiments (see Section 4.3.1).

only used for texture work where they can offer specific advantages over x-rays. Such situations are as follows:

- Where the greater depth of penetration is an advantage, for example, irregularly shaped or large-grained specimens and small volume fractions of second phases or porous materials
- Where the speed of measurement related to the greater volume is an advantage, for example, for dynamic studies of texture development
- Where low crystal symmetry or multiphase systems is present
- Where the sample must not be prepared for x-ray texture analysis, that is, nondestructive texture analysis

The elastic scattering of thermal neutrons by crystalline matter consists of two contributions, nuclear and magnetic scattering. *Nuclear scattering* is caused by the interaction between the neutrons and the atomic nuclei. These interactions lead to diffraction effects akin to the diffraction of x-rays and are utilized for analysis of the crystallographic texture. However, there are some characteristic differences between x-ray and neutron diffraction, which warrant mentioning at this stage. Since neutrons are mainly scattered at the atomic nucleus, interaction is much weaker than the interaction of x-rays at the shell electrons, leading to a much lower absorption and much higher penetration depth (Section 3.5). Furthermore, for x-rays the scattering amplitude f_X increases proportionally with the atomic number Z (Figure 3.14). However for neutrons, no such simple dependency exists, but the nuclear scattering consists of the potential scattering (scattering of a planar wave at a rigid sphere), which increases with $Z^{1/3}$, and the resonance scattering, which yields an irregular dependence on Z (Figure 3.14). Because of the resonance scattering, different isotopes of the same

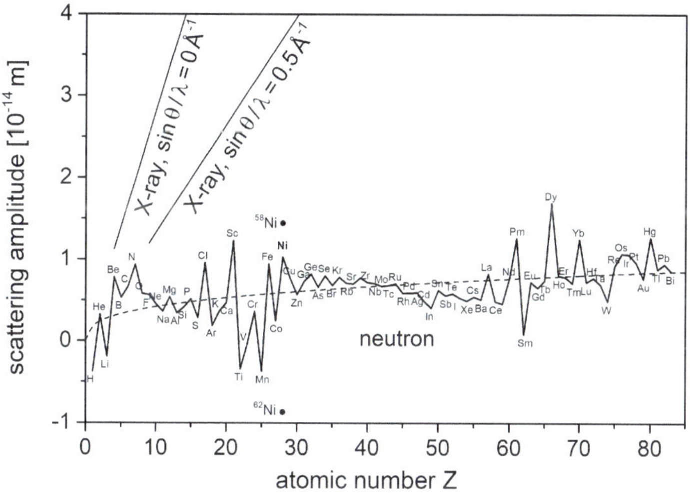

FIGURE 3.14 Variation of the neutron scattering amplitude f with atomic number Z due to the superposition of resonance scattering and potential scattering. The regular increase for X-rays is shown for comparison. The scattering amplitude for X-rays at 0° is given by $(e^2/mc^2)\cdot f_0$ or $(0.282\cdot 10^{-12}$ cm × Z). Values for neutrons are for the elements in their natural isotopic abundances and are taken from Sears (1992).

elements may have different scattering amplitudes and can consequently be separated with neutron diffraction. Moreover, signals from light elements are similar in magnitude to those from heavy ones, so that they can also be detected by neutron diffraction.

The scattering amplitude of x-rays strongly depends on the diffraction angle θ, whereas that of neutrons is virtually independent of θ (Figure 3.2), though the overall intensity still decreases somewhat with θ because of thermal vibration and other effects. Therefore, peaks at higher diffraction angles can be evaluated better by neutron diffraction. Furthermore, most neutron diffraction instruments provide better spectral resolution than x-rays, which is advantageous for deconvoluting complex diffraction patterns with overlapping peaks, such as in multiphase systems or low-symmetry materials (Section 4.5).

Magnetic scattering results from the classical dipole-dipole interaction of the magnetic momentum of the neutrons with the electrons of the atomic shell when these have a resulting magnetic momentum (Bacon, 1975; Szpunar, 1984; Bunge, 1989b). In general, the intensity of the magnetic scattering is much weaker than that of the nuclear scattering and can be neglected for texture analysis. However, in some magnetic materials—Fe and Mn—magnetic scattering can give rise to peaks in the diffraction pattern, which can be exploited to derive magnetic pole figures, that is, the distribution of the magnetic dipoles in the sample. However, the magnetic contribution to diffracted neutrons is always superimposed by the contribution of the nuclear scattering, that is, the crystallographic texture, which means that the magnetic pole figures have to be separated from the crystallographic pole figures (Zink et al., 1994; Birsan et al., 1996; Chateigner et al., 2010).

3.6.3 Electrons

Electron diffraction is employed in a fundamentally different way to x-rays or neutrons for studying texture. This is a consequence of first the fact that electrons are the only radiation listed in Table 3.1 to carry a charge, and second their penetration depth and interaction volume is small enough to allow diffraction from *individual* grains rather than volumes containing many grains as for x-ray/neutron diffraction. The charge on electrons allows them to be deflected by magnetic lenses and scanned in a raster and therefore to be used to produce highly magnified images in TEM and scanning electron microscopy (SEM), respectively. This means that diffraction information from discrete sample volumes that are submicron in size can be obtained concurrently with images of the same region. Hence, electron diffraction is used to obtain the orientations of individual grains, that is, microtexture, rather than the macrotextures as obtained by x-ray/neutron diffraction experiments.

The wavelength λ of the electron radiation is controlled by the accelerating voltage U of the electron microscope; it can be computed from the de Broglie relation (Equation 3.1) considering the relativistic mass increase of the electrons:

$$\lambda = \frac{h}{\sqrt{2meU}} = \frac{h}{\sqrt{2m_0eU\left(1+eU/\left(2m_0c^2\right)\right)}} \tag{3.16}$$

where h = Planck's constant, e = charge on an electron, m = electron mass, m_0 = electron rest mass, and c = velocity of light.

For typical accelerating voltages of U = 20 kV in an SEM and U = 200 kV in a TEM, the wavelengths are λ = 0.00859 and λ = 0.00251 nm, respectively (Table 3.1). Because of these small wavelengths—and the strong dependence of the diffracted intensity on the scattering angle (Section 3.2; Figure 3.2)—the Bragg angles θ_B for reflection of electrons (Equation 3.3) are typically very small (Section 6.1.1). Even in the case of high-index planes, the angles reach only approximately 2°. This means that the reflecting lattice planes are always situated approximately parallel to the primary electron beam.

When the electron beam in an electron microscope is directed onto a metallic sample, different scattering events and interactions with the sample material give rise to a variety of different electron and x-ray signals (Figure 3.15). These signals are used to produce the images in electron microscopes; they also provide information on the chemical composition and, with regard to the subject of this book, on the crystallography and crystallographic orientation of the sampled volume. Part of the primary electron signal is scattered elastically without losing energy. Elastically backscattered electrons are used for imaging purposes, as their intensity yields information on the atomic number Z (Z-contrast) of the sample material and thus on the chemical composition of the sampled region, for example, of particles. In transmission, coherent elastic scatter of electrons at crystalline objects gives rise to selected area diffraction (SAD) patterns that can be used for orientation determination (Section 6.2.1).

Electrons that are inelastically scattered at the atomic shell are subject to energy loss and absorption or they induce emission of secondary electrons, Auger electrons,

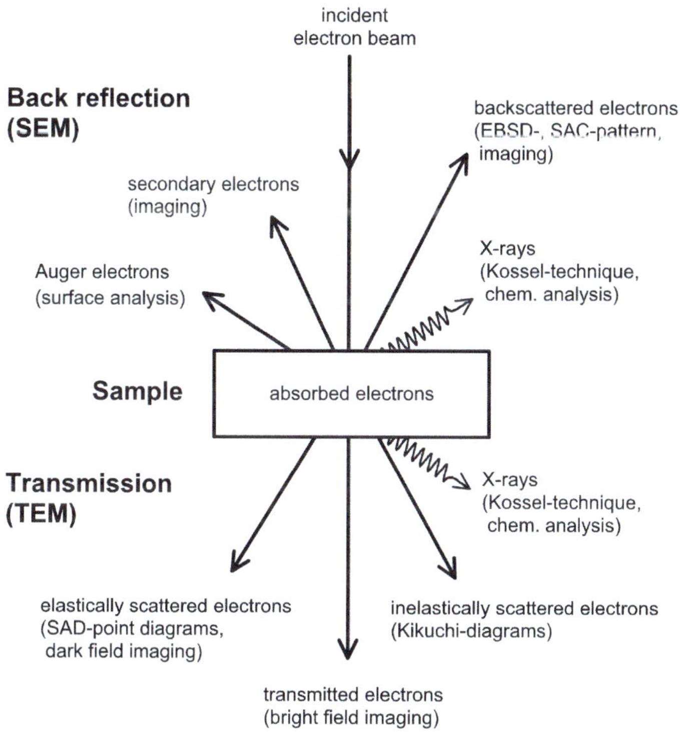

FIGURE 3.15 Summary of the various signals obtained by interaction of electrons with matter in an electron microscope.

or x-rays. Secondary electrons with an energy less than 50 eV come from a depth of only a few nanometers and are consequently very well suited for imaging of the sample topography in an SEM with high spatial resolution. Auger electrons with energies up to 2 keV are formed in the furthest surface layers (~1 nm). As their energy depends on the emitting element, they can be used for surface analysis. X-rays are typically used for chemical analysis, but they also yield information on the crystallographic orientation since they can be diffracted at the crystal lattice to give Kossel diffraction (Section 6.3.5). Multiple scatter of electrons in somewhat thicker sample regions gives rise to Kikuchi diagrams, which are very well suited for orientation determination in the TEM, as discussed in Section 6.2.6. In the SEM, similar diagrams can be produced by electron channeling patterns (ECP; Section 6.3.4) and electron backscatter diffraction (EBSD; Section 6.3.1).

3.7 SUMMATION

Most techniques for texture analysis rely on diffraction and Bragg's law. Since the angle between the incident and diffracted beam cannot exceed 90°, it essentially follows from Bragg's law that the wavelength λ of the incident radiation must be

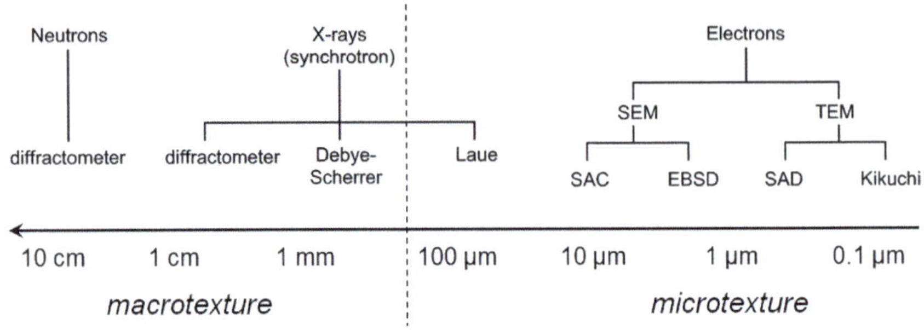

FIGURE 3.16 Categorization of the mainstream techniques for texture determination according to the radiation used as a probe. Macrotexture and microtexture methods are on the left- and right-hand parts of the diagram, respectively.

of the same or smaller size order as the lattice spacing d or, more precisely, $\lambda \leq 2d$ (Equation 3.3). Comparing the wavelengths of different radiations given in Table 3.1, it is clear that x-rays, neutrons, and electrons are diffracted by lattice planes, whereas light is not.

Besides categorization of the techniques for texture determination with respect to the type of radiation used as a probe—x-rays, neutrons, or electrons—the techniques can also be classified on the basis of whether it is a macrotexture (averaged orientation data from many grains; Section 1.1) or a microtexture (single orientations; Section 1.2) method. Figure 3.16 shows all the mainstream techniques for texture measurement by diffraction. A few peripheral techniques such as acoustic, ultrasonic, magnetic, or optical methods also exist, some of which are described in Chapter 13. With regard to the usage of the techniques featured in Figure 3.16, x-ray goniometry and EBSD are by far the most popular choices for macrotexture and microtexture analysis, respectively. All the techniques in Figure 3.16 are discussed in this book, with the extent of the coverage depending on the relevance to modern texture analysis. The choice of method(s) used to obtain orientation information will depend ultimately on factors such as the scale and resolution of the inquiry, the nature of the information sought, and the equipment available.

Part II

Macrotexture Analysis

4 Macrotexture Measurements

4.1 INTRODUCTION

This chapter introduces the common techniques to derive the distribution of crystallographic orientations present in a crystalline sample (its *macrotexture*) by means of x-ray and neutron diffraction. As mentioned in Section 3.4, x-ray diffraction was first employed in 1924 by Wever to investigate preferred orientations in metals by evaluating the inhomogeneous intensity distribution along the Debye–Scherrer rings (Figure 4.1). With the introduction of texture goniometer and the use of Geiger counters by Decker et al. (1948) and Norton (1948), pole figures could directly be recorded, and in particular the fundamental work by Schulz (1949a,b) initiated modern quantitative x-ray texture analysis. The principles of pole figure analysis by diffraction methods and the techniques and procedures to obtain quantitative texture data by means of x-ray diffraction are described in Sections 4.2 and 4.3. Since the 1960s, neutron diffraction has also been employed regularly to determine crystallographic textures, and the corresponding methods are discussed in Section 4.4.

The principles of texture analysis by x-ray or neutron diffraction apply for all materials, independent of their crystal structure. However, when investigating materials with low crystal symmetry or multiphase systems, some specific additional problems arise, which are addressed in Section 4.5.

4.2 PRINCIPLE OF POLE FIGURE MEASUREMENT

The principle of pole figure measurement by means of diffraction techniques is based on Bragg's law for reflection of radiation, x-rays, or neutrons, at the crystal lattice planes (Equation 3.3). Chapter 3 discusses the diffraction of monochromatic radiation at the lattice planes in a nontextured polycrystalline material. Now we will focus on the use of diffraction to determine the orientation of an individual crystal. To derive the crystallographic orientation of a given crystallite, the arrangement, with respect to an external reference frame, of a set of lattice planes (hkl) has to be determined. As each set of lattice planes has a different lattice spacing d, reflections from various sets of lattice planes can be distinguished by setting the detector to the corresponding angle 2θ with respect to the incident radiation (Section 3.2). First, we consider reflection at a single crystallite as schematically shown in Figure 4.2. The crystal is irradiated with monochromatic radiation at the proper Bragg angle for reflection at the lattice planes (hkl), and the detector is set at the angle 2θ with respect to the primary beam. Of course, a reflected intensity is only measured if the corresponding lattice planes are arranged such that they lie parallel to the sample surface, that is, their normal is the bisector of the angle between

DOI: 10.1201/9781003258339-6

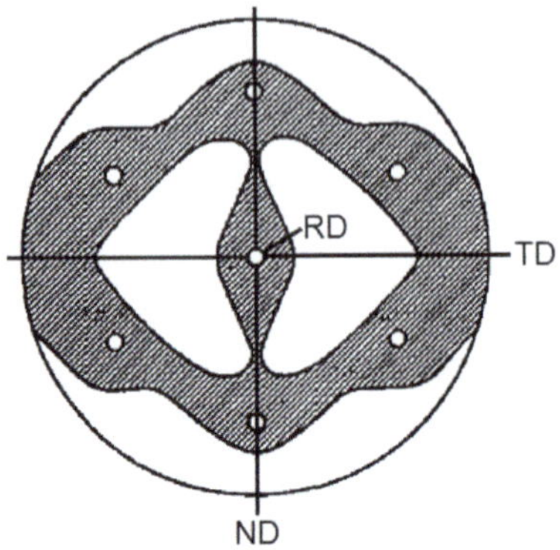

FIGURE 4.1 First pole figure, determined by Wever (1924) through evaluation of the inhomogeneous intensity distribution along the Debye–Scherrer rings in rolled aluminum.

the incident and reflected beams, as shown in Figure 4.2a, which means that in most cases *no* reflection is obtained (Figure 4.2b). To ensure reflection, the sample has to be rotated or tilted until the lattice planes are in reflection condition, that is, parallel to the sample surface (Figure 4.2c). The necessary rotation and tilt angles are a measure for the arrangement of the lattice planes within the crystal, which means

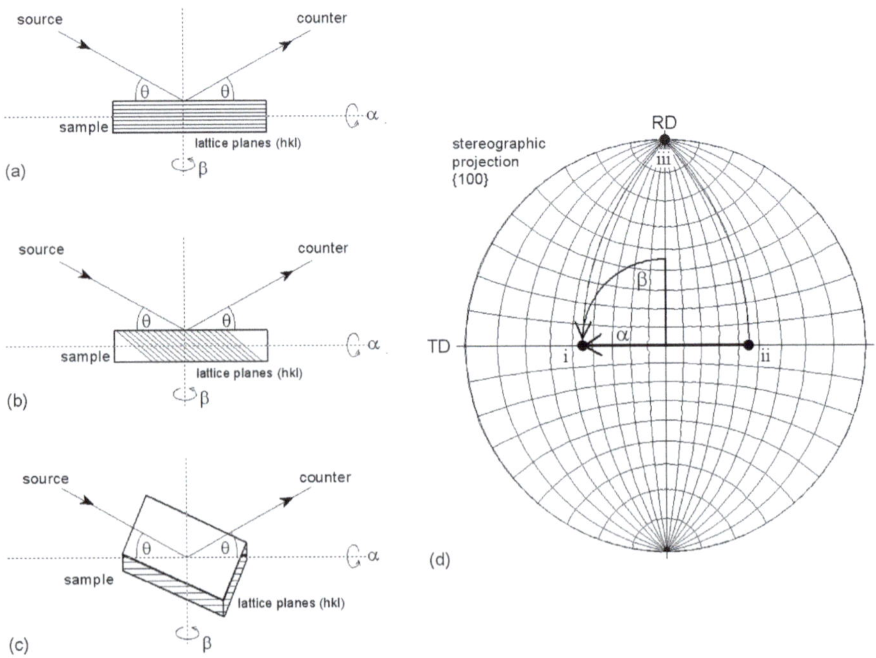

FIGURE 4.2 Sketch to illustrate the effect of sample rotation on the arrangement of the lattice planes and the resulting reflection conditions in a texture goniometer (see text for details).

they are characteristic of the orientation of the crystal with respect to the external sample frame. For example, when we irradiate a Goss-oriented crystal—{110}<001> ("cube-on-edge"; see Figure 2.9)—at the Bragg angle for reflection from the (200) planes, reflected intensities will be obtained for a sample tilt of 35° plus an additional rotation by either 90° or 270° (Figure 4.2d). A third peak would be expected at a sample tilt of 90°, but in that case the incident beam is parallel to the sample surface so that no reflection can occur (Section 4.3.2).

To determine an unknown crystal orientation in practical applications, the sample is systematically rotated in a texture goniometer about well-defined angles in such a way that all possible lattice planes are successively brought into the reflection condition and the reflected intensities are recorded as a function of these rotation angles. As we will see later, the rotation angles are directly related to the pole figure angles α (radial) and β (azimuthal), so that the reflected intensities can readily be represented in a pole figure (Section 2.5.1).

For polycrystalline samples, which are composed of an array of single crystallites, the intensity recorded at a certain sample orientation is directly proportional to the number of crystallites—more precisely, to the volume fraction of crystallites—which currently is in a reflection condition (this statement is qualified later in this section). Thus, the pole figure of a polycrystalline sample reflects the distribution of orientations in that sample, that is, its crystallographic texture. As an example, Figure 4.3 shows the {111} pole figure of aluminum, after rolling to 97% thickness reduction, plotted in the stereographic projection together with the definition of the pole figure angles α and β (Section 2.5.1).

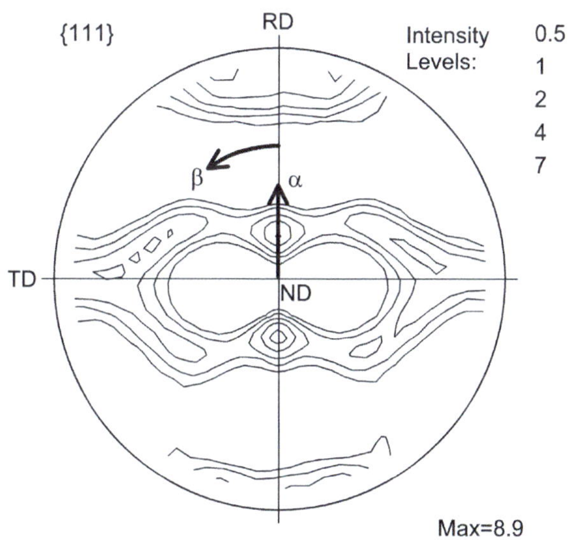

FIGURE 4.3 {111} Pole figure of 97% cold-rolled aluminum with the definition of the pole figure angles α and β. The orientation densities are given by iso-intensity lines in multiples of a random orientation distribution.

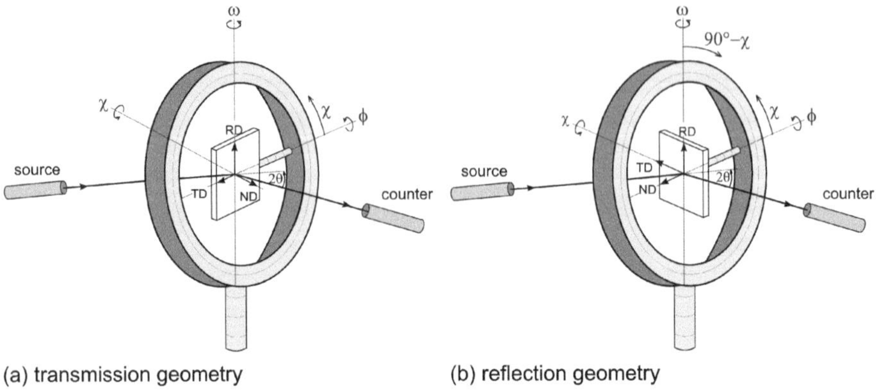

FIGURE 4.4 Diffraction in four-circle texture goniometers with definition of the instrument angles. (a) Transmission geometry; (b) reflection geometry.

Two different geometrical setups of texture goniometers are conceivable: transmission and reflection geometry. Determination of texture can be performed on thin samples that are penetrated by the neutrons or x-rays; this *transmission geometry* is shown in Figure 4.4a (Decker et al., 1948; Schulz, 1949b). Note that in Figure 4.4 we use a nomenclature for the rotation axes that is different from that in Figures 4.2 and 4.3. The nomenclature with the axes θ, φ, χ, and ω is standard in single-crystal diffractometry and marked on most diffractometers. Section 4.3.2 describes how these angles are related to the pole figure angles α and β. As discussed in Section 3.5, neutrons are readily able to transmit samples of several centimeters' thickness, and therefore texture analysis by neutron diffraction is generally performed in transmission geometry. X-rays, in contrast, are strongly absorbed by matter, and so the transmission method is generally restricted to materials with low absorption (e.g., magnesium or aluminum) and to extremely thin samples with thicknesses below approximately 100 μm. However, with a sufficiently strong x-ray source, which has been utilized for online texture control of steel sheets in industrial applications (Kopinek, 1994), sheets with thicknesses up to several millimeters can be penetrated by x-rays. Besides such exceptions, however, x-ray texture analysis is mostly performed on thick bulk samples with a plane surface in *reflection geometry*, as shown in Figures 4.2 and 4.4b (Schulz, 1949a).

In the preceding discussion of the principle of pole figure measurements, it was stated that the intensity recorded at a certain sample position is proportional to the volume fraction of crystallites in a reflection condition. This statement is oversimplified, however, and needs three supplements:

- The pole figure data need several corrections, which are described in Section 4.3.6.
- The positions of at least two poles are required (Section 2.5.1) to characterize unambiguously a given orientation in a pole figure. In turn, this means that a pole figure does not provide the complete *orientation* distribution but only the distribution of the *poles*.

- The reflections *hkl* and $\bar{h}\ \bar{k}\ \bar{l}$ from the opposites sides of a given lattice plane have equal intensities. This is caused by the fact that the diffraction averages over the volume, and it is irrelevant whether the beam impinges a lattice plane from the front or the back side, which is known as Friedel's law (Friedel, 1913). Therefore, pole figures measured with diffraction techniques are always centrosymmetric, that is, the symmetry contains an inversion center, even if the crystals are noncentric (Section 2.2). As we will see later, this also causes the "ghost error" during computation of the orientation distribution functions (ODFs) from pole figure data (Section 5.3.3).

4.3 X-RAY DIFFRACTION METHODS

A complete x-ray texture goniometer system is composed of the following main units:

- X-ray generator and x-ray tube
- Goniometer with a sample stage
- Detection system

Electronics and computer controls to control the rotations and read the recorded intensities are also required. Figure 4.5 shows a photograph of the setup that has been developed at the Institut für Metallkunde und Metallphysik, RWTH Aachen

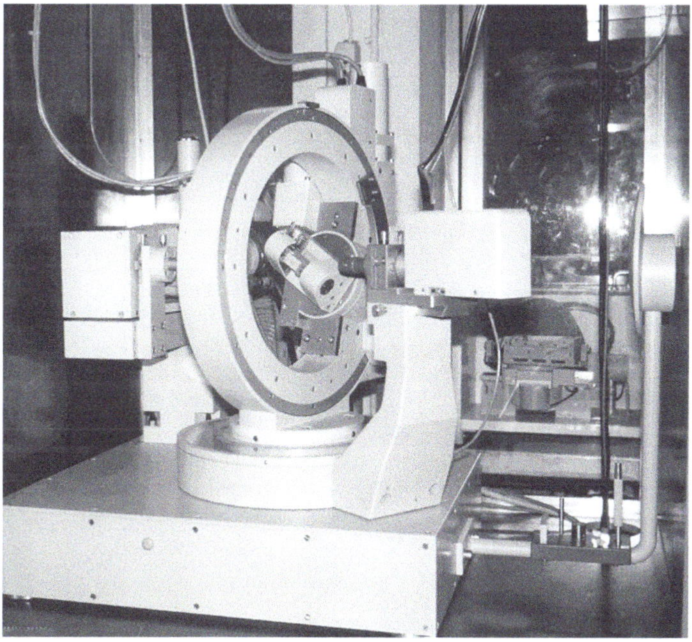

FIGURE 4.5 Texture goniometer at the Institut für Metallkunde und Metallphysik, RWTH Aachen University.

(Hirsch et al., 1986). Nowadays, fully functional texture systems are commercially available from several vendors. Sections 4.3.1 through 4.3.7 address the various units. As already mentioned, x-ray pole figure measurements are mostly performed in reflection geometry today. Accordingly, we will mainly focus on reflection geometry, and transmission geometry is only occasionally addressed for completeness.

4.3.1 Generation of X-Rays

In most texture applications, conventional sealed x-ray tubes with a particular target metal are used as an x-ray source. A high-voltage power source is connected across the cathode and anode of the tube, leading to the emission of electrons from the cathode (Figure 4.6a). When these electrons hit the anode, they are slowed down, producing x-rays. However, x-ray production is very inefficient—99% of the incident energy is lost in the form of heat—such that the dissipation of heat at the focal spot is one of the main limitations on the power that can be applied. By rotating the anode

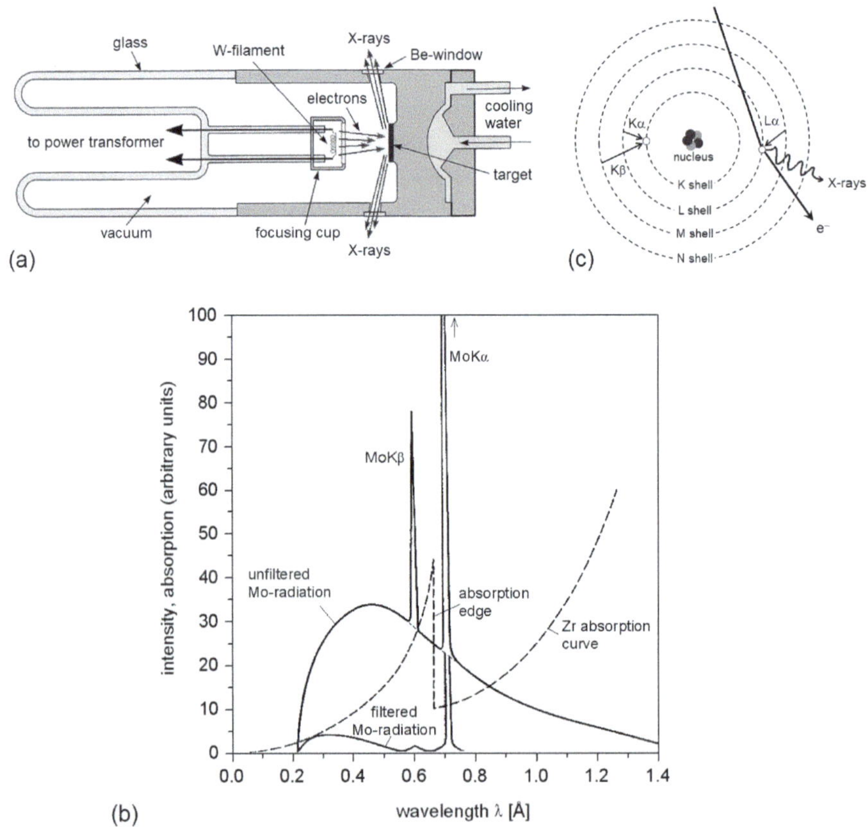

FIGURE 4.6 (a) Scheme of an x-ray tube for diffraction experiments, (b) emission spectrum of a molybdenum x-ray tube operating at 55 kV and absorption spectrum of a zirconium filter (see Figure 3.12), and (c) Siegbahn notation of characteristic x-radiation.

past the focal spot, the heat load can be spread over a larger area, enabling five to ten times higher intensities than in ordinary vacuum tubes. However, such *rotating anode tubes* are much more expensive and require considerable maintenance. Hence, they are not often used for diffraction experiments.

X-ray tubes produce a continuous wavelength spectrum (also called white radiation or *Bremsstrahlung*) that, at sufficiently high accelerating voltages of several tens of kilovolts, is superimposed by several sharp intensity maxima. As an example, Figure 4.6b shows the spectrum of a molybdenum-tube operating at 55 kV. The *continuous spectrum* originates when a highly accelerated electron encounters an atom in the target, here molybdenum. If the entire kinetic energy E of the incident electron is converted into a single X-ray photon the frequency of the radiation ν is given by the quantum relation

$$E = eV = h\nu \tag{4.1}$$

where e is the electron charge, V is the accelerating voltage of the tube, and h is Planck's constant. This frequency ν corresponds to the maximum energy of x-rays that can be emitted by the tube. In turn, the maximum energy of the x-rays defines the short-wavelength limit of the x-ray spectrum λ_{min} which, for a certain accelerating voltage V (in volt) is given by (in angstroms)

$$\lambda_{min} = \frac{12398}{V} \tag{4.2}$$

While all incoming electrons hit the target with the same energy eV, only rarely will they transfer this energy in a single event. Rather, in most cases, the electrons will dissipate their energy in a series of encounters with several target atoms and, as such, will produce either heat or x-ray quanta with lower frequency than the maximum. This results in the continuous spectrum extending from the above maximum frequency (minimum wavelength) down to very low frequencies or large wavelengths. The maximum in the intensity of the continuous spectrum occurs at a wavelength of roughly 1.5–2 times λ_{min}.

As soon as the accelerating voltage is increased beyond a critical value the continuous spectrum is superimposed by several characteristic sharp peaks. Since these peaks are extremely narrow and since their wavelengths are characteristic of the particular anode material, but do *not* depend on the accelerating voltage, they are called *characteristic lines*. Characteristic radiation appears when the bombarding electrons have sufficient energy to knock out electrons of an inner shell of the target material. Then electrons from higher states drop down to fill the vacancy, emitting x-ray photons with precise energies determined by the electron energy levels of the actual target material. For x-ray diffraction, the most important characteristic lines are $K\alpha 1$, $K\alpha 2$, and $K\beta$. In the *Siegbahn notation* of characteristic spectral lines, when an electron drops from the L electron shell down to the K shell, the emitted x-ray is called $K\alpha$ (Figure 4.6c). Similarly, $K\beta$ means that an electron falls from the M shell to the K shell. The difference between the $K\alpha 1$ and $K\alpha 2$ doublet lines is that they come from the two different subshells 2*s* and 2*p*. Accordingly, $K\alpha 1$ and $K\alpha 2$

TABLE 4.1
Characteristics of Various X-Ray Tubes and Appropriate Filters

	Wavelength (nm)		Kβ Filter			
Anode Material	**Kα**	**Kβ**	**Material**	**Edge Wavelength (nm)**	**Thickness for Kβ/Kα = 1/500 (μm)**	**Loss in Kα (%)**
Cr	0.22909	0.20848	V	0.22690	17	51
Fe	0.19373	0.17565	Mn	0.18964	18	53
Co	0.17902	0.16208	Fe	0.17433	19	54
Cu	0.15418	0.13922	Ni	0.14880	23	60
Mo	0.07107	0.06323	Zr	0.06888	120	71
Ag	0.05609	0.04970	Rh	0.05338	92	73
			Pd	0.05092	90	74

Source: Data taken from Hahn (2005).

have very close wavelengths which are often referred to as a single Kα line. Table 4.1 lists the wavelengths of the characteristic Kα and Kβ radiations of some typical anode materials. The characteristic Kα peak has an intensity that is about 100 times higher than the wavelengths of the adjacent continuous spectrum and about five to ten times higher than the Kβ peak, and this obviously favors the Kα radiation for x-ray diffraction experiments.

Table 4.1 indicates that among the typical tube materials, chromium has the longest characteristic wavelength and, as such, is well suited for diffraction experiments with materials with large unit cells as well as measurements of residual stresses with high resolution. Iron and cobalt tubes are used for analysis of ferrous alloys and steels to reduce the fluorescence in iron and are well suited for stress analysis. Copper is the most common tube material for diffraction experiments, including standard powder diffraction and measurements of residual stress and texture. Short wavelengths as provided, e.g., by molybdenum and, especially, silver contract the diffraction pattern toward low Bragg angles, leading to lower accuracy and resolution. Nevertheless, Mo tubes are commonly used for chemical crystallography due to their combination of high intensity and efficiency with relatively low absorption. Mo radiation is also applied for texture analysis of steels and other metals and alloys with comparably small unit cells. By way of contrast with the application of characteristic radiation for diffraction experiments, tungsten tubes operated at high accelerating voltages of 40–60 kV may be used when an intensive white x-ray spectrum is of more interest than the characteristic radiation, for instance for analysis of Laue diagrams (Section 3.4) or energy-dispersive measurements (Section 4.3.5).

For most techniques of texture analysis, monochromatic radiation is required; but, besides the strong Kα peak, the x-ray spectrum also contains the Kβ peak as well as the continuous spectrum. The Kβ peak can substantially be reduced by applying a filter in front of the detector, which is made of a material whose K-absorption edge (Section 3.6.1; Figure 3.13) lies just between the Kα and the Kβ peaks of the tube material. For a molybdenum tube, zirconium is the right filter material. The absorption edge of zirconium as well as the resulting filtered x-ray spectrum are shown in Figure 4.6b. Suitable filter materials for other tubes are given in Table 4.1. As an example, following the discussion in Section 3.5, for the commonly used copper tube, the intensity of the Kβ radiation is reduced to less than 1% if radiation passes through a 20-μm nickel foil (μ/ρ = 283 cm^2 g^{-1}) whereas Kα is only attenuated to 44%. Considering further that the intensity of Kβ has only 10%–20% of the intensity of the Kα peak in the first place, it means that an appropriate Kβ filter may substantially reduce the ratio of Kβ to Kα to values that are appropriate for texture analysis. The sixth column in Table 4.1 displays the thicknesses of the various filter materials leading to a reduction of the ratio of Kβ to Kα to approximately 1:500, which gives sufficiently "monochromatic" radiation for most texture applications. Thicker filters would further improve the ratio of Kβ to Kα, but of course they would also attenuate the intensity of the desired Kα radiation.

Continuous contributions to the x-ray spectrum remain at wavelengths larger than Kα and at very short wavelengths (Figure 4.6b). Although these intensities are relatively small compared to the Kα intensity, they may become significant in the case of very sharp textures, for example, for diffraction from single crystals or from thin films that are superposed on those from a substrate crystal. This is illustrated in Figure 4.7 for a 400-nm thin film of the high-temperature superconductor yttrium–barium–cuprate (YBaCuO) that had been deposited on a lanthanum-aluminate ($LaAlO_3$) substrate crystal (Wenk, 1992). The pole figure measured with truly monochromatic CuKα radiation (Figure 4.7b) contains much less peaks than the pole figure of the same sample measured with nickel-filtered CuKα radiation (Figure 4.7a). The additional poles arise from diffraction by the substrate with some wavelengths of the continuous spectrum, and thus they are artifacts. Figure 4.8 shows the {200}

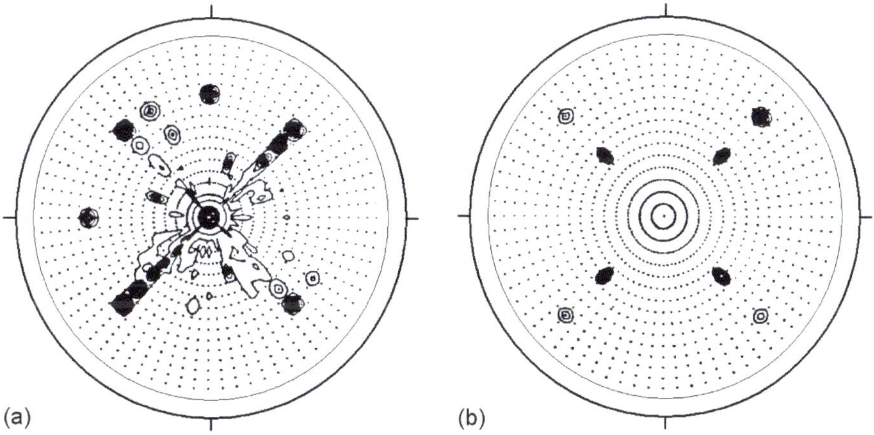

FIGURE 4.7 {102} Pole figure of a 400-nm YBaCuO film on a $LaAlO_3$ substrate measured with (a) Kβ-filtered CuKα radiation and (b) truly monochromatic CuKα radiation. Note additional peaks in (a) that are due to diffraction with different wavelengths on the substrate. (Courtesy of H.R. Wenk.)

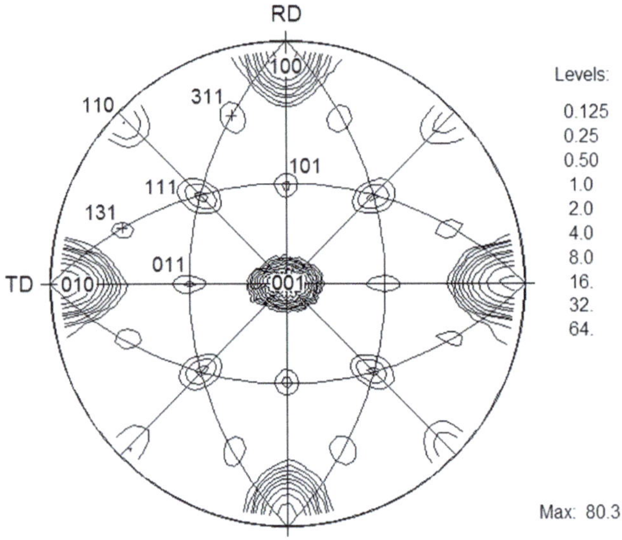

FIGURE 4.8 {200} Pole figure of highly textured aluminum foil, showing additional weak reflection peaks of nonmonochromatic radiation at different lattice planes.

pole figure of aluminum foil for high-capacity condenser applications. This material is produced to develop a very sharp cube texture (001)[100] so as to yield maximum capacitance (Section 5.5.4). It is seen that the nonmonochromatic parts of a Kβ-filtered x-ray spectrum are sufficiently strong to produce weak yet noticeable pole figure intensities for reflection at the {111}, {220}, and {311} planes.

Truly monochromatic radiation can be obtained by using single-crystal monochromators, which yield radiation of a certain wavelength in accordance with Bragg's law (Equation 3.3). The crystal type in a monochromator is selected based on the performance required in terms of intensity and resolution. Crystals such as Si, Ge, and quartz have small rocking angles, resulting in high resolution but low intensities. Graphite and LiF crystals, in contrast, produce high intensity but lower resolution due to larger mosaic spreads. With a mosaic graphite monochromator, intensities of the effective beam are only slightly lower than those for Kβ-filtered x-rays. As the monochromator also reduces the background intensity, the signal-to-background ratio is even better than that for filtered radiation. A flat monochromator crystal produces a parallel x-ray beam, whereas curved monochromators may be used to focus x-rays (Figure 4.9). The monochromator is best inserted into the primary beam between the x-ray tube and the collimator, since such a geometry causes no problems with defocusing during pole figure measurements. If the monochromator is inserted into the reflected beam, that is, between the sample and the detector, the x-ray path must be controlled by Soller slits that reduce the intensity.

As already addressed in Section 3.5, x-ray texture analysis is strongly influenced by the high absorption of x-rays in matter because this accounts for the penetration depth of x-rays in the sample and also influences the choice of an appropriate anode material of the x-ray tube. In general, absorption increases with rising atomic

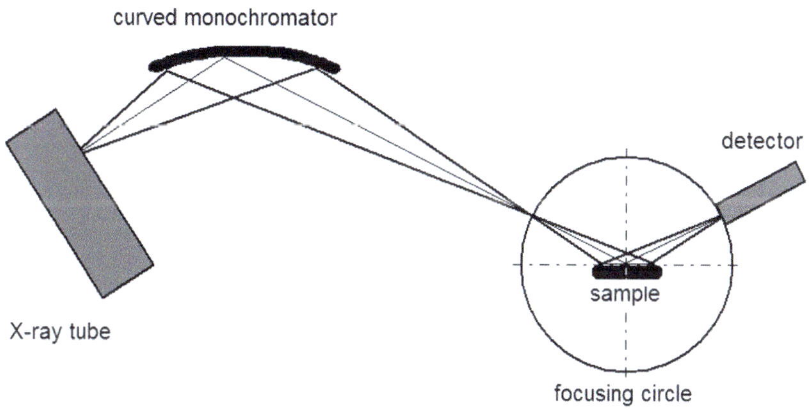

FIGURE 4.9 Principle of a curved monochromator in the primary beam path to produce monochromatic radiation.

number Z of the analyzed sample material (Table 3.4). Furthermore, it increases with the wavelength of the x-ray tube used, which means that x-rays with short wavelengths (silver, molybdenum, or copper anode) are typically less absorbed than those with large wavelengths (cobalt, iron, or chromium anode). However, there are some combinations of sample and target materials that yield anomalously high absorption because in these instances the x-ray photons have a sufficiently high energy to yield fluorescence (Section 3.6.1). Since high absorption is generally undesirable for texture analysis, such unfavorable combinations of x-ray tubes and sample material should be avoided. For example, steel or materials containing cobalt and manganese should not be analyzed with CuKα radiation (Table 3.4).

Another important aspect of absorption is the depth of x-ray penetration. The mass absorption coefficients for x-rays are typically of the order of 100 $cm^2\ g^{-1}$ (Table 3.4), and for transmission geometry it has already been discussed in Section 3.5 how the intensity of x-rays passing through a material with a given absorption coefficient can be derived (Equations 3.13 and 3.14). In reflection geometry, the situation is more complicated. Here the question is, from what depth of the specimen does the diffracted information come? For normal incidence, that is, a nontilted sample ($\alpha = 0°$), the fraction G_x of the total diffracted intensity that is contributed by the surface layer of depth x can be estimated from (Cullity, 1978)

$$G_x = 1 - \exp\left(\frac{-2\mu x}{\sin\theta}\right) \tag{4.3}$$

(The factor 2 considers that the radiation has to pass through the layer twice, namely, before and after the reflection.) If we now arbitrarily define that a contribution from the surface layer of 95% (or 99%) of the total diffracted intensity is sufficient, then x is the depth of penetration. It is seen that the penetration depth increases with Bragg angle θ, and it would also decrease with increasing sample tilt α. Figure 4.10 shows the evolution of G_x with depth x in aluminum for three different

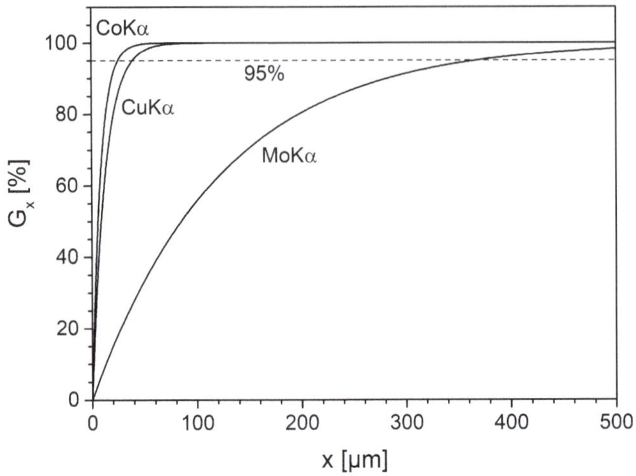

FIGURE 4.10 Evolution of G_x with depth x in aluminum for three different x-ray tubes (reflection angle $\theta = 20°$, normal incidence).

x-ray tubes—three different absorption coefficients (Table 3.4)—for a Bragg angle $\theta = 20°$ and a nontilted sample ($\alpha = 0°$). If we consider CoKα or CuKα radiation, it turns out that 95% of the reflected intensity comes from a layer with a thickness of 25 or 39 μm, respectively, which typically in recrystallized samples would mean that only one or two grain layers are recorded. Furthermore, this information is again heavily weighted; when 95% of the recorded information stems from a depth of 39 μm, then 50% of *that* information originates in the first 8.4 μm. Some further examples of sample thicknesses calculated for G values of 95% are listed in Table 4.2. Note that fluorescence leads to very low G values of only 2 μm if iron is analyzed with CuKα radiation.

4.3.2 Pole Figure Diffractometry in the Texture Goniometer

A pole figure goniometer essentially consists of a four-axis single-crystal diffractometer. The goniometer positions the detector with respect to the incident x-ray beam

TABLE 4.2
Depth x (in μm) That Leads to G_x Values of 95% for Different X-Ray Tubes and Materials Analyzed in Reflection Geometry at a Bragg Angle $\theta = 20°$

	CoKα	CuKα	MoKα
Al	25.4	39.1	368.0
Fe	12.3	2.1	16.9
Cu	7.0	10.8	11.2

at the proper Bragg angle 2θ. In the widely used *Eulerian cradle* (Figures 4.4 and 4.5), the sample is positioned relative to the x-ray beam by rotations about the three perpendicular axes φ, χ, and ω; the ω axis coincides with θ (Figure 4.4). Stepper motors, controlled by a personal computer, allow positioning of the goniometer to any arbitrary angular position of the four axes 2θ, ω, χ, and φ (within a certain range to avoid mechanical collisions).

In *reflection geometry*, a sample with a flat surface is mounted on the sample holder with its normal direction parallel to the axis of the φ-rotation (Figure 4.4b). Then, it is rotated in its plane about the φ axis, so that the angle φ corresponds to the azimuth β of a pole in the pole figure. After one full rotation, the sample is tilted about the χ axis, but as χ and the pole figure radial angle α are defined in the opposite direction, the relation is $\alpha = 90° - \chi$.

For most applications ω is kept constant at 0°, so that the χ circle is symmetrical between incoming and diffracted beams, that is, it is positioned at the Bragg angle θ. This arrangement is referred to as the Bragg–Brentano focusing condition for reflection geometry. ω-Scans (rocking curves) can be used to assess the quality of crystals and derive exact peak shapes (Heidelbach et al., 1992; Tizliouine et al., 1996; Sous et al., 1997; Borbély et al., 2008; Masiello et al., 2014). For special applications, such as texture measurements in thin films, it may be advantageous to use a low incident beam angle to increase the path length of the x-rays in the specimen (grazing beam technique; e.g., Bowen and Wormington, 1992; Szpunar et al., 1993; Van Acker et al., 1994; Tizliouine et al., 1996; Groves et al., 2001; Baker et al., 2010). In that case, the χ circle is no longer symmetrical and ω is the deviation from the bisecting position.

Although the limiting value for α in the Bragg–Brentano focusing condition is obtained when both the incident and the reflected beams are parallel to the sample surface, so that $\alpha_{max} \leq 90°$, there are two aspects that usually limit α_{max} to values of 60°–85°. Even when we analyze a sample with a perfectly uniform "random" texture, the intensity of the reflected peak strongly decreases with α. With increasing tilting angle α, the intersection of the incident x-ray beam with the sample—the projection of the beam onto the sample surface—becomes increasingly elongated; furthermore, different incident angles θ add an additional distortion of the projected beam (Figure 4.11a). Figure 4.11b shows the area irradiated by an originally square-shaped x-ray beam on the sample surface for different angles α and θ. Owing to the penetration depth of the x-rays into the sample, the true geometry of the reflecting volume is even more complicated. The projected area on the sample surface, and therewith the region from where x-rays are reflected, increases with increasing sample tilt α. It also increases with an increasing deviation of the Bragg angle θ from the $\theta = 90°$ position, and therefore pole figures should not be measured at $\theta < 10°$. To achieve optimum angular resolution, diffractometers are constructed such that reflection of the—originally divergent—incident x-ray beam at the plane surface of an *untilted* sample ($\alpha = 0°$) brings the diffracted beam back into a focal point that coincides with the receiving slits of the detector. Tilting of the sample to angles $\alpha > 0°$ has the effect that half of the sample is behind and half of it is in front of its original plane. For those sample regions, however, the focusing condition is no longer exactly fulfilled, which leads to a decrease in intensity simultaneously with a broadening of

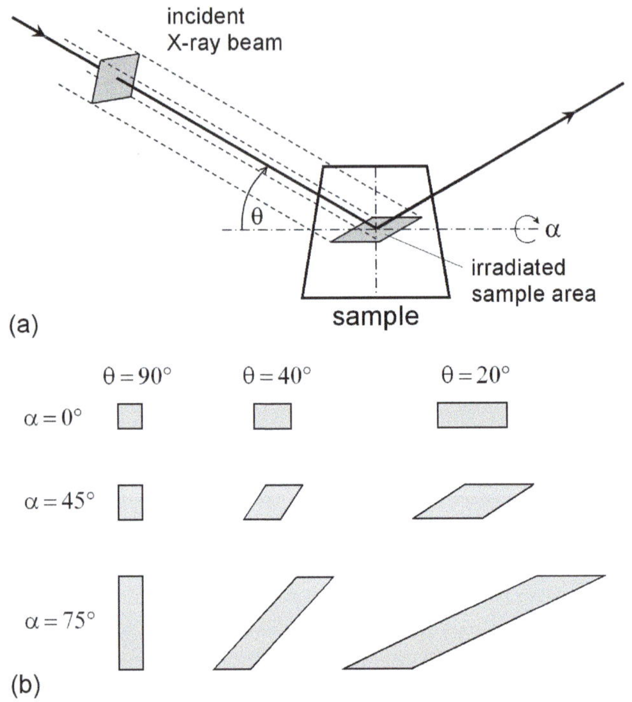

FIGURE 4.11 (a) Sketch to visualize the distortion of the irradiated spot on the sample surface. (b) Shape of the irradiated spot in dependence on tilting angle α and Bragg angle θ for a square-shaped incident beam.

the reflected peaks. Accordingly, this effect is called the *defocusing error* (Huijser-Gerits and Rieck, 1974).

Figure 4.12a shows the peak broadening for two diffraction peaks (111 and 200) of a randomly oriented copper powder sample for three different tilting angles α. Although the *integral* intensity of the peak remains constant, the detector, which is equipped with a system of receiving slits, records only the intensities diffracted from a small, constant area of the sample surface, so that the recorded intensity decreases with increasing sample tilt (Figure 4.12b). Measures to correct for the intensity drop due to defocusing are given in Section 4.3.6. Besides the defocusing error, absorption of the x-rays within the sample also has to be considered; with increasing sample tilting, the path length of the x-rays in the sample increases, which means that the reflected intensity decreases due to absorption. However, for thick samples this increase in absorption is exactly balanced by the increase in irradiated volume that contributes to the diffracted intensity (Schulz, 1949a). Note that "thick" samples mean that the sample is thicker than the x-ray penetration depth, which, as we have seen earlier, is usually only of the order of 100 μm. However, this does not hold for very thin samples; for example, for thin films, and in such cases, absorption has to be corrected for as well (Section 4.3.6).

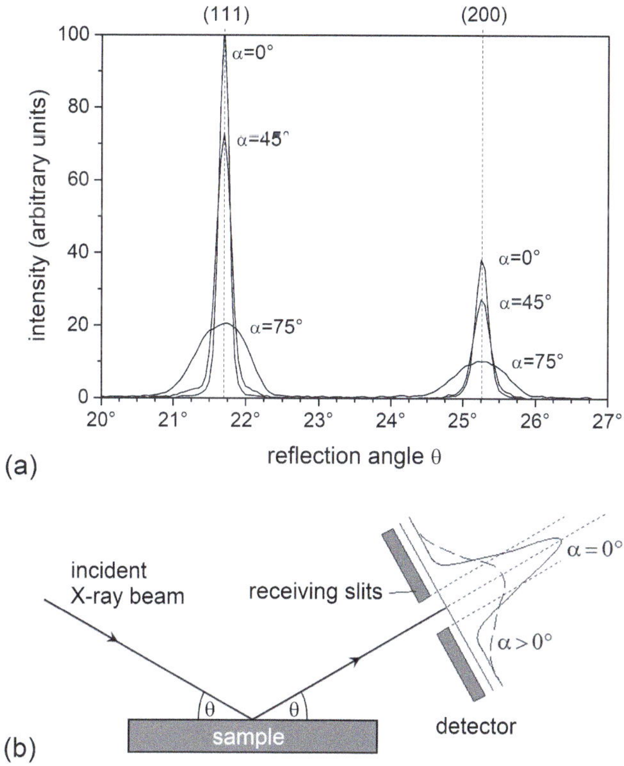

FIGURE 4.12 (a) Peak broadening with increasing tilting angle α (copper powder sample, CuKα radiation). (b) Sketch to visualize the defocusing error.

In *transmission geometry*, the sample holder is turned 90° such that the sample, a thin sheet, is mounted normal to the χ axis (Figure 4.4a). Here, the angle φ is kept constant at 0° and the sample is tilted about the ω axis, which corresponds to the complement of the pole distance (90° – α) in the pole figure. The azimuthal angle β in the pole figure is obtained from a rotation of the sample about the χ axis. In transmission geometry, the outer parts of the pole figure with $\alpha_{min} \le \alpha \le 90°$ are covered. As described for reflection geometry, the recorded intensity strongly decreases with sample tilt ω, so that the practical minimum of α is reached before the theoretical limit, which is given when the incident or reflected beam is parallel to the sample surface, so that $\omega < 90° - \theta$ or $\alpha_{min} = \theta$. In transmission geometry, the increase in absorption with sample tilt is not compensated for by the increase in the reflecting volume (Decker et al., 1948; Schulz, 1949b), so that an absorption correction is always necessary (Section 4.3.6).

We have seen that for pole figures measured in reflection geometry, the errors strongly increase with α, so that the inner circles of a pole figure (e.g., $\alpha < 45°$) can practically be obtained without corrections, whereas the outer parts need strong corrections ($\alpha > 45°$) or are even not accessible at all ($\alpha \approx 90°$). In transmission geometry vice versa, the outer parts of a pole figure can readily be determined and the inner circles are missing. Accordingly, *complete pole figures* can be obtained by combination of two pole figures, one of which is measured in reflection and the other in transmission geometry. Subsequently, the pole figure data are overlapped, which often requires some intensity adaptations to bring the two measurements to the same intensity scale. However, because of the laborious preparation of samples for transmission texture measurements, nowadays in most cases only incomplete pole figures are measured in reflection with tilting angles up to typically 60°–85°. If necessary, complete pole figures are derived by recalculation from the 3D orientation distribution functions (Section 5.3.2).

Besides the rotations to scan the pole figure described so far, the sample can be subjected to an additional translation in its plane to increase the sampling area (Bunge and Puch, 1984; Section 4.6).

4.3.3 Principles of Pole Figure Scanning

Older texture goniometers were driven by only one motor, and the pole figures were scanned by changing both angles α and β simultaneously on a spiral (Figure 4.13a). In modern stepper motor-driven goniometers, the pole figures are mostly scanned on equidistant concentric circles of constant α in steps of typically 5° in both α and β (Figure 4.13b). For a complete pole figure, this results in (α × β) = 19 × 72 = 1,368 points. At counting times of typically 1 s per point, this yields a total measuring time of less than 30 min per pole figure. For very sharp textures, the standard 5° × 5° grid may be too coarse and finer grids, for example, 2.5° × 2.5°, are required. Also, for calculating ODFs according to the series expansion method (Section 5.3.2) with high series expansion coefficients l_{max}, finer grids may be necessary to derive a sufficient number of data points.

For each point (α,β), the reflected intensity is measured by an electrical counter (Section 4.3.4) that "sees" a certain angular range of the pole figure. This *pole figure window* not only is defined by the aperture systems of x-ray tube and counter, and especially by the receiving slits of the latter, but also depends on the geometry of

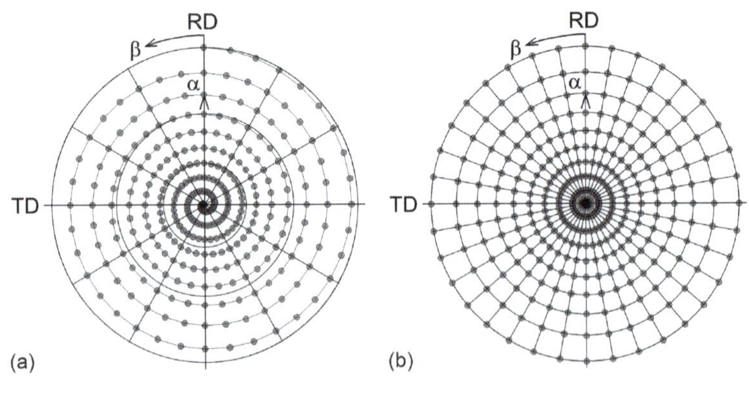

FIGURE 4.13 (a) Spiral and (b) circular, equal-angle scanning grids for pole figure measurements (note that for clarity both scanning schemes are shown in steps of 20°).

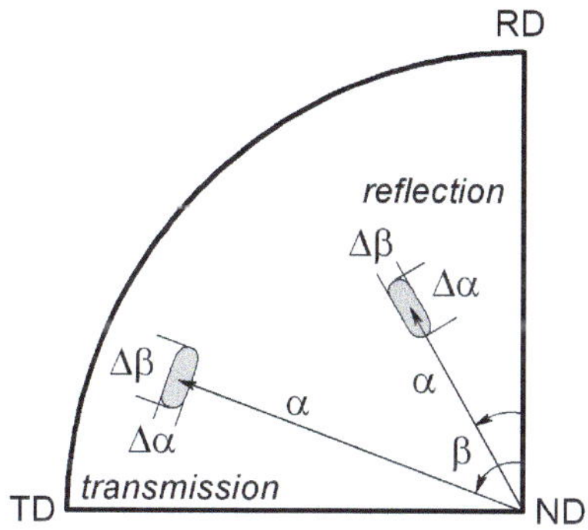

FIGURE 4.14 Shape of the pole figure window for reflection and transmission geometry.

the setup, that is, the reflection or transmission geometry, and on the settings of the Bragg angle θ and sample tilt α. Usually, the pole figure window has an elongated shape, as shown schematically in Figure 4.14.

With regard to pole figure scanning, two different modes are possible: step scanning and continuous scanning (Bunge and Puch, 1984). In the step mode, the sample is rotated by a certain angle Δβ, typically 5°, either on a spiral or on concentric circles, and then at this position the reflected intensity is determined by counting for a given time, for example, 1 s. In the continuous mode, the intensities are recorded continuously, thus integrating over an angular interval Δβ, and subsequently the measured intensity is attributed to the center point β of that interval. In modern texture goniometers, the rotations are performed by stepper motors, which can be moved in small steps of say 0.01° so that with a suitable computer control both scanning modes can readily be realized.

It should be noted that in the step mode, intensities are missed and sharp peaks may even be skipped entirely, as the step width is typically larger than the pole figure window Δβ (Figure 4.14). Continuous scanning, however, records all intensities in a certain range of Δβ. As the length Δα of the pole figure window concurrently integrates over a range Δα, continuous scanning ensures a rather comprehensive coverage of the entire pole figure. However, continuous scans are restricted to homogeneous scanning grids like spirals or concentric circles, whereas more complex scanning grids, such as equal area scans (see the following text), require step scanning.

In the standard *equal angle* (α × β) 5° × 5° grid, the density of measuring points per area is very inhomogeneous, as the pole figure center is covered with a much higher density in β than the outer circles (Figure 4.13b). To cover the pole figure more uniformly, an *equal area* scan can be applied, and various schemes to achieve equal area coverage have been proposed in the literature (Welch, 1986; Will et al., 1989; Matthies and Wenk, 1992; Morris and Hook, 1992; Helming, 1996). All these schemes have the following in common: They are close to a true equal area coverage (Appendix IV) and reduce the number of measurement points by about one-half (Figure 4.15). Thus, for a given angular resolution, the recording time can be halved, which is particularly interesting if measuring time is of concern (e.g., in neutron diffraction; Section 4.4), or, by allowing the same total measuring time, the angular

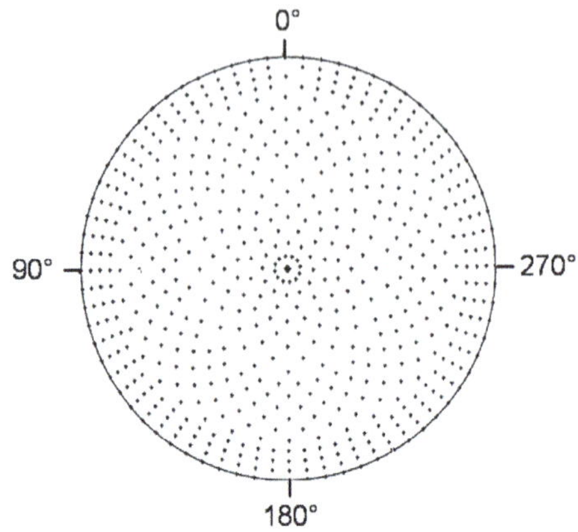

FIGURE 4.15 Equal area scanning grid for pole figure measurements. (Adapted from Brokmeier, 1989.)

resolution can be doubled. If necessary, the data for the conventional 5° × 5° grid, or for any other grid, are subsequently obtained by interpolation.

More recently, the idea of a uniform coverage of pole figures has been further advanced by applying a locally refined grid with the refinement depending on the pattern of preferred crystallographic orientation (Schaeben et al., 2007). This is achieved by an adaptive successive refinement of an initial coarse uniform grid as a function of the texture being measured. The prerequisite of this automated refinement is a very fast pole figure inversion that does not rely on a predefined grid during the course of the texture measurement.

4.3.4 X-Ray Detectors

In modern computer-controlled texture goniometers, the reflected x-ray intensities are recorded by means of an appropriate electrical counter. There are four different types of counters that are suitable for x-ray diffraction experiments (e.g. Prince, 2004):

- Geiger counters
- Scintillation counters
- Proportional counters
- Semiconductor counters

All these counters depend on the power of the x-rays to ionize atoms, either of a gas (Geiger and proportional counters) or of a solid (scintillation and semiconductor counters). Appendix VI describes the working principles and characteristics of these four types of x-ray counters.

Geiger counters were the first electrical counters used to detect radiation, but these are now virtually obsolete in diffractometry. In all counters that are commonly used today—scintillation, proportional, and semiconductor—the output voltage is proportional to the energy of the triggering x-ray quanta. Therefore, electrical units that distinguish between pulses of different sizes, *pulse-height analyzers*, can distinguish between x-rays of different energies. As discussed in more detail in Appendix VI, with the use of pulse-height analyzers, background and fluorescent radiation with different wavelengths can efficiently be reduced, so that the peak-to-background ratio can markedly be improved.

When we analyze the diffraction spectra of materials with high crystal symmetry, cubic or hexagonal, the diffraction lines are usually widely spaced (Figures 3.6 and 3.8). Thus, even when considering some peak broadening at large tilting angles (defocusing error; Sections 4.3.2 and 4.3.6), the reflected peaks can generally be separated easily (Figure 4.12a). In some cases, finer receiving slits may be necessary to improve angular resolution (e.g., to separate the 0002 and 1011 reflections in Figure 3.8). However, this is not necessarily true in samples with line-rich diffraction patterns—materials with low crystal symmetry or multiphase alloys, where peaks may partially, or even completely, overlap (Section 4.5.1). In such cases, the peaks can no longer be separated, and peak profile analysis becomes necessary to attribute unambiguously the recorded intensities to the individual *hkl* reflections. To register the peak profiles with conventional counters, the $\theta/2\theta$ spectrum must be scanned successively with a small receiving slit for each pole figure point (α,β) to achieve sufficiently high angular resolution. The resulting peaks are then separated by an appropriate deconvolution procedure, for example, by fitting the diffraction spectra with Gauss, Lorentz, or other functions, which finally yields the exact integral intensity of the various peaks for each pole figure point. This is an extremely time-consuming procedure, however, which can be accomplished much more easily with the help of a location-sensitive counter that records the entire peak profile simultaneously (Heizmann and Laruelle, 1986; Wcislak et al., 1993; Tarkowski et al., 2005).

A typical one-dimensional *position-sensitive detector* (PSD) suitable for texture analysis consists of a location-sensitive proportional counter with a resistance wire anode (Figure 4.16a). An incoming x-ray quantum generates an electrical charge Q at the point x of the anode. This electrical charge Q has the possibility to flow off in both directions to the left and right preamplifiers. However, due to the resistivity of the wire, the partial charges Q_l and Q_r flowing to either side generate voltage pulses U_l and U_r with the ratio $U_l/U_r = x/L{-}x$, where L is the active length of the resistance wire. Thus, the location x of the incoming x-ray quantum can be detected electronically and the electrical pulse is accumulated in the corresponding channel of a multichannel analyzer (MCA). So, eventually, the contents of the various channels of the MCA reflect the intensity distribution of the diffraction peak at different locations of the PSD. Alternatively, a linear silicon strip detector has been developed that consists of 128 Si strips, each 100 μm wide, so that the total width is 12.8 mm (Dabrowski et al., 2003).

In most applications, short linear detectors are used, which are mounted tangential to the 2θ-goniometer circle (a setup where the PSD records intensities along the α circle is described in Section 4.4.1). Such PSDs typically provide a viewing angle of 5°–10° with an angular resolution of the order of 0.01°, which is sufficient to record individual peaks, inclusive of peak broadening at high tilting angles, and this also permits the separation of overlapping peaks. Long, curved detectors with viewing

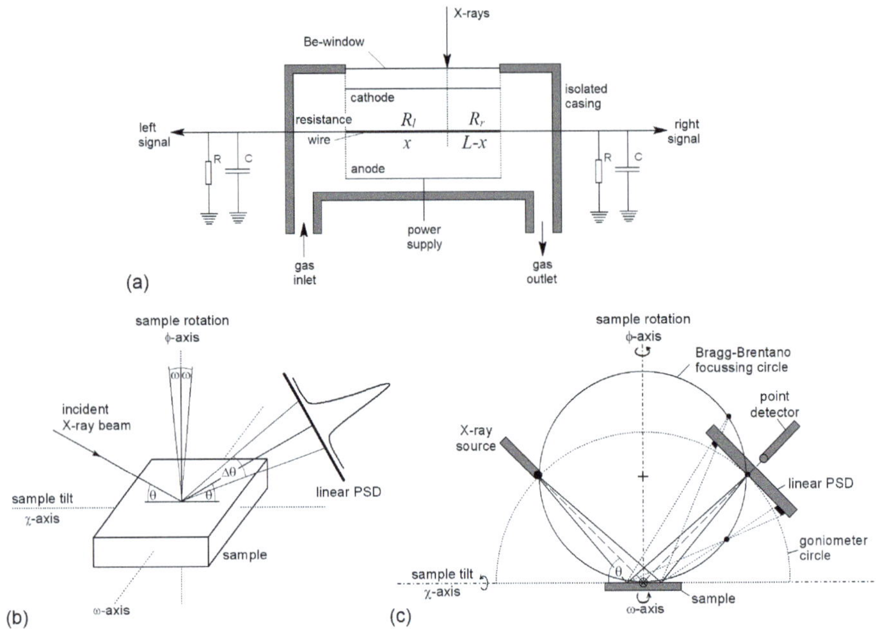

FIGURE 4.16 (a) Principle of a one-dimensional location-sensitive proportional overflow, overpressure gas ionization counter. (b) Pole figure measurements with a position-sensitive detector (PSD). (c) Deviation from the Bragg–Brentano focusing condition for a linear PSD. (Adapted from Wcislak and Bunge, 1996.)

angles of up to 120° are also in use (Wölfel, 1983; Bessières et al., 1991). However, these are mainly used to record diffraction spectra for phase analysis rather than for texture measurements, since with increasing deviation from the Bragg–Brentano diffraction condition two problems arise, which are described in the following.

In the Bragg–Brentano diffraction condition, the Eulerian cradle is set in such a way that the χ circle coincides with the bisector of the angle between the incident and diffracted beams, and any direction on the pole figure can be brought parallel to this direction by two rotations, 90°-χ and φ, corresponding to the pole figure coordinates α and β (see the preceding text). With a PSD this is true only for one position that is usually the middle of the detector, whereas for all other positions the χ circle is inclined by ω to the bisector (Figure 4.16b) (in the PSD geometry, ω is equal to $\Delta\theta$, the deviation of the χ circle from the bisecting position). As illustrated in Figure 4.16c, this means that the focusing condition is violated, leading to some peak broadening, and this effect increases with increasing distance from the exact Bragg–Brentano reflection condition (Heizmann and Laruelle, 1986; Wcislak and Bunge, 1991). Furthermore, when several pole figures are measured simultaneously, a given setting (ω,χ,φ) of the texture goniometer corresponds to different pole figure angles (α,β). Thus, the (ω,χ,φ) coordinates need to be transformed into the pole figure angles (α,β), and the corresponding transformation laws have been derived for transmission geometry (Bunge et al., 1982; Heizmann and Laruelle, 1986) and reflection geometry (Wcislak and Bunge, 1991). However, poles with $\alpha < \omega$ cannot be brought into coincidence with the bisector, which results in a nonmeasured "blind" area in the center of the pole figure. With increasing deviation from the exact Bragg–Brentano condition (with increasing ω), the size of this blind area increases, which means that the pole figures become less meaningful. Thus, for these two reasons—the violation of the focusing condition and the blind area—too large deviations from the exact Bragg–Brentano condition should be avoided.

A one-dimensional PSD records the intensities linearly, mostly along a circle in the equator plane, whereas a two-dimensional (area) detector would simultaneously register a certain part of the sphere. Accordingly, a conventional single detector can be considered as a zero-dimensional point counter. As area detectors yield a high angular resolution in two dimensions, they enable measuring sharp textures with high accuracy. If materials with peak-rich diffraction spectra are investigated, peak profile analysis can be performed and, under certain circumstances, the high angular resolution even allows determination of the orientations of individual crystals in a polycrystalline aggregate (Bunge and Klein, 1996; Rodriguez-Navarro, 2006).

In principle, a *photographic film* positioned so as to record part of the diffraction sphere is a 2D area detector (Wassermann and Grewen, 1962). As a matter of fact, area detection using photographic film was essential to the historic first observation of x-ray diffraction by Laue and colleagues (Friedrich et al., 1912). In the 1990s, *image plates* have been introduced for texture analysis (Brokmeier and Ermrich, 1994; Ermrich et al., 1997), which are far superior to films in terms of dynamic range, sensibility, and efficiency. However, they also require separate registration and readout periods of several minutes, which prevent an interactive, online analysis of the results. To overcome this shortcoming, charged-coupled device (CCD)-based detector systems have been developed for recording diffraction patterns generated by synchrotron radiation (Bourgeois et al., 1994; Garbe et al., 1996; Section 13.2). For real-time texture analysis in conventional x-ray texture goniometers, Siemens developed a 2D *multiwire proportional counter* filled with a high-pressure xenon gas mixture (Smith and Ortega, 1993; He, 2003). The detector is equipped with a concave beryllium window with a diameter of 11.5 cm. The reflected intensity data are received as a 1024 × 1024 pixel frame that can be transferred into a computer for storage and evaluation. Such a system is available from Bruker AXS as "general area detector diffraction system" (GADDS) with the detector HI-STAR and its more recent successor VÅNTEC-500 (He, 2012). Figure 4.17a shows a schematic of a standard x-ray diffractometer equipped with a 2D area detector, which simultaneously records parts of two Debye–Scherrer rings, whereas Figure 4.17b gives an example of the {111} and {200} rings of a heavily textured aluminum sample.

For an area detector all diffraction vectors belonging to one Debye sector form an arc (bisector) on the pole sphere, with its center being given by the direction of the primary beam (Bunge and Klein, 1996; Helming and Prechwinkel, 2005; Rodriguez-Navarro, 2007). Thus, for a fixed position of the pole figure angles ω and α, a large portion of the pole figure can be scanned with one β scan, that is, sample rotation. The exact coverage of the pole sphere naturally depends on many geometrical details of the entire setup, including the preset diffraction angle 2θ and, most notably, the distance between the sample and the detector. In general, a shorter sample-detector

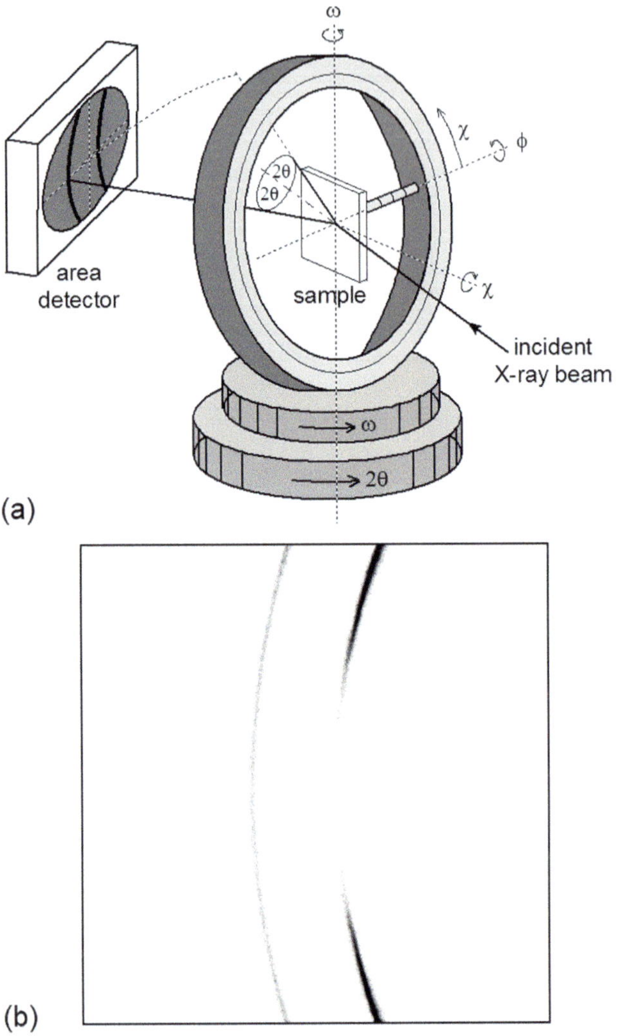

(a)

(b)

FIGURE 4.17 (a) Pole figure measurement with an area detector; (b) {111} and {200} Debye–Scherrer rings of a strongly textured aluminum sample. (Courtesy of I. Lischewski.)

distance leads to a greater field of view, but at the expenses of angular resolution. Figure 4.18 shows an example of the pole figure coverage ({111} and {200} pole figures simultaneously), which may be achieved with two β scans for α angles 40° and 80°, respectively (black symbols). This may be combined with an ω scan performed at constant α and β (gray symbols). However, the maximum angle ω and hence the covered pole figure area is restricted by the Bragg angle θ, since 0° < ω < θ. Thus, in the present example, three scans with the area detector give approximately the same information as 38 scans with a conventional point detector. Similarly, as discussed

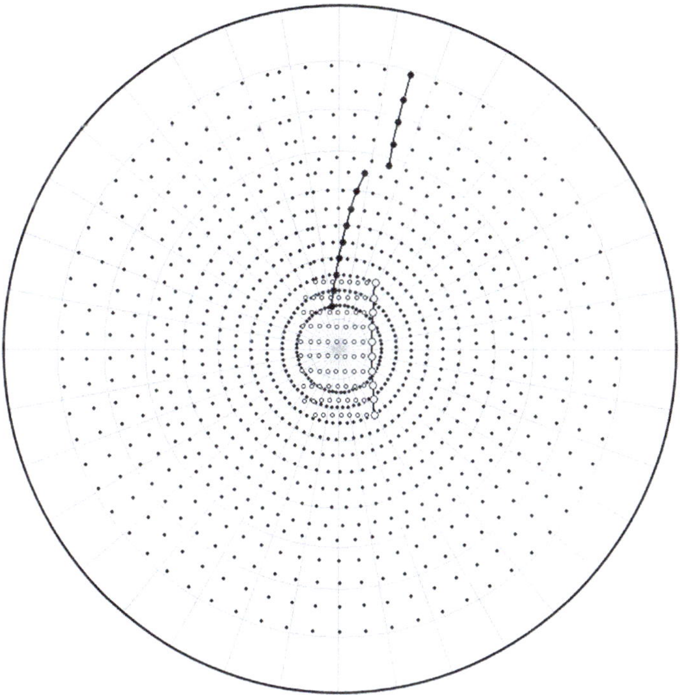

FIGURE 4.18 Pole figure coverage with two β scans (rotations) for $\alpha = 40°$ and $80°$ (black symbols) and one ω scan at $\alpha = \beta = 0°$ (gray symbols). In this example, the diffraction angle was $2\theta = 45°$, enabling simultaneous analysis of {111} and {200} pole figures in an aluminum sample.

for the one-dimensional PSD, the results obtained by an area detector have finally to be transformed into the standard $\alpha \times \beta$ pole figure grid for further evaluation and ODF computation.

4.3.5 Energy-Dispersive Diffractometry

In all applications mentioned so far, the counters were used in an *angular-dispersive* way—the diffracted intensities for monochromatic radiation were recorded in dependence on different angular settings of the sample, and energy discrimination is only used to reduce background and fluorescence by setting a narrow window of the pulse-height analyzer (Appendix VI). Figure 4.19 is a representation of Bragg's law $\theta = f(\lambda)$ for reflection at the various lattice planes (hkl) in copper. To determine the reflection spectrum in conventional angular-dispersive experiments with monochromatic radiation, $\theta/2\theta$ scans are performed with x-rays of a given wavelength, for example, $\lambda = 0.15$ nm (line B in Figure 4.19), and diffraction peaks are detected at each intersection of line B with the Bragg curves for the lattice planes (hkl).

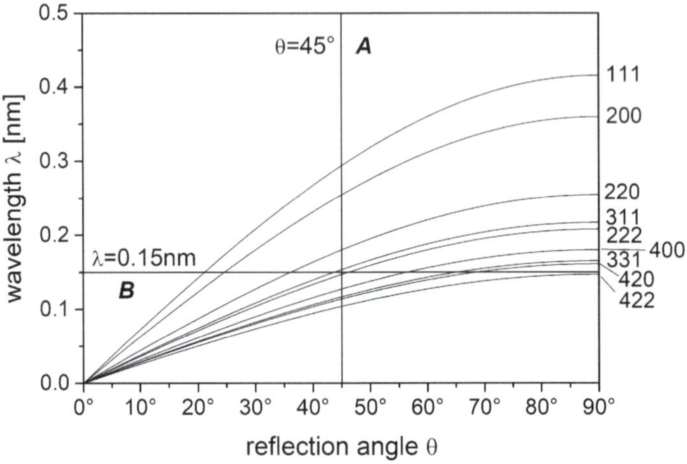

FIGURE 4.19 Representation of Bragg's law for reflection at the various lattice planes (*hkl*) in copper. The vertical line *A* represents the energy-dispersive and the horizontal line *B* the angular-dispersive diffraction experiments.

Alternatively, the energy or wavelength of the reflected radiation can be exploited as well (*energy-dispersive diffractometry*; Giessen and Gordon, 1968). If we introduce the energy of the radiation into Equation 3.3, we can write Bragg's law in a different way:

$$E = h\nu = h\frac{c}{\lambda} = \frac{hc}{2\sin\theta} \cdot \frac{1}{d_{hkl}} \tag{4.4}$$

It is seen that at a constant setting with Bragg angle θ, the energy of the diffracted radiation E is a measure for the lattice spacing d_{hkl} of the reflecting crystal planes (*hkl*). Thus, for polychromatic radiation covering a whole range of wavelengths, a detector at a fixed scattering angle, for example, $\theta = 45°$ (line *A* in Figure 4.19), would again record the various *hkl* diffractions that now are distinguished by different wavelengths. To perform energy-dispersive diffractometry, the diffracted intensities of polychromatic radiation—e.g. produced with a tungsten tube—are recorded by an energy-dispersive detector and then analyzed by an MCA that transforms the pulse-height spectrum into a wavelength (or energy) spectrum. To achieve maximum energy resolution, use of a semiconductor counter is recommended, but nonetheless, the resolution of neighboring diffraction lines is inferior to that of conventional angular-dispersive techniques. The diffraction spectrum is obtained by plotting the contents of the various channels of the MCA and, as an example, Figure 4.20 shows a diffraction spectrum obtained from a polycrystalline steel sheet measured with a chromium tube at a constant Bragg angle $\theta = 15°$. The fluorescence lines at the low-energy side of the spectrum are characteristic of the sample material, whereas the diffraction lines disclose its crystallographic structure.

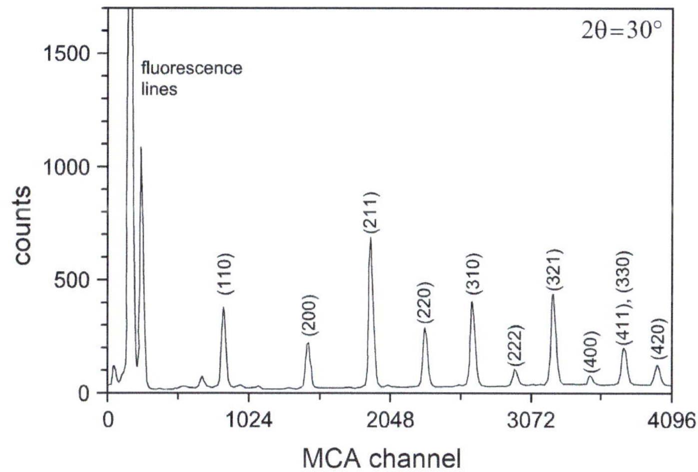

FIGURE 4.20 Diffraction spectrum of a steel sheet with random texture, obtained by energy-dispersive diffractometry. During the measurement, the sample was rotated about its normal direction; the time necessary to acquire the diffraction spectrum was 200 s ($2\theta = 30°$, Cr tube, liquid nitrogen-cooled Ge-semiconductor counter). (Adapted from Shimizu et al., 1986.)

To utilize the results obtained by energy-dispersive diffractometry for deriving the texture of the sampled volume, the wavelength spectrum must be convoluted into the contributions from the different reflector planes. By rotating the sample in a goniometer as in conventional angular-dispersive applications, a whole range of lattice planes can be registered simultaneously (Szpunar and Gerward, 1980). This means that the measuring time is virtually independent of the number of pole figures recorded, which makes this method particularly attractive for low-symmetry materials with peak-rich diffraction spectra. Furthermore, as mentioned for the PSDs, overlapping peaks can be separated by peak profile analysis.

Alternatively, the diffraction spectra can be exploited for a given, fixed sample position, which means that the sample does not have to be moved during texture analysis. The resulting texture information corresponds to an inverse pole figure of the sample surface (Section 4.3.7). Texture data from four inverse pole figures have been used to calculate a 3D ODF with a reasonable accuracy (Szpunar, 1990). Texture analysis at a fixed sample position is also of great interest for in-situ experiments (Gerward et al., 1976) as well as for online quality control in production plants. In the arrangement described by Kopinek (1994) and Klinkenberg et al. (1999), a system for online texture analysis with energy-dispersive detectors has been installed at the exit of a continuous annealing line of steel sheets. Four detectors, each recording the reflected intensities of five lattice planes, provide enough information to compute the low-order C coefficients that can be used to calculate average physical properties of the final sheet product, including *r* value, Young's modulus, and magnetic properties (Bunge, 1993).

4.3.6 Correction and Normalization of Pole Figure Data

Before the pole figures can be evaluated, or used for subsequent ODF calculation, several corrections have to be applied (Humbert, 1986). This section discusses measures to correct for background errors and errors introduced by defocusing and absorption and also addresses the influence of counting statistics. Finally, the texture data that are given as counts per pole figure point need to be normalized to standard units—multiples of a random orientation distribution (m.r.d.)—that are independent of any experimental parameters.

The first error that has to be corrected for is the *background error* (Figure 4.21). Background intensities are caused by incoherent scattering and fluorescence in the sample. Minor contributions may also result from interaction of the x-ray beam with any material in the path of the x-rays—collimator, beam stop, and air—as well as from electronic noise. Especially when the experiments are performed with "monochromatic" x-radiation produced by a Kβ-absorption filter, reflection of x-rays with other wavelengths of the continuous spectrum strongly contributes to the overall background intensity. Several measures to improve the peak-to-background ratio have been addressed in earlier sections: to prevent fluorescent radiation an appropriate x-ray tube with wavelength larger than the K-absorption edge of the sample material should be used (Section 4.3.1), but the short-wave components of the continuous spectrum may still excite fluorescence. A monochromator (Section 4.3.1) or an energy-dispersive detector (Section 4.3.5) substantially reduces background intensities with different wavelengths. With conventional counters, the background can be reduced by choosing a small window of the pulse-height analyzer (Appendix VI). Background is also diminished by using smaller receiving slits, but this would enhance the "defocusing error," that is, the drop in peak intensity with increasing tilting angle α (see the following text).

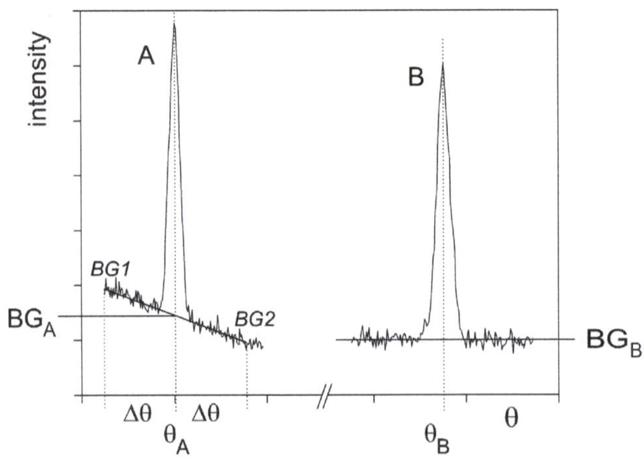

FIGURE 4.21 Determination of the background intensity *BG* in a θ/2θ-diffraction spectrum. Case A illustrates a strong dependence of the reflected intensity on θ, whereas in case B the background is almost independent of θ.

Although background intensities can—and of course should—be reduced by such measures, they cannot completely be eliminated in practice, and therefore the diffraction data have to be corrected by subtracting the background from the measured intensity (Figure 4.21). For correcting pole figure data, the background intensity BG is determined and subtracted from the measured pole figure intensity I_{meas}:

$$I_{corr} = I_{meas}(\alpha, \beta) - BG(\alpha) \tag{4.5}$$

Within a given pole figure, the background changes with increasing tilting angle α, but—for sufficiently large samples—does not depend on the sample rotation β. Thus, a background correction curve $BG(\alpha)$ can be obtained by measuring a complete pole figure away from the diffraction peak θ and integrating over the pole circles, that is, over β. At small angles θ the background typically strongly decreases with θ, and hence it is necessary to measure the background at either side of the Bragg peak and subsequently interpolate the results (case A in Figure 4.21). Measuring at one side may suffice if the background intensities remain more or less constant with θ (case B in Figure 4.21). Of course, in both cases, it must be ensured that the distance from the Bragg peak is large enough to allow for peak broadening at large sample tilts α but that it does not interfere with neighboring peaks. The decrease in the background intensity with α is fairly independent of θ. Therefore, it is sufficient to measure the background intensity only at one angle α, typically $\alpha = 0°$, and then derive the background values for other α from a predetermined correction curve.

Not only the background intensity but also the reflected peak intensity strongly decreases with α. Increasing sample tilt leads to a broadening of the reflected peaks (Figure 4.12a). The detector, however, is equipped with a system of receiving slits and hence "sees" only a small, constant area of the sample surface, so that peak broadening results in a marked decrease in recorded intensity (Figure 4.12b). A typical intensity drop obtained for copper with increasing α, normalized to 100% for $\alpha = 0°$, is shown in Figure 4.22a for two different Bragg angles θ, that is, two different pole figures.

To correct for this *defocusing error*, a correction function $U(\alpha)$ must be applied, which for any value of α normalizes the intensity of a random sample to the values at $\alpha = 0°$:

$$I_{corr} = \frac{I_{meas}(\alpha, \beta) - BG(\alpha)}{U(\alpha)} \tag{4.6}$$

As described in Section 4.3.2, the defocusing error is mainly determined by the sample position in terms of the two angles α and θ. Therefore, it appears reasonable to obtain the correction curve from geometrical considerations (Gale and Griffiths, 1960; Tenckhoff, 1970; Ortiz and Hermida, 1981; Saleh et al., 2016). However, the intensity drop also depends on the alignment of sample and goniometer as well as the size of the collimator and the receiving slit. The smaller the collimator size and the larger the receiving slit, the smaller is the defocusing error, though large receiving slits decrease the angular resolution. Thus, empirical correction curves $U(\alpha)$, as derived from measurements of a random sample with the particular experimental

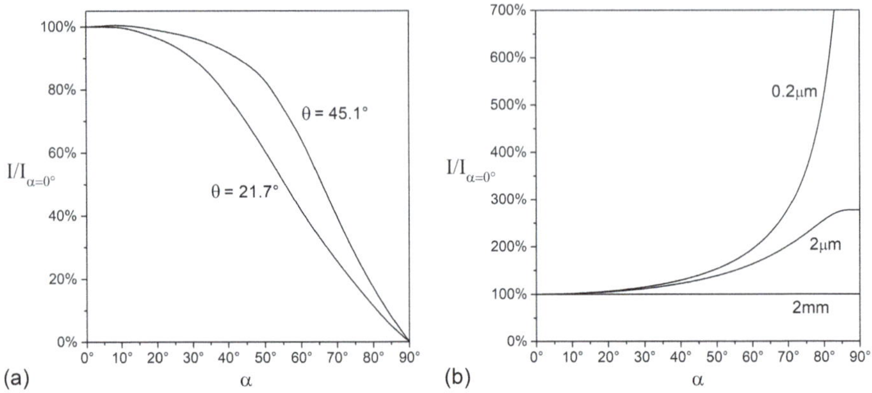

FIGURE 4.22 (a) Intensity drop with increasing sample tilt α, normalized to 100% for α = 0°. Data were obtained experimentally by recording the {111} (θ = 21.7°) and {311} (θ = 45.1°) pole figures of a copper powder sample with CuKα radiation. (b) Evolution of *I* with α for three different sample thicknesses, computed according to Equation 4.8 and normalized to 100% for α = 0°. For the 2-mm-thick sample the increase in absorption is exactly balanced by the increase in diffracting volume, whereas for the thinner samples an absorption correction is necessary ({200} pole figure).

setup, usually give better results. For that purpose, pole figures of a sample with a random texture are measured at the Bragg peak, integrated over the rotation angle β, and normalized to unity, that is, $U(0°) = 1$ (Figure 4.22a). It should be mentioned here that the preparation of a standard sample with a completely random texture is not trivial. In most cases, powder compacts are used, which may be slightly compacted or embedded, as in epoxy. Isostatically pressed powders can reveal too pronounced textures when the stress was not purely hydrostatic, whereas in too loosely bonded powder compacts, the reflected intensity may deviate from the curves obtained for 100% dense bulk samples. However, it is not necessary to have powder samples for all materials measured. As the defocusing correction $U(\alpha)$ only slightly depends on the Bragg angle θ, correction curves of other materials with similar Bragg angles can be used if no powder sample is available. Thus, textures obtained in different steel or aluminum alloys can generally be corrected with the correction curves derived from a powder sample from pure iron and aluminum, respectively.

Another error that has to be considered is caused by *absorption*. When a sample analyzed in transmission geometry is tilted, the path length of the x-rays within the sample increases much more than the increase in the diffracting volume, resulting in a strong decrease in diffracted intensity. According to Schulz (1949b) and Cullity (1978), the intensity of the reflected x-ray, in dependence on the Bragg angle θ and the tilting angle ω (α = 90° – ω), can be derived from

$$\frac{I_\omega}{I_{\omega=0°}} = \frac{\cos\theta}{\mu t} \cdot \frac{\exp\{-\mu t/\cos(\theta-\omega)\} - \exp\{-\mu t/\cos(\theta+\omega)\}}{\exp\{-\mu t/\cos\theta\} \cdot \{\cos(\theta-\omega)/\cos(\theta+\omega) - 1\}} \tag{4.7}$$

where μ is the linear absorption coefficient (Section 3.5). In the case of reflection of x-rays at an "infinitely" thick sample, the increase in absorption is exactly balanced by the increase in diffracting volume, such that the reflected integrated intensity remains constant and a special correction is not necessary. In the case of very thin samples (thin films), however, the volume increase is dominating and an absorption correction becomes necessary. The ratio between the reflected intensities from a sample of thickness t and an infinite thick sample is given by (Schulz, 1949a)

$$\frac{I(t)}{I_\infty} = 1 - \exp\left(-\frac{2\mu t}{\sin\theta\cos\alpha}\right) \tag{4.8}$$

In Figure 4.22b, the evolution of $I(t)/I_\infty$ with α is shown for copper sheet with three different thicknesses, again normalized to 100% for $\alpha = 0°$ (Figure 4.22a). For the 2-mm "thick" sample (or a material with a very high absorption coefficient) the correction factor stays constant at 1. With decreasing sample thickness, and especially for the 0.1-μm thin film, the intensity increases greatly with sample tilt, however. Generally speaking, when the sample thickness is below the penetration depth of the x-radiation, a correction of the absorption/volume effect (Equation 4.8) is necessary (Chateigner et al., 1994).

Finally, it should be mentioned that any pole figure measurement is influenced by *counting statistics* that imposes an uncertainty on each measurement with a standard error of the net intensity (Cullity, 1978; Humbert, 1986). The number of counts N derived per time unit at a given sample position is subject to statistical scatter that, for sufficiently large N, follows a Gauss law with the average value N_0 and a standard deviation $\sigma = \sqrt{N_0}$. In the presence of a background, the situation is slightly more complicated (Figure 4.21). If N_P and N_{BG} are the number of counts obtained in the peak and the background, respectively, then the relative standard deviation σ is

$$\sigma = \frac{(N_P + N_{BG})^{1/2}}{N_P - N_{BG}} \tag{4.9}$$

For instance, for $N_P = 1000$ and $N_{BG} = 100$, Equation 4.9 yields a relative standard deviation σ of 3.7%. Note that the error only depends on the *number* of pulses counted and not on the *rate*. Thus, the accuracy of the measurements can be improved by prolonging the counting times. In the preceding example, doubling the measuring time would yield $N_P = 2000$ and $N_{BG} = 200$, and this would reduce the relative error to 2.6%.

After pole figure measurement and subsequent correction of the data with respect to background intensity, defocusing error, and, if necessary, absorption, the pole figure data are available as number of counts, or counts per second, per pole figure point (α,β). For representation of the pole figures and for a subsequent evaluation, however, it is necessary to normalize the intensities to standard units that are not dependent on the experimental parameters. The commonly used convention is to express the data in m.r.d. For that purpose, a normalization factor N is derived by integrating over the measured and corrected intensities $I_{corr}(\alpha,\beta)$ over the full pole figure and weighting the intensities with regard to their areal contribution in the pole figure:

$$I_{norm}(\alpha,\beta) = \frac{1}{N} \cdot I_{corr}(\alpha,\beta) \tag{4.10}$$

where $N = \int_i I_{corr}(\alpha,\beta)\ \sin\alpha_i \,/ \int_i \sin\alpha_i$. Therewith, the integral over the full pole figure range becomes 1, which means that the pole figure of a randomly oriented standard sample would be 1 at all points. Pole figure regions with densities larger than 1 indicate that more lattice planes are aligned in those directions than in a random sample and vice versa.

For incomplete pole figures a true normalization is not possible, but a pseudo-normalization factor N' can be derived by summation over the measured pole figure range:

$$N' = \frac{\sum_{\alpha_{min}}^{\alpha_{max}} \sum_{\beta=0}^{360} I_{corr}(\alpha,\beta)\cdot\left(\cos(\alpha-\Delta\alpha/2)-\cos(\alpha+\Delta\alpha/2)\right)\cdot\Delta\beta}{2\pi\cdot\left(\cos(\alpha_{min}-\Delta\alpha/2)-\cos(\alpha_{max}+\Delta\alpha/2)\right)} \tag{4.11}$$

A more accurate normalization factor can be derived during a subsequent ODF analysis (Dahms and Bunge, 1986).

4.3.7 Inverse Pole Figures

In most cases, inverse pole figures—the presentation of the sample axes with respect to the crystal axes (Section 2.5.2)—are recalculated from the 3D ODF that has been derived from conventional pole figure data of the corresponding sample (Section 5.3.2). However, inverse pole figures (not to be confused with "pole figure inversion," Section 5.3.1) can also directly be measured by means of diffraction methods (Harris, 1952; Morris, 1959; Bunge and Roberts, 1969). The sample is mounted in reflection geometry with its normal parallel to the direction of interest—usually rolling direction (RD), transverse direction (TD), or direction normal (ND)—and the reflected intensities are measured in a $\theta/2\theta$ scan. The resulting peak intensities are normalized to the intensities of a standard sample with random texture and then the inverse pole figure can be constructed.

Thus, inverse pole figures can be derived with fixed sample geometry; no rotations in χ or φ are necessary, so that a standard powder diffractometer without Eulerian cradle can be used. In particular, use of a PSD (Section 4.3.4) or energy-dispersive detector (Section 4.3.5) allows rapid acquisition of the necessary data without any moving parts.

However, this technique only yields the intensities of reflections *hkl* that are in the accessible θ range, that is, θ is smaller than 90°, and for which no extinction rules apply. Furthermore, most experimental points lie on symmetry lines so that this method is much inferior to the conventional pole figures with a continuous coverage of the pole sphere. As an example, Figure 4.23 shows the inverse pole figure of a rolled steel sheet that consists of only eight reflections. The results are superimposed on the inverse pole figure that has been recalculated from the ODF of the same sample, indicating fairly good agreement between both methods. It is seen that although for high-symmetry materials only a couple of diffraction peaks can be considered, the coverage is better for materials of low crystal symmetry with line-rich diffraction patterns. Application of an energy-dispersive detector is advantageous, as this allows one to consider high-index planes that correspond to the short-wavelength components of the x-ray spectrum. However, the decreasing intensity of high-index

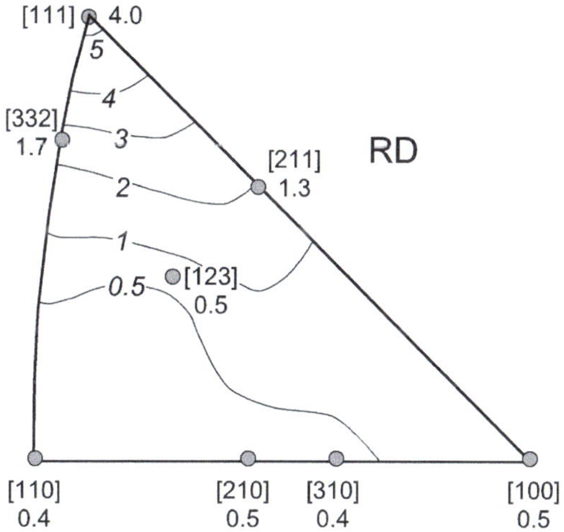

FIGURE 4.23 Inverse pole figure of the sheet normal direction ND obtained from a commercially produced stabilized steel sheet in the temper-rolled condition. Measured points—inverse pole figure measured as described in the text; intensity lines—inverse pole figure recalculated from the three-dimensional ODF. (Adapted from Bunge and Roberts, 1969.)

reflections and problems with peak separation still limit the application of this method for texture analysis.

Finally, it should be mentioned that from several inverse pole figures—obtained for different sample directions—the ODFs can be calculated in a similar way as that used for pole figures (Section 5.3.2).

4.4 NEUTRON DIFFRACTION METHODS

The first texture measurements by means of neutron diffraction were performed in 1953 by Brockhouse at Chalk River National Laboratory in an attempt to determine magnetic structures in steel (Brockhouse, 1953). In the 1960s and 1970s, this technique was further developed (Bunge and Tobisch, 1968; Kleinstück and Tobisch, 1968; Tobisch and Bunge, 1972; Kleinstück et al., 1976) and the advantages of neutron diffraction for texture analysis have been established (Section 3.6.2). Nowadays, the application of neutron diffraction for texture analysis is well established; see the reviews by Szpunar (1984), Welch (1986), Bunge (1989a), Wenk (2006), Brokmeier et al. (2011), and Vasin (2022). In principle, pole figure analysis by neutrons is equivalent to x-ray pole figure measurements (Section 4.4.1). However, more sophisticated methods like time-of-flight (TOF) measurements (Section 4.4.2) offer additional advantages over x-ray diffraction techniques.

4.4.1 Pole Figure Measurement by Neutron Diffraction

Texture investigations by means of neutron diffraction are mostly performed in research reactors where nuclear fission provides a supply of fast neutron. The neutrons are slowed down by interaction with appropriate moderators, until they

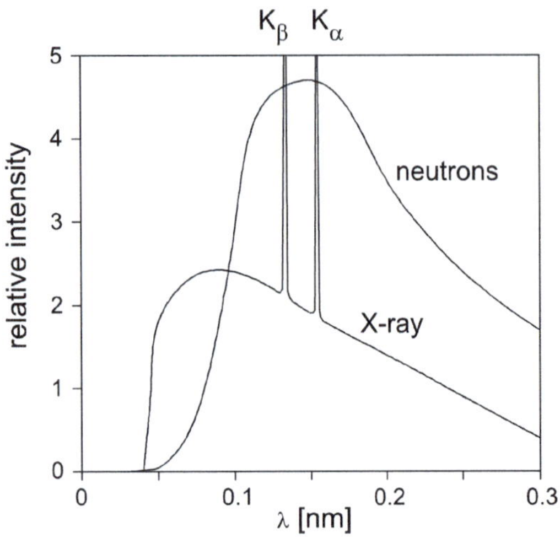

FIGURE 4.24 Wavelength spectrum of a thermal neutron source, compared with the spectrum of a copper x-ray tube with characteristic lines.

approach an approximate Maxwell velocity distribution with a maximum wavelength in the desired range from 0.05 to 0.3 nm, which is suitable for diffraction experiments (Figure 4.24). However, no pronounced maxima analogous to the characteristic x-ray peaks form, so that application of a monochromator is required to receive monochromatic radiation. Often copper (111) or graphite (0002) monochromator crystals are used.

The advantages of TOF measurements (Section 4.4.2) have encouraged interest in pulsed neutron sources, and this has drawn attention to using linear accelerators for generating neutron beams. When pulses of electrons accelerated to energies of 100 MeV fall on a target of a heavy element, through the intermediary of γ rays, pulses of neutrons are generated that can be used for diffraction experiments (Bacon, 1975). A more efficient way is to use high-energy proton beams that produce neutrons by a process called *spallation* (Lisowski and Schoenberg, 2006; Bauer, 2010; Zanini et al., 2019). When high-energy protons with energies of several 100 MeV hit the atomic nuclei of a given target material, neutrons and a variety of other spallation products are produced. The spallation products may react with other target nuclei, producing a cascade of particles with approximately 10–50 times more neutrons emitted than incident protons. To utilize these processes for generating a neutron beam, a high-energy (approximately 1 GeV) proton beam is generated in a particle accelerator and focused on a heavy metal target, for example, lead or tungsten. The energy of the resulting neutrons is too high (i.e., the wavelength is too small) for diffraction experiments, however, which means that moderation to thermal Maxwell distribution is necessary.

The principles of pole figure analysis by neutron diffraction are equivalent to x-ray pole figure measurements. The neutron detector is set to the corresponding Bragg angle 2θ for the selected lattice planes $\{hkl\}$ (Figure 4.25). The pole densities of that

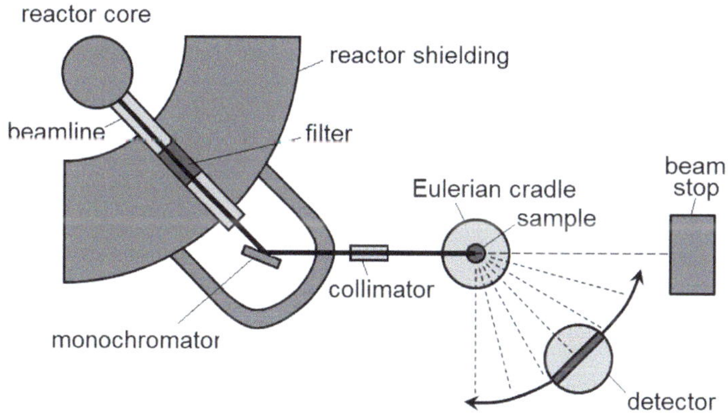

FIGURE 4.25 Scheme of pole figure measurements with a neutron source.

lattice plane in different sample directions are scanned with a goniometer by rotating the sample around two axes, φ and χ, corresponding to the pole figure angles β and α, to cover the entire orientation range (Section 4.3.2). The main difference between neutron and x-ray diffraction is the much lower, almost negligible, absorption of neutrons by matter (Table 3.4). Therefore, large samples, usually several centimeters in size, can be analyzed in transmission geometry. As shown by Tobisch and Bunge (1972), roughly spherical samples are completely sufficient for neutron experiments. In contrast to x-ray experiments, where the x-ray beam must not leave the sample surface, the sample must not leave the neutron beam for neutron diffraction experiments. Then, no intensity corrections are necessary and complete pole figures (up to $\alpha = 90°$) can be obtained in one scan. Absorption corrections for non-spherical samples are generally straightforward (e.g. Luzin and Brokmeier, 2002; Prince, 2004). The number of grains encountered is much larger, usually, by four to five orders of magnitude, which means that grain statistics is much better. Because of the low absorption, environmental stages can be used for in-situ investigations during heating, cooling, and straining of the samples (Hansen et al., 1981; Elf et al., 1990).

One drawback of neutron diffraction is that even in high flux reactors, the intensity is much lower compared to x-ray experiments, so that counting statistics (Equation 4.9) become an important consideration. A conventional pole figure scan takes between 6 and 48 h, compared to approximately 30 min in x-ray diffraction (Section 4.3.3). Since further access to neutron beam time is generally limited, techniques to accelerate the pole figure measurement are of great interest.

To speed up neutron diffraction experiments, the faster equal area scanning grids are frequently used (Section 4.3.3). Furthermore, measuring time can be cut down very efficiently by using PSDs that record the diffracted intensities of several pole figures simultaneously. In most applications, a curved one-dimensional PSD is mounted to record a certain range of the 2θ spectrum (Bunge et al., 1982; Will et al., 1989), as discussed in Section 4.3.4 for x-ray texture analysis. Unlike in x-ray diffractometry, however, the samples analyzed by neutrons are typically small compared to the goniometer dimensions, and therefore the focusing condition is still fulfilled for larger deviations from the exact Bragg–Brentano condition. This allows determination of continuous diffraction profiles from which integrated intensities can be extracted and used for pole figure determination, peak separation, etc.

An alternative arrangement with a vertical detector fixed at the position $2\theta = 90°$ was realized at the 10-MW neutron source DR3 at Risø National Laboratory, Denmark (Juul Jensen and Kjems, 1983) before the reactor was decommissioned in 2000. This unique arrangement simultaneously recorded the diffracted intensities corresponding to a range of positions within a given pole figure, rather than one point of several pole figures, so that fast texture analysis was possible. With recording times of about 15 min per pole figure, this system could be used for in-situ investigations, for example, in recrystallization kinetics or phase transformations (Juul Jensen and Leffers, 1989). To measure different pole figures, the wavelength of the incident neutrons had to be varied, which was realized by a special setup that allowed continuous rotation of the entire detection unit—the monochromator, the beam path between monochromator and sample, and the Eulerian cradle—about the center of the monochromator. Finally, McIntyre (2015) discusses the application of area detectors in neutron diffraction experiments, including film and, more recently, image plates and CCD-based detectors at neutron reactors, or scintillation detectors at spallation sources.

4.4.2 Time-of-Flight Measurements

In analogy to the energy-dispersive x-ray diffractometry described in Section 4.3.5, an alternative method to measure large numbers of pole figures simultaneously utilizes the *polychromatic* nature of the neutron source. However, as no energy-sensitive detectors for neutrons are available, another way had to be found. The method is to determine the flight time of the neutrons, which in a given experimental setup is a measure for their energy or wavelength, giving rise to the term TOF (Szpunar, 1984; Feldmann, 1989; Wenk, 1994). It follows from the de Broglie relation (Equation 3.1) that energy and wavelength of the neutrons are related to their velocity through

$$E = \frac{1}{2} m_N v_N^2 = \frac{h^2}{2\lambda^2 m_N} \quad \text{or} \quad v_N = \frac{h}{\lambda m_N} \tag{4.12}$$

where m_N is the mass and v_N the velocity of the neutrons. Thus, for neutrons with an energy spectrum between 10^{-3} and 10 eV (corresponding to wavelengths between $\lambda = 0.5$ and 0.01 nm), the velocities would range from $v_N = 10^3$ to 4×10^4 m s^{-1}. In a given experimental arrangement, such velocities can be determined in terms of the flight times of the neutrons. Figure 4.26 shows a schematic arrangement to perform TOF measurements. The total flight time T of the neutrons is given by their velocity v_N and the distance they travel, which is the length L_1 between source and sample plus the length L_2 between sample and detector. Thus, the relation between TOF, t_{TOF}, and the energy E—or wavelength λ—is given by

$$t_{TOF} = \frac{(L_1 + L_2)}{v_N} = \frac{(L_1 + L_2)}{(2E/m_N)^{1/2}} = \frac{(L_1 + L_2) \cdot h}{\lambda m_N} \tag{4.13}$$

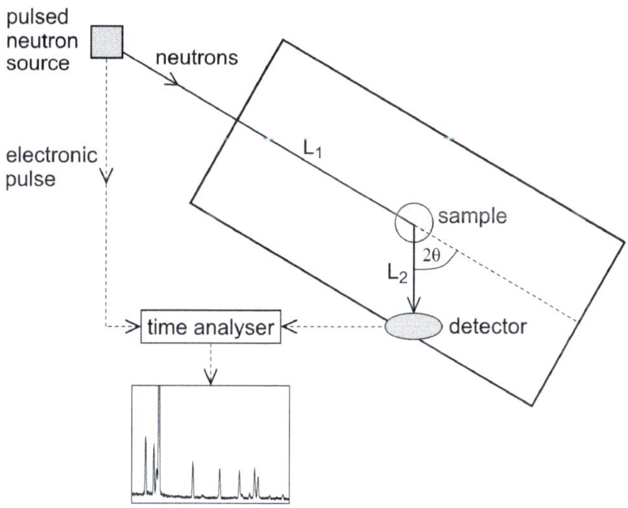

FIGURE 4.26 Schematic arrangement to perform TOF measurements. The total flight time T of the neutrons is determined by their velocity and the length L_1 between source and sample and l_2 between sample and detector.

This means that in a typical setup with flight lengths $L_1 + L_2$ of approximately 10–50 m, time differences of the order of milliseconds have to be resolved. To apply TOF in practice for determination of diffraction spectra, a pulsed neutron source is required, that is, TOF experiments have to be conducted either at a pulsed reactor (Feldmann et al., 1980; Helming et al., 1992; Walther et al., 1995; Conrad et al., 2008) or at a spallation source (Wenk et al., 1991; Jansen et al., 1998). The detector signal is stored in a multichannel time analyzer that is started synchronously with each neutron pulse. The recorded spectrum is successively built up in dependence on the neutron flight time, which means that the first signals that correspond to the fastest (i.e., high energy/low wavelength) neutrons are stored in the first channels and so on.

The number of measurements required to provide sufficient data to compute an ODF is given by the number of pole figures (hkl) times the number of sample orientations (α,β) (Section 5.3). Considering the possibilities of PSDs and TOF, it appears that a combination of both would substantially speed up pole figure measurements, and this has indeed been realized with 2D position sensitive TOF detectors at the Intense Pulsed Neutron Source (IPNS) at Argonne National Laboratory and the Los Alamos Neutron Science Center (LANSCE) (Wenk et al., 1991). With the installed setups, coverage of the entire hemisphere of reciprocal space—that is, complete pole figures for a number of different reflections hkl plus background data—requires about 27 different sample settings as opposed to the 1,368 points for *one* pole figure measured in conventional angular-dispersive diffractometry (Section 4.3.3).

The concept of combining PSDs with TOF measurements was advanced by building instruments that comprise multiple TOF detectors so as to increase the coverage of the pole sphere (Wenk, 2006), as on the SKAT neutron diffractometer in

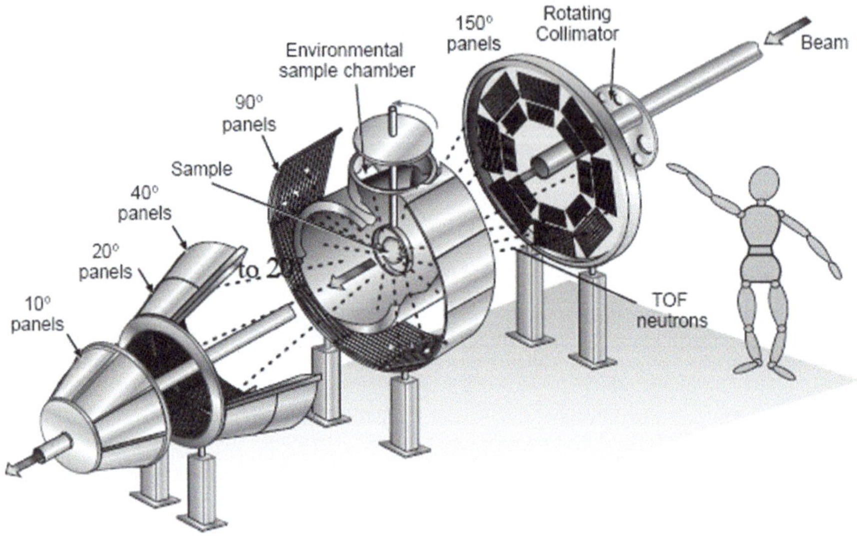

FIGURE 4.27 Schematic view of the neutron TOF diffractometer HIPPO with five banks of detector panels arranged on rings of constant diffraction angle. The sample is located in a sample chamber. For scale a sketch of a fairly large person is added; the distance from the 150° panels to the 10° panels is 3 m. (Courtesy of S. Vogel.)

Dubna (24 detectors, Ullemeyer et al., 1998) and the IPNS (14 detectors, Xie et al. 2003). In 1998–2002, the TOF diffractometer HIPPO (high-pressure preferred orientation) was designed and installed at LANSCE (Wenk et al., 2003; Vogel et al., 2004; Matthies et al., 2005). HIPPO achieves very high neutron count rates by virtue of a short (9-m) initial flight path and a novel 3D arrangement of detector banks with 1,360 ^{3}He tubes situated on five conical rings with scattering angles ranging from $2\theta = 10°$ to 150° around the incident beam direction (Figure 4.27). The large detector coverage of 4.8 m^2 allows texture measurements with a very small number of changes of sample orientation, typically between four and eight. Furthermore, rotation around a single axis is sufficient, as contrasted with the standard two-circle goniometers employed on conventional neutron and x-ray texture instruments. This provides the unique capability of rapid texture measurements under various sample conditions, such as temperature and uniaxial stress. The exceptionally high data rates of HIPPO also make it useful for time-resolved studies. In addition to the standard ancillary equipment (100-specimen sample changer, closed-cycle He refrigerator, furnace), a 500-ton toroidal anvil press TAP-98 was incorporated into the HIPPO diffractometer which is capable of achieving pressures of 30 GPa at ambient and high (2000 K) temperatures with samples up to 500 mm^3 in volume (Zhao et al., 2010). A low-temperature gas cell allows studies in the pressure range up to 5 kbar and temperatures between 10 K and ambient temperature. The rotatable stress-rig CRATES, built by the Technical University Hamburg-Harburg, Germany, allows the study of crystal lattice strains and texture changes in situ during uniaxial deformation in tension or compression up to stresses of 100 kN (Hartig et al., 2006).

A similar concept was pursued at the pulsed neutron source ISIS at the Rutherford Appleton Laboratory in the UK. The *general materials diffractometer* GEM is equipped with six detector banks, housing a total of more than 7,000 individual detector elements in 86 modules, covering a scattering angle range from 1° to 169° and hence providing considerable coverage in orientation space. For the analysis of texture, the detectors are gathered in 164 groups, each covering approximately 10° × 10°. With this configuration, it is possible to obtaining quantitative bulk texture data in a "single shot" within a matter of minutes without any sample reorientations (Kockelmann et al., 2006). As with HIPPO, the high temperature furnace facility on the GEM enables in-situ texture measurements as a function of temperature to monitor and quantify texture variation during phase transformation (Obasi et al., 2012).

4.5 TEXTURE MEASUREMENTS IN LOW-SYMMETRY AND MULTIPHASE MATERIALS

The principles of texture analysis by x-ray or neutron diffraction and subsequent ODF calculation described so far apply for all materials, independent of their crystal structure. However, in the case of multiphase materials or structures with low crystal symmetry, there are some specific additional problems that have to be considered (Bunge, 1985; Brokmeier, 1989; Lazik and Esling, 1993):

- Problems with peak separation caused by peak-rich diffraction spectra in both multiphase and low-symmetry materials (Section 4.5.1)
- Volume fraction of the various phases in multiphase materials (Section 4.5.2)
- Anisotropic absorption of the reflected x-rays in multiphase materials (Section 4.5.2)

In general, neutron diffraction offers some advantages over x-rays in analyzing both low-symmetry and multiphase materials. It should be emphasized here that the problems mentioned do not occur when the samples are studied by means of single-grain orientation methods (e.g., EBSD), where the textures of all phases can be derived separately (Section 9.4.4).

4.5.1 Peak Separation

Both multiphase materials and low-symmetry materials are characterized by line-rich diffraction spectra; Figure 4.28a and b shows examples for two-phase α/β brass with 20 vol% β phase and for calcite with trigonal crystal structure, respectively. In such line-rich diffraction spectra, the various diffraction peaks may be situated very close together, which can cause the problem of a proper separation of the contributions of the corresponding reflections/phases. In Figure 4.28a the {111} peak of the α brass is situated very close (0.5°) to the {011}-β-peak. In particular, with peak broadening at high sample tilting during pole figure analysis, this can lead to a substantial overlapping of such closely spaced diffraction peaks, which makes texture analysis in complex crystal structures—e.g., high-temperature superconductors—a very challenging task (Wenk et al., 1996).

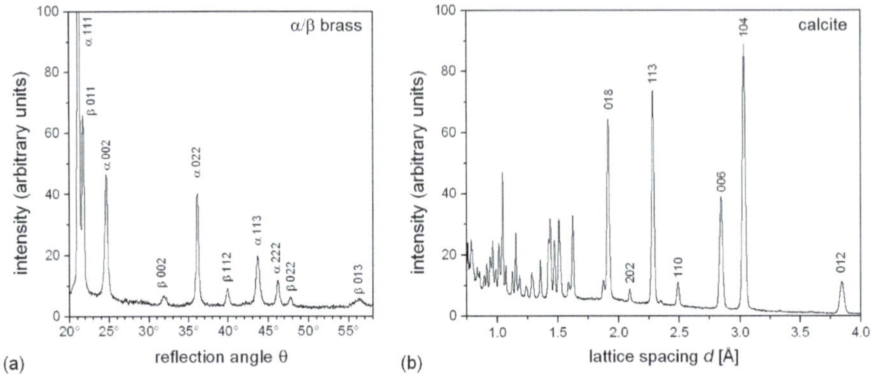

FIGURE 4.28 Examples of line-rich diffraction spectra. (a) Two-phase brass (Cu–40%Zn, x-ray θ/2θ spectrum). (From Engler and Juul Jensen, 1994. With permission.) (b) Calcite (neutron TOF spectrum.) (Data from Lutterotti et al., 1997.)

In contrast to the *partial coincidence* of reflection peaks in the examples in Figure 4.28, diffraction peaks can also *completely* overlap. For instance, in cubic metals the two sets of lattice planes {330} and {411} have the same lattice spacing *d*, namely, $d_{330} = d_{411} = a/\sqrt{18}$ (*a*: lattice constant). Consequently, the Bragg angles are identical too, which means that these two diffraction peaks completely coincide (Figure 4.20). Although this is not really a problem in the cubic structures—the example was chosen only as a simple illustration of the effect—there are more severe cases where all diffraction peaks of a given phase are systematically overlapped by another phase that is usually related by a phase transformation. Examples include ferrite and martensite in dualphase steels (Schwarzer and Weiland, 1988) and the primary α phase and the martensitic α″ phase in two-phase titanium alloys (Dunst and Mecking, 1996). Enantiomorphic materials, that is, materials consisting of left-handed and right-handed crystals, can be considered as special cases of two-phase materials with complete coincidence of *all* peaks so that the textures of the two phases cannot be determined from macrotexture pole figures (Bunge, 1985). In such cases of complete peak coincidence, texture analysis has to be performed by electron diffraction (microdiffraction) techniques that determine the textures of the various phases independently.

In the case of a *partial coincidence*, peak separation can slightly be improved by using a smaller receiving slit of the detector (Section 4.3.4). Use of an x-ray tube producing x-rays of a longer wavelength (Table 4.1) would shift the reflections toward larger Bragg angles θ and, in particular, would increase the distance Δθ between neighboring peaks. However, this can eventually result in the loss of diffraction peaks at too high reflection angles (θ ≥ 90°) and, furthermore, the problem of fluorescence must be considered (Section 4.3.1). Likewise, second-order peaks at larger θ are more widely spaced, but they suffer from poor signal-to-noise ratio (Figure 4.28a). Neutron diffraction is superior to x-ray diffraction as neutron diffractometers in general offer a higher spectral resolution, especially at large Bragg angles (Section 3.6.2). However, in many cases only a few peaks are affected; for example, in α/γ-duplex steels only the {111} peak of the face-centered cubic (fcc)

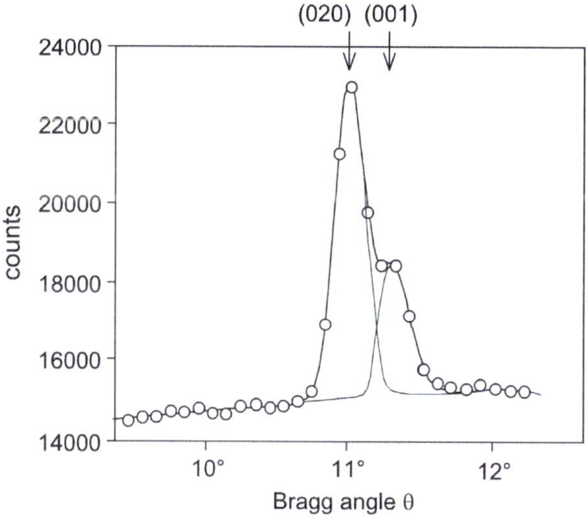

FIGURE 4.29 Separation of two partially overlapping feldspar diffraction peaks recorded by a linear PSD. (Adapted from Brokmeier, 1989.)

γ austenite and the {011} peak of the body-centered cubic (bcc) α ferrite overlap (similar to that in Figure 4.28a). In such cases usually enough nonoverlapping pole figures are available to compute the ODF; if necessary, the overlapping pole figures can eventually be recalculated (Section 5.3.2).

A very efficient way to separate partially overlapping peaks is to record the entire diffraction profile of the overlapping peaks by means of a PSD or area detector (Section 4.3.4) or an energy-dispersive detector (Section 4.3.5). The peaks are fitted with appropriate functions that allow derivation of the integral intensities of the corresponding pole figures (Jansen et al., 1988; Antoniadis et al., 1990; Wenk and Pannetier, 1990; Rodriguez-Navarro, 2006). Figure 4.29 shows an example of separation of the (020) and (001) peaks in feldspar, which are less than 0.5° apart.

Dahms and Bunge (1989) developed an analytical deconvolution scheme that achieves peak separation by an iterative procedure of ODF calculation and recalculation of the corresponding pole figures until a proper peak separation is obtained. This method even allows determination of the textures of two phases in cases where some peaks *completely* overlap, and where all other methods described so far would necessarily fail. Examples of successful peak separation have been given for low-symmetry materials (natural quartz; Dahms, 1992b) and for multiphase systems, including α/γ-titanium aluminides (Dahms et al., 1994) and two-phase titanium alloys (Dunst and Mecking, 1996). Most notably, proper peak separation was vital in the texture analysis of paragneiss, a natural mineral consisting of the three low-symmetry phases quartz, feldspar, and mica (Heinicke et al., 1991). A similar scheme to separate overlapping peaks based on the direct WIMV method (Section 5.3.4) has been developed by Kallend et al. (1991b).

Rather than extracting information from single diffraction peaks, it is more efficient to use the whole diffraction spectrum and extract texture information in a way similar to that by which crystallographers extract structural information from a powder pattern with the Rietveld technique (Rietveld, 1969). A method proposed by Wenk et al. (1994) relies on an iterative combination of crystallographic Rietveld profile analysis and quantitative ODF calculation (Rietveld texture analysis, RITA). Meanwhile, this method has been applied to several synthetic test cases (Matthies et al., 1997; Wang et al., 1997) as well as to biological samples (Lonardelli et al., 2005), geological materials (Lutterotti et al., 1997; Von Dreele, 1997; Wenk et al., 2001), and complex oxide ceramics (Jones et al., 2004; Guilmeau et al., 2005). In all cases resolution was very good, even for the highly overlapped regions of the diffraction spectrum. Thus, the RITA method can be expected to improve quantitative texture analysis of low-symmetry compounds and multiphase materials with overlapping peaks, to enable efficient data collection by requiring only a small part of the pole figure, and to make maximal use of data obtained by means of energy-dispersive methods that record complete spectra with many diffraction peaks, especially at neutron TOF facilities (Section 4.4.2).

4.5.2 Multiphase Materials

When analyzing the textures of multiphase materials, two further problems have to be considered besides the problem of overlapping peaks treated in the preceding section. Firstly, the reflected intensity is proportional to the volume fraction of the considered phase and hence it becomes difficult to determine properly pole figures of phases with small volume fractions. In Figure 4.28a, it is obvious that in the case of a duplex brass containing 20% β phase, the diffraction peaks of the minor component are much smaller than those of the α phase, resulting in a rather poor signal-to-background ratio for the former (though in this case the textures of both phases could properly be determined; Engler and Juul Jensen, 1994). In general, 5 vol% can be estimated to be the lower limit for texture analysis by x-ray diffraction techniques. Because of the much higher penetration depth of neutrons, a much larger volume contributes to the reflected intensities in neutron diffraction, yielding sufficient counting statistics also for phases with small volume fractions (Bolmaro and Brokmeier, 1999). As an example, Brokmeier (1993) analyzed the texture of an 8–10-μm galvanized zinc coating on a 0.7-mm steel sheet, and even the diffraction peaks of as little as 0.05% copper in aluminum could be recorded (Brokmeier, 1989).

Moreover, one (or several) phases of the multiphase compound may be strongly elongated, e.g. in eutectic structures or in plastically deformed materials, where the various phases are typically highly aligned along the deformation direction (Figure 4.30a). Differences in the absorption of x-rays in the different phases can cause errors in x-ray texture measurements (Bolmaro et al., 1994). Depending on the sample setting—Bragg angle θ and pole figure angles α and β—the intensity reflected in a given phase has to pass through different lengths of the other phase, and hence it is differently attenuated by absorption (Figure 4.30b). That is to say, the measured X-ray intensity does not only depend on the texture of a given phase, but also on the alignment of the other phases (Bunge et al., 1987). In the case of large differences in absorption coefficients in the different phases, this may cause large errors in the intensity distribution in the pole figures obtained. For neutron

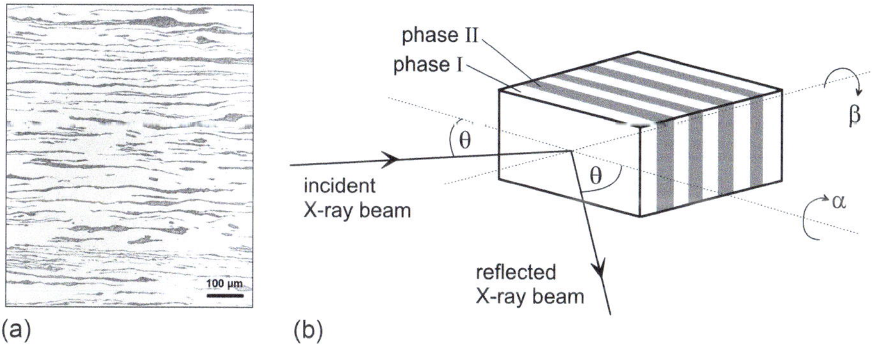

FIGURE 4.30 (a) Microstructure of cold-rolled two-phase Cu-40%Zn illustrating the alignment of the phases with deformation. (b) Schematic representation of the reflection of x-rays during pole figure measurements, illustrating the occurrence of anisotropic absorption in layered two-phase structures.

diffraction, absorption is in general much lower (Section 3.6.2) and—independent of the sample setting—the neutrons are scattered from the total sampled volume so that neutron absorption is virtually not affected by a possible alignment of the various phases (Bunge, 1985; Brokmeier, 1994; Engler and Juul Jensen, 1994).

In conclusion, for texture measurements in multiphase materials, neutron diffraction is clearly advantageous compared to x-ray diffraction because of the much higher penetration depth in matter of the former. Smallest volume fractions can be determined with sufficient counting statistics and the anisotropic absorption as described for x-rays can be neglected in the case of neutron diffraction.

4.6 SAMPLE PREPARATION

One of the advantages of neutron diffraction experiments is the flexibility in the sample geometry suitable for neutron texture analysis, so that in most cases only a minimum of sample preparation is required. Because of the large penetration depth of neutrons, bulk samples of centimeter size can readily be investigated in transmission geometry (Section 4.4.1). Sample holders made of vanadium are preferable since vanadium has an extremely low coherent scattering length (Bacon, 1975). Neutron texture analysis was greatly advanced by the spherical sample method (Figure 4.31) since in this geometry no absorption correction is necessary, yielding highest accuracy of the texture (Tobisch and Bunge, 1972). The deviations from the exact sphere may be quite large, however, so that cylindrical or even cubic samples, maybe with rounded edges, are usually sufficient (Brokmeier and Ostwaldt, 1996; Welch, 1986). Thus, roughly spherical bulk samples can generally be analyzed without special sample preparation, which makes neutron diffraction attractive for geological studies. For analysis of too thin specimens, such as sheet material, a cylindrical composite sample may be produced by stacking several disks cut from the sheet and tied together with screws or a wire (preferably of a material with different diffraction peaks), as shown in Figure 4.31e. Gluing is less suitable since the hydrogen in the

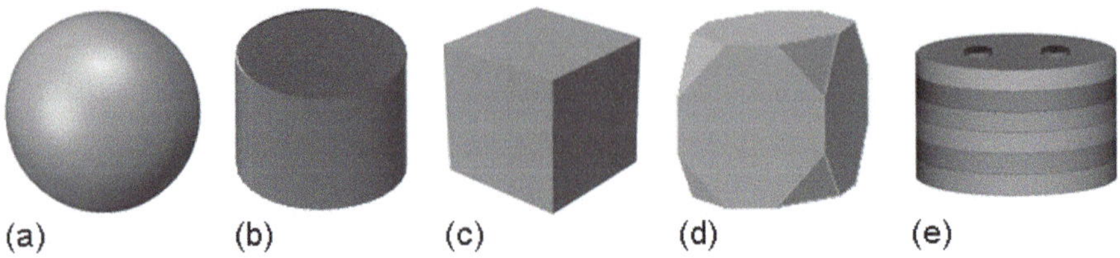

FIGURE 4.31 Samples for texture analysis through neutron diffraction. (a) Spherical sample; (b) cylindrical sample; (c) cubic sample; (d) cubic sample with rounded edges; (e) cylindrical composite sample produced by stacking several disks cut from a thin sheet.

glue produces a significant amount of neutron scattering. One may wish to first electropolish or etch the sheets so as to reduce the influence of a possible surface texture (Juul Jensen and Kjems, 1983).

In contrast, the low penetration depth of x-rays in most materials usually entails a certain amount of sample preparation. Texture measurements in transmission geometry require very thin parallel samples with thickness typically below 100 μm, so that this method is restricted to foils or thin wires that can be penetrated by the x-rays. Thicker samples may be thinned by grinding and etching. Today, a vast majority of texture measurements are performed in reflection geometry, and in the following text, we will address some ways for an appropriate preparation of the samples for x-ray texture analysis in reflection geometry.

To obtain the textures of wires, a bundle of parallel wires is mounted on a flat holder (Wenk, 1998). In most cases, production of the wires results in very heterogeneous textures with the mantle of the wire being different from the core. Provided the penetration depth of the x-rays used is not too large, the surface texture can be measured from the wires as they are, that is, without any preparation. Subsequently, the surface can be removed by chemical etching and the texture of the core determined (Montesin and Heizmann, 1992). When only the average texture is of interest, texture measurements can be conducted on the cross section of a bundle of wires. However, for analysis of axial-symmetric textures in general, an arrangement with the symmetry axis being parallel to the sample surface, rather than perpendicular to it, is advantageous with a view to separation of crystallographic texture and artifacts caused by geometrical corrections (Wenk and Phillips, 1992).

Preparation of bulk samples, including sheet materials, for x-ray pole figure measurements in reflection geometry is usually easy and can be performed with standard metallographic techniques. Samples can be round or rectangular shaped, with their size typically ranging from 10 to 30 mm and a thickness of 0.5 mm up to several millimeters. The sample surface must be flat, and the samples should carefully be ground down to 1,000 grit. Finer grinding or additional polishing is generally not necessary. However, the sample surface must not show any deformation as this would alter the texture of the affected layers and, therewith, the results obtained. Therefore, chemical etching or, if appropriate, short electropolishing of the surface is essential. For most metals, standard etching reagents that homogeneously attack the surface yield satisfactory results, for example,

- For *aluminum* and many *aluminum alloys*, best results are obtained by etching for 10–20 min in approximately 50% NaOH. Heating of the solution to ~60°C ensures a more homogeneous etching, but should be avoided for deformed samples to prevent recovery. After the etch, many aluminum alloys (in particular, copper-containing ones) have a gray or black layer that can readily be removed in ~10% HNO_3.
- *Copper* can be etched in diluted HNO_3. For copper alloys too high concentrations (e.g., exceeding 50%) should be avoided as they might selectively etch different phases.
- *Ferritic steels*, including low-carbon steel and iron-silicon transformer steel, can be etched for 1–2 min in a solution of 100-mL H_2O_2 and 5–15 mL HF. In alloys with high chromium contents, 10-mL HNO_3 and 15-mL HCl can be added and the etching time should be increased to 10–20 min (Raabe, 1992).
- Etchants suitable for some other bcc metals are 100-mL HCl, 90-mL HF, 100-mL HNO_3, 30-mL H_2O_2 (*tungsten*); 100-mL H_2SO_4, 40-mL HNO_3, 40-mL HF (*tantalum*); 100-mL HNO_3, 35-mL HF, 1–5-mL H_2SO_4 (*niobium*); and 100-mL HNO_3, 8-mL HF or 100-mL HCl, 30-mL H_2O_2 (*molybdenum*). Etching time of 20–30 min removes 5 μm (Raabe, 1992).
- Kelly et al. (1996) developed an etching technique for *tantalum* and *tantalum alloys*, which can be used for preparation of samples for x-ray analysis. After standard mechanical grinding and polishing, the specimens are swab etched with a solution consisting of lactic acid, HNO_3, and HF. It turned out that this technique is suitable for a surprising variety of materials, including other bcc metals and alloys as well as *titanium* and *titanium alloys*.

Of course, when working with strong acids or bases, the relevant health and safety precautions must be taken. For ceramics and geological materials, sawing and mechanical grinding are generally adequate. The bulk samples are usually cut with a diamond saw and then further ground with silicon carbide, alumina, or diamond abrasive powders down to a grit size of 10 μm.

Owing to the small penetration depth of x-rays, grain statistics often is a problem for x-ray texture analysis. If we assume a penetration depth of 0.1 mm and an area of 10×10 mm^2 being irradiated by the x-ray beam, the recorded volume is of the order of 10 mm^3. Then, for a material with a grain size of 25 μm, more than 10^6 grains contribute to the reflected intensity that definitely yields good statistical evaluation. For a grain size of 250 μm, in contrast, only slightly more than 1,000 grains are recorded, which is anticipated to be the lower limit for sufficient grain statistics (Section 10.3.3). The number of grains that contribute to the reflected intensity can be enlarged by a translation movement of, for example, 10 mm in the plane of measurement (Bunge and Puch, 1984), provided the sample is sufficiently large. For samples with elongated grains (deformed materials), it is advantageous to perform this translation perpendicular to the direction of the grain elongation (i.e., parallel TD for rolled sheets) as this enlarges the number of encountered bands.

To enlarge further the sampling area, it has been proposed to perform a meander-shaped scan of the sample surface (Hirsch et al., 1986). By this, a sampling area of as

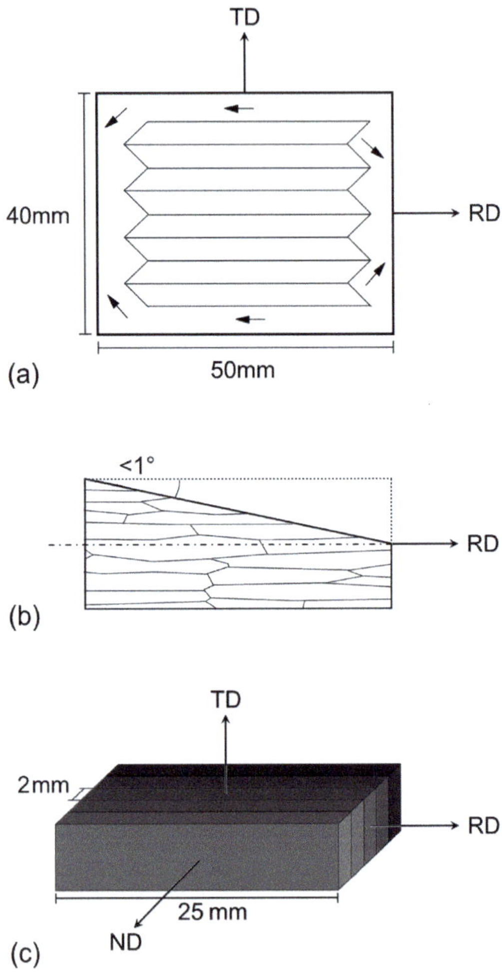

FIGURE 4.32 Measures to increase the sampling area during x-ray texture analysis. (a) Meander scan: the sample is translated in x- and y-direction, and at the end of each x-stroke of 42 mm, a lateral movement of 4 mm is carried out. After eight of these lateral shifts, the direction is reversed. (b) Sample with slightly slanted surface averaging over several through-thickness layers (not drawn to scale). (c) "Sandwich" sample stacked from several rotated pieces.

much as 50×40 mm^2 can be completely scanned (Figure 4.32a), so that textures of coarse-grained materials can be determined. However, this method requires a rather complicated control of the goniometer head with the possibility for an *x*/*y*-sample positioning. Rather, it is much easier to measure pole figures of several standard-sized samples and then average the resulting pole figure data (Engler et al., 1996b). This procedure is easy to perform and the sampling size is (theoretically) unlimited. It must be emphasized that to enlarge the sampling area these methods require a homogeneous texture; in other words no pronounced texture gradients should exist throughout the positions from which the different scans are taken.

For rolled sheets with strongly elongated but very thin deformed bands, the measuring statistics can also markedly be improved by preparing the sample with a slightly slanted surface (Figure 4.32b). If we allow for a maximum angle of 1°, which is tolerable in most cases, at a sheet sample of length 25 mm, a depth profile of 450 μm would be sampled, which for most materials is significantly more than the average penetration depth of x-rays (Table 4.2). The "extreme case" of a sample with a slanted surface would be a sample that is prepared perpendicular to the reference direction—rotated by 90° about one of the reference directions. For example, in the case of hot-rolled bands, preparation of a sandwich sample from five pieces stacked together gives very satisfactory results (Figure 4.32c). Note that in these cases, the resulting textures integrate over the sample thickness; that is, they reveal an average over the different layers of the sheet.

In any case, some preliminary knowledge on the microstructure and, if possible, on the texture of the material investigated is valuable to select the best way of sample preparation and, therewith, to achieve optimum experimental results.

4.7 SUMMATION

Measurements of pole figures by means of x-ray diffraction started in the 1940s by introducing texture goniometers to scan the pole figures and electrical counters to record the reflected intensities. Since the 1960s, pole figures are also regularly determined by neutron diffraction. Today, automated computer-controlled systems are available and the corresponding measuring techniques and the necessary correction methods that have been introduced in this chapter are very well established, so that diffraction experiments can routinely be performed to yield reliable, quantitative results for macrotexture.

Because of the much easier access to x-ray goniometers than to neutron sources, a vast majority of pole figure measurements are performed by means of x-ray diffraction. Neutron diffraction offers some advantages, however, which are mainly attributed to the much lower absorption of neutrons by matter. The much greater penetration depth of neutrons in the sample material allows investigation of spherical samples of centimeter size in transmission geometry, which yields the opportunity to perform nondestructive texture measurements or in-situ experiments. Complete pole figures can be obtained without intensity corrections. The grain statistics are better, which is of particular interest in coarse-grained samples or for detection of low volume fractions of a second phase.

Texture analysis can be greatly advanced by the use of PSDs or energy-dispersive detectors (in neutron diffraction: TOF experiments). In contrast to the conventional point detectors, these detectors record a part of the diffraction spectrum, so that the intensities of several pole figures, or large parts of one pole figure, can be measured simultaneously. This can be used for fast texture analysis—in online applications or in-situ experiments. Furthermore, PSDs and energy-dispersive detectors greatly facilitate texture analysis in low-symmetry and multiphase materials with line-rich diffraction patterns, where overlapping peaks can be separated by peak profile analysis.

5 Evaluation and Representation of Macrotexture Data

5.1 INTRODUCTION

Chapter 4 introduced the common techniques used to determine macrotextures experimentally by means of x-ray and neutron diffraction. The direct output of such texture measurements is the reflected intensity in the form of two pole figure angles α and β, that is, the distribution of the poles in the traditional *pole figure*. Hence the pole figure is the obvious choice for a critical assessment of experimental texture results. This method is discussed in some detail in Section 5.2.

The projection of the 3D orientation distribution onto a 2D pole figure causes a loss of information, so it is not possible to determine the orientation density of crystallites in a polycrystalline sample, that is, its *texture*, from the experimental pole figures without a certain ambiguity. Thus, for quantitative analysis of macrotexture, a 3D description—in terms of the 3D ODF—is required. However, unlike pole figures, orientation distribution functions (ODFs) cannot be directly measured by means of diffraction techniques; they need to be calculated from the pole figure data, typically from a set of several pole figures obtained from a given sample. Section 5.3 introduces the most common approaches to accomplishing this *pole figure inversion* and addresses their main advantages and disadvantages.

To display the 3D ODF, a representation in an appropriate 3D frame is required. Chapter 2 introduced several suitable orientation spaces, namely, Euler space (Section 2.6), cylindrical angle/axis space (Section 2.7), and Rodrigues space (Section 2.8). Among them, the Euler space is most often used to represent macrotexture data. Section 5.4 addresses the properties of Euler space and describes suitable techniques for displaying textures in Euler space. A "statistical" misorientation distribution function (MODF) can be derived from the ODF and displayed in Euler space. However, since the topic of misorientation distributions is more relevant to microtexture, the statistical MODF is discussed in Section 10.7 rather than in this chapter. The cylindrical angle/axis space and Rodrigues space are mostly used to represent microtexture data (Chapter 10), but have not found wide applications in conventional macrotexture analysis so far.

Section 5.5 presents a number of examples of the most typical rolling and recrystallization textures of fcc, bcc, and hexagonal metals. This section is intended to provide a guide to the characteristic texture types that are frequently encountered in metallic materials.

DOI: 10.1201/9781003258339-7

5.2 POLE FIGURE AND INVERSE POLE FIGURE

The present section introduces the principles of representing textures in pole figures (Section 2.5.1) and inverse pole figures (Section 2.5.2), which form the backbone of macrotexture analysis. We will address means of normalizing and contouring the orientation densities in a pole figure and illustrate use of the inverse pole figure, which finds applications in the case of axial symmetric textures (e.g., in tensile tests).

5.2.1 Normalization and Contouring of Pole Figures

For a comparison of pole figures of different samples, it is necessary to normalize the orientation densities to standard units that are not dependent on the experimental setup. The commonly used convention is to express the data in multiples of a random distribution (m.r.d.) of orientations, which means that the pole figure of a random standard sample would be "×1" at all points (Section 4.3.6). Pole figure regions with intensities higher than this indicate that more lattice planes are aligned in those directions than in a sample with random texture and vice versa.

The most commonly used way of contouring the orientation density in a pole figure is to draw lines of equal pole density, or equal intensity, as known from topographic or weather maps (Figure 5.1). Iso-intensity lines, also called contour lines or isopleths, are curves that connect contiguous points of the same pole density. As an

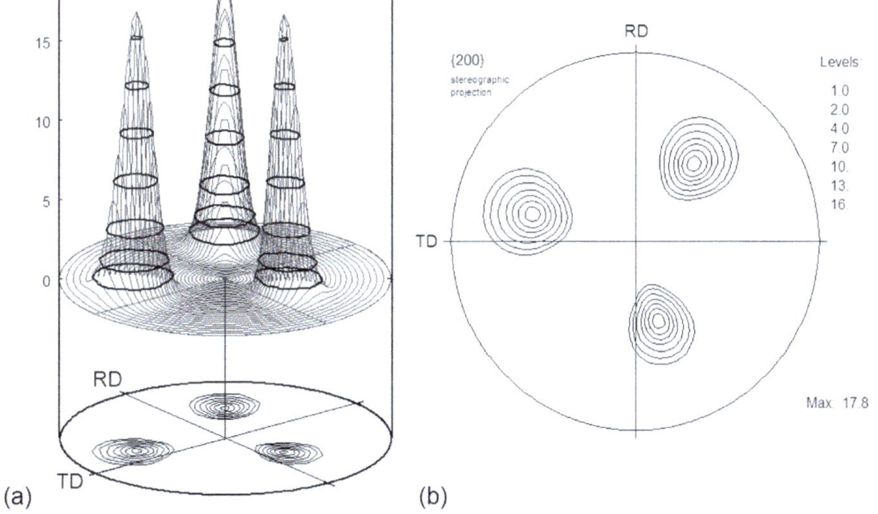

FIGURE 5.1 (a) Contouring of a 3D pole density distribution to derive the standard 2D pole figure presentation shown in (b) ({200} pole figure, stereographic projection).

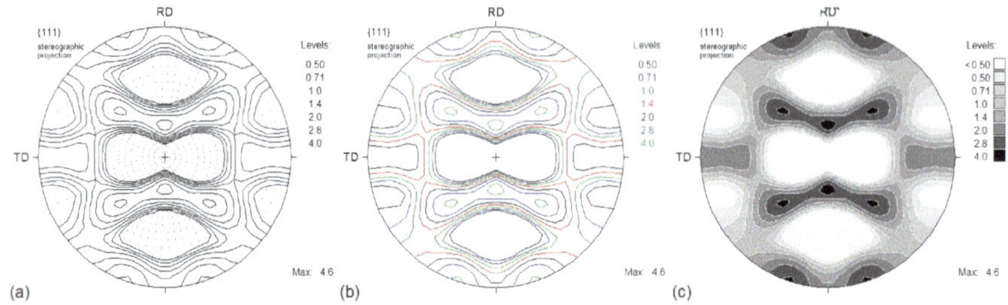

FIGURE 5.2 {111} Pole figure plotted with contour lines (logarithmic progression; regions with intensity lower than 0.5 are dotted); (b) same pole figure as in (a) plotted with colored contour lines; (c) same pole figure as in (a) plotted in gray shades (aluminum alloy AA 5005-H22, partially recrystallized).

example, every point along a line marked "×2" indicates a pole density of two times random. Note that iso-intensity lines must not touch or cross, nor do they branch or fork.

The values of the various contour lines do not need to be linearly spaced. Often, it is more instructive to use logarithmic progression to reveal texture details (Figure 5.2a). For an easy visual comparison of different pole figures, it is of considerable help to use standard distances, preferably starting at the same minimum level. In other words, levels in a given percentage, say 10%, 30%, 50%, and 75%, of the maximum intensity should be avoided. In some cases, in particular for weak textures, contour lines may be ambiguous, which means that intensity variations may not unmistakably be resolved. In such cases the individual contour lines need to be labeled with their values (Figure 5.3) or clearly marked by different (minimum 3) line styles or colors (Figure 5.2b). In many cases, those ambiguities can be avoided by simply marking pole figure regions below the minimum or above the maximum level by shadings or by dots (Figures 5.2a and 5.3). Alternatives to the contour lines are different gray shadings or colors in accordance with the intensity of the corresponding pole figure regions (Figure 5.2c). Such pole figures are more complicated to reproduce, however.

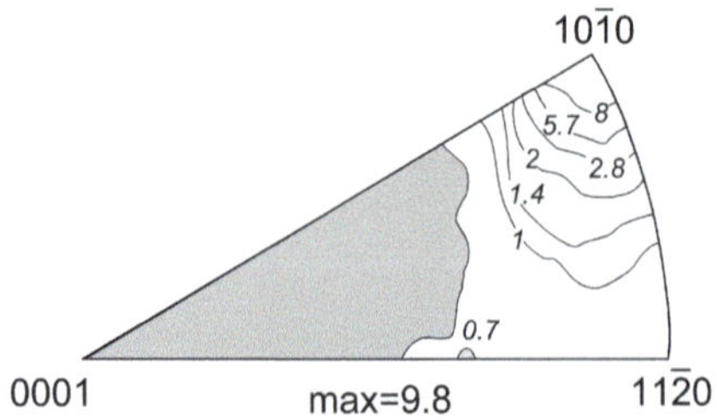

FIGURE 5.3 Inverse pole figure of titanium deformed by extrusion to a von Mises equivalent strain of 1.75 (inverse pole figure of the extrusion axis). (Adapted from Rollett and Wright, 1998.)

5.2.2 Representation of Orientations in the Inverse Pole Figure

As addressed in more detail in Section 2.5.2, the inverse pole figure is a representation of the orientations of the specimen coordinates in the crystal coordinate system. As an example, Figure 5.3 shows the inverse pole figure of the extrusion axes of titanium, which was plotted with iso-intensity lines in the hexagonal standard triangle formed by the axes (0001), $(11\bar{2}0)$, and $(10\bar{1}0)$. The texture displays a concentration of the extrusion axes in the vicinity of $(10\bar{1}0)$. This means that the basal plane normals are perpendicular to the extrusion axis, which is the common texture of hexagonal materials deformed by extrusion (Rollett and Wright, 1998). Examples of inverse pole figures of various fcc metals will be presented in Figure 5.22.

Although it is generally anticipated that in uniaxially deformed samples a fiber texture is present, this information cannot be derived from Figure 5.3 since the inverse pole figure only shows the orientation of one reference axis—here the extrusion axis—whereas rotations about this axis are not considered. Thus, for a complete representation of the 3D orientation distribution, the orientations of *two* reference axes would be required. Although for analysis of the macrotexture of fiber textures this may not be necessary, for the evaluation of misorientations the full orientation information must be available. Section 10.6 addresses the implications of this in more detail. In conclusion, for the representation of macrotexture data, inverse pole figures are usually less well suited than the pole figures.

5.3 DETERMINATION OF THE ORIENTATION DISTRIBUTION FUNCTION FROM POLE FIGURE DATA

5.3.1 The Orientation Distribution Function

The projection of the 3D orientation distribution onto a 2D projection plane results in a loss in information, so that in turn the 3D orientation distribution (the texture) cannot be derived from a given pole figure without some uncertainty. Only in the case of very simple textures consisting of a small number of orientations—textures of single crystals or the cube texture in recrystallized fcc metals—may the pole figures be sufficient to yield unambiguously the entire texture information in a quantitative manner. In general, however, the intensities of the individual poles that form a given orientation cannot clearly be assigned to that orientation, in particular when the poles of different orientations overlap, which renders a clear separation of the volumes associated with the individual orientations complicated or even impossible. Furthermore, most experimentally obtained pole figures are *incomplete pole figures*, which means that not the entire pole figure range has been recorded (Section 4.3.2). This further reduces the texture information available in such pole figures.

To overcome these ambiguities, and thus to permit a quantitative evaluation of the textures, it is necessary to describe the orientation density of grains in a polycrystal in an appropriate 3D representation, that is, in terms of its ODF. The ODF is defined as a probability density function of orientations g, expressed, for example, in the form of the Euler angles φ_1, Φ, and φ_2. Let us assume a microstructure consisting of different grains i of volume V_i with different orientations g_i (Figure 5.4). Grains of

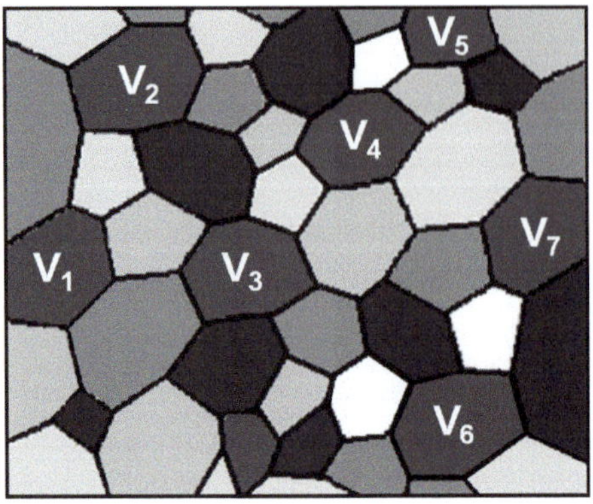

FIGURE 5.4 Schematic representation of a microstructure consisting of different grains *i* of volume V_i with different orientations. Similar orientations *g* within an orientation range *dg* are color coded with the same gray value.

similar orientations *g* within an orientation range *dg* are color-coded with the same gray value. The ODF $f(g)$ is defined by the following relationship:

$$\frac{dV}{V} = f(g)dg, \quad \text{with} \quad dg = \frac{1}{8\pi^2} \sin\Phi \, d\varphi_1 \, d\Phi \, d\varphi_2 \quad \text{and} \quad \oint f(g)dg = 1 \tag{5.1}$$

where *V* is the sample volume and *dV* is the volume of all crystallites *i* with the orientation *g* in the angular element *dg*. Note that the volume element *dg* is heavily distorted in orientation space, which is related to the definition of the Euler angles (Section 2.6.1; see also Section 5.4.1).

Experimental input for ODF computation is usually provided by macrotexture pole figure measurements (Sections 4.3 and 4.4). Mathematically, the pole figure is defined by the volume fraction *dV/V* of crystals having their crystal direction *h* parallel to the sample direction *y*—typically given by the two pole figure angles α and β (see Section 2.5.1):

$$\frac{dV}{V} = \frac{1}{4\pi} P_h(y)\,dy \tag{5.2}$$

where y = {α,β} and $dy = \sin\alpha \, d\alpha \, d\beta$.

Note that this pole figure definition is based on directions $h = [h_1,h_2,h_3]$ in the *direct* crystal lattice. Experimental pole figures, however, are commonly expressed in terms of the reflecting lattice plane normals, whose Miller (or Miller–Bravais) indices—e.g., {111} or {1120}—correspond to the coordinates of the crystal direction *h* in the *reciprocal* lattice. This distinction is not important for cubic structures (Appendix I), but it must be taken into account for the noncubic crystal classes.

As discussed in the preceding section, pole figures are usually normalized such that the pole densities are given in m.r.d. For that purpose, the pole figure data expressed in arbitrary units (e.g. in counts per seconds) are normalized such that

$$\frac{1}{4\pi} \cdot \oint P_h(y)\,dy = 1 \tag{5.3}$$

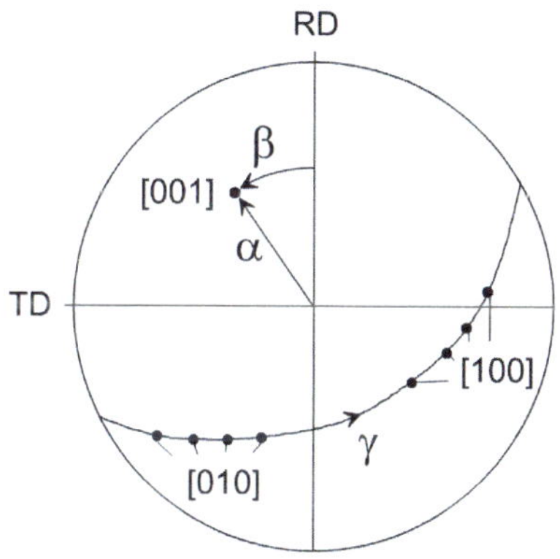

FIGURE 5.5 Representation of the fundamental equation of pole figure inversion in a {100} pole figure.

This expression can be used to normalize the pole figure, as described in Section 4.3.6.

As shown schematically in Figure 5.5, a pole that is defined by the direction y in a given 2D pole figure $P_h(y)$ corresponds to a region in the 3D ODF $f(g)$ that contains all possible rotations with angle γ about this direction y in the pole figure. Written in a mathematical manner, this means

$$P_h(y) = \frac{1}{2\pi} \cdot \int_{\gamma=0}^{2\pi} f(g)\, d\gamma \tag{5.4}$$

where $y = \{\alpha,\beta\}$ and $g = \{\varphi_1,\Phi,\varphi_2\}$.

The factor $1/2\pi$ is necessary because of the two normalization conditions of ODF (Equation 5.1) and pole figure (Equation 5.3). Equation 5.4 provides a relationship between the functional values of the pole figure $P_h(y)$ and the orientation distribution function $f(g)$. Hence, Equation 5.4 represents the *fundamental equation of ODF computation*, which has to be solved to calculate the ODF. This is called *pole figure inversion*. In general, the ODF is not completely determined by one pole figure, and the missing information must be provided by additional pole figures. An analytical solution to the problem of pole figure inversion is not possible, however. In the literature, several mathematical approaches to solving the integral in Equation 5.4 have been proposed. In the following sections, we will briefly introduce some of the most common methods for pole figure inversion. In Section 5.3.2 the well-known *series expansion method*, also known as *harmonic method*, is introduced, while their main errors—*truncation error* and *ghost error*—will be treated in Section 5.3.3. Section 5.3.4 will provide an overview about several *direct methods* which were designed to derive the ODF values $f(g)$ directly from the pole figure data in an iterative procedure. These sections are not intended to provide a rigorous mathematical description of the various approaches, but they try to outline their underlying ideas and compare their main advantages and disadvantages.

5.3.2 The Series Expansion Method

The basic premise of the *series expansion method* is the assumption that both the measured pole figures $P_h(y)$ and the ensuing ODF $f(g)$ can be fitted by a series expansion with suitable mathematical functions. Appropriate functions to use in a spherical coordinate system are the "spherical harmonic functions," and therefore this method has also been called the *harmonic method*. A complete scheme for texture analysis by series expansion has been developed independently by Bunge (1965) and Roe (1965). The following text outlining the series expansion method is based on the nomenclature of Bunge (1982) and Bunge and Esling (1982, 1985); for more details, see Jura and Pospiech (1978), Kallend (1998), Schaeben and van den Boogaart (2003), Chateigner et al. (2019), and Man (2022).

The series expansion method is based on the fact that an analytical solution to Equation 5.4 can be given for certain specific distribution functions—the generalized harmonic functions. For these two functions, $k_l^n(y)$ and $T_l^{mn}(g)$, Equation 5.4 becomes

$$\frac{1}{2\pi}\cdot\int_0^{2\pi} T_l^{mn}(g)\,d\gamma = \frac{2}{2l+1}\cdot k_l^{*m}(h)\cdot k_l^n(y) \tag{5.5}$$

where the asterisk indicates the complex-conjugate quantity and $k_l^{*m}(h)$ and $k_l^n(y)$ are the surface spherical harmonic functions defined on a sphere fixed to the crystal frame and the sample frame, respectively. The distribution described by the latter is the special pole figure belonging to the special ODF $T_l^{mn}(g)$. Thus, the pole figure $P_h(y)$ and the ODF $f(g)$ can be described by functions for which the solution is known. There is a specific generalized harmonic function $T_l^{mn}(g)$ for every possible combination of the three indices l, m, and n. Functions $T_l^{mn}(g)$ with low values of l are long-wave functions, which means they define large angular distances between neighboring maxima and minima of the functions. Conversely, the short-wave functions $T_l^{mn}(g)$ with high values of l describe small angular distances between neighboring maxima or minima. The generalized harmonic function $T_l^{mn}(g)$ are well-known functions of the Euler angles φ_1,Φ,φ_2, which is one of the main reasons why these angles were adopted in texture analysis.

In the series expansion method, the ODF $f(g)$ is expanded in a series of (symmetrized) generalized spherical harmonic functions $T_l^{mn}(g)$:

$$f(g) = \sum_{l=0}^{\infty}\sum_{m=-l}^{l}\sum_{n=-l}^{l} C_l^{mn}\cdot T_l^{mn}(g) \tag{5.6}$$

Any ODF $f(g)$ can be composed by adding functions $T_l^{mn}(g)$ in appropriate amounts, that is, each function $T_l^{mn}(g)$ is multiplied by an appropriate factor C_l^{mn}. Thus, the ODF is completely described by the *series expansion coefficients* C_l^{mn}. This is the general principle of a series expansion, which is well known in the form of Fourier analysis, where the special functions are sine or cosine functions. The harmonic functions $T_l^{mn}(g)$ are more complex 3D functions in three variables, but the principle is the same. Numerical values of the harmonic functions $T_l^{mn}(g)$, $k_l^{*m}(h)$, and $k_l^n(y)$, which are independent of the experimental data, are usually precalculated as a function of crystal and sample symmetry and stored in numerical libraries (Pospiech and Jura, 1975; Bunge, 1982; Wagner et al., 1982).

The pole figures can also be expanded in a series expansion. As the pole figure is characterized by the two angles α and β, an expansion in (symmetrized) spherical harmonic functions $k_l^n(y)$ is sufficient:

$$P_h(y) = \sum_{l=0}^{\infty}\sum_{n=-l}^{l} F_l^n(h)\cdot k_l^n(y) \tag{5.7}$$

Substitution of Equations 5.6 and 5.7 into Equation 5.4 under consideration of Equation 5.5 yields an expression relating the 2D series expansion coefficients $F_l^n(h)$ of the pole figure $P_h(y)$ to the C_l^{mn} coefficients of the ODF $f(g)$:

$$F_l^n(h) = \frac{4\pi}{2l+1}\cdot\sum_{m=-l}^{l} C_l^{mn}\cdot k_l^{*m}(h) \tag{5.8}$$

If the coefficients $F_l^n(h)$ and C_l^{mn} fulfill Equation 5.8, then the functions $f(g)$ and $P_h(y)$ fulfill the fundamental Equation 5.4. Therefore, the solution of Equation 5.8 yields the solution of Equation 5.4. In other words, the fundamental integral relation between the functional values of the pole figure and the orientation distribution function (Equation 5.4) is replaced by a system of linear equations between the various series expansion coefficients (Equation 5.8), which can be solved much more easily. This is the main advantage of the series expansion method.

Thus, Equation 5.8 represents a system of linear equations with the series expansion coefficients C_l^{mn} as unknowns. There is one such system for every combination of the indices l and n; each of these equations contains as many unknown C_l^{mn} coefficients as there are different values of the third index m. Given that m runs from $-l$ to $+l$ in the summations, there are $2l + 1$ unknown C_l^{mn} coefficients that have to be determined. In order to solve a system of linear equations with $2l + 1$ unknowns, there must be at least $2l + 1$ equations in that system containing the same values of l and n, that is, $2l + 1$ pole figures $P_h(y)$. Thus, the C_l^{mn} coefficients can be determined unambiguously up to $l = L$ if the coefficients $F_l^n(h)$ are known for $2L + 1$ pole figures. It is noted that the indices m in the functions $T_l^{mn}(g)$ correspond to the different crystal symmetries. For triclinic crystals, all values of $-l \leq m \leq +l$ are required in the series expansion, whereas for higher crystal symmetries, only a subset of m values is needed. From Figure 5.6, it appears that this greatly reduces the number of unknowns in Equation 5.8 and, accordingly, the number of pole figures necessary to compute an ODF.

In order to solve Equation 5.8 for the unknown series expansion coefficients C_l^{mn}, the coefficients $F_l^n(h)$ must be known for several pole figures $P_h(y)$. The $F_l^n(h)$

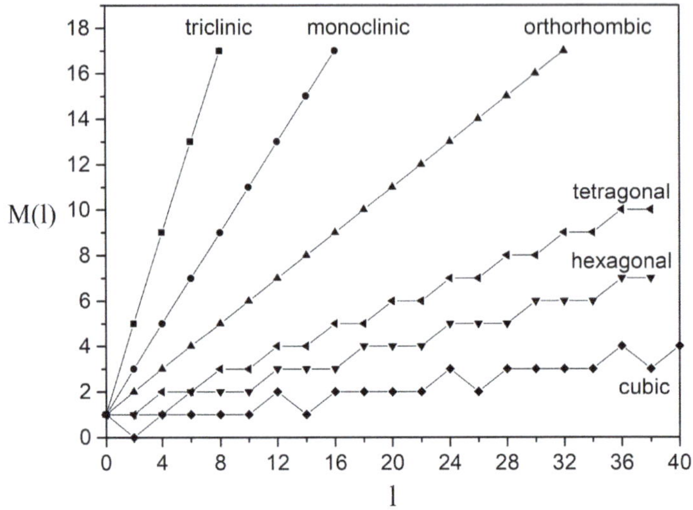

FIGURE 5.6 Number of linearly independent harmonic functions $M(l)$ as a function of the degree of series expansion l for different crystal symmetries (only for even-order l). (Data taken from Bunge, 1982.)

coefficients can be obtained by multiplying Equation 5.7 by the term $k_{l'}^{*n'}(y)$ and integration over the whole sphere y:

$$\int P_h(y)\cdot k_{l'}^{*n'}(y)\,dy = \sum_{l=0}^{\infty}\sum_{n=-l}^{+l} F_l^n(h)\int k_l^n(y)\cdot k_{l'}^{*n'}(y)\,dy \tag{5.9}$$

Making use of the orthonormality relation of the spherical harmonics then yields

$$F_l^n(h) = \int P_h(y)\,k_l^{*n}(y)\,dy \tag{5.10}$$

Equation 5.10 allows computing the coefficients $F_l^n(h)$ for every combination of the indices l and n and for every measured pole figure h. The solution to the fundamental equation 5.4 is achieved by solving Equations 5.10, 5.8, and 5.6, in turn. Equation 5.10 is used to expand the experimental pole figure $P_h(y)$ into a series with the coefficients $F_l^n(h)$. In analogy to the Fourier analysis this step is called *analysis*. Then, Equation 5.8 is solved for the unknown C_l^{mn} coefficients, which is called *coefficient transformation*. Finally, with Equation 5.6 the ODF $f(g)$ can be generated for any desired orientation g, which is called *synthesis*.

Whenever possible, a larger number of pole figures are used to calculate an ODF than is mathematically necessary. For instance, for cubic and hexagonal materials, respectively, three to four and five to six pole figures are usually determined. The experimental data are then averaged by a least-square scheme that reduces the impact of measuring errors and, in particular, allows the use of incomplete pole figures for the ODF calculation. Strictly speaking, the above derivation is only valid if complete pole figures are used as an input. It is possible to apply the harmonic method to computation of ODFs from incomplete pole figures, but without making use of the orthogonality relation (Pospiech and Jura, 1974; Morris, 1975; Humbert and Bergmann, 1980; Van Houtte, 1980, 1984; Dahms and Bunge, 1986). This implies that the terms of different indices l can no longer be treated separately.

For I incomplete pole figures available in a range Z, the C_l^{mn} coefficients are determined from the condition that the deviation between the experimental pole figures $P_h(y)_{\text{exp}}$ and the pole figures recalculated from the C_l^{mn} coefficients $P_h(y)_{\text{calc}}$ is minimized by a least-square method:

$$\Delta = \sum_{i=1}^{I} w_i \cdot \int_Z \left[P_{hi}(y)_{\text{exp}} - P_{hi}(y)_{\text{calc}} \right]^2 dy \overset{!}{=} \min \tag{5.11}$$

where $P_{hi}(y)$ is the intensity distribution measured in the ith pole figure ($h_i = hkl$) in the y direction, w_i is a weighting factor of pole figure i, often simply chosen to unity, and Z is the measured part of the pole figures. The recalculated pole figures $P_h(y)_{\text{calc}}$ are derived from the C_l^{mn} coefficients by substituting $F_l^n(h)$ in Equation 5.7 by Equation 5.8:

$$P_h(y) = \sum_{l=0}^{\infty} \sum_{m=-l}^{l} \sum_{n=-l}^{l} \frac{4\pi}{2l+1} \cdot C_l^{mn} \cdot k_l^{*m}(h) \cdot k_l^n(y) \tag{5.12}$$

Note that it is also possible to recalculate pole figures that cannot be measured, either because of the extinction rules in the specific material (Section 3.3) or because of overlapping diffraction peaks in line-rich materials (Section 4.5).

The integral in Equation 5.11 is commonly substituted by a summation over the entire measured range Z of the pole figures. This leads to a set of simultaneous linear equations from which the unknown C_l^{mn} coefficients can be derived. However, this system is much larger than the system set up for complete pole figures (Equation 5.8) and hence determination of the C_l^{mn} coefficients is rather time-consuming.

Another method which is easier to implement involves an initial extrapolation from the measured to the missing part of the pole figures, e.g. with a third order polynomial (Kallend et al., 1991a). Then, the above procedure of pole figure inversion based on complete pole figures is applied to provide a first estimate of the C_l^{mn} coefficients, which are used to re-calculate the missing parts of the pole figures (Equation 5.12). Since these recalculated pole figure parts contain information from the "real" experimental data, they are much better estimates than the original extrapolated first-guess values and hence are substituted for them. This process is repeated iteratively to improve the reliability of the missing parts, until a satisfactory self-consistency of the data is achieved. Dahms and Bunge (1987) suggested to improve this procedure by adding data from additional, non-measured, pole figures into the iteration.

Instead of the pole figure $P_h(y)$, a general axis distribution function $A(h,y)$ is often used. If the sample direction y is varied for a constant crystal direction h—e.g. {111} or {200} poles—the general axis distribution function $A(h,y)$ reduces to the pole figure $P_h(y)$ (Equation 5.12). However, the general axis distribution function $A(h,y)$ can also be specified in the opposite way, that is, the sample direction y is kept constant while the crystal direction h is varied. Then, the general axis distribution function $A(h,y)$ becomes the *inverse* pole figure $R_y(h)$:

$$R_y(h) = \sum_{l=0}^{\infty} \sum_{m=-l}^{l} H_l^m(y) \cdot k_l^{*m}(h) \tag{5.13a}$$

with the coefficients

$$H_l^m(y) = \frac{4\pi}{2l+1} \sum_{n=-l}^{l} \cdot C_l^{mn} \cdot k_l^n(y) \tag{5.13b}$$

Thus, from an ODF, inverse pole figures can be calculated and, vice versa, inverse pole figures can likewise be used to derive the ODF.

5.3.3 Truncation Error and Ghost Correction

An ODF computed with the series expansion method as described in the previous section inevitably contains several errors, even under the assumption of ideal pole figure data. One error is caused by the necessity to truncate the series expansion of pole figure and ODF expansion. The summations in Equations 5.6 and 5.7 are controlled by the index l, which is formally extended to infinity. This is not possible for numerical calculations, however, but the series must be terminated at a finite maximum value l_{max}. The resultant *truncation error* leads to broadening of the texture peaks as well as to minor artifact intensities in the vicinity of strong texture components. In a series expansion the degree l_{max} of the expansion generally depends on the sharpness of the function, in that sharper textures require a higher expansion degree l_{max}. Vice versa, a weak, smooth texture can be obtained with lower l_{max} which, in turn, often requires a lower number of pole figures. According to Matthies (1982), the value of l_{max} at which truncation errors can be neglected can be related to the orientation resolution $\Delta\omega$ through $l_{max} = 360° / 2\Delta\omega$.

In order to illustrate the impact of the truncation error on the resulting ODFs, Figure 5.7a through c shows ideal ODFs composed of two orientations—the Bs orientation {011}⟨211⟩ at $(\varphi_1,\Phi,\varphi_2)$ = (35°,45°,0°) and the S orientation {123}⟨634⟩ at (60°,35°,65°)—computed with l_{max} = 12, 22, and 34, respectively; while Figure 5.7d shows the resulting orientation density profiles of the Bs peak (dashed line in Figure 5.7c). Apparently, all three ODFs qualitatively display the same orientation distributions, with approximately similar intensity ratios among the two main components. The texture sharpness strongly increases with l_{max}, however, by more than doubling between l_{max} = 12 and 34. Simultaneously, the orientation peaks become narrower (Figure 5.7d). This reflects the better resolution of sharp peaks, which is achieved with higher degrees of l_{max}. It is also seen that largest differences arise between l_{max} = 12 and 22, whereas the differences between 22 and 34 are much less pronounced, mainly concerning the maximum intensity.

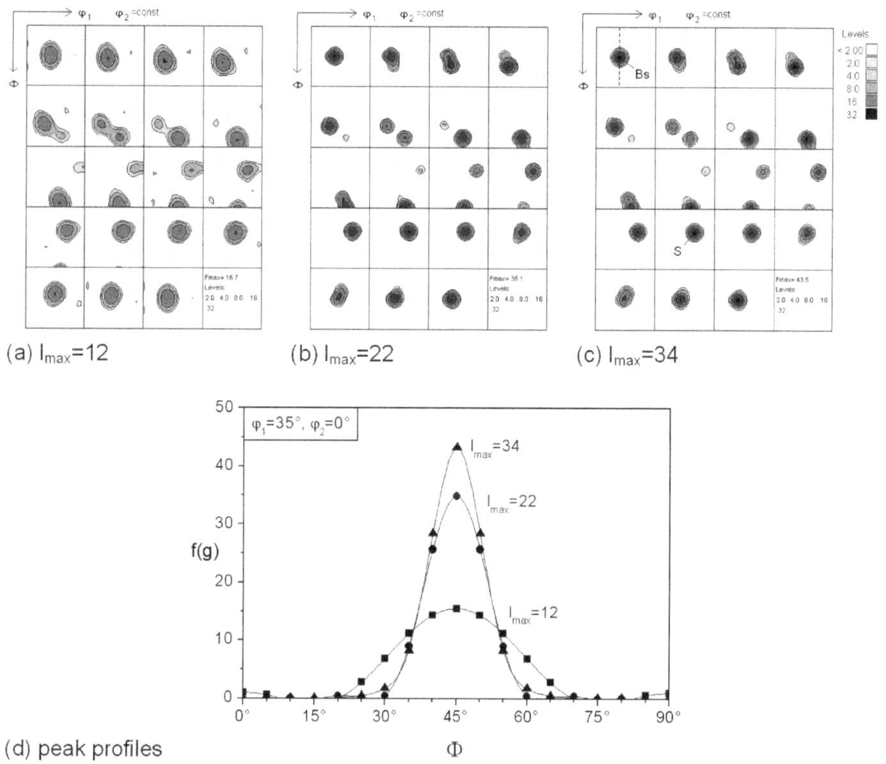

FIGURE 5.7 ODFs composed of two orientations—The Bs orientation {011}⟨211⟩ at $(\varphi_1,\Phi,\varphi_2) = (35°,45°,0°)$ and the S orientation {123}⟨634⟩ at (60°,35°,65°)—computed with different degrees of series expansion l_{max}: (a) $l_{max} = 12$; (b) $l_{max} = 22$; (c) $l_{max} = 34$. (d) Orientation density profile through the Bs peak (dashed line in [c]).

The quantitative error introduced by truncation can principally be determined by comparison of an ODF computed with a certain value of l_{max}, $f_{l_{max}}$, with the "real" texture, f_0. Two ODFs can quantitatively be compared by computing the relative mean square deviation between the two according to

$$\rho = \frac{\oint \left(f_0(g) - f_{l_{max}}(g)\right)^2 dg}{\oint \left(f_0(g)\right)^2 dg} \cdot 100\% \tag{5.14}$$

If the two ODFs, $f_0(g)$ and $f_{l_{max}}(g)$, are identical, the parameter ρ becomes zero. Figure 5.8 shows the evolution of the ρ parameter for three different textures of a cubic material as a function of the maximum series expansion coefficient l_{max} up to a value of $l_{max} = 34$. The reference ODFs f_0 were synthesized from a set of individual orientations directly in Euler space, i.e. without making use of a series expansion, such that the truncation error does not play a role. The three textures are mainly

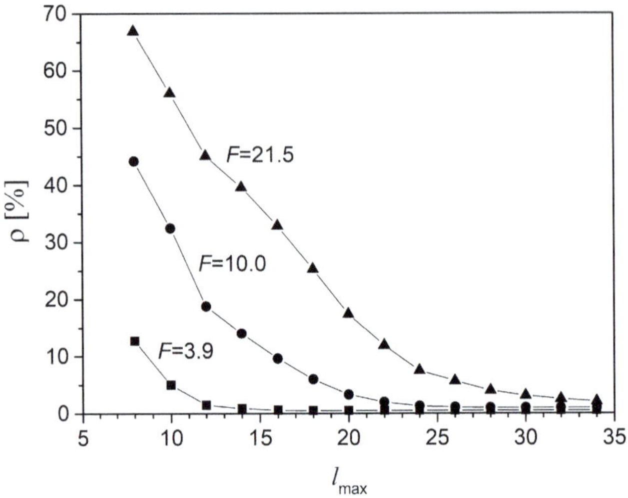

FIGURE 5.8 Reduction of the truncation error as indicated by a decreasing parameter ρ with increasing l_{max} for three ODFs differentiated by texture sharpness, which is specified by the texture index *F*.

differentiated by their sharpness which is indicated by the values of the *texture index F* (see Section 5.3.6). It is seen that ρ generally decreases with increasing l_{max}. For weak to medium textures ($F < 10$) the truncation error almost vanishes for series expansion degrees l_{max} in excess of 20. For sharper textures the truncation error is larger, and reasonably low ρ values below 5% are achieved for larger values of l_{max} only. Accordingly, for cubic materials the order of the series expansion is often set to $l_{max} = 22$ or, for very sharp texture, to $l_{max} = 34$, which usually yields sufficiently accurate results at an acceptable number of pole figures (Figure 5.6).

Another, more severe, problem of ODF calculation is the so-called *ghost error*. In the 1970s, Lücke and coworkers observed that intensities of the ODF may be missing (negative ghosts), or wrong intensities may appear (positive ghosts) (Lücke et al., 1981). Matthies (1979) proved that these ghosts are caused by the lack of the odd-order series expansion coefficients C_l^{mn}. The definition of the pole figure in Equation 5.2 is based on an individual crystal direction $[h_1,h_2,h_3]$, for example, $[1\bar{1}1]$, as opposed to a family of crystallographically equivalent directions $\langle h_1,h_2,h_3\rangle$, for example, ⟨111⟩. Although this distinction is irrelevant for directions related to each other by rotational symmetry elements, it is important for inversion centers in that a pole figure $h = [h_1,h_2,h_3]$ needs to be distinguished from its counterpart $-h = [\bar{h}_1, \bar{h}_2, \bar{h}_3]$. Thus, because of the centrosymmetry of the experimental pole figures derived by diffraction experiments (Friedel's law; see Section 2.2), they contain less information than necessary to reproduce unambiguously an ODF. The spherical harmonics functions $k_l^n(y)$ are symmetric with respect to an inversion center when l is even, but antisymmetric when l is odd. Therefore, fitting Equation 5.7 to centrosymmetric pole figures demands that the odd-order coefficients F_l^n are zero. Originally it was assumed that the corresponding odd-order ODF coefficients C_l^{mn}

also had to be zero, until Matthies (1979) showed that there is no such restriction, as long as the C_l^{mn} fulfill certain constraints such that the F_l^n in Equation 5.8 are zero. Thus, as long as these constraints are met, the odd-order C_l^{mn} coefficients have no influence on the pole figures $P_h(y)$ and, in consequence, pole figures obtained by diffraction experiments cannot provide information on the odd-order C_l^{mn} coefficients (this is different if orientations obtained by single-grain orientation measurements are analyzed, Section 10.3.2).

Written in a formal manner, the complete ODF $f(g)$ is composed of two parts (Matthies, 1979):

$$f(g) = \tilde{f}(g) + \tilde{\tilde{f}}(g) \tag{5.15}$$

The first part, $\tilde{f}(g)$, which is defined by the even-order C coefficients, can be derived from diffraction pole figures, whereas the second part, $\tilde{\tilde{f}}(g)$, which is given by the odd-order C coefficients, cannot. Because of the missing odd part $\tilde{\tilde{f}}(g)$, it may seem that as much as one-half of the information is missing. However, since the odd-order C coefficients are zero for low l—e.g. for cubic structures the first non-zero odd terms are obtained for $l = 9$—and quite small for large l, the impact of the odd-order C coefficients is limited in practice and the errors are typically approximately 20%. To visualize the impact of the ghost error on the ODFs, Figure 5.9a shows a typical fcc rolling texture, plotted in sections through the 3D Euler space (for details of ODF representation in Euler space see Section 5.4). Since this texture was not measured by diffraction techniques, but synthesized from a set of individual orientations, it principally comprises both the even and the odd part of the ODF (see the following text). Figure 5.9b shows the respective even part $\tilde{f}(g)$, which was computed only with the even-order C coefficients; similarly, the odd part $\tilde{\tilde{f}}(g)$, merely based on the odd-order C coefficients, is displayed in Figure 5.9c. Although the general texture features of Figure 5.9a and b are similar, it is obvious that the overall sharpness of the ghost-free complete ODF is appreciably higher while the even-order ODF contains some small extra peaks, which means that these are the errors attributable to the ghosts. It is noted that the odd-order ODF $\tilde{\tilde{f}}(g)$ in Figure 5.9c contains some intensities in the range up to about 4 (dark shadings)—mainly concerning the main fcc rolling texture orientations—together with significant regions with intensities below zero (bright shadings). Clearly, addition of the two parts shown in Figure 5.9b and c in accordance with Equation 5.15 gives the complete ODF which is presented in Figure 5.9a.

Thus, to derive complete, ghost-corrected textures by means of the series expansion method, the part $\tilde{\tilde{f}}(g)$ given by the odd-order C coefficients must be determined under certain assumptions. Most procedures to derive the odd-order C coefficients make use of the condition that the ODF must be nonnegative for all orientations, that is, $f(g) > 0$.

The first approach to determine the missing part $\tilde{\tilde{f}}(g)$, viz. the *zero-range* method proposed by Bunge and Esling (1979) and Esling et al. (1981), was based on the assumption that ranges in the pole figures with zero intensity yield an approximation

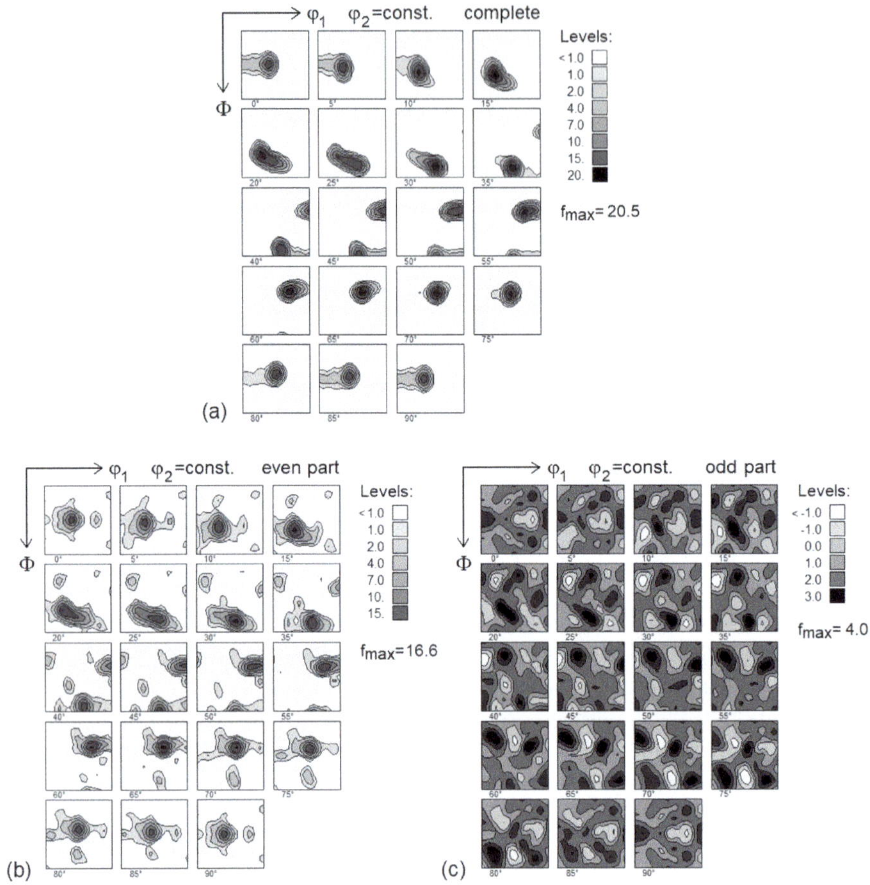

FIGURE 5.9 Ideal rolling texture of an fcc material, presented (a) as complete ODF $f(g)$, computed with both even and odd-order C coefficients (l_{max} = 22), (b) even part $\tilde{f}(g)$ and (c) odd part $\tilde{\tilde{f}}(g)$ of the ODF. (Note different intensity levels in (c).)

to the indeterminable part $\tilde{\tilde{f}}(g)$ of the ODF. From the pole figure zero ranges, the corresponding zero ranges in the orientation space can be derived by geometrical considerations. For these ranges, it readily follows from Equation 5.15:

$$\tilde{\tilde{f}}(g) = -\tilde{f}(g) \tag{5.16}$$

Thus, in the zero ranges of the orientation space, the function $\tilde{\tilde{f}}(g)$ is known and can be expressed in terms of the series expansion. The corresponding C coefficients are now considered as an approximate solution of $\tilde{\tilde{f}}(g)$ in the *entire* orientation space. It is obvious that this method can only yield satisfactory results for pole figures with well-defined zero ranges, that is to say, in the case of sharp textures.

In the *quadratic method* (Van Houtte, 1983), the ODF $f(g)$ is expressed in a quadratic form $h^2(g)$, or alternatively in an exponential form (Van Houtte, 1991), which cannot be negative. The function $h(g)$, which produces a satisfactory description of the true ODF $f(g)$, is then found in an iterative procedure. This method disadvantageously alters the (known) even-order C coefficients, which introduces some residual errors. As in the case of the zero-range method, the quadratic method is only applicable if well-defined zero ranges exist, i.e., for sharp textures.

The *positivity method* by Dahms and Bunge (1988, 1989) is a generalization of the zero-range method. At the nth step of this iterative procedure, the ODF $f_n(g)$ is given by

$$f_n(g) = f_{n-1}(g) + \tilde{\tilde{f}}_n'(g) \tag{5.17}$$

where the (complete) correction function $f'(g)$ approximates the negative parts of the $(n-1)$th function $f_{n-1}(g)$ (Figure 5.10):

$$f'(g) = \begin{cases} -f_{n-1}(g), & \text{if} \quad f_{n-1}(g) < 0 \\ 0, & \text{if} \quad f_{n-1}(g) \geq 0 \end{cases} \tag{5.18}$$

The initial solution ($n = 0$) is the experimental ODF, that is, the ODF directly calculated from the pole figure data. Note that only the odd part of the correction function is added to the previous solution in Equation 5.17, so that the (known) even part of the ODF $\tilde{f}(g)$ is never modified. This procedure is repeated until the entire ODF is positive or until the remaining negative regions are sufficiently small. This method was improved by introducing a small positive offset r in Equation 5.18 (Dahms and Bunge, 1989; Wagner et al., 1990). By choosing different values of r, the width of the range of possible solutions can be estimated, and the solution with maximum r

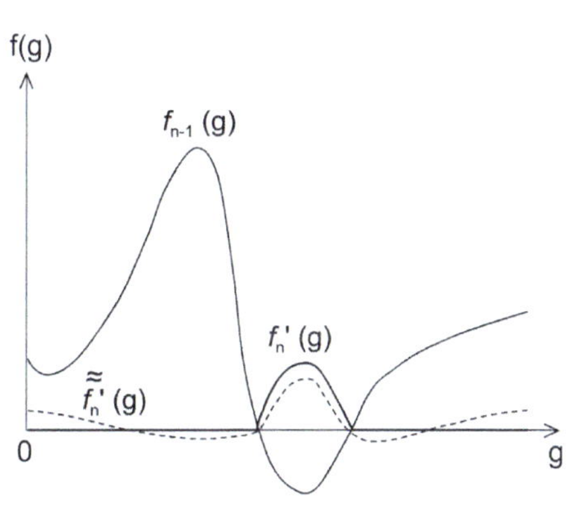

FIGURE 5.10 Ghost correction according to the method by Dahms and Bunge (1988, 1989) at the nth iteration step in a 1D representation. $f_{n-1}(g)$: ODF at the preceding iteration step; $f_n'(g)$: (complete) correction function to approximate the negative ranges of $f_{n-1}(g)$; $\tilde{\tilde{f}}_n'(g)$: odd part of the correction function $f_n'(g)$. (Adapted from Wagner and Dahms, 1991.)

corresponds to the solution with maximum isotropic background, or "phon" (Dahms, 1992a). This generalized positivity method offers a fast, efficient way to obtain the complete ODF and thus to correct for the ghost error.

Along similar lines, the concept of *maximum entropy* has been developed (Schaeben, 1988, 1991; Wang et al., 1992, 2014; Böhlke, 2005). In the present context the term "entropy" is associated with probability distributions, uncertainty, and information rather than being meant in a thermodynamic sense (Section 5.3.6). Out of the possible range of complete ODFs, the solution with maximum entropy has the smoothest distribution with minimum additional information to avoid artifacts, for example, ghosts, for which there is no evidence in the pole figures. This method has the advantage of providing ghost correction also in cases where no negative ODF ranges occur.

In a completely different framework, Lücke et al. (1981, 1986) proposed to deconvolute the experimental ODF into a set of several—typically five to ten—individual orientations. Each orientation is associated with a Gauss-type distribution in Euler angle space with the (half) scatter width ψ_0. Thus, the distribution of the orientation density S as function of the angular distance ω from the center orientation can be computed according to:

$$S(\omega) = S_0 \cdot \exp\left[-\left(\frac{\omega}{\psi_0}\right)^2\right] \tag{5.19}$$

where S_0 is the height of the Gauss peak. Figure 5.11 shows an example of a Gauss-shaped orientation in Euler space with a peak height $S_0 = 10$ and a scatter width of $\psi_0 = 8°$, which is the angular distance at which the intensity of the peak has dropped

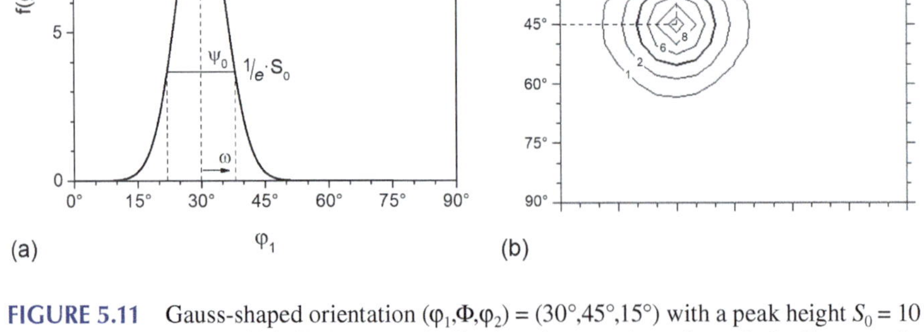

FIGURE 5.11 Gauss-shaped orientation $(\varphi_1,\Phi,\varphi_2) = (30°,45°,15°)$ with a peak height $S_0 = 10$ and a scatter width $\psi_0 = 8°$ in Euler space. (a) Section along angle φ_1 through the Gauss-peak; (b) presentation of the Gauss-peak in the section $\varphi_2 = 15°$ of the Euler space.

to $1/e$ of its maximum value (Bunge, 1982). From the set of i orientations g_i, a model ODF $f_{\text{mod}}(g)$ can be derived from a superposition of the individual Gauss peaks

$$f_{\text{mod}}(g) = \sum_i f_i(g) = \sum_i S_i \cdot \exp\left[-\left(\frac{\omega}{\psi_i}\right)^2\right] \quad (5.20)$$

Here, ω denotes the angular distance from a given point in the ODF to the center orientation g_i. The volume fractions M_i of the texture components are derived from integration of Equation 5.19:

$$M_i = \frac{\Delta V_i}{V} = \frac{1}{2\sqrt{\pi}} \cdot S_i \cdot Z_i \cdot \psi_i \left(1 - \exp\left(-\frac{\psi_i^2}{4}\right)\right) \quad \text{with} \quad \sum_i M_i = 1 \quad (5.21)$$

where Z is the symmetrical multiplicity of the center orientation g (Hansen et al., 1978).

With Equation 5.6, the model ODF $f_{\text{mod}}(g)$ can also be expressed in a series expansion:

$$f_{\text{mod}}(g) = \sum_{l=0}^{\infty} \sum_{m=-l}^{l} \sum_{n=-l}^{l} C_l^{mn} \cdot T_l^{mn}(g) \quad (5.22)$$

The C_l^{mn} coefficients describing the model ODF in Equation 5.22 can be expressed in dependence on the volume fractions M_i and scatter widths ψ_i of the set of individual orientations (Pospiech and Lücke, 1975):

$$C_l^{mn} = \sum_i M_i \cdot \frac{\exp\left(-l^2\psi_0^2/4\right) - \exp\left(-(l+1)^2\psi_0^2/4\right)}{1 - \exp\left(-\psi_0^2/4\right)} \cdot T_l^{mn}(g_i) \quad (5.23)$$

Note that this equation, which is also used to compute ODFs from single grain orientation measurements (Section 10.3.2), yields a *complete* ODF; i.e. both odd- and even-order C coefficients are obtained. By subtracting the even-order part of the model ODF from the experimental ODF, a difference ODF is obtained that can be used to improve the fit between the model ODF and the experimental ODF by shifting the position of the model components or by adding new ones. The complete ODF is finally derived by combining the even-order C coefficients of the experimental and the odd-order C coefficients of the model ODF (see Equation 5.15):

$$f(g) = \tilde{f}_{\text{exp}}(g) + \tilde{\tilde{f}}_{\text{mod}}(g) \quad (5.24)$$

By this method, the entire experimental information is retained but still a substantial reduction of the ghost error is achieved, even if the model fit is poor. As a rule of thumb, a fit of the difference ODF to 25% (which in most cases is easy to achieve) or, better, to 10% would, respectively, reduce the ghost error to 5% or even as little as 2%. It should be emphasized that although the assumption of Gauss-type orientations—in case of weak textures or pronounced fiber textures—is not necessarily justified, this method yields very satisfactory results with regard to a reduction of the ghost error. This method has successfully been applied to a huge amount of experimental data by the Aachen texture group. Furthermore, the information of the volume fractions M_i and scatter widths ψ_i of the main Gauss components can be used for a very condensed, quantitative description of the ODFs (Engler et al., 1994b). The major disadvantage of this method is that the iterative derivation of the model ODF has not been automated so far; thus the method is rather tedious to apply.

5.3.4 Direct Methods

Besides calculating the ODF by means of the harmonic method in a Fourier-space, other algorithms have been developed to derive the ODF values $f(g)$ directly from the pole figure data. These methods are summarized under the head *direct methods*. Such methods consider the pole figures $P_h(y)$ at a finite number of individual points y_i and, accordingly, the ODF $f(g)$ at a finite number of individual orientations g_j. Typically, both the pole figures and the orientation space are subdivided into a regular grid with a spacing of 2.5° or 5°. The relation between the individual pole figure points and the corresponding cells of the orientation space is established under consideration of the crystal geometry. Each pole figure cell corresponds to cells along one or more projection lines through the ODF, depending on sample and crystal symmetry. Thus, the integral in the fundamental Equation 5.4 can be replaced by a finite summation for each pole figure cell i:

$$P_h(y) = \frac{1}{N}\sum_{i=1}^{N} f(y \Leftarrow g_i) \tag{5.25}$$

Equation 5.25 defines a set of linear equations that, under appropriate conditions, can be solved to yield the ODF $f(g)$. After an initial estimate of the ODF, the fit between the ODF values and the value of the associated pole figure points is improved by means of an iterative procedure. In all direct methods, the positivity condition is directly taken into account.

In the following text, the principles of the so-called *WIMV method* are outlined, which is implemented in the Los Alamos texture software package popLA (Kallend et al., 1991a) as well as the Berkeley texture package BEARTEX (Wenk et al., 1998). The method goes back to Williams (1968) and Imhof (1977, 1982) and has since been improved by Matthies and Vinel (1982), which led to the acronym WIMV. For a more comprehensive review, the reader is referred to Matthies and Wenk (1985), Kallend (1998), and Matthies (2002).

The initial $f(g)$ values of each ODF cell (e.g., in a 5° × 5° × 5° grid) are estimated by the geometric mean of the values in the associated cells in the experimental pole figures, $P_{h_i}^{\exp}$:

$$f_0(g) = N_0 \cdot \prod_{i=1}^{I}\prod_{m_i=1}^{M_i} P_{h_i}^{\exp}\left(y_{m_i}\right)^{1/I \cdot M_i} \tag{5.26}$$

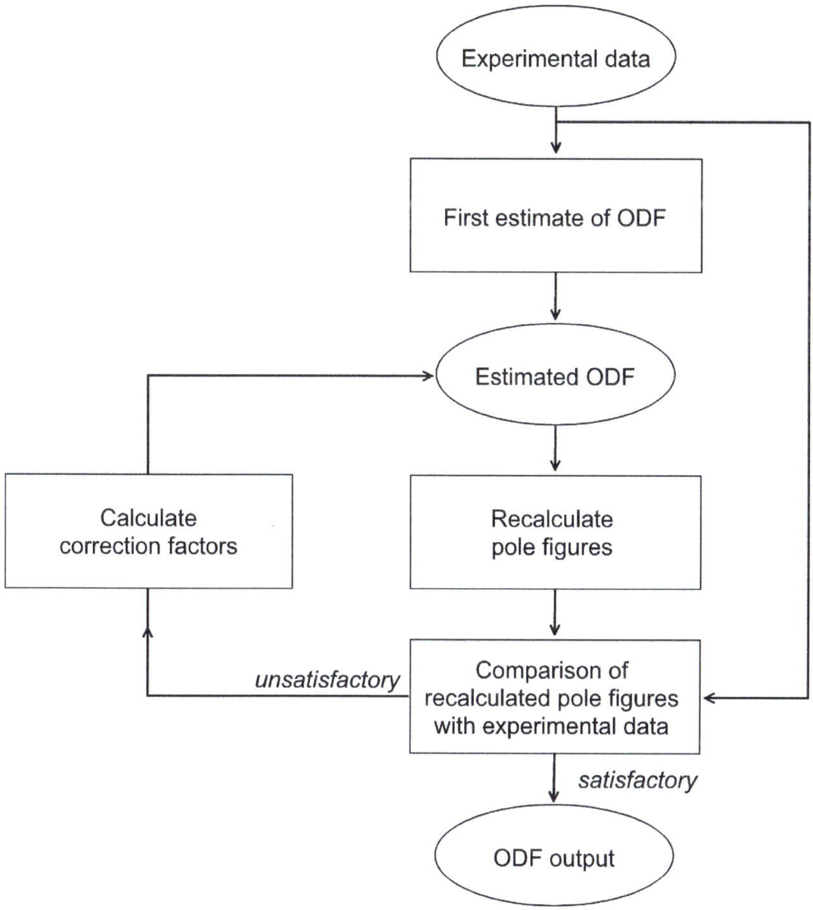

FIGURE 5.12 Flow chart of ODF calculation according to the WIMV algorithm. (Adapted from Kallend, 1998.).

where I is the number of measured pole figures, M_i is the multiplicity of pole i, and N_0 is the normalization.

Thus, as required, zero values in the pole figure automatically lead to zero values in the orientation space. The ODF is then refined by a series of so-called inner iterations (Figure 5.12). If $P_{h_i}^n$ denotes the recalculated pole figure after the nth iteration step, for each cell of the ODF a correction factor is derived as the ratio of the geometric mean of the corresponding cells in the experimental pole figure to the one in the recalculated pole figure $P_{h_i}^n$. The next estimate is then derived by multiplying the value of each ODF cell with this correction factor:

$$f_{n+1}(g) = N_n \cdot f_n(g) \cdot \frac{f_0(g)}{\prod_{i=1}^{I} \prod_{m_i=1}^{M_i} \left[P_{h_i}^n(y_{m_i})\right]^{1/I \cdot M_i}} \tag{5.27}$$

If, for example, the ODF value in a given cell is too large, then the corresponding recalculated pole figure values would also be too large. Consequently, the correction factor would be less than 1 so that the corresponding ODF value is reduced in the next step.

Additionally, an outer iteration can be applied to limit the minimum value of the ODF to $f(g) \geq f(g)_{min}$, which means that the phon is raised. By this any potential negative ghost will be filled, which in turn reduces the corresponding positive ghosts as well. In practice, the algorithm converges rapidly, and typically a satisfactory solution is obtained after approximately ten iterations.

Besides the WIMV method, there are several further direct methods for ODF computation that are listed here for completeness:

- Based on the WIMV method, Pawlik developed the *arbitrary defined cells* (ADC) method (Pawlik, 1986, 1994; Pawlik et al., 1991). Although most direct methods consider projection lines in relating the pole figures to the ODF cells, ADC uses projection *tubes*. The cross section of these tubes is determined by the shape of the corresponding pole figure cells. Computation of the intersections of the tubes with the ODF cells allows to derive the volume fraction of an ODF cell that contributes to the pole figure and leads to a better smoothing in the final result. This method is implemented in the commercial texture software package *LaboTex*.
- In the *vector method*, both the pole figures and the orientation space are subdivided into small subregions, which are considered as components of vectors of high dimensions (Ruer and Baro, 1977; Ruer and Vadon, 1982; Schaeben et al., 1985; Vadon and Heizmann, 1991). The two vectors are coupled by a huge, though sparse, matrix that contains the geometrical relations between the pole figures and the orientation space. The texture vector representing the final ODF is derived from the pole figure vectors through an iterative fitting procedure. The method is straightforward and easy to apply, but handling of the large matrices causes problems with computation efficiency, which reduces the resolution of this method.
- In the *component method* put forward by Helming and colleagues (Helming and Eschner, 1990; Helming et al., 1994), the ODF is described by a set of a finite number (mostly less than 30) of generalized Gauss- (or Lorentz-) based model functions. Similarly, as described earlier (Section 5.3.3), each of these functions has an individual height and range (full width at half maximum) as a measure for the strength and scatter of the texture component in orientation space. The approaches of Eschner and Fundenberger (1997) and Helming (1998) even allow using asymmetric texture functions with different scatter in different angular directions. The orientations and several parameters to describe the generalized Gauss components are directly derived from the pole figures by an iterative nonlinear optimization procedure. Though this method is rather tedious to apply, it provides a very condensed description of textures with physically meaningful quantities. Furthermore, this method is well suited to derive ODFs from incomplete pole figures with uncommon non-measured areas, such as selected area

electron diffraction (SAD) pole figures (Helming and Schwarzer, 1994; see Section 6.2.2).

- Hielscher and Schaeben (2008) introduced the *MTEX algorithm* for ODF computation from diffraction pole figures which is especially well suited for sharp textures and high-resolution pole figures measured with respect to arbitrarily scattered specimen directions, e.g. by area detectors. The estimated ODF is computed as the solution of a nonlinear minimization problem which simultaneously considers the orientation densities and normalization coefficients. The minimization procedure applies discretization by radially symmetric functions and fast Fourier techniques to guarantee smooth approximation and high performance. The algorithm is implemented as part of the freely available MATLAB toolbox for quantitative texture analysis (Bachmann et al., 2010) and satisfies the following key requirements: (i) the estimated ODF is non-negative; (ii) the method allows for arbitrary crystal and diffraction geometries; (iii) the method allows for superposed pole figures; and (iv) the unknown normalization coefficients of the pole figures are determined simultaneously with the ODF.

5.3.5 Comparison of Series Expansion and Direct Methods

Comparing the series expansion method, along with its different approaches to derive the odd part of the ODF, with the various direct methods reveals some characteristic advantages of each, which are summarized in the following text.

The series expansion method is a fast, reliable method that is rather insusceptible to experimental scatter in the pole figures. Its main output (the C coefficients) represents a condensed manner with which to characterize the texture, which is superior to storing the entire ODF data. The C coefficients also disclose valuable additional information on texture-related properties—elastic modulus and elastic and plastic anisotropy. Furthermore, changes in the reference coordinate system can be accommodated by manipulation of the C coefficients, so that texture changes during phase transformations (Davies et al., 1976; Inoue et al., 1992; Park and Bunge, 1994) or recrystallization (Pospiech and Lücke, 1979; Lücke and Engler, 1990; Engler, 1997) can readily be analyzed. Rotations in the sample frame are also possible (Inoue and Takasugi, 2001). Finally, normalization of the experimental pole figures (Section 4.3.6) is easier.

In the direct methods, the condition of nonnegativity and the handling of the ghost problem are usually implicitly fulfilled, and no truncation error arises. Analysis of incomplete pole figures as well as different sample or crystal symmetries is easier to accomplish than in the harmonic method. In general, direct methods require fewer pole figures to obtain satisfactory solutions, which is of particular interest when low-symmetry materials are analyzed.

All current approaches to compute ODFs from pole figure data have now been developed to a state that, provided good experimental data are available, they are able to yield routinely reliable and reproducible quantitative texture results (Matthies et al., 1988; Wenk et al., 1988; Bacroix et al., 1994; Caleyo et al., 2001). Nonetheless, because of the systematic differences between different methods, it is highly recommended not to change the method within a series of experiments and especially not to switch between series expansion and direct methods.

5.3.6 Texture Parameters

Once the ODF *f(g)* has been obtained, the overall texture strength can be assessed with the help of the so-called *texture index F* (sometimes labeled *J*) (Bunge, 1982). The texture index *F*, expressed in units of m.r.d., is defined as

$$F^2 = \frac{1}{8\pi^2}\sum_i \left[f(g_i)\right]^2 \Delta g_i \tag{5.28}$$

where Δg_i is the volume of the ODF cells. Thus, the texture index F represents the mean-square value of the ODF, which varies from unity for a random texture to infinity for a perfect texture as e.g. a single crystal. Using harmonic functions, the texture index F can be calculated from the C_l^{mn} series expansion coefficients through

$$F^2 = 1 + \sum_{l=2}^{l_{max}} \left(\frac{1}{2l+1}\right) \sum_{m=-l}^{l} \sum_{n=-l}^{l} \left|C_l^{mn}\right|^2 \tag{5.29}$$

The texture index is useful to compare the textures of samples with the same crystal symmetries and similar texture components.

As an alternative strength parameter the *texture disorder* may be evaluated, which led to the introduction of the *texture entropy S*:

$$S = \frac{1}{8\pi^2}\sum_i f(g_i)\ \ln f(g_i)\Delta g_i \tag{5.30}$$

Maximization of the texture entropy forms the backbone of the entropy method advanced by Schaeben (1988, 1991) described in Section 5.3.3.

5.4 REPRESENTATION AND DISPLAY OF TEXTURE IN EULER SPACE

The quantitative description of a given texture in terms of its ODF, that is, the orientation density in terms of the three Euler angles, necessitates an appropriate 3D orientation space for its representation. The first orientation space was introduced by Perlwitz et al. (1969) by defining three angles ρ_1, ρ_2, and ρ_3 to represent the orientation distributions of rolled copper and brass, which were determined by electron diffraction in a TEM (Pitsch et al., 1964). This space was chosen so that the most important rolling texture components all came to lie in a 2D section with $\rho_3 = 0°$ through the 3D space, which strongly facilitated the visualization and evaluation of textures in this space.

In Section 5.4.1, we focus on the orientation space—the Euler angle space—that is most frequently used today. This orientation space is formed by the three Euler angles such that an orientation given by a triple of Euler angles corresponds to a distinct point in Euler space whose coordinates are defined by these three angles (Figure 2.13).

Because of the necessity to represent the orientation data in a 3D orientation space, texture analysis is often considered a complicated problem that requires a large ability to visualize things. This task can greatly be facilitated by suitable ways to display and, if possible, condense the texture information. In Section 5.4.2, we address the common methods used to represent textures in the Euler space as well as appropriate means to condense textures and thus to facilitate texture analysis.

5.4.1 Properties of Euler Space

The Euler space is a 3D orientation space whose axes are formed by the three Euler angles. In most cases, the three Euler angles are chosen to form an orthogonal coordinate system resulting in a Cartesian Euler angle space (Morris and Heckler, 1968; Van Houtte, 1987). With regard to the axes of this space, Section 2.6.1 introduced three different sets of the Euler angles, namely, the angles φ_1, Φ, φ_2 (Bunge); Ψ, Θ, Φ (Roe); or Ψ, Θ, ϕ (Kocks). Although all these conventions likewise define an orientation space, in the remainder of this section the discussion is confined to Bunge's angles φ_1, Φ, φ_2, as these are by far most commonly used in the literature. The principal considerations hold for all conventions, however.

As already pointed out in Section 2.6.2, an orientation given by a triple of Euler angles corresponds to a distinct point in Euler space whose coordinates are unequivocally given by these three angles (Figure 2.13). We have also seen how sample and crystal symmetry influence the size of the Euler angle space necessary to represent all possible orientations in an unequivocal manner (Table 2.3), and it turns out that the widely used method of representing textures in a cubic Euler space with $0° \leq (\varphi_1,\Phi,\varphi_2) \leq 90°$ inevitably assumes orthonormal sample symmetry. It should be emphasized that even in cases where orthonormal sample symmetry applies in principle—in samples deformed in plane strain—this reduction can lead to loss in information and can even cause errors. It is clear that asymmetries in the textures that point to deviations from the idealized plane strain state cannot be resolved in such reduced ODFs (Sections 5.4.2). More severely, in all cases where the initial texture of a sample did not depict orthonormal symmetry before deformation, reduction of the Euler angles under the assumption of orthonormal sample symmetry is misleading, although orthonormal sample symmetry does apply. For instance, single crystals in most cases have no orthonormal symmetry (with the exception of crystals with highly symmetrical orientations like cube {001} ⟨100⟩ and Goss {011}⟨100⟩) so that important texture features such as direction and amount of rotations cannot properly be resolved.

In accordance with the definition of the Euler angles (Section 2.6.1), for $\Phi = 0°$ (and $\Phi = 180°$), the two remaining angles φ_1 and φ_2 describe rotations about the same axis, namely, the normal direction, ND, which gives rise to strong distortions in the Euler angle space. Figure 5.13a shows the position of the three equivalent variants of the orientation $(123)[63\bar{4}]$ in the reduced Euler angle space. For each central orientation (marked by a dot), an isotropic scatter of ~10° is marked. It is obvious that the resulting spheres in the Euler angle space are far from being isotropic, but

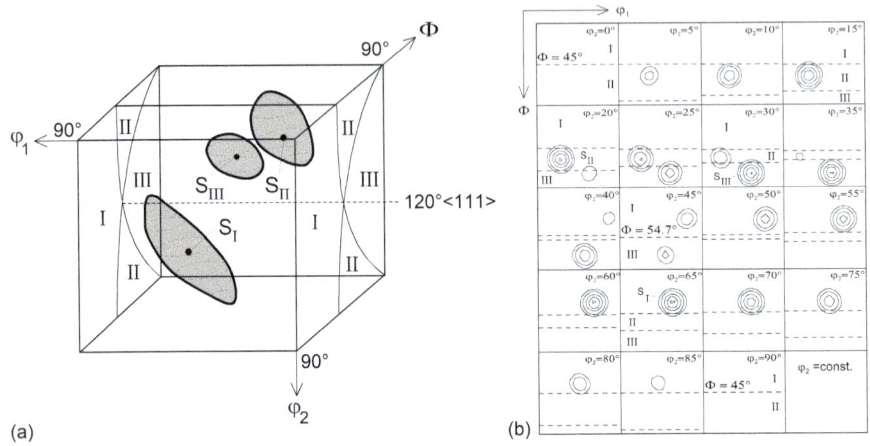

FIGURE 5.13 Spatial arrangement of the three components of the orientation (123)[634] with isotropic scatter of 10° in the reduced Euler angle space showing strong distortions at small angles Φ. (b) Same texture as in (a) plotted with iso-intensity lines in φ_2 sections ($\Delta\varphi_2 = 5°$) through the Euler space.

rather show strong distortions that become most pronounced for small angles Φ. An orientation with Φ = 0° even deteriorates to a line with the condition $\varphi_1 + \varphi_2 =$ constant (see Section 2.6.1). These distortions become most evident by representing the texture in the (unusual) Φ sections; an example is shown in Section 10.6.1. With regard to the cube recrystallization texture, the distortions of Euler space result in the fact that rolling direction (RD) rotations of the cube orientation {001}⟨100⟩ can be found in the $\varphi_2 = 0°$ section by following the angle Φ at $\varphi_1 = 0°$, whereas transverse direction (TD) rotations are visible in the same section along Φ at $\varphi_1 = 90°$. In the following section, we will see that these distortions can be reduced by representing the textures in polar rather than Cartesian coordinates.

5.4.2 Representation and Display of Textures

A representation of an ODF in a 3D space as in Figure 5.13a is generally not suitable for publication on the printed page. Therefore, it is common practice to represent ODFs in the form of sections through the orientation space (Figure 5.13b). As already discussed for the pole figures, the intensity distribution in the individual sections can be displayed by different colors, different gray values, or, most commonly, by iso-intensity lines (Figure 5.14). Usually, equal-distance sections along one of the angles in 5° steps (sometimes 10° steps) are used. In principle, sections of all three Euler angles can be produced and evaluated alike, though surveying the literature shows that ODFs of fcc materials are usually presented in φ_2 sections, whereas the textures of bcc materials are traditionally shown in φ_1 sections, though φ_2 sections offer some advantages (see the following text). For hexagonal materials both φ_1 and φ_2 sections are found. Φ sections are hardly ever used, in contrast.

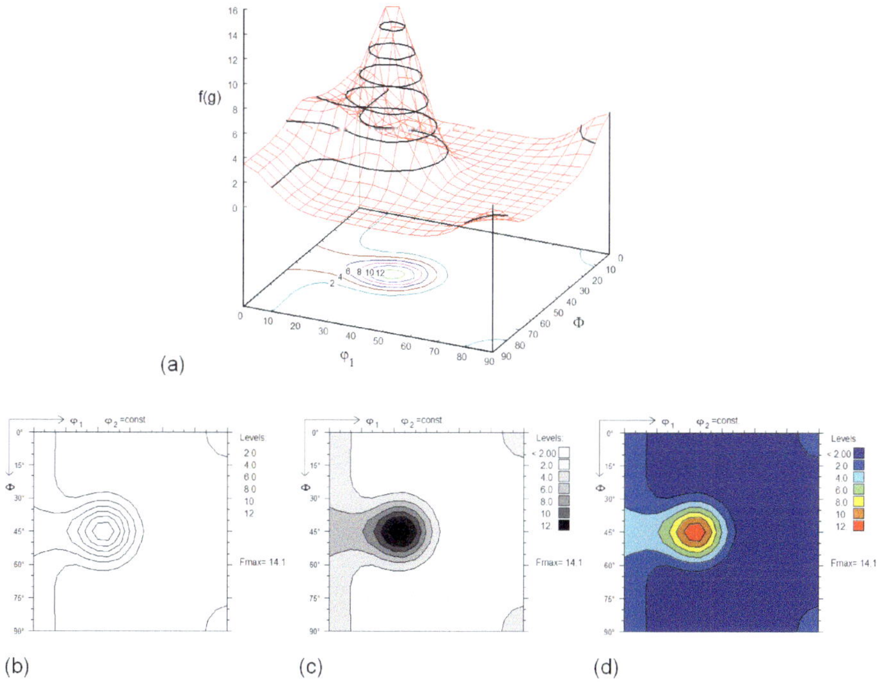

FIGURE 5.14 (a) Contouring of a 3D orientation distribution to derive the standard 2D presentation of ODF sections; (b) orientation distribution of (a), plotted with iso-intensity lines; (c) orientation distribution of (a), plotted with gray shades; (d) orientation distribution of (a), plotted with colors (ODF $\varphi_2 = 0°$ section).

As mentioned in Section 5.4.1, the distortions in the Cartesian Euler space at small angles Φ can be greatly reduced by representing texture data in polar rather than the usual Cartesian coordinates (Williams, 1968; Hansen et al., 1978; Wenk and Kocks, 1987). For that purpose, the angles φ_1 and Φ are plotted in polar coordinates—just as in a pole figure—and φ_2 is represented as the axis of a cylinder perpendicular to the polar plot, as shown in Figure 5.15a. Analogous to the φ_2 sections through the Cartesian Euler space (Figure 5.13b), a set of sections through the resulting cylindrical space at constant values of φ_2 is adequate to represent texture in a 2D image (Figure 5.16). Following the definition of the Euler angles (Section 2.6.1), for fixed values of φ_1 and Φ, the angle φ_2 defines a rotation about a distinct [001] axis. Thus, the projection of the ODF in the plane with $\varphi_2 = 0°$ corresponds to a {100} pole figure (Figures 5.15a and 5.16). φ_2 sections have been labeled *crystal orientation distribution* (COD) to indicate that in this representation the crystal orientations are expressed in a sample coordinate system (Wenk and Kocks, 1987). Alternatively, one may wish to represent the angles φ_2 and Φ in polar coordinates with φ_1 as cylinder axis (Figure 5.15b). As for fixed angles φ_2 and Φ, φ_1 indicates a rotation about the sample ND, a projection of the ODF in the plane with $\varphi_1 = 0°$ corresponds to an inverse pole figure of the ND. For that reason, φ_1 sections may be referred to as *sample orientation distribution* (SOD) (Wenk and Kocks, 1987).

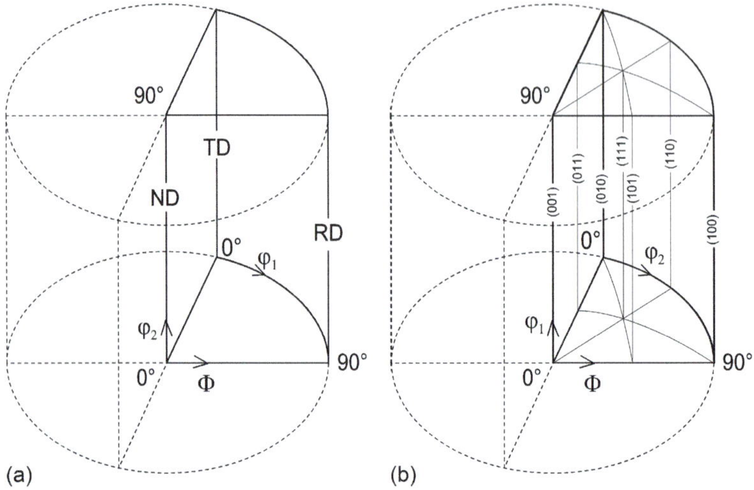

FIGURE 5.15 Representation of the orientation space in a cylindrical Euler space. (a) Cylinder axis parallel to φ_2; the projection of texture on the base ($\varphi_2 = 0°$) corresponds to a pole figure (COD). (b) Cylinder axis parallel to φ; the projection of texture on the base ($\varphi = 0°$) corresponds to an inverse pole figure (SOD).

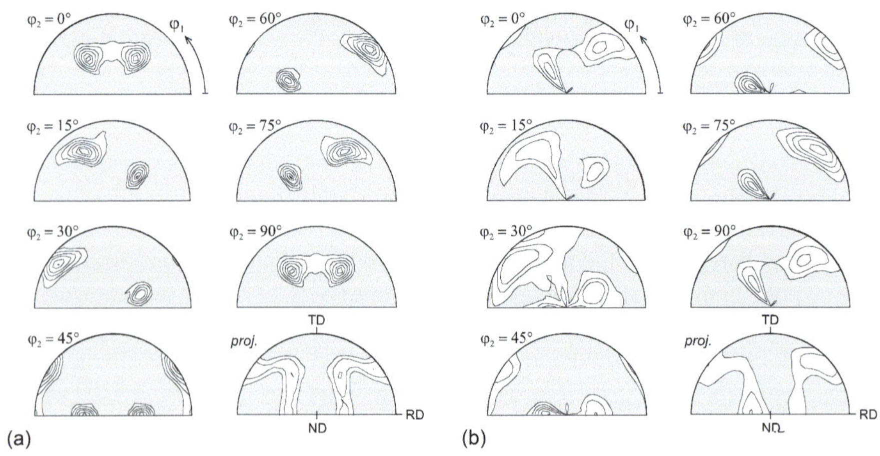

FIGURE 5.16 Texture of hot-rolled aluminum AA 6016 represented in the cylindrical Euler space (φ_2 sections, COD). (a) Rolling texture at the center layer of the sheet, depicting orthonormal sample symmetry. The sections $\varphi_2 = 0°$, 45°, and 90° as well as the (projected) pole figure are symmetrical with respect to TD. Also, note symmetries between sections $\varphi_2 = 15°$ and 75° and $\varphi_2 = 30°$ and 60°. (b) Shear texture at the sheet surface, depicting monoclinic sample symmetry. The various semicircles show no symmetry with respect to TD, representing the deviation of the strain state from orthonormal sample symmetry.

The texture presentation in a cylindrical space in sections φ_2 = constant is particularly suitable for demonstration of the influence of sample symmetry on the necessary range of orientation space. As outlined in Section 2.6.2, sample symmetry affects the range of φ_1. Thus, for orthonormal, monoclinic, and triclinic sample symmetry, respectively, a quarter circle $\left(\varphi_1^{max} = 90°\right)$, a semicircle $\left(\varphi_1^{max} = 180°\right)$, and a full circle $\left(\varphi_1^{max} = 360°\right)$ are necessary to present the texture. As an example, Figure 5.16a and b shows the texture of a hot band of an aluminum alloy at the band center and at the surface, respectively. In both cases, the textures are represented in the cylindrical Euler space in φ_2 sections (i.e., CODs). At the center layers of the sheet, the assumption of orthonormal sample symmetry is very well fulfilled, which results in symmetric semicircles representing the different φ_2 sections as well as in a symmetric projection, that is, pole figure (Figure 5.16a). At the band surface, in contrast, the different semicircles evidently are not symmetric (Figure 5.16b), representing the deviation of the strain state from orthonormal sample symmetry.

It is seen in Figure 5.16 that the representation of ODFs in polar coordinates still reveals some problems at small Φ angles—at the center of the circular sections—which may particularly affect representation of the cube orientation. These distortions can further be refined by introducing *oblique σ sections* (with $\sigma = (\varphi_1 + \varphi_2)/2$; Matthies et al., 1990; Kocks, 1998b), which, however, is not discussed here. The main disadvantage of the representation of textures in the cylindrical Euler space is that determination of the orientations is more complicated than in an orthogonal space.

In many studies, a series of comparable textures is analyzed, for example, rolling textures at different levels of deformation or recrystallization textures obtained after annealing at different temperatures. To compare the—usually rather similar—textures in such cases, it may be useful to confine the representation to the sections of the ODF that display the characteristic texture changes, which typically affect only one or two sections rather than the entire Euler angle space. For instance, most texture aspects of bcc materials can be found in the ODF section with $\varphi_2 = 45°$. Therefore, despite the fact that bcc textures are traditionally represented in φ_1 sections (see the preceding text), analysis of the section $\varphi_2 = 45°$ offers a very condensed way of analyzing bcc rolling and recrystallization textures, examples of which are given in Sections 5.5.2 and 5.5.5. Many interesting features of recrystallization textures in fcc materials are visible in the $\varphi_2 = 0°$ section (Engler et al., 1995a).

Alternatively, instead of displaying the entire ODFs, many textures can be represented in a very condensed way by plotting the orientation intensity along certain characteristic paths or distinct crystallographic *fibers* through the orientation space versus an angle that defines the position along this path or fiber (Bunge and Tobisch, 1968; Lücke et al., 1985). For instance, the characteristic texture evolution during rolling can readily be visualized by plotting the orientation density along the corresponding rolling texture fibers, examples of which are presented in Sections 5.5.1, 5.5.2, and 5.5.5. Provided that the texture information not covered by these fibers is of minor importance, such diagrams evidently represent a very convenient way of displaying characteristic textural features compared to the ODF plots.

5.5 EXAMPLES OF TYPICAL TEXTURES IN METALS

A number of examples of the characteristic deformation and annealing textures of fcc, bcc, and hexagonal metals are described in this section. Given the large body of publications on textures, it is not practicable to address all possible textures or present an exhaustive survey of the existing literature. Rather, the intention is to provide the reader with a guide to the most typical texture types that are frequently encountered in metallic materials and address the underlying formation mechanisms during deformation and recrystallization.

5.5.1 Deformation Textures of fcc Metals

In the earlier days of texture analysis by pole figure measurements, most effort was made to analyze the rolling textures of fcc metals and alloys, most notably, of copper and Cu alloys. In those days, rolling textures were subdivided into two categories, namely, textures of *pure metals*—copper, nickel, and aluminum, and *alloy-type* textures—typically observed in brass and austenitic steel (Hu et al., 1952; Wassermann and Grewen, 1962; Dillamore and Roberts, 1965; Leffers and Ray, 2008). In many fcc metals and alloys, there is a characteristic dependence of the rolling texture on the materials' *stacking fault energy* (SFE). The SFE depends on the sample material and is usually strongly affected by alloying elements. For example, pure metals such as nickel and aluminum have very high SFEs in excess of 200 mJ m^{-2}, whereas copper and gold have an intermediate value of 50–60 mJ m^{-2} (Gallagher, 1970). Most alloy elements appreciably reduce the SFE to very low values of below 20 mJ m^{-2}, for example, in Cu alloys such as brass or bronze or in austenitic steel. Interestingly, pure silver has a very low SFE as well and, in consequence, reveals typical "alloy"-type rolling textures. The influence of SFE on the rolling texture can be understood in terms of its impact on the deformation mechanisms, most notably, deformation twinning. In metals with medium to high SFE deformation, twinning is not observed; hence, deformation is solely attained through dislocation slip. Low SFE enables deformation twinning and, subsequently, massive shear band formation, which results in the altered deformation texture evolution (Hutchinson et al., 1979; Hirsch et al., 1988; Leffers and Bilde-Sørensen, 1990). As an example, Figure 5.17 shows the {111} pole figures of cold-rolled pure copper and brass (Cu–37%Zn), whereas the corresponding ODFs are given in Figure 5.18. The texture shown in Figures 5.17a and 5.18a is a typical example of the characteristic "pure-metal" (or *copper-type*) rolling texture. This texture is commonly described in terms of three main texture components, the copper (or Cu) orientation {112}⟨111⟩, the S orientation {123}⟨634⟩, and the brass (or Bs) orientation {011}⟨211⟩ (Table 5.1).

However, a more detailed analysis of the pole figure (Figure 5.17a) and, especially, the ODF (Figure 5.18a) reveals that a description with ideal components is inadequate. Rather, the textures can be described as a spread of the orientations from the Cu orientation through the S orientation to the Bs orientation. That is to say, the typical copper-type rolling texture is best described by a continuous *orientation tube* running through orientation space. Figure 5.19a shows a schematic representation of this orientation tube in Euler space (for clarity only onc of the symmetrically

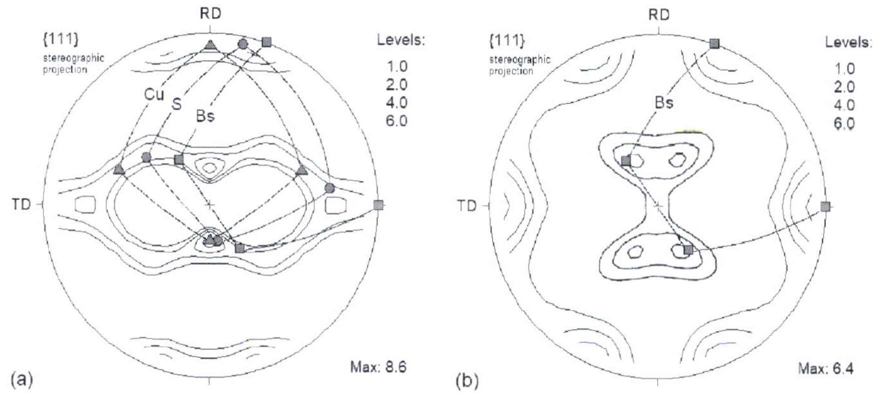

FIGURE 5.17 Rolling texture ({111} pole figure) of (a) pure copper and (b) brass (Cu–37%Zn).

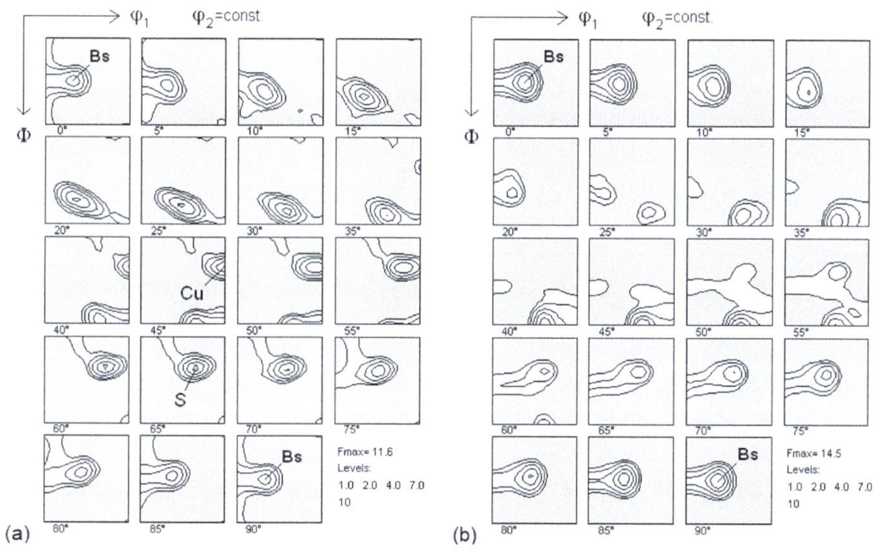

FIGURE 5.18 Rolling texture (ODF) of (a) pure copper and (b) brass (Cu–37%Zn).

TABLE 5.1
Miller Indices and Euler Angles of the Most Important Orientations of Al and Al Alloys after Rolling and after Recrystallization (Approximated)

Designation	Miller Indices {*hkl*}⟨*uvw*⟩	Euler Angles φ_1 Φ φ_2
Copper (Cu)	{112}⟨111⟩	90° 30° 45°
S	{123}⟨634⟩	59° 34° 65°
Brass (Bs)	{011}⟨211⟩	35° 45° 0°/90°
Goss	{011}⟨100⟩	0° 45° 0°/90°

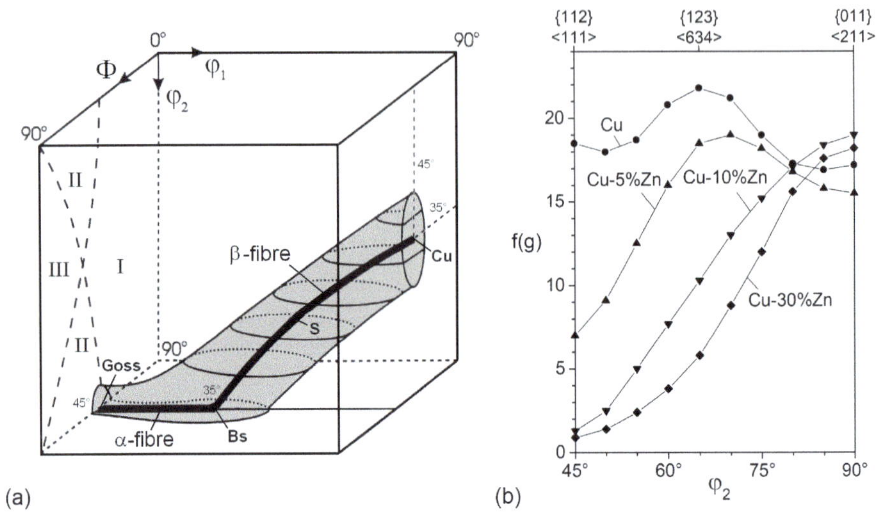

FIGURE 5.19 (a) Sketch of the position of α and β fibers in Euler space (for clarity, only one of the symmetrical variants is shown); (b) β fibers of pure copper and three different copper-zinc alloys, showing the influence of SFE on the rolling textures (95% rolling reduction). (Adapted from Hirsch and Lücke, 1988.)

equivalent branches in the range 45° ≤ Φ ≤ 90° is shown). The orientation tube is usually described in terms of its central fiber. In fcc metals, the most important orientation fibers are the α and the β fibers: at low deformation degrees the α fiber with {011}||RD is observed, which runs from the Goss orientation {011}⟨100⟩ to the Bs orientation {011}⟨211⟩. At higher deformation degrees the β fiber prevails, which runs from the Cu orientation {112}⟨111⟩ through the S orientation {123}⟨634⟩ to the Bs orientation {011}⟨211⟩ where it meets the α fiber. Table 5.2 summarizes the characteristics of these two fibers together with several other fibers that are less frequently observed in the textures of fcc materials.

TABLE 5.2
Characteristic Fibers in fcc Metals and Alloys

Fiber	Fiber Axis	Euler Angles[a]
A	⟨011⟩||ND	0°,45°,0°—90°,45°,0°
Γ	⟨111⟩||ND	60°,54.7°,45°—90°,54.7°,45°
T	⟨011⟩||TD	90°,0°,45°—90°,90°,45°
B	[b]	90°,35°,45°—35°,45°,90°

Notes:

[a] Typical values without symmetry considerations.

[b] Defined by the maximum intensity rather than by exact crystallographic position.

Thus, specific differences in the rolling textures of different fcc materials can conveniently be displayed by plotting the intensity distribution along the α and β fibers. Figure 5.19b shows the β fibers of rolled pure copper and copper with different amounts of zinc after 95% rolling reduction. Note that—unlike most other texture fibers—the path of the β fiber is not fixed in Euler space; the β fiber rather connects the relative intensity maxima in the corresponding ODF sections along the fiber, giving rise to the alternative term *skeleton line*. Therefore, besides analyzing the intensity distribution along the β fiber, subtle details in the position of the β fiber in Euler space can be evaluated, which may yield valuable information with regard to the underlying deformation mechanisms (Hirsch and Lücke, 1988; Engler et al., 1989).

The alloy, or brass, type texture displayed in Figures 5.17b and 5.18b is dominated by a quite pronounced Bs orientation {011}⟨211⟩. Figure 5.19b shows that there is a systematic decrease in the intensities close to the Cu and S orientations to the advantage of the Bs orientation with increasing zinc content, that is, with decreasing SFE. This reflects the characteristic rolling texture transition from the copper or pure metal type to the brass or alloy type.

Rather than changing the SFE, alloy additions may precipitate in the form of second-phase particles, and different precipitation states affect the deformation behavior by interaction of the slip dislocations with the particles in dependence on particle size and volume. Figure 5.20 shows the rolling textures obtained in an aluminum-copper alloy after different preaging treatments, where it turned out that the

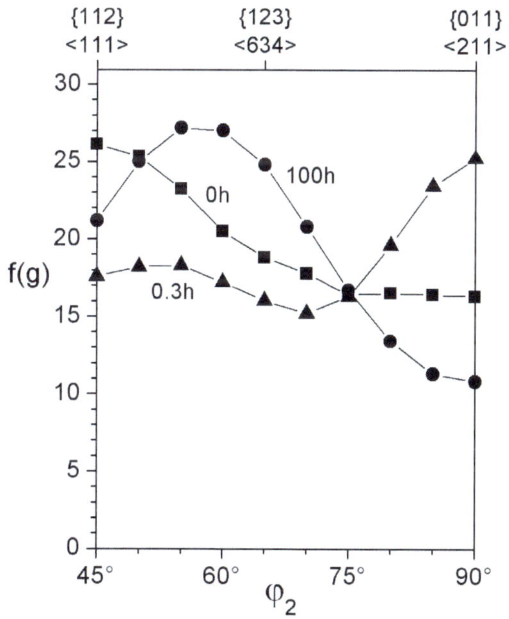

FIGURE 5.20 β fibers of Al–1.8%Cu cold rolled after different aging treatments, showing the influence of precipitates on rolling textures (95% rolling reduction). (Data taken from Engler et al., 1989.)

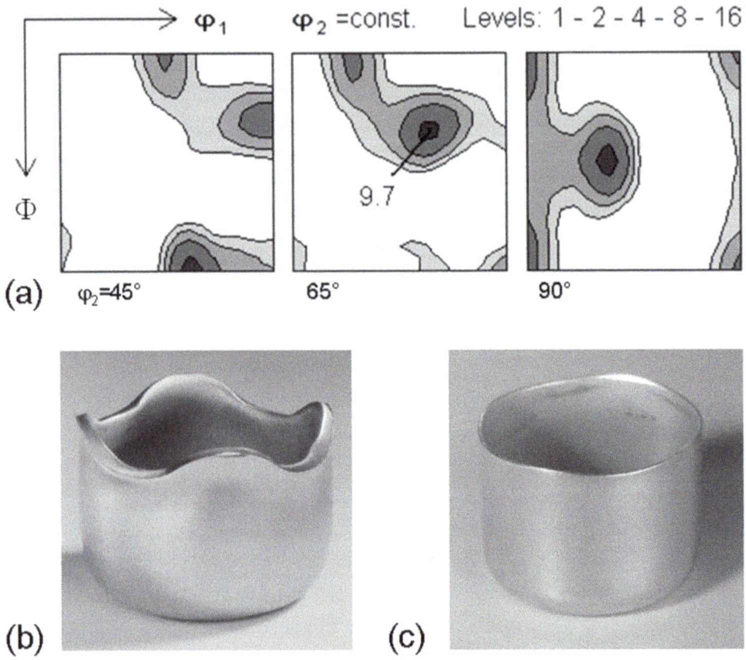

FIGURE 5.21 Correlation of texture and earing in AA 3104 can body stock. (a) Texture at final gauge (H19), showing a rolling texture superimposed on a cube texture retained from the recrystallized hot strip; (b) cup drawn from hot strip, showing pronounced 0°/90° earing; (c) cup drawn from final gauge sheet, showing very weak earing with mixed 0°/90° and 45° ears. (Adapted from Engler and Hirsch, 2007.)

resulting different precipitation states significantly influenced the intensity distribution along the β fiber (Engler et al., 1989; Lücke and Engler, 1990). Other parameters including initial grain size (Engler, 1995), deformation temperature (Maurice and Driver, 1997), and initial texture (Leffers and Juul Jensen, 1986; Mao et al., 1988; Engler et al., 2005a) have also been shown to affect the intensity distribution along the various fibers, though generally less dramatically than alloy composition.

With a view to the latter point—impact of initial texture—Figure 5.21a shows an example of the rolling texture obtained in Al beverage can stock (alloy AA 3104, AlMn1Mg1). This example is of great importance for the aluminum industry because of the correlation of texture and mechanical anisotropy, especially, the *earing* behavior. When a cylindrical cup is deep drawn from a circular blank of a textured sheet, the rim of this cup will not be level but will depict a number of distinct high points—usually called ears—and an equal number of low points, called troughs, as shown in Figure 5.21b and c. This earing is very undesirable, as it causes inhomogeneous wall thickness reduction and may interfere with automated deep-drawing operations.

For Al beverage cans, highly cold-rolled states (H19) are used to meet the strength requirements. Thus, the sheets comprise a strong β-fiber rolling texture, which is inevitably accompanied by the formation of ears at the four positions ±45° to the RD. Upon the production of Al can stock, it is aimed at offsetting these 45° ears by

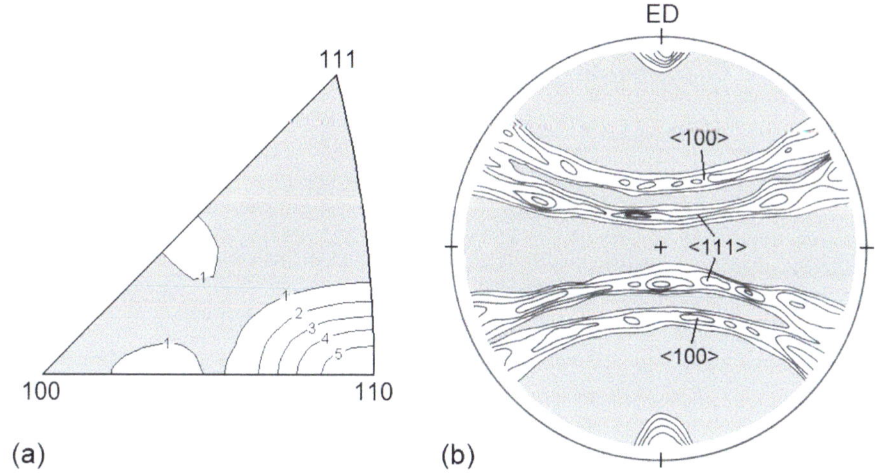

FIGURE 5.22 (a) Inverse pole figure of the compression direction of copper after compression to a strain of ~1; (b) pole figure of the extrusion texture of an aluminum alloy (ED: extrusion direction). (Courtesy of J. Hirsch.)

a proportion of {001}⟨100⟩ cube texture that is retained from the recrystallized hot strip (Section 5.5.4). Due to the slow rotation rate of the cube orientation, the resultant 0°/90° ears (Figure 5.21b) are able to balance the newly forming 45° ears up to rather high rolling reductions, typically ranging between 80% and 90% (Hutchinson et al., 1989; Engler et al., 2007). Figure 5.21a shows an example of the texture of final gauge can stock, which consists of a mixture of a retained cube texture and the newly formed rolling texture components. The resulting cup height profile reveals a characteristic six-ear profile with ears under 0°, 45°, 135°, 180°, 225°, and 315° to the RD (Figure 5.21c).

Uniaxial deformation—tension, compression, wire drawing, or extrusion—usually leads to simpler textures than rolling. Since only one deformation axis is prescribed, invariably fiber textures form, which are often represented as inverse pole figures. Figure 5.22a shows the compression texture of a copper alloy, which consists of a ⟨110⟩ fiber with some scatter toward ⟨221⟩ (Engler, 2000). In low-SFE materials, ⟨111⟩ components may also form. After tension or uniaxial extrusion of fcc materials, the textures can be described by two fibers with ⟨111⟩ and ⟨100⟩ parallel to the deformation axis. Figure 5.22b shows an example of this double fiber texture in an extruded Al alloy. The relative proportion of the two fibers may vary with initial texture, deformation mode, and SFE (English and Chin, 1965; Stout and O'Rourke, 1989).

5.5.2 Deformation Textures of bcc Metals

Despite the importance of steel industry, deformation textures in bcc metals and alloys have been less extensively investigated. In general, bcc metals and alloys tend to form fiber textures, which means that most orientations are assembled along two characteristic fibers through orientation space. These are summarized in Table 5.3 (Hutchinson, 1984; Von Schlippenbach et al., 1986). Figure 5.23a shows a typical

TABLE 5.3
Characteristic Fibers in bcc Metals and Alloys

Fiber	Fiber Axis	Euler Angles[a]
A	⟨011⟩‖RD	0°,0°,45°—0°,90°,45°
γ	⟨111⟩‖ND	60°,54.7°,45°—90°,54.7°,45°
H	⟨001⟩‖RD	0°,0°,0°—0°,45°,0°
Z	⟨011⟩‖ND	0°,45°,0°—90°,45°,0°
E	⟨110⟩‖TD	90°,0°,45°—90°,90°,45°
B	[b]	0°,35°,45°—90°,54.7°,45°

Notes:
[a] Typical values without symmetry considerations.
[b] Defined by the maximum intensity rather than by exact crystallographic position.

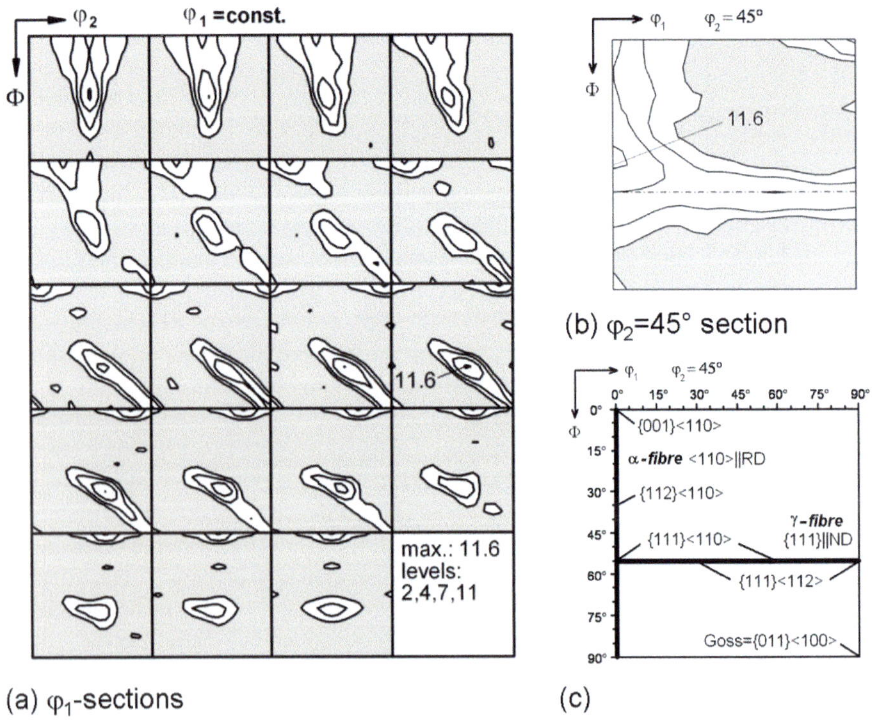

FIGURE 5.23 Rolling texture of a Ti and Nb microalloyed interstitial-free steel in the Euler space (90% rolling reduction). (a) Conventional representation of the ODF in φ_1 sections; (b) $\varphi_2 = 45°$ section displaying the intensity distribution of the α- and γ-fiber orientations; (c) schematic representation of the most important orientations in bcc materials in the $\varphi_2 = 45°$ section. (Courtesy of M.Y. Huh.)

cold-rolling texture of a microalloyed interstitial-free (IF) steel, plotted in φ_1 sections. The, mostly partial, α fiber comprises the orientations with a common ⟨110⟩ crystal axis parallel to the RD, that is, the orientations {*hkl*}⟨110⟩, which include the orientations {001} ⟨110⟩, {112}⟨110⟩, and {111}⟨110⟩. In the Euler space, the α fiber can be found between $(\varphi_1, \Phi, \varphi_2) = (0°, 0°, 45°)$, (0°, 35°, 45°), and (0°, 54.7°, 45°) (note that, because of the cubic crystal symmetry, these orientations also appear at other, symmetrically equivalent locations in the Euler space). The γ fiber comprises the orientations with {111} parallel to the ND, that is, the orientations {111}⟨*uvw*⟩, including {111}⟨110⟩ and {111}⟨112⟩. This fiber runs from (0°, 54.7°, 45°) to (90°, 54.7°, 45°) in Euler space but, because of the cubic crystal symmetry, the range $60° \leq \varphi_1 \leq 90°$ is sufficient for a representation of the full γ fiber. As already alluded to in Section 5.4.2, the characteristic texture changes during rolling can readily be visualized by plotting the orientation density along the most important texture fibers. Figure 5.24 displays the typical texture evolution upon cold rolling of a low-carbon steel in the form of the intensity density along the α and γ fibers (Hölscher et al., 1991). It is evident that, starting from the almost random hot strip texture (see the following text), there is a significant texture sharpening with increasing strain. Furthermore, increasing rolling deformation tends to shift the texture maxima from {001}⟨110⟩ along the α fiber over {112}⟨110⟩ toward {111}⟨110⟩ and {111}⟨112⟩ in the γ fiber.

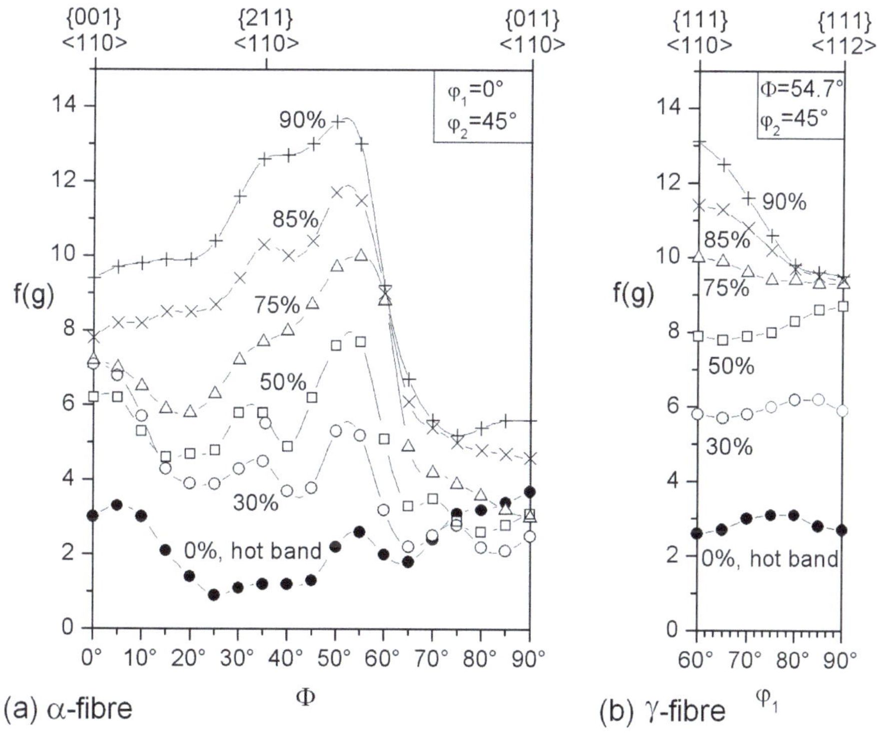

FIGURE 5.24 Development of the rolling texture fibers of a low-carbon steel with increasing rolling reduction; (a) α fiber, (b) γ fiber. (Data from Hölscher et al., 1991.)

As already mentioned, the textures of bcc materials are often represented in φ_1 sections, though φ_2 sections offer some advantages. In Figure 5.23b, the texture of Figure 5.23a is represented in the $\varphi_2 = 45°$ section, and it is seen that the most important texture features—including both the α fiber ⟨011⟩||RD and the γ fiber {111}||ND—can all be found in this section (Figure 5.23c).

Unlike in fcc metals, the rolling texture evolution in iron and low-carbon steels is largely independent of composition and processing (except the described general texture sharpening with strain). During hot rolling and subsequent coiling, recrystallization and the phase transformation from γ to α iron take place, leading to grain refinement and a randomization of the texture (Ray et al., 1994). As a consequence, the hot bands of low-carbon steels comprise fairly small grains with a random texture, which constitutes quite constant starting conditions for the subsequent cold rolling and recrystallization.

This is different in highly alloyed steel grades, like Fe–Si transformer steel or Cr-containing ferritic stainless steel (Seidel et al., 1989; Raabe, 1996; Huh and Engler, 2001). Such steel undergo no or merely a limited phase transformation during hot rolling, which restricts the opportunities for a texture randomization through a γ- to-α-phase transformation during the thermomechanical processing. The as-cast ingots usually comprise a columnar grain structure with the fast-growing {001} crystal directions aligned parallel to the ND of the ingot. Accordingly, the hot bands of ferritic stainless steels commonly display pronounced textures and strong through-thickness gradients in both microstructure and texture. Small recrystallized grains with a weak shear texture form at the outer surface layers, and highly elongated very large pancake-shaped grains with {001}⟨110⟩ are often observed in the sheet center (Figure 5.25a). This inhomogeneous hot strip has a marked impact on the evolution of texture and properties during the subsequent cold rolling and recrystallization. Because of the preferred {001}⟨110⟩ component in the hot strip, the cold-rolling textures usually comprise higher α-fiber intensities, whereas the γ fiber is less well developed than in plain low-carbon steel (Figure 5.25b).

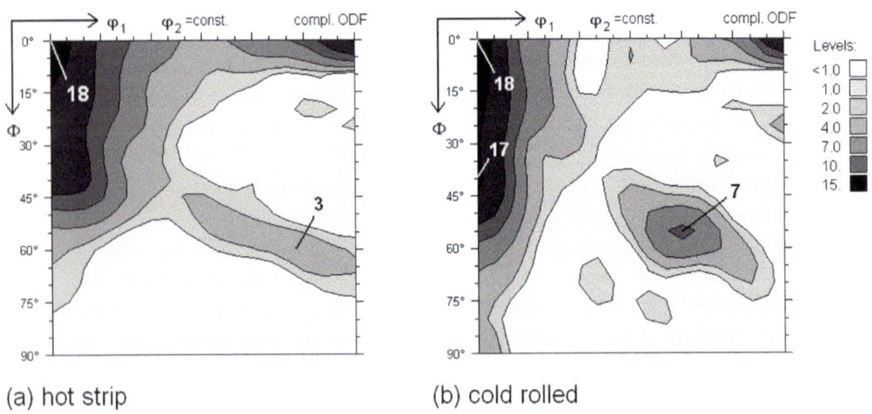

(a) hot strip (b) cold rolled

FIGURE 5.25 Texture of (a) hot strip and (b) cold strip of ferritic stainless steel Fe–17%Cr (ODF $\varphi_2 = 45°$ section, sheet center). (Data from Huh and Engler, 2001.)

Other bcc materials, like tantalum, molybdenum, and niobium, show similar α- and γ-fiber rolling textures as steel. As discussed for the highly alloyed Fe–Si and Fe–Cr steel grades, usually the component {001}⟨110⟩, which is retained from the hot strip texture, is quite strong (Raabe and Lücke, 1994; Raabe et al., 1994). Moreover, refractory metals have been reported to develop pronounced through-thickness texture gradients (Vandermeer and Ogle, 1969; Klein et al., 1994; Wright et al., 1994).

In uniaxial tension or wire drawing, bcc materials form (110) fiber textures, as fcc materials deformed in compression. Vice versa, in compression of bcc metals, mixed ⟨100⟩ and ⟨111⟩ fiber texture form, which resemble the typical fcc tension textures (Section 5.5.1). For the resemblance of fcc and bcc deformation textures due to reversing slip plane and slip direction, see Hölscher et al. (1994) and Rollett and Wright (1998). A notable difference between tensile deformation of fcc and bcc structures, however, is the phenomenon of *grain curling* (Hosford, 1964; Gil Sevillano et al., 2007). Upon drawing of bcc wires, grains tend to form ribbons and wrap around one another, rather than deforming uniaxially. This is attributed to the formation of a ⟨110⟩ fiber texture, which—at variance to the macroscopic uniaxial strain state—favors a plane strain deformation on the grain level.

5.5.3 Deformation Textures of Hexagonal Metals

Compared to cubic metals (both fcc and bcc), studies on the texture evolution in hexagonal materials are very limited. Nevertheless, hexagonal textures are of interest because of the use of zirconium alloys *Zircaloy* in nuclear reactors and titanium and magnesium sheet in transportation applications (see reviews by Tenckhoff, 1988; Philippe, 1994; Wang and Huang, 2003). Note that for hexagonal structures, two different conventions for the definition of crystal axes are used in the literature (Section 2.3.1; Appendix I). All examples in this chapter are based on the definition in Equation 2.2a.

With a view to the evolution of deformation textures, hexagonal materials are different from cubic metals in that, besides dislocation slip, twinning is almost always required to accommodate the external plastic strain state (Kocks and Westlake, 1967). Fcc materials have 12 equivalent slip systems of type {111}/⟨110⟩ (i.e., slip plane/slip direction) that can be activated, so that there will always be enough slip systems to accommodate plastic deformation. In bcc materials the slip direction is ⟨111⟩, whereas the slip planes are not so clearly defined. In most bcc materials, {110} planes are activated, whereas in some cases, additional slip occurs on {112} or {123} planes. Thus, bcc metals have 12, 24, or even 48 possible slip systems to accommodate plastic strain. In hexagonal structures, basal slip {0002}/(1120) is the easy glide system and, as such, activated most often. However, there are only three slip systems of this type (and only two mutually independent ones), which in most cases is not sufficient to accommodate plastic deformation. Additionally, the prismatic slip systems {1010}/⟨1120⟩ and the pyramidal slip systems {1011}/⟨1120⟩ and {1012}/⟨1120⟩ may be activated, but they have the same slip direction as the basal glide (cross slip). Pyramidal slip {1122}/⟨1123⟩ is the only slip system that has a slip direction *not* parallel to the basal plane. That is to say, pyramidal $(c + a)$ glide is the only mechanism that permits the accommodation of deformation along the c axis. Because of

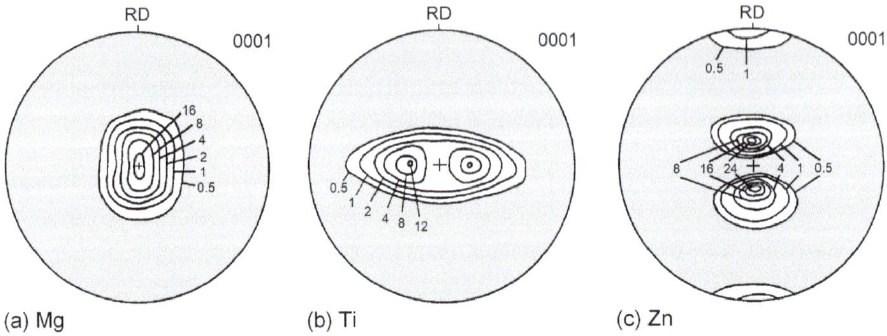

FIGURE 5.26 {0001} Pole figures for rolled hexagonal metals as a function of the *c/a* ratio; (a) magnesium (*c/a* = 1.624), (b) titanium (*c/a* = 1.587), and (c) zinc (*c/a* = 1.856). (Adapted from Grewen, 1973.)

this general lack of slip systems, slip cannot continue as the only deformation mode but twinning becomes significant already at rather low strains. The most important twinning system is {1012}/⟨1011⟩, but in materials with a low *c/a* ratio, other twinning systems may be activated as well. The shears achieved by twinning are quite low, typically between 0.1 and 0.2 (Grewen, 1973). Therefore, hexagonal materials usually contain a large volume of twins, which reorient the structure and hence facilitate further dislocation slip.

The deformation textures of hexagonal materials can roughly be subdivided into three groups in accordance with their *c/a* ratio (Wassermann and Grewen, 1962; Grewen, 1973; Barrett and Massalski, 1980; Philippe et al., 1988). For metals like magnesium or cobalt with a close-to-ideal *c/a* ratio of ~1.633, the rolling textures are dominated by a pronounced {0001} fiber texture (Figure 5.26a). This is a direct consequence of the dominant basal slip that leads to a rapid rotation of the grain orientations toward {0001} and, therewith, readily exhausts the possibility to activate other slip systems or twinning systems. This severely limits the possibility for both further cold rolling and subsequent forming operations of Mg sheet (Agnew et al., 2003; Barnett et al., 2004; Gehrmann et al., 2005; Yang et al., 2005). Upon hot deformation, other slip systems (prismatic slip as well as pyramidal $c + a$ slip) are activated much more easily; furthermore, recovery or recrystallization are enhanced (Ion et al., 1982; Del Valle et al., 2003; Backx et al., 2004; Walde and Riedel, 2007; Al-Samman and Gottstein, 2008). Figure 5.27 shows the ODF of hot-rolled Mg alloy AZ31 sheet (Mg–3%Al–1%Zn). Note that because of crystal symmetry, the angular range of φ_2 is limited to 60°, whereas there is no threefold symmetry as in the case of cubic structures (Section 2.6.2; Table 2.3). The ODF shows a rather weak {0001} fiber texture with some scatter toward {0112}⟨2110⟩. Sheet formability is likewise greatly enhanced at elevated temperature (Doege and Dröder, 2001).

For metals with a lower *c/a* ratio than ideal (titanium, zirconium), the basal planes are rotated by ±30–40° about the RD (Figure 5.26b). Figure 5.28a and b shows the evolution of the cold-rolling texture of titanium sheet for two different rolling degrees. The textures are displayed in two important φ_2 sections of the ODF,

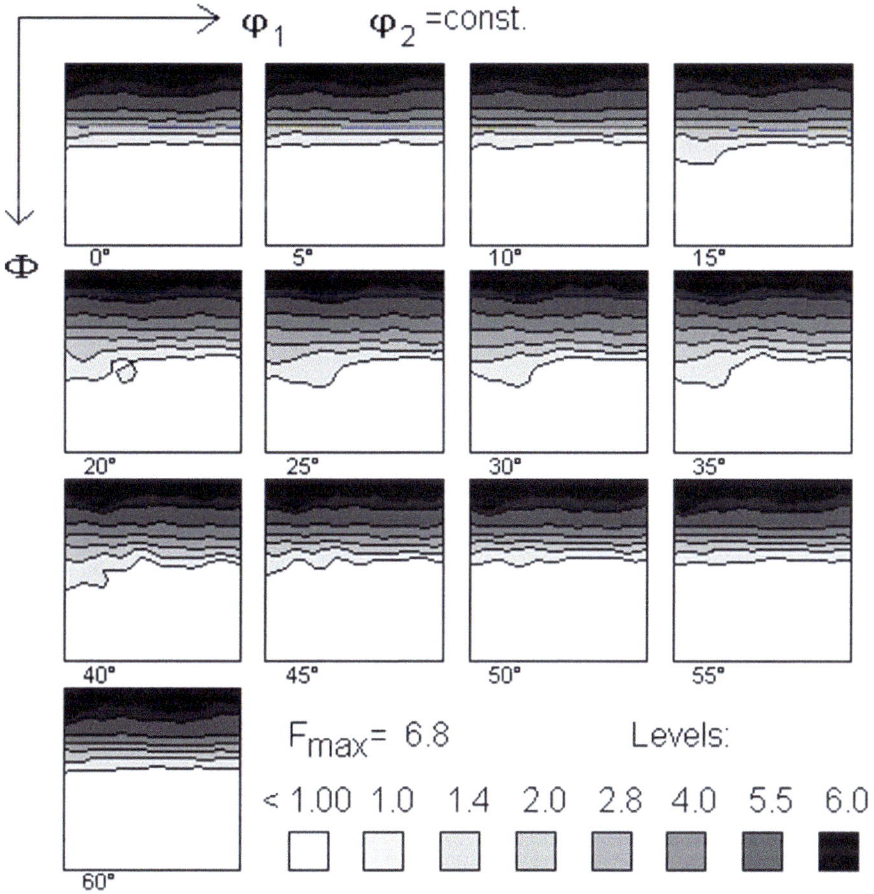

FIGURE 5.27 Hot-rolling texture of Mg alloy AZ31. (Courtesy of T. Al-Samman.)

$\varphi_2 = 0°$ and $\varphi_2 = 30°$ (Figure 5.28c). After medium rolling degrees, the orientations tend to align along a fiber along Φ at $\varphi_1 = \varphi_2 = 0°$, which are all orientations having a common ⟨1010⟩ direction parallel to the RD, plus minor intensities of {0001}⟨uvt0⟩ (Figure 5.28a). At higher rolling degrees, the texture sharpens whereas the orientations concentrate at $\Phi \sim 35°–40°$, that is, at {1214}⟨1010⟩ (Figure 5.28b) (Inoue and Inakazu, 1988; Lee et al., 1988; Singh and Schwarzer, 2000; Wagner et al., 2002). Such textures form through prevalent {1010}/⟨1120⟩ prismatic slip, whereas pyramidal and basal cross slip as well as pyramidal (*c* + *a*) slip are less important. Furthermore, there is significant contribution from two twinning systems, viz. {1012} tension twinning and {1122} compression twinning (Tomé et al., 1991).

In zinc and cadmium with higher *c*/*a* ratios, the basal orientation is split up in that the basal planes are tilted symmetrically by ±20–30° about the TD, toward the RD (Figure 5.26c); the ⟨1010⟩ directions are aligned with the TD (Dillamore and Roberts, 1965; Philippe et al., 1994; Diot et al., 1998). Figure 5.29 shows an example

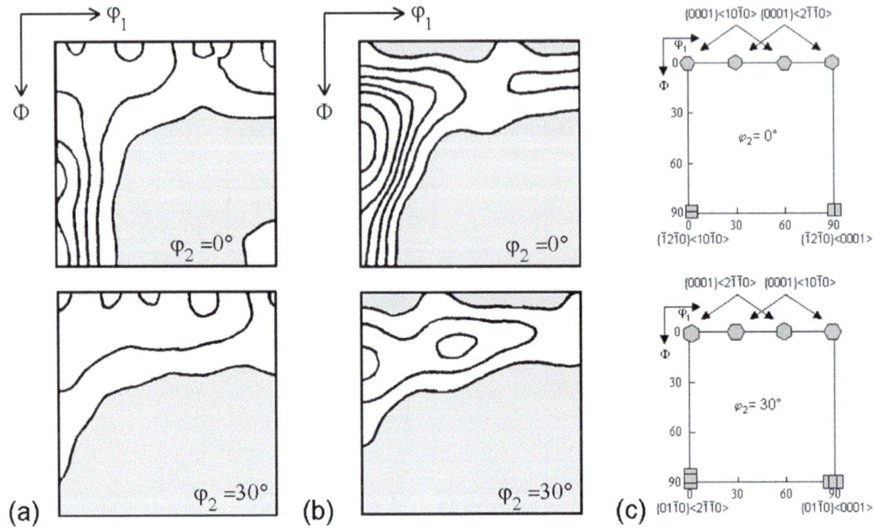

FIGURE 5.28 Rolling texture of titanium sheet: (a) 50% cold rolled, (b) 90% cold rolled, and (c) position of important orientations of hexagonal materials in Euler space (ODF $\varphi_2 = 0°$ and 30° sections, intensity levels 1–2–3–4–5–7–9). (Adapted from Inoue and Inakazu, 1988.)

of the ODF of a ZnCuTi alloy after 80% cold rolling, plotted in the section $\varphi_1 = 90°$. The texture is composed of a fiber along φ_2 at $\varphi_1 = 90°$ and $\Phi = 20°$, with a texture maximum at $\varphi_2 = 45°$, corresponding to crystallites with the c axes tilted by ~20° from ND in the ND-RD plane. This texture can be interpreted by a combination of basal slip and significant contributions of prismatic and pyramidal ($c + a$) slip systems, whereas {1012} twinning, though present, is less significant.

Of course the categorization by the *c/a* ratio is quite crude, and other factors, including initial texture, strain rate, deformation temperature, and chemical

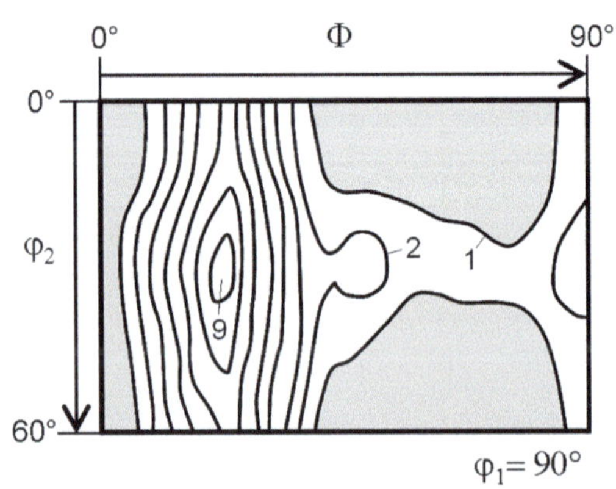

FIGURE 5.29 Rolling texture of 80% cold-rolled ZnCuTi alloy sheet (ODF $\varphi_1 = 90°$ section). (Adapted from Philippe et al., 1994.)

composition—both intentional alloying and impurities—will also impact texture development of hexagonal metals. For instance, the oxygen content has been reported to have a strong effect on the rolling texture evolution through changes of the critical resolved shear stress of the various slip systems and, most notably, through suppression of twin activities (Philippe et al., 1988; Zaefferer, 2003).

After wire drawing or extrusion, Zr and Ti usually form ⟨1010⟩ fiber textures; an example for this has already been given in Section 5.2.2 (Figure 5.3). Extruded Mg develops ⟨1010⟩ and ⟨1120⟩ double fiber textures, where the intensity ratio between the two fibers depends on the deformation conditions (Müller et al., 2006). Zn wires have a so-called "spiral texture," with ⟨0001⟩ approximately 70° from the wire axis. This difference is attributed to twinning, which leads to significant reorientations of material from a position in which the basal plane is parallel to the wire axis (Dillamore and Roberts, 1965; Barrett and Massalski, 1980).

5.5.4 Recrystallization Textures of fcc Metals

Recrystallization of deformed metallic materials is accomplished by the nucleation of new undeformed grains and their subsequent growth into the deformed matrix (Humphreys et al., 2017). These processes are characterized by the formation and the movement of large-angle grain boundaries. Therefore, recrystallization gives rise to a change in the distribution of the crystallographic orientations and, consequently, to a change in the crystallographic texture of a metallic material.

The rolling textures of fcc materials are mainly governed by two fibers, the α fiber and, most pronounced, the β fiber. It has been shown in Section 5.5.1 that the addition of alloy elements mainly alters the intensity distribution along these fibers, most notably through affecting the materials' SFE. While the rolling textures of Cu alloys show a rather smooth, continuous transition as a function of alloy composition (Figure 5.19b), the recrystallization textures change more drastically, however, and several points have been established on how alloying elements may potentially affect recrystallization textures. In correspondence to the two fundamental steps of recrystallization, nucleation, and growth, the texture changes during recrystallization are traditionally interpreted in terms of the two rival theories of oriented nucleation and oriented growth (Doherty, 1997).

In the case of the *oriented nucleation* argument, it is assumed that the preferred formation of special orientations determines the final recrystallization texture. With the emergence of the techniques for local texture analysis available today—analysis of Kikuchi diffraction patterns in the TEM and electron backscatter diffraction (EBSD)—nucleus orientations can be readily studied and, therewith, the occurrence of oriented nucleation in the sense of the preferred formation of special orientations at each nucleation site has been established.

In the case of *oriented growth* or *growth selection* arguments, it is assumed that, starting from a broad spectrum of nucleus orientations, those with the best growth conditions with respect to the deformed matrix grow fastest and, therefore, dominate the recrystallization texture. The well-known classical work by Liebmann et al. (1956) and, more recently, detailed bicrystal experiments performed by Molodov

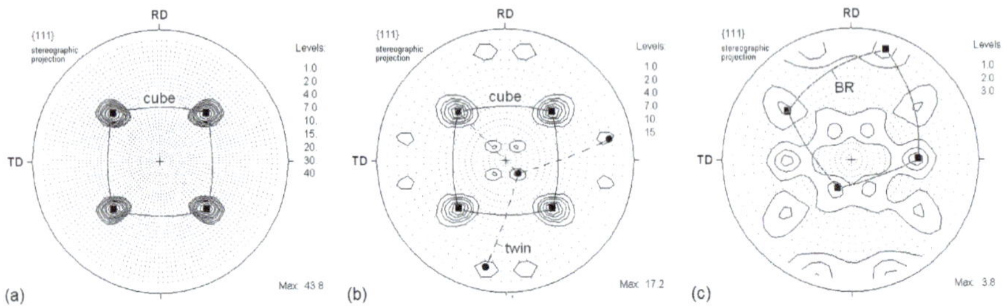

FIGURE 5.30 Recrystallization texture of cold-rolled fcc metals: (a) aluminum, (b) copper, and (c) brass (Cu–37%Zn) ({111} pole figure).

et al. (1995) have proven that in aluminum, grain boundaries with a rotation relationship of approximately 40° about a ⟨111⟩ axis depict maximum mobility.

It must be emphasized here that each of these two theories describes only one aspect of recrystallization as a limiting case, such that a discussion solely based on either of these two theories will be incomplete. Instead, in the majority of cases, a combination of both theories in the sense of a 40° ⟨111⟩ growth selection out of a limited spectrum of preferentially formed nucleus orientations takes place, which explains well the recrystallization textures of many fcc metals and alloys.

Materials with a high SFE like Al, Ni, and Cu typically form a cube recrystallization texture {001}⟨100⟩; an example is shown in Figure 5.30a. The mechanisms leading to the cube recrystallization texture have been subject to debate for more than six decades; a systematic review of this research is given by Hu (1986). Only with the advent of techniques for microtexture analysis, the nucleation and growth mechanisms of cube-oriented grains have been understood quite comprehensively. In brief, cube-oriented nuclei emerge from bandlike structures, the so-called cube bands, which are part of the deformed microstructure (Ridha and Hutchinson, 1982; Dons and Nes, 1986; Hjelen et al., 1991). These cube bands are either transition bands (Dillamore and Katoh, 1974) or deformed grains with retained cube orientation (Vatne et al., 1996). During the subsequent growth, grains with the cube orientation prevail, which is due to their optimum growth conditions into the rolling texture (Lücke, 1974).

With an increasing amount of alloying elements, the SFE of most Cu alloys decreases. Because of the decrease in SFE, recrystallization twinning may become an important mechanism for providing new orientations, thus leading to recrystallization texture components that are not present in the as-deformed state. For instance, copper has a lower SFE than Al. Although the SFE of Cu is not low enough to allow for deformation twinning (Section 5.5.1), growth twins may emerge. Accordingly, the recrystallization texture of pure Cu is composed of the cube orientation and its first-generation twins (Figure 5.30b).

Usually very weak recrystallization textures are observed in Cu alloys with medium contents of Zn, Al, or Ge, which contain twins of the cube and Goss orientation as well as a few new orientations, including the transition or U orientation {258}⟨121⟩ and the P orientation ~{011}⟨122⟩ (Figure 5.31a; Table 5.4). In such

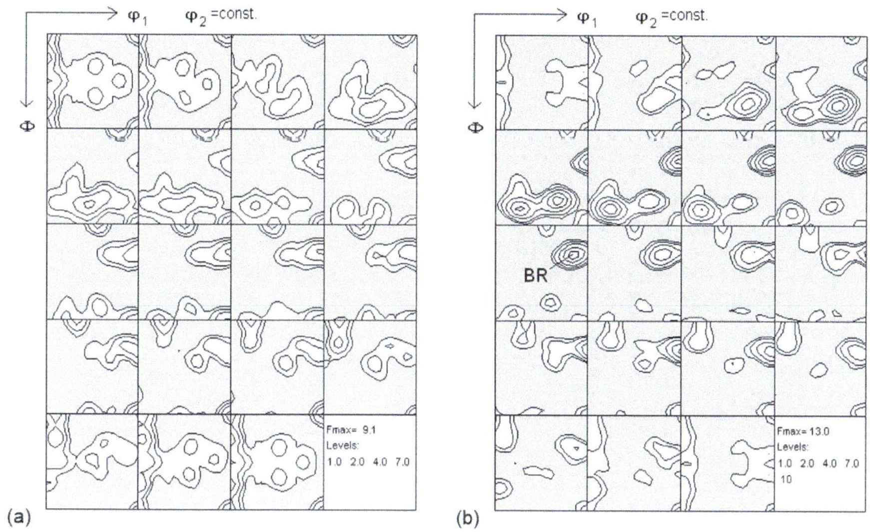

FIGURE 5.31 Recrystallization texture of cold-rolled copper alloys. (a) Cu–8%Mn and (b) Cu–37%Zn.

materials, the SFE is not sufficiently reduced so as to promote mechanical twinning and subsequent shear band formation and, consequently, provide additional nucleation sites. However, SFE is small enough to facilitate recrystallization twinning, which explains the increased occurrence of the twin orientations in the corresponding textures. Moreover, the alloying elements in solid solution interfere with the grain boundary motion, which retards the growth of the recrystallized grains, slows down recrystallization kinetics, and counteracts the formation of a well-defined

TABLE 5.4
Miller Indices and Euler Angles of the Most Important Recrystallization Texture Orientations of fcc Metals and Alloys (Approximated)

Designation	Miller Indices {*hkl*}⟨*uvw*⟩	Euler Angles φ_1 Φ φ_2
Cube	{001}⟨100⟩	0° 0° 0°/90°
$Cube_{RD}$	(013)⟨100⟩	0° 22° 0°/90°
$Cube_{ND}$	{001}⟨310⟩	22° 0° 0°/90°
Cube-twin (first generation)	{122}⟨212⟩	27° 48° 27° or 63° 48° 63°
Goss	{011}⟨100⟩	0° 45° 0°/90°
BR	{236}⟨385⟩	80° 31° 35°
U (transition)	{258}⟨121⟩	45° 35° 20°
R	{124}⟨211⟩	53° 36° 60°
P	{011}⟨122⟩	65° 45° 0°/90°
Q	{013}⟨231⟩	45° 15° 10°

recrystallization texture (Schmidt and Lücke, 1979; Eichelkraut et al., 1984; Savoie and Lücke, 1991).

A further decrease in the SFE leads to changes in the as-deformed microstructure, especially by promoting profuse shear band formation. Upon subsequent annealing, this gives rise to nucleation of recrystallization at the shear bands, which can substantially alter the resulting recrystallization textures. As an example, the recrystallization texture of brass is dominated by the so-called brass recrystallization texture component "BR" {236}⟨385⟩, as displayed in Figures 5.30c and 5.31b. The BR orientation is the characteristic recrystallization texture component of materials with low SFE, like silver or Cu alloys with high Zn, Al, or Ge contents (Schmidt and Lücke, 1979; Engler et al., 1995b). Grains with the BR orientation emerge from shear bands through subsequent twinning. Furthermore, the BR orientation has an approximate 40° ⟨111⟩ orientation relationship to the Bs orientation, the main rolling texture component in such low-SFE alloys (Figure 5.18b), which favors it for a successful growth. Similar weak BR textures have been reported in recrystallized sheets of austenitic stainless steels (Chowdhury et al., 2007). Interestingly, this is not the case in the austenitic phase in duplex steels, which is attributed to the absence of shear bands in the as-deformed austenite (Keichel et al., 2003; Zaid Ahmed and Bhattacharjee, 2014).

In the field of Al alloys, texture investigations are of great interest since the texture determines the plastic anisotropy that is important for practical applications, for example, for good deep-drawing properties. Furthermore, with a view to recrystallization, the occurrence of specific texture components yields viable information about the underlying mechanisms of recrystallization, namely, formation and growth of the nuclei. In Al alloys, the SFE is too high to allow for significant twin formation. Nonetheless, there is quite a variety of recrystallization textures, which is caused by the interplay of preferred nucleation and growth effects (Engler, 1996).

Hot-rolled Al usually shows a more or less strong cube recrystallization texture (Daaland and Nes, 1996). Figure 5.32 shows an example of the hot strip texture of aluminum alloy AA 3104 can stock, where a strong cube recrystallization texture is key to the formation of Al sheet suited for the production of beverage cans (Section 5.5.1). Pronounced cube recrystallization textures are also observed in cold-rolled high-purity Al (Figure 5.30a; see also Figure 4.8), for example, in Al foil

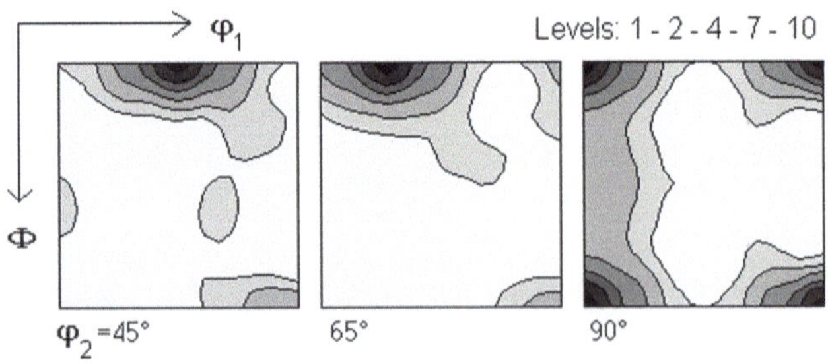

FIGURE 5.32 Cube-recrystallization texture of hot-rolled aluminum (can stock alloy AA 3104, recrystallized hot strip, ODF φ_2 = 45°, 65°, 90° sections).

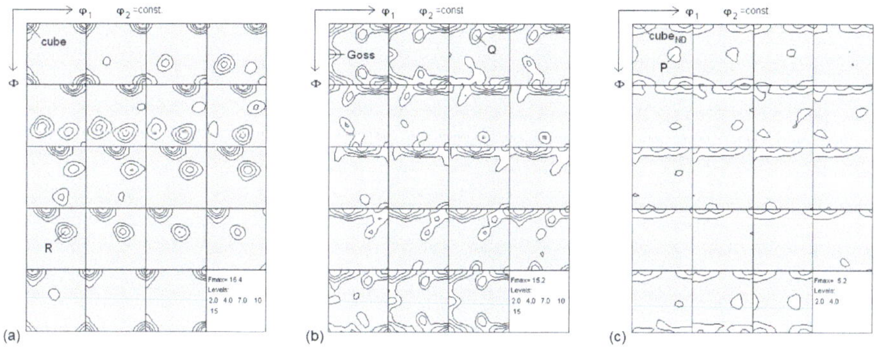

FIGURE 5.33 Characteristic recrystallization texture of cold-rolled aluminum alloys. (a) Cube + r texture of commercial purity al; (b) Goss + Q texture of Al-3%Mg; (c) weak PSN recrystallization texture of AlFeSi with large second-phase particles.

for high-voltage electrolytic capacitors (Engler and Huh, 1999; Takata et al., 2004; Boutin, 2010).

In commercial purity Al, the cube component is frequently accompanied by the R orientation, which is quite similar to the rolling texture S orientation (Figure 5.33a). R-oriented grains presumably form through nucleation at the preexisting large-angle grain boundaries and subsequent growth into the as-deformed matrix (Engler, 1999).

On cold rolling of Al alloys with high amounts of solute, Cu or Mg, shear bands may form (Wagner et al., 1995). On subsequent recrystallization annealing, these shear bands are viable nucleation sites because of their strong local lattice rotations and highly localized strains. As shown in Figure 5.33b, nucleation at shear bands leads to the formation of two new orientations, the Goss orientation {011}⟨100⟩ and the Q orientation {013}⟨231⟩ (Lücke and Engler, 1990; Engler, 2001). During the growth of the nuclei, both Q and Goss-oriented grains are regarded as being preferred owing to a 40° ⟨111⟩ orientation relationship to the main rolling texture orientations.

In Al alloys containing large particles (>1 μm), nucleation of recrystallization takes place in the deformation zones that form around these large particles during deformation, which is commonly referred to as *particle stimulated nucleation* (PSN) (Humphreys, 1977). PSN typically results in rather weak recrystallization textures with some preference of the P orientation {011}⟨122⟩ and a characteristic scatter of the cube orientation about the sheet ND toward {001}⟨310⟩ (Figure 5.33c).

5.5.5 Recrystallization Textures of bcc Metals

Analysis of recrystallization textures in bcc steels is of major industrial importance because of the correlation of texture and formability, especially, deep drawability. Compared to fcc materials, bcc metals usually exhibit a narrower range of recrystallization textures. Figure 5.34 shows the recrystallization texture of an IF steel optimized for deep-drawing applications. Evidently, the grain orientations assemble along the characteristic α and γ fibers and, as such, the recrystallization texture of ferritic steel sheets resembles the corresponding rolling textures (Figures 5.23 and 5.24).

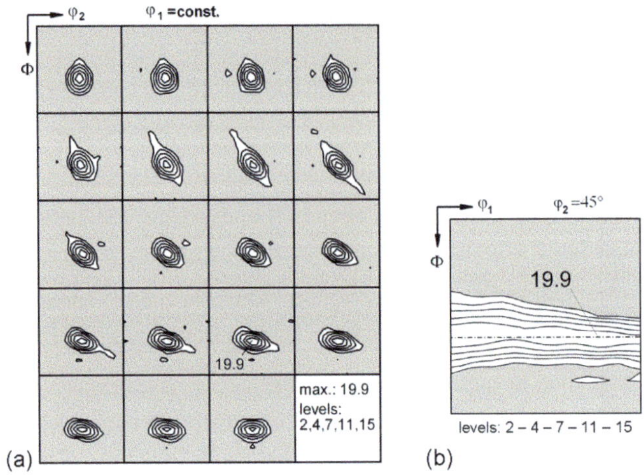

FIGURE 5.34 Recrystallization texture of a Ti-alloyed interstitial-free steel showing a pronounced γ fiber: (a) φ_1 sections; (b) $\varphi_2 = 45°$ section (90% cold rolled, annealed for 1 h at 700°C). (Courtesy of M.Y. Huh.)

The main effect of recrystallization is sharpening of the γ fiber at the expenses of the α fiber (Emren et al., 1986; Kern et al., 1986; Därmann-Nowak and Engl, 1991; Samajdar et al., 1997). For other bcc materials, only limited information is available in the literature, but, generally, similar recrystallization textures consisting of α and γ fibers with varying ratios have been reported (Raabe et al., 1994; Briant et al., 2000; Hünsche et al., 2004; Lobanov et al., 2016).

The typical γ-fiber recrystallization texture of bcc materials is attributed to the preferred formation of recrystallized nuclei at the preexisting large-angle grain boundaries between the deformed grains, where optimum conditions for nucleation prevail (Hutchinson, 1989; Inagaki, 1994; Tse et al., 2000). Because of the nucleation within the as-deformed microstructure, the new grains will retain orientations of the former rolling texture. However, the different rolling texture components differ in density and local arrangement of the dislocations, and, thus, in driving force for recrystallization. The γ-fiber orientations between {111}⟨110⟩ and {111}⟨112⟩ are commonly believed to have higher stored energy than the competing rolling texture orientations (Rajmohan et al., 1997; Samajdar et al., 1997). Therefore, nuclei form preferably in grains with γ-fiber orientations, which then grow at the expense of other orientations, including the α-fiber orientations.

It has further been argued that because of deformation inhomogeneities and local strain gradients that persist in the vicinity of the grain boundaries, the nuclei forming in those regions may be subjected to substantial rotations about the ND with regard to the matrix at a distance from the grain boundary (Hutchinson, 1989; Tse et al., 2000). This means that regions close to the grain boundaries between two symmetrically equivalent components of {111}⟨110⟩ may actually comprise {111}⟨112⟩ and vice versa. Thus, in the presence of a dominant {111}⟨110⟩ rolling texture, {111}⟨112⟩ will nucleate preferably and, consequently, eventually prevail in the recrystallization

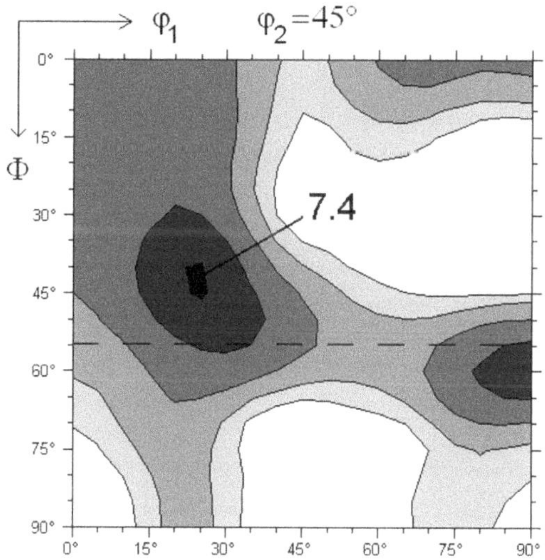

FIGURE 5.35 Texture of recrystallization annealed ferritic stainless steel Fe–17%Cr (ODF $\varphi_2 = 45°$ section, sheet center). (Data from Huh and Engler, 2001.)

texture. It is noted that good deep-drawing properties demand both high intensities of the γ-fiber orientations and a uniform intensity distribution along the γ fiber. Thus, optimization of deep drawability entails control of texture and microstructure during rolling and subsequent recrystallization so as to achieve a uniform intensity distribution along the γ fiber (Figure 5.34b).

For ferritic stainless steels, once again, different recrystallization textures are found. Figure 5.35 displays the texture of 80% rolled Fe–17%Cr sheet after the final recrystallization anneal (see Figure 5.25b). The recrystallization texture is quite weak, consisting of γ-fiber orientations with a fairly uneven intensity distribution. The texture maximum is notably shifted by approximately 8° from {111}⟨$\bar{1}$12⟩ on the γ fiber to an orientation {334}⟨4$\bar{8}$3⟩ with the Euler angles $(\varphi_1,\Phi,\varphi_2) = (26°,48°,45°)$. The occurrence of this orientation is explained in terms of a growth preference into {112}⟨110⟩ through a mechanism termed "selective particle drag" (Raabe and Lücke, 1992). Finely dispersed microcarbides exert a retarding force on grain boundary motion (Sinclair et al., 2005). The amount of this so-called Zener drag depends, among others, on the specific energy of a given grain boundary. Thus, in materials with a fine particle dispersion, coincidence site lattice (CSL) boundaries with a low specific energy experience less Zener drag than ordinary high-energy boundaries, which, in turn, favors them for growth on recrystallization. A similar solitary orientation {113}⟨3$\bar{6}$1⟩ at (20°,25°,45°) was recently reported in recrystallized tungsten sheet which contained 1 wt% of La_2O_3 particles (Zhang et al., 2020b).

A direct consequence of the limited γ → α-phase transformation during the thermomechanical processing of ferritic stainless steel sheets (Section 5.5.2) is the propensity of some grades of chromium steel to develop an undesirable surface phenomenon known as *ridging* (or *roping*) when the sheets are deformed in tension or deep drawing. During hot and cold rolling, the columnar grain structure of the as-cast slab is converted into highly elongated bands with a sharp texture. On the final recrystallization annealing, these bands transform into grain colonies of similar orientations (Brochu et al., 1997; Park et al., 2002; Huh et al., 2005). Figure 1.4 shows the resulting orientation clusters with an example of an EBSD orientation map. During the subsequent forming operation, these orientation clusters are likely to show different plastic behavior than their surroundings. At the free sample surface, these variations result in different strains in the sheet ND and, therewith, in ridging (Engler et al., 2005b).

An important effect of annealing of samples with low rolling strains or annealing at low temperatures is the occurrence of extended recovery, which is of particular importance for the orientations around {001}⟨110⟩. Compared to the other rolling texture orientations, {001}⟨110⟩ is distinguished by minimum stored energy as well as a very homogeneous microstructure. Accordingly, both nucleation and growth of {001}⟨110⟩ are retarded with respect to the competing α- and γ-fiber orientations (Vandermeer and Snyder, 1979; Raabe, 1995). This effect of extended recovery is of particular importance for bcc transition metals, for example, tantalum and tungsten. Figure 5.36a shows an example of a tantalum sheet, which was 80% cold rolled and then annealed for 1 h at 1000°C. Apparently, besides the sharpened γ-fiber orientations, significant intensities of {001}⟨110⟩ are retained from the rolling texture (Raabe et al., 1994). In specimens with stronger deformation or higher annealing temperatures, a pronounced {111} γ-fiber texture is formed through discontinuous recrystallization (Figure 5.36b).

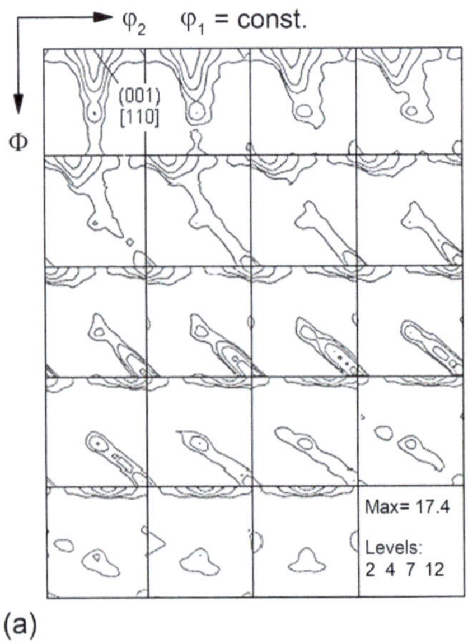

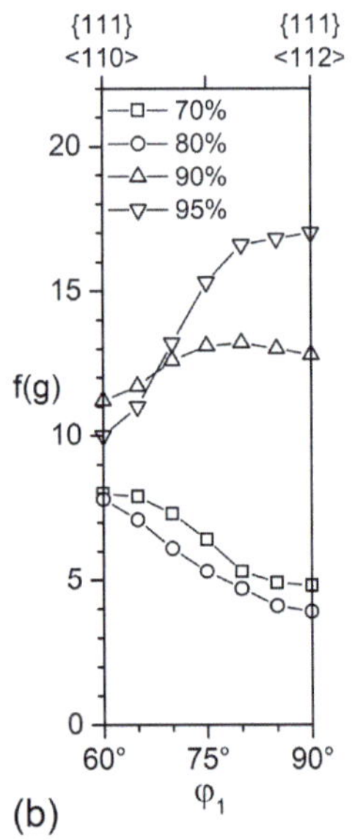

FIGURE 5.36 Annealing texture of cold-rolled tantalum showing strong recovery. (a) 80% rolled, annealed for 1 h at 1000°C (ODF, φ_1 sections); (b) rolled by various rolling degrees, annealed for 1 h at 1000°C (γ-fiber plots). (Courtesy of D. Raabe.)

5.5.6 Recrystallization Textures of Hexagonal Metals

In contrast to the quite well-established recrystallization textures of cubic metals, recrystallization of hexagonal material has been studied much less extensively. Early studies on recrystallization textures of hexagonal metals reported that the former rolling textures were often retained (Wassermann and Grewen, 1962; Barrett and Massalski, 1980). More detailed ODF analysis proves that at high strains and high annealing temperatures, there are distinct texture changes. Often, however, recrystallization and rolling textures are related through a 30°⟨0001⟩ orientation relationship that—for hexagonal crystal symmetry—is not easy to discern, in particular in {0001} fiber textures like those commonly found in Mg alloys (Figure 5.27).

At high strains and high annealing temperatures in both Ti and Zr alloy sheets, a preferred orientation ~{1013}⟨1210⟩ (φ_1, Φ, φ_2) = (0°,30°,30°) has been described (Inoue and Inakazu, 1988; Singh and Schwarzer, 2000; Zhu et al., 2009; Ghosh et al., 2017). This orientation has a 30°⟨0001⟩ orientation relationship to the rolling texture orientation ~{1214}⟨1010⟩ (φ_1, Φ, φ_2) = (0°,35°···40°,0°), as exemplified in Figure 5.37 by way of the rolling and recrystallization textures of 80% cold-rolled Zr sheet. There is some evidence that nucleation of recrystallization starts in the deformed grains where twinning during the preceding rolling deformation is important. A large fraction of the microstructure (approximately 75%) recrystallizes without any orientation change, however, suggesting extended recovery or in-situ recrystallization. More significant texture changes toward the described annealing textures then only occur after finalization of (primary) recrystallization through a selective grain growth that is influenced by the inheritance of the texture and microstructure developed during the first stage of recrystallization annealing (Wagner et al., 2002). After subsequent grain growth, a broad peak centered around (0°,35°,30°) has been reported, similar as in the associated rolling texture (Bozzolo et al., 2005).

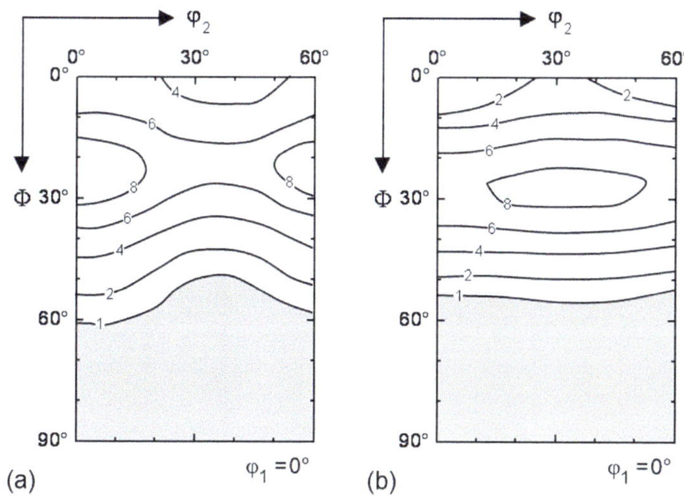

FIGURE 5.37 (a) Rolling and (b) recrystallization texture of zirconium (90% cold rolled, annealed for 350 min at 650°C, ODF φ_1 = 0° section). (Adapted from Zhu et al., 2009.)

5.6 SUMMATION

In this chapter, the most common methods to represent and evaluate macrotexture data have been introduced and their main advantages and disadvantages addressed.

The most well-known way of representing macrotextures is to project the crystal orientations into the sample coordinate system in a *pole figure*. Since the density distribution in a pole figure is the direct output from most techniques to determine textures experimentally, pole figures are well suited to a critical assessment of the experimental data. *Inverse pole figures*, the projection of the sample coordinate system into the crystal coordinate system, are used frequently to represent the orientation distribution of the main sample axis in the case of uniaxial sample symmetry. Although for analysis of fiber textures representation of one reference direction may suffice, for the evaluation of misorientations the full orientation information must be considered, however, which renders inverse pole figures inferior to the pole figures for the representation of texture data.

The projection of the 3D orientation onto a 2D pole figure results in a loss in information, so that for a quantitative evaluation of macrotextures a 3D representation (the ODF) is required. However, ODFs cannot be measured directly in x-ray or neutron diffraction experiments but have to be calculated from several pole figures, known as *pole figure inversion*. An analytical solution to this problem of pole figure inversion is unfortunately not possible. A variety of numerical approaches to derive the ODF have been developed, which can basically be subdivided into two groups.

- The harmonic method is based on a series expansion of ODF and pole figures. This method inherently suffers from the so-called ghost error, which is caused by the impossibility to derive the odd-order C coefficients from diffraction experiments. This means that the odd-order C coefficients must be determined based on additional assumptions, for example, the condition that the ODF must be nonnegative for all orientations. In combination with an appropriate method to correct for the ghost error, the series expansion method is a fast, reliable method that is rather insusceptible to experimental scatter in the pole figures. The resulting C coefficients represent a condensed manner to characterize the texture; furthermore, they disclose information on texture-related properties.
- In direct methods, the ODF is computed in an iterative procedure directly in the orientation space; not in a Fourier space as in the harmonic method. The condition of nonnegativity and the handling of the ghost problem are usually implicitly fulfilled and no truncation error arises. Analysis of incomplete pole figures as well as different sample or crystal symmetries is straightforward to accomplish.

A comparison of the various methods to compute ODFs from pole figure data proves that nowadays all methods are able to yield reliable and reproducible quantitative results, but because of the systematic differences between methods, it is suggested to stay with a given method within a series of experiments.

To represent the orientation density in dependence on the three Euler angles, that is, the ODF, an appropriate 3D orientation space is necessary. Traditionally, textures are represented in sections through an orthogonal orientation space whose axes are formed by the three Euler angles, that is, the *Euler space*. However, this Cartesian Euler space depicts strong distortions at small angles Φ, which renders a visualization of the orientations and textures complicated. These distortions can be reduced by representing the textures in polar rather than orthogonal coordinates—similar as in pole figures—which results in a cylindrically shaped Euler space.

Texture analysis can greatly be facilitated by suitable ways to condense the data. In many cases, texture representation can be confined to characteristic regions of the Euler space, which contain the texture information of interest rather than showing the entire ODF. For instance, if material processing affects only a few sections of the Euler space, the representation of these sections may be sufficient to visualize important texture changes. Many textures can be represented by plotting the orientation intensity along certain characteristic paths or distinct crystallographic fibers through the orientation space versus an angle defining the position along this path or fiber.

The chapter concludes with a guide to the most typical texture types of fcc, bcc, and hexagonal metallic materials as well as the underlying formation mechanisms during deformation and recrystallization.

Part III

Microtexture Analysis

6 Diffraction Techniques in TEM and SEM

6.1 INTRODUCTION

There is a basic divide between macrotexture techniques (based on x-ray or neutron diffraction; Chapters 4 and 5) and microtexture techniques (based usually on electron diffraction; Chapters 6 through 12), which relates to the scale of the inquiry and hence affects the nature of the primary output data. For macrotexture techniques, the primary output is a profile of diffracted intensities, which is characteristic of a large, usually in the order of cubic millimeters, contiguous sample volume, whereas for microtexture—individual grain—techniques, the primary output is a diffraction pattern from each sampled volume, usually of the size of approximately $100 \times 100 \times 10$ nm³ in scanning electron microscopy (SEM) and $10 \times 10 \times 100$ nm³ in transmission electron microscopy (TEM) (orders of magnitude of surface area times depth of probing for one pattern), i.e. about 10^{13} to 10^{14} times (!) smaller than the x-ray-measured data. By automated collection of many of these individual patterns a total volume between $10.000 \times 10.000 \times 0.001$ μm³ for SEM and $10 \times 10 \times 0.1$ μm³ for TEM can be probed (total probed area times depth of probing). Note, however, that the probed volume is not fully measured but only at individual measuring points, separated by the mapping step size, see Section 9.3.1. The SEM values may at the end, if smart probing is applied, be competitive with those of macroscopic techniques. Provided the probed volume is single-crystalline (potentially including crystal defects) then every individual diffraction pattern embodies the complete crystallographic information, including phase and orientation but also, to a certain extent, defect density, and elastic strain.

Most microtexture investigations are nowadays performed by electron backscatter diffraction (EBSD) in the SEM because it is by far the most highly automated and versatile technique as will be discussed in Section 8.2. EBSD is, in principle, only limited by its spatial resolution, which is in the order of few 10 to a few 100 nm (see Sections 6.3.3 and 8.2.9 for more details) and by its demand for a relatively high beam current, which makes it in some cases unsuitable for beam-sensitive samples. In case highest spatial resolution is required TEM diffraction techniques, in particular spot diffraction, may play a role, though the TEM techniques fall short of the EBSD technique because all require thin foils and the thickness of the foil is in this case the resolution-limiting factor. In contrast, EBSD is a bulk sample technique which allows much larger sample areas to be observed and requires much easier sample preparation. Besides TEM and SEM-based electron diffraction techniques, very rarely and only for special purposes, in particular accurate local measurement of elastic strains, also the micro-Kossel technique, an x-ray diffraction technique in the SEM, may be applied (Section 6.3.5). In the present chapter we will first introduce

DOI: 10.1201/9781003258339-9

some important basic concepts of electron diffraction, namely the reciprocal lattice (which is, indeed, important for any form of diffraction, also x-rays, for example) and the basic electron scattering mechanisms. We then introduce the individual diffraction techniques in detail starting with an overview on the general forms of electron diffraction in both SEM and TEM. Only techniques which allow crystal orientation measurements will be considered. We then continue first with those techniques applied in TEM before going into details on the most used technique for orientation, texture and microstructure measurements, namely the EBSD technique in the SEM.

6.1.1 Description of Bragg Diffraction in Reciprocal Space

Before describing the details of different electron diffraction geometries, it appears necessary to introduce the concept of reciprocal space which is used to explain Bragg diffraction in a very general way. Note that this concept will first appear rather abstract. However, it will quickly turn out to be very helpful in understanding and interpreting diffraction patterns.

The reciprocal space is an $\mathbb{R}3$ space with all dimensions being reciprocal to the real space, i.e. all length metrics of real space, ℓ, change to reciprocal lengths, $1/\ell$. The directions of the coordinate axes of reciprocal space are orthogonal to pairs of coordinate axes of real space. This is described in the basic geometric transformation equation of real to reciprocal space coordinates:

$$\mathbf{a}_i^* = \frac{\mathbf{a}_j \times \mathbf{a}_k}{\mathbf{a}_i \cdot (\mathbf{a}_j \times \mathbf{a}_k)}, \quad i,j,k = 1,2,3 \tag{6.1}$$

where the vectors with asterisks indicate the base vectors of reciprocal space and those without are those of real space. For example, the vector $\mathbf{a}_1^*$ of reciprocal space stands perpendicular to the vectors $\mathbf{a}_2$ and $\mathbf{a}_3$ of real space and its length is reciprocal to the lattice plane spacing of the planes spanned by $\mathbf{a}_2$ and $\mathbf{a}_3$ vectors:

$$\mathbf{a}_1^* = \frac{\mathbf{a}_2 \times \mathbf{a}_3}{\mathbf{a}_1 \cdot (\mathbf{a}_2 \times \mathbf{a}_3)} \tag{6.2}$$

Similar to the atomic *real lattice* of the crystal which is built by the base vectors $\mathbf{a}_i$, the base vectors, $\mathbf{a}_i^*$ of the reciprocal space form a lattice consisting of *reciprocal vectors* which is called the *reciprocal lattice*:

$$\mathbf{h}_{(hkl)} = h\ \mathbf{a}_1^* + k\ \mathbf{a}_2^* + l\ \mathbf{a}_3^* \tag{6.3}$$

It can be shown that these vectors $\mathbf{h}_{hkl}$ are normal vectors on the lattice planes (*hkl*) of the real lattice, with the property that their vector lengths are reciprocal to the plane distances of the lattice planes, i.e. $|\mathbf{h}_{(hkl)}| = 1/d_{(hkl)}$.

The same concept of reciprocal vectors can be applied to plane waves, for example of electrons or x-rays traveling in real space: these waves can be described by *wave vectors*, **k**, whose direction is perpendicular to the (plane) wave front and whose length is reciprocal to the wavelength, i.e. $|\boldsymbol{k}| = 1/\lambda$.

Mathematically spoken the reciprocal space is the *Fourier transform* of real space. The Fourier transform of a *crystal lattice* in real space leads to the *reciprocal lattice* and to *plane waves* of electrons or x-rays traveling in real space The concept of reciprocal space and Fourier transforms is of great importance for electron and x-ray diffraction and is therefore described in great detail in all textbooks on diffraction techniques. The reason is that the electron or x-ray intensity distribution that is obtained by an electron or x-ray plane wave traveling through a lattice of atoms (or generally of scattering centers) is described by the Fourier transforms of the spatial atom arrangement. The Fourier transform of a set of lattice planes (*hkl*) in a crystal with *n* atoms at coordinates $(xyz)_i$ is called the *structure factor* F_{hkl} and is calculated according to:

$$F_{hkl} = \sum_{i=1}^{n} f_i\left(\lambda/\sin\theta/2\right)\exp\left[2\pi i\left(hx_i + ky_i + lz_i\right)\right] \tag{6.4a}$$

The structure factor is a complex number, which indicates phase and amplitude of the resulting wave. To obtain the diffracted wave amplitude the absolute value of the structure factor is calculated according to:

$$\left|F_{hkl}\right| = \sqrt{2\pi}\sqrt{\left[\sum_i f_i\cos\left(hx_i + ky_i + lz_i\right)\right]^2 + \left[\sum_i f_i\sin\left(hx_i + ky_i + lz_i\right)\right]^2} \tag{6.4b}$$

here *fi* is the *atomic scattering amplitude* of atom *i* in the unit cell, calculated for the current wavelength and diffraction angle. The atomic scattering amplitude can be understood to be the Fourier transform of every single atom in the unit cell. The calculation of *f* is described in detail for x-ray diffraction in the International Tables of x-ray crystallography (see Section 3.3). The atomic scattering amplitude of a given atom for electrons, f_e, is different from that of x-rays, f_x, but can be calculated from it:

$$f_e\left(\lambda/\sin\theta/2\right) = \frac{2\mathrm{me}^2}{\mathrm{h}/2\pi}\left(\frac{\lambda}{\sin\frac{\theta}{2}}\right)^2\left(Z - \frac{\mathrm{mc}^2}{\mathrm{e}^2}f_x\right) \tag{6.5}$$

where *m* and *e* are mass and charge of the electron, *c* is the speed of light, and *h* is Planck's constant.

The structure factor describes, in principle, the basic intensity of a diffraction spot or a Kikuchi line. Besides this the intensity of a spot or line depends on the shape of the diffracting crystal (as briefly described in Section 6.2.1). It is also modified by the dynamic nature of electron diffraction that is discussed in Section 6.3.2.

The best (and most simple) illustration of the importance of reciprocal space for diffraction processes is the description of the *Bragg equation* in this concept. Although the Bragg equation has already been introduced in Section 3.2, we will repeat it here shortly. Coherent scattering of a plane wave by the regularly arranged atoms of a crystal leads to constructive and destructive interference of the elementary waves in different directions in space according to Huygens principle. The process of

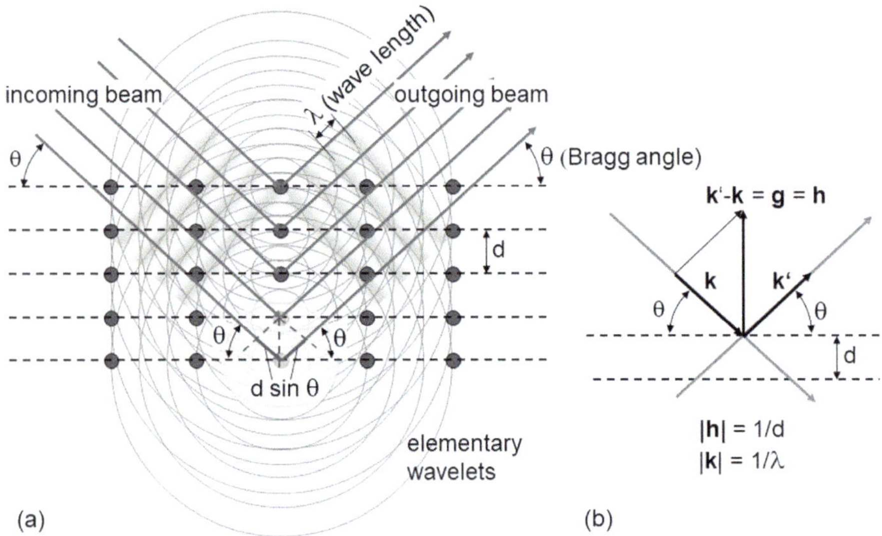

FIGURE 6.1 (a) Sketch of the Bragg equation by superposition of elementary wavelets, emitted from every atom of the real lattice. (b) Reciprocal space display of the Bragg equation **g** = **h**. **k** and **k′** indicate the wave vectors of the incoming and outgoing radiation. **h** indicates the reciprocal lattice vector of the lattice planes with interplanar spacing d.

interference of coherent waves is called diffraction, and the resulting intensity distribution is a diffraction pattern. The directions of those plane waves with maximum electron intensities (i.e. the directions of constructive interference) are described by the Bragg equation

$$n\lambda = 2d_{hkl} \sin\theta_{hkl} \tag{6.6}$$

Here θ_{hkl} is the *Bragg angle* which is the angle between the set of diffracting lattice planes *hkl* and the wave vector of the incoming and outgoing electron wave, d_{hkl}, is the lattice plane distance of the planes *hkl*, λ is the electron wavelength and *n* is a positive integer number. The Bragg equation—illustrated in Figure 6.1a—describes the position of the maxima in electron (and x-ray) diffraction patterns and is therefore of utmost importance for understanding the geometry of all kinds of diffraction patterns. The Bragg equation is one of the corner stones of kinematic electron diffraction theory. This theory claims that every incoming electron wave is scattered only once before leaving the crystal again. This assumption is given up in the dynamical electron diffraction theory. This equation can be conveniently described using the reciprocal lattice: for this the incident wave vector of the primary beam will be named $\mathbf{k}_0$ and that of the wave vector caused by Bragg diffraction on a set of lattice planes $\mathbf{h}_{(hkl)}$ will be called $\mathbf{k}_g$ (the Bragg "reflection"). The difference between these vectors is called the *diffraction vector*, **g**:

$$\mathbf{g} = \mathbf{k}_d - \mathbf{k}_0 \tag{6.7}$$

The diffraction vector indicates the change of the electron wave vector during the diffraction process and is extremely important and appears in probably all equations on diffraction theory.

We note that Bragg scattering is an elastic scattering process, i.e. the electrons do not lose energy. That means that the lengths of the two wave vectors $\mathbf{k}_0$ and $\mathbf{k}_g$ are both equal to $1/\lambda$. Now we show that constructive interference of coherently scattered elementary waves occurs, as displayed in Figure 6.1b, if the diffraction vector, **g**, is equal to a vector of the reciprocal lattice, **h**. As seen in Figure 6.1b the length of **g** is in this case $|\mathbf{g}_{hkl}| = 2 \sin \theta\, |\mathbf{k}| = 2 \sin \theta\, 1/\lambda$. With $|\mathbf{h}_{hkl}| = 1/d_{hkl}$ we obtain the Bragg equation. The Bragg equation in reciprocal space is thus

$$\mathbf{g} = \mathbf{h}_{hkl} \tag{6.8}$$

which is also called the Laue condition. Often the diffraction vector that fulfils the Bragg condition is denoted as $\mathbf{g}_{hkl}$. Note that also vectors with lengths $g_{n(hkl)} = n \cdot \frac{1}{d_{hkl}}$ with n being an integer number will be allowed.

6.1.2 Forms of Electron Diffraction—Overview

We will treat here only such forms of electron diffraction which have some importance for orientation microscopy (ORM), i.e. which can be used for the determination of crystal orientations and phase. Orientation determination is in most cases based on an analysis of the pattern geometry; this will therefore be the focus of the treatment. Nevertheless, also the intensity distribution in the patterns may be of importance (in some cases it is, in fact, essential) and is thus regarded as well.

Physically, three basic forms of electron diffraction can be distinguished, the parallel-beam type patterns, the convergent-beam type patterns, and the "point-source" or Kikuchi-type patterns. They differ in the way how the primary electron wave for diffraction is created. Geometrically, different types may look alike, while patterns of the same type may appear quite different. All these pattern types will be introduced in the subsequent chapters in detail. Here we will give only a brief overview.

The parallel-beam type patterns are formed when a parallel primary electron beam (i.e. a beam in which all electron wave vectors are parallel) is diffracted. The created diffracted beams again consist of parallel electron wave vectors with directions given by the Bragg equation. A parallel-beam type diffraction pattern is best observed by means of a lens which focuses parallel beams into sharp spots in its focal plane. In TEM the diffraction lens system fulfills this task. To the group of parallel-beam type patterns belong the selected area diffraction patterns (SADP) and the Debye–Scherrer ring patterns (DSP), which are formed by the overlap of SADP from many, differently oriented crystals.

Convergent beam-type patterns are formed by a convergent electron beam illuminating a sample. The plane electron waves of each incident electron are diffracted according to the Bragg equation but since the incident electron waves come from a cone of different directions, the diffracted electron waves also show a certain directional spread. Consequently, in the focal plane of an objective lens—a lens is indispensable here—no sharp spots but disks are visible, the diameter of which varies with the angle of convergence. The disks show a complex intensity distribution because each position inside the disk is created by diffraction under a different incident angle of the electron beam and thus under a different excitation error, s, which will be introduced in Section 6.2.1. Additionally to the disks, convergent beam-type patterns may show structures of dark lines inside of the disk. These lines are so-called higher-order Laue zone (HOLZ) lines. They are deficiency Kikuchi lines (see below) stemming from HOLZ diffraction vectors. Pattern types belonging to this class are the convergent beam electron diffraction (CBED) patterns and the large-angle convergence beam electron diffraction (LACBED) patterns. The small convergence beam electron diffraction (SCBED) patterns display an intermediate form between the parallel and convergent-beam type. Also the electron channeling patterns, ECP, in the SEM belong, from their basic formation principle, to the convergent beam-type patterns, although their way of recording is different from the others and their appearance is more like that of Kikuchi-type patterns.

The third pattern type, the point-source or Kikuchi-type patterns are formed from a small electron source inside of the crystal. This electron source is created by incoherent scattering of the electrons of the finely focused primary electron beam inside of the material close to its surface. The scattered electrons which correspond to plane waves with a wide range of propagation directions are diffracted according to the Bragg equation and it can be easily shown that the locus of the diffracted electrons consists of pairs of cones with a wide opening angle, symmetrically positioned about the diffracting lattice planes. Since diffraction occurs from a "point" source no lens system is required to obtain sharp patterns. In the original work, Kikuchi used a simple lens-free diffraction apparatus and observed transmission Kikuchi patterns (TKP) (Kikuchi, 1928). In TEM, TKP can be observed using the diffraction lens system. In SEM again, no lens system is used to obtain backscatter Kikuchi patterns (BKP) (which are more frequently called EBSD patterns). To the Kikuchi-type patterns also belong the Kossel patterns obtained by x-ray diffraction. Here a small x-ray source is created inside of the sample by exciting a characteristic x-ray fluorescence line of the material with some primary radiation.

Kikuchi-like patterns are created without any coherency relation between the primary beam electrons and the lattice. The direction of incidence of the primary beam onto the sample has therefore no influence on the geometry of the observed pattern (this is, however, not true for the intensity distribution in these patterns). In principle, every lattice plane reflection may therefore be visible under all beam incident conditions and it is only the position of the lattice plane which determines the position of a reflection in the pattern. This is different for the parallel and convergent beam pattern types, which are created due to a coherency relationship between the primary beam electrons and the lattice. In this case, the direction of the primary beam determines which lattice plane reflections are visible in a diffraction pattern.

Due to the smaller amount of geometric constraints, Kikuchi-like patterns are usually easier to use for orientation determination than the other type of patterns. It will be shown, however, that the spot-like patterns have some significant advantages for ORM in the TEM.

In the following the different pattern types will be described in more detail. The focus will thereby be put on the suitability of the different types for orientation and phase determination.

6.2 DIFFRACTION IN TEM

6.2.1 Selected Area Diffraction Patterns (SADP) and Ewald Construction

The classical and probably still most widely used form of diffraction pattern in TEM is the selected area diffraction patterns, SADP, named because an aperture is used to select the area of a sample which contributes to the formation of the diffraction pattern. In the simplest approach these patterns are imagined to be created by Bragg reflection of the primary beam on the lattice planes as displayed in Figure 6.2. The diffraction angle between the primary beam and the diffracted beam is equal to $2\theta_B$, caused both the diffracted and the primary beam include an angle of θ_B with the lattice planes. Because the Bragg angle is very small the lattice planes that contribute to the pattern are almost parallel to the primary beam. Furthermore, Figure 6.2 shows that the distance of a diffracted spot to the primary beam spot, R, is $R = \frac{\lambda L}{d}$ where λ is the wavelength, L is the distance between sample and detector (the so-called camera length), and d is the interplanar distance of the diffracting lattice planes, which means R is reciprocal to d. *The vector connecting the primary beam spot with a diffracted beam spot thus stands perpendicular to the lattice planes and its*

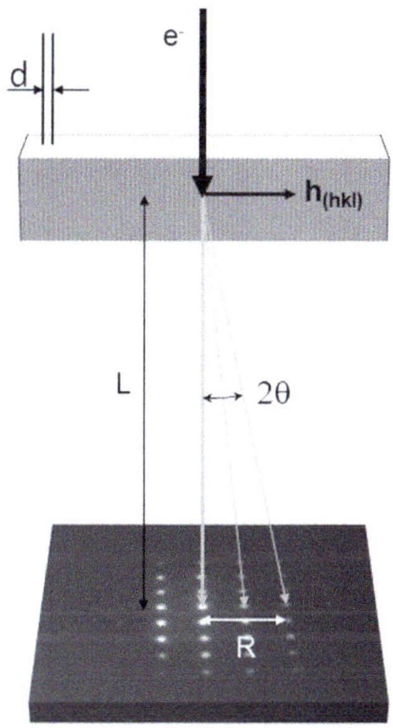

FIGURE 6.2 Sketch of Bragg diffraction in TEM. Note that the wavelength λ is in the order of 2 pm, and the lattice plane distance, d, in the order of 0.2 nm. As a consequence the Bragg angle is in the order of 1° (here shown too large). The camera length, L, is in the order of 0.1–1 m. That means, the vector, R, has a (real space) length of approximately 3–30 mm. The vector R is (almost) parallel to the vector h and its length is proportional to $1/d$, thus it is, in principle, a reciprocal lattice vector projected into real space.

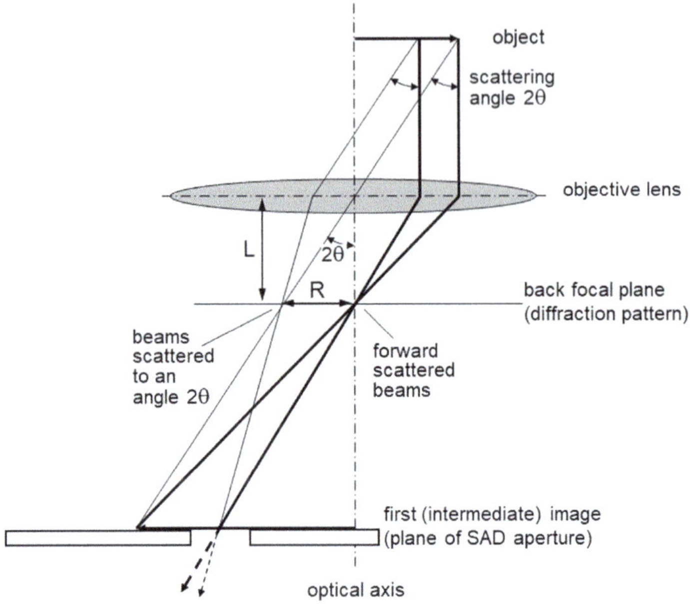

FIGURE 6.3 The role of the objective lens for the formation of diffraction patterns in the back focal plane and the first intermediate image of the sample. Rays of similar direction (thick or thin lines) intersect in the back focal plane and form the diffraction pattern. Rays from similar object points intersect in the image plane and form the first intermediate image. The selected area aperture is inserted into this plane to select the area contributing to a diffraction pattern.

length; R, is reciprocal to the interplanar spacing, d; this means it is a reciprocal lattice vector! We will see shortly, that indeed, the diffraction pattern is a projection of the reciprocal lattice.

Before discussing this, it is advisable to recall the operation of the TEM objective lens in terms of simple ray-optics: as displayed in Figure 6.3 the rays leaving an object (the specimen) are deflected by the lens such that parallel rays (plane waves) are focused into the back focal plane of the lens. The position of focus in the plane depends on the direction of the wave vector. Diffracted rays are plane waves (rays leaving the sample in the direction defined by the wave vector). Consequently, these rays are focused into the back focal plane of the objective lens and built up the diffraction pattern there. Further below the diffraction plane is the image plane where a magnified image of the object is created. In a TEM one uses the intermediate lens (behind the objective lens) to image either the back focal plane of the objective lens—this makes the diffraction pattern visible—or to image the first image plane, which makes the image of the sample visible. An SAD pattern is produced by illumination of the thin foil sample with a parallel electron beam (i.e. a plane wave front). The parallel diffracted beams are focused into sharp spots in the back focal plane of the objective lens. In order to limit the part of the sample that contributes to the diffraction pattern an aperture is introduced into the intermediate image plane as indicated at the bottom part of Figure 6.3. This aperture which selects the area of the sample that contributes to the diffraction gives the name to the technique "selected area diffraction pattern." If the size of the aperture is smaller than approximately 500 nm. diffraction on the aperture makes the assignment of diffraction pattern and selected area impossible. This size, therefore, corresponds to the lateral resolution limit of the technique.

As mentioned above the vectors between primary beam spot and the diffracted spots can be interpreted as reciprocal lattice vectors. This fact can be understood and efficiently used by means of the Ewald construction. The Ewald construction named after the physicist and crystallographer P.P. Ewald (1921) is a very convenient and powerful way to derive the geometry and, to a certain extent, also the intensity of diffraction spots. The Ewald construction starts by plotting the reciprocal lattice, i.e. the lattice of all the vectors **h**, for a given lattice. This is sketched by the gray disks (or ellipses, as will be explained further down) in Figure 6.4, together with a few $\mathbf{h}_{hkl}$-vectors. The reciprocal lattice is a 3D lattice but in Figure 6.4a we plot only a 2D section. Next, the wave vector of the incident wave, $\mathbf{k}_0$, will be plotted such that it ends on the origin of the reciprocal lattice. Note again that the wave vector points in the direction of electron wave propagation and has a length of the inverse wavelength, $|\mathbf{k}| = 1/\lambda$. Now the diffracted wave vector, $\mathbf{k}_d$, is plotted at the origin of $\mathbf{k}_0$. As this vector also has the length $1/\lambda$, all potential vectors of diffracted waves end on the surface of a sphere with the radius $1/\lambda$ which is plotted as a circle in Figure 6.4a. This sphere is called the *Ewald sphere*. The diffraction vector **g** which points from the vector $\mathbf{k}_0$ to the vector $\mathbf{k}_d$ always ends on the Ewald sphere.

As discussed in Section 6.1.1, the Bragg equation is fulfilled when the diffraction vector is equal to a reciprocal lattice vector, i.e. $\mathbf{g} = \mathbf{h}_{hkl}$. Graphically this means that *a Bragg reflection always occurs when the Ewald sphere intersects a reciprocal lattice point* $\mathbf{h}_{hkl}$ as this is the case for the vector marked in the Figure 6.4a. The Ewald construction therefore allows very conveniently to see which lattice planes (i.e. which reciprocal vectors) contribute to the diffraction pattern. Note that in Figure 6.4a the Ewald sphere is drawn with strong curvature. In real electron diffraction, however, the curvature of the Ewald sphere is very small, because the wavelength is very small compared to the reciprocal lattice plane distances: for an acceleration voltage of 200 kV the radius of the Ewald sphere is about $1/0.00273\ \text{nm}^{-1}$, while the distance between the reciprocal lattice vectors is only in the order of $1/0.3\ \text{nm}^{-1}$. Thus, the Ewald sphere is about 100 times larger than the reciprocal vector length and appears almost flat. As a consequence it intersects almost only the reciprocal spots of the so-called zero-order Laue zone, ZOLZ (see Figure 6.4a).

In Figure 6.4a we finally use a trick to illustrate the relation between the Ewald sphere construction and the real diffraction pattern. On the top of the figure we display a crystal with the lattice planes, the incoming beam and the Bragg-diffracted beam, under an angle 2θ. The middle part displays the Ewald sphere construction in reciprocal space. The bottom part then displays the resulting diffraction peaks again in real space in a (two-dimensional) cut through a diffraction pattern. Note that this schematic is in principle not allowed because we plot the crystal diffraction, which occurs in real space—with dimensions ℓ—together with the Ewald sphere construction in reciprocal space—with dimension $1/\ell$—together in one image.

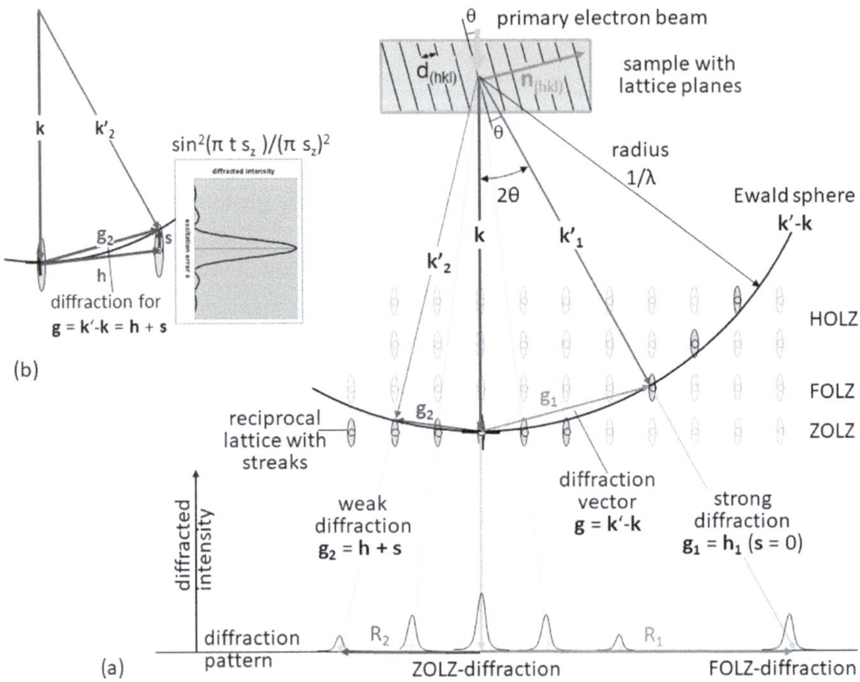

FIGURE 6.4 The Ewald sphere construction and its relation to the diffraction pattern. The sketch shows reciprocal space (the reciprocal lattice, the Ewald sphere, and the electron wave vectors) and real space (the crystal lattice, the diffraction pattern, and the electron beams) in the same graph. This is, of course, in principle not allowed, since the dimensions of both spaces are different (1/length and length). (a) The wave vector of the primary beam is **k**, the one of a diffracted beam **k**′. The Ewald sphere indicates all loci of possible diffraction vectors **g** = **k**′ − **k** in reciprocal space. Constructive interference occurs whenever the Ewald sphere intersects a reciprocal lattice point **h**, i.e. **g** = **h** (Bragg equation is satisfied). On a thin foil sample for TEM the reciprocal lattice points degenerate into streaks elongated into the direction of the thinnest sample direction. Thus, diffraction also occurs when the Bragg equation is not exactly satisfied. This situation is indicated in (b). Here **g** = **h** + **s**, where the deviation vector (or excitation error) **s** indicates the deviation from the exact Bragg conditions. The diffracted intensity strongly depends on the value of **s** (the component of **s** normal to the foil surface), as indicated in the graph. In order to obtain the diffraction pattern, the directions of the wave vectors of diffracted waves, **k**, are interpreted as diffracted beams in real space. A strongly excited beam is obtained when the Bragg equation is exactly satisfied $\mathbf{g}_1 = \mathbf{h}$. For weakly excited beams, the Bragg equation is not exactly satisfied and the wave vector equation is $\mathbf{g}_2 = \mathbf{h} + \mathbf{s}$. The reciprocal lattice vectors most closely perpendicular to the primary wave vector **k** are the members of the zero-order Laue zone (ZOLZ). Their diffraction spots are close to the primary beam spot. Reciprocal lattice vectors who make the next larger angular deviation to **k** are the members of the first-order Laue zone (FOLZ). All others belong to HOLZ. (Adapted modified from Zaefferer, 2011.)

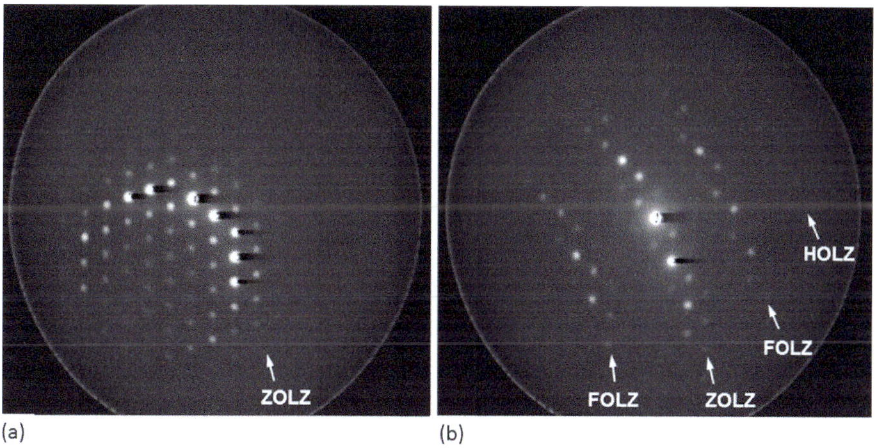

FIGURE 6.5 Two typical TEM spot diffraction patterns, obtained from fcc Fe–36% Ni alloy at 200 kV and a camera length of 0.2 m. (a) The intersection of the Ewald sphere with the reciprocal lattice. All spots stem from zero-order Laue zone. (b) Zero-, first- and higher order are visible.

To show that the Ewald construction is, indeed, extremely helpful we finally demonstrate two typical spot diffraction patterns from an austenitic steel sample in Figure 6.5. Figure 6.5a shows that many spots arise which form a regular pattern, as from a regular lattice (the reciprocal lattice), but the spherical cut of the Ewald sphere through the ZOLZ lattice points is clearly visible. In Figure 6.5b the reciprocal lattice is cut such that ZOLZ and FOLZ spots are visible. Each of the Laue zones indicates the cut of the Ewald sphere with the regular lattice.

Once the reciprocal lattice is known for a given lattice structure, then the reflections appearing in a diffraction pattern can be easily calculated. If the reciprocal lattice is calculated as Fourier-transform of the real crystal then the Ewald construction also allows, in a kinematic diffraction approximation, to calculate the intensities of the appearing Bragg reflections (see the abovementioned textbooks on electron microscopy and, for example, Zaefferer, 2000, 2002).

Now we finally have to address the reason for the drawing of the reciprocal lattice points as elongated ellipses in Figure 6.4. In fact, as shortly mentioned in Section 6.1.1, the reciprocal lattice is the Fourier transform of the real lattice. It is beyond the scope of this text to discuss the nature of the Fourier transform in detail but we note that features which are large in real space are small in reciprocal space and vice versa. The fact that a TEM sample is very thin in the direction of the electron beam leads to the fact that the reciprocal lattice points degenerate into streaks elongated into this direction. As a consequence, the reciprocal lattice streaks are also cut by the Ewald sphere if the sphere does not pass exactly through the reciprocal lattice point. This can also be interpreted in the way that diffraction also occurs when the Bragg equation is not exactly satisfied. This situation is indicated in Figure 6.4b. Here is $\mathbf{g} = \mathbf{h} + \mathbf{s}$, where the deviation vector (or excitation error) $\mathbf{s}$ indicates the deviation from the exact Bragg condition. The intensity I of a spot $\mathbf{g}$ in dependence of s ($=|\mathbf{s}|$) is calculated, in the simplest case, by the two-beam dynamical theory according to the following equation where I_0 is the primary beam intensity and A the transmitted area:

$$I_g(s,t) = \frac{I_0}{A}\left(\frac{\pi}{\xi_g}\right)^2\left(\frac{\sin(\pi t s_{eff})}{\pi\ s_{eff}}\right)^2 \quad \text{with} \quad s_{eff} = \sqrt{s^2 + 1/\xi_g{}^2} \tag{6.9}$$

The intensity I depends on the foil thickness, t, and on the excitation error, s, which in the schematics of the Ewald construction is equal to the distance of the respective reciprocal lattice vector to the Ewald sphere. The extinction distance, ξ_g, is a constant for a given diffraction vector **g** and a given material:

$$\xi_g = \frac{\pi V_e \cos\theta_g}{\lambda|F_g|} \approx \frac{\pi V_e}{\lambda|F_g|} \quad (\text{because } \theta_g \text{ is small}) \tag{6.10}$$

V_e is the volume of the crystal lattice unit cell, λ the electron wavelength, θ the Bragg angle for the diffraction vector **g**, and F the structure factor of **g**.

Note that in the case of x-ray diffraction the reciprocal lattice points are not degenerated into streaks because the diffracting volume is large and, thus, destructive interference leads to extinction of the x-ray intensity for all directions where the Bragg equation is not accurately satisfied. As a consequence only very few Bragg reflexes are visible in a single crystal Bragg-diffraction experiment with x-rays. (In the Ewald-sphere image this means that the Ewald sphere only rarely intersects a reciprocal lattice point.) This is the reason why either a powder, a rotating sample, or polychromatic radiation is required to obtain a reasonable amount of Bragg reflections. In contrast, the thin foil nature of TEM electron diffraction is responsible for the large amount of reflections visible in a TEM diffraction experiment. Additionally, a convergent beam may be used to increase the number of reflections as it will be discussed in the following.

In conclusion, the SAD pattern has to be understood as the central projection of a spherical cut through the reciprocal lattice into real space. This statement—which is in contrast to the simplifying textbook assumption that an SADP is a planar cut through the reciprocal lattice—has two important consequences for the use of spot patterns for orientation determination. First, as shown in Figure 6.5b, for example, the curvature of the Ewald sphere leads to the occurrence of spots which do not all origin from the same zone axis or Laue zone (i.e. from only coplanar diffraction vectors). Rather, also spots from FOLZ or HOLZ may be visible. This enables the determination of a unique orientation from just one spot pattern. In contrast, in the case of a planar cut only one zone axis would be available and at least two crystal orientations, related by a 180° rotation about the zone axis, would fit the pattern. The second consequence is that indexing of spot patterns is more complicated than generally thought: for the correct analysis of spot diffraction patterns one needs to consider the geometrical relation between diffraction vector and corresponding diffraction point as displayed in Figure 6.4b. In the exact Bragg case, $|s| = 0$ (which is a rare case) the diffraction vector **g** is exactly equal to a reciprocal lattice vector $\mathbf{h}_{hkl}$. The pattern coordinates of $\mathbf{h}_{hkl}$ can then be directly calculated from the position of the spot in the pattern and the know Ewald sphere curvature and subsequently used for indexing. In all other cases, the vectors **g** and $\mathbf{h}_{hkl}$ differ by the deviation vector, **s**, which can only be determined (or estimated) from the intensity of the spot. This points out the importance of spot intensity analysis for accurate orientation determination which has been discussed in detail by Zaefferer (2002).

Although SADP are classically used for orientation determination they are not well suited for ORM. One reason is the already mentioned bad spatial resolution of the technique of about 500 nm. The second reason is the limited angular resolution at least for the case of the classical "textbook" approach where spot patterns are described as planar cuts through the reciprocal lattice. If only the geometry of the pattern is used and only patterns from low-indexed zone axes are considered, the orientation can only be determined to some degree (±3°) close. The reason for this bad angular resolution lies in the fact that the diffracting angle is rather badly defined in thin foils (diffraction occurs for a large s-range) and consequently the positions of the reciprocal lattice vectors $\mathbf{g}_{hkl}$ are not well determinable. Second, the solid angle of SADP is rather small (a few degrees around the primary beam)[1] which introduces a relatively large error for the determination of the reciprocal lattice direction perpendicular to the pattern plane. The spatial resolution achievable with spot patterns can be significantly improved by using another technique for pattern creation than SADP, i.e. small-angle convergent beam diffraction (SCBED) patterns (Section 6.2.3) or reconstructed spot patterns created by the conical dark field scanning (CDFS) technique (Section 6.2.5). Furthermore, the angular resolution can be significantly improved using not only spot positions but also their intensities. This is further discussed in Section 7.2.4. With these improvements, the use of spot patterns becomes the basis for the most powerful techniques for TEM-based ORM (see Section 7.4.2).

Nevertheless, there are important applications also for the classical SADP. One is the precise analysis of the reciprocal space (and therefore of the crystallography of the illuminated area) based on the fact that SADP permit a direct look onto a cut through the reciprocal lattice. A lot of details about the observed crystal, for example symmetry, orientation, morphology, and orientation relation of very small, oriented precipitations (by occurrence of streaks) are revealed. Since spot patterns only cover a small solid angle (the maximum scattering angle for elastic scattering is about 5–6°), tilt experiments are often required for exploration of a sufficiently large section of reciprocal space. Tilt experiments are, however, quite difficult to carry out with spot patterns. Here Kikuchi patterns are better suited. Another important application of SAD patterns is the selection of well-defined diffraction conditions for the observation of dark-field images. Dark-field images are important for defect observation but can also be used for ORM as it will be discussed in Section 6.2.5.

6.2.2 Debye–Scherrer Ring Patterns (DSP)

DSP will be only briefly discussed here because, different from all other discussed pattern types, DSP are not single crystal patterns but are formed by the overlap of SADP from many, differently oriented crystals. Figure 6.6 shows a true ring pattern from a nanocrystalline material, together with a schematic of pattern formation. DSP cannot be used for individual crystal orientation determination, nevertheless, they can be used to measure the orientation distribution of small areas in the sample.

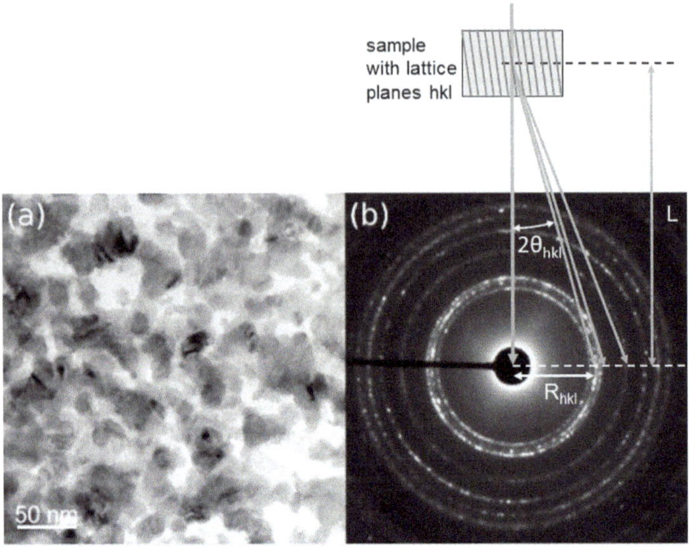

FIGURE 6.6 A Debye–Scherrer ring diffraction pattern produced from a nanocrystalline sample of ZnO at 200-kV acceleration voltage. (a) Bright field image and (b) ring diffraction pattern and a schematic of its geometry. (Adapted from Straumal et al., 2013.)

DSP have a lot in common with the SAD patterns. They are created using an SAD aperture to select a certain, polycrystalline area of the sample. In reciprocal space the reciprocal lattices of many crystals overlap, leading to the formation of concentric spheres of diffracted intensity. The radii of the spheres correspond to the lengths of all possible diffraction vectors occurring in the reciprocal lattice of the individual crystals. Using the Ewald sphere construction, the diffraction pattern, thus, consists of concentric circles where each circle belongs to one set $\{hkl\}$ of lattice planes. The radii of the circles are calculated from the Bragg equation. The intensity of the rings depends on the crystal structure, the radiation (accelerating voltage), the beam incidence direction, the sample conditions (foil thickness), and the orientation distribution function. As an example, Figure 6.7 shows the SAD patterns obtained in two aluminum samples with different textures. A random texture yields uniform intensities along the Debye–Scherrer rings (Figure 6.7a), while in the case of a pronounced texture—as in the case of a sample deformed by rolling—the intensities are distributed heterogeneously (Figure 6.7b). Thus, the intensity distribution along the individual Debye–Scherrer rings can be exploited for determination of the local texture of the irradiated volume. Such techniques for measuring local pole figures have been developed independently by Schwarzer (1983) and Humphreys (1983). For experimental details, see also Schwarzer and Weiland (1986) and Weiland and Panchanadeeswaran (1993).

The diffraction rings only reveal the spread of crystal planes perpendicular to the incident beam (because the Bragg angle is only in the order of 1°), and hence one ring diffraction pattern only comprises information corresponding to one large circle in a pole figure (Figure 6.8a and b). To obtain larger coverage of the pole figure, the sample has to be tilted similarly to conventional x-ray texture analysis (Section 4.3), as shown in Figure 6.8a and c. To derive the complete orientation distribution, for

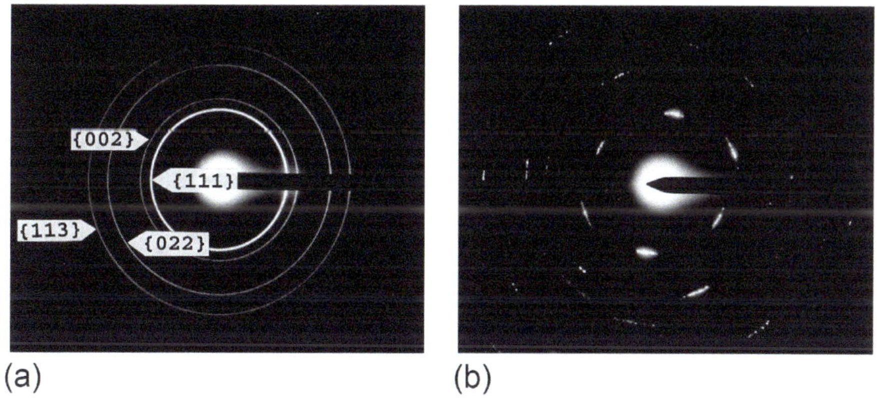

FIGURE 6.7 SAD ring patterns in polycrystalline assemblies. (a) SAD diffraction pattern of evaporated aluminum with random texture and (b) SAD diffraction pattern of cold-rolled aluminum with strong texture. (Courtesy of H. Weiland.)

each pole figure point, the intensity along the corresponding Debye–Scherrer ring has to be recorded. This is done using a wide-angle digital camera fixed either above or below the viewing chamber of the TEM. This camera records the full pattern including several complete *hkl* rings for every sample tilt α (Schwarzer et al., 1996a; Schäfer and Schwarzer, 1998). The total recording time is then as short as approximately 1 min. In most TEMs, the tilting angle is limited to a range of 0° to approximately ±60°, so that only incomplete pole figures with a large non-measured region

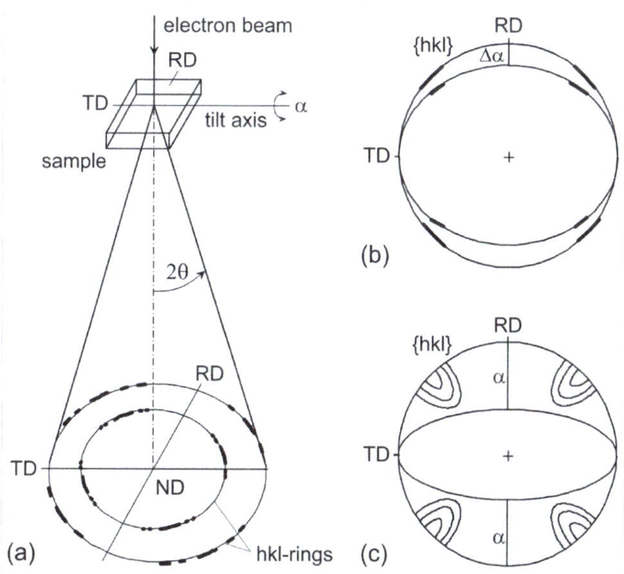

FIGURE 6.8 (a) Formation of SAD pole figures in a TEM; (b) and (c) coverage of the pole figure as the angle α is gradually increased.

in the pole figure center are obtained (Figure 6.8c). To receive complete information about the texture, several incomplete pole figures can be determined and, similarly to x-ray texture analysis (Section 5.3), from these an ODF is computed (Weiland and Panchanadeeswaran, 1993; Helming and Schwarzer, 1994). At the time, a few application examples demonstrated the applicability of the technique, including measurements of the microtextures in deformation zones around particles (Humphreys, 1977) or within deformation or shear bands (Schwarzer, 1988).

The technique has a number of drawbacks. One concerns the quantitative evaluation of the SAD pole figures which is complicated by the fact that the intensity distribution along the Debye–Scherrer rings is not directly proportional to the pole density in the pole figures, and thus requires relatively complicated additional experimental or analytical corrections, including background intensity and an increase in diffracting volume and absorption with increasing sample tilt (Schwarzer, 1983, 1993; Humphreys, 1984). A second concerns the observable volume which is limited on the small side by the spatial resolution of the SADP technique to about 1 μm and on the large side by the largest field of view achievable in a TEM image, of about 100 μm diameter. Furthermore, the technique only measures the texture without any microstructural information. With the advent of automated spot pattern acquisition and analysis, the DSP technique has become largely outdated. Pattern mapping techniques allow to measure orientations directly (and not via pole figure inversion) and to obtain the microstructure at the same time. We describe these TEM-based techniques in Section 8.3.

6.2.3 Convergent Beam Electron Diffraction Patterns (CBEDP)

CBED patterns are created using a convergent electron beam for illumination of the sample. The convergence is achieved by focusing the beam onto the specimen. The angle of convergence can be controlled by the size of the probe on the sample and the size of the condenser aperture. For conventional CBED patterns, the angle of convergence, i.e. the half cone opening angle, should be about the value of the Bragg angle of the observed diffraction spots.

As in the SADP case, CBED patterns are formed by elastic, coherent scattering of the primary beam. However, since the primary electrons enter the sample under a range of different directions, the diffraction pattern does not consist of sharp diffraction spots but of disks whose diameter is proportional to the convergence angle. Each disk in a CBED pattern represents a two-dimensional rocking curve over the crystal for the respective diffracted beam. Figuratively, it can be understood as an array of individual spot patterns, each one for another beam incidence direction, laid side by side within the borders of the convergence angle of the primary beam (quoted after Deininger et al., 1994). Therefore, the interior of these disks is, if their diameter is large enough, not of uniform brightness but show numerous details given by the different beam incidence directions for every position in the disk. The details in the disks can be used for quantitative characterization of many properties of the illuminated area. It is, for example, possible to determine the sample thickness, the character of dislocations and stacking faults, the crystallographic symmetry, and lattice constants of the material. CBED is a very powerful technique and is described in detail in several monographs and textbooks on TEM as for example Tanaka et al. (1988, 1994), Mansfield (1984), Williams and Carter (2009), and Steeds and Morniroli (1992). It is not the aim of this text to describe details of these techniques; rather the use of a special form of CBEDP for orientation determination, an aspect that is disregarded in the literature, will be discussed here in more detail.

If the convergence of the primary beam is small (i.e. if a small condenser aperture is used and the beam is condensed to a small spot onto the sample)—convergent angles are typically 1/5 to 1/10 of the Bragg angle—the usual CBED disks shrink together and become well-confined spots of uniform intensities as in a typical SADP; Figure 6.9a shows a typical example. We refer to these patterns in the following as SCBED patterns. It was probably the present author, who introduced these patterns for the first time as a powerful tool for the determination of crystal orientations with high spatial resolution although Riecke (1962) described a similar form of diffraction patterns. Basis for orientation determination from SCBED is the Ewald

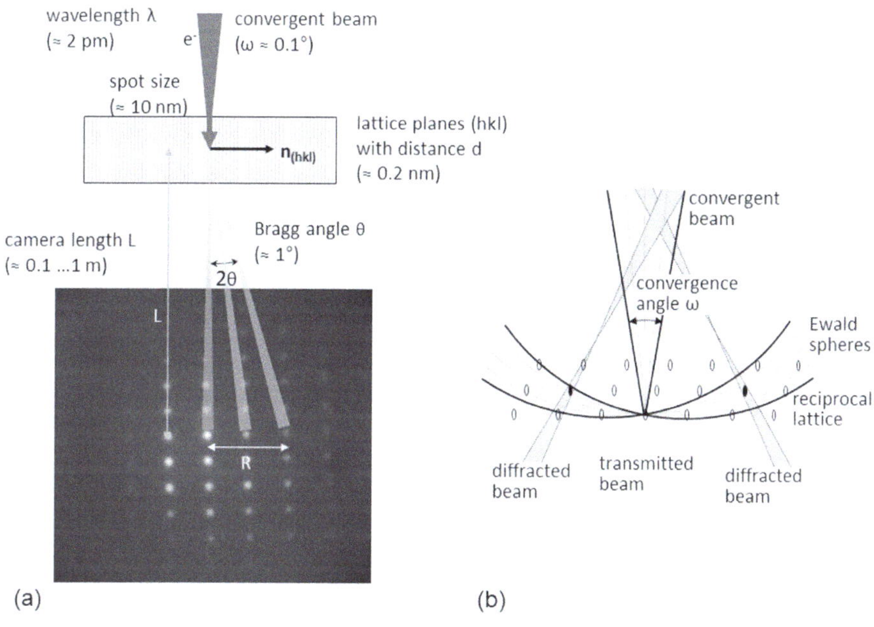

FIGURE 6.9 (a) A typical SCBED pattern obtained from the ⟨011⟩ zone axis of iron. The pattern is created using a convergent electron beam rather than parallel illumination. The spot size on the sample can be as small as 10 nm. In order to obtain a small angle of convergence, ω (on the order of 0.1–0.2°), a small condenser aperture has to be selected. The resulting pattern looks very much like a classical SAED pattern but the diffraction spots are wider due to the primary beam convergence. (b) Ewald construction for convergent beam electron diffraction (the same figure holds for precession electron diffraction). Note that the convergence angle is displayed much larger than it is in reality. All reciprocal lattice vectors that are positioned within the enveloping Ewald spheres lead to diffraction spots. The spot intensity is given by the integral over that part of the reciprocal lattice streak that is inside the enveloping Ewald spheres. For two vectors of the reciprocal lattice (shown with their reciprocal lattice streaks), the appropriate diffraction geometry is sketched. (Adapted from Zaefferer, 2011.)

sphere construction that can still be applied for the geometric interpretation of the pattern (Figure 6.9b). The spot intensities are in this case integral intensities over all incidence angles covered by the primary beam, i.e. over a certain range of the deviation parameter s. For a constant thickness t of the crystal the intensity of a spot depends monotonically on s (Equation 6.9) because the higher-order maxima for $I(s) \sim (\sin(s)/s)^2$ are too small to create appreciable intensity. Therefore, the intensity of a spot can be used, after it has been normalized by the appropriate ξ_g (cf. Equation 6.9) to estimate the distances of the corresponding reciprocal lattice point to the Ewald sphere. As it will be shown in Section 7.2.4, this point is quite important for automatic orientation determination from these patterns.

With respect to orientation determination, SCBED as compared to SADP have the strong advantage that they are created by a finely focused beam which allows reaching a much higher spatial resolution of <10 nm, compared to 500 nm reachable with SADP. Compared to the orientation determination from Kikuchi patterns (see Section 6.2.6), SCBED patterns have the great advantage of being relatively insensitive to lattice defects and can thus be used to measure orientations in highly deformed metals or other crystals with high defect densities. The reason for this is sketched in Figure 6.10: a slight lattice rotation due to crystal defects leads to a slight

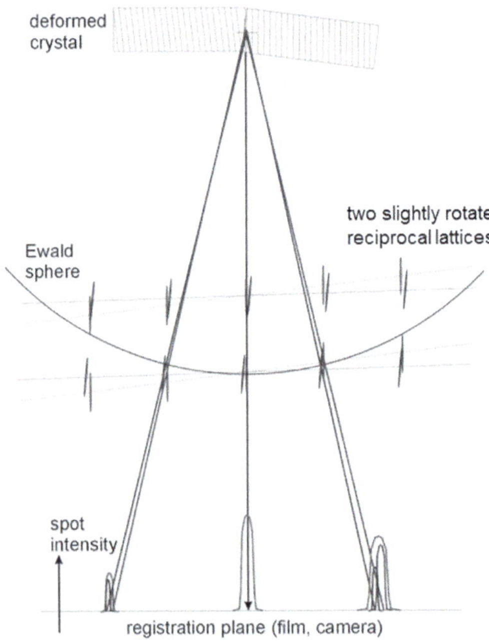

FIGURE 6.10 The effect of lattice deformation by dislocations on spot diffraction patterns. The sketch uses the same combination of reciprocal and real space as Figure 6.4. The deformed crystal is represented by two slightly misoriented crystals. The reciprocal lattices of both crystals show the same misorientation as the crystal lattices. The spot *intensities* from both crystals vary but their *positions* almost do not. As a consequence, the pattern is still visible and analyzable. (Adapted from Zaefferer, 2011.)

rotation of the Ewald sphere but to the formation of still clearly visible diffraction spots. The reduced sensitivity to small lattice rotations is, of course, also the reason for the reduced angular resolution of this technique. Finally, SCBED patterns are also available on very thin samples, when incoherent scattering is not sufficient to make Kikuchi patterns clearly visible.

6.2.4 Spot Diffraction with Electron Beam Precession (Precession Electron Diffraction, PED)

This section closely follows a text published earlier in Zaefferer (2011). Besides a direct convergence of the primary beam as it is employed in SCBED it is also possible to obtain focused convergent beam diffraction patterns by precession of the primary electron beam. This technique has been developed by Vincent and Midgley (1994). Figure 6.11a describes the principle of the technique: the primary beam is moved along a hollow cone with the opening angle ρ and the pivot point positioned on the sample. In a second step the diffracted beams exited by the sample

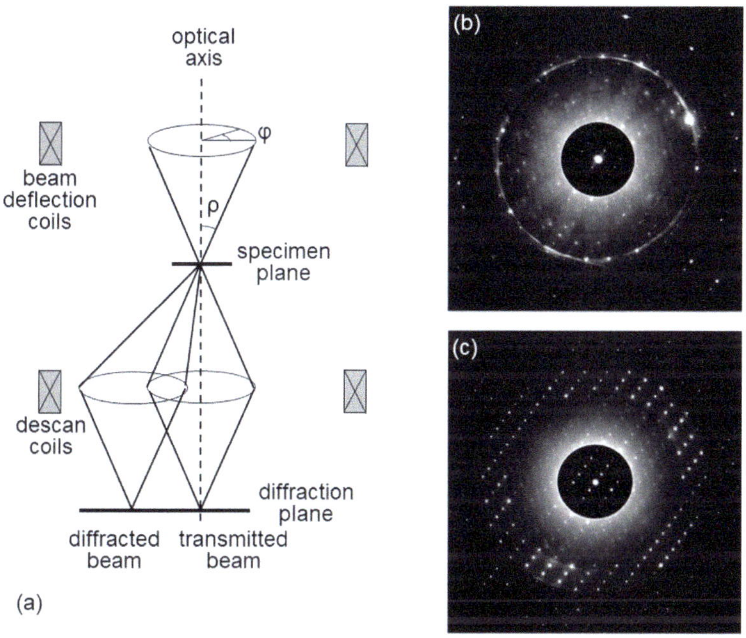

FIGURE 6.11 Precession electron diffraction (PED). (a) Schematic drawing of PED in a TEM. The beam is tilted off zone by angle ρ using the beam tilt coils and serially processed through an angle $\varphi = 2\pi$. A complementary de-tilt is provided below the specimen by de-scan coils to restore the zone axis pattern. (b) Selected area DP of the [532] zone axis of magnesium orthovanadate ($Mg_3V_2O_8$). (c) Processed SAEDP of the same crystal using a moderate precession angle of 5.2 mrad. Images (b) and (c) have identical exposure times, digitizing conditions, and have received the same digital image processing, so they can be directly compared. (Adapted from Own, 2005.)

are de-scanned with a second set of scan coils, thus resulting in a diffraction pattern with focused diffraction spots. In contrast to the SCBED technique the angle of convergence can be controlled by the precession angle without enlarging the diffraction spots to disks. Compared to SADP and SCBED the PED shows a significantly larger amount of diffraction spots as indicated in Figure 6.11b and c (Own, 2005). While Figure 6.11 shows the special case of a pattern taken along a low-indexed zone axis, it should be noted that a similarly significant increase of the amount of spots is also observed in off-zone axis patterns. Figure 6.9b, which was displayed for SCBED, can also be employed to understand the features of PED: first, all reciprocal lattice points included within the enveloping Ewald spheres lead to diffracted intensities that means a large amount of spots is visible in the pattern. This means in particular that HOLZ are enhanced in their visibility thus allowing unequivocal orientation determination. Second, the intensities of the spots are generated by integration over a large s interval, which largely leads to the disappearance of dynamical diffraction effects, as discussed above. In fact, it has been shown in a number of publications (e.g. Barnard et al., 2010; Eggeman et al., 2010) that precession diffraction patterns can be treated with the kinematical theory. The large amount of spots and their kinematical intensity distribution makes PED an ideal candidate for orientation determination by template matching which is described in Section 7.4.2.

For ORM electron precession has to be coupled with beam scanning. This complex beam movement is realized by the ASTAR system produced by NanoMegas (www.nanomegas.com). A detailed overview on the technique is found in Rauch and Veron (2014). In this paper, the authors claim the spatial resolution of the technique to be close to the diameter of the primary beam, i.e. in the order of 1 nm. This is, however, only true for columnar grain structures with grain boundaries parallel to the electron beam. For all other microstructures, it is the sample thickness which is the resolution-limiting factor (see also Section 8.3.5).

6.2.5 Conical Scanning

Closely related to the precession scanning described in the previous section is the CDFS method, developed by Wright and Dingley (1998), partly based on former ideas of Humphreys (1983) and Schwarzer (1983). To the author's knowledge, the technique has not been developed further, although the method has some interesting and powerful features. In particular, it might be superior to other TEM ORM techniques in terms of spatial resolution. In terms of angular resolution of the orientation determination, the technique appears to be behind other techniques. Concerning speed of measurements the technique seems to be comparable to the STEM techniques.

CDFS is based on the consecutive recording of centered dark-field images obtained under different beam incidence angles. To obtain a centered dark-field image, the sample is illuminated in normal imaging mode with a parallel primary beam. The diffraction aperture is positioned on the optical axis, and the primary beam is inclined until a diffracted beam travels along the optical axis. In imaging mode, then a dark-field image is visible where only those parts of the sample light up, that fulfill the Bragg condition for the particular beam inclination. As shown in Figure 6.12a the primary beam is then scanned by an angle ϕ around cones with

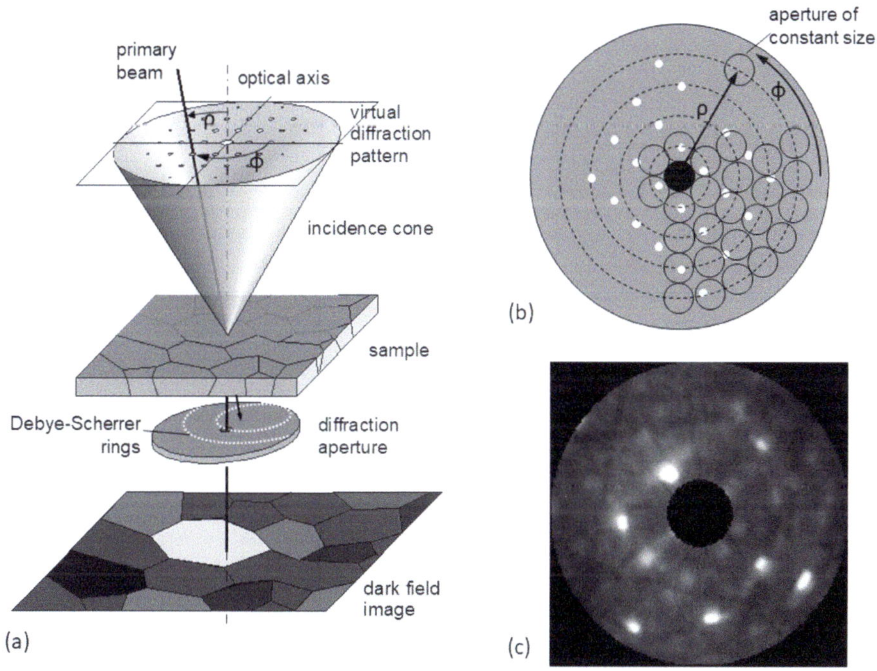

FIGURE 6.12 (a) Schematic drawing of the CDFS principle. The primary beam is rocked about the angles *r* and *w* and the appropriate centered dark field images are recorded. In order to construct the virtual diffraction pattern for every pixel in the image, the intensities of that pixel are projected to the ρ-ω plane indicated in the graph. (b) Demonstration of the changing angular resolution inside of each reconstructed spot pattern due to the constant aperture size that is used to record the patterns. The inner circles are the most inaccurate ones. The larger is ρ the smaller is the angular resolution. (c) A typical reconstructed spot pattern obtained from copper with an angular step size of 2° for ω and 0.1° for ρ. (Adapted from Zaefferer, 2011.)

successively larger opening angles ρ up to a maximum angle of approximately 4°. For every beam incidence (ϕ, ρ) those positions, (x, y), in the dark field image light up, which fulfill the Bragg conditions for this particular incidence.

After recording all dark-field images for all incidence angles, the intensities at every position (x, y) for the different inclinations are collected and placed together in a polar coordinate system (ϕ, ρ) as shown in Figure 6.12b. As a result for the point (x, y) a diffraction pattern as the one displayed in Figure 6.12c is reconstructed (indicated as "virtual diffraction pattern" in Figure 6.12a). In this way, a diffraction pattern is reconstructed for every pixel in the dark-field images. These diffraction patterns can then be indexed in a conventional manner or by template matching, as it is described in Sections 7.3 and 7.4, and an orientation map can be created (see Chapter 9).

At the time the technique was developed the data acquisition time for one complete scan with sufficiently high angular resolution was between 20 minutes and more than 2 hours for dark-field images of 512 × 512 pixels. The postprocessing time was in the order of several hours. The spatial resolution of the technique is, first of all, determined by pixel resolution of the camera and the chosen magnification of the image. It can, in principle, be as high as the resolution of the dark-field image. In practice, the resolution is deteriorated by the misalignment of the microscope or by sample drift, so that the pixel positions in the recorded images do no longer all correspond to exactly the same position in the sample. The fact that all pixel of an orientation map is recorded in parallel allows applying image cross-correlation to align the different images and correct to a certain extent for sample drift and misalignment. This is another advantage over the STEM techniques, where a similar correction is not possible.

To the authors knowledge the technique has not been developed further, mainly because the STEM techniques developed by Rauch and Midgley (see the previous section) became more powerful in automation and speed of acquisition. The intriguing properties of precession electron patterns together with the powerful analysis by image matching have made the PED technique the most powerful technique for orientation mapping in TEM.

6.2.6 Transmission Kikuchi Diffraction Patterns (TKP)

Kikuchi patterns are named after the famous Japanese nuclear scientist Seishi Kikuchi (in Japanese Kikuchi Seishi 菊地 正士), who discovered these patterns in the year 1928 (Kikuchi (1928). Kikuchi patterns are created when a focused electron beam enters a sample, sufficiently thick to allow a significant amount of electrons being incoherently scattered. These incoherently scattered electrons can be imagined to emerge from a point source inside of the sample emitting electrons in all directions of the surrounding crystal. Subsequently the first incoherently scattered electrons undergo coherent scattering by the lattice planes, thus forming the typical line-geometry of the Kikuchi patterns, as visible in the two examples shown in Figure 6.13. Kikuchi patterns offer a precise and straightforward way of determining

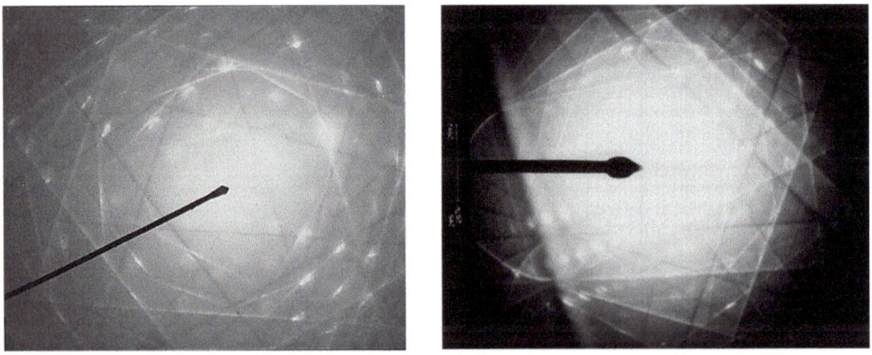

(a) α-titanium (b) γ-iron

FIGURE 6.13 Typical transmission Kikuchi pattern obtained at 200 keV from (a) α-titanium and (b) γ iron. Note that Kikuchi lines are rather weak compared to the strong background which is due to the small amount of coherently scattered electrons with respect to the incoherently scattered ones. In (b), a blurred low-indexed (111) band is visible.

crystal orientations from small areas. Furthermore, they are excellently suited to perform crystal tilt experiments required for defect analysis in TEM and SEM. If highly indexed lines are measured, they can also be used to determine—with medium precision—lattice constants or accelerating voltage.

As already mentioned in the introduction, Kikuchi patterns form a whole class of diffraction patterns which differ from the others by two important facts. First, the primary beam electrons are first incoherently scattered before they are in a second step coherently scattered, i.e. diffracted, by the lattice. The diffracted electrons thus do not observe a coherency relation with the primary beam electrons. Second, the incoherent scattering which is limited to a rather small volume in the sample forms a point-like source of electrons inside of the crystal emitting electrons in a wide range of directions, which is responsible for the Kikuchi pattern geometry. Both points are discussed in the following in more detail.

The main source of incoherent scattering contributing to the formation of TKP is thermal diffuse scattering, TDS. As will be addressed in Section 6.3.2.1, the average energy loss per TDS even is in the order of 0.1 eV which means that this scattering process may be considered "quasi-elastic". Energy filtering of diffraction patterns actually shows that sharp Kikuchi patterns are obtained if only the electrons of the zero-loss peak are used which have lost at maximum a few tenth of an eV in energy. A wide angular distribution of electrons in Kikuchi patterns is only observed when a sufficient amount of incoherent scattering occurs. This is only the case for samples that are thick enough (a thick TEM sample or a bulk sample in SEM) or for sufficiently low electron energies. The effect of TDS on Kikuchi patterns will be discussed in more detail in conjunction with the spatial resolution of the EBSD technique in Section 6.3.3. In TKP the maximum scattering angle (with respect to the incident beam direction) is in the order of 10–15°.

The electrons emerging from incoherent scattering may be subsequently coherently scattered at the crystal lattice leading to diffraction under angles according to the Bragg law. Due to their large angular distribution they "illuminate" the crystal lattice planes from all directions, thus for every set of lattice plane some electrons arrive under Bragg conditions and are diffracted. The locus of these electrons are pairs of cones, so-called *Kikuchi or Kossel cones*, around the normal of the diffracting lattice planes as is shown in a two-dimensional sketch in Figure 6.14a and in three dimensions in Figure 6.14b. The opening angle of the cones is $180°–2\theta$, which means that the angle between the cones belonging to one set of lattice planes is 2θ. Since θ has a relatively small value for high energy electrons (in the order of 1 degree) the cones intersect a flat detector plane in almost straight, parallel line pairs, the so-called *Kikuchi lines*. The pair of lines often envelops an area of higher intensity, which is called the *Kikuchi band*. As indicated in Figure 6.14a, the center line between these pairs corresponds to the *gnomonic projection* of the diffracting lattice planes onto the observation plane. The Kikuchi pattern essentially embodies all the angular relationships in a crystal—both the interzonal and interplanar angles—and hence implicitly contains the *crystal symmetry*. Note that also the *atomistic information* about the crystal is encoded in the diffraction pattern, albeit not in its symmetry but in its intensity distribution, as will be shown in Section 6.3.2. As will be shown later in Figure 7.2 and discussed in Section 7.2.1, if the origin of the projection, the

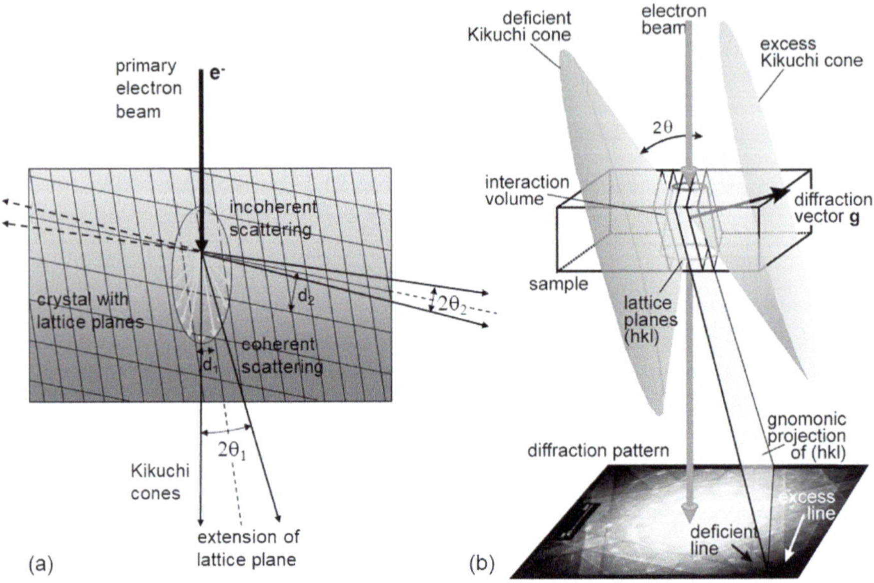

FIGURE 6.14 (a) 2D-schematic of Kikuchi cone formation on a crystal lattice. In a first incoherent scattering step a point source is created in the crystal. This point source leads to Bragg reflection on all lattice planes, creating pairs of cones with an opening angle of 180°–2θ. The cones thus have an angular distance of 2θ. (b) 3D schematic drawing of the geometry of a TKP. The diffraction vector **g** stands perpendicular on the plane of the gnomonic projection of the lattice planes. Note that in reality the Kikuchi cones have a much larger opening angle, such that they intersect the pattern along the respective Kikuchi lines

so-called *pattern center* is known then it is straightforward to determine from the position of the Kikuchi band on the detector, the spatial position of the diffracting lattice plane or its normal vector, the diffraction vector $\mathbf{g} = \mathbf{h}_{hkl}$, of these planes. The length of the diffraction vector, which is equal to $1/d_{hkl}$, can also directly be calculated from the distance between the two Kikuchi lines and their position on the screen, as indicated in Figure 7.2, taking into account the angular difference between the lines proportional to 2θ. Once the positions and lengths of several $\mathbf{h}_{hkl}$ of a given crystal are known it is easy to index the pattern (i.e. assign Miller indices to the visible Kikuchi bands) and calculate the orientation from it (see Section 7.2).

The intensities of the two Kikuchi cones are not equal, which is due to the anisotropic intensity distribution of the incoherently scattered electrons. In thin foils the intensity distribution can be approximated by a Gaussian distribution (Reimer, 2013b):

$$I = I_0 \exp\left(-\left(\Theta/\Theta_0\right)^2\right) \tag{6.11}$$

with I_0 being the primary intensity, Θ the angle of observation, and Θ_0 a constant. This steep distribution is responsible for the occurrence of an excess and defect Kikuchi cone in each pair of cones, as indicated in Figure 6.15: the electron intensity

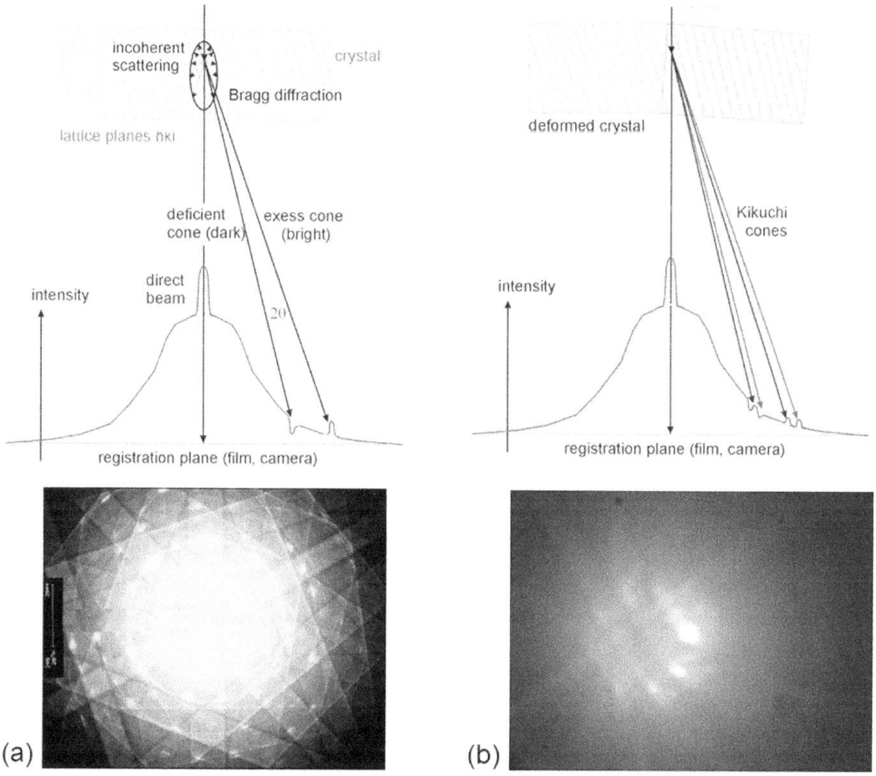

FIGURE 6.15 The effect of lattice deformation by dislocations on Kikuchi diffraction patterns. The defected crystal is represented by two slightly misoriented crystals. (a) Kikuchi diffraction on a defect free crystal, developing sharp Kikuchi lines. (b) Kikuchi diffraction on a defected crystal, simulated as two slightly misoriented crystals, leading to two displaced Kikuchi patterns. The low Kikuchi line intensity is distributed over two, slightly misoriented Kikuchi cones. The Kikuchi lines become blurred and finally disappear. (Adapted from Zaefferer, 2002.)

in the cone further away from the primary beam is higher than the surrounding background because it is created by diffraction from a direction of high electron density into a direction of low electron density. This cone is called the excess cone and the appropriate excess Kikuchi line appears brighter than the background. The electrons diffracted into the excess cone are missing in the background intensity, creating, in principle, a symmetrical cone consisting of a deficit of electrons. This deficit is not completely filled by the electrons diffracted into the second Kikuchi cone because this one is due to diffraction from a direction of low electron density into a direction of high electron density. This cone of lower electron density than the surrounding background is the deficiency cone and creates a dark line, the deficiency line. In TKP the deficiency line is closer to the primary beam, the excess line further away. On thicker samples or for low-index reflections the Kikuchi line intensities become more controlled by dynamic diffraction effects which leads to the appearance of Kikuchi bands enveloped by the corresponding Kikuchi lines which may show, depending on sample thickness and orientation, lower or higher intensity than the background (*deficiency* or *excess band*). It may also be that the Kikuchi lines vanish and only the band is visible. Finally, the lines may also change from the deficiency to the excess type and vice versa. A couple of examples are found in Figure 6.13b.

In contrast to the rather simple geometry of Kikuchi patterns, describing the intensity of Kikuchi lines requires more complicated considerations. The reason is that Kikuchi lines only occur under conditions where incoherent scattering of primary electrons is a frequent event, like in thick samples or at low accelerating voltages. These are, however, the conditions where the dynamical theory of electron diffraction has to be taken into account for the description of diffracted intensities.

Neither the general features nor the details of the Kikuchi band intensity distribution can be described by simple models. Rather, dynamical electron diffraction theory including anomalous absorption has to be applied (for an extensive overview on literature see, e.g. Wang, 1992; Omoto et al., 2002; Reimer, 2013a, 2013b). The dynamical theory of electron diffraction will be further discussed in Section 6.3.2.3 in conjunction with EBSD, where dynamic effects are much stronger visible than for TKP. Note that the intensity distribution observed in Kikuchi patterns is not only due to dynamical electron diffraction but also due to incoherent and inelastic electron scattering.

The occurrence of Kikuchi bands of varying contrast depending on sample thickness and accelerating voltage has consequences for orientation determination. One is that for low indexed reflections frequently only blurred excess Kikuchi bands and no Kikuchi lines are visible. Furthermore, the center of a Kikuchi band may not precisely correspond to the plane position as the band contrast may be asymmetric. Consequently, neither the Bragg angle i.e. the length of **g** nor the position of **g** can be determined accurately for these reflections. The higher is the (hkl) index of a reflection, the sharper are the corresponding Kikuchi lines, thus, the more precise **g** can be determined. In contrast, the intensity of these Kikuchi lines becomes increasingly weak with larger (hkl) index. Whether Kikuchi lines appear with or without an excess band can be estimated, according to Reimer (2013b), from the profile width, $\Delta\theta$, of a Kikuchi line:

$$\Delta\theta = \frac{\Delta w}{|g_{hkl}|\ \xi_g} \tag{6.12}$$

$\Delta\theta$ must be roughly larger than the Kikuchi band width (2θ) to create a visible band, i.e. $\Delta\theta \geq 2\theta$. Δw is a constant which is, in fact, not easy to determine. Reimer proposes a value of approximately two. For high-order reflections Equation 6.12 may also be used to estimate the sharpness of Kikuchi lines.

TKP have several important advantages for orientation determination. First, using sharp, higher-order Kikuchi lines the accuracy of orientation determination may be better than 0.1° which is one order of magnitude higher than for spot patterns. Second, the visible solid angle of the reciprocal space can be rather large in a TKP (more than 20° compared to about 5° for spot patterns). This normally allows observation of at least two zone axes in one pattern which is sufficient to determine the orientation of a crystal unequivocally. Also tilt experiments can be performed with great ease and 3D crystallographic symmetries may be directly observed. Third, for orientation determination with TKP only the geometry of the pattern (i.e. the line positions) has to be taken into account while the intensity of the lines is not of great importance. This makes experiments and calculations easier and more straightforward.

The high angular sensitivity also implies a serious disadvantage of TKP, which is their sensitivity to lattice defects like dislocations, stacking faults or elastic strain. In fact, with increasing density of defects TKP blur out and can thus not be used for orientation determination in highly strained lattices. The reason for this is illustrated in Figure 6.15b, where a lattice defect is indicated as a (very) small misorientation between neighboring crystal areas. The existence of this boundary leads to the distribution of the anyway relatively small diffracted intensity into several closely spaced lines which, with sufficient disturbance, combine into very weak, blurred bands. This is in contrast to BKP (Section 6.3.2) where small misorientations just lead to broader, but still well-recognizable bands. Another disadvantage of TKP for orientation measurements is the difficulty to automate the detection of line positions by means of software algorithms. The reason for this lies in the large variations of contrast in dependence of local sample conditions and crystal orientation. Automatic orientation determination, however, is an important issue in TEM-based ORM where crystal orientations are to be determined as quick and as reliable as possible. The available algorithms and their difficulties are discussed in Section 7.2.3.

6.3 DIFFRACTION IN SEM

Over the last 30 years ORM based on electron backscatter diffraction (EBSD) in SEM has become a very popular and powerful tool for materials and geological science (Adams et al., 1993; Schwarzer et al., 2009). The EBSD technique is not only the basis of the most comprehensive and most quantitative tool for characterization of microstructures, including phase distribution, defect density characterization and, to a certain extent, residual stress measurements, but it is also the most frequently used tool for microtexture analysis. Combined with smart probing it is even much used to perform macrotexture measurements.

Although diffraction effects in SEM are known already since the work of Coates (1967), only the advent of sensitive detectors (low light cameras) and powerful computers made the enormous development of the technique possible. Two forms of diffraction patterns, electron backscatter diffraction (EBSD) patterns, also called backscatter Kikuchi patterns (BKP), and electron channeling patterns (ECP) are useful for orientation microscopy (ORM). Both result from diffraction on bulk samples and both have the same geometry as transmission Kikuchi patterns (TKP). The automation of the analysis of EBSD patterns for automatic crystal orientation mapping (ACOM, OIM) has made EBSD one of the most powerful techniques for the study of microstructures and microtextures of crystalline materials. EC patterns are not much used nowadays mainly because their generation is more complicated than that of EBSD patterns and because their spatial resolution is significantly worse. Nevertheless, for special applications, they may have advantages over EBSD patterns. One reason to include them into this discussion is that their intensity distribution is responsible for orientation contrast observed with a backscatter electron (BSE) detector on crystalline samples, namely electron channeling contrast imaging, ECCI. This is a technique for microstructure observation which is included because it is increasingly used in conjunction with orientation measurements by EBSD. Also, ECCI is the basis of a new technique, eChord, for orientation mapping, possibly used as alternative to EBSD. Finally, also the micro-Kossel technique, which was one of the forerunners of EBSD, is included here for completeness, although it is used nowadays only for some special applications.

Wilkinson and Hirsch (1997) give a comprehensive overview on the EBSD and ECP techniques in terms of theory and praxis. Also Reimer (2013a) treat both types of diffraction patterns in common. Indeed, as already pointed out by Booker et al. (1967), both types are related by the reciprocity principle: the intensity distribution in an ECP is due to the angular variation of the primary beam which creates a Bloch wave field Ψ_{in}, inside of the crystal; the outgoing wave field Ψ_{out} is incoherent and not of large importance. In contrast, in BKP the intensity is due to the outgoing wave field Ψ_{out}, formed inside of the crystal, while the input wave field is incoherent and of minor importance (but, in fact, more important than Ψ_{out} for ECP). Also in terms of ray-geometry, a reciprocity relationship between both patterns can be established, as shown in Figure 6.16a and b: for ECP the incoming electrons arrive with a conical direction distribution, while the outgoing electrons leave in all directions (but are registered only at the detector position). For BKP, the situation is inversed: the primary electrons travel in all directions (created by incoherent scattering of the primary beam) but they leave with a conical intensity distribution toward the large detector.

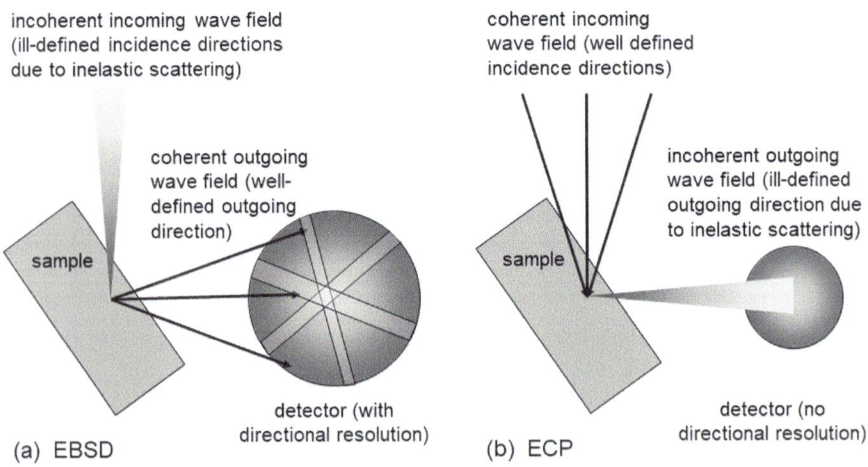

FIGURE 6.16 The reciprocal relationship between (a) EBSD pattern and (b) EC pattern. In EBSD an incoherent source creates a coherent signal (the EBSD pattern). The angular crystallographic information is encoded in the outgoing signal. In ECP a coherent source creates an incoherent signal. The angular information is created by the angular scanning of the primary beam.

6.3.1 Backscatter Kikuchi Patterns (BKP) or Electron Backscatter Diffraction (EBSD) Patterns

Backscatter or reflection Kikuchi patterns have been observed already shortly after the discovery of TKP by Nishikawa and Kikuchi (1928). In the 1930s, von Meibom and Rupp (1933) and Boersch (1937) produced Kikuchi patterns from backscattered electrons, which eventually led to the development of EBSD in the SEM. In 1953 Alam and co-workers published a very detailed study on the intensity distribution of BKP in dependence of beam incidence angle, observation angle, beam energy, and material properties and drew important conclusions on the physical formation mechanisms of the patterns (see further below) (Alam et al., 1954). While Alam et al. used a simple diffraction apparatus for their examinations, Venables and coworkers observed BKP for the first time in a scanning electron microscope (Venables and Harland, 1973; Venables and Bin-Jaya, 1977). As already described for TKP, crystal orientation and crystallographic phase can be easily determined from the patterns. The application of the technique in the SEM therefore became the starting point for the development of "orientation microscopy" by Dingley (1984) and later by Adams et al. (1993), as will be further described in Sections 8.2, 9.3, and 9.4.

In order to obtain BKP, the surface of a flat sample is illuminated under a shallow angle by the primary electron beam. For optimum observation conditions, a beam incidence angle of approximately 20° with respect to the sample surface is selected and electrons scattered into angles between 30° and 120° with respect to the primary beam are recorded. In practice, the sample in the SEM chamber is tilted to 70° from the horizontal position while the EBSD detector faces the sample surface from the chamber side wall. The appropriate geometry is sketched in Figure 6.17a and c. Detection of the patterns occurs by a phosphor screen combined with a highly light sensitive CCD camera, which is positioned close to the sample (usually in the order of some centimeters). Figure 6.17b shows a typical pattern, obtained on a CdTe crystal with 15-kV electrons. The geometry of the pattern, displayed in Figure 6.17c, is exactly that of TKP with the difference, that the angular range is usually much larger and that the projection center, which is the point where the electron beam hits the sample, is not visible (in TKP the pattern center is marked by the primary beam spot). The pattern formation may be understood—similar to that of TKP—in a two-step process as indicated in Figure 6.18. The first step consists of incoherent scattering of the primary electrons, producing a point-like electron source inside of the crystal. In the second step, these electrons are coherently scattered, creating Kikuchi cones. The scattering angles obtained by the incoherent scattering process range from 0 to 180° with an intensity peak approximately in the direction of reflection of the primary beam on the sample surface. The exact direction of this intensity maximum depends on the acceleration voltage and the average atomic number of the specimen.

Good understanding of the physical processes that lead to formation of the patterns allows for optimum application of the technique and potential further technical development. As mentioned above, the formation of Kikuchi patterns is assumed to occur in two steps, an incoherent, quasi-elastic, large-angle scattering process, which leads to the formation of an "electron source" in the crystal and a coherent

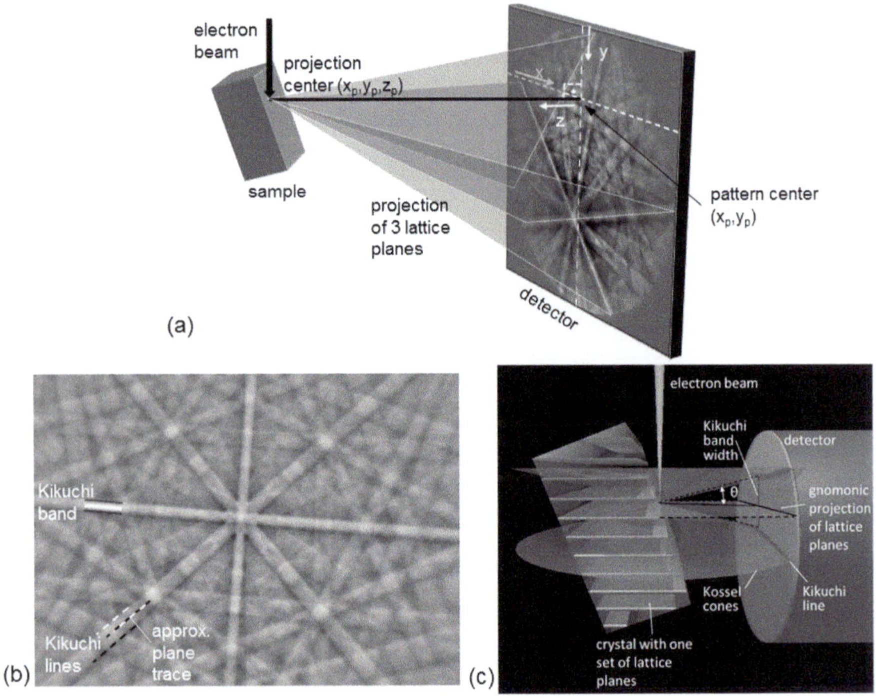

FIGURE 6.17 (a) Experimental setup for measurement of EBSD patterns in a scanning electron microscope. The position of the electron beam on the sample marks the projection centre. Three gnomonic projections of lattice planes are plotted for 3 Kikuchi bands. (b) An EBSD pattern of CdTe, obtained at 15-kV acceleration voltage. (c) Schematic drawing explaining the formation of Kikuchi bands and Kikuchi lines from Kossel cones.

small-angle scattering process, which creates the Kikuchi cones. While the first step determines the spatial resolution of the EBSD technique. i.e. the interaction volume from where the pattern originates (see Section 6.3.3), the second step is responsible for the creation of crystallographic information in the pattern. A good understanding will allow for improvements in the field of phase determination, in particular for cases of pseudo-symmetry which will be discussed in Section 7.4.4. Incoherent and coherent scattering mechanisms will be discussed in the subsequent sections.

6.3.2 Physics of Backscatter Kikuchi Pattern Formation—Dynamical Theory of Electron Diffraction

In the previous section, it was explained that the Kikuchi lines can be imagined as the intersection of the Kossel cones of Bragg-diffracted electrons with the flat detector. Because for typical electron energies in electron microscopy (~10–200 keV), the Bragg angles in electron diffraction are rather small (in the order of 1–5°) and the intersections of the Kossel cones with the detector are almost straight Kikuchi lines (in fact, they are slightly curved hyperbolae). This simple *kinematic* (geometric) construction of the Kikuchi bands would yield an intensity distribution consisting of two

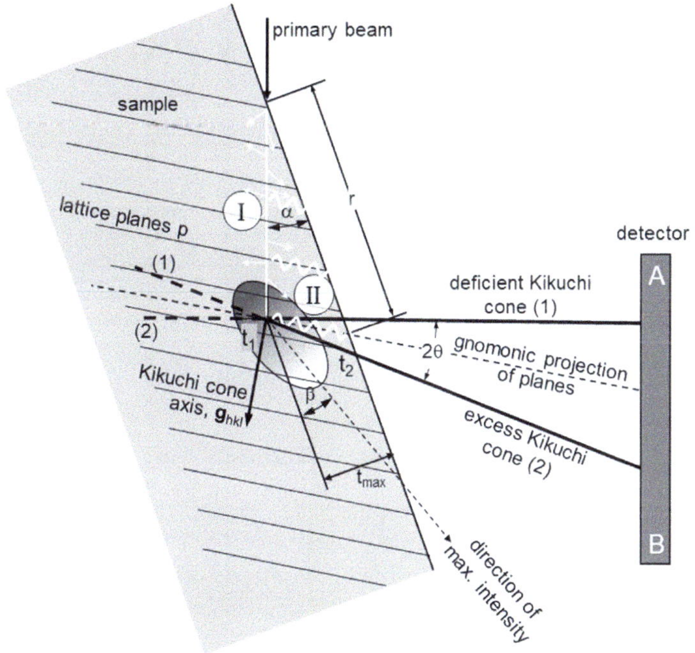

FIGURE 6.18 Schematic display of geometry and scattering mechanisms of backscatter Kikuchi diffraction. The primary beam enters the material at an angle α. During the traveling through the material electrons are continuously inelastically scattered (I). These electrons form the source of the pattern with anisotropic distribution of electrons. The highest intensity is found approximately in reflection direction of the primary beam, $\beta \approx \alpha$, but varies with acceleration voltage and atomic number of the sample. The total path length determines the lateral resolution of the technique, r, and the depth resolution, t_{max}. On the way out the electrons undergo coherent diffraction (II), creating two Kikuchi cones. The excess Kikuchi cone is closer to the direction of maximum intensity. Integration of the wave field, as described in Equation 6.15, occurs from t_1 to t_2. Electrons that arrive on different points on the detector (A) and (B) have traveled along different path lengths through the material.

bright Kikuchi lines (lines of Bragg-diffraction) without any significant intensity between them. This is, indeed, what is frequently observed in TEM Kikuchi patterns, where white excess lines and dark deficiency Kikuchi lines are explained in a straightforward manner with kinematic electron diffraction theory (Section 6.2.6). However, in EBSD (Figures 6.17 and 6.19) and ECP (Figures 6.22b and 6.26b) in the SEM—and also, more faintly, in TEM Kikuchi patterns (Figure 6.13)—the intensity distribution is very different: dark Kikuchi lines envelope a bright Kikuchi band with various internal fine structures. This band intensity profile can only be understood taking into account *dynamical* electron diffraction theory.

The main difference between kinematic and dynamical theory is that in the former the electron wave field is assumed to be scattered only once, leading to the well-known Bragg reflection. In dynamical theory, in contrast, the scattered wave field itself is re-scattered multiple times before it finally leaves the sample. This

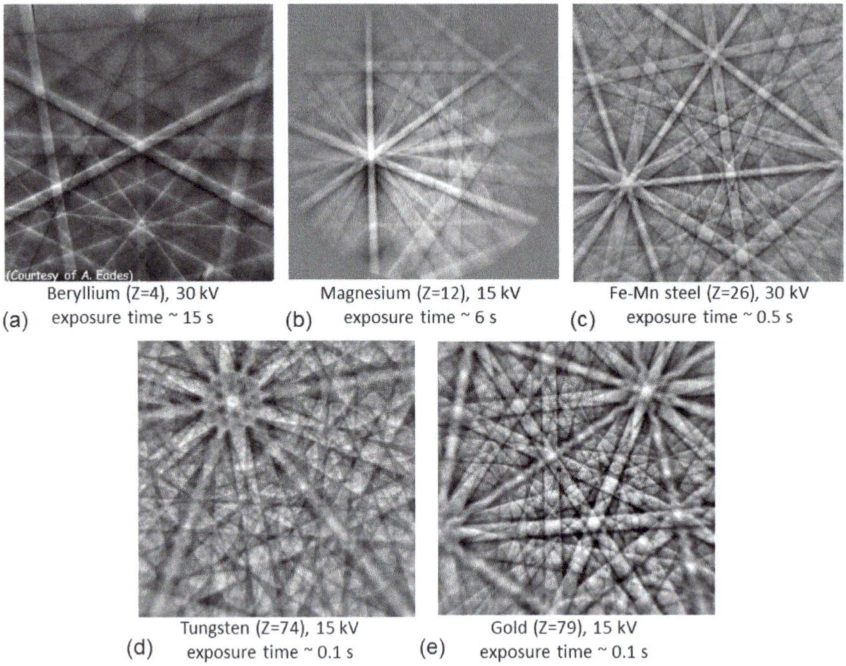

FIGURE 6.19 EBSD patterns of five materials with different atomic numbers. The different backscattering yield of the different materials is reflected in the different exposition times. In (a) (courtesy of A. Eades), most reflections show only Kikuchi lines (kinematic and elastic scattering). (b), (c) With increasing atomic number the intensity of bands becomes stronger (dynamic and inelastic scattering), that of lines weaker. (d) EBSD pattern of tungsten with high amount of fine structure (high indexed lattice plane reflections). (e) EBSD pattern of gold (courtesy G. Nolze) shows less fine structures compared to tungsten because of its internal deformation (Au pattern in (e) Courtesy G. Nolze).

dynamical process leads to a characteristic re-distribution of intensities between the individual Bragg diffraction peaks and also to intensity in between the Bragg peaks. For illustration, Figure 6.19 shows a couple of typical EBSD patterns from different materials: Figure 6.19a is rare example of a pattern from beryllium. Be has a very small atomic number ($Z = 4$) and strong atomic bonds. Its low Z-number is responsible for its weak scattering power and, therefore, for an almost kinematic diffraction behavior: Kikuchi line couples show the dark-bright contrast typical for TEM TKP patterns and Kikuchi bands are mostly absent. The strong bonding leads to very sharp bands. Figure 6.19b from Mg ($Z = 12$) shows already much more dynamical diffraction behavior, visible by bright Kikuchi bands partly, however, still enveloped by dark-bright Kikuchi lines. Figure 6.19c is a pattern from an austenitic steel. The great uniformity of the pattern in terms of brightness and the clear appearance of separated Kikuchi bands is ideal for automatic detection with a Hough transform (Section 7.2.2) and shows how well EBSD is suited for the investigation of materials in this range of atomic numbers. Figure 6.19d, from W with $Z = 74$, shows exclusively dynamic diffraction with lots of fine structures (higher-order diffraction vectors) and bright Kikuchi bands. Note also the appearance of HOLZ rings. Finally, Figure 6.19e displays a diffraction pattern of gold ($Z = 79$) which should show a similar amount of fine structures as tungsten in case of a perfect crystal. However, the fact that gold is very soft and has a low stacking fault energy (<30 mJ m^{-2}) commonly leads to a high amount of defects in the material. These defects extinct most of the fine diffraction structures as this was also shown by Nolze and Winkelmann (2012). Nevertheless, HOLZ rings are still clearly visible.

The basic principles of dynamical diffraction theory will be laid out here below. For deeper understanding we refer to the chapters on dynamical electron diffraction theory in the well-known textbooks on TEM by Fultz and Howe (2012), Carter and Williams (2016), and Reimer (2013b). For the case of EBSD it appears that the textbook of Reimer (2013a) on SEM is the best reference. The text by Winkelmann (2009a) dedicated to the formation of EBSD patterns requires a good basic understanding of the principles of dynamical diffraction theory.

6.3.2.1 Electron Scattering Mechanisms

In order to understand the physics of electron diffraction one first has to realize that electrons interact with matter in a number of ways. Coherent and elastic scattering ("Bragg" or Rutherford scattering) preserves the energy and the phase of the electron wave. Inelastic scattering processes, on the other hand, are processes where the electron wave loses energy and therefore also the phase relation. The inelastic processes can be further divided into two: the first is thermal diffuse scattering (TDS) where only a very small fraction of the electron energy is lost (below 1 eV) but usually the direction of the wave is strongly changed (up to 10°). The process is also called phonon scattering because it is assumed to be due to the thermal vibration of the atoms. If several of the TDS processes occur one after the other the electron traveling direction can change significantly (up to 180°) almost without losing any energy (Wang, 2003, 2013; Zaefferer, 2007). The second inelastic scattering process is core loss scattering, where the electrons lose a significant amount of energy. This process is responsible for continuous energy loss of the electrons, for example by formation of x-rays. The combination of Bragg scattering and TDS builds the basic mechanism of EBSD.

As a side note, it should be remarked that the TDS process is responsible for spatial resolution and angular distribution, while TDS and core loss scattering determine the energy distribution of backscattered Kikuchi patterns (BKP) in the SEM and TKP in the TEM. This is illustrated in Figure 6.18. Based on the observation that Kikuchi lines are relatively sharp, Zaefferer (2007) suggested that TKP and BKP have undergone only small energy losses, which are quantified in Table 6.1. The energy loss per TDS event is in the order of 0.1 eV. The average scattering angle is in the order of about 10° for 15-kV electrons and 2.5° for 200-kV electrons (Wang, 2013). From TKP it is known that the total energy loss is about 1.5 eV (the width of the zero-loss peak in TEM), which corresponds to about ten scattering events. This would lead to a maximum scattering angle of 25°, which is very close to what is actually observed. If one assumes a maximum scattering angle ($\alpha + \beta$ in Figure 6.18) of 100° one also obtains an average of about ten scattering events required for the

TABLE 6.1
Energy Losses and Scattering Events upon the Formation of TKP (TEM) and BKP (SEM)

Pattern	Accelerating Voltage (kV)	Zero Loss Peak Width (eV)	Number of TDS Events	Scattering Angle per Event	Total Scattering Angle
TKP	200	~1.5	10	2.5°	0 … 25°
BKP	15	? (1)	10	10°	0 … 100°

Note: (1) the question mark for the energy loss in BKP indicates that this value is still an issue of debate in literature.

electrons to change their direction toward the detector screen. This would lead to an equally low energy loss in BKP as in TKP, meaning the appearance of a zero-loss peak in the electron energy spectrum of BKP. This was roughly confirmed by Eades et al. (2009) and Deal and Eades (2005), who showed by energy filtering that EBSD patterns were most clearly visible when only energies with less than 500-eV energy loss were allowed on patterns created with 15-keV energy for different materials. They also showed that patterns with very small energy losses (in the order of 100 eV) appeared more blurred than those with slightly larger losses. The authors explained this with the small penetration depths of those electrons that had lost only very little energy. For these small penetration depths, the diffraction profile (see below) is strongly depth-dependent, leading to a strong apparent blurring. In contrast to these observations, Ram and De Graef (2018) suggested, using Monte-Carlo simulations assuming a continuous energy loss approach (caused by core-loss scattering), that EBSD patterns may show different electron energies, depending on the angle of scattering. For a 15-keV primary electron energy, they suggested the occurrence of energies between 10 keV for electrons scattered through an angle of about 140° to the top of the detector (point A in Figure 6.18) and of 13 keV for a scattering angle of 40° (point B in Figure 6.18). In view of the experimental results of Eades et al. (2009), these simulated values clearly appear unrealistic. This was later also pointed out by Winkelmann et al. (2019). They estimated the average energy loss of the electrons in an EBSD pattern of silicon to be in the order 1–1.5 keV with maximum differences between different parts of an EBSD detector in the order of 500 eV (which is still more than the values found in the experimental examinations of Eades et al., 2009). Based on energy loss measurements, Vos and Winkelmann (2019) estimated the mean free path of electrons (i.e. the distance traveled between to scattering events) contributing to EBSD patterns to be in the order of 30 nm for 15-keV electrons in silicon. Together with the estimation of Zaefferer (2007) that about ten TDS events are required to create EBSD patterns this leads to a spatial resolution of EBSD of 300 nm for 15-keV electrons on Silicon ($Z = 14$). This value agrees very well with the experimental measurements by Tripathi and Zaefferer (2019), who measured about 300-nm resolution for 15-keV electrons on magnesium ($Z = 12$). The interaction volume, i.e. the physical spatial resolution, will be discussed further in Section 6.3.3.

6.3.2.2 Theory of Electron Channeling

EBSD patterns look very similar to ECP (cf., for example, the Figures 6.19b and 6.27b). The reason for this is that the physics of pattern formation is identical, with the only difference that the sequence of the contributing scattering processes (Bragg scattering and TDS) is inverted (e.g. Wilkinson and Hirsch, 1997; Zaefferer and Elhami, 2014). For ECP the primary electron wave is first Bragg scattered, i.e. the primary electron wave is *channeling into* the lattice, and then backscattered, mainly by TDS. For EBSD the primary electron wave is first incoherently scattered by TDS and it is then *channeling out* of lattice by Bragg scattering. The close relation of the two processes is illustrated in Figure 6.20 (which will be discussed in more detail further below). Since the process of electron channeling is more intuitively understandable it is reasonable to resort to this process first in order to understand the physics of EBSD.

In *electron channeling* (also called *channeling-in*), displayed in Figure 6.20a, a plane electron wave (corresponding to a primary beam electron) enters into a crystal and forms, by means of Bragg scattering, a standing electron density wave field, coherent (i.e. in a particular phase relationship) with the atomic lattice. This wave field is schematically sketched in Figure 6.21 as a wave and color pattern. Wave maxima, red-colored, indicate high electron densities, blue-colored minima low densities. The electron density distribution changes in a characteristic manner with the incidence angle of the primary electron wave onto the crystal. This distribution may be calculated using dynamical electron diffraction theory as described in a bit more detail further below. When the incident electron beam travels along a

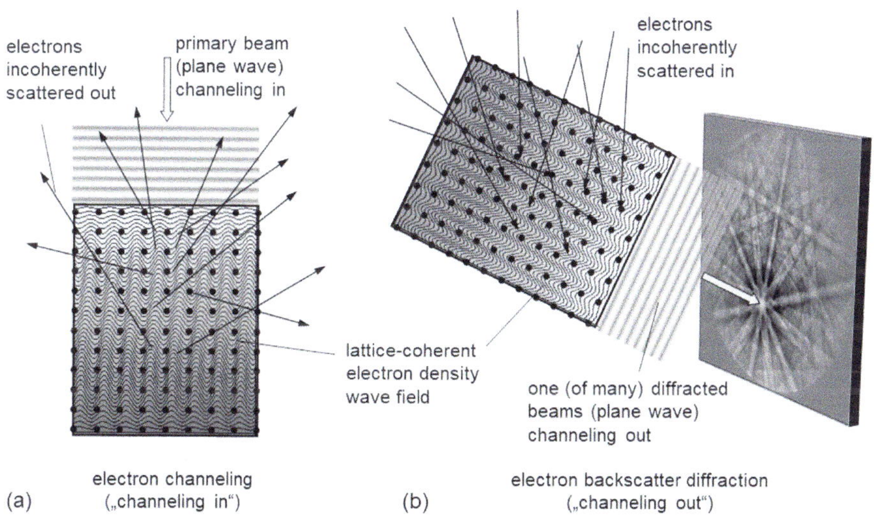

FIGURE 6.20 Comparison of (a) electron channeling and (b) electron backscatter diffraction. Bright gray features indicate plane electron waves traveling in vacuum. The wave features indicate the electron density wave field inside of the crystal. The black arrows indicate incoherently scattered electrons.

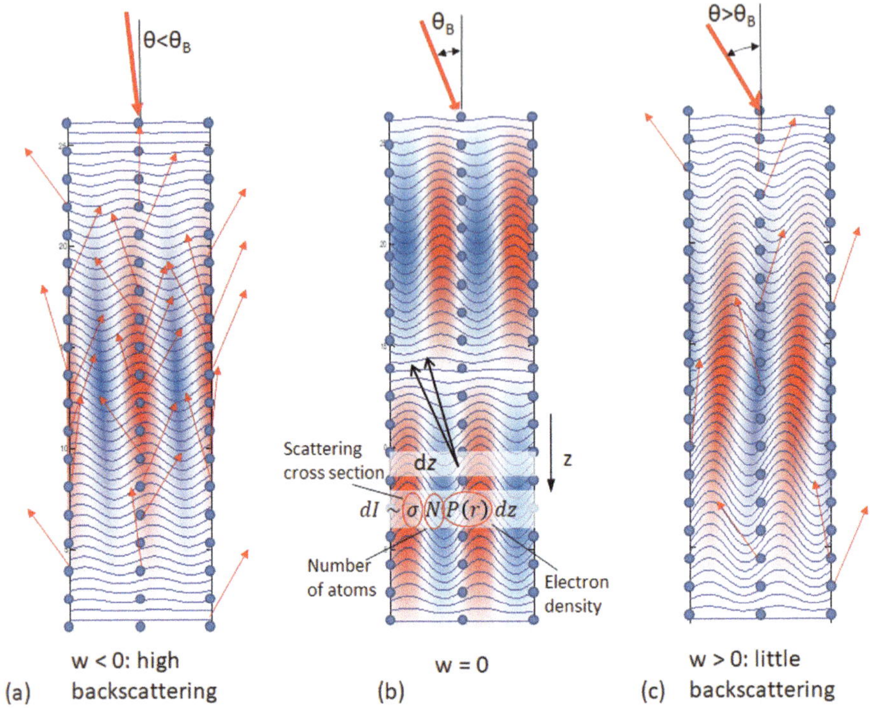

FIGURE 6.21 Electron wave function for (a) the backscattering case, (b) the exact Bragg condition, and (c) the channeling case. Red indicates a high value of the wave function (i.e. a high electron density), blue a low density. The red arrows symbolize the electrons being scattered out of the coherent wave field by thermal diffuse scattering or another scattering process. They finally form the (diffuse) BSE signal used for ECCI. In (b) the integration is plotted into the graph. Integration occurs over thin slices *dz* parallel to the beam entry surface.

direction corresponding almost exactly to the Bragg angle of a set of lattice planes (Figure 6.21b and c), the highest electron density is found *in between* the atomic columns (case "C" (channeling)). In contrast, for the case that the electron beam travels at angles smaller than the Bragg angle of a set of lattice planes (parallel or close to parallel to the lattice planes) the highest electron density is found *at* the atomic columns (case "B" (backscattering)) (Figure 6.21c). For the exact Bragg case the distribution of maxima is exactly balanced. Next, one has to realize that the electrons in the crystal lattice do not only interact with the atoms by coherent Bragg diffraction, but also by TDS (i.e. they are scattered by the thermally vibrating atoms) whereby they lose a bit of energy but, more importantly, they lose their phase relation with the incoming electron wave and significantly change their direction (Wang, 2003; Zaefferer, 2007). The amount of this incoherent scattering is proportional to the density of electrons at the atomic nuclei and, additionally, depends on the scattering power of the respective atoms where usually heavy atoms have stronger scattering power than lighter ones. For abovementioned case C, where the highest electron density is in between the atoms, very little incoherent scattering occurs. The electron wave field thus propagates almost un-attenuated through the crystal. This condition is called *electron channeling*. Because the amount of backscattered electrons is low in this case, a backscatter detector placed somewhere outside of the sample would record only a low signal. In case B, in contrast, intense *backscattering* will occur and a strong signal will be recorded.

The electrons that are recorded on the detector are not the electrons that channel through the crystal, but those that have been scattered out of the channeling wave field by some sort of incoherent scattering process. These electrons, therefore, may show a very broad energy distribution (from 0 to almost the primary beam energy) and their spatial distribution is simply a cos-φ distribution (φ being the deviation from the surface normal), typical for point source illumination below the surface. They do not form a diffraction pattern, in particular. The appearance of a diffraction pattern in an ECP (which is a pattern constructed out of many individual intensity measurements for different beam incidences) is because every point in this pattern corresponds to a different beam incidence angle, i.e. to a different channeling condition. For every point the total backscattered intensity is recorded. This situation is sketched in Figure 6.22, where the backscattered intensity is plotted over the beam

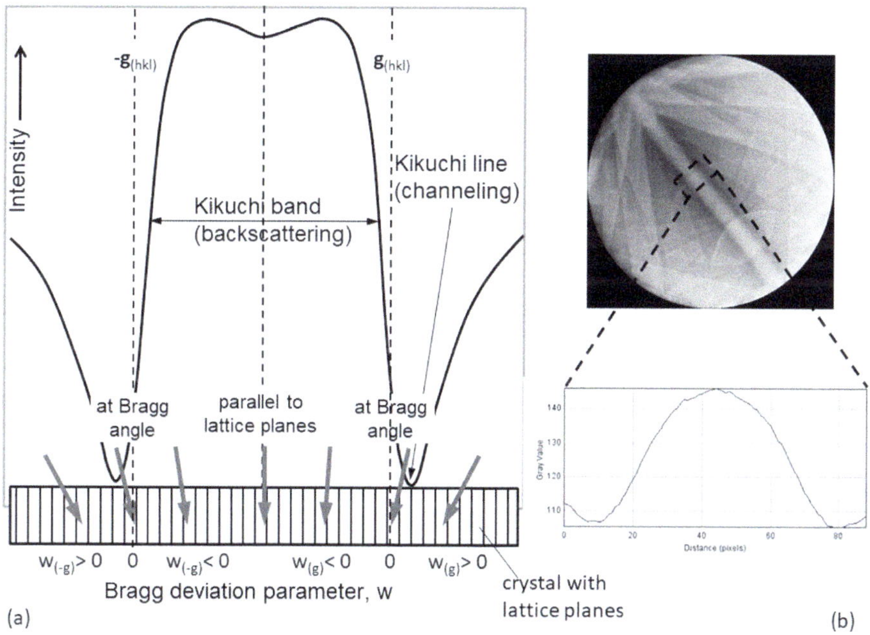

FIGURE 6.22 Kikuchi band profile according to Equation 6.13 calculated from integration of the equation indicated in Figure 6.21. (a) Calculated band profile for different beam incident directions indicated by the Bragg deviation parameter, w. Note that two parameters are plotted, one for the positive, one for the negative reciprocal vector of a set of lattice planes. The two curves overlap in the middle for negative values of w (for situations when the primary beam is close to parallel to the lattice planes). The gray arrows indicate the beam direction relative to the crystal lattice planes. (b) Experimental electron channeling pattern with one typical beam profile being displayed. Note the bright Kikuchi band enveloped by the dark Kikuchi lines.

incidence angle. In Figure 6.22a a one-dimensional profile through a diffraction pattern is plotted, together with a sketch at the bottom of the figure indicating the electron beam inclination relative to the crystal lattice. In Figure 6.22b the two-dimensional distribution in a real ECP is shown. Every dark line in the image corresponds to the mentioned cases of type C, the bright bands to cases of type B. In the next section the calculation approach for the here displayed intensity profile will be outlined.

In order to understand the dependence of the backscattered intensity on the primary beam inclination on a quantitative basis, one first needs to calculate the distribution of the electron wave field in the crystal. Details of the mathematical treatment are given by Reimer (2013a). Zaefferer and Elhami (2014) follow exactly these lines but give a number of further intuitive explanations. Here we will not repeat the mathematical treatment but only describe the principle steps and the final result. The basic principle of this calculation is to find an electron wave function that solves the Schrödinger equation for high energy electrons (i.e. the primary beam electrons) propagating in the charge field of a crystal. The Schrödinger equation is a partial differential equation describing the energy balance between kinetic and potential energy of electrons in the crystal. A suitable solution of this equation is found by a Bloch-wave Ansatz with elementary wavelet ("Bloch") functions (see Section 6.3.2.3 for more mathematical details). The final solution of the Schrödinger equation is then given as a summation over the individual Bloch wavelets with different weight factors. For the simplest case, when only one set of lattice planes contributes to the wave field formation, one obtains an analytical solution for the Schrödinger equation. In TEM this case is called the *two-beam case*, because only the forward-diffracted beam with wave vector $\mathbf{k}_0$ (the "primary" beam) and one further deflected beam (Bragg-diffracted on a particular set of lattice planes) with wave vector $\mathbf{k}_g$ appear in the TEM diffraction pattern. The mathematical description of the electron wave function will not be given here in order to keep the text simple. Instead, we display in Figure 6.21, the wave fields, calculated for different incidence directions close to two-beam conditions. Usually, many lattice planes contribute at the same time to the diffraction process and the calculation of intensities requires to solve a rather large interaction matrix.

From the calculated electron density field, the backscattered intensity is obtained assuming that the local backscattered intensity is proportional to the local electron density, the local number of atoms, and the scattering power of the local atom cores (sketched in Figure 6.21b). The total backscattered intensity, $\Delta\eta$, is then calculated in dependence of the beam direction, indicated by a parameter w (see further below), by integration of these local incoherent backscattering events over the total penetration depth of the primary electron wave into the crystal. For the two-beam case the result of these calculations is a surprisingly simple equation:

$$\Delta\eta(w_g) = \frac{N\sigma_B}{2\pi}\xi_0'\left(-\frac{(w_g + \xi_0'/\xi_g')}{1 + w_g^2 - (\xi_0'/\xi_g')^2} + \frac{w_g}{1 + w_g^2 + \left[(1 + w_g)(\xi_0'/\xi_g)\right]^2}\right) \tag{6.13}$$

Here N is the number of atoms per unit volume, and σ_B is the cross section for backscattering of the atoms in the volume. ξ_0' and ξ_g' are the absorption distances, whose values are tabulated or have to be determined based on quantum mechanical calculations.

Note that the calculations are not done directly in terms of incident angles but, instead, in dependence of the Bragg-deviation parameter, w, which is zero at the exact Bragg angle, negative for values smaller than the Bragg angle (inside of a Kikuchi band) and positive for larger values. The parameter w is calculated according to $w_g = |\mathbf{s}_g| \cdot \xi_g$, where ξ_g is the extinction distance which depends on the respective material, the active diffraction vector, $\mathbf{g}$, and electron energy. $\mathbf{s}_g$ is a reciprocal space vector and indicates the deviation of the primary beam direction from the Bragg diffraction vector $\mathbf{g}_{(hkl)}$. Understanding its meaning requires to resort to Figure 6.23, displaying the Ewald construction (see Section 6.2.1) for a case close to the Bragg diffraction case. The conventional diffraction geometry is drawn in thick solid lines for the exact Bragg case: the primary beam direction, $\mathbf{k}_0$, and the diffracted outgoing beam, $\mathbf{k}_g$, are related according to $\mathbf{k}_g - \mathbf{k}_0 = \mathbf{g}_{(hkl)}$. The situation for a slightly deviating incoming beam direction, $\mathbf{k}_0'$, is plotted by thin broken lines (note that the deviation angles in Figure 6.23 are drawn much larger than what they are in reality): the rotation of the incidence direction leads to a similar rotation of the Ewald

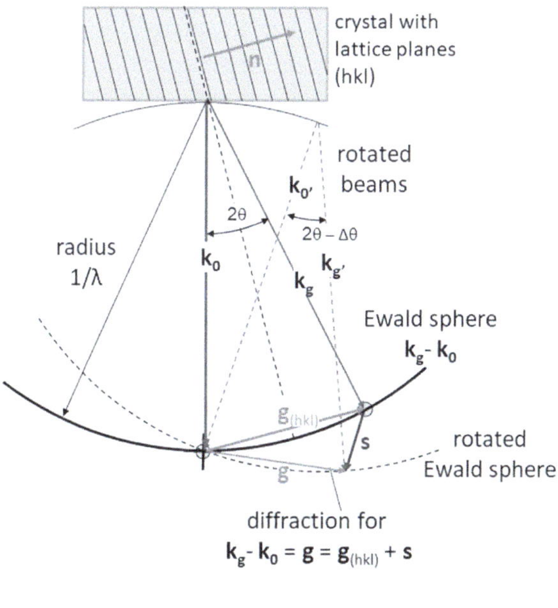

FIGURE 6.23 Ewald sphere construction for the profile across a Kikuchi band. For different positions of the profile the electron beam, $\mathbf{k}_0$, travels along different incident directions. Together with the primary beam the Ewald sphere rotates. For the exact Bragg position (close to the dark line of a Kikuchi band) the diffraction vector $\mathbf{g}$ superimposes with the reciprocal lattice vector $\mathbf{g}_{(hkl)}$. This situation is indicated by bold lines. For small deviations from the Bragg position, given by the deviation vector, $\mathbf{s}$, $\mathbf{g}$ is different from $\mathbf{g}_{(hkl)}$. Note that $\mathbf{g}$ is still on the rotated Ewald sphere. This situation is indicated by thin, broken lines.

sphere about its center $\mathbf{g} = 0$. The active diffraction vector, $\mathbf{g}$, continuous to reside on the Ewald sphere, as indicated in the construction. It now does no longer coincide with $\mathbf{g}_{(hkl)}$, meaning it is no longer exactly in Bragg diffraction. The difference vector between $\mathbf{g}$ and $\mathbf{g}_{(hkl)}$ is the abovementioned deviation vector, $\mathbf{s}_g$, pointing in the direction of the current primary beam direction $\mathbf{k}_0'$. It is plotted as a blue vector in Figure 6.23. $\mathbf{s}$ is considered positive when it points parallel to $\mathbf{k}_0'$ and negative when it is anti-parallel.

An example curve $\Delta\eta\left(w_g\right)$ is plotted in Figure 6.22a. This curve is actually an overlap of two mirror-symmetrical curves, one for $-\mathbf{g}_{(hkl)}$ on the left side and one for $+\mathbf{g}_{(hkl)}$ on the right side. The total curve corresponds to a typical intensity profile of a Kikuchi band in an ECP. A typical experimental band profile is shown in Figure 6.22b. The intensity is minimum at incident angles close to the Bragg angle ($w = 0$, $\mathbf{s} = 0$), while it is very high for angles almost parallel to the lattice planes ($w < 0$, $\mathbf{s}$ pointing in negative direction). This constitutes the typical Kikuchi band described in the introduction to this chapter, consisting of dark edges (close to the Bragg angle) and a bright central band.

6.3.2.3 Dynamic Theory of Electron Backscatter Diffraction

Based on the ECP formation, it is now straightforward to understand the formation of EBSD patterns by applying the reciprocity theorem. This theorem suggests that the physics of all electron interactions remains the same when the propagation direction of the electron wave is inverted. This is illustrated in Figure 6.20a and b: in electron channeling the primary beam electrons arrive on the surface of the crystal as a plane-wave. These electrons then channel into the crystal in form of a standing wave pattern and are finally backscattered by TDS; for EBSD the electron scattering processes are exchanged in order. First the electrons are scattered by TDS, before they channel out into well-defined directions. For every direction a plane wave is leaving the crystal and arrives at one position on the EBSD detector. Each of these plane waves is associated with a lattice-coherent electron density wave field formed inside of the crystal. These wave fields are "supplied" with the TDS electrons, which are incoherently and randomly scattered primary beam electrons. The critical point to understand is that the reciprocity theorem claims that *the probability that an electron is scattered into one of the individual coherent wave fields in EBSD formation is the same as the probability that an electron is scattered out of the wave field in electron channeling: it depends on the scattering power of the atoms that scatter the primary electron wave and on the probability (or intensity) of the outgoing wave field at the respective scattering position.* The total intensity obtained for a given exit direction, i.e. for a given position on the EBSD detector, is obtained by integration of the wave field over the total exit depth of the electrons. The intensity profile that is created by this *channeling-out* process is, in principle, identical to the one created by the channeling-in process, displayed in Figure 6.22a: for angles close to the Bragg angle of a lattice plane one obtains low intensities, the Kikuchi lines, while for angles smaller than the Bragg angle high intensities are obtained. These latter constitute the bright Kikuchi bands.

The quantitative description of this process requires, as for the channeling-in case, to calculate the electron wave field which resides in the crystal using a Bloch-wave approach. In the case of channeling-in there is only one wave field to calculate which fits with the primary beam direction. For EBSD, in contrast, there is a wave field for every outgoing beam direction **u**, i.e. for every pixel on the EBSD detector. The wave field that forms for every direction **u** in the crystal is supplied by electrons through TDS of the primary electron beam. The integration of the wave field is, therefore, over the ingoing wave, not over the outgoing as is the case for electron channeling. The mathematical description of the channeling-out process has been worked out in a number of papers by Allen, Rossouw and co-workers (see Rossouw et al., 1994, and the literature cited therein). The work of Rossouw was taken further by Winkelmann and applied to the backscattering process (Winkelmann et al., 2007). From this latter work we reproduce only the final equations, describing the intensity scattered in a crystal direction **u**. Understanding these equations requires to get deeply familiar with the mathematical details of dynamical diffraction theory (see the textbooks on electron diffraction theory cited at the beginning of this chapter), which we will not attempt here. Nevertheless, we try to briefly line out the background of these equations. First, the electron wave function at every position **r** in the crystal, $\Psi(\mathbf{r})$, is determined by solving the Schrödinger equation for high energy electrons traveling in the periodic potential of the crystal. This is done using the Bloch wave approach, which results in the description of the wave function by a sum of Bloch waves. The so-determined wave function has the following form:

$$\Psi(\mathbf{r}) = \sum_j c_j \exp\left(i\mathbf{k}^{(j)} \cdot \mathbf{r}\right) \sum_g C_g^{(j)} \exp(i\mathbf{g} \cdot \mathbf{r}) \tag{6.14}$$

with $\mathbf{k}^{(j)} = \mathbf{K} + \lambda^{(j)}\mathbf{n}$ being the wave vector of the Bloch wave (j) and $\mathbf{K}$ being the wave vector of the high energy electron inside of the crystal (which is slightly different from that in vacuum because of the presence of an average crystal potential). **n** is the surface normal vector.

Once the Schrödinger equation has been solved the weights of the individual Bloch wave components c_j and $C_{\mathbf{g}}^{(j)}$, as well as the values $\lambda^{(j)}$ are known. Using this fundamental solution one now needs to consider that the electron wave function is supplied with electrons, randomly scattered from the primary electron source into the wave field. The probability of scattering into the wave field is, applying the principle of reciprocity, proportional to the electron density of the electron wave at the location of scattering. To obtain finally the total amount of scattering into the electron wave one has to integrate over the total scattering depth in the crystal ranging from t_1 to t_2. The total intensity is the given as the product of the electron wave function with its complex conjugate, $\Psi\Psi^*$:

$$I_{dyn,\,u} \propto \sum_{n,i,j} Z_n^2 B^{ij}(t_1,t_2) \sum_{\mathbf{g},\mathbf{h}} C_{\mathbf{g}}^{(i)} C_{\mathbf{h}}^{(j)} \exp\left(-M_{\mathbf{g}-\mathbf{h}}^n\right) \exp\left(2\pi i (\mathbf{g}-\mathbf{h})\mathbf{r}_n\right) \tag{6.15a}$$

with

$$B^{ij}(t_1,t_2) = c_i c_j^* \frac{\exp\left[2\pi i\left(\lambda^{(i)} - \lambda^{(j)^*}\right)t_2\right] - \exp\left[2\pi i\left(\lambda^{(i)} - \lambda^{(j)^*}\right)t_1\right]}{2\pi i\left(\lambda^{(i)} - \lambda^{(j)^*}\right)} \tag{6.15b}$$

In these equations the (double) sum $\Sigma_{\mathbf{g},\mathbf{h}}\ldots$ runs over all pairs of wave vectors **g** and **h**, which are activated for the beam direction **u** (determined, for example, through an Ewald-sphere construction). The Bloch waves only have contributions at the atom sites $\mathbf{r}_n$ of the n atoms in the unit cell. The double sum indicates the exchange of electrons between the wave vectors **g** and **h** for all wave vector pairs, as it is the basis of the dynamical coupling concept. The Debye–Waller factor $M^n_{\mathbf{g}-\mathbf{h}}$ indicates the blurring of the diffraction conditions by thermal vibrations of the different atoms. Each of the Bloch waves (i) and (j) is multiplied by an individual weight, $B^{ij}(t_1,t_2)$, which represents the amount of electrons scattered into the Bloch wave fields by TDS. As already mentioned, the calculation of these weight factors assumes that the scattering into the Bloch wave field by an atom is proportional to the wave function at that position. Integration over the total scattering depth in the crystal ranging from t_1 to t_2 then leads to the weight factor B.

Based on the equations outlined above, Winkelmann (2009b) developed a computer program for simulation of EBSPs. Figure 6.24a shows a simulated EBSD pattern of aluminum and compares it with a real pattern in Figure 6.24b; the similarities are striking. Note, however, that some characteristic differences exist. One is band asymmetry, which is due to anisotropic illumination on a steeply tilted crystal in the experimental case: almost all Kikuchi bands in the experimental pattern show, similar to TEM Kikuchi patterns, a bright line and a dark line. One example is marked with arrows in the pattern. In contrast, the simulated pattern shows perfect band symmetry with two dark lines.

A computer program for even more realistic pattern simulation has been developed by Callahan and De Graef (2013), who coupled the coherent scattering calculations with incoherent scattering, creating an asymmetric background intensity distribution. The realistic pattern simulations have triggered a number of highly

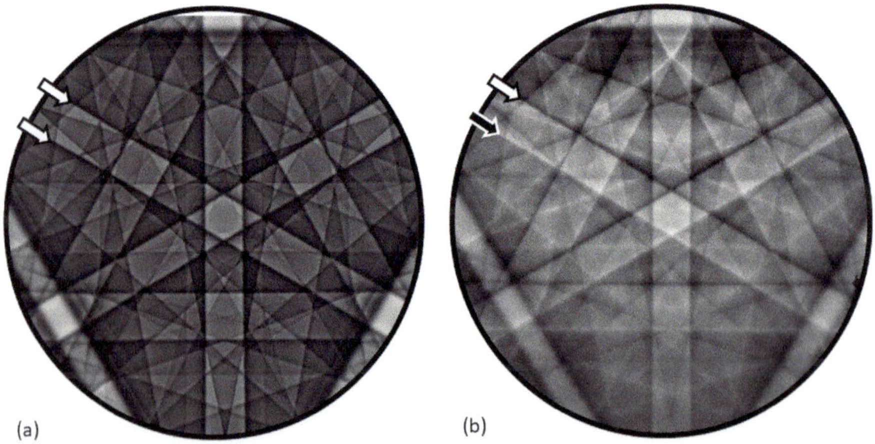

(a) (b)

FIGURE 6.24 (a) Simulated and (b) experimental EBSD pattern of aluminum at 15-kV acceleration voltage. The center pole is the [111] zone axis. Arrows mark differences between the simulated and experimental patterns. The simulation has been done with the program Esprit-S (Bruker) programmed by A. Winkelmann.

interesting applications, including phase determination and orientation determination for crystals without centrosymmetry. For the latter a new indexing method, called dictionary indexing (De Graef, 2020), is used which compares experimental patterns with realistically simulated ones. These applications will be introduced and discussed in Section 7.4.3.

6.3.3 Physical Spatial Resolution of EBSD

The most serious limitation of EBSD-based ORM is its spatial resolution, i.e. its lateral resolution measured on the surface of a sample and its depth resolution concerning the distance under the sample surface from which the patterns originate. The resolution depends, among other things, on the atomic number, density and constitution of the investigated material, on the acceleration voltage of the primary electrons, on the beam profile, and on the software used for pattern analysis. A number of authors have studied the lateral resolution but confusion has arisen due to the fact that two different definitions have been used. The *physical resolution* measures the size of the *interaction volume* from which the pattern is actually generated. This indicates, for example, how far away from a large-angle grain boundary one obtains diffracted intensities from both crystals. The *effective resolution*, on the other hand, indicates how accurately an ORM system may resolve a large-angle grain boundary using software algorithms to deconvolve overlapping patterns. The effective resolution depends on the physical one but is usually better (i.e. smaller). Nevertheless, for heavily deformed or nanocrystalline materials or for small-angle grain boundaries, where the crystals on either side create very similar patterns which cannot be deconvolved by software, the physical resolution is the relevant one. In the present section we will discuss the physical resolution; the effective resolution will be addressed in Section 8.2.9.

When defining the lateral resolution one has to realize that the interaction volume is not spherical but—due to the significant tilt of the sample—highly anisotropic. This point is illustrated in Figure 6.25a: the interaction volume of those electrons leading to EBSD formation has a shape strongly elongated into the direction of surface inclination. The resolution thus takes significantly different values in horizontal and vertical direction of the tilted specimen. The vertical values are about a factor three larger than the horizontal ones. A number of papers deal with the determination of the physical, particularly the *lateral resolution*, by both experiments and simulations, e.g. Ren et al. (1998) and Isabell and Dravid (1997). The authors of these papers also mention values on the *depth resolution* (i.e. depth below the surface from where electrons contribute to pattern formation) in the order of 20–50 nm. More recently, Tripathi and Zaefferer (2019) again performed measurements of the physical resolution and compiled the resolution data available in literature in dependence of atomic number and acceleration voltage. Their main results are reproduced here in Figure 6.25b and c. It shows that for high acceleration voltage (30 kV) and low atomic number (Mg, $Z = 12$) the lateral resolution can be as bad as 3000 nm. For small acceleration voltage (5 kV) the resolution on Mg may be a small as 250 nm. For tungsten, in contrast, with $Z = 74$, the authors found that the resolution does not vary much over the investigated voltage range and amounts to approximately 50 nm.

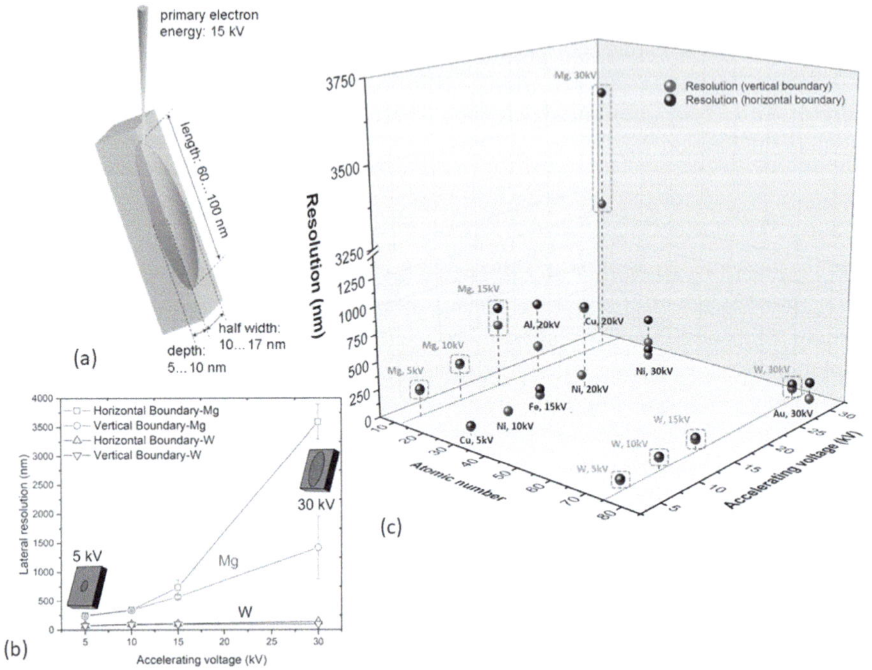

FIGURE 6.25 Physical spatial resolution of conventional EBSD. (a) Sketch of the shape of the electron-matter interaction volume. Values are for Fe. (b) Results of resolution measurements on Mg and W for different acceleration voltages. The little three-dimensional sketches represent the shape of the interaction volume. (c) Literature values on spatial resolution as a function of atomic number and acceleration voltage. Maximum values are ~50 nm (W, 5 kV) ··· 3 μm (Mg, 30 kV). (Adapted from Tripathi and Zaefferer, 2019.)

There have been a number of attempts to use classical Monte Carlo (MC) electron trajectory simulations to understand the origin and properties, particularly the lateral and depth resolution and their Z-dependence, of EBSD (Dingley and Field, 1997; Ren et al., 1998; Pouchou, 2004). Careful analysis of their results leads to the conclusion, however, that BKP are formed by mechanisms different from those assumed in classical MC simulations. Indeed, these simulations all do not take into account interaction cross sections for TDS, which is, however, the most important inelastic scattering mechanism that contributes to EBSD pattern formation.

The discussion above shows that a serious improvement of the physical spatial resolution can be obtained by reduction of the energy of the primary electrons, thus reducing the size of the interaction volume. For ORM, the reduction of the primary electron energy leads to a few problems that have to be solved: first, the currently used phosphors for electron detection are optimised for energies >15 keV and are, therefore, not sensitive enough to record low energy patterns with a sufficient signal to noise ratio. Also contamination and beam stability disturb increasingly. Finally, with decreasing voltage the Kikuchi lines become wider and show more dynamic effects, meaning that the band intensities are reduced and more smeared out. Solutions around these problems appear to be low-voltage sensitive direct electron detection cameras (see Section 8.2.5) and pattern analysis by a pattern matching approach (Section 7.4.3).

6.3.4 Electron Channeling Techniques

As explained in the section on the dynamical theory of electron diffraction (Section 6.3.2), electrons may efficiently channel through a crystal when their traveling direction is close to the Bragg angle of a set of lattice planes. This case is usually called "Bragg diffraction" when working in transmission and "Channeling" when observing the BSE signal. When strong channeling occurs the amount of backscattered electrons is very low. In contrast, when the primary electron beam travels along a direction exactly or closely parallel to a set of the lattice plane or parallel to a zone axis of the crystal lattice where several lattice planes intersect, the backscattering signal is particularly high. This dependence of the intensity of BSE on crystal orientation can be observed and exploited in an SEM in two ways, namely ECP and ECCI. Both modes are discussed separately in the next sections.

6.3.4.1 Electron Channeling Patterns (ECP)

To produce ECP a flat crystalline sample is placed horizontal or close to this in the SEM and the BSE signal is recorded. The geometric setup is displayed in Figure 6.26a. The incident beam is now rocked about a pivot point on the sample and

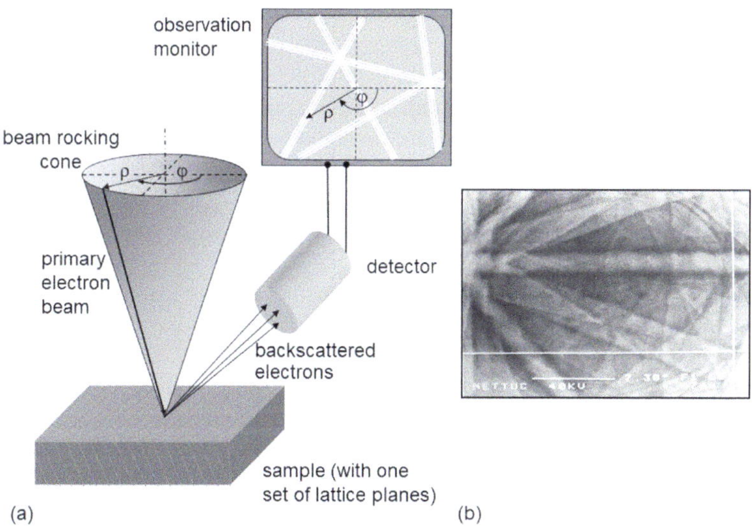

FIGURE 6.26 An ECP pattern and its recording principle. (a) Schematic drawing of the technical formation of an ECP: the beam is rocked about a relatively large solid angle (<10°) with the pivot point being on the sample surface. To each beam incidence the intensity is recorded and displayed on the screen in polar coordinates. (b) An ECP recorded from copper. (From Schwarzer and Zaefferer, 1994b.)

the BSE intensities are displayed in dependence of polar coordinates of the beam rocking. As a result, a Kikuchi-diffraction type pattern (cf. Section 6.1.2) is produced on the screen, as shown in Figure 6.26b. Such patterns have been called electron channelling patterns (ECP) by Coates (1967). The dark lines correspond to channeling lines, where the beam runs parallel to the Bragg angle of a set of lattice planes. The angular distance between two almost parallel lines (they are, in fact not exactly parallel because they are conical sections) is twice the Bragg angle. The centers of the bright bands, bands of strong backscattering, can be considered as the gnomonic projection lines of the different lattice planes.

Although the geometry of ECP and BKP is rather similar due to their reciprocity relationship (see Section 6.3.2), ECP patterns show more details than BKP, because the energy spread in ECP is only due to energy spread of the primary beam while in BKP it is due to energy loss during TDS (e.g. Wilkinson, 1997). Also, the angular intensity distribution in ECP is more homogeneous because it is determined by the primary beam intensity. Wilkinson and Hirsch (1997), for example, mention the occurrence of [12 12 0] lines in an ECP of silicon while, at maximum, [6 6 0] lines are visible in an EBSD pattern. Due to the fine details visible in ECP, they are very well suited for the determination of lattice constants from the width of the Kikuchi lines or for the precise determination of crystal orientations. On the other hand, the sharpness of the lines in conjunction with their relatively low contrast makes ECP very sensitive to lattice defects: each change of orientation inside of the illuminated area, for example due to strain fields of dislocations, leads to a blurring of the lines and to their rapid disappearance.

ECP patterns are not much in use nowadays, simply because most modern microscopes do not allow large-angle beam tilts and because EBSD is a much more convenient method to obtain diffraction information in the SEM. Also the spatial resolution of ECP, which is given by the size of the pivot point on the sample, is usually at least in the order of 1 μm which is at least an order of magnitude worse than that of EBSD. The relatively low spatial resolution also leads to the fact that only materials that have low defect densities produce analyzable diffraction patterns. In those cases where a microscope allows angular scanning the scanning angle is usually limited to about 10°, leading to a total angular range of 20° for the diffraction pattern which makes the pattern difficult to analyze by automatic means. Interested readers find a detailed survey on ECP in Joy et al. (1982).

6.3.4.2 Electron Channeling Contrast Imaging

While ECP is little used nowadays, ECCI is a technique that receives increasingly more attention (Simkin et al., 2003; Zaefferer and Elhami, 2014; Gianola et al., 2019; Gutierrez-Urrutia, 2019). For ECCI the beam is scanned in a conventional manner over the sample and the orientation-dependent BSE signal is recorded. As mentioned above, and discussed in depth in Section 6.3.2, a crystal allows deep electron channeling if the electron beam impinges at the Bragg angle of a set of lattice planes onto the crystal. In this case the backscattering signal is low and the respective area appears darker than differently oriented areas or grains. For lattice planes that have a high structure factor the channeling effect is usually particularly intense and, thus, creates particularly low intensity minima. These planes are often (but not always)

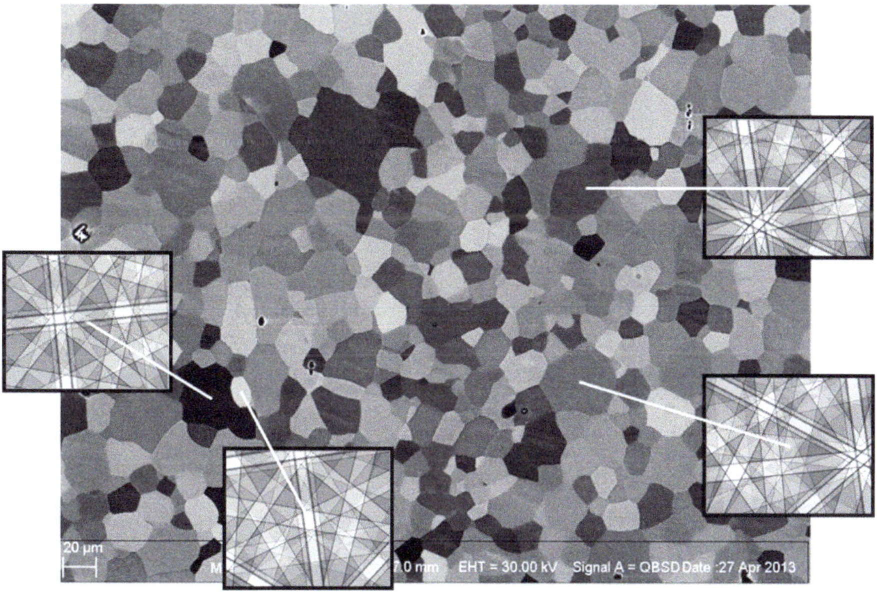

FIGURE 6.27 An electron channeling contrast image of an Fe-Si sample with bcc crystal structure. Simulated Kikuchi patterns indicate the illumination conditions for selected grains. For very bright grains the primary beam direction corresponds to the direction of a low-indexed pole. For very dark grains the beam direction runs parallel to a low-indexed Kikuchi line (i.e. it follows the Bragg angle of a low-indexed lattice plane).

low-indexed planes. Figure 6.27 shows an ECC image obtained from a recrystallized grain structure of a stainless steel. Different grains appear in different brightness, in accordance with the EC patterns displayed. Note that these patterns have been simulated from the crystal orientations obtained for the grains at the different positions using EBSD mapping.

If a crystallite or an area of it is oriented such that it fulfills channeling conditions, i.e. it appears dark in the BSE image, then every defect that changes the atomic order or the angle of the lattice planes relative to the primary beam (e.g. dislocations or stacking faults) will lead to a deviation from channeling to backscattering and the defect will become visible as bright feature. A stacking fault, for example, is a plane in the lattice where the stacking order changes as indicated in Figure 6.28. The part of the crystal that is below the stacking fault is shifted by a certain fraction of the lattice cell distance. It thus shows a phase shift of the channeling wave field and is therefore in backscattering mode. As a consequence, the stacking fault appears as a bright area in the BSE image as shown as an example in Figure 6.29. This figure shows four different stacking faults in a crystal of a Co-based superalloy (Makineni et al., 2018). The ECP in the upper corner indicates that the (002) lattice plane is the channeling plane. This ECP is simulated using the crystal orientation known from an EBSD map. Two features of the image should be pointed out here. First, the stacking fault image is very bright on one side and fades away to the other. This is due to the

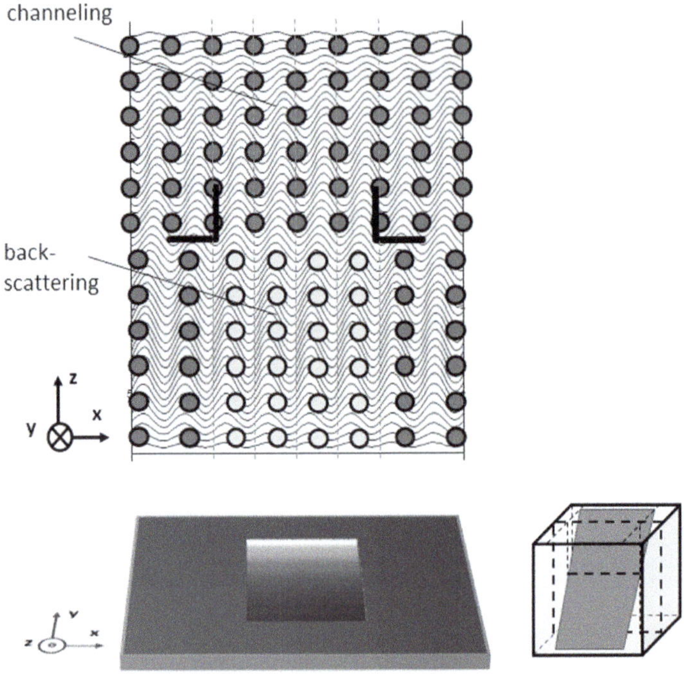

FIGURE 6.28 Schematic representation of the Bloch wave theory for the formation of defect contrast for a stacking fault bound by two Shockley partial dislocations. The contrast of the defect is observed on the top surface (z-normal). The position of the defect in the sample is indicated in gray in the overview box. The atoms below the stacking fault, indicated by bright gray shading, are shifted by half an atom column distance in x-direction with respect to the atoms in the perfect crystal (dark). The Bloch wave field enters the perfect crystal in channeling conditions. Below the stacking fault the conditions change to strong backscattering. (Adapted from Zaefferer and Elhami, 2014.)

increasing depths of the stacking fault below the surface, which reduces the number of backscattered electrons. Second, the SF image does not fade continuously but with intense intensity oscillations. These oscillations, also called extinction fringes and well known from TEM, are a beating effect of the electron wave in the lattice. Detailed explanations are found in all textbooks on TEM but also in Zaefferer and Elhami (2014) for the ECCI case. Figure 6.29 also shows the stereographic projection (SP) of the {111} and {100} planes for the given crystal orientation. The stacking fault traces and the corresponding perpendicular large circles are plotted into this SP (see Section 11.2.2 on trace analysis). From trace analysis, it is found that all four {111}-type SFs are present in the imaged crystal. The SP also allows, for example, determination of the inclination angles under which these SF stand to the surface. In contrast to TEM the inclination angle is unambiguous because top and bottom of the feature can be clearly distinguished. Besides SFs, the image also shows dislocations which pile up at the γ–γ' boundaries, though they are not so well visible.

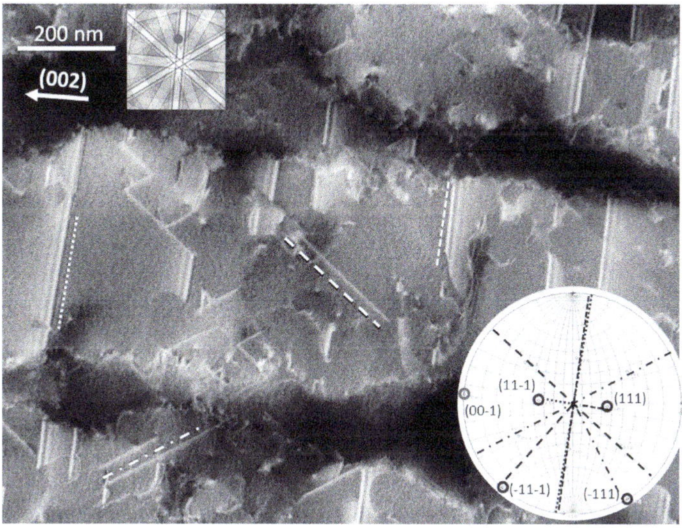

FIGURE 6.29 Stacking faults observed by ECCI in a Co-based superalloy. The imaging diffraction vector is **g** = (002). All four possible {111} stacking faults are visible at different inclinations that can be read out from the stereographic projection. The SF traces are plotted as broken lines into the stereographic projection. (Courtesy of S. Makineni.)

To obtain the required channeling conditions for defect observation, as in TEM the crystal must be tilted such that a suitable set of lattice planes is illuminated under Bragg angle. This is achieved by measuring the orientation, usually by EBSD at a high sample tilt angle (see Section 6.3.1), then tilting the sample back to close to horizontal position and using a pattern simulation software (e.g. TOCA by Zaefferer, 2000) or a software for manipulation of stereographic projections to determine the correct tilt angles. If EBSD is used to determine the crystal orientation, then the use of a software is imperative because, in contrast to TEM, the diffraction pattern is not visible at the tilt angle of microstructure observation.

The short example above illustrates that ECCI allows observation of extended defects on bulk samples in a very similar manner as TEM, though with reduced contrast and resolution. Table 6.2 concludes this section by a comparison of some aspects of ECCI and TEM.

6.3.4.3 Electron Channeling Orientation Mapping

Based on ECCI, Lafond and Langlois and co-workers developed a new method for orientation determination and orientation mapping, named electron channeling orientation mapping, eChord (Lafond et al., 2018). Although this method has not yet received much attention, it has the potential to measure large orientation maps in a much shorter time than EBSD (Lafond et al., 2020), and it may allow measurement of high-resolution orientation maps on light materials.

The technique is closely related to the conical scanning technique in TEM, described in Section 6.2.5. The basic principle of this technique is the recording of

TABLE 6.2
Comparison of Some Key Aspects of ECCI and TEM Imaging

	ECCI	TEM
Lateral resolution	8 ··· 10 nm	1 nm
Depth of observation	50 ··· 100 nm	100 ··· 200 nm
Observable area	10^8 μm²	10^4 μm²
Sample	Bulk, defects in a potentially more natural surrounding	Thin foil, defects may be modified due to the mechanically weak structure
Sample handling (control of diffraction conditions)	Tilt-rotation stage, reaches a circle of 25° diameter around sample normal	Double tilt stage reaches a circle of 25° diameter around sample normal (identical to ECCI)
Possibility of in-situ experiments	Easy: deformation (tensile, bending compression, shear, monotonic and cyclic), heating, cooling, oxidation and other gas reactions, possibility to perform interrupted ex-situ experiments (e.g. corrosion)	Usually difficult: deformation (tensile, bending monotonic), heating, cooling, oxidation, and other gas reactions Ex-situ experiments are hardly possible because of delicate sample handling

ECC images of a fixed area on the sample for different sample tilt and rotation angles. Note that this is different from the conical scanning technique: for the latter technique, the beam is tilted, for eChord the sample is tilted and rotated. For every point of these images the intensities vary according to the respective channeling conditions. From all collected ECC images, for example for one tilt angle of 10° and 360° rotations in 1° steps around the sample normal, all intensities along one circle through the (virtual) diffraction pattern are recorded for all pixels simultaneously as this is indicated in Figure 6.30. For orientation determination these intensities are then compared with pre-calculated intensities for all potential orientations until a match is found. Before the intensities can be calculated it is clear that the individual images need to be aligned and corrected for distortions. If all this is done, the technique results in orientation maps, very similar to those obtained by EBSD, as shown by Lafond et al. (2018). These authors claim an angular resolution of the technique better than or equal to 0.5°. At the same time the ECC images, potentially, show defects, imaged under different channeling conditions, which can directly be used for defect analysis, provided the images are of high enough quality. Alternatively, the images directly indicate the correct tilt angles for good defect analysis.

The technique has the advantage that rather large areas can be recorded with relatively little distortion and in short time. The authors mentioned acquisition rates of 1 hour for 1,000,000 data points (1° rotation step size, 10 s recording time for one image). Furthermore interesting is that no EBSD equipment is necessary. However, a sufficiently accurate stage is required. Most recently Lafond et al. (2021) demonstrated acquisition of highly resolved orientation maps of an Al_2O_3 sample, recorded at 1 kV, which currently appears impossible to do with EBSD.

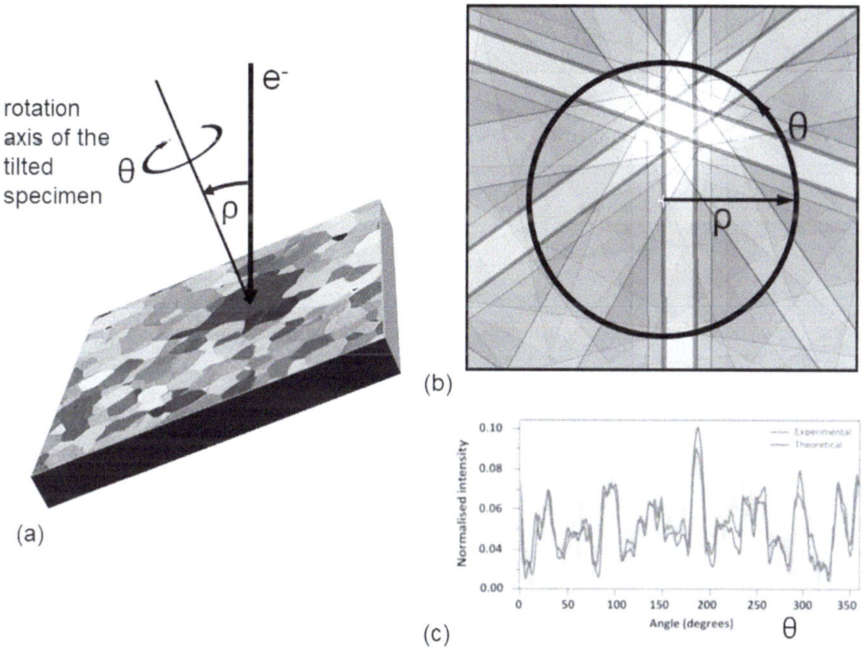

FIGURE 6.30 Schematic of the electron channeling orientation mapping eChord: (a) a sample in the SEM is tilted and then rotated in steps about the tilted axis. For every rotation step an ECC image is taken. (b) For every pixel in the ECC images one virtual diffraction pattern exists (unknown before the investigation). The movement of the sample about ρ and θ probes one circle in this diffraction pattern. (c) The intensity distribution along the probed circle as an example taken from Lafond et al. (2018). By comparison of the experimental profile with precalculated profiles for all possible orientations, the correct orientation is determined.

6.3.5 Micro-Kossel Technique

Although the Kossel technique is slightly exotic and not much used nowadays and although it is not an electron diffraction technique but is based on x-ray diffraction we include this technique here, first for completeness and second because it may have some very interesting applications, namely the absolute measurement of lattice constants, i.e. residual stresses, with an accuracy and lateral that cannot be reached by any other technique.

The basis of the Kossel technique (Kossel et al., 1935) is the diffraction of x-rays at the crystal lattice, where the x-ray source is created by the electron beam and is situated *within* the sample under investigation. For crystallographic studies—including orientation determination—by means of this technique an electron beam is focused on the sample. The electrons excite characteristic x-rays along the beam path. The so-created "point source" inside the crystal emits monochromatic x-ray radiation in all directions of space. In a simple kinematic model of x-ray diffraction, similar to that invoked to describe Kikuchi-type diffraction (see Section 6.2.6), those x-rays

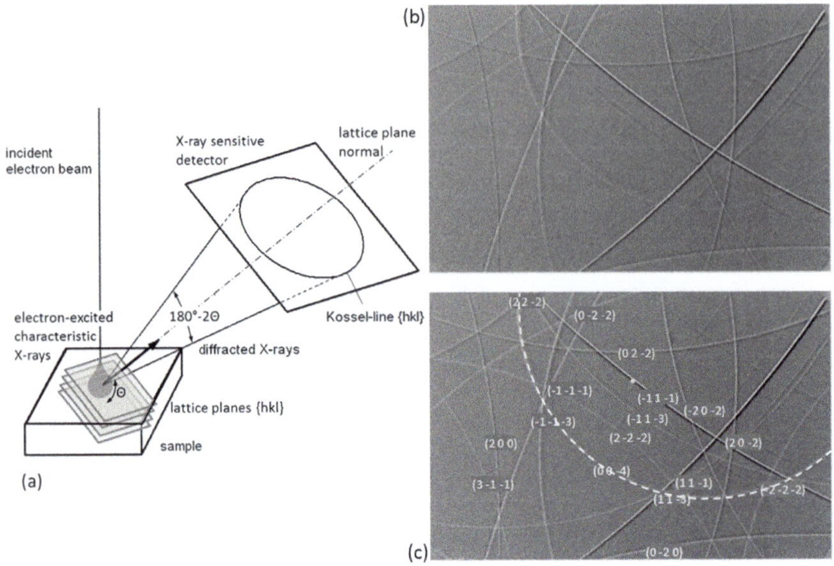

FIGURE 6.31 (a) Schematic of the formation of Kossel cones on one set of lattice planes. The conical section on the detector is selected to be circular. (b) A true Kossel pattern of a Ni-base superalloy created by Ni Kα radiation. (c) same as (b) but with plane indices at the Kossel lines. A circular Kossel cone section is marked as dotted line. (Adapted from Bouscaud et al., 2014.)

impinging on a set of planes under the Bragg angle, θ_B, are diffracted according to Bragg's law and continue to travel in directions, which enclose the Bragg angle with the set of lattice planes. The locus of all Bragg-reflected x-ray beams from one set of lattice planes is a pair of cones with the opening angle 180°–2θ_B as displayed in Figure 6.31a. Compared to electron-diffraction-based Kikuchi patterns, the diffraction angles of x-ray diffraction are usually much larger and, therefore, no almost straight pairs of line appear on the detector, but largely hyperbolic conical sections. An example is shown in Figures 6.31b and c, which are taken from one of the most recent treatments of the technique by Bouscaud et al. (2014).

The diffracted x-ray intensity along these cones can be recorded on an x-ray sensitive film or on an x-ray sensitive camera (e.g. an electron-protected EBSD camera), which is usually placed several centimeters before (reflection geometry) or behind (transmission geometry) the sample within the microscope chamber. In comparison to Kikuchi electron diffraction (Sections 6.2.6 and 6.3.1), the wavelength of the diffracting x-rays used in the Kossel technique is much larger and much more accurately defined. Therefore, all Bragg angles θ between 0° and 90° may occur, so that the resulting projection lines on the detector plane are strongly curved (Figure 6.31b) and significantly sharper than the Kikuchi lines created by electron diffraction.

Kossel et al. (1935) first used this method to achieve diffraction patterns for macroscopically large crystals. To investigate local orientations in microscopic regions (micro-Kossel x-ray diffraction, MKXD), Kossel measurements are usually performed in a microprobe (Peters and Ogilvie, 1965; Bevis and Swindells, 1967; Inokuti and Doherty, 1977) or an SEM (Dingley and Steeds, 1974; Dingley, 1978).

In order to obtain Kossel patterns, the accelerating voltage of the electrons must be about two to three times larger than the critical voltage to produce characteristic x-ray radiation (Section 4.3.1). For some elements Kα radiation may be used, for others Lα or even Mα radiation may be suitable. In some cases, however, the wavelength of the excited characteristic radiation of the sample material is too large to yield reflection according to Bragg's law. The wavelength λ of characteristic radiation is calculated according to

$$\lambda = \frac{hc}{E} \tag{6.16}$$

with h being Planck's constant and c being the speed of light. Inserting for λ the Bragg equation yields:

$$\frac{nhc}{2dE} = \sin\theta \tag{6.17}$$

From this, since the arcsine function cannot take arguments larger than one, it follows:

$$E > \frac{nhc}{2d} \tag{6.18}$$

or, with values for h and c and with $n = 1$:

$$E[\mathrm{eV}] > \frac{619.8}{d\,[\mathrm{nm}]} \tag{6.19}$$

With this, one obtains for a lattice plane spacing of, for example, 0.2 nm that the minimum energy of the characteristic radiation has to be ~3 keV to be able to create a Kossel line. For many light elements, e.g. Al and Si, the Kα radiation is thus not energetic enough to create any Kossel line. But also for heavier elements Kossel lines may be not created. This is because an x-ray line is only excited strongly if the excitation energy is 2–3 times higher than the line energy. That means, in an SEM with a maximum electron energy of 30 keV only lines up to max. 15 keV can be strongly excited. For a number of elements, the Kα line is, thus, too energetic to be excited, while the Lα or Mα radiation is still not energetic enough to fulfill Equation 6.19.

In those cases where no suitable internal characteristic radiation is available, x-radiation of a suitable wavelength can be produced by an external x-ray source that is placed as close as possible to the surface of the sample (pseudo-Kossel technique). For example, a thin layer of a different material can be deposited on the sample surface (Peters and Ogilvie, 1965; Ferran et al., 1971) or the sample can be covered with fine particles of another metal (Ullrich and Schulze, 1972). These layers of particles are then used to create suitable characteristic x-rays.

Although Kossel patterns have been achieved from volumes of only a few microns in diameter (Dingley and Steeds, 1974), the spatial resolution is, in most cases, limited to 5–10 µm, so that only rather coarse-grained samples can be analyzed. In case of the transmission technique, the spatial resolution is correlated with the transmitted sample thickness, which generally results in poorer spatial resolution (Ferran et al., 1971). Only with extremely thin samples a spatial resolution of 10–20 μm has been realized (Bellier and Doherty, 1977). Furthermore, Kossel patterns have a very high background and, consequently, a low signal-to-noise ratio, caused by the non-diffracted x-rays as well as the Bremsstrahlung (continuous radiation). For orientation determination the Kossel technique is therefore not competitive to EBSD. However, the technique allows the determination of lattice constants with unrivaled accuracy at reasonably good spatial resolution. That is why Kossel patterns are highly interesting for analysis of unknown crystal structures (Tixier and Wache, 1970; Ullrich et al., 1992) and for high-precision internal stress measurements (Ellis et al., 1964; Brechbühl et al., 1999), as lattice constants a can be derived with an accuracy of better than $\Delta a/a = 10^{-5}$ (Maurice and Fortunier, 2008; Bouscaud et al., 2014; Morawiec, 2016).

It is noted that since the 1990s, Kossel experiments are also performed using the fluorescent radiation excited by the highly collimated x-ray beam of synchrotron facilities (Section 13.2). Compared to the Kossel technique with electron sources, the synchrotron-based Kossel technique stands out by lower sample heating, larger penetration depth, reduction of the exposure time by several orders of magnitude, and the possibility of measuring in air (Ullrich et al., 1994). Especially, the high beam intensity together with the absence of the continuous radiation leads to an appreciably better signal-to-noise ratio; hence, synchrotron Kossel patterns can be recorded with image plates (Bauch et al., 1999).

6.4 SUMMATION

Electron diffraction in TEM and SEM is the basis of microtexture and microstructure investigations. Over the years a large amount of different techniques have been developed, which are differentiated mainly by the fact whether the diffraction occurs based on a *parallel* primary beam (TEM spot diffraction techniques), on a *convergent* beam (TEM convergent beam patterns, ECP), or on a *point source* in the crystal (transmission and BKP in the TEM and SEM and Kossel patterns in the SEM).

Understanding the geometry and intensity distribution in these different patterns is essential for a comprehensive analysis of them in terms of orientation, phase, defect density, and lattice parameters of the diffracting volume. One of the basics for this is the understanding of reciprocal lattice and the Ewald sphere construction. Furthermore, a basic understanding of the dynamical theory of diffraction is important to be able to deal with the many potentially confusing features that are visible in diffraction patterns.

With respect to their suitability to orientation determination—which is the basis of texture measurements—the techniques differ considerably. In TEM classical SAD patterns fall short of spatial resolution, while Kikuchi patterns allow very accurate and straightforward orientation analysis but quickly blur out in the presence of lattice defects and on too thin samples. The most promising method in TEM is the formation of small-angle convergent beam patterns with or without precession of the primary beam. In Section 7.4 it will be shown that there is a powerful way of automated analysis of these patterns, namely *template matching*. In SEM it is the EBSD or backscatter Kikuchi pattern (BKP) technique which is by far the most powerful and versatile technique, as will be discussed in Chapters 7 and 8. The main limitation of the EBSD technique is its spatial resolution and its demand for relatively high beam current. There is one interesting competing technique, the eChord technique which is based on recording of electron channeling images. This technique has the potential to work at much lower beam currents and at even higher acquisition rates than EBSD. However, it is inferior in terms of spatial resolution, phase analysis capabilities, and measurement of elastic strains and defect densities. The Kossel technique, finally, is an x-ray diffraction technique which is rarely used but offers very accurate local elastic strain measurements.

NOTE

1. The rather small angular distribution of electrons in SADP is due to the strong attenuation of electron scattering with increasing scattering angle. In fact, the Mott formula shows a decay of the scattering factor with $1/\sin^2\theta$, i.e. a decay of the intensity proportional to $1/\sin^4\theta$.

7 Procedures for Orientation Determination from Electron Diffraction Patterns

7.1 INTRODUCTION

In the previous chapter various forms of electron diffraction patterns have been introduced. With the exception of Debye–Scherrer ring patterns these patterns allow the determination of individual crystal orientations which are, in turn, the basis of any quantitative crystallographic analysis, interpretation of diffraction contrast images in transmission electron microscopy (TEM) and scanning electron microscopy (SEM), and, most importantly, of orientation microscopy. In this chapter, algorithms used to determine crystal orientations from diffraction patterns are introduced.

All textbooks on electron microscopy describe the manual procedures for indexing diffraction patterns and extracting orientations from these. These procedures are in principle based on the extraction of interplanar distances and angles from diffraction patterns and the subsequent comparison of these values with values calculated and stored beforehand for the considered crystal structure. For simple crystal structures and highly symmetric orientations (e.g. low-indexed zone-axis-patterns), these procedures are straightforward. They will be shortly discussed in Section 7.2.1. The second part of this chapter will be devoted to a more general approach, which is able to deal with any crystal structure and any type of diffraction pattern. Finally, in the third part, methods for phase and orientation determination will be discussed which are based on comparison of simulated diffraction patterns with experimental. These methods require the realistic simulation of diffraction patterns. These algorithms are called template matching (Section 7.4.2) or dictionary indexing (Section 7.4.3). These "brute force" methods require large computer power for the processing of large amount of images but may be superior to classical methods in terms of crystallographic analysis. A very new, more refined version of dictionary indexing is the so-called spherical indexing which will be shortly treated at the end of this book.

Generally, orientation determination is based on the comparison of the theoretical reciprocal lattice for a given crystal lattice with a set of reciprocal lattice vectors determined from a diffraction pattern. It shall be reiterated here that the reciprocal lattice vectors describe the lattice planes in a crystal lattice (see Section 6.1.1). The

DOI: 10.1201/9781003258339-10

reciprocal vectors stand perpendicular on the lattice planes normal, and their lengths are reciprocal to the lattice plane distances. The comparison may be accomplished vector by vector, which requires that the experimental vectors are extracted from the patterns first. Algorithms for extracting diffraction vectors from diffraction patterns will be discussed in Section 7.2. The comparison procedures, also called indexing procedures, are discussed in Section 7.3. All comparison algorithms require a pre-calculation of the reciprocal lattice for the crystal lattices under consideration. The points that have to be considered there are discussed in Section 7.1.2.

Before going into details of indexing and orientation determination it needs to be repeated that the term "crystal orientation" means that the position of the crystal coordinate system is described with respect to the coordinate system of the sample. The sample itself is described in the microscope coordinate system while the diffraction process is described in the detector coordinate system. The relationships between these different coordinate systems are discussed in the next section.

7.1.1 Coordinate Systems

For a first introduction into coordinate systems see also Section 2.3.1. The orientation of a crystal describes the relationship between a crystal-related coordinate system and a sample-related one. In order to define a crystal orientation we therefore need to define the appropriate Cartesian coordinate systems.

The sample coordinate system, S, is fixed by distinguished directions in the sample which are often prescribed by the production process or the desired application of the material. On a rolled metal sheet, for example, the coordinate system is usually defined by the rolling direction (RD) as x-axis and the normal on the rolling plane (normal direction, ND) as z-axis. The y-axis is fixed appropriately and is in the example parallel to the transverse direction (TD). In the following text matrices are described by bold capital letters, **M**:

$$\mathbf{M} = \begin{pmatrix} m_{xx} & m_{xy} & m_{xz} \\ m_{yx} & m_{yy} & m_{yz} \\ m_{zx} & m_{zy} & m_{zz} \end{pmatrix} \tag{7.1}$$

Vectors are described by bold lower case letters, **u**. Real space vectors, e.g. direction vectors in crystals and samples, are column vectors, i.e.

$$\mathbf{u} = \begin{bmatrix} u_x \\ u_y \\ u_z \end{bmatrix} \tag{7.2}$$

while reciprocal space vectors (i.e. diffraction vectors or lattice plane indices) are line vectors:

$$\mathbf{h} = \begin{pmatrix} h_x & h_y & h_z \end{pmatrix} \tag{7.3}$$

The crystal coordinate system, **C**, is fixed by the crystallographic axes, **a**, **b**, **c**, of the crystal. Naturally, these axes do not set up a Cartesian coordinate system. A Cartesian system, $\mathbf{C}_0$, fixed to **C** is therefore introduced by the usual convention that the x-axis is set parallel to the a-direction and the z-axis perpendicular to *a* and *b*. The transformation matrix between **C** and $\mathbf{C}_0$ is the crystal matrix, **A** (see Section 2.3.1):

$$\mathbf{A} = \begin{pmatrix} a & b\cos\gamma & c\cos\beta \\ 0 & b\sin\gamma & c(\cos\alpha - \cos\beta\cos\gamma)/\sin\gamma \\ 0 & 0 & c\sqrt{1+2\cos\alpha\cos\beta\cos\gamma - (\cos^2\alpha + \cos^2\beta + \cos^2\gamma)}/\sin\gamma \end{pmatrix} \tag{7.4}$$

where *a*, *b*, and *c* are the length of the base vectors and α, β, γ are the angles between *b* and *c*, *c* and *a*, and *a* and *b*, respectively. The transformation occurs according to

$$\mathbf{C}_0 = \mathbf{A}\ \mathbf{C} \tag{7.5}$$

Vectors **h** of the reciprocal lattice (e.g. crystal planes) are transformed into Cartesian coordinates by

$$\mathbf{h}_0 = \mathbf{h}\ \mathbf{A}^{-1} \tag{7.6}$$

Vectors **u** of the real lattice (e.g. crystal directions) are transformed by

$$\mathbf{u}_0 = \mathbf{A}\ \mathbf{u} \tag{7.7}$$

The different coordinate systems involved in orientation measurement and description are displayed in Figure 7.1. Orientations are determined from diffraction patterns, which appear on some kind of observation screen. The position of this measurement system is defined by the detector coordinate system, **D**. The x-y plane of the detector is put parallel to the observation plane while the z-axis stands perpendicular on this plane and points toward the electron source.

The microscope coordinate system, **M**, which is fixed at the sample holder in the microscope describes the position of the sample relative to the microscope and to the pattern measurement system which is fixed at the microscope.

In order to describe a crystal orientation with respect to the sample the orientation is first determined with respect to **D**, subsequently transferred into **M** and then into **S** by appropriate matrix operations. See Nolze (2007) for a good and comprehensive description of the appropriate matrix operations.

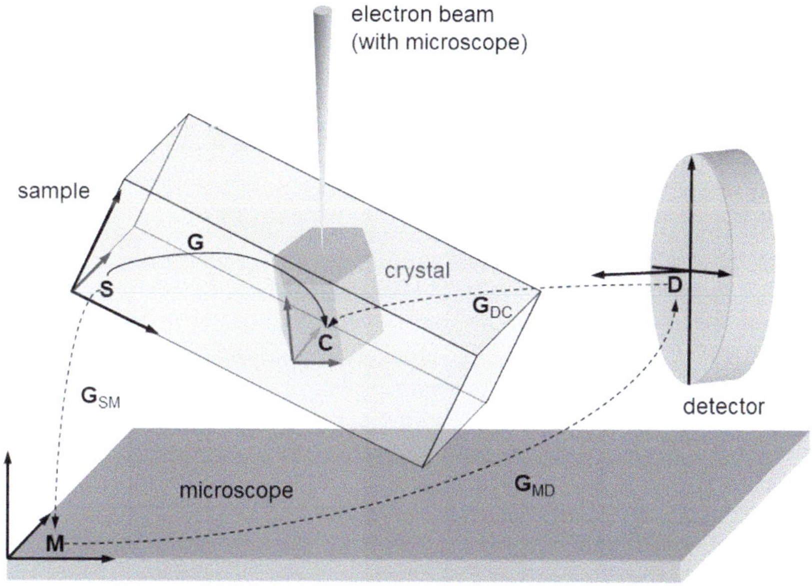

FIGURE 7.1 Definition of coordinate systems required for orientation determination. The orientation of a crystal is given by the rotation matrix the rotates the sample coordinate system S into that of the crystal coordinate system C. In order to determine G = C → S, one measures first the orientation of a crystal with respect to the detector system **D**, $\mathbf{G}_{DC} = \mathbf{C} \rightarrow \mathbf{D}$. The detector system is related to the microscope system $\mathbf{G}_{MD} = \mathbf{D} \rightarrow \mathbf{M}$. Finally the sample is related to microscope system as well $\mathbf{G}_{SM} = \mathbf{M} \rightarrow \mathbf{S}$. Combining all rotations and inverting the resulting matrix results in the orientation matrix.

7.1.2 Calculation of the Reciprocal Lattice

Orientation analysis from diffraction patterns is based on a comparison of measured diffraction vectors with the vectors of the reciprocal lattice, calculated for the crystal structure under consideration. We remind the reader here that the reciprocal lattice vectors are nothing else but the normal vectors on the lattice planes of the crystal, with a length reciprocal to the lattice plane distance. Their vector indices are the Miller indices of the lattice planes. In many computer programs for orientation analysis the list of diffraction vectors is called a "reflector table." The comparison might be based only on the geometry of the vector arrangement or also on their related structure factors, which is determined from the intensity of the diffraction features. In any case, one needs a pre-calculated reciprocal lattice. Most conveniently, the reciprocal lattice vectors are calculated beforehand and stored in a look-up table, which is organized according to the symmetry of the reciprocal lattice.

The length of a reciprocal lattice vector, **h** (which is the reciprocal of the lattice plane distances) is

$$|\mathbf{h}| = |\mathbf{h_o}| = \sqrt{h_{o,x}^2 + h_{o,y}^2 + h_{o,z}^2} \tag{7.8}$$

The angles, α, between two different reciprocal vectors (i.e. between different lattice planes with indices hkl_1 and hkl_2) are calculated in a straightforward manner for every crystal symmetry by

$$\cos \alpha_{hkl_1,hkl_2} = \frac{\mathbf{h}_{o1} \cdot \mathbf{h}_{o2}}{|\mathbf{h}_{o1}||\mathbf{h}_{o2}|} \tag{7.9}$$

As has been described in Section 6.1.1 the reciprocal lattice is the Fourier transform of the real lattice. The Fourier transform conserves the rotational symmetry of the lattice while every maximum in Fourier transform, i.e. every reciprocal lattice point reflects the translational symmetry of the lattice for the respective lattice planes. When calculating the total reciprocal lattice one therefore needs to calculate only the vectors belonging to the fundamental zone of the crystal structure and apply the rotational symmetry of the lattice to these. Every reciprocal vector not only has a position, but also a structure factor assigned to it, which approximately determines the intensity of the respective diffraction spot, line, or band. The calculation of the structure factor is done employing Equation 6.4.

The rotational symmetry of the lattice is, in principal, represented by one of the 32-point groups. However, the diffraction process introduces in most cases a center of symmetry into the Fourier transform, thus, it is in most cases sufficient to use, also for non-centrosymmetric crystals, one of the 11 Laue groups which are those point groups that contain a center of symmetry. In the case of kinematic diffraction conditions the use of the Laue symmetry is justified by Friedel's law which states for these conditions that the diffracted intensities from planes *hkl* are the same as for the planes $\bar{h}\ \bar{k}\ \bar{l}$. For parallel illumination, e.g. for selected area electron diffraction (SAD) patterns, this is even approximately valid in the case of dynamic diffraction as shown by von Laue (1948). Von Laue also showed that in case of Kikuchi patterns the fact that the electron source is situated inside of the crystal leads to differentiation of the line intensities for planes *hkl* and $\bar{h}\ \bar{k}\ \bar{l}$. Also dynamic scattering effects contribute to this differentiation as discussed by Baba-Kishi (1991) and Marthinsen and Hoier (1988). It is, therefore, in principle possible to distinguish point groups with and without center of symmetry or to distinguish enantiomorphic crystals by measuring the intensities of dynamic reflections. Two examples for this will be presented later in Section 7.4.4. The pattern geometry, however, does not show any differences. Convergent beam diffraction patterns are well suited to study the point group of crystals (e.g. Tanaka et al., 1983) but also backscatter Kikuchi patterns (BKPs) have been used for this sake (Baba-Kishi and Dingley, 1989). Nevertheless, in BKP and transmission Kikuchi patterns (TKP) and small angle convergent beam diffraction pattern (SCBED) the differences in intensities of (*hkl*) and –(*hkl*) are very small and difficult to detect. Concluding, for orientation determination from electron diffraction patterns, in particular for automatic orientation microscopy, the limitation to the Laue group symmetry is in most cases sufficient: in SAD patterns there is anyway no distinction possible. Orientation microscopy based on Kikuchi-type patterns, up to know, is in most cases based on the analysis of the pattern geometry only and not on detailed line intensities. However, the development of so-called dictionary approaches, which use image matching between simulated and experimental patterns, allows overcoming this limitation, as will be shown in Section 7.4.3.

The point or Laue-group symmetry describes the geometry of diffraction patterns comprehensively. Symmetrically equivalent reciprocal lattice vectors, called members of the same reflector family, have the same structure factor, F, whose value has to be known for the calculation of the diffracted intensity of a reflection in the diffraction pattern. F is calculated according to Equation 6.4, which requires the knowledge of all atoms in the unit cell with their positions and atomic scattering amplitudes. The atom positions in the unit cell are created from a basic unit of atoms (the basis) by applying the space group symmetry of the structure.

Pattern intensities can, of course, only be calculated if the atom positions in the unit cell are known. Sometimes, however, cases arise where this information is not available. Nevertheless, even in these situations, patterns can usually be indexed uniquely if at least the crystal system, better the Laue group, and the approximate lattice constants are known, because the pattern symmetry is still correct. In turn, missing crystallographic information may be determined from the indexed diffraction patterns.

7.2 EXTRACTION OF LATTICE PLANE (REFLECTOR) POSITIONS FROM DIFFRACTION PATTERNS

All modern techniques for orientation microscopy use a digital pattern recording system. The digital images are transferred to a computer where the coordinates of diffraction spots or lines may be determined either manually by marking selected positions or automatically by digital image processing (DIP) algorithms. Although manual procedures are widely outdated and replaced by sophisticated DIP algorithms there are still niches for them. One of these is the determination of Kikuchi lines from TKP of thin or heavily distorted samples. Also electron backscatter diffraction (EBSD) patterns are sometimes so faint that the human eye and brain are superior to the currently used computer algorithms in finding them. Manual algorithms are, of course, very slow and are therefore only of use when few orientations are to be determined as this is frequently the case in TEM-related work. For orientation microscopy, however, hundreds of thousands to millions of orientations are to be determined. For these applications, automation of spot, line, or band detection is indispensable. Also for precise orientation determination from spot patterns, which requires the measurement of spot positions and intensities, automatic detection is necessary. We will present the algorithms for the determination of Kikuchi lines or bands and of diffraction spots separately in the following.

7.2.1 Manual Determination of Diffraction Plane Vectors

In a spot diffraction pattern the vector, R, which points from the primary beam spot to a diffracted beam spot (as displayed in Figure 6.2), corresponds almost exactly to the reciprocal vector as it was discussed in Section 6.2.1. The vector R just has to be expressed in detector coordinates to obtain the reciprocal vector. Note, however, that the assumption made here about the position of R is not very accurate, as it will

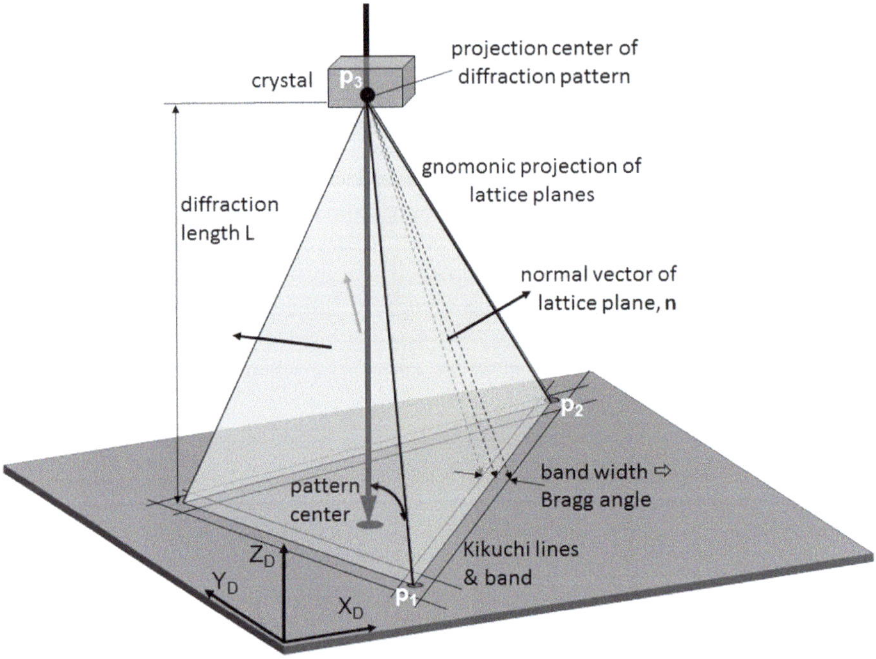

FIGURE 7.2 Geometry of transmission Kikuchi diffraction and determination of reciprocal vectors from a TKP. Three points, marked, p_1, p_2 and p_3, on a lattice plane need to be measured in the detector coordinate system, **D**, to determine the plane normal vector. One of the points is the projection center of the pattern, the other two are just any two points along the center of a Kikuchi band. The projection center is measured from the pattern center (the primary beam spot) in Z-direction of **D**. From the measured band width the Bragg angle and from this the interplanar spacing can be determined.

be discussed further in Section 7.2.4. For a transmission Kikuchi pattern Figure 7.2 describes the relation between the diffraction vector and a Kikuchi line. The two circles, p_1 and p_2, marked in the detector plane indicate two points on the center line of a Kikuchi band measured in detector coordinates XYZ_D. Alternatively also two positions on each corresponding Kikuchi line can be measured. The third point, p_3, marks the pattern (or projection) center which is that point where the beam hits the sample. In the case of TKD patterns this point has the XY_D coordinates of the primary beam on the detector plane and the diffraction length, L, as Z_D coordinate. Note that in the case of an EBSD pattern there is no primary beam spot on the detector, which is why it is more difficult in this case to determine the projection center and, thus, to determine the correct reciprocal vector coordinates. From the measured three points it is straightforward to calculate the normal vector of the respective diffraction plane as the cross product of any two vectors connecting the three points (e.g. the two ones marked in Figure 7.2):

$$\mathbf{n} = (\mathbf{p}_2 - \mathbf{p}_1) \times (\mathbf{p}_3 - \mathbf{p}_2) \tag{7.10}$$

From the width of the Kikuchi band, it is furthermore easy to calculate the opening angle of the Kikuchi cone and, thus, to determine the Bragg angle and the length of the reciprocal vector.

7.2.2 Automatic Kikuchi Pattern Analysis: The Hough Transform for EBSD Pattern Analysis

The automatic detection of Kikuchi lines by DIP has been originally developed for the evaluation of BKP in the SEM. For the DIP a Hough transform (HT) (Hough, 1962) has been most successfully applied and will be the only algorithm discussed here. Before the use of the HT the so-called Burns algorithm was tried out for a short time (Wright and Adams, 1992). The use of a HT for Kikuchi line detection was first introduced by Krieger-Lassen et al. (1992) (see also Krieger-Lassen, 1994), and shortly after implemented in a system for orientation microscopy ("orientation imaging microscopy," OIM) by Adams et al. (1993).

The HT is a discrete form of the Radon transform (Radon, 1917), which determines the line integral along a given parameterized curve $f(x,y)$ on a surface (x,y). The integral corresponds to the sum of the values encountered along the curve. In image processing the (x,y) surface corresponds to an image which consists of discrete pixels, each with a brightness value. The HT then follows a parameterizable line through the image, e.g. a straight line, an ellipse or a circle and adds up all the intensities found along this line and saves this value in a collector space. For the detection of lines the normal parameterization of a line is used in the form (cf. Figure 7.3a)

$$\rho = x\cos\omega + y\sin\omega \;\; \rho = xcos\omega + ysin\omega \quad \text{with} \quad -R > \rho > R \quad \text{and} \quad 0 > \omega > \pi \tag{7.11}$$

where ρ is the distance of the line from the origin, R the radius of the image, and ω the angle of the normal of the line with the x-axis. The collector space is a Cartesian coordinate system ρ over ω, called the Hough space. In this space the intensity sums for all possible lines in an image, each one represented unequivocally by a set of (ρ, ω), are collected. The resulting Hough space is indicated in Figure 7.3b. In this way bright or dark lines in the original image correspond to bright or dark spots in the Hough space. By application of a HT the problem of detecting lines in the real image is thus simplified to the problem of the detection of points in the Hough space. In fact, since a BKP image does not show really sharp Kikuchi lines but rather bright bands the HT does not create sharp points in the Hough space but "butterfly"-shaped spots as shown in Figure 7.3b. These may be detected by convolution of the Hough image with an appropriate mask (Krieger Lassen, 1994). Here it is important to note that size of the convolution mask must be adapted to the size of the HT peaks for optimum peak detection. In Figure 7.3c, for example the mask is well suited for convolution of the peaks 1, 2, and 3, but it is too small for the peaks 4, 5, and 6. The peaks 1, 2, and 3 belong to narrow bright bands in the pattern, while the other peaks belong to wider bands, which consist almost only of Kikuchi lines. If these peaks are to be properly determined (which is recommendable because the accuracy and confidence of orientation determination increases with the number of correctly determined bands), the HT

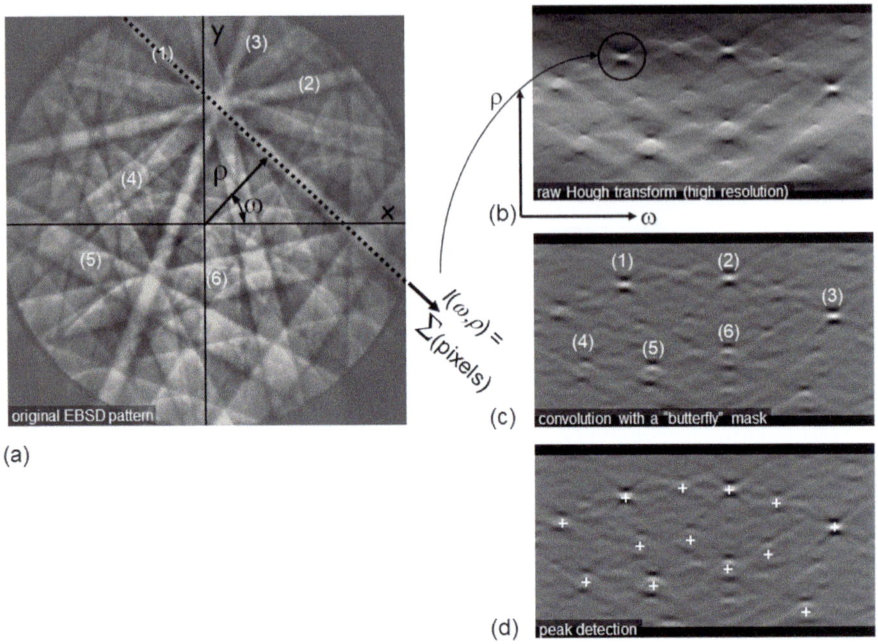

FIGURE 7.3 Illustration of the Hough transform (HT) for detection of Kikuchi bands in an EBSD pattern. (a) A typical diffraction pattern. (b) The raw high-resolution HT (ρ = 240 pixel, $\Delta\omega$ = 0.5°). (c) HT after convolution with a filter mask to enhance characteristic peaks. (d) Crosses indicate the detected maxima. Numbers indicate a few characteristic bands and corresponding HT peak. (1), (2), and (3) are correctly convoluted, for (4), (5), and (6), the used convolution mask is not large enough.

either has to be reduced in resolution (which increases speed and decreases accuracy) or the mask size has to be increased (which decreases speed and increases accuracy). It should be noted that the brightness of the Hough peaks, averaged over a whole pattern, may be used as a very important metric for the quality of the diffraction pattern. In orientation maps it is frequently used as "image quality" or "pattern quality".

In order to reach high calculation speed the HT algorithm has been highly optimized for the evaluation of BKP. In particular, the always identical sine and cosine operations in Equation 7.11 are pre-calculated and stored. The HT can be significantly speed up by reducing the resolution of the Hough space. A resolution of 1° in ω and 100 pixel in ρ is sufficient to determine a crystal orientation with an accuracy of 0.5° to 1° (the subject orientation accuracy is addressed in some more detail in Section 8.2.10). In this way a pattern analysis rate of up to ~3000 s^{-1} (on a 16-core processor PC, year 2023) can be reached. On the other hand, in order to reach high orientation accuracy, the Hough space resolution has to be increased. With a resolution of ρ = 240 pixels and ω = 0.5° (HT 240 × 0.5) an orientation accuracy in the order of 0.1° can be reached (see Section 8.2.10). Such a HT is calculated, however, at a 100 times lower rate.

The above-described HT procedure which uses a fixed-size convolution mask works well in many standard cases of EBSD-based orientation microscopy. However, with improving camera performance new applications may come into reach which require more flexible detection algorithms. Such applications are on the one hand the reliable phase determination which requires precise band width measurements. On the other hand, techniques with improved spatial and angular resolution by decrease of acceleration voltage or a retraction of the camera will become available. In these cases, the detection of wide bands is required. A well-suited algorithm which allows the reliable detection of most kinds of lines and bands is the Hough backmapping technique introduced by Krieger-Lassen (1998). The algorithm requires that a conventional HT has been calculated. It then selects for each pixel (x_i, y_i) the most probable accumulator in the Hough space as the maximum or minimum (ρ_m, ω_m) along the $\rho(\omega)$ curve (Equation 7.11). Each pixel of the real image is thus just added once into the Hough space but most probably at the position of an existing bright or dark line (ρ, ω). As a result, the Hough space is filled much more sparsely than in a conventional HT, the maxima are more easily detectable and lines can be easily separated into dark and bright ones depending on whether the minimum or maximum (ρ_m, ω_m) is taken. Furthermore, the backmapping HT is calculated more quickly than the filter-convolution of a normal HT.

There are further ideas of improving the evaluation accuracy of a HT, for example the detection of the hyperbolic shape of the Kikuchi line by employing a three-dimensional HT (to detect three parameters of the Kikuchi conical sections) (Maurice and Fortunier, 2008) or the application of a line-refinement approach (which is not fully published) proposed by Thomsen et al. (2013) which will, however, not be discussed here further.

Before applying the HT to a BKP, it is usually necessary to eliminate the strong background by subtracting an "empty" image which has been created by integration over a large number of different patterns. This approach may fail when multiphase materials consisting of phases of very different densities and thus backscattering efficiency are investigated. This may lead to very different background intensities. Also different surface inclination due to a strong surface relief may change the background of a pattern so strongly that a static background image may not be suitable for background removal. Finally, when investigating single crystals, it might also be difficult to obtain a homogeneous background. In all these cases the application of a low-pass filter may be reasonable to remove the background. Some strategies for background removal will be discussed later in Section 8.2.6.

7.2.3 The Hough Transform for TKP

Already soon after the introduction of the HT for BKP evaluation efforts were made to use the same algorithm also for the evaluation of TKP (Zaefferer and Schwarzer, 1994; Krieger-Lassen, 1995; Schwarzer and Sukkau, 1998). These investigations did not lead to reliably working algorithms but pointed out the difficulties that have to be faced when dealing with TKP. The basic problem is that the density of crystallographic information (the number of Bragg-diffracted electrons over all electrons) is much lower in a TKP than in a BKP (because of thinner samples and higher acceleration voltage) and the detection algorithms therefore must be more sensitive. Figure 7.4 shows a typical TKP and a BKP to point out the difficulties in detail. The first difference between both kinds of patterns is the strongly varying background in

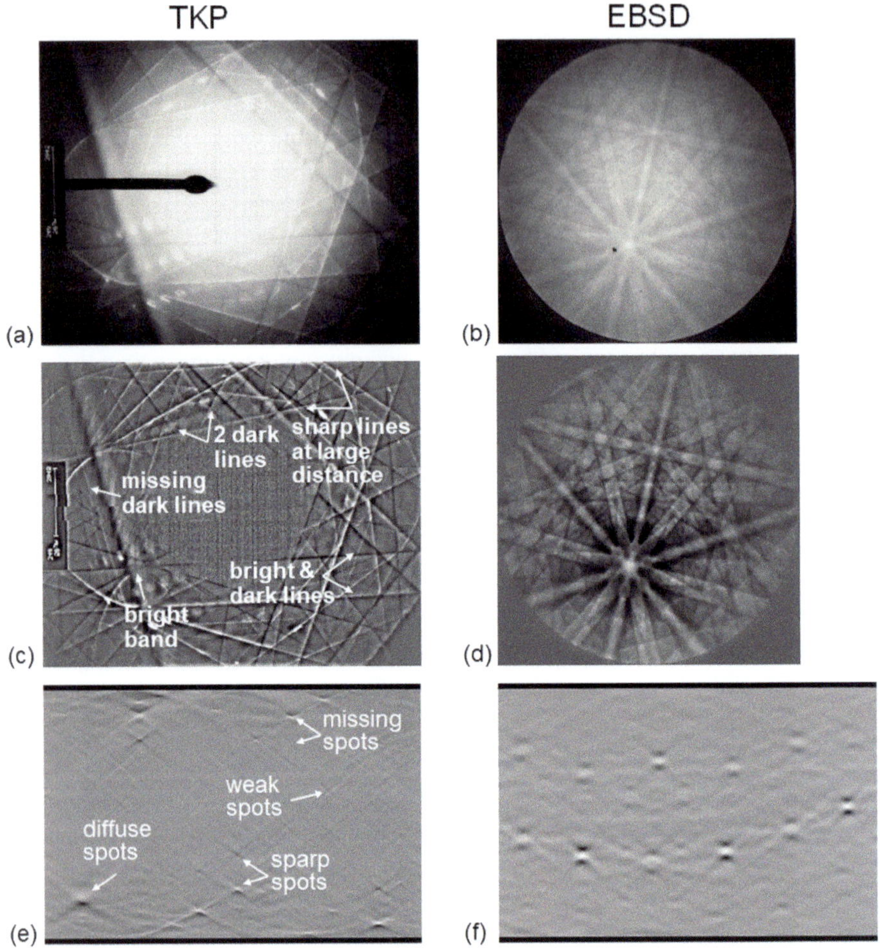

FIGURE 7.4 Comparison of the application of a HT for analyzing TKP and EBSD patterns. (a) and (c) TKP without and with background subtraction. (b) and (d) EBSD pattern without and with background subtraction. In (c) the difficulties of applying a HT to TKP are marked. (e) and (f) resulting HT for the two sorts of pattern. In (e) the difficult features are marked.

TKP. The contrast may vary so much that it becomes necessary to acquire the image in two parts—an inner part with short exposition time and an outer one with a longer time. This procedure has been applied by Fundenberger et al. (2003). The removal of the background by subtraction of an empty image is not possible with TKP because the pattern background may change drastically with sample thickness. Instead, the background may be eliminated by the use of a high-pass filter which allows only the high frequency Kikuchi lines to pass through into the filtered image. A typical result is shown in Figure 7.4c. The use of such a software-based filter increases, of course, the calculation time for automatic pattern evaluation.

A second problem for automatic TKP analysis is the variable appearance of TKP Kikuchi lines as shown in Figure 7.4a and c. Couples of sharp dark and bright lines may appear as well as couples of only dark lines or bright diffuse bands. Furthermore, the lines of a couple may have an appreciable distance. In contrast, BKP mainly show bright, relatively narrow bands. The application of a backmapping HT for the evaluation of TKP is therefore an obvious choice.

A further problem for the application of a HT for TKP is the appearance of aligned spots in thin or in highly deformed sample areas. Aligned spots may be misinterpreted by the HT algorithm as bright lines. There are two ways around this problem. One is the separate detection and subsequent elimination of the spots from the image, the other is the use of a robust indexing algorithm which is capable of sorting out lines which do not fit to the others. A last difficulty that frequently arises in TKP is faint contrast of the defect line of Kikuchi line pairs. However, when only the excess line is available indexing becomes more difficult because neither the length nor the exact position of the diffraction vector can be determined.

In the light of these difficulties and due to the increasing power of the automated SEM techniques it is clear that only comparatively little effort has been put into the development of the fully automated evaluation of TKP (see Section 9.6 for a comparison of the SEM and TEM techniques). Nevertheless, Fundenberger et al. (2003) published a fully working system for orientation microscopy. Another example, where a high-resolution HT is successfully applied is the detection of higher-order Laue zone (HOLZ) lines in convergent beam electron diffraction (CBED) patterns for the determination of lattice strain (Krämer and Mayer, 1999). Here, however, only clear lines with constant contrast (dark lines on a white background) are to be detected which makes the procedure significantly easier. Finally, also the present author is working on algorithms using a Hough backmapping approach for TKP analysis. It should be mentioned, however, that we see the future of TEM orientation microscopy in the use of some sort of spot diffraction pattern (see the following section and Section 7.4).

7.2.4 Extraction of Diffraction Vectors from Spot Patterns

While in Kikuchi patterns the position of the lines or bands is sufficient to accurately determine the crystal orientation, in spot patterns it is necessary to determine spot positions and intensities. The reason for the required intensity is as follows: in spot patterns, the diffraction vector, **g**, is related to the spot position observed in the pattern via the Ewald sphere construction as shown in Figure 6.4. In the pattern coordinate system, the x- and y- coordinates of **g** are identical with the x- and y-coordinates of the appropriate spot divided by the camera constant λL. In contrast, the determination of the z-coordinate requires to take into account the curvature of the Ewald sphere, $1/\lambda$.[1] However, the reciprocal lattice vector corresponding to a diffraction spot is not necessarily exactly on the Ewald sphere but may deviate from it by the vector **s**, as was shown in Figure 6.4b. Thus, the precise determination of $\mathbf{g}_{hkl}$ requires knowledge on **s**. The value of **s** is related to the intensity of a diffraction spot. For simple crystal structures like fcc and bcc, where the structure factor of all visible reflections is identical, the value of **s**, i.e. the distance of the reciprocal lattice

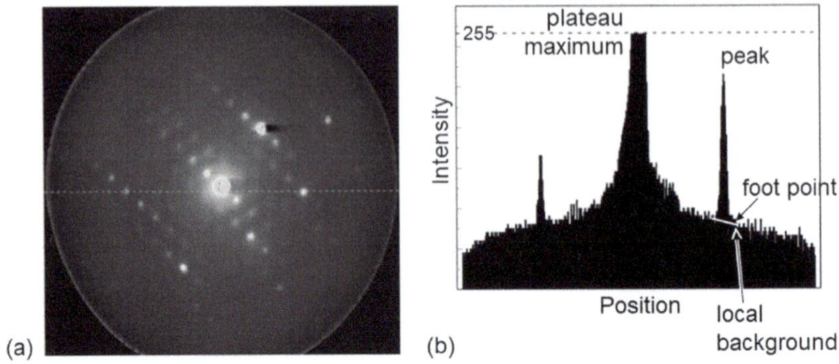

FIGURE 7.5 Image processing for automated detection of diffraction spots in SAD and SCBED patterns. (a) A typical diffraction pattern. (b) An intensity cross section through the pattern along the marked line in (a). See text for further details.

vectors to the Ewald sphere, is inversely proportional to the intensity obtained from Equation 6.9 normalized by the extinction distance of the respective spot. Those diffraction vectors **g** with the largest normalized intensity are closest to their appropriate $\mathbf{g}_{hkl}$ and therefore best suited for a precise orientation determination. Only those are used for indexing.

For materials with more complex crystal structures the structure factor differs for different reflectors which makes application of the above procedure impossible. A better way, which does not require any assumption on the position of the Ewald sphere, is the use of the template matching technique, which will be introduced in Section 7.4.2.

Zaefferer (2000) and (2002) proposed an image processing procedure for position and intensity determination from SAD or SCBED patterns. The procedure works without any preceding background subtraction procedure on images with a size in the order of 300 × 300 pixel. Larger images are usually binned to half the original size. The procedure consists of three steps. First the spot positions are searched, second the exact spot maximum is determined, and third the spot intensity for each spot is calculated by a numerical integration over the whole peak area followed by a subtraction of the local background. The different steps are illustrated in Figure 7.5. Details can be found in Zaefferer (2000) and (2002).

7.3 INDEXING AND ORIENTATION CALCULATION

Indexing of diffraction patterns, i.e. consistent assignment of Miller indices (*hkl*) to the measured diffraction vectors, is one of the complex problems in the process of orientation determination. Basically, one has two sets of vectors which have to be brought to a best match, the vectors measured in the pattern, $\mathbf{g}_P$—which are subject to measurement errors—and the pre-calculated reciprocal lattice vectors $\mathbf{h}_C$. The two sets of vectors are illustrated in Figure 7.7. The largest complication for indexing arises due to the occurrence of vectors in the data set which do not belong to the same crystal. This case occurs due to the overlap of diffraction patterns from several crystals and is more common, for example, in orientation microscopy on nanocrystalline materials. When performing EBSD investigations on these materials—if patterns are visible at all—the patterns from neighboring crystals commonly overlap. More serious, however, is the fact that the crystals may be so small that even in a thin TEM sample several crystals may occur along the beam path and form several overlapping patterns. It is therefore clear that with decreasing grain size the indexing procedures become increasingly complicated (see Section 8.3).

The vector matching procedure for indexing can either be performed serially—by matching one vector after the other—or parallel, by matching all vectors simultaneously. In the following we describe and discuss two approaches to indexing. We start with the most obvious, serial algorithm that has been implemented in many computer programs for pattern indexing (e.g., Young et al., 1973; Rhoades, 1976; Schwarzer and Weiland, 1984; Engler et al., 1996a; Zaefferer, 2000, 2002). Representative for most algorithms developed we describe in the following the algorithm applied in the interactive software TOCA (tools for orientation determination and crystallographic analysis) (Zaefferer, 2002). The algorithm was developed for indexing any kind of TEM diffraction patterns. In Section 7.3.2, parallel algorithms are discussed.

7.3.1 Serial Indexing

From the diffraction pattern a set of diffraction vectors $\mathbf{g}_{D,i}$, described with respect to the detector coordinate system **D**, is extracted. Basis of serial indexing is a serial comparison of the angles between these vectors and, if available, their lengths with the reciprocal lattice vectors, $\mathbf{h}_C$, calculated for possible crystal structures. To this end first all inter-vectorial angles are determined for the experimental vectors using Equation 7.9 for the experimental reciprocal vectors $\mathbf{g}_{D,i}$. Now one of the experimental vectors, $\mathbf{g}_{D,1}$ (which is, for example, selected to be the shortest vector or the vector corresponding to highest diffracted intensity) is assigned in a systematic manner a vector $\mathbf{h}_{C,1}$ from the look-up table for the reciprocal lattice. A pre-selection of this vector may be done on the basis of a vector length comparison. The next experimental vector $\mathbf{g}_{D,2}$ is assigned a second vector $\mathbf{h}_{C,2}$ and the angles between $\mathbf{h}_{C,1}$ and $\mathbf{h}_{C,2}$ are compared to those between the experimental vectors. The vector $\mathbf{h}_{C,2}$ is varied until one is found which gives a similar angle. The procedure is continued until all $\mathbf{g}_D$ vectors are assigned consistently to different $\mathbf{h}_C$ vectors and all angles and lengths fit together. If no consistent assignment is found the assignment of $\mathbf{g}_{D,1}$ is changed and the procedure started again.

This relatively simple "trial-and-error" algorithm is complicated by the fact that the measured vectors are subject to experimental errors and that vectors may occur in the measured data set that do not belong to the same reciprocal lattice, for example due to overlapping patterns or false detection of spots or bands. One can deal with these issues by comparing the vectors within a certain tolerance and by systematic exclusion of one or more of the measured vectors until consistent indexing is achieved. In order to reduce the number of vector combinations to be checked, the pattern normal, $\mathbf{d}_z$, can be calculated for each pair of indexed vectors (using Equation 7.12) and compared for consistency with that calculated from the other pairs.

Once consistent indexing is found the orientation with respect to **D** is calculated by solving the following equation system for each pair of diffraction vectors $\mathbf{h}_{C,1}$, $\mathbf{h}_{C,2}$ assigned to the experimental vectors, $\mathbf{g}_{D,1}$, $\mathbf{g}_{D,2}$:

$$\cos\measuredangle\left(\mathbf{g}_{D,1},\mathbf{d}_i\right)=\frac{\mathbf{h}_{C,1}\cdot\mathbf{d}_i}{\left|\mathbf{h}_{C,1}\right|}$$
$$\cos\measuredangle\left(\mathbf{g}_{D,2},\mathbf{d}_i\right)=\frac{\mathbf{h}_{C,2}\cdot\mathbf{d}_i}{\left|\mathbf{h}_{C,2}\right|} \quad (7.12)$$
$$\left|\mathbf{d}_i\right|=1$$

where $\mathbf{d}_i$ denominate the unknown column vectors of the orientation matrix in **D**, either the detector normal $\mathbf{d}_z$ or a reference direction in the detector plane $\mathbf{d}_x$. The cosines of the angles between the indexed diffraction vectors $\mathbf{h}_C$ and either of the two reference directions **d** correspond to the coordinates of the measured diffraction vectors $\mathbf{g}_D$. Equation 7.12 has two possible solutions for **d** for each pair of vectors $\mathbf{g}_1$, $\mathbf{g}_2$. By comparison of all solutions for **d** the correct direction is chosen. The final direction is calculated as the average over all solutions. The determined vectors $\mathbf{d}_x$ and $\mathbf{d}_z$ are not perfectly perpendicular to each other, therefore the resulting orientation is subsequently determined such that the orthogonal matrix vectors have the smallest angular difference to the determined vectors. Detailed mathematical considerations about this problem have been made by Krieger-Lassen (1996).

The first solution found by serial indexing is not necessarily the best match even when it is calculated with smallest tolerances and smallest amount of excluded vectors. It is therefore necessary to determine a number of (in principle all) possible solutions, calculate for each solution a confidence value, and then select the most confident solution. In the interactive software TOCA (Zaefferer, 2000) solutions are sorted by several parameters which include an indexing confidence index, given by the ratio of the number of consistently indexed diffraction vectors over all measured vectors and an accuracy value calculated as the average angular difference between measured and recalculated diffraction vectors. Other software solutions use similar metrics.

The described procedure becomes lengthy and prone to errors in cases where a significant fraction of vectors is not belonging to the same crystal. In this case parallel indexing described in the next section is much more powerful.

7.3.2 Parallel Indexing

A form of parallel indexing—where all measured diffraction vectors are matched to theoretical ones at the same time—has been proposed by Wright (1992) and is used in the commercial software package OIM for the automated orientation determination from EBSD patterns (Adams et al., 1993). The algorithm is based on the selection of all possible subsets of three non-coplanar vectors $\mathbf{g}_P$ out of all measured diffraction vectors. Note that in EBSD patterns such triples of vectors correspond to triangles of Kikuchi bands, as displayed in Figure 7.6. Each triple is indexed separately, as described in the previous section, usually resulting in a large number of possible

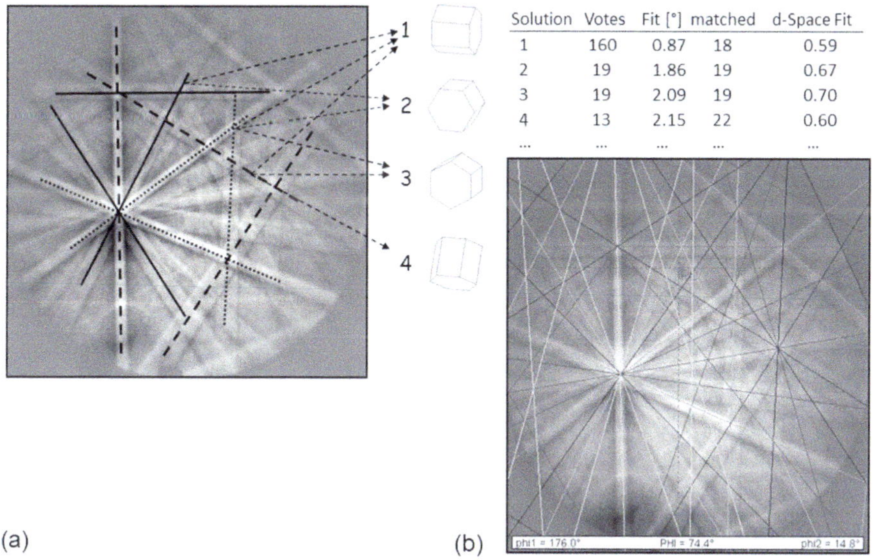

Solution	Votes	Fit [°]	matched	d-Space Fit
1	160	0.87	18	0.59
2	19	1.86	19	0.67
3	19	2.09	19	0.70
4	13	2.15	22	0.60
...	...	...	...	...

FIGURE 7.6 Triplet indexing algorithm for EBSD pattern indexing on the example of a Mg pattern. (a) In an EBSD pattern all detected bands that form triangles (non-coplanar diffraction vectors) are selected and indexed separately. The resulting orientations are displayed as lattice unit cells at the side. The solution list shows which orientation obtains how many votes. (b) The original pattern, overlayed with the simulated bands for the solution with the highest number of votes. All lines are correctly positioned, indicating the correct solution.

orientation solutions. These orientations are then accumulated and it is counted how often each orientation has been calculated, i.e. how many votes it got. In order to select the most confident orientation from this set of solutions a confidence index CI is calculated according to

$$\mathrm{CI} = (V_1 - V_2)/V_{\mathrm{IDEAL}} \tag{7.13}$$

where V_1 and V_2 are the number of votes for the first and second solutions and V_{IDEAL} is the total possible number of votes for the detected bands. In Figure 7.6b the solution with the highest number of votes is overlaid to the original pattern.

The described algorithm works very robustly and reliably in the case of most EBSD patterns. Problems may arise, however, if the measured data set $\mathbf{g}_\mathrm{D}$ is small, if the vectors are closely spaced (low symmetry structures) or if a large set of vectors $\mathbf{h}_\mathrm{C}$ has to be checked (complicated crystal structures). These cases arise particularly in measurements taken from TEM diffraction patterns: (i) an unknown deviation vector $\mathbf{s}$ may introduce measurement errors (in spot patterns); (ii) the small solid angle of the pattern allows measurement of only few, closely spaced $\mathbf{g}_\mathrm{D}$; (iii) also highly indexed diffraction vectors may be visible in the pattern and have to be taken into account for indexing which requires a large set of pre-calculated lattice vectors. It is thus clear that there is—particularly for the case of overlapping TEM diffraction

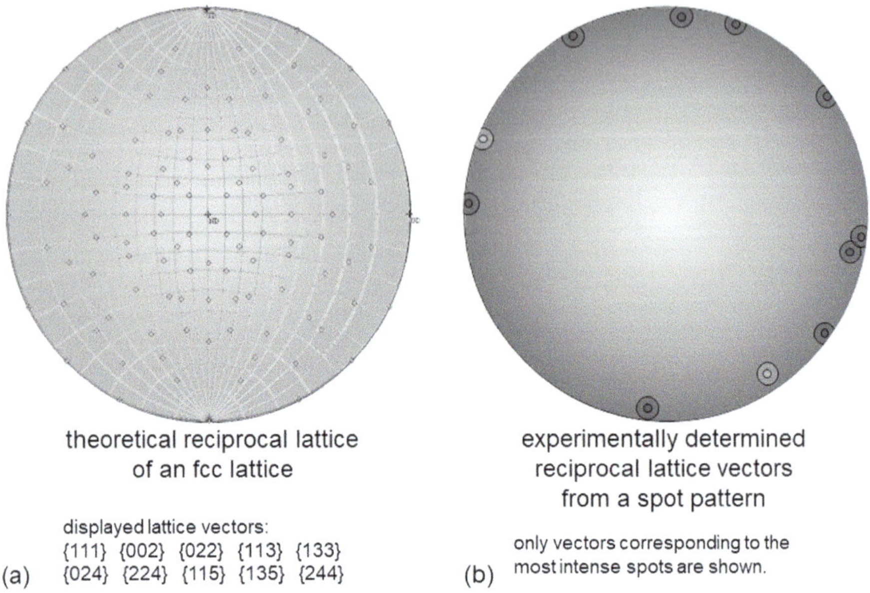

FIGURE 7.7 The basic idea for a general indexing algorithm: positioning of a theoretical set of diffraction vectors such that it superimposes with a measured set vectors in the limits of measurement errors and possibly wrongly detected vectors. (a) The theoretical set of lattice vectors for an fcc crystal structure in a stereographic projection (all vectors are normalized to a length of 1). The graph contains only the indicated diffraction vectors (i.e. up to (h k l) = (2 4 4) although the experimental pattern may contain many more reflector families (up to (11 11 11) has been observed). (b) The determined diffraction vectors obtained from an experimental spot pattern. The small circles indicate the actually detected vector positions, the large circles indicate the allowed error cones.

patterns—still a need for alternative, parallel indexing algorithms. This issue has been addressed in a formal and general way by Morawiec and Bieda (2005). The principal idea is that the two sets of vectors, the measured and the theoretical ones as indicated in Figure 7.7, are rotated such that the average distance between the vectors of both sets becomes minimum. Though we do not propose a general solution to this problem here, a practical algorithm for indexing spot patterns is the template matching process presented in the following section.

7.4 PATTERN MATCHING AND DICTIONARY INDEXING

7.4.1 Introduction

It has been discussed in the previous sections that the spot pattern technique is the most promising way for orientation microscopy in TEM on nanocrystalline materials. Nevertheless, with the classical indexing procedures, significant problems arise, particularly the determination of the correct excitation error, *s*, for unknown structure factors and the treatment of overlapping patterns that appear regularly when attempting orientation measurements on truly nanocrystalline materials. An efficient way to solve these problems is offered by the template matching technique developed by Rauch and Dupuy (2005). The technique has since developed amazingly; the most recent state-of-the-art is described in Rauch and Veron (2014). The principle of this technique is the comparison of experimental diffraction patterns with a large set of pre-calculated simulated diffraction patterns, in order to find the best fitting one. The set of possible pre-calculated diffraction patterns is called a pattern dictionary which gives the name "*dictionary indexing*" (DI) to the respective EBSD-technique. Indeed, the DI technique is not only well suited for TEM spot pattern indexing but is also an attractive, alternative way for the analysis of EBSD patterns. The advantage of this template matching approach is that it uses the full intensity information of the diffraction pattern, in contrast to the HT, which analyzes mainly geometric information on position and width of diffractions bands. In this way the DI approach is not only able to deal more easily with complicated crystal structures but also with more noisy patterns generated with lower beam current or wider Kikuchi bands caused by lower acceleration voltage. So far, the dictionary indexing method for analysis of EBSD patterns has been mainly developed by the group of De Graef (Chen et al., 2015; Singh and De Graef, 2016; De Graef, 2020)—inspired, certainly, by the earlier work of Rauch and Dupuy (2005) on the TEM spot pattern template matching. The pattern matching method is particularly interesting when crystals with "pseudo-symmetries" or phases with very similar diffraction pattern geometries but different intensity distributions are investigated. This has been shown in a number of important publications by Nolze et al. (2016) and Winkelmann and Nolze (2015a, 2015b) (see Section 7.4.4 for a discussion of their work). Furthermore, the robustness of the method with respect to indexing of patterns of very low quality appears to be higher than that of the HT-based analysis. This method has become feasible because of the ever-increasing computation power and the significant advances in realistic pattern analysis.

7.4.2 Template Matching for Spot Pattern Indexing in TEM

The technique is based on a direct comparison of experimental diffraction patterns with simulated ones, using an image cross-correlation technique. A good match between a pattern simulated for a certain orientation and the experimental pattern immediately leads to the orientation of the related crystal. The technique therefore requires neither the determination of individual spot positions nor any discrete indexing algorithm. It requires, however, realistically calculated template patterns.

Template matching is performed by calculating the cross-correlation between the experimental and the simulated pattern:

$$f_{CI} = \frac{\sum_{x,y} P_{x,y} \cdot T_{x,y}}{\sqrt{\sum_{x,y} P_{x,y}^2} \cdot \sqrt{\sum_{x,y} T_{x,y}^2}} \tag{7.14}$$

where P is the experimental pattern, T is the template, x and y indicate all x-y pixels in the two images. f_{CI} becomes maximum for the best fit of the two patterns. For orientation determination the experimental pattern is compared with templates calculated in small angular steps for all possible orientations. The maximum of f_{CI}

will then yield—with some exceptions—the correct orientation. This basic procedure is quite calculation intensive because a large number of orientation templates and a large number of pixels in each pattern have to be considered: for cubic crystal symmetry and 1°-resolution about 650,000 templates, each composed of about 100 × 100 pixels would have to be compared. It should be kept in mind, however, that the pattern analysis can be performed off-line, on a separate computer, therefore not affecting on-site TEM occupation. By focusing the correlation calculation only to those pixels in the in the template and the pattern that contain information the number of pixels per template is reduced to 50 to 100. Furthermore, reduction of calculation is obtained by reducing the templates to the beam directions in the standard orientation triangle of the respective crystal structure. All patterns having the same pattern normal are just rotated with respect to each other. This reduces the template space to just 1,800 for cubic crystals. Furthermore, as proposed by Wu and Zaefferer (2009), the procedure can be made more robust and speed up drastically by a transformation of pattern and template into in polar coordinates ρ and ω with respect to the pattern center. A pattern rotation about the center then corresponds to a simple shift along ω as this is indicated in Figure 7.8. Additionally, in the same paper Wu and Zaefferer introduced the circular projection as a pre-selection

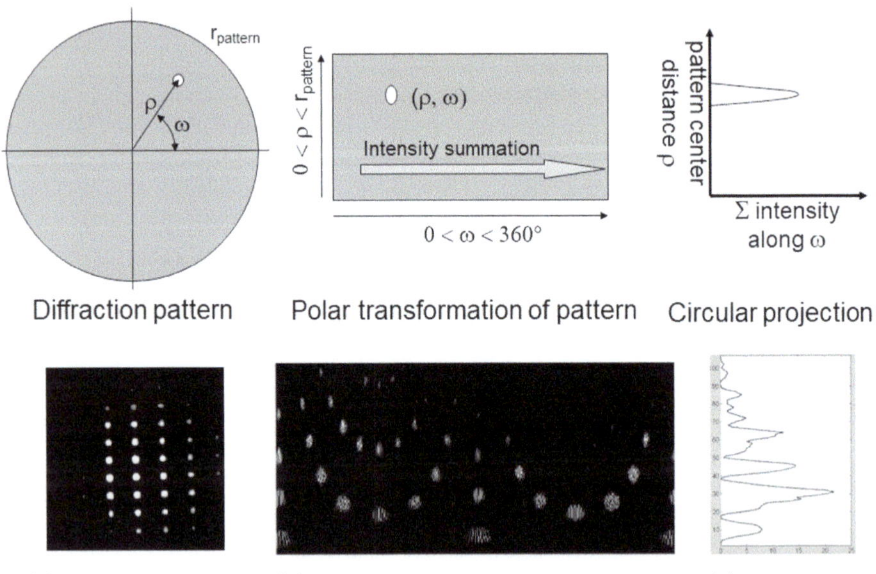

FIGURE 7.8 The two principal steps of spot-pattern template matching. A spot pattern (a) is converted into a polar coordinate system (b) such that the rotation about the center pole which is necessary for template matching can be expressed by a simple shift operation along ω. (c) In order to accelerate the search for possible templates the circular projection is calculated for every template and for the experimental pattern. The circular projection consists in adding all intensities along lines of ρ = const. In the template matching process only those templates are checked which have a circular projection fitting to that of the experimental pattern.

of possible templates. The circular projection consists in a summation of all pixel intensities along ω in the polar transform. When comparing the circular projections of the experimental pattern with those of the templates only 20–40 templates are left over for the final cross-correlation calculation.

The template matching technique for orientation determination from spot patterns has three important advantages compared to the individual spot matching techniques. First, the method implements in principle an ideally parallel indexing algorithm—all spots in the pattern are matched simultaneously. Furthermore, the template matching comprises all steps for orientation determination (reflector determination, indexing, orientation determination) in one single, in principle simple procedure. The probably biggest advantage of the template matching procedure is its ability to deal with overlapping patterns as they frequently occur in nanocrystalline or heavily strained material. An example is shown in Figure 7.9, where the particularly difficult case of overlapping patterns with double diffraction effects is displayed: the figure shows diffraction patterns obtained from a twin. Figure 7.9a and b shows the patterns of the individual matrix and twin crystals and Figure 7.9c

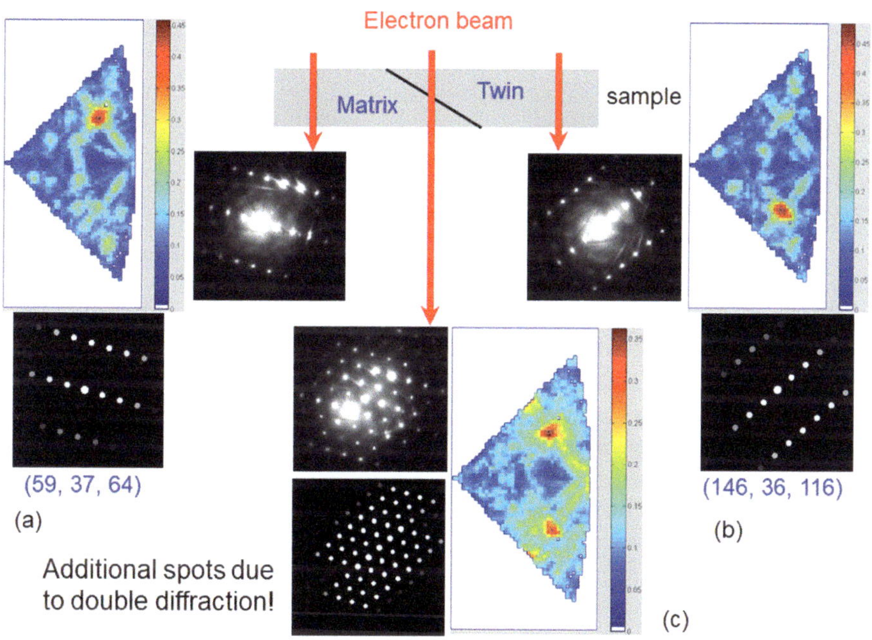

FIGURE 7.9 Template matching applied to solve the overlapping patterns of twin and matrix in a copper sample. (a) Spot diffraction pattern of the matrix and the solution by template matching. (b) Same as (a) for the twin. (c) Spot diffraction pattern of overlapping twin and matrix patterns at the twin boundary. Note the occurrence of double diffraction, i.e. of diffraction spots which appear neither in the matrix nor in the twin pattern. These spots arise due to a diffracted beam from the first crystal acting as primary beam for the second one. The template matching method is anyway able to determine the correct orientation of both overlapping crystals.

that from the boundary where both crystals overlap. The cross-correlations for the individual patterns are shown in Figure 7.9d and e as a projection into the standard orientation triangle. Both show sharp peaks. The cross-correlation for the boundary pattern shows exactly two peaks at the positions of the individual crystal orientations. The strong double diffraction peaks (which occur because one of the diffracted beams of the first crystal acts as "primary beam" for the second crystal) do not influence the result at all.

Finally, a very significant improvement is obtained when combining template matching with precession electron diffraction, as introduced in Section 6.2.4. The precession makes more diffraction spots visible, particularly those at larger scattering angles, which renders orientation determination more accurate and more robust. The combination of precession electron diffraction and template matching thus developed into the most successful system for orientation mapping in TEM. A number of recent publications have documented the strength of the system, e.g. Cautaerts et al. (2021) and Rauch et al. (2021).

7.4.3 Dictionary Indexing for EBSD Pattern Analysis

Similar to the TEM spot pattern method, template matching can also be applied to EBSD patterns. The approach again consists of two important parts: (i) calculation of a dictionary of realistic input patterns and (ii) a rapid and reliable comparison strategy. The realistic calculation of diffraction patterns requires the application of dynamical theory of coherent electron diffraction, potentially coupled with a model for incoherent scattering.

(i) For creation of the pattern dictionary it is first necessary to calculate a master pattern. This master pattern is most conveniently a spherical diffraction pattern as displayed in Figure 7.10 (Day, 2008). The diffracting crystal is put into the middle of a sphere and the pattern center (i.e. the electron point source) is placed into the center of the crystal. The simulated diffraction pattern of the whole crystal can then be displayed on the surface of this sphere. Figure 7.10 displays furthermore that the experimentally observed diffraction patterns are simply gnomonic projections of this pattern to a tangential plane of the sphere. The detector reference system, **D** (introduced in Section 7.1.1), is fixed to the point where the detector plane touches the sphere. Note that this point is sometimes called the *pattern center*, not to be confused with the *projection center* which is the center of the sphere. De Graef (2020) proposes a square Lambert projection for the storage of the spherical master pattern in (Cartesian) computer memory. It is, furthermore, a mathematically demanding question how the orientation space is efficiently and uniformly sampled in order to obtain equally spaced diffraction patterns for a given angular resolution of orientation space. Singh and De Graef (2016) discuss the use of the *cubochoric orientation space*. This space allows, in contrast to Euler space or Rodrigues space, a fully uniform mapping of orientations in a 3D space (Roşca et al., 2014). From the spherical master pattern, every potential diffraction pattern, including different positions of the pattern center, can be rapidly calculated by central projection onto the tangential detector plane. The calculation of realistic pattern intensities is done using the dynamical theory of electron diffraction that has been introduced in Section 6.3.2.3.

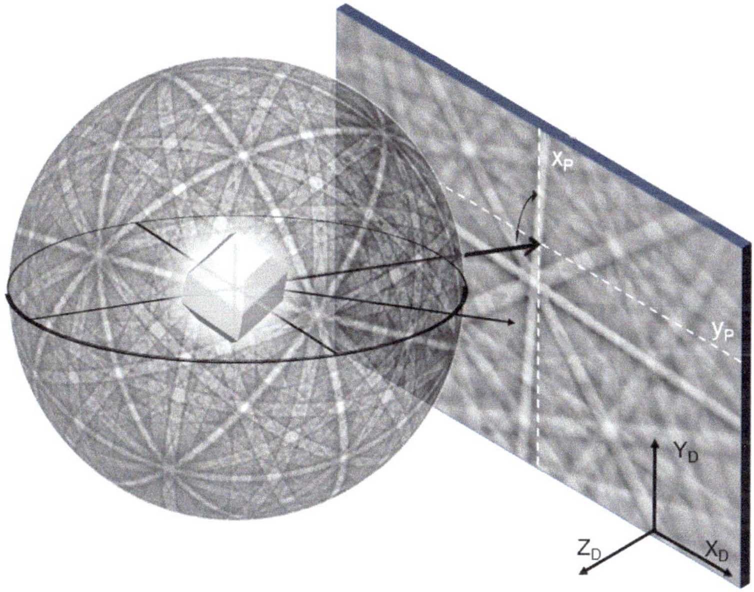

FIGURE 7.10 A spherical EBSD master pattern and its projection onto a detector plane. The diffracting crystal is positioned in the middle of the spherical pattern. The detector plane stands tangential to the sphere or at a certain distance to it. The center of the sphere is the pattern *projection center*. It needs to be measured in detector coordinates (subscript D). The point X_p, Y_p on the detector indicates the *pattern center*. (The master pattern image has been adapted from De Graef, 2020.)

For every crystal structure, only a limited section of the sphere, corresponding to the standard orientation triangle, is relevant. If the orientation space is sampled in steps of 1.4° then 333,227 patterns are required for cubic crystal symmetry and 8,000,000 patterns for triclinic sample symmetry (De Graef, 2020). All other areas are identical through symmetry.

Although the currently software-implemented approaches (Winkelmann, 2009b; Callahan and De Graef, 2013) yield very realistic patterns, it is still possible to improve these by convoluting the coherent patterns with incoherent electron scattering, calculated by a Monte-Carlo electron trajectory model using thermal diffuse scattering cross sections and scattering angles. This leads to the typical asymmetric background intensity of EBSD patterns. This approach, however, has the disadvantage that no master pattern can be used because every beam incidence angle leads to a different distribution of deficiency and excess lines.

For comparison between the experimental pattern and the simulated patterns in the dictionary, various metrics can be taken. For example, a full cross-correlation can be calculated. It turns out, however, that simpler and significantly quicker approach is sufficient. For this both patterns are considered as a one-dimensional vector, **e** for the experimental and **d** for dictionary patterns, with the number of image pixels as number of components. The conventional normalized vectorial dot product between the two vectors than measures the degree of similarity between a pattern, ρ (Chen et al., 2015):

$$\rho = \frac{1}{N_{max}} \sum_{n=1}^{N_{\max}} \frac{\mathbf{e} \cdot \mathbf{d}_n}{|\mathbf{e}||\mathbf{d}_n|} \tag{7.15}$$

Here $N_{\max}$ is the total amount of entries in the dictionary.

Depending on the particular application, two different comparison strategies have been employed. The first "brute force" approach simply compares the experimental pattern with all patterns in the dictionary and selects the one with the largest dot product. This requires a large number of calculations and is, therefore, only suited for off-line indexing, after all patterns of a map have been stored. De Graef (2020) realized that even if the patterns are significantly binned, to sizes as small as 15 × 15 pixels, they can still be indexed with good accuracy. Binning may, thus, speed up the indexing process significantly. De Graef and co-workers (Singh et al., 2018) showed that the robustness of dictionary indexing is comparable or even significantly higher than that of HT-indexing. Even on patterns where the HT approach yields no reliable results (low acceleration voltage, low electron dose, highly defected crystals) the dictionary indexing still resulted in very good orientation maps. Their results are reproduced here in Figure 7.11. While the dictionary indexing rate was almost unaffected by the pattern quality, the HT approach varied between 80 and only 4% of correctly detected orientations. It should be mentioned that since shortly a new indexing algorithm, named "spherical indexing" has been developed (Lenthe

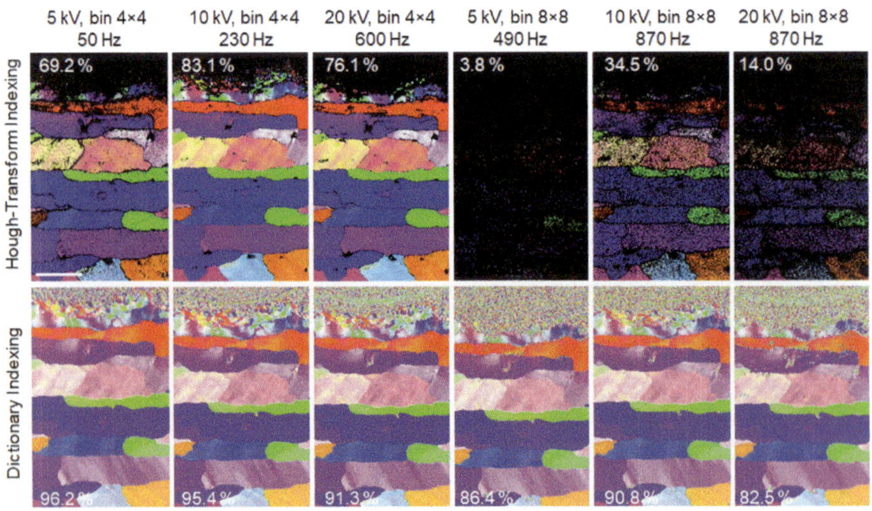

FIGURE 7.11 Indexing of a deformed aluminum sample with a classical HT approach, including serial indexing (top row) and with dictionary indexing at different measurement conditions (bottom row). On top of every column the acceleration voltage, the camera binning, and the measurement rate are given. (Adapted from Singh et al., 2018.)

et al. 2019), which is even more powerful than dictionary indexing. It is superior to dictionary indexing in terms of robustness, orientation accuracy, and indexing speed. A short chapter about this new technique can be found at the end of this book.

The second approach to pattern matching is mainly applied in the case of pseudo-symmetry resolution. In this case, a preliminary indexing of the diffraction pattern is performed by a traditional HT-approach using the Laue group of the point group in question. This results in a crystal orientation with pseudo-symmetry. To resolve this the respective point-group diffraction patterns are selected from the dictionary and high-resolution pattern matching then results in the correct solution. Since the case of solving pseudo-symmetries is becoming increasingly important, we dedicate the following section to this topic.

7.4.4 Pseudo-Symmetries

Pseudo-symmetries are symmetries that appear in diffraction patterns but do not reflect the true symmetry of the diffracting crystal. An informative overview on the appearance of pseudo-symmetries in EBSD patterns can be found in Nolze et al. (2016). Usually the symmetry that is detected is higher than the true symmetry of the crystal. For example trigonal crystals may appear hexagonal or tetragonal ones as cubic. Some so-called pseudo-symmetries are only caused by an insufficient number of diffraction vectors ("reflectors") being taken into account during indexing. A prominent case is the occasional false indexing of bcc crystals. In electron diffraction patterns pseudo-symmetries arise due to two reasons, the first being the physical nature of the diffraction process, the second being the analysis of EBSD patterns with the HT algorithm. We first address the physical background.

In x-ray diffraction, as well as in kinematic electron diffraction, e.g. in spot diffraction patterns obtained from thin crystals, Friedel's law applies, saying that the intensities of reflections of (*hkl*) and ($\bar{h}\bar{k}\bar{l}$) are identical. Consequently, only the Laue group rather than the full point group of a crystal is visible to these techniques. Non-centrosymmetric crystal structures therefore *appear* to have a symmetry center, i.e. they show pseudo-symmetry. With kinematic diffraction techniques, the correct orientation of a non-centrosymmetric crystal can principally not be determined unambiguously. This is a problem because in such crystals positive and negative planes or directions may have different physical properties and it may be important to measure their correct signs and their distributions in a polycrystalline sample. Under conditions of dynamic electron diffraction as they prevail in CBED patterns (e.g. Tanaka et al., 1983; Steeds and Morniroli, 1992) and EBSD patterns; however, Friedel's law is invalid and reflections of (*hkl*) and ($\bar{h}\bar{k}\bar{l}$) will show different intensities, even though these differences may be subtle (see the remarks in the example further below). Thus, in principle, EBSD patterns can be used to distinguish the correct signs of polar axes and planes in non-centrosymmetric crystals, provided the reflector intensities are taken into account. This has been shown as early as in 1989 by Baba-Kishi and Dingley (1989).

The second source of pseudo-symmetry lies in the EBSD pattern analysis process, which classically consists, as shown above, of a combination of band detection via HT and subsequent indexing of the detected Kikuchi bands (related to lattice planes), by an appropriate algorithm. The HT is a very robust algorithm that works also on very low-quality EBSD patterns. However, the HT together with band indexing is mainly based on the geometry of the pattern, i.e. on position and, potentially, width of the Kikuchi-bands. The intensity of bands or their internal structure, in contrast, is largely disregarded. Intensities are only considered by setting intensity thresholds for experimental and simulated bands. All bands which overcome this threshold, however, are considered without any intensity information. This approach is efficient, as long as crystal structures have relatively few strong Kikuchi bands, like many metals (fcc, bcc structures). It may fail, however, when many Kikuchi bands of closely similar intensities are present (this may be already the case with hcp crystal structures). The HT approach fails, in particular, in cases of pseudo-symmetry from non-centrosymmetric structures.

In the following we present two examples illustrating two different cases of pseudo-symmetry, namely GaP crystals without centrosymmetry and enatiomorphic quartz, and their solutions using pattern matching. They are taken from works of Winkelmann and Nolze (2015a, 2015b). A third example—recognition of cubic-tetragonal pseudo-symmetry in γ-TiAl—will be described in Section 9.6.3.

Note that the next two sections closely follow a text published earlier by Zaefferer and Habler (2017). One of the most prominent and difficult-to-solve examples of pseudo-symmetries is caused by the abovementioned absence of a center of symmetry in certain crystal structures. This applies, for example for crystals of ZnS (Zinkblende) prototype with space group $F\bar{4}3m$, no 216. Besides ZnS, several important semiconducting materials crystalize in this structure type, e.g. GaAs, CdTe, or GaP. Note that if all atom positions in the ZnS structure are occupied with the same atoms the diamond structure is formed (Si, Ge, diamond), space group $Fd\bar{3}m$, no. 227, which, in contrast, has a centre of symmetry. The absence of a center of symmetry results in polar crystallographic directions, where physical properties in a direction $[uvw]$ are different from those in the opposite direction $[\bar{u}\bar{v}\bar{w}]$. The members of the ZnS structure, for example, show polar piezoelectric behavior, and it may be important to measure the direction of the polar axis and its distribution in a polycrystalline sample correctly. Winkelmann and Nolze (2015a) demonstrated that the use of realistic diffraction pattern simulations allows determination of this asymmetry: for the GaP with Zincblende structure they demonstrated that the direction of the polar [111] axis can be distinguished, i.e. that EBSD allows the distinction of the [111] and the $[\bar{1}\bar{1}\bar{1}]$ crystal direction. The major results from this work are summarized in Figure 7.12. The principle of orientation distinction is the fact that the {111}-type Kikuchi bands show an intensity anisotropy, which depends on the direction of illumination of the crystal. In Figure 7.12a an experimental pattern is shown. This diffraction pattern may fit to either of the two simulated diffraction patterns in Figure 7.12b and c, which show the mentioned intensity anisotropy of the {111} bands. By image matching of the patterns with the simulated patterns the intensity anisotropy was clearly detected and the correct orientation was determined as it is shown for complete orientation maps in Figure 7.12d, measured without pattern matching, and e, determined with pattern matching. Note that the same distinction is not possible for all ZnS structures: in the case of GaP, the atomic scattering amplitudes of Ga (Z = 31) and P (Z = 15) atoms are very different and the contrast

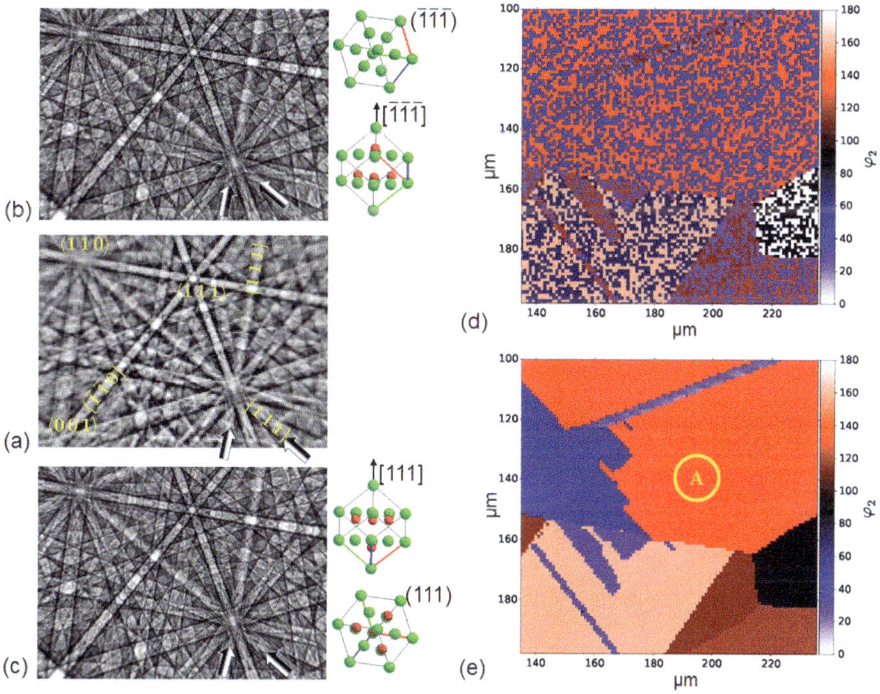

FIGURE 7.12 Use of pattern matching for indexing of non-centrosymmetric GaP crystals. (a) is the experimental EBSP obtained from crystal a in (e). (b) and (c) are the two possible simulated diffraction patterns solving (a). The orientations just differ by the sense of the polar [111] axis as indicated by the lattice unit cells. Note that the (111)-plane views correspond to the orientation of the pattern whereas the additional lattice cells display the view from the side onto the unit cell to reveal the different order of the Ga and P atom layers more clearly. The arrows in (b) and (c) indicate those Kikuchi bands which show a pronounced asymmetry that can be used to distinguish the two orientations. It is clear that (c) fits better to (a) than (b). (d) Orientation map from a polycrystal of GaP indexed taking into account only the geometry of the pattern: the two possible polar directions cannot be distinguished and lead to random scatter. (e) Same map as (d) but indexed taking into account the intensity distribution. The polar directions are correctly distinguished. (Adapted from Winkelmann and Nolze, 2015b.)

in the {111} bands is therefore large. In contrast, for GaAs (with Z = 33 for As) or CdTe (Cd 48, Te 52) the scattering differences of the atoms are probably too small to create detectable contrast.

In a second study Winkelmann and Nolze (2015b) showed that it is, in principle, possible to determine the chirality of trigonal α-quartz. α-quartz crystallizes in the truly chiral point group 32, which means that it can form two enantiomorphic crystals with space groups $P3_121$ and $P3_221$ (containing a left-rotating or right-rotating threefold screw axis). As displayed in Figure 7.13a and b, these crystal structure varieties are mirror-symmetric similar to the left- and right-hand pair. This enantiomorphism of α-quartz has important consequences for its piezoelectric properties, as the two enantiomorphic crystals show opposite polarity. Note that properties

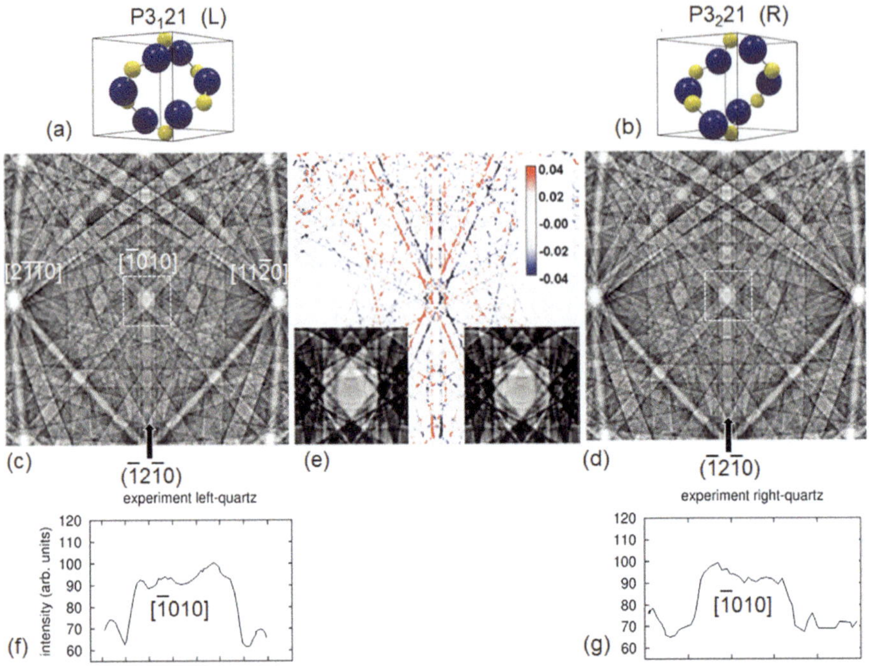

FIGURE 7.13 Use of EBSD to distinguish left-handed (left side) and right-handed (right side) quartz. (a) and (b) are sketches of the atomic configuration and their appropriate space groups. (c) and (d) are EBSD patterns simulated for both enantiomorphic crystals centered about the [$\bar{1}$010] pole. (e) The differences between the patterns in (c) and (d) calculated as Ac (see text). The inlets in this figure display the details of the two different [$\bar{1}$010] poles marked by a white square in (c) and (d). (f) and (g) display the experimentally determined profiles, measured through the center of the [$\bar{1}$010] poles. Note that the original graphs also contained data from Dauphiné twins which are left away here for simplification. (Adapted from Winkelmann and Nolze, 2015a.)

which are by themselves centrosymmetric, e.g. thermal conductivity or elasticity, are not affected by the enantiomorphism. Using dynamically simulated EBSD patterns the authors showed that the diffraction patterns of left-handed and right-handed quartz crystals show, indeed, weak but characteristic differences. Their results are reproduced in Figure 7.13c and d which show the two simulated diffraction patterns, whereas Figure 7.13e displays the normalized difference between both which is calculated by $A_C = \dfrac{I_R - I_L}{I_R + I_L}$ where I_R and I_L correspond to the signal intensities in the patterns of right- and left-handed quartz, respectively. The insets in these figures zoom into the [$\bar{1}$010] zone axes which show clear left- and right-hand intensity differences. When analyzing true electron backscatter patterns (EBSPs) from left- and right-handed crystals these differences are, however, for two reasons very difficult to spot. The first reason is that experimental patterns usually show a certain amount of blur which is due to the detection system and due to the energy variation of the electrons, caused by the inelastic scattering events. The second reason is the existence of excess and deficiency lines in true patterns which are, similar to the TEM case, due to the anisotropic intensity distribution in the illumination source inside the crystal. Of the two Kikuchi lines enveloping a Kikuchi band, the line which is closer to the direction of highest backscatter intensity (which is approximately the direction of reflection of the primary beam, see Figure 6.17) appears darker while the other line is brighter than their average intensities. This illumination contrast superimposes onto the crystallographic contrast and needs to be carefully separated from the latter. The excess-deficiency contrast is zero for those bands, which run vertically through the pattern center. Therefore, such bands are most reasonably selected for the investigation. Furthermore, although R-L differences may be found all over the pattern as shown in Figure 7.13e, it seems that the differences are strongest in the $(\bar{1}2\bar{1}0)$ band which corresponds to the pseudo-mirror plane. The authors selected a $(\bar{1}2\bar{1}0)$ band running exactly vertically across the screen and showed that it is, indeed, possible to determine sufficient details to distinguish the two crystal structure varieties. Figure 7.13f and g reproduces the intensity profiles obtained at a cross section through the $[\bar{1}010]$ zone axis across the $(\bar{1}2\bar{1}0)$ band. The contrast is clearly asymmetric which characterizes the right- or left-handed crystal. Note that it is important to observe the profile into a well-defined direction, for example the positive c-axis direction, in order not to confuse the left and right side.

Our explanations here above show that the R-L distinction is in principle possible. In practice, however, this is very difficult and, at least at the current moment not a generally applicable technique. Nevertheless, the investigations described above show that EBSD has great potential, not only for determination of microstructure and texture but also for crystal structure analysis, including crystallographically challenging structures. In combination with dynamical theory simulations it may become possible to determine even unknown crystal structures, although this is certainly still a way to go.

7.4.5 Cross-Correlation EBSD for the Measurement of Elastic Strain Fields

Note that the following sections closely follow a text published earlier by the author (Zaefferer and Habler, 2017). As will be discussed in Section 8.2.10, the angular resolution of HT-based EBSD analysis is in the range of about 0.5–1°. The reasons for this relatively large value are manifold: first, the HT itself is of limited angular resolution; next, the conventional HT can detect only straight features though Kikuchi-lines are actually curved; finally, the pattern projection center is only approximately known. Another related shortcoming of HT-based EBSD analysis is the very inaccurate determination of interplanar spacings of lattice planes, i.e. of Bragg angles, as they are used in x-ray diffraction for the measurement of lattice constants and elastics strains. The reasons for this are, first, the rather small values of typical Bragg angles in electron microscopy. Second, the dynamic nature of the diffraction process leads to rather blurred diffraction patterns as compared to x-ray diffraction patterns. As a consequence—and in contrast to x-ray diffraction—absolute lattice constants or an elastic (*hydrostatic*) *expansion or compression* of the lattice cannot be measured accurately. In contrast, a *deviatoric strain* (i.e. a strain which does not homogeneously expand or compress the lattice but somehow shears the lattice) leads to the change of the interplanar angles, i.e. the angles between different sets of lattice planes or diffraction vectors. By careful comparison of an EBSP from a measurement position with a very similar EBSP from an *assumingly strain- and rotation-free reference* position using image cross-correlation (ICC) techniques, the accuracy of EBSP analysis can be improved significantly, so that very small rotations and even elastic lattice distortions can be measured. This technique has been developed mainly by Wilkinson et al. (2006), based on earlier works which are referenced in this paper. Recent overviews can be found in Wilkinson et al. (2014) and Gianola et al. (2019). We will refer to it here as *cross-correlation (CC) EBSD*. It can also be found under the name *high resolution (HR) EBSD*, although this may lead to confusions with spatial resolution. The technique is based on the fact that lattice distortions as well as small misorientations cause characteristic shifts of different areas within an EBSP, as schematically illustrated in Figure 7.14. These shifts can be determined by comparing the measured pattern with a reference pattern of a crystal of almost identical orientation and without lattice distortions. To this end, the pattern is segmented into reasonably sized regions of interest (ROI). For each of these ROIs the shift is determined by ICC. Note that elastic strains and small lattice rotations yield very small shifts in the diffraction pattern. A large elastic strain of 0.2% (which is approximately the largest elastic strain reachable in metals before they deform plastically) produces an angular shift of about 0.1° in an EBSD pattern. For usual

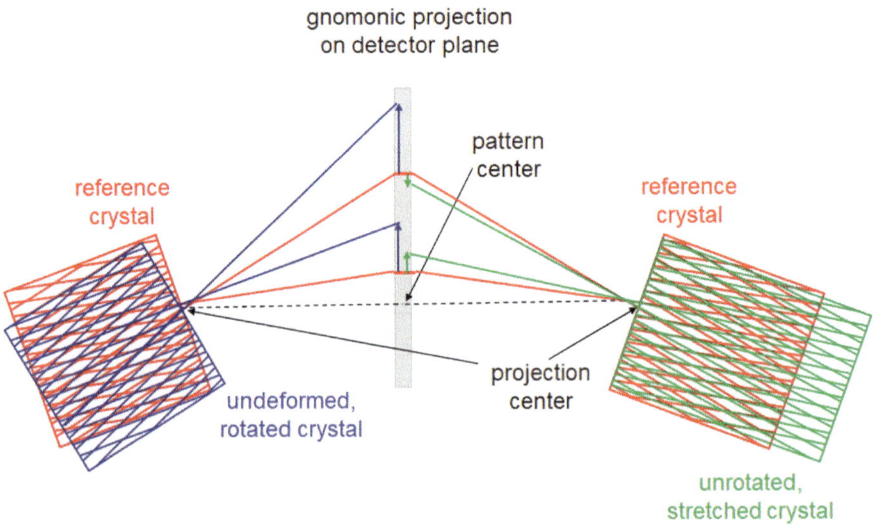

FIGURE 7.14 Schematic display of the shift of Kikuchi bands or their respective intersections (crystallographic poles) by a rotation or stretch of the crystal lattice. The arrows on the detector plane indicate the observed shifts with respect to the reference position. The red crystal displays the unrotated and stretch-free reference crystal, the blue crystal shows the effect of a small rotation, the green crystal that of a small stretch. For all crystals the pattern center relative to the detector plane, indicated by the black dotted line, is the same.

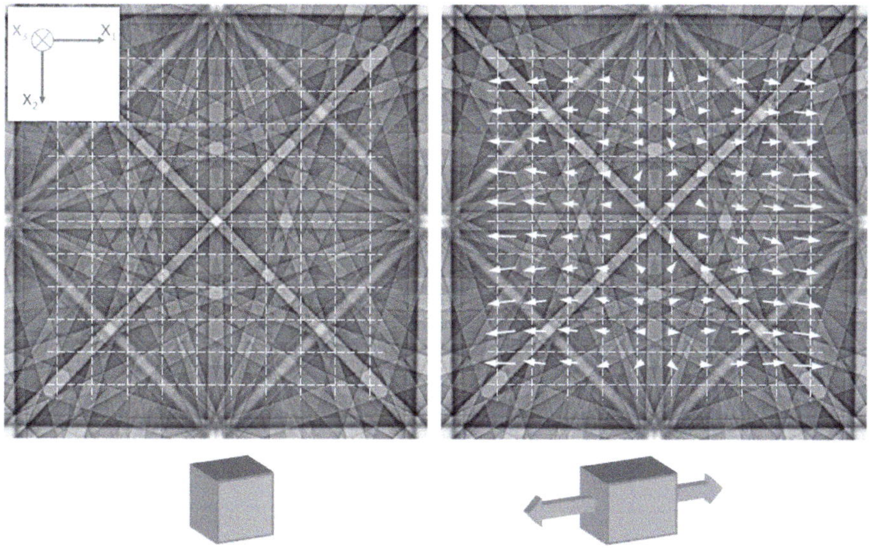

(a) unstrained (b) strained in x_1 direction

FIGURE 7.15 Illustration of pattern cross-correlation for determination of small crystal lattice distortions or rotations using simulated diffraction patterns from a crystal with known lattice distortion. The white squares in (a) indicate the regions of interest in the unstrained (simulated) pattern. The arrows in (b) indicate the shifts of the ROIs after an extension of the lattice in x_1 direction. (Adapted from Jaepel, 2015.)

detection geometries, this corresponds to a shift of about one pixel across a detector width of about 1,000 pixels (note that all these values are just rough estimates!). Thus, if reasonable elastic strain resolution is requested it is necessary to achieve sub-pixel resolution for band shift detection. This is accomplished in the ICC by fitting of the cross-correlation peaks with a smoothing function, for example a cubic spline. In this way shifts as small as a few hundredths of a pixel can be detected from high-quality EBSPs. Figure 7.15 shows an example for the determination of lattice distortion: for each ROI, marked by white squares in the unstrained image, a shift vector is determined, which is displayed as a green arrow on the strained image. From the derived shift vector field, the deformation gradient tensor can be calculated and from this, via multiplicative decomposition, the rotation and distortion tensors inferred, using the assumption of infinitesimal strains. The shifts determined for the individual ROIs correspond to changes of the angles between crystallographic directions. This means that only lattice rotations and the deviatoric part of the strain tensor can be determined directly. The missing three normal components can be determined additionally presuming that the measurement is performed at the surface of the material and that there is no stress component perpendicular to this surface. A detailed mathematical description of the analysis of shift vectors is presented by Wilkinson et al. (2006). Currently a strain accuracy in the order of 10^{-4}, and of rotations in the order of 0.01°, can be reached. It should be mentioned, however, that elastic strains (or residual stresses) can, currently, only be measured relative to a reference point in the same crystal, i.e. only third kind residual stresses are accessible. For the measurement of absolute lattice constants the pattern projection center would have to be determined very accurately (on about 1 μm close), which is currently impossible. Current techniques for the determination of the projection center will be introduced in Section 8.2.8.

One important precondition for reliable application of the cross-correlation technique is that the features compared in the two images are only shifted with respect to each other, but not distorted. When comparing EBSD patterns, this condition is fulfilled as long as the lattice rotations stay small. Larger lattice rotations, exceeding about 3°, however, lead to severe distortions caused by the gnomonic projection of the pattern and lead to increasingly large errors in stress and strain determination. The cross-correlation technique has, therefore, originally only been employed to measure residual stresses in semiconductor materials (Wilkinson, 2000; Wilkinson et al., 2014) which contain very few dislocations and, consequently, only very small lattice rotations. In contrast, the application to plastically deformed metals (e.g. Vaudin et al., 2008; Dingley et al., 2010; Britton and Wilkinson, 2011) led to less reliable results because they create stronger rotations caused by geometrically necessary dislocations. In the meantime the problem posed by large lattice rotations has been largely solved by employing so-called pattern remapping techniques (Britton and Wilkinson, 2012; Maurice et al., 2012). In this technique the lattice rotation between measured and reference orientation is determined using a standard HT-based approach. Subsequently, one of the patterns is remapped onto the orientation of the other one by a pixel-by-pixel operation. Finally, the remaining differences are determined using the cross-correlation approach. With this approach misorientation up to about 10° can be compensated satisfactorily and the accuracy of strain measurement be significantly improved. A few examples on the application of the cross-correlation EBSD technique for mapping of elastic strains and small rotations will be presented in Sections 9.4.7 and 9.4.8.

7.5 SUMMATION

In Chapter 7, we have introduced the principles of geometric and intensity analysis of electron diffraction patterns and methods for crystal orientation determination from the patterns. For spot diffraction patterns in TEM it is important to extract position and intensities of spots to calculate an accurate orientation. This process makes use of the fact that a spot diffraction pattern is the intersection of the curved Ewald sphere with the regular reciprocal lattice. From TEM spot diffraction patterns, orientations can be determined with an accuracy of about 0.5°.

For Kikuchi-type diffraction patterns, the HT has been established for extraction of diffraction vectors (or lattice plane positions). The HT detects straight lines or bands in an image with a freely definable resolution. Nevertheless, the orientation accuracy determined through the process is not better than about 0.2° at best, because Kikuchi lines and bands are not perfectly straight features. The HT approach works satisfactorily for a large amount of EBSD patterns. It fails or becomes inaccurate, however, for EBSD patterns taken at low acceleration voltage or from crystals with very particular intensity distributions or for TEM Kikuchi patterns.

Instead using a HT or a spot detection algorithm, TEM and SEM diffraction patterns can also be analyzed by comparing them with a dictionary of precalculated template images. This calculation-intensive approach has the great advantage of automatically taking into account the intensity distribution in diffraction patterns. It may, therefore, potentially, detect subtle intensity differences between different phases more sensitively. Also the approach appears to work on patterns of much lower quality—i.e. lower beam current or shorter exposure times—than the conventional approaches. Also EBSD patterns measured with low acceleration voltage (and therefore at highest spatial resolution) can be analyzed without limitation.

NOTE

1. In many textbooks the z-coordinate is assumed to be 0 (flat Ewald sphere surface) which, however, does not give an accurate orientation determination.

8 Practice of Orientation Measurement and Orientation Microscopy

8.1 INTRODUCTION

In the previous chapters we have introduced the physics and the principles of analysis of electron diffraction patterns for orientation measurements. These techniques form the basis of "orientation microscopy" (ORM), which allows, based on spatially resolved determination of individual crystal orientations and crystallographic phases, the reconstruction of microstructures and the determination of microtexture (see Chapter 9 for a more detailed discussion). This chapter is dedicated to the practice of these techniques. The focus will be on the electron backscatter diffraction (EBSD) technique (Section 8.2) because it is the by far most used and probably also most versatile technique. We start the chapter with a few hints on specimen preparation, although this wide field cannot be covered completely and requires a lot of manual experience. We will then introduce hardware issues, like geometry of the microscope setup, including the classical backscattering geometry and the newer transmission geometry and different detector types. Furthermore, we will discuss the determination of optimum microscope settings and pattern pre-processing methods. A very important issue is the calibration of the technique, as only the correct calibration guarantees correct results. All this finally influences spatial and angular resolution of the technique which is currently also the most important limit of EBSD-based ORM.

The spatial resolution limit of EBSD can be overcome by transmission electron microscopy (TEM)-based techniques, though on the account of much smaller investigated areas which may compromise the representativeness of such results (see Chapter 9). In Section 8.3 a few technical hints to the TEM techniques will be given, but the most important topic will be a comparison of resolution limits of TEM, transmission Kikuchi diffraction (TKD) in scanning electron microscopy (SEM) and EBSD. This discussion will then be guided further in Chapter 9 where a few typical cases of SEM vs. TEM ORM will be presented.

8.2 EBSD-BASED ORIENTATION MICROSCOPY IN THE SEM

The basic geometric setup of EBSD-based ORM in a SEM is sketched in Figure 8.1 together with a few of its variants. The basic setup (Section 8.2.4), displayed in Figure 8.1a, consists of a bulk sample (sample conditions and preparation are found in Section 8.2.1 and 8.2.2), steeply tilted on the SEM stage, facing toward an EBSD

 DOI: 10.1201/9781003258339-11

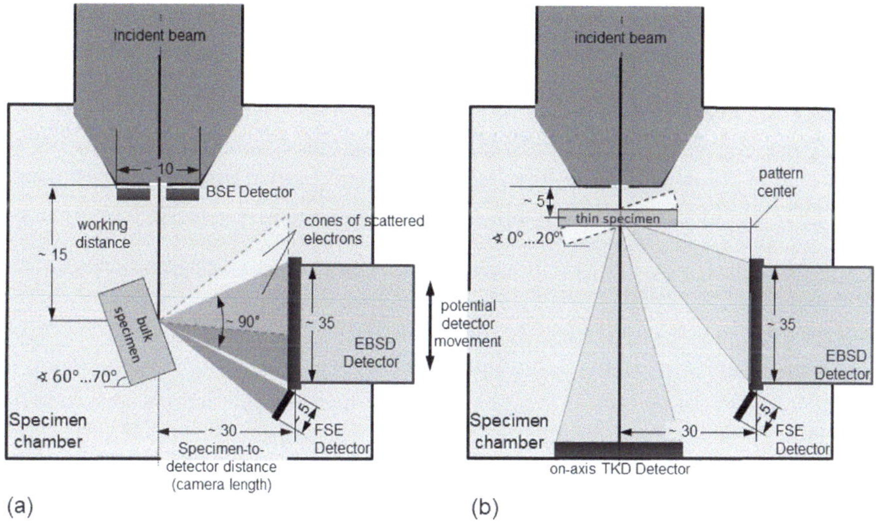

FIGURE 8.1 Sketch of the geometric setup of EBSD and TKD in an SEM chamber. All approximate measures given in millimeters. (a) Setup of conventional EBSD on bulk specimens. The cone sketched by a broken line indicates the situation of lower acceleration voltage, and higher atomic number of the specimen. Placing this cone on the detector requires to move the detector upwards or the sample downwards. (b) Setup of transmission Kikuchi diffraction on a thin specimen. The detector is either a conventional EBSD detector or an on-axis detector which has needs to be able to cope with the high contrast of diffraction patterns at that position.

detector (Section 8.2.5) which is mounted at a side flange of the SEM chamber. The detector, in most cases still a retractable transparent phosphor screen observed by a highly light-sensitive camera, is positioned close to the tilted sample (usually between 20 and 40 mm away from the beam-sample intersection). The sample is illuminated with a stationary electron beam (beam conditions in Section 8.2.3) while the EBSD pattern formed on the detector is recorded. The recording time may vary between less than a millisecond for high-speed mapping to several seconds for high-quality pattern recording (Section 8.2.6). Then the beam may be moved to the next point and mapped step-wise over the surface. A similar process can be performed on thin electron transparent specimens mounted below the objective lens as shown in Figure 8.1b and discussed in Sections 8.2.4 and 8.2.9 and in Section 9.6.2.

It should be mentioned that EBSD-ORM is—compared to other scanning techniques in the SEM—a rather slow technique. Although the speed of acquisition reaches currently up to several thousand measurements per second, maps which usually consist of several 100,000 pixels still require measurement times in the order of half an hour or more.

The quality of an EBSD measurement and the capabilities of an EBSD system are measured by means of a number of parameters. These are the *spatial resolution*, the *orientation or angular resolution*, the *correctness of indexing*, the *acquisition and processing speed* of the system, and the amount of information available in the diffraction patterns. These parameters are controlled by a number of factors, including, under others, the *sample* (sample preparation, material, microstructure), the *microscope* (emitter type, accelerating voltage, beam size and current, mechanical and electronic stability), the *EBSD camera* (light sensitivity, pixel resolution, read-out time), and the EBSD software (processing speed, robustness). In the following some of these factors are discussed. A further detailed (though a bit outdated) treatment can be found, for example, in Humphreys (2001).

8.2.1 Specimen Requirements

EBSD has a few characteristics that dictate certain sample requirements. (i) EBSD is a surface-sensitive technique where the signal stems from the first 10–50 nm of the sample. (ii) EBSD is done at high tilt in order to expose the sample surface to the detector. (iii) EBSD requires relatively high beam currents in order to obtain patterns with an acceptable noise level.

Point (i) requires that that the sample surface is free of any deformation layer created by sample preparation, e.g. grinding and mechanical polishing. Best polishing results are usually obtained by chemical or electrochemical procedures. Very successful, particularly on metallic materials, is also mechano-chemical polishing with a suspension of oxide particles (OPS). Also ion polishing either with a broad ion beam (BIB) of Ar^+ ions or with a focused ion beam (FIB) of Ga^+ or Xe^+ can give very good sample preparation results. In the next section a few common recipes will be mentioned.

Point (ii) claims that the sample, or at least the investigated part of it, must be flat and it must be possible to expose the flat surface at an approximately 70° angle toward the detector. Flatness guarantees constant incidence and take-off angles across the sample surface which is important for EBSD contrast. As shown in Figure 6.18, the intensity center of an EBSD pattern lies approximately in reflection direction. Changing the sample surface inclination for example due to a surface relief thus leads to a relatively strong shift of the intensity distribution across the pattern as schematically indicated in Figure 8.2. This is particularly disturbing if a constant background is subtracted from the patterns for contrast enhancement (see Section 8.2.6): in case of a shift of the intensity center the background does no longer fit to the current pattern. There are a number of cases where a surface relief cannot be avoided. Examples may be multi-phase materials where different phases may show different removal rates during mechanical or chemical sample preparation. Another example is the in-situ observations of deformation processes, where surface roughness develops due to differently deforming crystals. Also, when measuring orientations of individual crystallites in a powder or on freely grown surfaces different surface inclination leads to very different pattern illumination conditions. In all these cases it is important to work with a dynamic background subtraction method (see Section 8.2.6).

The critical dependence of the EBSD technique on surface conditions is also reflected in the sensitivity to contamination: the relatively high beam current used to produce EBSD patterns leads to the quick polymerization of residual carbohydrates on the surface. The growing contamination layer may even disable the formation of diffraction patterns. It is therefore of great importance to use clean samples (rinse

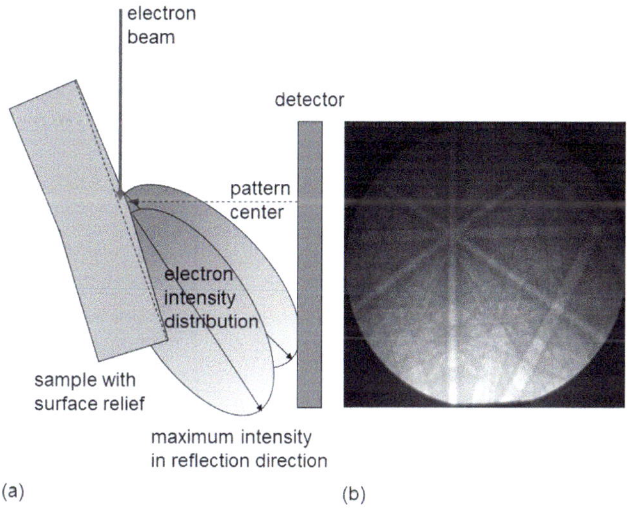

FIGURE 8.2 Influence of different surface inclination on pattern intensity distribution. (a) Schematics of intensity distribution in patterns: the intensity maximum is found approximately in reflection position of the primary beam with respect to the surface. A changing surface inclination thus leads to a significant change of the intensity distribution at the detector. Note that a static background subtraction procedure will not work properly and result in bad pattern quality. (b) Experimental diffraction pattern of a TiC precipitation without background correction. The pattern center is positioned at the top third of the pattern, the intensity center is at its bottom.

several times with clean alcohol or acetone and dry the samples well afterwards, do not touch samples with the hand) and a clean vacuum.

Point (iii), finally, requires good electrical and thermal conductivity of the sample. On metals, usually no problem arises as long as the sample is well contacted to the sample stage. Nevertheless, resin-mounted samples, even if they are well conductive, often lead to sample drift during longer measurements, probably due to their reduced thermal conductivity. Non-conductive materials have to be coated with a thin layer of carbon (up to few tens of nanometers) or an even thinner layer (~2 nm) of amorphous metal; Cr has been found to give good results for this case.

The quality of an EBSD pattern (i.e. the signal-to-noise ratio, SNR) and the sharpness of the Kikuchi lines) also depends on the material observed. Generally, the quality increases with increasing atomic weight and increasing bonding forces of the material. Usually, heavy metals or minerals show better patterns than light ones and covalently bonded materials show sharper patterns than those with metal bonding. In Figure 6.19 the patterns of five metals with significantly different atomic numbers and different bonding strength are compared; note the exposure times required to obtain the patterns. Finally, it has to be mentioned that lattice defects (e.g. dislocations) disturb the formation of clear patterns. The higher is the lattice defect density of the sample, the lower is the pattern quality. This relationship, on the other hand, can be used to map the density of lattice defects in the sample (see Chapter 9).

8.2.2 EBSD Specimen Preparation

One of the attractions of EBSD is that specimen preparation is straightforward, often similar to that for optical microscopy, though the surface finish is usually more critical. This is in contrast to the requirements for TEM-based electron diffraction techniques, which require complex thin-foil preparation procedures.

The specimen preparation objective for EBSD can be stated very simply: the top 10–50 nm of the specimen should be representative of the region from which crystallographic information is sought, since it is only from this surface region that diffraction occurs. The shallowness of the diffraction volume results from the highly tilted specimen position in the SEM chamber and the consequent reduction of the mean free path length for electrons (Figure 6.17). The practical upshot of this, and indeed the "only" crucial aspect of specimen preparation for EBSD, is that the specimen surface must not be obscured in any way—by mechanical damage (e.g., grinding), surface layers (e.g., oxides and most coatings), or contamination.

The standard metallographic preparation route for most specimens, especially metals and alloys, is mounting, grinding, and polishing (Katrakova and Mücklich, 2001). The adaptation of this route for EBSD is discussed first, since metallic materials are still the most common group to which EBSD is applied. Mounting the specimens in a conducting medium is clearly advantageous for SEM work; otherwise, electrical contact with the specimen can be established by using silver or carbon paint or conductive tape, or simply by cutting the specimen from the mount after the preliminary preparation stages. Note that the use of silver or carbon paint may significantly increase the tendency to form contamination layers during the EBSD mapping. Note also that different conductive resins exist. Most preferable is one with carbon filling because of its chemical inertness. Fillings with iron or copper powder may lead to unwanted electrochemical reactions with the sample during wet-metallographic preparation.

It is the final preparation step that ensures suitability for EBSD. Diamond polishing is not an appropriate final stage because of the remnant mechanical damage entailed (e.g. Wynick and Boehlert, 2005). However, many common electrolytes or etchants that are used to reveal the microstructure for optical microscopy do so by attacking the specimen surface. This has the concurrent effect of removing polishing damage. Hence, for many metals and alloys the route used for optical metallography, with minimized diamond polishing to avoid surface damage and a subsequent chemical or electrochemical step, is all that is needed for EBSD. A useful guide is that the same electrolytes that are used to prepare TEM thin foils can often also be used to prepare the surfaces of bulk specimens for EBSD. A comprehensive collection of TEM sample preparation recipes is found in Thompson-Russell and Edington (1977). Non-metallic specimens that cannot be electropolished (typically, minerals, ceramics, and semiconductors) may be fractured, cleaved, or mounted and mechanically polished for EBSD. Specimens that contaminate or oxidize rapidly may need to be prepared immediately before SEM examination or alternatively ultrasonically cleaned and stored under solvent (e.g., white tin). Particularly difficult specimens can be ion milled for a few hours (Michael and Kotula, 2008; Chen et al., 2012; Nowakowski et al., 2016).

A few examples of preparation methods used for EBSD are given here, to illustrate some of the points made in this section.

- *Commercially pure aluminum, titanium alloys.* Electropolish in 5% perchloric acid in ethanol at –25°C. Etchants that deposit a film on aluminum are unsuitable for EBSD.
- *Aluminum-lithium alloys.* Immerse for several seconds in Keller's reagent, slightly warmed.
- *Mild steel.* Swab with 2% nital for several seconds.
- *Many rocks and minerals.* Diamond polish block specimens followed by colloidal silica polishing for several hours, particularly on a vibration polisher.
- *Polysilicon.* Wash in detergent and immerse for 1 min in 10% hydrofluoric acid in a plastic beaker.

A highly recommended method for preparing a variety of specimens for EBSD is final polishing in colloidal silica, since this medium does not introduce the harsh mechanical damage associated with diamond polishing (Nowell et al., 2005; Chen et al., 2012). Unlike standard abrasive slurries, colloidal silica or OPS (oxide polishing suspensions) use a combination of a chemical surface reaction based on the pH and oxidation ability of the colloidal base together with a mechanical removal of the reaction products by silica particles of <50 nm diameter. Large particles in metallic matrices (e.g. intermetallics in aluminum alloys or carbides and nitrides in steels), low-melting-point metals and alloys, ceramics, composites, and minerals are all amenable to polishing for up to several hours in colloidal silica (Katrakova and Mücklich, 2002).

Difficulties almost always arise when multi-phase samples are to be prepared as different phases have different chemical, electrochemical, mechanical, and ion sputtering behavior. In these cases, often different polishing quality is obtained on different phases and surface relieves may develop that disturb the acquisition of patterns. No universal preparation method can be given for these cases but OPS polishing is frequently a good compromise. Collections of sample preparation methods can be found in Katrakova and Mücklich (2001) and Vander Voort (2011a and b), as well as on the websites of the EBSD-system producers (EDAX-TSL and Oxford-HKL Technology).

A powerful way for preparation of multi-phase materials or materials that do not easily polish chemically or electrochemically is grazing incidence BIB polishing (with incidence angles of the BIB as shallow as 10°). There are a number of producers of ion beam milling instruments. The usual geometric setup of BIB milling for surface polishing is sketched in Figure 8.3. A very informative text on the fundamentals and practice of this technique has been written by Brodusch et al. (2015). Figure 8.3b–e taken from their work shows the influence of beam energy and specimen inclination on beam damage and sputtering yield, i.e. milling efficiency for different metals. These graphics give hints on the correct sputtering parameters for different materials. The authors show that for milling of lithium metal, which can hardly be prepared by any other means, they obtain very good results for 15 minutes

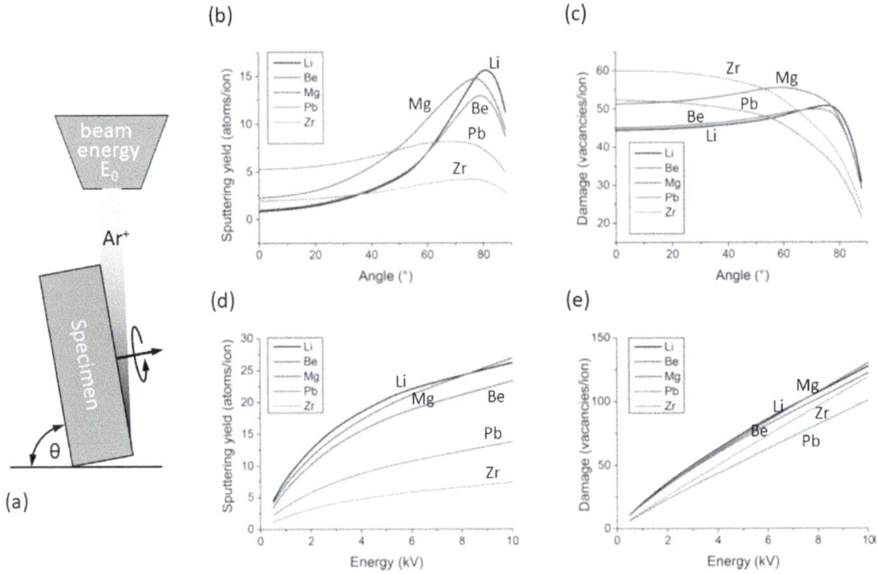

FIGURE 8.3 Specimen surface preparation by broad ion beam milling. (a) Geometric setup. A strongly tilted, rotating specimen is illuminated with a broad Ar^+ ion beam of a given beam energy E_0 (given as acceleration voltage of the Ar beam). (b) to (e) Variation of the sputtering yield and damage as a function of the ion accelerating voltage E_0 and incident angle θ for Li, Be, Mg, Zr, and Pb. (b) Sputtering yield and (c) damage at $E_0 = 3$ kV, (d) sputtering yield, and (e) damage at θ = 80°. Simulations were performed with the SRIM software (Ziegler et al., 2010). (Adapted from Brodusch et al., 2015.)

of milling at 80° of sample tilt using a beam energy of 3 kV and 80-μA current. Mamtani et al. (2020) investigated the use of BIB polishing for preparation of multiphase rocks and minerals and could show that the technique has great benefits, both in terms of surface quality and preparation time. One important disadvantage of the BIB method is the thermal energy that is introduced into the sample. Brodusch et al. (2015) discuss this issue and claim, from other literature, that the specimen temperature may rise to a few hundred degrees if the beam current and the beam energy are selected high. In their case they realized that the temperature of the investigated lithium stayed below its melting point of 180°C. A good device to cope with sample heating is a cryo-BIB instrument. A further disadvantage of ion milling is the appearance of dimples on the surface when hard (sputter-resistant) and soft (easily sputtering) phases are found in one sample. A reduction of milling time and increase of sputtering angle may be a way to cope with this problem. Finally, the prepared sample surfaces may be limited to a few mm in diameter, in contrast to samples of traditional metallographic sample preparation which may reach centimeters in diameter.

Finally, small areas of a specimen can also be prepared by an FIB either of Ga^+ ions or Xe^+ ions (using a so-called plasma-FIB). FIB preparation is particularly used for 3D EBSD, described in Section 11.4.1. Besides the relatively small size of the prepared area (usually between 20 and 100 μm in diameter) and surface roughness caused by the parallel beam inclination, a serious problem of FIB polishing is the introduction of damage into the crystal lattice. The damage is usually not very dramatic on metals (probably because they recover sufficiently quick), but it may lead even to amorphization on materials like silicon, intermetallics, and other structures with ionic or covalent bonding. Michael and Kotula (2008) showed that the beam damage on Si is only a few nanometer thick when using acceleration voltages as low as 2–5 kV, for both Ar^+ and Ga^+ beams. In contrast, Kato et al. (1999) showed that the amorphous layer can be several 10 nm thick at higher voltages which is too thick to obtain EBSD patterns.

Inadequate or inappropriate specimen preparation could give rise to misleading data interpretation. For example, the diffuseness of an EBSD pattern is a guide to the amount of plastic strain in the lattice; however, if a specimen contained diamond polish damage prior to etching, the pattern might be interpreted erroneously as arising from a deformed specimen. Furthermore, the absence of an EBSD pattern in a sampled region is evidence either that it is noncrystalline, or highly strained, or that the grain size is smaller than the interaction volume. However, these interpretations are only valid after correct specimen preparation.

Non-conductive samples are often coated with a thin layer of a conductive material like carbon or chromium. The thickness of this coating layer that still allows to obtain patterns from the underlying material is a function of both the accelerating voltage and the atomic number of specimen and coating. Higher accelerating voltages allow deeper penetration of the electron beam, whereas higher atomic numbers are associated with lower elastic mean free paths of electrons and thus shallower penetration. The sensitivity of EBSD patterns to coating thickness has been explored by coating single crystals of silicon with various thicknesses (from 5 to 40 nm) of aluminum, nickel, or gold (Michael and Goehner, 1994). Figure 8.4 illustrates the effect of coating silicon with 5 nm of nickel. An EBSD pattern from the underlying silicon is easily observed when an accelerating voltage of 40 kV is used, yet for an accelerating voltage of 10 kV, the beam hardly penetrates through 5 nm of nickel. It was confirmed that the interaction thickness is approximately twice the elastic mean free path (Joy and Newbury, 1998), giving critical interaction depths—the maximum thickness of surface material that can be penetrated by the electron beam—at 40 kV accelerating voltage of 100, 20, and 10 nm for aluminum, nickel, and gold, respectively.

Another way to obtain EBSD patterns from non-conductive samples is to use a gas inlet to produce a gas-stream, e.g. air, nitrogen or argon, in the SEM chamber close to the specimen surface. The gas allows, by ionization through the electron beam, sufficient conductivity to discharge the sample surface without broadening the electron beam too much. Alternatively, a low-vacuum SEM may be used, although the long beam path through reduced vacuum reduces spatial resolution considerably.

In summary, since the EBSD signal appears from the volume immediately below the surface of a specimen, it is highly important to prepare good, artefact-free specimen surfaces. In some cases, particularly for single phase metallic materials of average

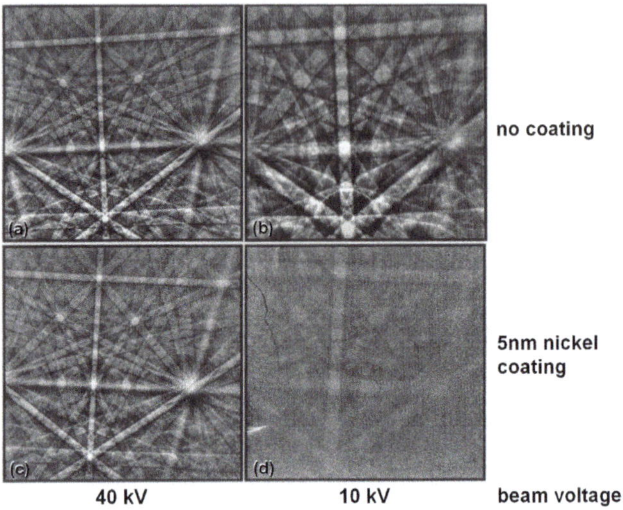

FIGURE 8.4 Illustration of the penetration depth of the electron beam in a silicon EBSD specimen. (A) No coating, 40-kV accelerating voltage; (b) no coating, 10-kV accelerating voltage; (c) coating with 5-nm nickel, 40-kV accelerating voltage; (d) coating with 5-nm nickel, 10-kV accelerating voltage. There is less beam penetration at 10 kV since the underlying silicon pattern is indistinct. (Courtesy of J.R. Michael.)

strength, this may be a straightforward task and almost as easy as the preparation of samples for light-microscopy. For many other materials, in particular for soft or multiphase materials, sample preparation for EBSD may, however, be a quite challenging task. Nevertheless, the very large possibilities given by EBSD-based ORM on quantitative materialographic investigations, certainly justify the necessary efforts.

8.2.3 Microscope Parameters

In principle, EBSD patterns can be obtained with every SEM. However, the quality of the obtained patterns and the achievable spatial resolution strongly depend on the microscope characteristics. Microscopes with *tungsten filament* deliver enough beam current but the beam spread is rather large when using a high beam current, thus leading to a reduced spatial resolution. On the other hand, instruments with *cold field emitter* allow for a small beam size but the relatively small beam current as well as the reduced long-term stability are not optimum for EBSD. The best overall EBSD performance is obtained from a microscope with a *thermal field emission gun (FEG)* and high beam current. A good compromise in terms of costs may be given by an SEM with *LaB_6 emitter*. There are surprisingly few and contradictory numbers available on the *probe current* at the specimen, necessary for optimum pattern creation. While Humphrey indicates probe currents for a FEGSEM in the order 50 to 100 nA and for a tungsten-filament SEM of 200 nA (Humphreys, 2001) for optimum EBSD performance, other authors quote beam currents between 0.75 nA to about 10 nA for a FEGSEM. It seems that Humphreys actually indicated emission rather than final probe current.

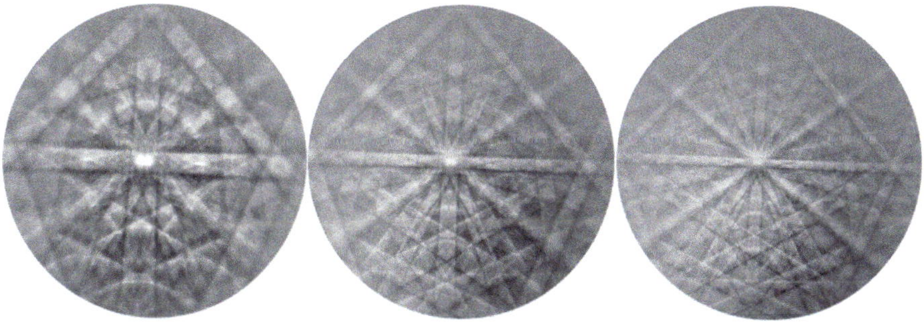

FIGURE 8.5 EBSD patterns obtained on a germanium single crystal at different acceleration voltages.

A small beam diameter is important to reach a high spatial resolution. In FEGSEMs, however, the beam diameter is so small that it is not the resolution-determining factor. In this case, the spatial resolution depends on the penetration depth of the beam into the material and is therefore influenced by the *accelerating voltage* and the investigated material (see Sections 6.3.3 and 8.2.9). Here we just mention that the spatial resolution in principle becomes better with decreasing accelerating voltage, however, currently an optimum is reached at about 15 kV. With lower voltages contamination of the sample starts playing a more important role. Also the sensitivity of the phosphor screen used for detection decreases. Finally, the Kikuchi band width increases with decreasing acceleration voltage which leads to problems with the currently used line detection algorithms. Figure 8.5 displays EBSD patterns that were taken on a Ge single crystal at three different acceleration voltages to illustrate the differences: with decreasing voltage the patterns become more noisy and the bands become wider and more blurry.

Two further points concerning the microscope should be mentioned: the *mechanical and electronic stability* and the *vacuum quality*. EBSD-based ORM requires rather long measurement times (up to 12 hours for very large maps) with the stage at a highly tilted position. During this measurement time the beam current has to be stable and its position must be perfectly controlled. Furthermore, for high resolution measurements with step sizes of few ten nanometers the stage drift must not exceed a few nm/min. Concerning the vacuum it was already mentioned that the technique requires a relatively high beam current and is therefore prone to the formation of contamination layers on the sample. For highest resolution measurements an oil-free vacuum, using a scroll pump for pre-vacuum pumping and a turbo-molecular pump and ion getter pump for high vacuum conditions seems to be beneficial, particularly in keeping the objective aperture clean. It should, however, be mentioned, that most of the contamination deposited on the sample during measurement stems from the sample itself: we observed that even a microscope with oil diffusion and rotary pump allows contamination-free ORM if the sample is clean. A plasma cleaning device mounted to the SEM chamber and employed before EBSD mapping may be helpful in the case of long or repeated measurements.

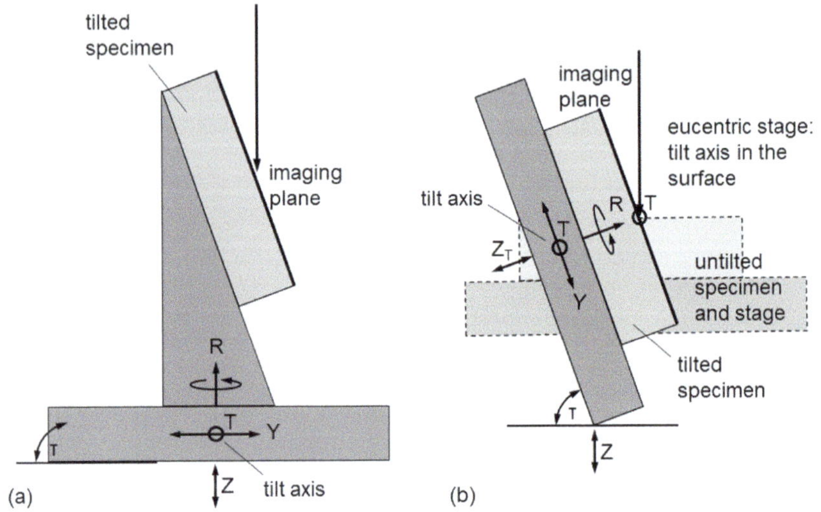

FIGURE 8.6 Sample holders for EBSD. (a) Pretilted stage mounted on a normal, non-tilted 5-axis stage, with axes X, Y, Z, T, and R (Y is out of the image plane). Sample movement in the imaging plane of the specimen requires movement of Y and Z or refocusing. (b) Tilted 5- or 6-axis (eucentric) stage, with axes X, Y, z, Z_T, T, and Rr. Sample movement in the imaging plane is accomplished simply by moving X or Y, no refocusing required. If the stage is eucentric (i.e. it has a Z_T axis), the Z_T axis can be used to move the tilt axis T into the imaging plane. Then the sample can be tilted from horizonal to EBSD position without losing the area of observation.

A final point to be mentioned concerns the *microscope stage* which holds and positions the sample in the SEM. In EBSD position the sample is usually tilted to 70°. This tilt angle can be realized in two ways, either by use of a pretilted holder on the SEM stage or by tilting the stage itself; both variants are sketched in Figure 8.6. The pretilted stage has the advantage that it is small and can be placed without the risk of collision rather close to the objective lens. It has, however, also a lot of disadvantages. In particular, the movement in the observation plane is rather complicated because the stage moves in its horizontal coordinate system. That means that every movement in Y-direction (in Figure 8.6) requires refocusing or change of sample height, Z. Also, the sample cannot be rotated about its normal or observed in horizontal position without changing the sample holder. The better choice is, thus, the conventional stage, tilted to 70° (with precaution of potential collisions). If the stage can tilt in a eucentric manner, i.e. it has either a well-defined eucentric sample height or even a Z_T axis to move the sample into the eucentric height position, then the move from horizontal to EBSD position becomes very convenient because the point of observation remains in the field of view and in focus. Any X-, Y-, and R (rotation) movement with the stage then occurs conveniently in the plane of observation and no large refocusing is required. Note that for both stage types it is important to align the tilt axis with the axis of (usually) horizontal scanning. This is done by performing a sample movement along the tilt axis (usually the X-axis) and controlling whether a point of interest remains on a horizontal line on the SEM observation screen. Potential deviations need to be corrected by a scan rotation (not by a sample rotation!). If scan direction and tilt axis are not aligned the high tilt position of the sample may lead to extreme distortions of the scan field on the surface and wrong texture results as has been shown by Nolze (2007).

In order to observe microstructure and orientation changes at temperatures above room temperature an SEM can be additionally equipped with a heating stage to perform in-situ annealing experiments under EBSD observation, In this way processes of recovery, recrystallization, and grain growth (Lens et al., 2005; Mirpuri et al., 2006) as well as phase transformations (Seward et al., 2004; Lischewski et al., 2008) can be observed. High-speed EBSD is of particular interest so as to allow for sufficiently high heating rates (e.g. Field et al., 2007). One of the potential problems with in-situ annealing experiments is that the free surface may influence the microstructural changes (Nakamichi et al., 2008), and hence it must be ascertained that the microstructure and texture effects observed at the surface are representative of the bulk behavior. Also, pattern quality may degrade at elevated temperature and emission of thermal electrons may interfere with the imaging of microstructure. It is furthermore possible to combine EBSD with in-situ deformation by uniaxial tension, compression, or bending (Han et al., 2003; Scheriau and Pippan, 2008). At present the applied deformation is halted in order to obtain an EBSD map (Karlsen et al., 2008). Combined in-situ EBSD annealing and deformation studies offer experimental advantages in the study of topics such as phase transformations and microstructure evolution.

8.2.4 Specimen/Microscope Geometry

The three main parameters that can be altered in the microscope/EBSD setup, displayed in Figure 8.1a, are the *specimen-to-screen distance* or *camera length*, the *specimen tilt*, and the *specimen height* or *working distance* in the microscope. Approximate values for all these parameters are necessary to calibrate the system (Section 8.2.8). Note that a modification of any of these values requires experience and attention as one risks collisions between sample and detectors or pole piece. It is recommendable to use a chamber camera when attempting a change of these values. Often for EBSD data collection, the specimen-to-screen distance or camera length has a fixed position at 20 to 30 mm, preset by the manufacturer of the system to obtain an optimum acceptance angle of approximately 90°. Nevertheless, it is usually possible to withdraw the camera from that position and thus decrease the solid angle of pattern acceptance. This creates a zooming-in into the pattern, similar to a change of the camera length in TEM. With larger camera length the pattern becomes darker but potentially more details are resolvable.

With regard to *specimen tilt*, EBSD patterns have been observed for specimen tilt angles >45°. However, as mentioned in Chapter 6, the path length of the backscattered electrons in the specimen decreases with increased specimen tilt, which leads to better contrast in the EBSD pattern with increased tilt angle (Venables and Harland, 1973). Tilt angles >80° are impracticable because of the excessive anisotropy of the sampled volume and distortion in the uncorrected image. A tilt angle of

70° represents a good compromise with regard to convenience and pattern contrast and it is used as a standard angle for EBSD. Note that with a change of specimen tilt also the position of maximum intensity on the detector changes. If, for example, the specimen tilt is reduced then the diffraction intensity moves upwards, as indicated by the shaded cone of diffracted electrons in Figure 8.2. In order to compensate for this either the camera has to be moved upwards or the specimen downwards (see also the next section on EBSD cameras). Note that for flat specimens the high specimen tilt distorts the interaction volume into an ellipsoid having an aspect ratio of tan70°:1 (3:1) on the surface as also discussed in Sections 6.3.3 and 8.2.9.

Although for standard SEM work best resolution and minimized focusing distortions are obtained from a short *working distance*, for EBSD the major consideration when choosing the working distance is to position the specimen such that the detector is homogeneously illuminated. This is often (but not always—see further below) the case when the pattern center is in the upper third of the detector. There are, however, further important factors: with increasing average atomic number of the specimen, decreasing acceleration voltage, and decreasing sample tilt angle the average total electron scattering angle increases and the maximum intensity of the pattern moves upwards. This is indicated by the shaded cones of electrons in Figure 8.1a. It is in this case optimum to move the camera upwards as well. Alternatively, but associated with a loss in spatial resolution, the sample can be moved downwards. A further limiting factor is the risk of collision with microscope hardware, particularly the pole piece, at small working distances. A further factor determining the optimum specimen position is, therefore, the geometry of the microscope. Taking all these factors into consideration, an optimum working distance is often in the range 15–25 mm; smaller values should be preferred, but optimum detector illumination should be given priority.

In recent times a second powerful geometry for ORM in SEM has been established, the use of transmission Kikuchi patterns, sometimes—confusingly—called "transmission-(t-)EBSD." This technique has been introduced by Keller and Geiss (2011) and since been refined by Trimby (2012) and Sneddon et al. (2016). It excels mainly at its superior spatial resolution which will be discussed in Section 8.2.9 but also by its ease of operation compared to TKP in TEM. The geometry of TKD is indicated in Figure 8.1b. A thin, electron transmittable specimen is placed directly below the objective lens. The patterns formed in the sample are recorded as transmitted electron signal below the sample, in most cases on the conventional EBSD detector. Examples of patterns obtained from a thin foil are displayed in Figure 8.7. In order to obtain optimum illumination of the detector the sample may be slightly (up to 20°) tilted away from the detector as indicated in Figure 8.1b. To this end it is necessary to use a dedicated sample holder (see the quoted literature). Because the projection center of the pattern is in the specimen which is in most cases above the detector, the pattern center is usually outside of the recorded pattern (see Figure 8.1b) and the diffraction patterns are relatively strongly distorted as this is visible in the diffraction patterns in Figure 8.7. Note that there are also attempts (Fundenberger et al., 2016) to mount a detector similar to the case of TEM, below the sample, on the optical axis. While this offers certain advantages in reduction of pattern distortion and pattern intensity, it suffers from the very strong brightness change (i.e. contrast)

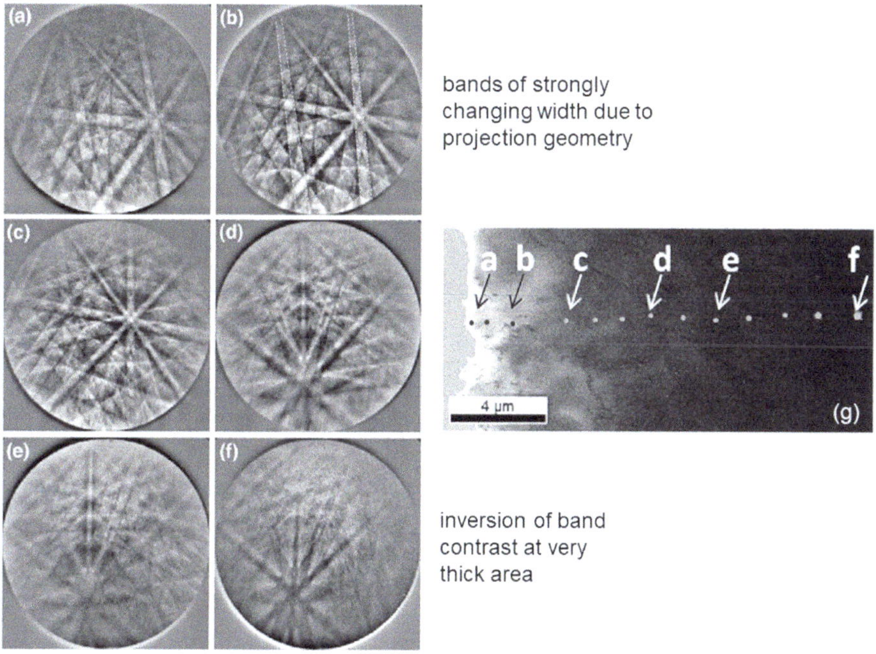

FIGURE 8.7 (a)–(f) TKD patterns obtained at different points in a thin foil of an 8Cr tempered martensite steel specimen at 25-kV acceleration voltage. (g) Pattern (a) is taken very close to the hole of the TEM sample. (Adapted from Suzuki, 2013.)

in the Kikuchi patterns around the primary electron beam, as it is also known from TEM TKP. While SEM TKP offers a very interesting option for high resolution orientation mapping (see Section 9.6.2) it sacrifices, at the same time, the important advantage of conventional EBSD, allowing observation of bulk samples.

8.2.5 EBSD Detector

The EBSD detector is probably the most essential part of an EBSD-based ORM system as it contributes importantly to spatial resolution, orientation resolution, speed of measurement, and resolution of pattern details for phase and stress determination. Figure 8.8a sketches a typical detector construction. It consists nowadays in most cases of a digital, highly light-sensitive Peltier-cooled CMOS (complementary metal oxide semiconductor) camera (some older models use CCD (charge coupled device) chips) which records the image on a phosphor screen in the specimen chamber. The phosphor screen is coated with a thin film of aluminum to ensure its conductivity. This layer is also important to block electrons of low energy, not carrying useful diffraction information. Note that the phosphor screen is easily damaged by accidental contact, e.g. with a sample or when handling the camera. The optical link between phosphor screen and camera is done either with a lens optics or via glass fibers. Most detectors can be used in several binning modes, thus enhancing

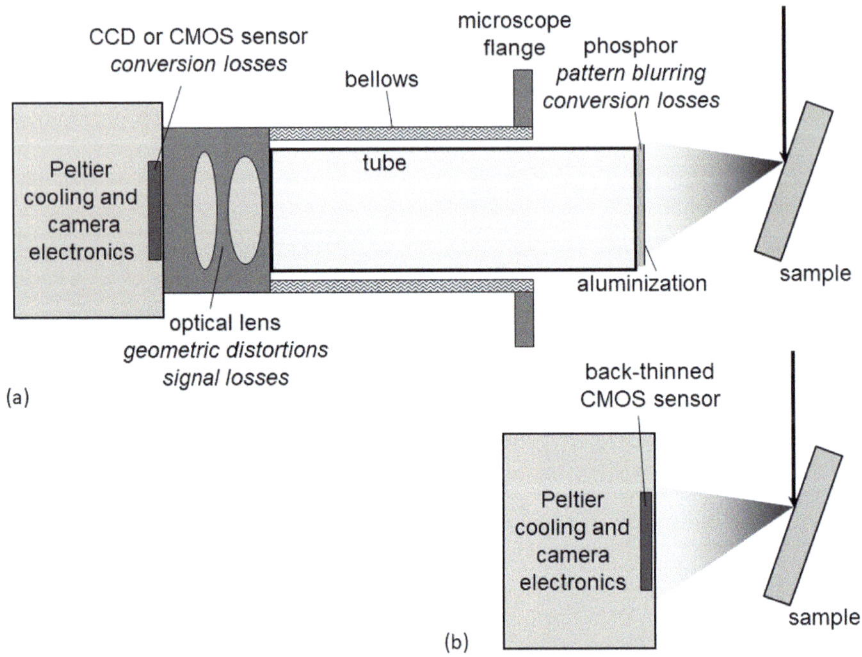

FIGURE 8.8 Design of EBSD cameras. (a) Typical camera with CCD or CMSOS sensor, lens, and phosphor screen. Potentially problematic features are marked in italic font. The sensor is mounted outside of the chamber. (b) EBSD camera for direct electron detection consisting of a back-thinned CMOS. The sensor is situated inside the chamber.

sensitivity and read out speed on cost of pixel resolution by comprising a certain number of pixels. Binning means that the content of several pixels is added together on the detector chip and read out collectively. A 4×4 binning, for example, means that 4×4 pixels are joint together, yielding 16 times higher intensity and equally shorter read-out time. In this way a detector may allow acquisition of highly resolved images ($\approx$1,000 × 1,000 pixel) at slow acquisition speed or of low-resolution images ($\approx$100 × 100 pixels) with high speed. In Figure 8.9 we reproduce patterns taken under different conditions. The technical development, however, is very quick and values for sensitivity and resolution are ever-increasing. Currently, some detector types are able to record and transmit more than 3,000 low-resolution images per second at a standard beam current.

At the present state, it is certainly improvements of the detector which will allow the most significant improvements in terms of spatial and angular resolution and applicability to phase and stress determination of an EBSD system. Phosphors with higher sensitivity for lower energy electrons, higher sensitivity of the chip, higher density of pixels, and a flexible detector positioning in the chamber have been developed. A flexible positioning of the camera in the chamber is a very useful feature as this allows placing the camera at an optimum observation position as discussed in the previous section.

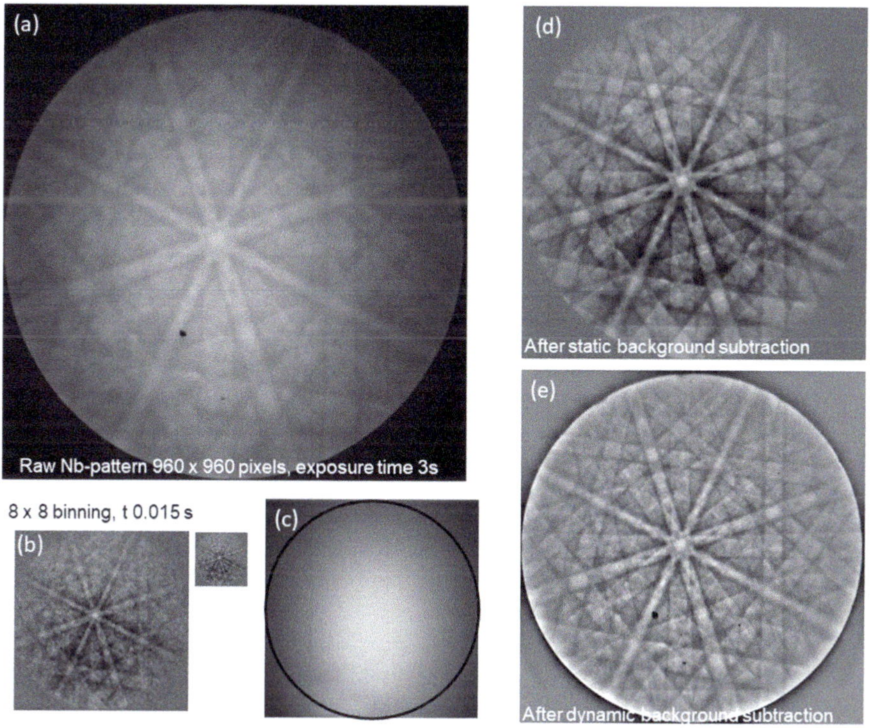

FIGURE 8.9 Recording and postprocessing of EBSD patterns from Nb, 15 kV acceleration voltage. (a) High-quality pattern recording, at full detector size (no binning) with long exposure time. (b) Pattern after 8 × 8 binning and background subtraction in original size and zoomed-in for better visibility. Exposure time 1/64 of unbinned pattern. (c) Example of a static background image. (d) Pattern from (a) after subtraction of a static background image. (e) Pattern from (a) after subtraction of a dynamic background image, obtained by a low-pass filter.

The currently most important development in area of detectors is the introduction of direct electron detection cameras for EBSD recording. These detectors use a back-thinned CMOS chip which is directly illuminated with electrons. Figure 8.8b illustrated that many sources of error drop off when optics and phosphor screen are removed. The move toward direct detection has a number of advantages, some of them are visible in Figure 8.10 (Wilkinson et al., 2013): first, the cameras show more details of the diffraction patterns. This may become important for phase determination, particularly when it comes to detection of pseudo-symmetries as discussed in Section 7.4.4. Second, they are significantly more sensitive to low voltage EBSD, which may be very important for high spatial and angular resolution EBSD. Third, they allow energy filtering of the electrons which contributes to lower noise and better sharpness of the patterns (Wang et al., 2020; Kacher et al., 2022). In addition, they have increased enormously in sensitivity over the last years such that exposure times as small as 0.5 ms are achievable. Finally, they may be positioned in a more

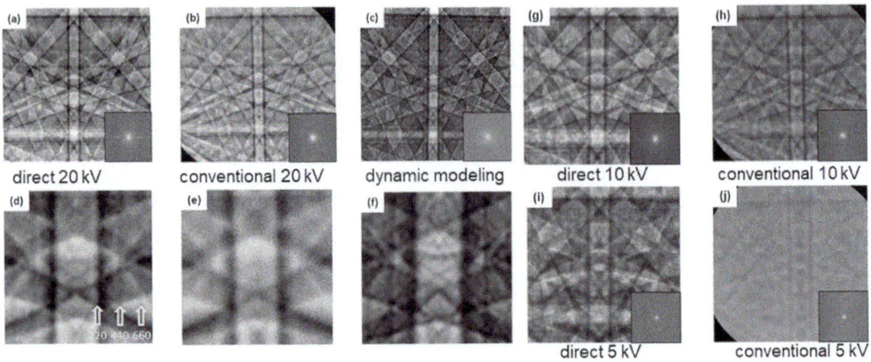

FIGURE 8.10 Comparison of patterns recorded from a Si single crystal with a classical EBSD detector and those recorded with a direct electron detector with a back-thinned CMOS sensor (sensor specifications: Pixel array 1024 × 1024, max. readout rate 28 full frames/s, pixel size 20 μm × 20 μm, dark noise 27 electrons rms (root mean square), analog-digital conversion 12 bits/pixel, linear dynamic range 60,000 electrons). (a)–(c) Comparison at conventional EBSD recording conditions, (d)–(f) Showing details of (a)–(c). (g)–(j) Comparison of performance at low-kV conditions. (Adapted from Wilkinson et al., 2013.)

flexible manner because they are smaller (e.g. Mingard et al., 2018). The only visible disadvantage is that the cameras are currently more expensive.

Finally, an interesting technique shall be mentioned that is possible, in principle, with all EBSD detectors, namely the use of the EBSD detector as a large and versatile general BSE detector. To this end freely defined regions of interest (ROI) on the EBSD detector are read out as one or several integral signals. Depending on the size and position of the ROI different sample information can be extracted. Nolze et al. (2017), for example, show that by placing the ROIs distributed horizontally across the detector and assigning each of these signals a different color, it is possible to create very informative pseudo-orientation maps. Also, because the total backscattering yield depends on the average atomic number of a phase, it is possible to sensitively visualize compositional differences in a sample. Furthermore, as the location of maximum intensity on a detector depends on the local surface inclination. This can be used to map, in a sensitive manner, a surface relief profile, as shown by Wright et al. (2015a).

8.2.6 Diffraction Pattern Enhancement

The real-time or "raw" diffraction pattern—obtained straight from the specimen—is typically rather noisy, blemished, uneven, blurred, or dim. Figure 8.9a shows an example of a full-size raw image. Lack of sharpness can be caused by material properties (e.g., lattice strain) or experimentally induced artifacts formed mainly by inappropriate specimen preparation. All EBSD systems include forms of diffraction pattern enhancement to improve the quality and so facilitate indexing on the processed pattern.

The diffuse background, plus minor screen blemishes, is stripped from a pattern by employing a *background correction* routine. This routine involves obtaining an image of the background intensity on the phosphor by integrating over the patterns obtained from a number of differently oriented grains such that the coherent diffraction information in the pattern averages out and only the continuous background remains. In a typical as-deformed or finely recrystallized microstructure, this can readily be accomplished by scanning over an area with a sufficient number of differently oriented grains while recording a diffraction pattern (or by averaging over individually recorded diffraction patterns)—if necessary at a lower magnification on the specimen. A typical background image is displayed in Figure 8.9c. After acquisition of the background it should be checked whether the background image is actually free of any faint diffraction bands, which would appear as "negative" bands in the background-corrected patterns and could lead to significant misindexing. This can happen, in particular, if the number of grains is too small to yield a uniform background, or, especially, on single-crystal materials. In this case a background can be obtained by rotating the sample around its surface normal during background acquisition (note that this may be difficult and dangerous because one risks collisions between sample and detector or objective lens). The background image is then electronically subtracted from the diffraction pattern, resulting in a pattern as displayed in Figure 8.9d. Alternatively, the diffraction pattern can also be divided by the background image, which results in a better correction of shading differences of the processed pattern. A new image of the background intensity is collected every time a new specimen is placed into the microscope, or when the operating conditions (voltage, beam current, working distance, specimen tilt-angle, material) are changed.

Instead of subtracting or dividing a static background image it is also possible to use a dynamic background image, which is calculated by a high band pass filter from the current experimental one. Although this filter may not function as ideal as an optimum fitted static background, it has a number of important advantages: it works when no suitable background image can be created, e.g. on single crystals, it works on samples with surface relief which creates different intensity distributions in the patterns (see Figure 8.2), it works on multi-phase materials where phases with different elemental composition and thus scattering power are present, and, finally, it does not have to be re-adapted when operating conditions change. Figure 8.9e shows a pattern after dynamic background subtraction.

After background removal it is advisable to spretch the contrast of the image for better visualization. Besides background subtraction the diffraction pattern may be *averaged* over a certain number of frames to reduce the amount of noise present. A long averaging time improves the pattern quality but it will noticeably slow down data acquisition.

8.2.7 Software

A couple of commercial software packages are available, which partly rely on quite different software algorithms. Two parts in the measurement software are most essential for a reliable and robust measurement. The first is the line detection algorithm, which is based on the Hough transform. Several variations have been developed since the original work of Krieger-Lassen (1998) as discussed in Section 7.2.2. Most of these are used for special cases, for example, for the precise detection of band width for the distinction of different phases or for indexing of complex crystal structures. A good control on the parameters of the Hough transform, i.e. the resolution of the Hough transform, the size of the filter mask and peak detection parameters is an important feature. The second part is the indexing algorithm (see Section 7.3). The most important feature here is how the algorithm deals with false or imprecisely detected bands. Furthermore, it is important how the algorithm deals with ambiguities (i.e. several possible orientation solutions). Values used to decide for the right solution have been discussed in Section 7.3. A good indexing algorithm improves in many cases the physical lateral resolution of an EBSD system as far as it allows separating, with high confidence, the solutions stemming from the overlapping patterns of two neighboring crystals. The so obtained higher resolution is called the effective lateral resolution, which will be the subject of Section 8.2.9.

Note that currently there is a strong move toward the use of image matching techniques for the orientation and phase analysis of diffraction patterns (see Section 7.4). With increasing computer power these "brute force" methods become more and more attractive in particular for the following areas: first, they ease the analysis of weak and noisy diffraction patterns which arise due to measurements at low beam current (e.g. on sensitive materials) or from heavily defected materials. Second, they allow comprehensive intensity analysis of diffraction patterns, which is important for determination of complicated phases or for the distinction of very similar patterns from pseudo-symmetric structures (see Section 7.4.4). Third, they enable the analysis of diffraction patterns obtained from low-voltage experiments. In these cases, the Kikuchi bands are very wide and the Hough-transform approach fails to detect them properly.

8.2.8 Calibration of an EBSD System

Calibration of an EBSD system means the determination of the projection center of a pattern measured in detector coordinates X_D, Y_D, Z_D. A sketch of the geometry of projection center and detector together with a number of aspects of calibration is plotted in Figure 8.11. Calibration is a very important process, because the correct calibration is required to obtain correct orientation results and even to be able to index diffraction patterns at all. Also, while scanning across a sample for ORM the pattern center changes constantly with the scan position as indicated in Figure 8.11a by two scan positions (1) and (2). Consideration of this movement is important at low magnification scans where the position of the beam on the sample changes significantly; at magnifications of 75 times the movement of the projection center when scanning once across an image amounts to several millimeters which corresponds to around 10% of the detector size (about 30 pixels on the detector). At 500 times the movement is still about 1%. For high magnification the projection center change is of no big importance for ORM. For the determination of absolute elastic strain values from EBSD patterns, in contrast, it would be essential to measure the projection center with higher accuracy (subpixel resolution) than what is possible today (see Basinger et al., 2011; Maurice et al., 2013; Pang et al., 2020). The projection center coordinates are most reasonably given in fractions of the detector

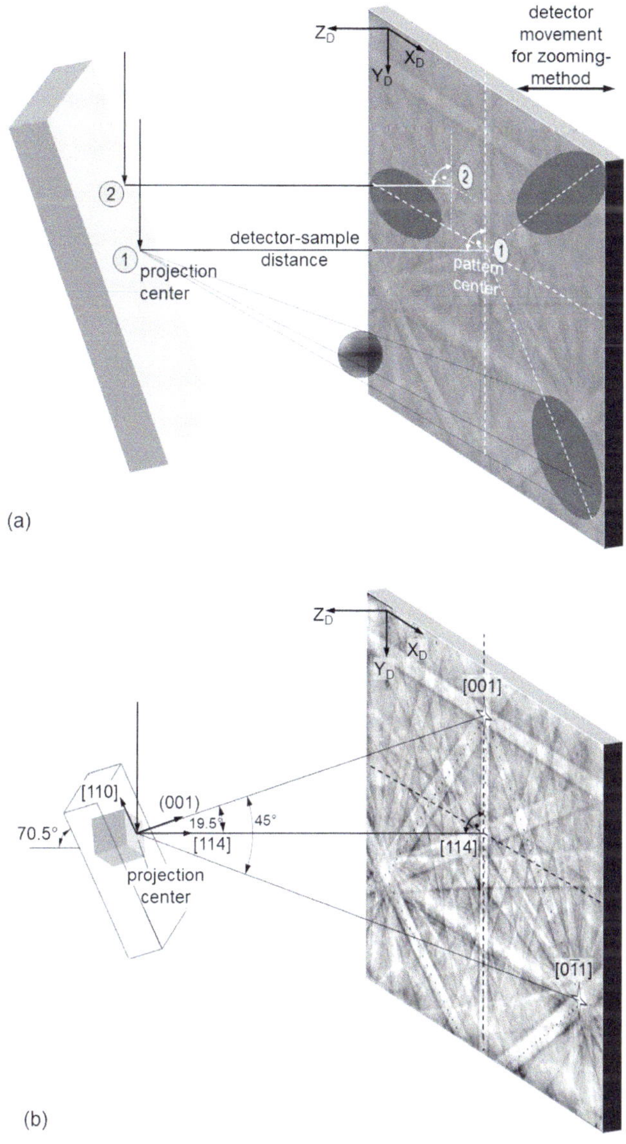

FIGURE 8.11 Various principles of EBSD projection center calibration. The projection center is the location where the electron beam hits the sample, measured in detector coordinates $X_DY_DZ_D$. Z_D is perpendicular on the detector plane. Usually the coordinates are measured in percent of the detector size in X-direction. The "pattern center" corresponds to the X_DY_D coordinates, the "detector-sample distance" to the Z_D coordinate of the projection center. When the location of the electron beam moves, e.g. from (1) to (2), the projection center coordinates move accordingly. (a) The shadow-casting method uses three balls, one of which is shown, casting shadows onto the detector. (b) The "known orientation" method places a (001)-oriented Si single crystal such that the pattern center is exactly at the [114] zone axis. To this end the crystal needs to be tilted 70.5° and the [110] direction must point perpendicular to the tilt axis.

size, e.g. its horizontal width. In some EBSD systems the projection center is separated into two units, the pattern center (PC, consisting of the coordinates X_D and Y_D), which indicates the x and y coordinates of the projection center on the detector and the specimen-to-detector distance (Z_D), which corresponds to the z-coordinate. All parameters are indicated in Figure 8.11a.

The problem with EBSD calibration is that the projection center is not directly visible in the pattern, but needs to be inferred indirectly. The calibration methods are as follows:

- Known crystal orientation
- Shadow casting
- Pattern zooming
- Iterative pattern fitting

For all methods it is recommendable to use good patterns—potentially better than those used during full-speed ORM pattern acquisition. It is therefore sensible to collect a pattern for calibration with a longer exposure or integration time than the patterns collected during the measurement. On materials with high defect densities it is furthermore recommended to move the beam around until a clear and sharp pattern is found. It is, however, important to perform this process for the electron beam being close to the center of the mapping area.

8.2.8.1 Known Crystal Orientation

The "known crystal orientation" method was developed as part of the original computer software that analyzed EBSD patterns online (Dingley et al., 1987). It survived for several years as the main method for EBSD calibration. Also now it may still be helpful for certain cases. Essentially, as illustrated in Figure 8.11b, a single crystal of silicon with (001) surface orientation is mounted in the microscope such that the [110] direction, which usually corresponds to the edges of commercial Si wafers, coincides with tilt axis of the microscope stage. The sample is then tilted by 70.5°. If this is done accurately the [114] pole (which is 19.5° (= 90° – 70.5°) away from the [001]) points exactly perpendicular to the electron beam, and thus perpendicular to the detector surface, provided that the detector is mounted with Y_D parallel to the beam. The position of the [114] pole thus marks the pattern center in the image. The projection distance Z_D is then obtained by measuring the distance in pixels on screen between two known zone axes and relating it to the known angle between them via the following relationship:

$$\cos\phi = \frac{(x_1 - X_D)(x_2 - X_D) + (y_1 - Y_D)(y_2 - Y_D) + Z_D^2}{\sqrt{(x_1 - X_D)^2 + (y_1 - Y_D)^2 + Z_D^2} \cdot \sqrt{(x_2 - X_D)^2 + (y_2 - Y_D)^2 + Z_D^2}}$$

and

$$\cos\phi = \frac{u_1 u_2 + v_1 v_2 + w_1 w_2}{\sqrt{(u_1^2 + v_1^2 + w_1^2)} \cdot \sqrt{(u_2^2 + v_2^2 + w_2^2)}} \tag{8.1}$$

where ϕ is the angle between the two chosen known zone axes $[uvw]_1$ and $[uvw]_2$, x_1, y_1, x_2, y_2 are the measured coordinates of these zone axes in detector coordinates, X_D, Y_D the measured coordinates of the PC, i.e. of the [114] pole, and Z_D is then the only unknown in Equation 8.1.

If the solid angle of the EBSD pattern is large enough it is easy, for example, to recognize the [001] and the $[0\bar{1}1]$ zone axis and measure their positions as indicated in Figure 8.11b. Their interzonal angle, Φ, is 45°. The operator then needs to measure the positions of the three detected known zones and solve Equation 8.1. It is important to mount the specimen with exactly the same geometry as the calibration crystal, in particular the focus (i.e. the working distance) must be kept similar. To this end, the SEM is focused on the calibration standard, the focus is then kept constant and the height of the specimen that is to be measured is adapted using the z-control of the stage until the specimen is in focus again. This may be facilitated by mounting sample and calibration standard at the same height together on the same sample holder.

{001}-oriented single-crystalline silicon is widely used for the known crystal calibration method since the material is readily available, stores well, and gives clear diffraction patterns. Germanium or gallium arsenide are also used, or lead sulfide that has the additional advantage of exhibiting natural {001} facets. Errors are introduced into the method if the single crystal is not exactly {001}. This can, however, be detected with electron channeling contrast imaging (ECCI) in the SEM or by independent means, such as x-ray methods.

The most serious *advantage* of the known orientation method is the possibility to obtain an absolute orientation calibration. The main *disadvantages* are that extreme care has to be taken when mounting the calibration crystal, including alignment of the sample with the microscope scan direction and tilt axis. Furthermore, the accuracy of the projection center determination depends strongly on the ability to determine the correct coordinates of known poles. Note that due to the gnomonic projection, the locations of known poles are not exactly in the geometric center of the Kikuchi band intersections. Finally, many SEM stages, nowadays, are limited to 70° of tilt, i.e. it may be impossible to reach the required 70.5° (and a misalignment of 0.5° may be inacceptable for exact pattern center determination).

8.2.8.2 Shadow Casting

For shadow casting (Dingley, 1981) three steel balls are mounted in front of the specimen which cast elliptical shadows onto the screen (sketched in Figure 8.11a). The major axes of the ellipses, when extended, intercept at the *PC*. Another shadow-casting method uses a fine square mesh attached to the phosphor screen (Schwarzer et al., 1996b). The shadow images of the mesh are trapezoids having their long axes pointing toward the *PC*. The determination of Z_D is then performed by measurement of the detector coordinates of two known poles and applying Equation 8.1.

The *advantages* of the shadow-casting method are that it is accurate and repeatable and may offer some protection from collision between the specimen and the screen; the *disadvantages* are that it requires additional, precision-made hardware and, unless the shadow-casting devices can be retracted, it significantly interferes with automatic pattern recognition.

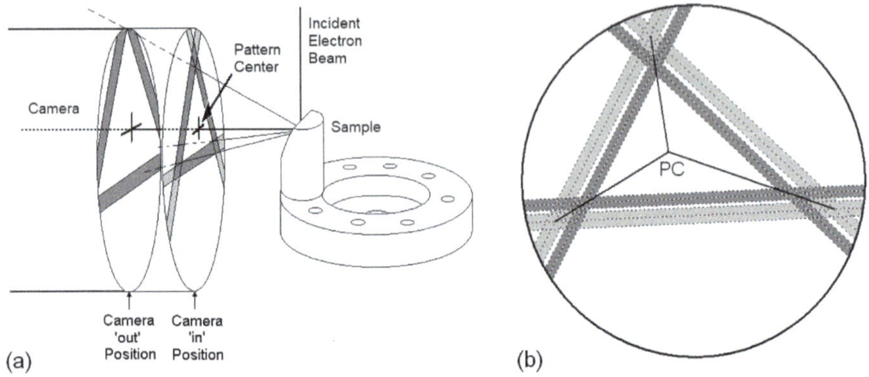

FIGURE 8.12 (a and b) Principle of the "pattern magnification" method of EBSD calibration. (Courtesy of K. Dicks.)

8.2.8.3 Pattern Zooming

When the EBSD detector is retracted from the specimen, Z_D is increased and the pattern is zoomed-in about the pattern center. The pattern center then remains the only point that does not change its position on the detector. Practically, the pattern is viewed at a minimum of two detector positions, which we call here the "in" and "out" positions. The easiest way to locate the pattern center is then to outline three Kikuchi bands that form a triangle in the "in" camera position and follow the *same* three bands in the "out" position on the magnified diffraction pattern, as shown in Figure 8.12a. The coordinates of each triangle corner for the "in" and "out" positions lie on a line on which the *PC* also lies. The intersection of the three lines from each triangle corner therefore defines the *PC* (Figure 8.12b; Hjelen et al., 1993). The coordinate Z_D is found again, as described for the known orientation method, employing Equation 8.1 for two known zone axes coordinates measured relative to the determined pattern center. The pattern magnification method does not require a standard crystal and can be applied to any well-defined diffraction pattern. However, determination of Z_D requires recognition of two zone axes. This can be done by computer software which attempts to index the pattern for different assumed camera lengths Z_D. It may also be convenient to keep a specimen with an easily identified EBSD pattern such as (001) silicon to facilitate this procedure.

The *advantage* of the pattern magnification method is that it requires no special equipment. The method may allow quite good accuracy of projection center determination if large zooming is applied. There are no great disadvantages, except that, as for the previous case, some pattern recognition capability may be needed to input the parameters to find Z_D.

8.2.8.4 Iterative Determination

Calibration via iterative pattern fitting is currently the most frequently used method in commercial software. It requires only an EBSD pattern of reasonable quality in which at least three bands or zone axes can be identified (usually automatically). Equation 8.1 can then be formulated $n(n-1)/2$ times, where n is the number of identified zone axes, substituting in the known indices of two of the poles in turn. These equations have to be solved numerically to obtain X_D, Y_D, and Z_D (Schmidt et al., 1991; Krieger Lassen et al., 1992; Krieger Lassen and Bilde-Sørensen, 1993). The numerical method needs a "first guess" for the values of the *PC* coordinates and Z_D, good enough to allow for automatic indexing of the diffraction pattern and, thus, automated identification of poles. The better is the first guess, the less iterations will be required to converge toward the right solution. This is usually obtained from a table of calibration values for different working distance stored when first setting up the EBSD system. Selecting the working distance closest to the current working distance, and if necessary, performing linear interpolation between projection center values for different working distances should always allow to determine a reasonable first guess.

The *advantages* of the iterative pattern fitting method are that no special attachments or standards are required and it is insensitive to the geometry of the setup. Moreover, iterative pattern fitting works for any good-quality pattern without operator interaction, which qualifies this method for recalibration during automatic EBSD runs. A *disadvantage* of this method is that it requires the knowledge of the lattice parameters of the phase under consideration. In cases where the phase is unknown (i.e. when phase determination is to be performed) the calibration must first be carried out on a sample of known parameters. The so-determined projection center must then be kept for the unknown sample (important is to also keep the focus, see Section 8.2.8.1). Furthermore, the technique is, at least currently, not able to obtain sufficiently high accuracy for absolute elastic stress determination (which is true, however, for all mentioned techniques). Also it does not allow to determine the absolute orientation of samples. This is only possible with the known-orientation method.

Table 8.1 summarizes the pertinent features of the calibration routines described earlier. The optimum approach to calibration is to exploit the best features from

TABLE 8.1
Summary of EBSD Calibration Options

Method	Advantages	Disadvantages
Known crystal orientation	Accurate and repeatable, absolute orientation determination possible; most basic procedure	Errors associated with crystal and mounting; some pattern recognition required, sufficient stage tilt (70.5°) required when using the (411) pole of Silicon
Shadow casting	Accurate and repeatable; some protection against screen collision	Requires precision hardware that interferes with automatic pattern recognition, works only if the lattice parameters of the sample are known
Pattern magnification	No special equipment or standards required	Some pattern recognition, i.e. "first guess" solution required, works only if the lattice parameters of the sample are known
Iterative pattern fitting	No special equipment or standards required; can be repeated any time; standard routine in all commercial EBSD programs	Requires "first guess" solution (no serious problem, though), no absolute orientation determination, works only if the lattice parameters of the sample are known

the different methods and form a combination procedure. In commercial programs the first calibration of a system is usually done by a known crystal orientation. The so-determined values are stored in a table and provided for later regular calibration using the iterative calibration routine.

8.2.9 Lateral Resolution

The *physical spatial* resolution, i.e. the size of the interaction volume for EBSD, has been discussed in Section 6.3.3 in some detail. We remind here again that the physical spatial resolution depends mainly on the average atomic number of the material and on the acceleration voltage. For a light material like magnesium ($Z = 12$) (this also concerns many minerals), this resolution is at best in the order of a few 100 nm, for a heavy one like tungsten it is about 50 nm. The general dependence is displayed in Figure 6.19. Note, for example, that to our experience, it is not possible to resolve Si- or Mg-rich particles in a 5000 series Al alloy, if they are smaller than about 500 nanometer in diameter.

Here, we add a few points on the *effective lateral* resolution, which may be defined, for example, as the distance of the electron beam to a large angle grain boundary where the EBSD system is still able to uniquely determine the crystal orientation of the current grain. As described by Humphreys et al. (1999) the effective lateral resolution can be determined by measuring the amount of points in an orientation map that are not indexed or indexed with a confidence index of 0 (i.e. two solutions are equally possible). Humphreys presents the following equation for the fraction of points being uniquely assigned to one grain, N_S

$$N_S = \frac{(D_A - \Lambda_A)(D_P - \Lambda_P)}{D_A D_P} \tag{8.2}$$

where D_A and D_P are the mean grain sizes parallel and perpendicular to the sample tilt axis, Λ_A and Λ_P are the effective parallel and perpendicular resolutions. For an equiaxed microstructure with $D_A = D_P$ and for $\Lambda_P = 3\Lambda_A$ (for 70° tilt) one obtains as resolution:

$$\Lambda_A = \frac{D(1 - N_S)}{4} \tag{8.3}$$

Using these equations or a similar approach, the best *effective* spatial resolution for a material such as brass is 25–50 nm for a tungsten-filament SEM and 9–22 nm for an FEGSEM, both measured parallel to the tilt axis (Harland et al., 1981; Dingley, 2000; Humphreys, 2001) have been claimed. An effective spatial resolution of 6–9 nm in a platinum sample has been quoted by Dingley (2004). Nevertheless, these numbers, which are significantly smaller than the physical resolution, have to be taken with great care: the resolution calculated by the equations above relies on the existence of *recognizable* grain or phase boundaries, that is, it depends on the ability of the EBSD system to deconvolve two overlaying diffraction patterns at a grain boundary. In microstructures of heavily deformed metals the diffraction

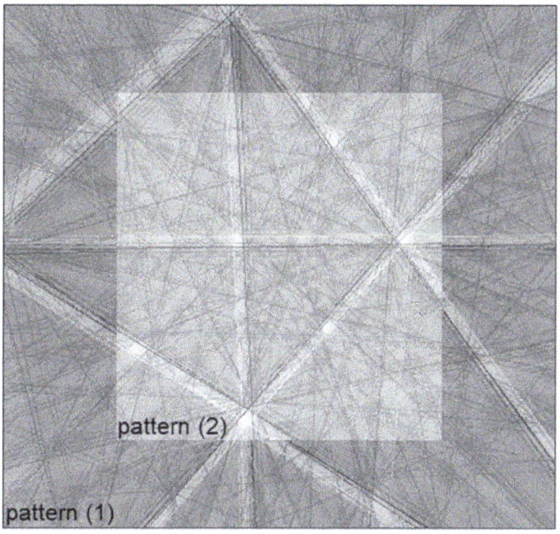

FIGURE 8.13 Demonstration of the effect of pattern overlap on lateral resolution in heavily deformed material: Two patterns, (1) and (2), taken from two sides of a 1° boundary are superimposed. Each of the patterns is blurred by the presence of a high dislocation density. The blurring of the patterns makes the distinction of the 1° misorientation impossible. Rather, automatic software determines an average orientation for each of the patterns. Conclusion: The effective resolution is not significant in this case. The physical resolution needs to be taken into account to determine whether the boundary can be recognized.

patterns are blurred due to the overlap of many slightly misoriented patterns. A simulation of this situation is shown in Figure 8.13. In this case pattern analysis software might not be able to recognize two different orientations, but it determines an average orientation between the two subgrains which will lead to the disappearance of the boundary. In the case of heavily deformed metals—where the resolution is of particular importance—the effective resolution is therefore not necessarily better than the physical one. Note that Figure 8.13 also shows that EBSD patterns remain analyzable even in case of highly deformed metals because the principle bands remain visible. This is in contrast to the case of transmission Kikuchi patterns discussed earlier in Section 6.2.6.

A way to "improve" the lateral resolution in such structures is the filtering of the microstructure map with a Kuwahara filter as proposed by Humphreys et al. (2001). The basic idea of this filter is to average over neighboring orientations, thus filtering out orientation noise. Although the application of the filter results in clean cellular microstructures (see also Brough et al., 2006) it is not clear to which extent the obtained microstructures correspond to the real ones and how many artifacts are created. Note that here ECCI in the SEM (Section 6.3.4.2) may help to resolve the true microstructure. A serious improvement of lateral resolution seems only to be possible by enhancement of the physical resolution, most probably by moving to smaller accelerating voltages as discussed briefly in Sections 8.2.3 and 8.2.5. Technically, this move requires in particular improvements on the camera (direct electron detection) and on the pattern analysis software (template matching).

8.2.10 Angular Resolution

The *angular resolution* of an EBSD-based ORM system indicates how *precise* an orientation can be determined, which means how large the angular scatter is if a crystal of a given orientation is measured several times. One may also ask how *accurate* this orientation measurement is, i.e. how close is the average of several orientation measurements to the true orientation. The angular resolution may, thus, be distinguished in the *relative (precise) and absolute (accurate) orientation resolution*. The absolute orientation resolution depends on a number of microscope parameters like accuracy of stage tilt and sample positioning, accuracy of camera positioning and, of course, on the accuracy of pattern analysis. The sample positioning, in particular, can have a very strong influence, as has been shown in a work of Nolze (2007). He worked out that the wrong alignment of the sample not only leads to distortion of the scan field but also to orientation deviations of up to 20°.

The relative angular resolution describes how precise a misorientation between two neighboring measurement positions can be determined. It may be defined as a statistical value, describing the orientation variation obtained when scanning across a single crystal of constant orientation. The relative angular resolution depends on the pattern acquisition and analysis conditions. As an example we show two values for the angular resolution measured under two different well-defined conditions: as sample served an electronic grade silicon single crystal with virtually no misorientation. For the measurement a high magnification of 5,000× was taken in order to avoid any influence of a change of pattern center during the scan. The measurements were carried out for different measurement conditions. For the case of high measurement speed a 100 × 100 pixels, 8 × 8 camera binning mode and a (HT 120 × 1) Hough transform (see Section 7.2.2 for this notation) was used. The results are shown in Figure 8.14. A maximum orientation deviation of 0.8° and an average of 0.35° were obtained. For the high-resolution case a 900 × 900 pixel 1 × 1 binning mode was used together with (HT 240 × 0.5) Hough transform. In this case a maximum deviation of 0.2° and an average deviation of 0.1° were obtained. The reasons for the angular imprecision of EBSD measurements were investigated by Ram et al. (2015). Most importantly it is due to an inaccurate detection of Kikuchi bands in a Hough transform. Kikuchi lines are actually slightly curved while the Hough transform only detects straight line features. This problem is even verstärkt when using a too small convolution mask. Also an inaccurate pattern center position plays a role.

The definition of the angular resolution given above takes only into account the misorientation angle, ω, but not the axis of misorientation, **u**. As Wilkinson (2001) and Wilkinson et al. (2006) pointed out, the axis of misorientation measured in the conventional way by a Hough transform becomes very imprecise as the angle approaches the angular resolution of the technique. In fact, the value of **u** becomes meaningless. Bate et al. (2005) developed an analytical relationship between the angular inaccuracy of the misorientation axis, ϕ, the angular resolution, δ, and the misorientation, ω:

$$\varphi = \arctan\left(\frac{\delta}{\omega}\right) \tag{8.4}$$

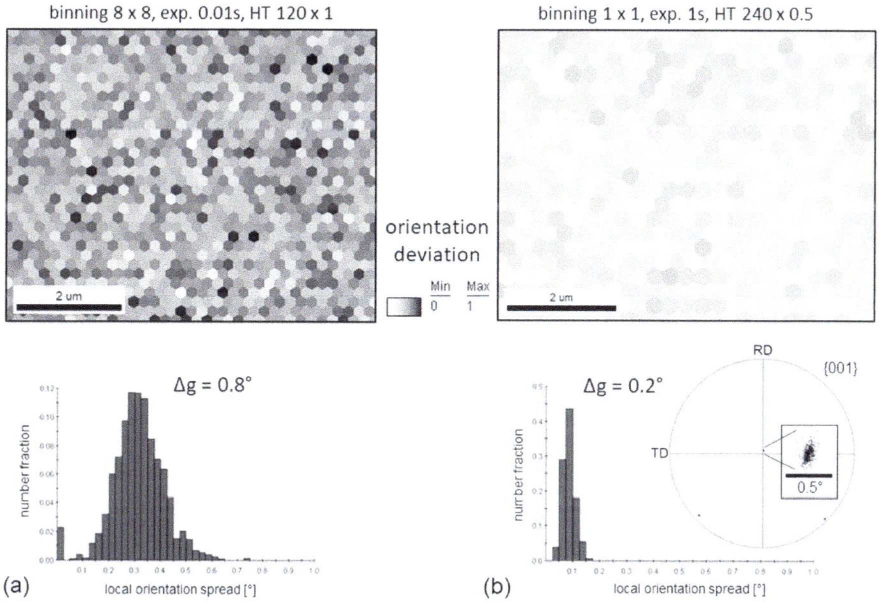

FIGURE 8.14 Angular resolution of an EBSD system in dependence of the measurement conditions measured by mapping a silicon single crystal of electronic grade. (a) Conventional high speed measurement conditions resulting in a poor orientation resolution Δg of about 0.8° maximum noise (0.3° most frequent noise value). (b) High Hough transform resolution measurements resulting in a small angular resolution of 0.2° maximum noise (0.1° most frequent noise) but also in low measurement rates. In (b) the orientation is additionally shown in the {001} pole figure, indicating that the absolute orientation measurement is approximately 2° wrong. The inset shows the distribution of the measured orientation. (Measurement with JEOL JSM 6500 F equipped with an EDAX-TSL EBSD system.)

For the measurement situation described above and in Figure 8.14 (δ = 0.1° or 0.3°) the misorientation axis across a 1° boundary could thus be determined with a precision of 5.7° or 16.7°, respectively.

A way to significantly improve the relative angular resolution for the case of small misorientations is cross-correlation analysis of EBSD patterns which has been introduced in Section 7.4.5. The technique allows reaching an angular resolution of 0.006° which might be precise enough to perform, for example, a dislocation analysis according to Nye's approach in two dimensions (Nye, 1953). An example for this application will be given in Section 9.4.6. While the technique is of significantly higher accuracy than the classical Hough transform technique, it is, on the other hand, much slower and therefore not well suitable for conventional ORM. Note also that CC EBSD does not allow to measure absolute orientations, but it only improves the measurement of misorientation of neighboring measurements, as long as the misorientations are not larger than a maximum of 3° without and 10° with application of pattern re-mapping.

8.2.11 In-Situ Investigations of Texture Formation Processes by EBSD

The large SEM chamber and the fact that for EBSD applications the working distance and detector distance are relatively large (both in the order of 10–30 mm) allows to place relatively bulky in-situ equipment into the SEM chamber in EBSD geometry. This equipment may be a deformation testing machine, e.g. for tensile, compression, bending or shear deformation, or a heating or cooling stage. The investigated samples can then have sizes in the order of several millimeters which makes their behavior already relevant for bulk specimens. Even in terms of texture, these samples may contain a sufficient amount of grains (see Section 9.4.8) to allow observation of the microscopic mechanisms of texture formation, be it during deformation, recrystallization, or phase transformation.

For annealing experiments the SEM may be equipped with a heating stage in order to study microstructure and orientation changes during annealing, most prominently, during recovery, recrystallization, and grain growth (Lens et al., 2005; Mirpuri et al., 2006; Sukhopar and Gottstein, 2016) or during phase transformations (Seward et al., 2004; Lischewski et al., 2008, Wakita and Suzuki, 2017). High-speed EBSD is of particular interest so as to allow for observation of rapidly changing microstructures (e.g. Field et al., 2007). A number of points have to be considered for proper in-situ heating experiments. First, external cooling of the stage must be arranged because energy removal is difficult in vacuum. It is important to keep the heat source as small and as homogenous as possible in order to reduce thermal drift (e.g. Bozzolo et al., 2012). Lischewski et al. (2008) report a dedicated heating system which uses a laser outside of the SEM coupled to the sample via light guide. This has the additional charm that no heating current is required at the imaging site and, thus, no magnetic field is induced at the sample. Finally, for high temperature experiments (>600°C) the EBSD detector has to be shielded against thermal radiation, as proposed, for example, by Wakita and Suzuki (2017) and Ubhi et al. (2014). This allows recording of high-quality diffraction pattern at high temperatures, also because defects may heal out during the annealing process. One of the potential problems with in-situ annealing experiments is that the free observation surface may influence the microstructural changes (Nakamichi et al., 2008), and hence it must be ascertained that the microstructure and texture effects observed at the surface are representative of the bulk behavior. A particular problem concerns the attempt to observe nucleation of recrystallization or phase transformation, which is, indeed, one of the still open problems in the related research: first, it is impossible to claim whether grains appearing on the surface are really new nuclei or whether they belong to larger grains that grew below the surface. The latter is, indeed, the more frequent case. Second, the presence of a free surface and the fact that all grain structures at the surface have been significantly modified in size, shape, and stress by the preparation of the free surface, strongly influences the physical conditions for nucleation. In-situ investigations of nucleation of recrystallization or phase transformations should, therefore, be regarded with great skepticism. A valuable alternative to SEM-based investigations of nucleation processes is, in principle, the in-situ investigation using non-destructive 4D (three spatial dimensions and time) observations by Laue or Debye–Scherrer methods with synchrotron radiation, as described in Section 13.2. These techniques, however, suffer from a lack of spatial resolution in the deformed region as well as for the nuclei. As a consequence, also these approaches are mainly good for the study of the growth stage of recrystallization rather than for the nucleation stage.

Very frequently EBSD-based in-situ observations are employed to study deformation processes. A good overview on the technique and its potential results can be found in Wright et al. (2016). Dedicated small deformation equipment is available for the SEM in EBSD geometry. It is important to note that the equipment needs to combine sufficiently high stiffness with a small and EBSD-suitable construction. This means, in particular, that the machine frame needs to be small enough to neither touch the EBSD detector nor any other part in the microscope and that it does not cast shadows onto the EBSD detector. For convenience the machine should allow symmetric straining to keep the area of observation close to the original field of view. In this regard, three-point bending is a very convenient option to perform in-situ deformation experiments because the center bending pole can be fixed with respect to the microscope. Furthermore, bending allows to study different degrees of compression and expansion in one sample. A shortcoming of three-point bending experiments is, however, that the sample surface does not stay flat when reaching larger strains (>10%). The tensile side sinks in while the compression side pops out. This problem does not occur with tensile samples (Han et al., 2003; Scheriau and Pippan, 2008). In order to obtain good diffraction patterns the deformation process is halted in order to obtain an EBSD map (Karlsen et al., 2008). In-situ experiments become particularly informative when the EBSD-based orientation information is combined, for example, with surface trace analysis of slip systems. In the case of non-cubic materials these traces and the measured orientations allow the determination of the activated slip systems in dependence of applied strain as it has been done, for example, in an investigation of active slip systems in Mg alloys by Cepeda-Jiménez et al. (2016). These observations can even be carried further if the measurements are carried out by cross-correlation EBSD (see Section 7.4.5) to measure the elastic strains and stresses (Ojima et al., 2011) or if they are combined with plastic strain measurements using digital image correlation (DIC) as shortly described in Section 9.4.5. A nice example for in-situ deformation measurements combining DIC and EBSD on aluminum alloys may be found in Zhang et al. (2020a). As for the heating experiments described above, but with less serious consequences, it is important to note that EBSD-based in-situ experiments are carried out on the surface of the material. In a deformation experiment there is, thus, no force perpendicular to the free surface, in contrast to the situation inside of the material. The results, e.g. type and density of activated slip systems and related lattice rotations, may, therefore, not be directly transferable to bulk material behavior. A correlation of experiments with crystal plasticity modeling may show, how trustworthy are the so-obtained results. This has, for example, been tested in some detail by Diehl et al. (2017).

8.3 TECHNIQUES OF TEM ORIENTATION MICROSCOPY

In contrast to the SEM-based ORM, where only one technique is of universal use, TEM-based ORM can be performed on the basis of a number of different techniques, including the use of Kikuchi patterns, spot patterns, and reconstructed spot patterns. The two first techniques are classical scanning techniques, where, similar to EBSD-based ORM, a finely focused electron probe is scanned across the sample and orientation and phase are extracted from the resulting diffraction patterns. The microscope is used as a scanning TEM (STEM). The third technique, conical dark-field scanning, works on a completely different basis as has been described in some detail in Section 6.2.5. In the last ten years the STEM technique using small-angle convergent beam electron diffraction (SCBED) patterns (Section 6.2.3), also including precession (Section 6.2.4), has turned out to be the most useful technique. We will shortly address some important practical issues concerning the STEM techniques. Examples for their application will be presented in Chapter 9.

8.3.1 Technical Aspects of TEM-Based Orientation Microscopy

In principle, TEM-based ORM can be performed on every conventional TEM. The main technical requirements for the described techniques concern the camera used for pattern or dark-field image recording and the appropriate scan generator. The latter is particularly important for the generation of distortion-free precession. Note that PED STEM requires a highly complex electron beam movement. First the beam has to be tilted before the sample and back-tilted below the sample which requires pairs of scan coils for x and for y tilt above and below the specimen. The precession of the tilted beam has to occur such as to move the beam on perfect circles while the precession center point must be well focused to one point on the sample in order to guarantee high resolution. Next, the focal point on the sample needs to be scanned across the sample thereby keeping the precession movement stable. Finally, one needs to make sure that the diffraction pattern formed on the back focal plane of the objective lens does not move significantly, in order to allow a similar angular field of view in the diffraction pattern. The complex process of beam alignment is described in great detail by Barnard et al. (2017). We reproduce one of their images in Figure 8.15 in which we point out the positions of critical alignment.

In contrast to scanning precession technique the Kikuchi and the conventional SCBED techniques are less demanding in alignment of the microscope. Here it is mainly the astigmatism and focus of the condenser lens for sample illumination and of the diffraction image that require critical control. A point that has to be taken into account when working with TEM Kikuchi patterns is that these patterns create intensities rather far off the optical axis (i.e. more than 10°). At these angles the focal distance is different than for positions close to the optical axis. This may lead to significant distortions of the patterns and to wrong angular measurements (see, e.g. Burton et al., 2020). We have therefore usually realized that the precise analysis of Kikuchi patterns requires a different diffraction focus setting than that of spot patterns. The correct focus is, however, not easy to detect. An iterative procedure is required where patterns at different focus settings are indexed until the smallest deviation between recalculated and experimental pattern is found.

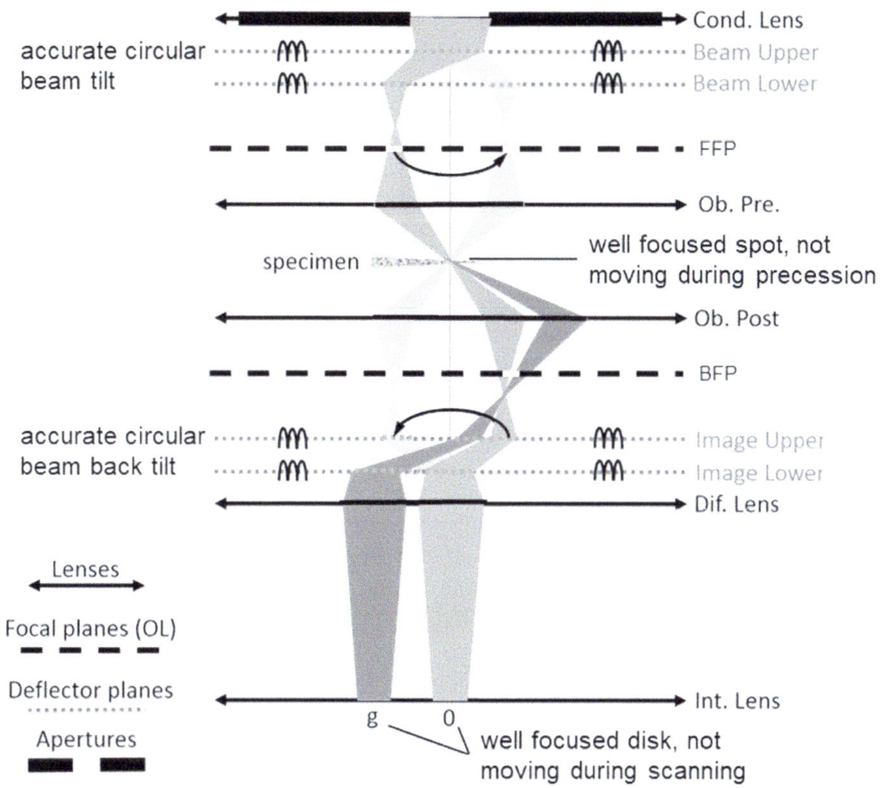

FIGURE 8.15 Schematic of the focused processing probe geometry for PED. Two primary beam inclinations (dark and bright) are depicted for the direct beam, **0**, with one diffracted beam, **g** (purple). The diffraction condition is only met at one particular primary beam inclination. The beam upper and lower deflector coils shift and tilt the beam above the sample, the image upper and lower deflector coils back-tilt diffraction pattern below it. Principal imaging and conjugate diffraction planes are indicated, as well as the principal focal planes of the objective lens (OL). Lenses are labeled as follows: condenser (Cond.), diffraction (Dif.), intermediate (Int.) and objective pre (Ob. Pre.) and post (Ob. Post) specimen lenses. FFP/BFP front/back focal plane. (Adapted from Barnard et al., 2017. Note that the original image contains further information that has been dropped out here for simplification.)

8.3.2 Calibration of Patterns

Once all alignments are done the system needs to be calibrated, i.e. the pattern projection center has to be determined, similar to what has been discuss for EBSD in Section 8.2.8. In contrast to EBSD, the calibration is much simpler in the case of TEM, simply because the primary beam position on the detector marks the pattern center ($X_D Y_D$). The primary beam spot needs to be measured accurately, which requires attention because the primary beam spot is usually a broad, uniformly illuminated disk rather than a well-focused point. As described for the EBSD calibration it is required to index the pattern correctly (i.e. a first guess for the camera length is required). Then Equation 8.1 is employed to calculate the camera length, Z_D. For calibration it is, of course, necessary to use a sample with known lattice structure.

8.3.3 Camera Considerations

With respect to the camera used for pattern recording neither pixel resolution nor sensitivity are of great importance for the application for orientation mapping. Rather, a high dynamic range and robustness against beam damage is required in order to record the large contrast occurring in diffraction patterns or in dark-field images. In terms of resolution a normal video format (800 × 600 pixels) or a medium CCD chip (1,000 × 1,000 pixel) is largely sufficient, the latter being particularly reasonable for the TKP technique. Concerning sensitivity, it should be remarked that Kikuchi diffraction patterns may be relatively weak, particular if one reduces the dose on the crystal in order to cope with radiation sensitive materials. In CCD or CMOS cameras for electron microscopy the transition of electrons into light is done either with a YAG (yttrium aluminum garnet) crystal or a phosphor scintillator. It appears that YAG scintillators are the better choice for diffraction work in terms of contrast and sensitivity to beam damage. The camera may be mounted either on the bottom flange of the microscope column, below the microscope chamber, or at the wide-angle port above it. The bottom mounted camera has the advantage of being easier to mount but the large disadvantage that it usually has a very small acceptance angle which is why the diffraction length has to be reduced strongly (≤100 mm) to record full diffraction patterns. Usually the performance of the microscope is less good at these small camera lengths. High-resolution CCD cameras mounted to the wide-angle port, in contrast, are able to record almost the whole image visible on the viewing screen of the microscope.

It should be mentioned that also simpler camera installations have been used to record patterns for ORM. For instance, Høier et al. (1994) and Schwarzer and Zaefferer (1994a) used a simple video camera directed through the front window of the microscope onto the tilted fluorescent screen in order to record Kikuchi patterns. Rauch and Dupuy (2005) used a similar geometry for recording SCBED patterns. In another setup, the chip of a Peltier-cooled slow-scan CCD camera was inserted into the electron beam slightly above the main viewing screen of the TEM (Krieger Lassen, 1995; Engler et al., 1996a); here the advantage is that the recorded Kikuchi pattern is undistorted. Alternatively, the camera was mounted inside the TEM below the main viewing screen (Weiland and Field, 1994). Fundenberger et al. (2003) used a wide-angle camera which was double exposed, once to obtain good illumination of the center part of a pattern, and once to expose the outer area correctly. We reproduce a typical pattern from their work in Figure 8.16.

8.3.4 Sample Preparation

Finally, we shortly address sample preparation. It is well known that sample preparation for TEM is a demanding and complicated task well beyond the scope of this book. A comprehensive text only on sample preparation is the book by Goodhew (2020). Since the ability to create suitable samples may severely limit the application

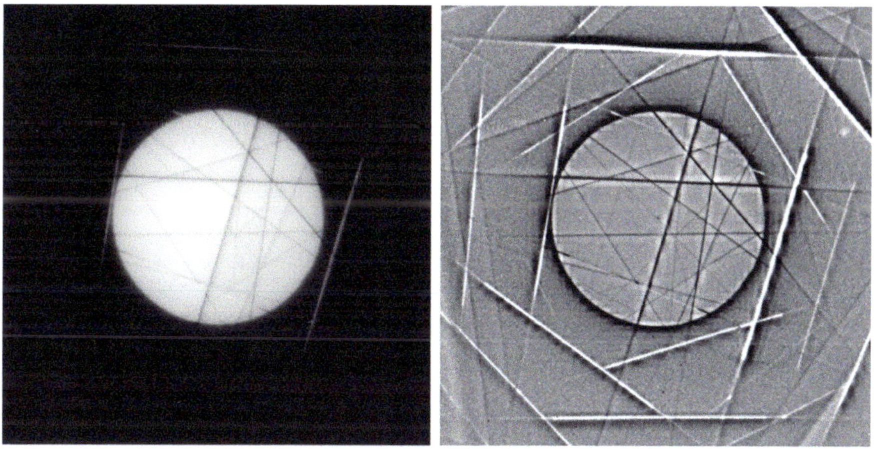

(a) (b)

FIGURE 8.16 (a) Raw Kikuchi diffraction pattern from a thick aluminum sample obtained by double exposure: outside areas are exposed longer than the inner area of the pattern. (b) Same pattern after background removal and brightness homogenization. (Adapted from Fundenberger et al., 2003.)

of TEM techniques for ORM we address here a few issues. Traditionally, TEM samples have been produced by electropolishing or dimple grinding and ion polishing (see the large collection of recipes for electropolishing in Thompson-Russell and Edington, 1977). The techniques create very well suited samples, but preparation of areas of interest is left to chance. TEM target sample preparation is nowadays frequently done by FIB milling which allows preparation of good samples, for example for high-resolution TEM observations. Observable areas, however, are often not larger than a few micrometers in diameter. For ORM these samples may be too small to reach the required amount of representativeness.

The most important sample preparation issue for ORM is the thickness of samples which strongly influences the appearance of TEM diffraction patterns. Very thin specimens only create spot patterns on a uniform low-intensity background as displayed in Figure 8.17a. Here a sample of an atomic number of $Z = 25$ may have a thickness of up to 100 nm. With increasing sample thickness, the amount of incoherent scattering becomes stronger. This leads to an increasing non-uniform background with appearance of Kikuchi patterns and spot patterns at the same time. Figure 8.17b shows mainly spot patterns with few Kikuchi lines while Figure 8.17c shows mainly Kikuchi patterns with weaker spot patterns. Here the sample may have a thickness of up to 200 nm. The strong background contrast leads to the fact that only small areas of the pattern may contain information while other areas are either under- or overexposed. Note that the pattern shown in Figure 8.17c has been recorded on film, which has a much larger contrast range than cameras and therefore shows significantly more features than a similar camera image. This image shows, furthermore, the presence of strong and aligned diffraction spots which may interfere with line detection algorithms, because they will be detected as lines, too.

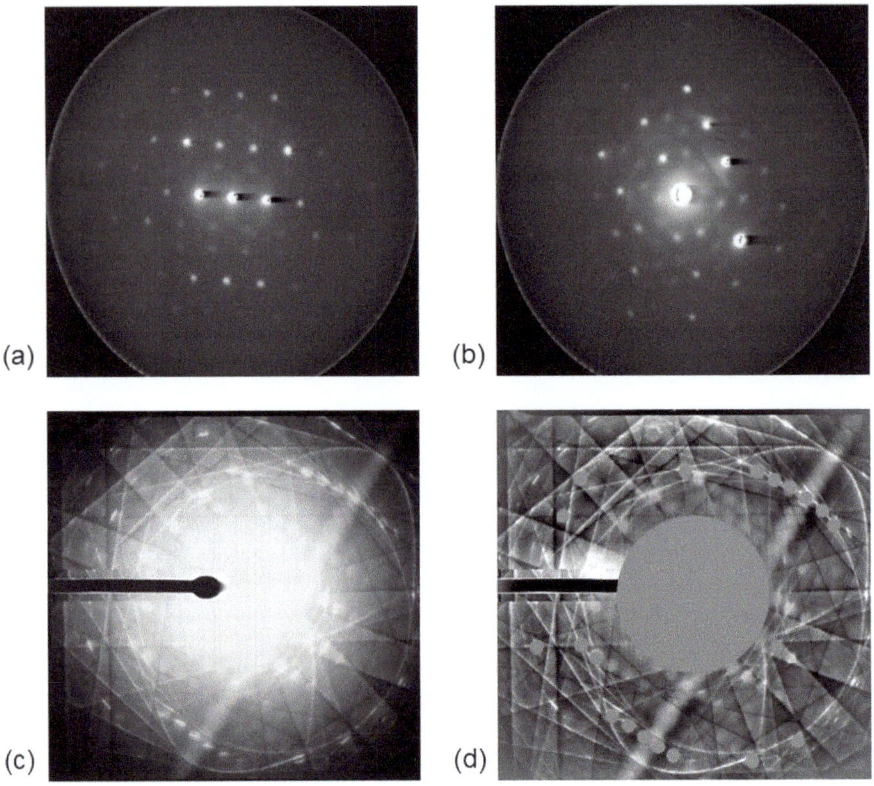

FIGURE 8.17 Diffraction patterns appearing from differently thick samples. (a) and (b) Spot patterns from an fcc Fe-Ni sample obtained on a simple CCD camera. (a) is a thin sample, (b) a slightly thicker one (see text). (c) Raw Kikuchi pattern from a thick bcc Fe-Cr sample, obtained on film. (d) Same sample after background and spot removal.

Proper image processing algorithms remove the spot patterns before the Kikuchi lines or bands are detected. Figure 8.17d shows the image, processed with the program TOCA (Zaefferer, 2002). For still thicker samples the background becomes more uniform, spots become weaker and Kikuchi lines and bands become strong (see Figure 6.13). As already mentioned before, the largest problems concerning different sample thicknesses arise on multi-phase samples where different phases may be thinned differently. Also in this case it appears, indeed, that spot pattern detection is the most reasonable approach to TEM ORM.

8.3.5 Spatial and Angular Resolution

In TEM ORM performed by STEM an electron transmissive sample is illuminated through its complete thickness by a high-energy electron beam. During this transmission the originally finely focused beam broadens slightly but observations show that the beam keeps a very small diameter in the range of few nanometers even

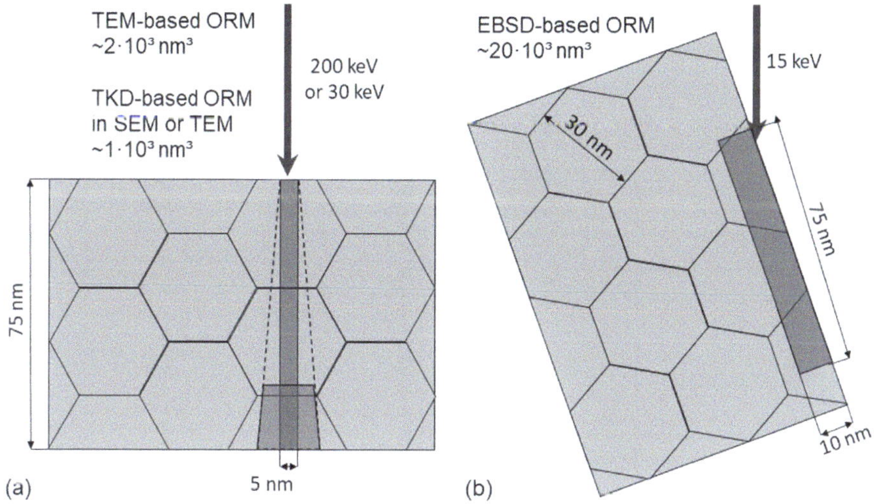

FIGURE 8.18 Comparison of spatial resolution of TEM and SEM-based orientation microscopy. As an example a schematic drawing of a nanocrystalline material with 30-nm grain size is sketched. The beam interaction volume is indicated in dark gray; its size is given in nanometers. (a) For TEM SCBED the diffracting volume is the full beam path. For TEM and SEM TKD only the bottom crystals contribute to the signal, leading to very small lateral (volume) resolution. (b) Note that for EBSD a reduction of voltage may significantly reduce the interaction volume.

on relatively thick samples. Some authors, therefore, claim a spatial resolution of STEM ORM, as good as 2 nm (Viladot et al., 2013). Nevertheless, this value has to be regarded critically as will become clear from Figure 8.18 which sketches the case that a finely focused beam transmits through a thin foil of, for example, 75-nm thickness. In this case, for SCBED patterns, all grains stacked along the beam path may contribute to the final signal and the resolution is, therefore, rather controlled by the sample thickness than by the beam size. In the case plotted in Figure 8.18 the resulting probed volume may be as large as about $1.5 \cdot 10^3$ nm³. Note that for Kikuchi diffraction it may be assumed that most of the signal only stems from the last grain in the stacked volume as the relatively subtle diffraction lines of one grain may blur out when transmitting through a further grain. It may, therefore, be assumed that Kikuchi diffraction in TEM may have a probed volume and resolution less than half of that of SCBED. For EBSD-based ORM, the spatial resolution is usually larger, as plotted in Figure 8.18b. For a material of a mean atomic number (e.g. iron) the probed volume obtained is in the order of $80 \times 30 \times 10$ nm³ = $2.4 \cdot 10^4$ nm³ about ten times larger than that of TEM ORM. Note, the resolution values of TEM ORM are, in contrast to EBSD, largely independent of the density of the sample. The TEM techniques are therefore clearly superior to the EBSD technique when working on low-Z materials ($<Z = 20$) where the EBSD interaction volume is rather large (~10^6 nm³) while the TEM interaction volume remains unaffected. Note, furthermore, that TKD in SEM (Section 8.2.4) may have spatial resolution comparable to that of TKD in TEM.

Finally we add a few words on the angular resolution of TEM-based ORM. It is obvious that the SCBED technique is less accurate than the TKD technique because the Kikuchi lines are sharper and the diffraction conditions are exactly fulfilled at a Kikuchi line while they are only approximately fulfilled on diffraction spots (discussed in Sections 6.2.3, 6.2.4, and 7.2.4). However, the diffraction vectors for spots at large Bragg angles are better defined than for those close to the optical axis. On SCBED the orientation accuracy, thus, becomes better with increasing maximum Bragg angle that is still visible in the pattern. Morawiec et al. (2014) compared SCBED and TKP techniques and came to the conclusion that SCBED results in an orientation resolution of about 1.1° while TKD may be as good as 0.3°.

8.4 SUMMATION

In this chapter we have addressed a number of aspects that have to be considered when performing orientation and phase measurements in the SEM or TEM. For optimum performance of the techniques in terms of spatial and angular resolution and quality of data it is essential to have optimum prepared samples and optimum imaging and recording parameters. For EBSD the largest improvements in terms of pattern quality and resolution are expected to result from optimization of the detector. A reduction of acceleration voltage may improve spatial resolution significantly if the sample is clean and the detector sensitive enough. For the TEM techniques a homogeneous sample thickness is a very important prerequisite for obtaining good results. It appears that the SCBED technique gives the most robust results even if the angular resolution is maybe inferior to that of the TKD technique.

9 Orientation Microscopy and Orientation Mapping

9.1 INTRODUCTION

The concepts of orientation microscopy (ORM) and orientation mapping were introduced in Section 1.2.3. *ORM* refers to the automated measurement and storage of crystallographic phases and orientations and potentially further information according to a predefined pattern of coordinates on the sampling plane of the specimen. Once this basic information (orientation plus position) is stored, it can be processed and output in a number of ways, which is the topic of this chapter. The pictorial output of these orientations with reference to the sampling coordinates provides an *orientation map* of the spatial orientation distribution; that is, it derives the "orientation topography."

Not only does an orientation map have great visual appeal, but it also gives access to fully quantified orientation statistics. In this chapter, we first consider the historical evolution of orientation mapping (Section 9.2) followed by some experimental aspects of ORM using electron backscatter diffraction (EBSD) (Section 9.3). The types of orientation mapping output and the information entailed are discussed in detail in Section 9.4, concentrating on application of orientation mapping to microtexture. Simple examples are used to show the basic types of microtexture parameters that are most commonly used in the maps to convey and depict aspects of the orientation topography. It should be emphasized that only a few examples, illustrating the main principles of orientation mapping to display microtexture, are shown. There are many more examples, of varying complexity, in the recent literature. The microtexture constituents in orientation maps are quantified and analyzed in the ways described in Chapter 10, and the examples show how the maps, or subsets of them, link to the statistical output.

ORM and orientation mapping are primarily based on EBSD in scanning electron microscopy (SEM). Some ORM techniques in transmission electron microscopy (TEM) are also available, which are described briefly in Section 9.5. Less effort has been put into the development of orientation mapping in TEM than in SEM because there are limitations with the TEM approach and there are fewer potential applications than there are for SEM. Section 9.6 finally provides a comparison of preferred applications of TEM- and SEM-based techniques.

9.2 HISTORICAL EVOLUTION OF ORIENTATION MICROSCOPY IN SEM AND TEM

The practice of linking microstructure and orientation in a map has existed ever since individual orientation measurements have been made, although it has not been reported frequently. A typical procedure would have been to observe or record the

DOI: 10.1201/9781003258339-12

grain structure, measure orientations of individual grains—usually one orientation per grain—then compile a map of the orientations (or grain boundary misorientations) superimposed on a diagram of the grain structure. Early examples of colored grain orientation maps derived by the Kossel technique (Section 6.3.5) were published by Inokuti et al. (1987). Although this method fulfilled the criterion of linking microstructure and orientation, the components of the microstructure and the orientations were first identified separately and then combined to produce a map. True orientation mapping takes a fundamentally different approach: diffraction patterns are collected according to a grid, and the spatial arrangement of the orientations thus obtained constitutes directly the map.

To the best of the authors' knowledge, the first time a true orientation map was published was in the 1960s; the technique was referred to as *orientation topography* (Haeßner et al., 1966). The maps, consisting of approximately 25 data points, were collected from rolled copper using selected area electron diffraction (SAD) in the TEM, with a square grid spacing of 1.8 μm. All in all, 900 SAD diffraction patterns were photographed from 14 different areas of the material. Figure 9.1a shows one of the larger contiguous maps composed of 100 individual orientations together with a pole figure (Figure 9.1b) which provides an orientation key to the symbols used (Gotthardt et al., 1972). The map, which relied on manual diffraction pattern evaluation, is certainly crude but it does embody the fundamental orientation mapping principle of constructing a diagram of the microstructure based on orientation measurements collected on a regular grid-sampling schedule. Furthermore, each measurement encompasses quantification of the orientation, which is conveyed in this case by a pole figure.

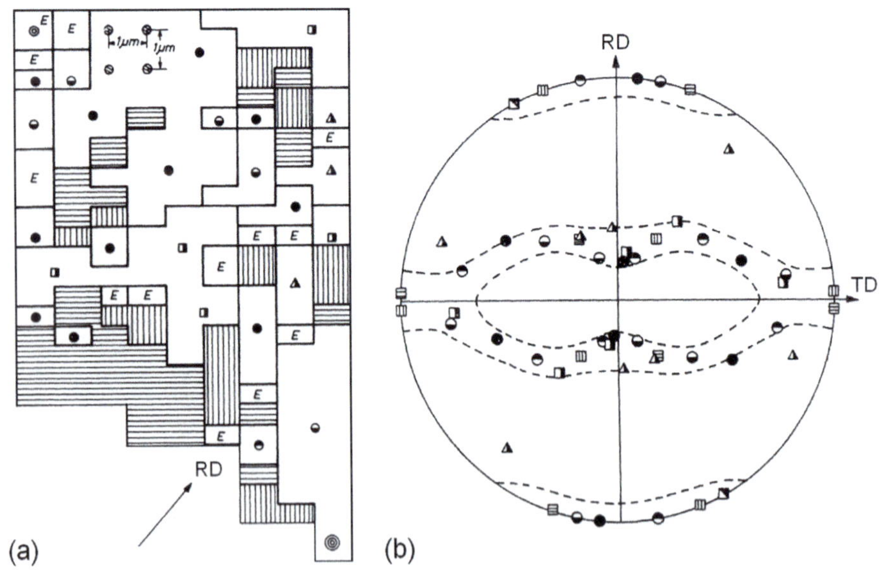

FIGURE 9.1 (a) Early orientation map, collected manually in the TEM from rolled copper using a square grid size of 1 μm. The symbols on the map refer to the pole figure in (b). (Adapted from Gotthardt et al., 1972.)

The technique did not receive much attention until Dingley (1984) developed a computer program for manual indexing of diffraction patterns and replaced the photographic film that was used for pattern registration by a setup of a phosphor screen and a TV camera. It should be mentioned that computer programs for analysis of EBSD patterns were written even earlier by Goehner (1975) and Goehner et al. (1976). In 1992 Krieger Lassen introduced the Hough transform (see Section 7.2.2) for the automatic detection of the Kikuchi bands in EBSD patterns (Krieger Lassen et al., 1992, 1994a). Together with an automatic indexing algorithm he created a computer program for the fully automated determination of crystal orientations from EBSD patterns. This development enabled Adams and co-workers in 1993 to develop a new way of scanning microscopy, the "orientation imaging microscopy," OIM (Adams et al., 1993). The publication of this work marks the breakthrough of the technique. The emergence of several commercial and non-commercial systems and an increasing number of users stimulated a strong development of both, software and hardware. In the course of the time several overviews have been written, e.g. Schwarzer (1997a, 1997b, 2006), Field (1997), Prior et al. (1999), Humphreys (2001), Baba-Kishi (2002), Dingley (2004), Randle (2007), and Wilkinson and Britton (2012). The most comprehensive overview is the collection of articles in the book of Schwartz et al. (2009). A second comprehensive book on the technique is available in French language (Pouchou, 2004). Besides the term OIM, which is a protected brand name of EDAX/TSL, a variety of other names have proposed, e.g. automated crystal orientation mapping (ACOM) (Schwarzer, 1997a, 1997b) or simply crystal orientation mapping, COM. We refer to the technique as EBSD-based ORM.

The evolution of orientation mapping techniques for TEM took a less straight pass, simply because of the many different techniques that are available, as it has been discussed in Chapters 6 and 7. Certainly, the investigations of Schwarzer (1983), Humphreys (1983), and Schwarzer and Sukkau (1998) were under the first. They all used Kikuchi pattern analysis because of its suitability for straightforward orientation determination. Zaefferer (2002) explored the possibilities of small-angle convergent beam diffraction (SCBED) patterns for TEM-based ORM in heavily deformed materials. This work motivated Rauch and Dupuy (2005) to develop the technique of template matching for fully automated orientation mapping. About at the same time Wright and Dingley (1998) and Dingley (2006) developed their conical scanning technique and Fundenberger et al. (2003) explored the possibilities of fully automated Kikuchi pattern analysis. Already earlier Vincent and Midgley (1994) showed that SCBED patterns could be improved by precession, leading to precession electron diffraction, PED. PED, SCBED, and template matching were then coupled by Rauch and Véron (2014) to a TEM ORM system. Over the last ten years or so, this technique has turned out to be the most powerful and robust way of dealing with TEM ORM, even if its orientation accuracy stays somewhat behind that of Kikuchi diffraction (Morawiec et al., 2014).

9.3 EBSD-BASED ORIENTATION MICROSCOPY

A typical result obtained by EBSD-based ORM is presented in Figure 9.2. The figure shows a microstructure of a partially recrystallized IF steel (Thomas et al., 2003)

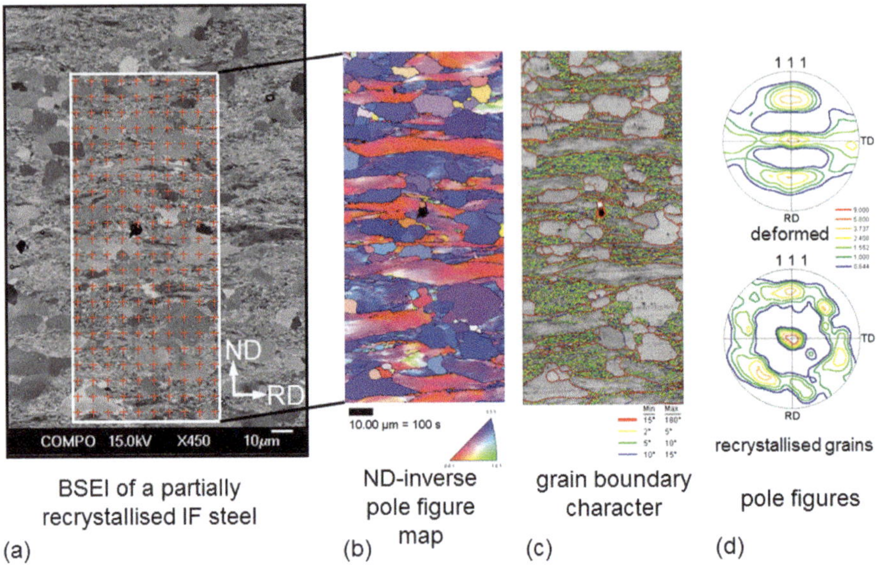

FIGURE 9.2 Illustration of the process of EBSD-based orientation microscopy. (a) A back-scattered electron image of a polished sample of a cold rolled and partially recrystallized IF steel. The measurement field for EBSD is indicated. (b) An orientation map in inverse pole figure coloring recalculated from the measured orientation data. The colors indicate the crystal direction pointing in the vertical direction of the map (the normal direction of the rolled sheet) (c) Grain boundaries of different misorientation angle superimposed to a pattern quality map. (d) (111) Pole figures calculated after microstructure partitioning for deformed and recrystallized grains (partitioning based on grain orientation spread and grain size).

as observed by electron channeling contrast and by ORM. In the crystal orientation map similar colors stand for similar crystallographic directions pointing toward the sample normal. Grains with uniform orientations (i.e. recrystallized grains) appear in uniform color while grains with internal orientation gradients show changing colors. In this way grain form and size, recrystallized and unrecrystallized areas, type of grain boundaries, and the local texture (here displayed in form of pole figures separated for deformed and recrystallized areas) can be easily recognized, partitioned, and quantified. Already this small example shows the strength of ORM: it is a truly quantitative microscopy technique because it first creates numbers for every scan position (crystal orientations and phases, quality data, strain, etc.) from which then microscopic images and statistical values can be calculated in large varieties. It allows creation of large and accurate data sets with more than 10^6 pixels, with 10^5 grains or other units can be measured in less than 4 hours. It reaches lateral spatial resolutions between 20 and 1000 nm (depending on the material), its angular resolution ranges between 0.5° and 0.01°, and it allows the observation of sample areas between square micrometer and square centimeters.

Before presenting an overview on the many applications of ORM we first discuss a few important points concerning the data collection strategies for ORM by EBSD (i.e., the actual collection of the raw data to produce a map) in Section 9.3.1 and ways to store data in Section 9.3.2.

9.3.1 Measurement Strategies

When designing the scanning strategy for an EBSD measurement it is important to consider two aspects. The first is the desired resolution of the measurement, which sets the step size of the map. The second is the requested statistical representativeness of the measurement which sets the total size of the probed area. Conflicts between both may arise when, for example, small precipitates are to be measured in a matrix of large grains. We will discuss all aspects and propose a strategy to solve also potential conflicts.

There is a choice of two basic approaches for location of the sampling spots on a specimen surface. Either the specimen is moved with respect to the stationary incident beam by means of a computer-controlled specimen stage, that is, *stage control* (Adams et al., 1993), or the electron beam itself is deflected across the stationary sample surface, similar to conventional SEM, that is, *beam control* (Schwarzer et al., 1995). Although stage control has a few interesting features, beam control is by far the most frequently used strategy for EBSD mapping due to the much higher speed, accuracy, and flexibility of the beam control.

Stage control can be realized by a high-precision specimen stage that positions the sample relative to the electron beam. The main advantage of stage control is that it allows to accomplish very large scans even in the order of square centimeters, which may be of interest for the observation of local texture variations in work pieces, for example. If the sample is flat and mounted on a stage as the one displayed in Figure 8.6b (a pretilted stage does as in Figure 8.6a is not suitable) the sampling position stays in focus and the calibration parameters stay constant. Stage control on its own is little used nowadays to locate sampling points. However, in conjunction with beam control it creates a very attractive large-area sampling mode which will be discussed further below.

Precise movement of the beam is achieved via algorithms in the EBSD software, which control the SEM deflection coils. In this way *high resolution maps at high magnifications* (about >2000 times magnification) can be quickly realized. It is important to understand that the resolution of a map (not the resolution of EBSD!) is defined by the step size of the map as long as the step size is larger than the physical or effective resolution (see Section 6.3.3 on the physical and Section 8.2.9 on the effective resolution). Step sizes smaller than the effective resolution make no sense and it is important to realize that in cases when small-angle grain boundaries are to be resolved the effective resolution is equal to the physical resolution. Resolution values have been discussed in the previous chapters (Section 8.2.3). As a rule of thumb step sizes down to about 200 nm make sense for materials with average atomic numbers smaller than about 20 (light metals, most minerals), 50 nm for those up to $Z = 50$ (first 2 lines of transition metals around Fe and Ru and there compounds) and down to 10 nm for even heavier materials. These values are true for SEMs with thermal field emission guns. For LaB6 or W-filament microscopes they are by a factor of 2 or 3 larger.

Frequently much larger step sizes than the physical resolution are recommendable. Humphreys (2001), for example, recommends selecting the step size such that the smallest well-visible grain has about ten steps in diameter. Note that if grain shapes and sizes are not important, e.g. when only the texture is relevant (see further below), even larger step sizes are reasonable. Maps with small step sizes may take long time and may, thus, be subject to all kinds of drift. A stage drift or a continuous beam drift due to sample charging or heating may lead to a significant macroscopic distortion of the originally rectangular scan field. Furthermore, short period drifts, for example due to heating-cooling cycles of the cooling water of the objective lens, may lead to regular sinusoidal drifts in the image. The large drifts mentioned first can be corrected by using a drift correction approach, where the software, after a certain measurement time regularly searches for a recognizable marker in the image and corrects for its position relative to its original position. This may reduce drift significantly. Further ideas for drift correction, e.g. by correlation of SE (secondary electron) or BSE (backscatter electron) images recorded without drift in a short time with the slowly measured EBSD map, are discussed by Winiarski et al. (2021). The sinusoidal drifts, however, cannot be corrected in this way. It appears necessary, instead, to optimize the SEM cooling device or other sources of short-term drift.

When performing *maps at magnifications smaller than approximately 2000 times*, it should be realized that the resolution of the SEM may not be independent of the magnification because with decreasing magnification the electron beam may vibrate around its expected position. This means that measurements with small step sizes performed at low magnifications, for example in order to cover a large area at high resolution, may not result in the desired resolution performance. For low magnification scans, it is also important to realize that the calibration may change significantly across the map (see Section 8.2.8). If this is not corrected, the measured orientations may be systematically wrong across the scan or indexing could even fail. To avoid this, it is necessary to update the calibration parameters at every scan position by interpolation of the parameters obtained for the start and end points of the scan. Note that this can be done automatically with all commercial EBSD systems.

When EBSD sampling is done at *magnifications below 500 times*, it becomes furthermore important to use the *dynamic focus* of the microscope to keep the focus across the steeply tilted sample. Because the specimen is highly tilted (e.g., 70°), the electron image of the sampling area is distorted and a dynamic focusing correction does not only keep the focus on the surface but may also account for the foreshortening effect and to produce an undistorted map.

In cases where high lateral resolution, e.g. to measure small features, is to be coupled with high statistical relevance it is optimum to combine stage and beam control in a multi-map approach. While the beam control operates at high magnification to measure small maps with high resolution the stage moves the sample from one map location to the next. If data acquired in this way are from contiguous regions on the specimen, the beam-scanned maps can subsequently be montaged in a post-processing step. By this procedure, many beam-scanned maps can be stitched together to cover a wide area of specimen surface. It is, however, also possible to measure non-contiguous maps, for example, by distributing small probing maps over a large area of the specimen (Davut and Zaefferer, 2010); an example will be given in Section 9.4.8. Every small map may show high-resolution details, while all maps together provide a statistically representative distribution of these details without measuring the whole surface. This may result in reduced measurement times and reasonable limitation of the amount of collected data. Such measurements can be very efficient to measure macrotextures, macroscopic grain size or shape distributions, or grain boundary distributions, to mention a few possible results. Examples for such measurements will be presented in Sections 9.4.8 and 10.3.3.

When using beam control all kinds of smart probing of a sample can be realized. Most importantly, some commercial systems allow deciding whether sampling is performed on a square or a hexagonal grid. The hexagonal grid has the advantage that each measurement point has six equal-distant nearest neighbors, whereas in a square grid four nearest neighbors plus four second nearest neighbors must be considered. Furthermore, the hexagonal grid facilitates delineation of interfaces. When probing large areas, particularly on materials, which show clear grain structures (e.g. recrystallized metals), it has been proposed to scan an area in a first step very coarsely and successively refine the scanning step size only between those points, which show a misorientation larger than a given bias. This can lead to a very significant reduction of the number of measured points and of measurement time while preserving a good resolution of the map.

It should be mentioned, finally, that EBSD and also TEM-based ORM only probes discrete points, of the size of the interaction volume of the respective technique, in the microstructure. Although the finally created maps appear contiguous, there may be a lot of unmeasured space in between the measured points, particularly when the measurement step size is much larger than the physical resolution of the technique. This is in contrast to the broad beam x-ray based techniques, which measure the total volume and therefore do not "overlook" features.

9.3.2 Data Storage

After diffraction pattern indexing each sampling point is stored with its phase identity, orientation (usually as Euler angles), spatial coordinates, measure of fit or confidence index, and diffraction pattern quality metric(s). The record file is usually stored in a proprietary file format of the EBSD system providers. Besides the measurement data this file also contains meta data like the geometry of sample set-up, measurement parameters (working distance, acceleration voltage) pattern center calibration values, crystal structures of the considered phases and more. Additionally, the file may contain x-ray count data data if simultaneous EDS measurements were carried out. Furthermore, also the determined Hough peak positions or even the complete diffraction patterns may be saved for every data point in order to allow re-indexing after the measurement. With all these data the data files may become very large, at the limit up to hundreds of Gigabyte when high-resolution patterns are saved. A certain amount of compression is possible by saving the numerical data in a binary format. These data, however, could, until recently, only be opened in dedicated software provided by the commercial EBSD system makers. To avoid this, all EBSD system makers also provide text-based output of their data, which, however, often do not allow access to all data stored in the binaries. In recent times efforts are made to store all measurement data in hierarchical data format, HDF, particulary in hdf5, as documented for example by Jackson et al (2014). This data file format allows universal reading and writing of measurement data and meta data. Furthermore, a recent initiative, started by Ånes and coauthors (Ånes et al, 2020), provides an open-source Python-based software (KikuchiPy, documentation) that is capable of reading and writing most proprietary file formats and transform them into each other.

It should be noted that saving diffraction patterns for every data point, although occupying large storage space, is essential for some applications and for others highly recommendable. Storing patterns is essential for cross correlation calculations as they were introduced in Section 7.4.5. For this application it is even important to store patterns at high pixel resolution (>500^2 pixels), where the required storage space increases quadratic with pattern resolution. Also for indexing of diffraction patterns with any form of pattern matching (template matching Section 7.4.2, dictionary indexing Section 7.4.3, or spherical indexing shortly described at the end of the book.), it is imperative to store the diffraction patterns, though they can be stored, at very low pixel resolution (down to 30^2 pixels) if angular resolution is not very important. There is also increasing demand to store patterns for post processing after mapping has been performed. Stored patterns, for example, allow re-indexing of the data set (or a part of it) with another set of crystallographic phases, or with different calibration or even with different image processing and analysis parameters. The first two tasks (re-indexing after change of phases or calibration parameters) can also be executed when only the Hough peaks have been stored (see Section 7.2.2), which is possible with some software systems. If one considers position, width, and brightness of Hough peaks to be stored as integer values for, for example, 12 detected Kikuchi bands, this requires 8×12 Byte to be stored additionally for every measured data point, in contrast to at least 30×30 Bytes if patterns are stored. In order to release storage volume it may be considered to delete or archive the patterns after final analysis of the data.

9.4 ORIENTATION MAPPING AND ITS APPLICATIONS

Depending on the information that is being sought, the main applications of orientation mapping to microtexture can be categorized as follows:

1. Spatial distribution of microtexture components (Section 9.4.1)
2. True grain size/shape distributions (Section 9.4.2)
3. Phase maps (Section 9.4.3 and Chapter 12)
4. Strain and dislocation density maps (Sections 9.4.4–9.4.6)
5. Texture maps (Section 9.4.7)

Display and quantification of grain boundaries and interfaces in terms of number, morphology, and crystallographic character is one of the strongest features of EBSD. Since grain boundaries are also important drivers of texture and microstructure evolution, e.g. as obstacles for dislocations during deformation, or as nucleation points for recrystallization or phase transformation we discuss the topic of interface characterization in detail in a separate Chapter 11.

More complex parameters, such as Taylor factors or other user-defined functions such as geometrically necessary dislocations, GNDs (El-Dasher et al., 2003) (see

Sections 9.4.5 and 9.4.6) or *r* values (Huh et al., 2005), can also be displayed in orientation maps.

Various noise filtering or "cleanup" algorithms are used to reduce the number of incorrectly indexed pixels in the map by extrapolating data into adjacent regions of poor pattern solve success or by replacing single orientations that are obviously wrong. A very important contribution to cleanup procedures is given by the fact that orientations that are closely related, e.g. that stem from the same grain should have similar orientations. This knowledge can be used to improve the reliability of data significantly. This may particularly help to improve the reliability of data close to grain boundaries. Use of cleanup routines should be carefully controlled to avoid introduction of artifacts into the data.

9.4.1 Spatial Distribution of Microtexture Components

Different texture components can be represented in a map using colors, gray scale, or patterns. This attribute is particularly useful for picking out the distribution of particular components with respect to grain size, shape, or position, for example, newly recrystallized grains in a deformed matrix or (111) grains with respect to their position within the steel sheet. Another area of interest since the advent of automated EBSD is the *connectivity* or grouping of orientations into spatially linked clusters (Larsen and Juul Jensen, 2003; Lewis et al., 2006). Orientation mapping indicates such clusters very graphically (Mason and Adams, 1994; Huh et al., 2005).

Color output is either linked directly to the orientation or user-defined. It is accompanied by a key to the use of colors. A popular output is based on assigning red, green, and blue to the 001, 011, and 111 corners, respectively, of the stereographic unit triangle for crystal directions that are parallel to a selected specimen direction. Note that, as for an inverse pole figure (IPF) (Sections 2.5.2 and 10.2), this representation displays only one direction of the 3D orientation information, and so rotations about this axis are not seen. For a complete interpretation, maps for one (or both) of the two other specimen reference directions can also be viewed. An example of this type of output is shown in Figure 9.3, which is an orientation map from commercially pure titanium in the specimen X, Y, and Z directions (Figure 9.3a, b, and c, respectively). The stereographic unit triangle color key for the maps is included in Figure 9.3d. The microtexture data from the maps are represented as IPFs from the X, Y, and Z direction maps in Figure 9.3e. Finally, Figure 9.3f shows the equivalent {0001} pole figure representation.

Euler angles (Section 2.6) or Rodrigues vectors (Section 2.8), each comprising three components, can be represented in maps by three colors. Only one map is necessary to display the full orientation distribution, rather than at least two maps, which is the case when the orientation color coding is linked to the specimen direction as in Figure 9.4. Figure 9.4a is the same map as Figure 9.3, but displayed with Euler angle coloring. The map is accompanied by a color key for the Euler angles (Figure 9.4b).

There is a very wide range of user-defined choices for depicting various microtexture subsets on maps. For example, choices can be based on a particular texture parameter, a grain size range, or both. These examples are illustrated by orientation maps obtained from silicon iron. Figure 9.5a shows an orientation map from an iron silicon alloy with grains colored blue if the specimen Z (normal) direction is

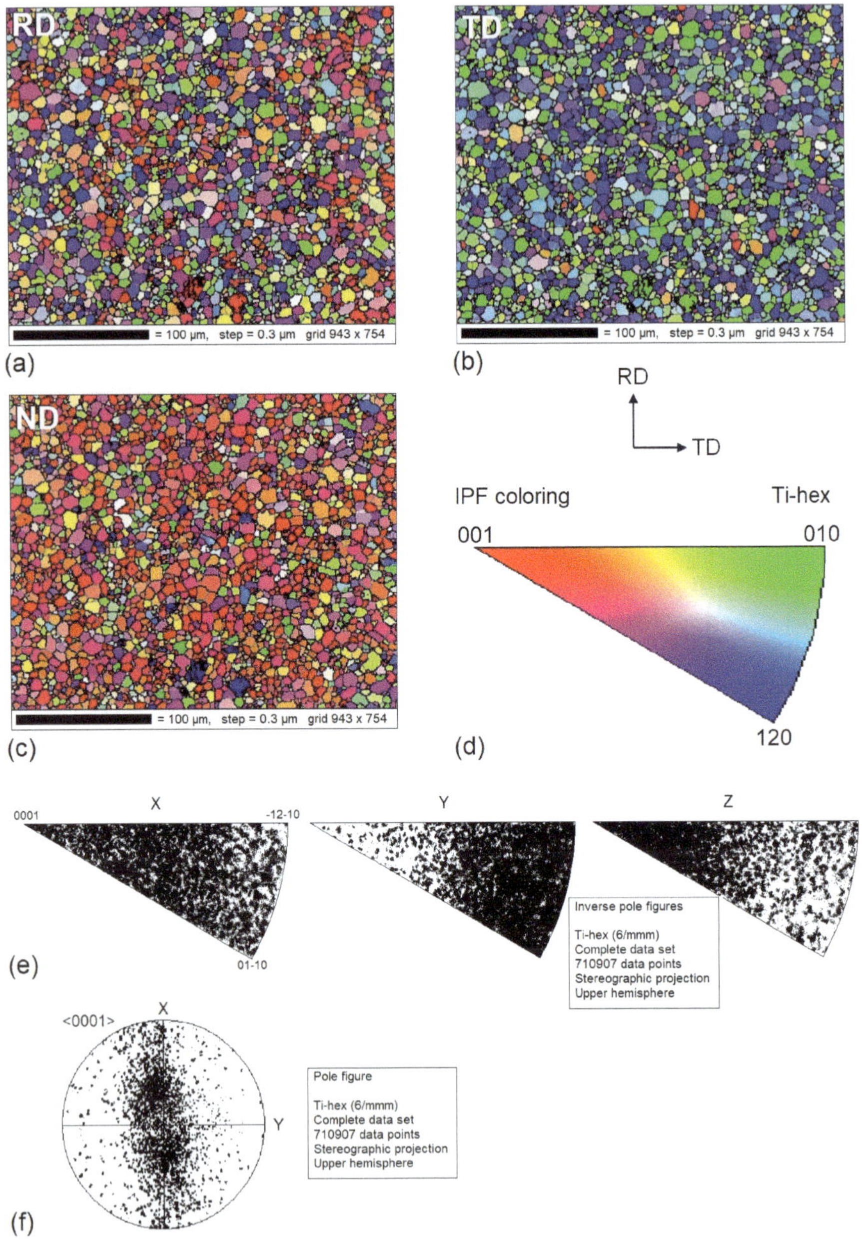

FIGURE 9.3 Texture of a sheet of recrystallized commercially pure titanium. (a), (b) (c), crystal orientation maps displayed in inverse pole figure colouring of (a) the sample rolling direction (RD || X), (b) the sample transverse direction (TD || Y) and (c) the sample normal direction (ND || Z). (d) is the inverse pole figure color code. (e) inverse pole figures from the X, Y, and Z directions; (f) (0001) normal pole figure.

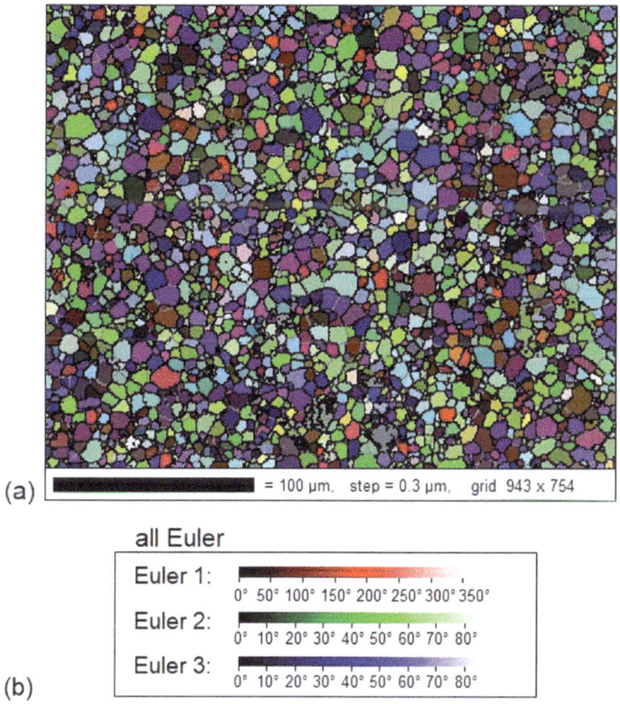

FIGURE 9.4 (a) Euler angle orientation map from commercially pure titanium. (b) Euler angle color key.

<20° from (111), which comprises 45% of the map area. The depth of color diminishes with increasing deviation from exact ⟨111⟩. Large-angle grain boundaries are depicted by black lines and small-angle grain boundaries by gray lines. Figure 9.5b is the {111} pole figure from the entire map in Figure 9.5a, shown as density contours (see Section 10.2.2).

Figure 9.6a shows another orientation map from an iron silicon alloy where large grains have arisen from abnormal grain growth. Here, the user-defined color key (Figure 9.6b) refers to proximity to the "α fiber," that is, (110) parallel to the direction of rolling during processing (Section 5.5.2). Red depicts the exact α fiber, changing to blue for orientations having the maximum deviation from the α fiber, that is, 45°. The {001} pole figure is shown in Figure 9.6c. The pole figure is linked to the map via numbering of the data points associated with large grains.

Orientation mapping has highlighted that, although a grain is defined as a unit of microstructure having a single orientation, in reality it has an *average* orientation since even undeformed materials may show some lattice curvature close to grain boundaries, and deformed grains exhibit a whole range of substructures that are accompanied by orientation shifts. Orientation mapping with a very small step size provides a means of studying these orientation perturbations (Adams et al., 1993; Bastos et al., 2006; Sandim et al., 2007). The results of such studies help to pinpoint

(a)

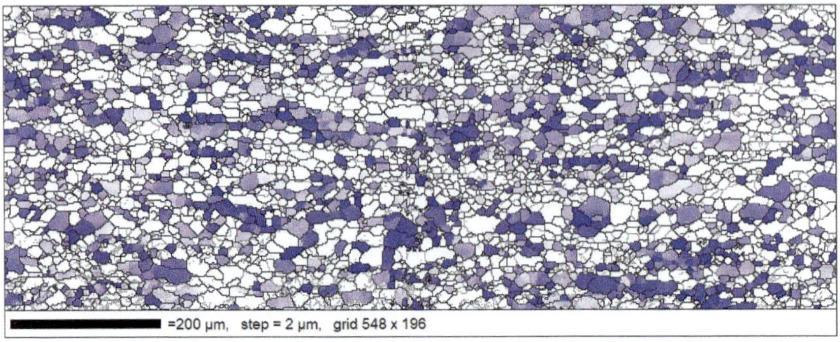

(b)

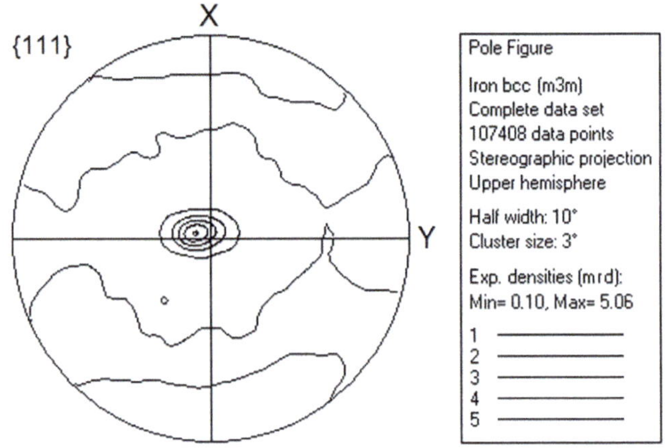

FIGURE 9.5 (a) Orientation map from an iron silicon alloy showing grains colored in blue if the specimen Z (normal) direction is <20° from (111). The depth of color diminishes with increasing deviation from exact (111). Large-angle grain boundaries are depicted by black lines and small-angle grain boundaries by gray lines. (b) {111} Pole figure from the map in (a), shown as density contours. The rolling direction is horizontal. (parallel to the specimen X direction.)

mechanisms of strain accommodation, which in turn have a large influence over a wide range of thermomechanical processes, including commercial fabrication of materials, and geological studies. For example, a new parameter, called the "modified crystal deformation," has been defined to quantify the spread of crystal orientation within individual grains arising due to dislocation accumulation during plastic deformation (Kamaya et al., 2006).

9.4.2 True Grain Size/Shape Distributions

Grain size distribution statistics can be readily computed by the EBSD software via the recognition of grain boundaries and hence grains. The map is digital and can therefore be input directly to various image analysis routines. Because it relies only on crystallographic characterization to determine a grain rather than etching or other traditional methods, the orientation map has the capacity to replace optical

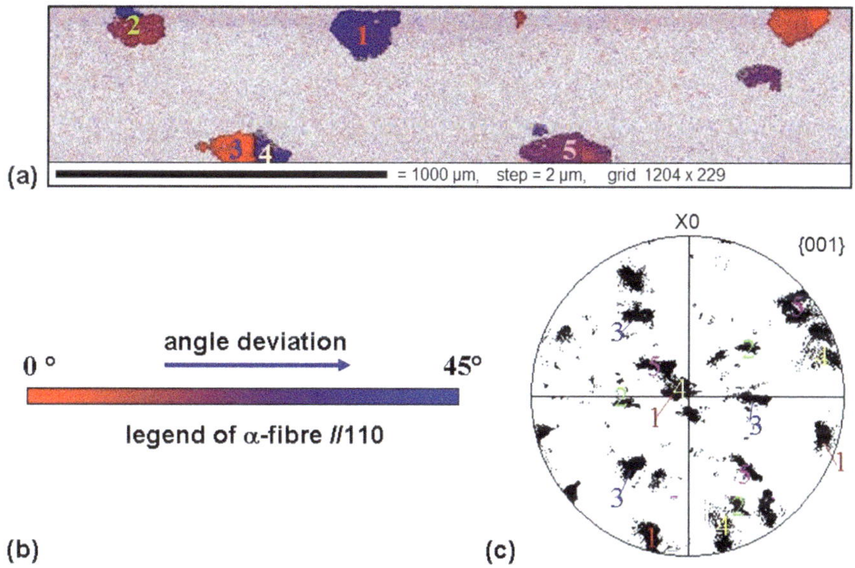

FIGURE 9.6 (a) Orientation map from an iron silicon alloy showing large grains (numbered) arising from abnormal grain growth. (b) User-defined color key for the map in (a); grains are colored according to proximity to the "α fiber," that is, (110) parallel to the rolling direction. Red depicts the exact α fiber, changing to blue for orientations having the maximum deviation from the α fiber (45°). (c) {001} Pole figure showing the orientations of the large grains, numbered accordingly, in (a). The rolling direction is horizontal (parallel to the specimen X direction).

microscopy for characterization of grain size. Microtexture measurements can hence also be linked to grain size. For example, the effect of second-phase particles on the rate of grain refinement has been studied by EBSD in an aluminum alloy (Apps et al., 2003). A weaker texture was found in the particle-containing alloy than in a similar alloy without particles. The link between grain size and texture has also been investigated by EBSD in a friction-stir processed Mg–Al–Zn alloy (Wang et al., 2006).

The misorientation used to define a large-angle and a small-angle grain boundary from an orientation map is user-defined and will affect the grain size calculation. Typically, a large-angle boundary has a misorientation of more than 15°. A more subtle consequence of use of orientation mapping to define grains is that the specification of a "grain" in the orientation map is based on the orientation of neighboring points being within a certain, small, tolerance of each other. However, the cumulative spread of orientation within a grain can be relatively large, especially for deformed materials (Kamaya et al., 2006).

9.4.3 Phase Maps

Microtexture can be determined simultaneously from known phases in multi-phase materials. Orientation mapping and microtexture determination when applied to multi-phase materials require some additional considerations regarding the simultaneous automatic pattern recognition of two or more phases (Field et al., 1996; Field, 1997). These issues include the following:

- The "background" diffraction signal may vary for each phase, which can be addressed by various hardware or software solutions.
- The computer algorithm needs to correctly recognize each phase. To search simply for the "best-fit" solution may have an unacceptable time penalty, and so a "confidence index" approach may be used. This requires some preliminary work to assess the confidence index parameters of each phase in the material.
- Multi-phase materials where the phases have the same crystal structures can only be analyzed with the help of other parameters such as pattern quality (Section 9.4.4).
- Specimen preparation of multi-phase materials, where the constituent phases polish at different rates, may be more complicated than that of a single-phase matrix.

Various steels and titanium alloys are typical examples of multi-phase systems that have been analyzed by EBSD orientation mapping (Schwarzer et al., 2000; Seward et al., 2004; Garzón and Tschiptschin, 2006). Not only is the microtexture of each phase obtained, but the output is a potent tool in the study of phase transformations (Gourgues-Lorenzon, 2009).

As an example, Figure 9.7 shows three different ORM maps obtained in a two-phase nickel-tungsten material, namely, an image quality map (Figure 9.7a; Section 9.4.4), an orientation map (Figure 9.7b), and a phase map (Figure 9.7c). As for the latter, orientations that could better be indexed for fcc crystal structure—those

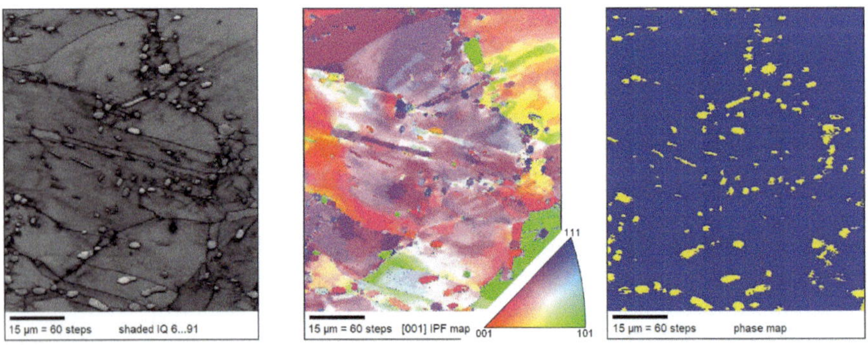

(a) IQ map (b) orientation map (c) phase map

FIGURE 9.7 EBSD analysis of a mildly deformed two-phase Nickel-tungsten specimen. (a) Image quality map (darker shading corresponds to poorer image quality; note that grain and phase boundaries are not specially marked, but show up due to their low image quality); (b) orientation map; (c) phase map (blue—nickel; yellow—tungsten). (Data taken from Sinclair et al., 2003.)

that are more likely to belong to the nickel—are colored blue, whereas the bcc orientations (i.e., tungsten) are colored yellow. Furthermore, large-angle boundaries are marked with black lines. This representation displays well the topographic arrangement of the two phases and estimates the respective volume fractions. In the present example, it turns out that the minority component, the tungsten, is situated preferentially at the nickel grain boundaries (Sinclair et al., 2003). The phase orientation relationship between the two phases can also be determined, which is addressed in Section 12.2.

9.4.4 Deformation Maps

One of the most important aspect of EBSD-based ORM is the measurement and characterization of microstructure and texture of plastically deformed materials and the quantification of plastic strain therein. Although this is a very large and quickly evolving application field—most studies appeared within the last ten years years—we will limit ourselves only to a few examples and refer the reader to overview papers from literature for further reading. Highly relevant papers with respect to the use of conventional EBSD-based ORM are the ones by Wright et al. (2011, 2015b), Kamaya (2011), Kamaya et al. (2006, 2020), Allain-Bonasso et al. (2012), and Field et al. (2010, 2012). References concerning the application of cross-correlation EBSD (CC EBSD) will be given in the next two sections.

Plastic strain, characterized by the presence of dislocations in a crystal lattice, can be observed in ORM maps in two characteristic manners. On the one side, dislocations in the lattice lead to a reduction of the diffraction pattern quality by blurring the diffraction bands and lines (see Chapters 6 and 8). On the other side, the presence of dislocations usually leads to local lattice rotations, which can, in turn, be used to quantify the amount of dislocations in the lattice. Dislocations that lead to a measurable lattice rotation are frequently called "geometrically necessary dislocations," GNDs, while those which do not create measurable rotations are summarized as "statistically stored dislocations," SSDs. SSDs only lead to a blurring of the diffraction pattern which becomes stronger with increasing amount of dislocations. The distinction of dislocations in GNDs and SSDs is rather artificial and more based on the way of measuring them than on a physical argument. Figure 9.8 gives a schematic overview on the two "types" of dislocations. For GNDs the long-range stress field associated with a dislocation line has relaxed in favor of a long-range lattice rotation. All those dislocations which are not associated with a long-range rotation field are classified as SSDs. This includes relaxed dislocation dipoles which consist of two closely spaced dislocation segments of opposite sign (e.g. created by dislocation loop), as well as unrelaxed individual dislocations. Figure 9.9 sketches the effect of GNDs and SSDs in an orientation map: a grain that contains a homogeneous distribution of GNDs (in the center of the image) shows a continuous orientation gradient, at subgrain boundaries the GNDs are clustered along lines, leading to local sudden orientation changes. If a grain would contain only SSDs (left side in the image), which is a theoretical situation, it would not show any orientation change, even if the dislocation density would be high.

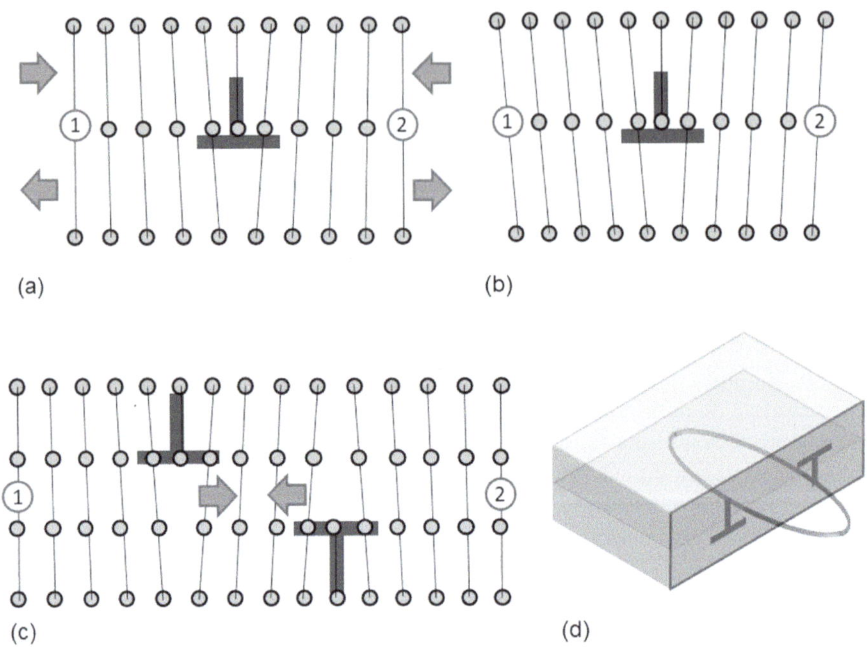

FIGURE 9.8 Illustration of different types of dislocations in strain fields. Rotations associated with the dislocations are measured between points (1) and (2). Stress fields are indicated by arrows. (a) Unrelaxed dislocation (considered to be a statistically stored dislocation, SSD) which does not create long-range lattice rotations but is associated with long-range stress fields. (b) Dislocation which has relaxed its long-range strain field into a rotation field and changed into a geometrically necessary dislocation (GND). (c) Dipole without long-range rotation field (SSD). (d) 3D-arrangement of a dipole, created from a dislocation loop.

Dislocations (not distinguishing SSDs or GNDs) are created during plastic deformation of a material. The dislocations originate in grain boundaries, at surfaces, or in Frank-Read dislocation sources, they move through the material and most of them leave it again. This "life cycle" of dislocations accommodates the strain the material is submitted to and rotates the grains, which eventually leads to the formation of a deformation texture. Only a small part of the dislocations remains in the material where they are responsible for the strain hardening of a material, for the stored deformation energy leading to recovery and recrystallization phenomena, and for strain damage. Of course, only those dislocations that remain in the material can be observed and measured by EBSD. Of these, only GND can be determined quantitatively.

In order to understand the deformation behavior of a material on the basis of EBSD measurements it is often assumed that the density of remaining dislocations reflects the density of moving dislocations. This may be true in the beginning of deformation but with increasing strain the dislocation evolution and the local plastic strain are no longer proportional but the dislocation density evolution roughly follows the Kocks-Mecking model (Kocks and Mecking, 2003). A more recent account

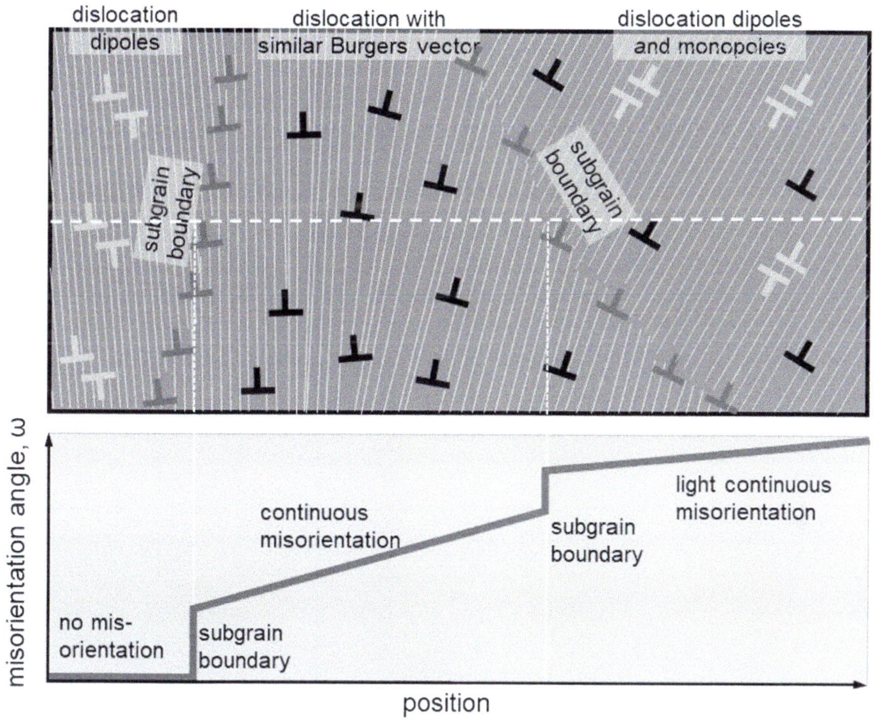

FIGURE 9.9 Relation between dislocation types and lattice rotations. Dislocations are drawn in different gray values to distinguish their different classification as SSDs, single GNDs, and GNDs organized in subgrain boundaries. Principally, however, they are all the same lattice defect. Geometrically necessary dislocations (GNDs) create measurable lattice rotations, statistically stored dislocations (SSDs) only lead to pattern blurring.

on this relationship is given, for example, by Davoudi and Vlassak (2018). It is furthermore assumed that the GND density, which can be quantified by EBSD maps as will be shown below, is proportional to the total stored dislocation density. Also this is, unfortunately, not true. It has been shown by a number of authors that the GND density is usually only a small and variable part of the total dislocation density. GNDs are only formed by heterogeneous deformation where the amount of positive and negative dislocations is imbalanced. Field et al. (2012) claimed a fraction of ρ_{GND}/ρ_{SSD} = 10%. By applying EBSD to measure the GND density and ECCI to measure the total dislocation density An and Zaefferer (2019) also showed a fraction of about 10%.

EBSD-based ORM offers a large number of misorientation descriptions that can be used for the quantification of strain. We mention a few:

- Grain average misorientation, which averages the misorientations between neighboring pixels for every grain.
- Grain orientation spread ("GOS"), which determines for every grain the largest misorientation.

- Grain reference orientation deviation ("GROD"), which determines for every pixel in a grain its misorientation with respect to a given reference orientation, for example the average orientation of a grain. The data can be expressed as angle and axis of misorientation. This is a particularly powerful way of exploring strain data, because it results in relatively large misorientations and shows their distribution in every grain.
- Kernel average misorientation ("KAM"). This measure is very much seen in literature and will therefore be described in some details in the following.

A *kernel* is a group of pixels that consists of a center pixels and all of its neighbors, potentially including also higher order next-nearest neighbors. A kernel is displayed in Figure 9.10a (here for the case of a hexagonal pixel array). For KAM all misorientations between the center pixel and its neighbors are measured and averaged. This value is then assigned as KAM value to the center pixel. The kernel may now be swept of an orientation map and the KAM value, ω_{KAM}, calculated

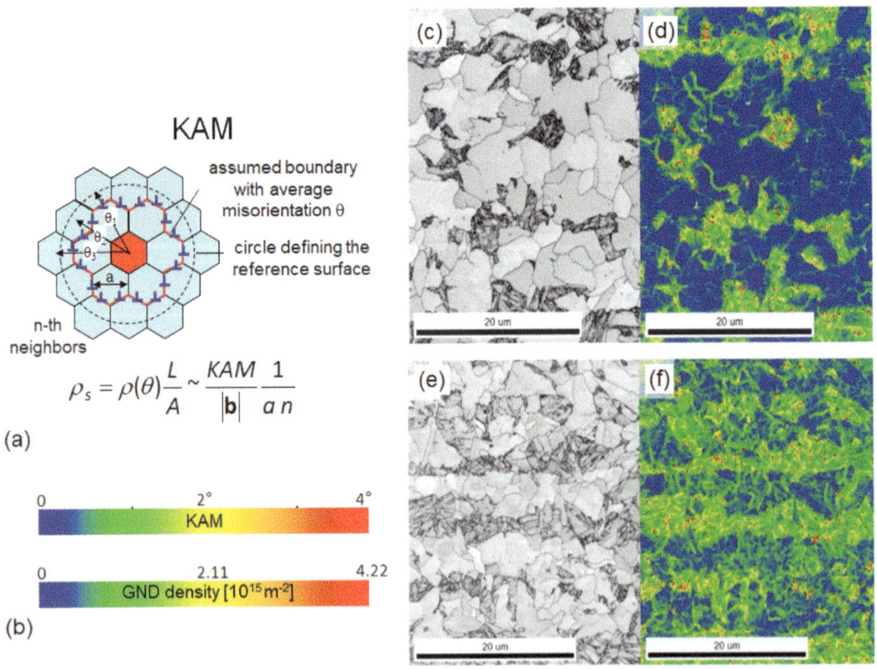

FIGURE 9.10 Illustration of the kernel average misorientation method. (a) Structure of a kernel in a hexagonal pixel grid. The red line in the kernel represents a virtual small-angle grain boundary between an inner and an outer grain, consisting of virtual dislocations. From the length of the boundary, L, the number of dislocations along it, ρ(θ), and the surface of the kernel, A, the dislocation density is calculated for a given KAM value. (b) Transformation of KAM values to GND density values applying the KAM equation. (c) and (d) Pattern quality map and KAM map of a DP (dual phase) steel annealed at 600°C for phase transformation (e) and (f), similar but annealed at 575°C. (Maps adapted from Schemmann, 2014.)

for every pixel in the map. Examples for the resulting KAM maps are displayed in Figure 9.10d and f which have been measured on two dual-phase steel samples, consisting of martensite and ferrite or bainite (Schemmann, 2014). The upper material has undergone a diffusive ferrite transformation while the lower underwent a partly displacive bainitic transformation. Figure 9.10c and e additionally displays the related pattern quality maps. These maps show the martensite as low-quality areas. Martensite has high dislocation densities and high lattice strain caused by the carbon incorporation into the lattice, both leading to low pattern quality. The material with ferrite transformation shows heavily strained martensite within an almost strain-free ferrite matrix. In contrast, the bainitically transformed material displayed in Figure 9.10e and f shows less heavily distorted martensite but more plastic strain in the bainitic matrix.

By assuming that the kernel consists of an inner grain and an outer grain, who have a misorientation corresponding to ω_{KAM}, which is accommodated by dislocations sitting along the grain boundary, it is possible to quantify the dislocation density is hosted in the kernel shown by Konijnenberg et al. (2015):

$$\rho_{GND} = \frac{c}{b\,a\,n}\,\omega_{KAM} \tag{9.1}$$

where b is the average length of the Burgers vector, a is the measurement step size, n the radius of the kernel in pixels, and c a geometric constant of the order of 3. ω_{KAM} is then given in radians. With this equation the KAM maps can directly be transformed into dislocation density maps, as shown in Figure 9.10b. Kamaya (2011) and Schemmann (2014) proposed a further improvement to the use of the KAM by employing increasingly larger kernels. From these, the orientation measurement noise and a noise-corrected KAM value can be extracted. This approach has been used very successfully for GND density quantification, for example in Godec et al. (2020).

Note that the KAM quantification only returns the total GND density but does not allow conclusions on the type of dislocations. In fact, as long as only "conventional" orientation data are measured (i.e. measured by a Hough transform analysis of EBSD patterns), the orientation data are too inaccurate to determine misorientation rotation axes (cf. Equation 8.4) which would be required for a more comprehensive dislocation determination. For this case, CC EBSD will deliver a sufficient accuracy of orientation data, such that the measured local lattice rotation tensor can be decomposed into individual dislocation tensors. The required mathematics for this decomposition will be briefly outlined in the next section. Wheeler et al. (2009) showed, however, that also with lower-resolution EBSD data reasonable dislocation information can be obtained, if the measured misorientations are integrated along larger paths through individual grains in orientation maps.

The above described techniques measure the amount of dislocations that remains in a microstructure after deformation. Note that it is also possible to measure the local distribution of strain that a material has accommodated during a deformation experiment. The required technique is called digital image correlation (DIC). For this technique it is required to place markers on the surface of a sample, record an image of the marker distribution, deform the sample, and record a second image. The local strain tensor at every pixel in the image is then determined by a comparison of the marker images before and after deformation. The high-end approach for this technique is the use of remodeled gold particles for nano-DIC which, at the same time, allows accurate EBSD map measurements and even ECCI observations. A highly informative text about this technique has been written by Di Gioacchino and Quinta da Fonseca (2013).

Finally it should be mentioned that all abovementioned strain measures, including pattern quality, can be used to partition ORM data into, for example, high-strain and low-strain areas in order to separate, for example, recrystallized on non-recrystallized materials fractions. This has been done for the example displayed in Figure 9.2 in order to determine the textures of recrystallized and non-recrystallized areas. Similarly, the recrystallized and unrecrystallized materials fractions were determined separately for various by materials by Džubinský et al. (2004) and Wagner et al. (2002). The pattern quality parameter is also used to separate out phases that have the same or similar crystal structures for subsequent microtexture determination. For example, bcc ferrite and bainite have been distinguished by measurement of image quality and confidence index in undeformed steel specimens (Petrov et al., 2007).

9.4.5 Quantification of Plastic Strain Fields by Cross-Correlation EBSD

Employment of the CC analysis of EBSD patterns as discussed in Section 7.4.5 makes it possible to achieve a much higher accuracy of local misorientation measurement than with the Hough transform-based analysis. The CC EBSD accuracy is sufficient to evaluate GNDs as proposed by Nye (1953), using his homogeneous strain field approach. Nye defined a dislocation (nowadays called "Nye") tensor, which describes the rotations caused by all dislocations in a local volume, as the curl of the lattice rotation field

$$\alpha_{ij} = \mathrm{curl}\left(g\right) = -e_{jkl} g_{il,k} \tag{9.2}$$

Here, α_{ij} is the dislocation (or Nye) tensor, **g** is the 3-dimensional orientation field (i.e. the orientation of a point and of all its neighbors in three dimensions), and e_{jkl} is the Levi-Civita permutation symbol. This symbol reads for example as follows:

$$\alpha_{12} = \frac{\partial g_{13}}{\partial x_2} - \frac{\partial g_{12}}{\partial x_3} \tag{9.3}$$

As shown by Nye (1953) and Kröner (1959), the Nye tensor is directly related to the total sum of the Burgers vectors of all GNDs in a given volume:

$$\alpha_{ij} = \sum_{k=1}^{K} \rho^k b_i^k \otimes t_i^j \tag{9.4}$$

This summation is carried out over all K different dislocation systems with their Burgers vector lengths b^k, their line elements t^k, and their densities ρ^k (the symbol $\otimes$ indicates the dyadic product). If the orientation field is known—for example, from EBSD measurements—Equation 9.2 can be used to calculate the dislocation tensor $\boldsymbol{\alpha}$. $\boldsymbol{\alpha}$ has nine components allowing determination of the densities of nine independent slip systems.

However, in most materials showing crystal-plastic behavior, more than nine slip systems are available. In the fcc lattice, for example, with $\langle 110 \rangle$ Burgers vectors and $\{111\}$ slip planes a total of 12 edge dislocations and six screw dislocations are possible. As a result the equation system is heavily under-determined and one may find $\binom{9}{18} = 48{,}620$ possible solutions to Equation 9.4. One way to select a unique solution out of these is to select the one which corresponds to the smallest total line energy. For a more detailed mathematical treatment and further references see, e.g., Maurice et al. (2012), Britton and Wilkinson (2012), and Konijnenberg et al. (2015).

An example of mapping of dislocations and residual stresses is displayed in Figure 9.11. The figure is based on investigation of An and Zaefferer (2019) on the fatigue loading behavior of a high Mn-steel. The material was shear-fatigued for

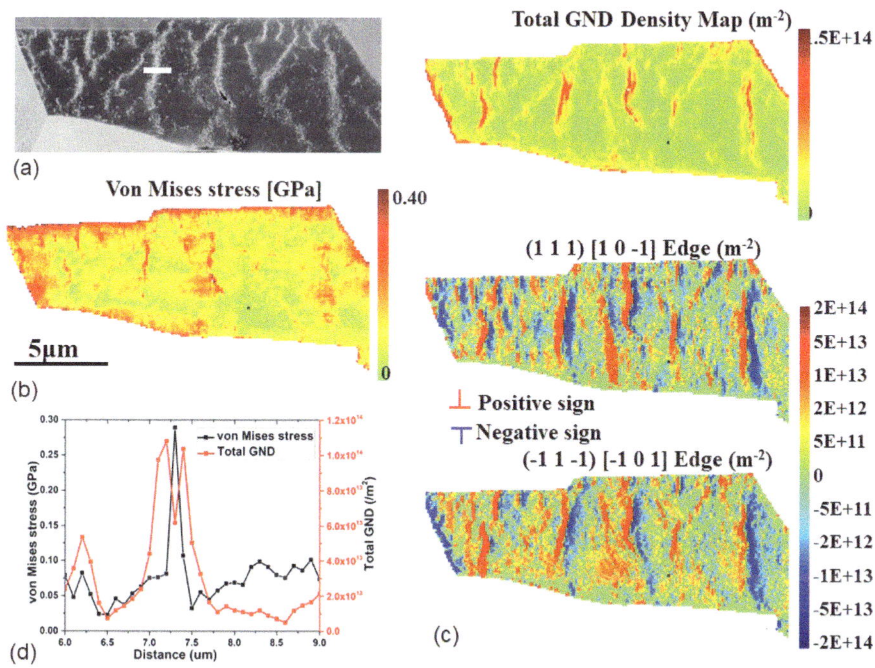

FIGURE 9.11 Residual stress maps and geometrically necessary dislocation (GND) density maps of a shear-fatigued high-Mn steel measured by cross-correlation EBSD. (a) Electron channeling contrast image of the deformed structure. (b) Von Mises stress. (c) Total GND density map and two maps of discrete slip system densities. (d) Graph of the stress and GND density along the white line in figure (a). (Adapted from An and Zaefferer, 2019.)

100 cycles with strain amplitudes of 0.5%. This led to the formation of fatigue dislocation walls, as visible in the ECC image in Figure 9.11a. Next, high-resolution, high-quality EBSD patterns (unbinned 900 × 900 pixels) were collected across the grain and from this the residual stresses and the GND densities were calculated. From these data the von Mises equivalent stress was calculated which is displayed in Figure 9.11b. The total GND density and GND density maps for two selected slip systems are plotted in Figure 9.11c. From these graphs it appears that the lattice rotations occur opposite on opposite sides of the dislocation walls (opposite dislocation signs). The detailed display in Figure 9.11d shows that the residual stresses are highest in the center of the dislocation wall while the GND densities are highest on the sides. Note that the amount of SSDs, which can be determined directly from the ECC images, is highest in the middle of the wall and their density is about ten times higher than that of the GNDs. As an example from the field of geomaterials we mention the work of Griffith et al. (2014) on the understanding of the role of dislocations for diffusion in garnet crystals. In this case the observed local, very small lattice rotations were used to estimate the density of dislocations which contributed to pipe diffusion in the material.

The examples discussed above illustrate the strength of CC EBSD for the measurement of small lattice rotations and distortions. However, the technique always requires a reference pattern from an area inside of the same crystal, and all strain and rotation measurements refer to this reference. Whether or not this reference itself is elastically distorted cannot be decided by the measurements. Note also that the technique is rather slow and very data-intensive, as for every measurement point a full high-resolution diffraction pattern needs to be recorded and stored.

9.4.6 Cross-Correlation EBSD for the Measurement of Elastic Strain Fields

In this section we present an example for the accurate measurements of elastic strains by CC EBSD. It was mentioned in Section 7.4.5 that one important precondition for reliable application of the CC technique is that the features compared in the patterns are only shifted with respect to each other, but not distorted, which limits the technique to materials with small plastic rotations. Minerals, due to their limited plasticity, usually develop only small plastic rotations and seem, therefore, well-suited candidates for application of the CC technique. However, they may pose other difficulties, as low pattern intensity, sample charging, beam damage etc., which need to be solved for every case. Schäffer et al. (2014) employed CC EBSD to determine lattice parameter changes and residual stresses during a phase transformation caused by interdiffusion of K and Na in alkali feldspar. For their investigations a Na-rich sanidine single crystal of gem quality was immersed in a melt of KCl at 850°C for 32 days and then cut and polished along the (001) and (010) planes as indicated in Figure 9.12a. During immersion K-Na diffusion profiles developed which showed significantly different compositional characteristics for directions perpendicular to the (001) and (010) planes. Figure 9.12b and c reproduces these results: while the diffusion profile parallel to the a-axis [100] is wide and smooth, the profile along the b-axis, [010] is very short and sharp. At the same time the crystal shows the development of characteristic cracks which indicate the presence of residual stresses.

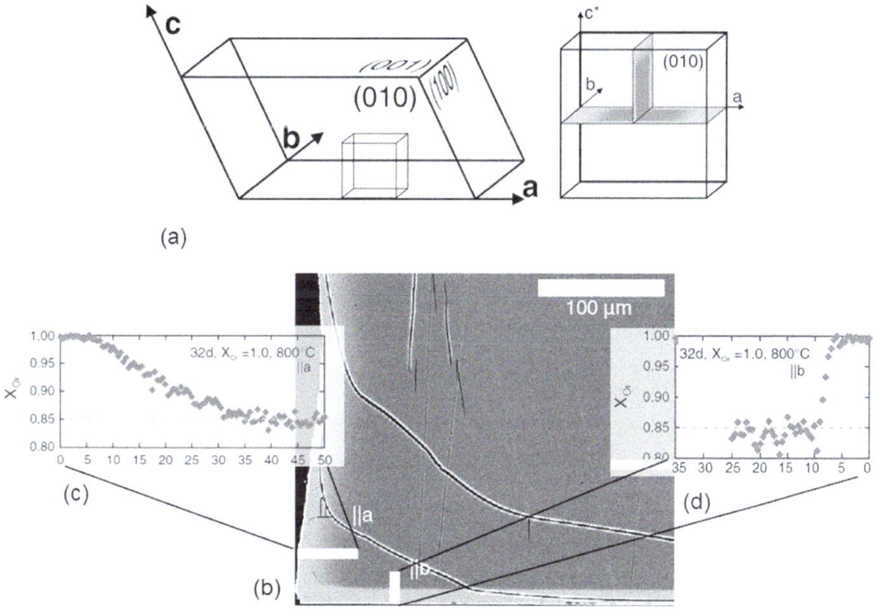

FIGURE 9.12 (a) Geometry of the prepared samples; *left* orientation of the sample within the unit cell of feldspar, *right* after the experiment the sample was cut into two mutually perpendicular directions (*gray planes*) to allow documentation of interdiffusion fronts in three different crystallographic directions. (b) BSE image of the (010) surfaces of the feldspar sample; bright areas along the grain surfaces indicate the K-rich area, darker areas correspond to unexchanged Na-rich feldspar. (c) and (d) FEG-EMP (electron microprobe) measurements of the cation ratio (K^+/Na^+) X_{or} along the profiles indicated in the BSE image. (Adapted from Schäffer et al., 2014.)

In order to measure these residual stresses CC EBSD was employed to analyze line scans along lines parallel to the **a** and **b** directions. The crystal distortion was analyzed using the software CrossCourt 3 by BLG Productions Ltd. (Bristol, UK). A pattern from the original, undistorted sanidine crystal was taken for reference. The results obtained on the sharp compositional boundary along the [010] direction are displayed in Figure 9.13 as strain tensor matrix. Interpretation of these results can be found in the original work by Schäffer et al. (2014) and in a short version in Zaefferer and Habler (2017).

9.4.7 Measurement of the Macroscopic Crystallographic Texture

In the present section we finally address the application of EBSD measurements for *macroscopic* texture investigations. An overview on the subject can be found at Wright et al. (2007). The technique is based on the fact that EBSD mapping can also be used to investigate rather large sample areas, in the order of several square centimeters. For these measurements, a rather large measurement step size is selected and a combination of sample stage movement and beam scanning is used. In this way, EBSD becomes a serious competitor for conventional x-ray diffraction (XRD)-based

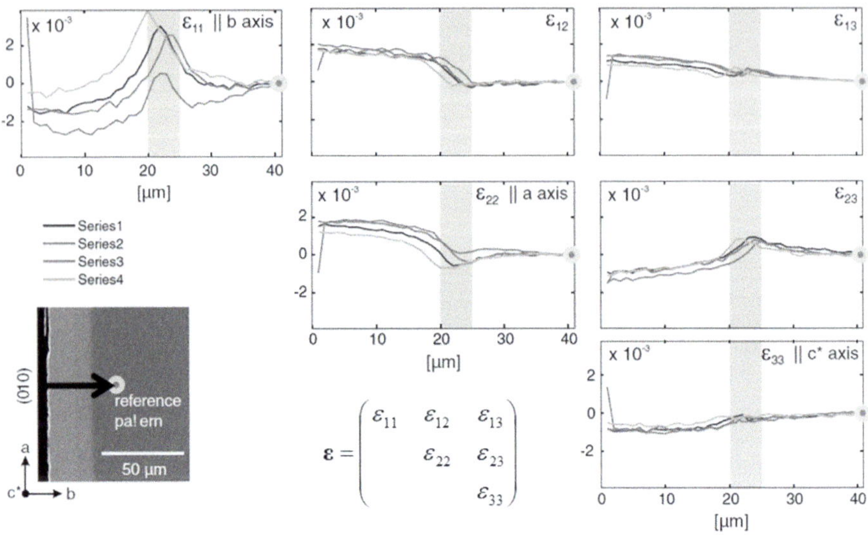

FIGURE 9.13 Line profiles of the different components of the strain tensor measured across a sharp diffusion front; the differently shaded profiles represent different measurements executed to test reproducibility, the reference pattern is located at the *right-hand side* as represented by the *gray dot* and the edge of the crystal is on the *left-hand side*. (Adapted from Schäffer et al., 2014.)

techniques for macroscopic texture measurements as EBSD-based measurements have a number of advantages over XRD-based ones:

i. The EBSD technique directly measures *orientations* and it is therefore in principle simple and straightforward to calculate the orientation distribution function (ODF) from these measurements. In contrast, the Bragg-type XRD-techniques only measure pole figures, i.e. crystal *directions*, and require some form of pole figure inversion to determine the ODF. Furthermore, as pole figures are, in most cases, measured only incompletely (usually, pole figures are measured to a maximum tilt angle of the primary beam of 60–80°) some form of pole figure extrapolation is required. Also, at high tilt angles necessary measurement corrections due to defocusing and absorption become important. In any case, when calculating the ODF using the harmonic series expansion according to Bunge (1993) these methods only determine the even coefficients, while the odd coefficients are determined from the assumption that the ODF must not be negative anywhere (positivity method). In contrast, odd and even coefficients are directly accessible through EBSD measurements. Thus, while the ODF is quasi directly *measured* by EBSD, complicated calculation and correction schemes are required by XRD.
ii. Complicated crystal structures are easily accessible for texture measurements with EBSD (provided that EBSD patterns can be obtained from the

sample). In contrast, in XRD frequently closely spaced diffraction peaks make individual pole figure measurements difficult.

iii. The same reason as (ii) applies to multi-phase materials. They are easily accessible by EBSD measurements because every phase is measured separately. Also, volume fractions of different phases can be easily determined by just counting the number of measurements for each phase. In XRD, again, one frequently struggles with overlapping peaks of different phases. Volume fraction determination requires the intensity integration over the full, recalculated pole figures.

iv. A further advantage is that EBSD-based texture measurements also yield the spatial distribution of textures, even if the measurement steps are so coarse that individual grains cannot be resolved; an example is shown in Figures 9.14 and 9.15. Figure 9.14 displays the texture distribution on the cross section of a hot-rolled Fe-Si transformer sheet. The measured area has a size of 2.2 × 8.8 mm², measured with a step size of 40 μm, resulting in about 12,000 data points. The measurement took about 10 minutes. Due to shear deformation the surface regions display a significantly different texture than the center area: while the surface shows a shear texture formed by inclination of a typical γ-fiber texture by ±26° in rolling direction. This

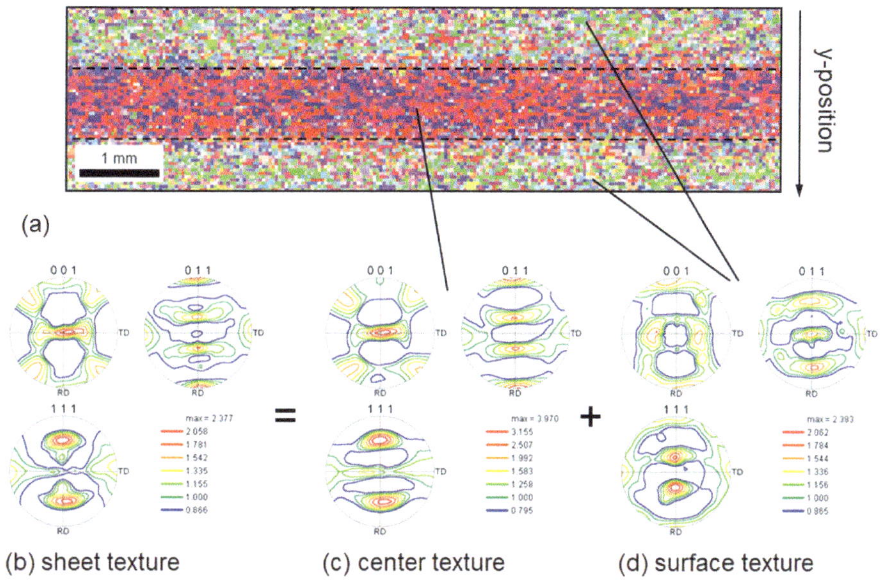

FIGURE 9.14 Large area EBSD map for macrotexture determination on the ND-TD plane of a hot-rolled Fe-3% Si sheet. Measurement parameters: area 2.2 × 8.8 mm², step size 40 μm, 12.000 data points, measurement time ~10 minutes, measurement by combined beam and stage scan. 15-kV tungsten filament SEM. (a) Orientation map in IPF-ND coloring code. Grains cannot be distinguished because the step size is approximately equal to the grain size. (b)–(d) Calculated pole figures of the marked areas: (b) total surface, (c) center area, and (d) both surface areas.

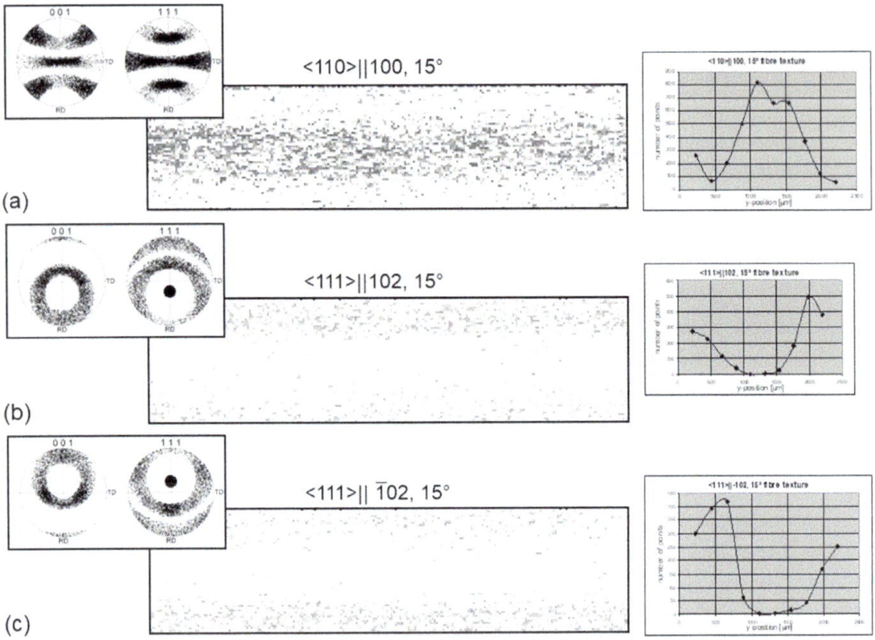

FIGURE 9.15 Distribution of selected texture components across the sheet thickness of the example shown in Figure 9.14. Fiber notation: (crystal direction ⟨uvw⟩) || (sample direction xyz), max. deviation angle. (a) symmetric plane strain texture component, (b) and (c) asymmetric shear texture components. Note that the shear texture components are different on the top and bottom of the sheet, indicating that the sheet was rolled unidirectionally. The maximum of the shear texture is on each side at about 1/10 to 1/8 of the sheet thickness. The maximum of the plane strain component (α-fiber) is in the sheet center.

fiber includes as a major component the Goss orientation (110)[001]. The center area shows a typical α-fiber texture. The evolution of the volume fraction of different texture components across the sheet cross section is easily calculated and displayed in Figures 9.14 and 9.15.

v. Finally, texture measurement by EBSD may also be significantly more rapid than that by XRD. In fact, as Wright et al. (2007) showed, the measurement of about 10.000 grains is largely sufficient to obtain a good ODF (in an earlier work Baudin et al. (1994 showed that, depending on texture strength and material, even less than 1,000 grains may be sufficient to estimate an ODF). The measurement of this number of grains takes about 1 minute if only one point per grain is measured (i.e. with a step size larger or equal to the grain size). If the time necessary for stage movement is added the measurement time goes up to 5–10 minutes for conventional measurements (see, however, the example further down). This time is not changing much even if several measurements per grain are carried out. On the other hand, XRD measurements, even if they are carried out with an area detector, usually take at least one hour.

Although the list above shows the strength of the EBSD technique, not everything can be measured in this way and for reliable EBSD-based texture measurements a number of points must be kept in mind: it is first of all important to make sure that the EBSD patterns are sufficiently good to be analyzed with good confidence. This is particularly important for those cases where only one measurement per grain is carried out. In these cases, neighboring measurement pixels do not allow assessing the correctness of data points. Also, it must be made sure that the quality of data is equally good for all texture or phase components; otherwise the texture results might be artificially biased toward the phase or texture component with the better patterns (for example when measuring textures on partly recrystallized materials). Of course, EBSD measurements cannot be used for materials which do not give diffraction patterns, like truly nanocrystalline materials or materials with very high defect densities.

Particularly important is, finally, the statistical aspect of EBSD-based texture measurements (see Section 10.3.3). While Baudin et al. (1994) mentioned that, depending on the texture strength at least 10^2 (for strong textures) to 10^3 grains have to be measured, measurements on a TRIP (transformation induced plasticity) steel with a relatively weak texture by Davut and Zaefferer (2010, 2012) suggested that rather about 10^4 grains are required. In this case, the number of grains sufficient to describe the texture was assessed by means of the perfection of sample symmetry reflected in pole figures. A number of 10^4 grains required for satisfactory texture description were also confirmed by Wright et al. (2007). Whether a sufficient amount of data has been measured depends on the number of grains that are included in the data set, the sharpness of the texture, and the spatial distribution of texture fields. A good proof of whether or not a sufficient amount of data has been measured is the visibility of sample symmetry in the texture data: many engineering materials show a symmetry in the distribution of their crystal orientations, which is created by the symmetry of the production process. A rolling process, for example, usually introduces an orthotropic symmetry which is frequently conserved also through a subsequent phase transformation or recrystallization process (see Section 5.5). It is a good proof of statistically significant data if this sample symmetry is well developed in the measured data.

For the example given above (Figures 9.14 and 9.15) we performed a symmetry check by an artificial symmetrization of the measured pole figure data. The symmetrized data were subsequently compared to the original, non-symmetrized data by calculation of the absolute difference between pole figure data points. Table 9.1 gives the differences for three calculated pole figures. It should be mentioned that this method fails when the sample symmetry of the specimen and its production process is unknown (for example in case of geological materials). In this case one might assess the reliability of a large master data set by subdividing it into smaller sets and comparing those to the master set. The scatter in the results might then be interpreted in terms of homogeneity of texture in the observed area.

A further example for successful EBSD-based macroscopic texture measurements is the measurement of the sharpness of the Goss texture of extremely coarse-grained electrical transformer sheets made from an Fe-3%Si alloy (Frommert et al., 2008). The sharpness of the Goss texture is an extremely crucial point for the magnetic performance of the material. Measuring the texture, however, displays a big

TABLE 9.1
Average Deviation of Triclinic and Orthotropic Pole Figures Points, Measured in Percent of the Orthotropic Intensity, for Different Sections of a Hot-Rolled Fe-3% Si Steel. The Orthotropic Data Are Calculated by Symmetrization of the Original Triclinic Raw Data

	(001) Pole Figure	(011) Pole Figure	(111) Pole Figure
Total sheet	4.9	3.7	5.1
Surface area	6.7	5.6	7.0
Center area	7.9	6.6	7.6

problem because of the enormous size of the grains of up to 2 cm in diameter in a sheet of a thickness of 0.2 mm. To get a sufficiently good statistics sheets of about 30 × 20 cm² surface area were cut into small stripes of 4 × 1 cm². As shown in Figure 9.16a the stripes were stacked together and the texture of the stacks was measured on the cross section using EBSD (see Section 4.6). In this way it was possible

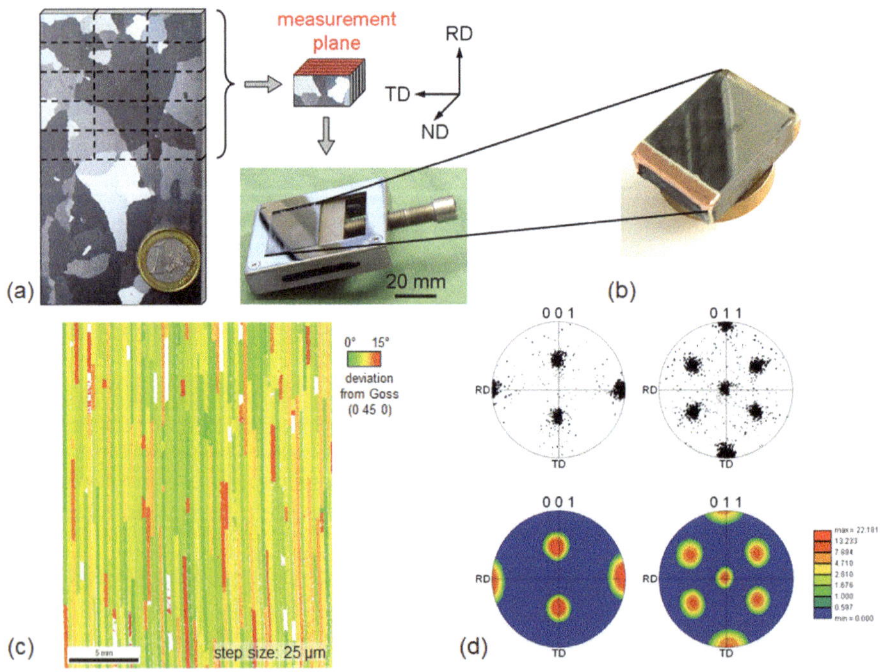

FIGURE 9.16 EBSD-based measurement of the sharpness of the Goss texture on a final annealed transformer sheet of Fe-3% Si (Frommert et al., 2008). (a) The microstructure of the very coarse-grained material (grain diameter ~ 10–30 mm) and the sample preparation approach. (b) The final sample ready for the SEM. (c) Very large orientation map measured at 15 kV in a tungsten filament SEM with a step size of 25 µm. (d) Discrete and density pole figures calculated from the maps. The clearly developed orthotropic sample symmetry of the orientation measurements indicates that sufficient data have been measured.

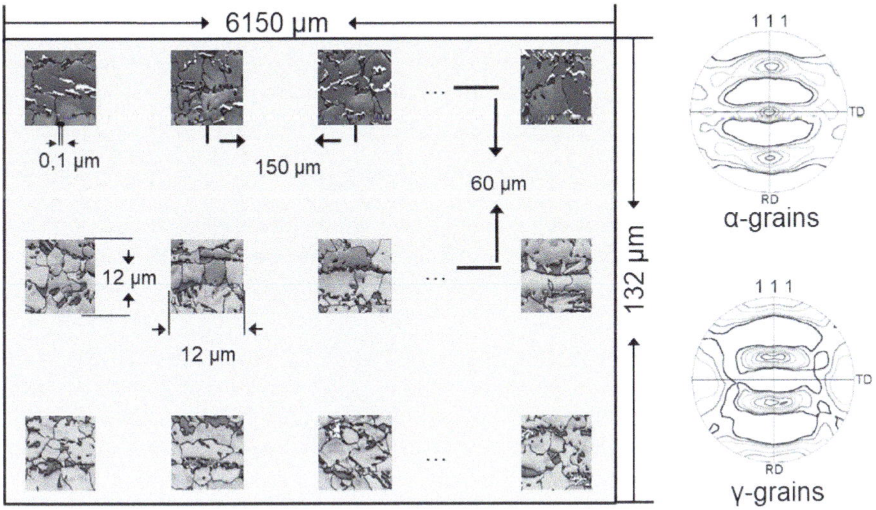

FIGURE 9.17 A new EBSD-based texture measurement scheme that allows acquisition of representative data by combination of measurement of small maps distributed over a large sample surface (adapted from Davut and Zaefferer, 2012). Example: cold-rolled and heat-treated low-alloyed TRIP steel consisting of a microstructure of ferrite (α-phase) and about 10 vol% of residual austenite (γ-phase). The well-developed orthotropic symmetry of the pole figures indicates that a sufficient amount of crystal orientations has been measured.

to measure about 1,000 grains on a conventionally grain oriented (CGO) steel sheet. Figure 9.16b shows one result from these measurements. The statistical significance of the data is once again proven by the good conservation of the rolling symmetry.

Davut and Zaefferer (2010) developed a new orientation mapping scheme which combines small-step size maps for local information taken at some mutual distance to cover large sample areas for statistically representative measurements, as displayed in Figure 9.17. This scheme requires approximately the same measurement time as XRD-based pole figure measurements to deliver texture information with the similar reliability, however, with all abovementioned advantages of EBSD-based analyzes. It should be concluded that for all EBSD-based texture measurements it is very important to select the probed sample volume carefully, considering the material grain size, texture strength, and texture homogeneity.

9.4.8 Analysis of Data Using MTex

We believe it has become clear until here that EBSD-based ORM (and ORM in general) is an extremely powerful and versatile technique for texture and microstructure characterization and research. All data shown so far have been analyzed with commercial software for plotting maps, orientation or pole distributions, or any kind of statistical data like grain sizes, shapes, or phase distributions. Although all these software tools are very comprehensive and flexible, they are finally limited to functions that are provided by the companies. In recent years an open source toolbox called MTex has been developed, which allows the user to freely program almost all kinds of analyzes of crystal orientations, misorientations, kernels, grains, or other units of discretely measured orientations. It also allows texture analysis based on x-ray pole figure measurements (see Section 5.3.4). The MTex toolbox library is based on the software platform MatLab and has been started by Bachmann, Hielscher, and Schaeben (2010). Since then it has been strongly promoted by a group of people around Ralf Hielscher, see e.g. Mainprice et al. (2015). Its webpage is currently found at this link: https://mtex-toolbox.github.io/index. Since it is an open-source project a large user community has developed and these people intensively contribute to the quantitative and qualitative growth of the software. Since we consider this software to be of very large interest for the texture and microstructure community we give here a very short "first steps" introduction into the programming with this software. The software functions as an interpreter of code. Very detailed tutorials and help files are found on the above-cited website. The example selected here stems from an investigation of a multi-phase coating on a Nickel-base superalloy (Wöllmer et al., 2003). For example only the bcc phase NiAl and the fcc phase Ni_3Al are considered. (The example has been provided by Shao-Pu Tsai; see Tsai et al., 2022.)

The program starts by cleaning up memory and screen:

```
close all
clear all
clc
```

Next, crystal and specimen symmetries are defined. Here three phases are specified: a class of non-indexed pixels, a bcc phase NiAl, and an fcc phase Ni_3Al. When defining the crystal symmetries not only the symmetry operators are invoked but also all basic matrix routines that have been described in Section 7.1.1 are properly taken into account automatically.

```
% crystal symmetry
CS = {...
'notIndexed',...
crystalSymmetry('m-3m', [0.36 0.36 0.36], 'mineral', 'Nickel
Aluminium', 'color', [1 0 0]),...
crystalSymmetry('m-3m', [0.286 0.286 0.286], 'mineral', 'Nickel
3Aluminium', 'color', [0 1 0])};
% plotting convention
setMTEXpref('xAxisDirection','north');
setMTEXpref('zAxisDirection','outOfPlane');
```

Now the file names containing the data are specified. MTex is able to read in a large amount of data formats. Here the ANG format of the company EDAX/Ametek is read in.

```
% path to files
pname = 'C:\Users\work\Desktop\SummerSchoolTrondheim_2022';
% which files to be imported
fname = [pname '\superalloy_group 1.ang'];
```

Here the data are read from the file and stored in a variable called "ebsd" which is actually the complete EBSD map.

```
% create an EBSD variable containing the data
ebsd = EBSD.load(fname,CS,'interface','ang',...'convertSpatial
2EulerReferenceFrame','setting 2');
```

Now, a first map is plotted which uses a couple of standard settings. By default a phase map is plotted as shown in Figure 9.18a.

```
% plot phases
figure; % to generate a new figure
plot(ebsd);
```

Next, two partitions of the original data set are created, containing the two crystal structures.

```
ebsd_NiAl = ebsd(ebsd.phase == 1); ebsd_Ni3Al = ebsd(ebsd.
phase == 2);
```

It is now possible to calculate data or plot maps for any of these partitions. For example, an IPF map for the Z sample direction for the NiAl phase is plotted in Figure 9.18b. White areas indicate the fcc phase, which is not plotted for this partition.

```
figure % create a new figure
ipfkey = ipfTSLKey(ebsd_NiAl);
ipfkey.inversePoleFigureDirection = zvector;
plot(ebsd_NiAl,ipfkey.orientation2color(ebsd_NiAl.
orientations));
mtexTitle('IPF wrt z direction');

% create a colour legend
figure
plot(ipfkey,'figSize','large');
mtexTitle('NiAl');
```

Now the code becomes a little more complex as grains are reconstructed by grouping together pixels of similar orientation. Here the threshold misorientation for pixels belonging to one grain is set to 10 degree. An IPF map with grain boundaries is plotted in Figure 9.18c.

```
[grains, ebsd_NiAl.grainId] = calcGrains(ebsd_NiAl,'angle',
10*degree);
ebsd_NiAl(grains(grains.grainSize < 5)) = [];
[grains, ebsd_NiAl.grainId, ebsd_NiAl.mis2mean]=
calcGrains(ebsd_NiAl,'threshold',[2*degree, 10*degree],
'boundary','tight');
```

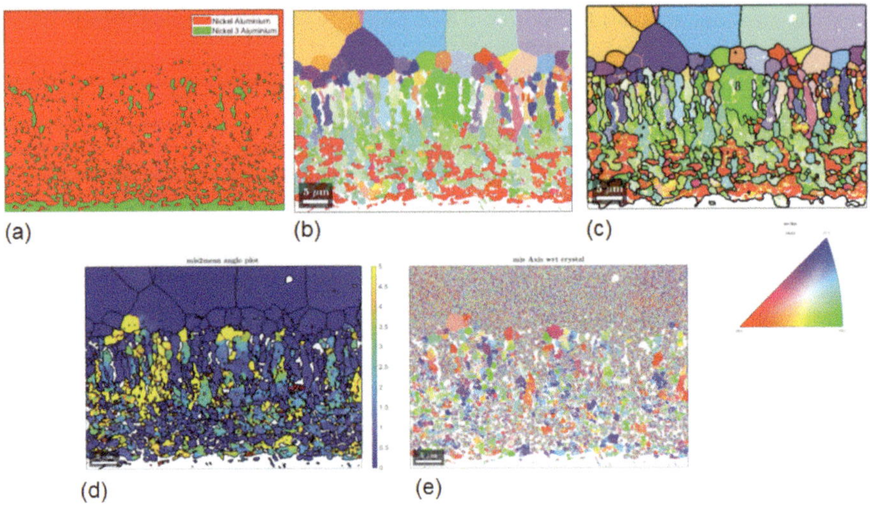

FIGURE 9.18 Examples of maps produced with MTex code for a Ni-base superalloy coating with a bcc NiAl phase and an fcc Ni_3Al phase. (a) Phase map, (b) inverse pole figure map of the z-axis, (c) IPF map and small- and large-angle grain boundaries, (d) grain reference misorientation angle plot, and (e) grain reference misorientation axis plot.

```
figure();
plot(ebsd_NiAl, ipfkey.orientation2color(ebsd_NiAl.
orientations));
hold on;
plot(grains.boundary,'linewidth',2,'linecolor','k');
plot(grains.innerBoundary,'linewidth',0.8,'linecolor',
'yellow');
hold off;
mtexTitle('noFilter');
```

Finally, we plot the misorientation angle in Figure 9.18d and, in Figure 9.18e, the misorientation axis of every pixel with respect to the average orientation of the grain it resides in.

```
figure % calculate misorientation angle to the mean
orientation of a grain
plot(ebsd_NiAl,ebsd_NiAl.mis2mean.angle / degree);
hold on
plot(grains.boundary,'linecolor',[0 0 0],'linewidth',2)
plot(grains.innerBoundary,'linecolor',[1 0 0],'linewidth',0.5)
caxis([0 5])
mtexColorbar
hold off
mtexTitle('mis2mean angle plot')
```

```
figure % calculate misorientation axis to the mean orientation
of a grain
aXisKey = TSLDirectionKey(ebsd_NiAl.mis2mean.axis);
plot(ebsd_NiAl,aXisKey.direction2color(ebsd_NiAl.mis2mean.
axis))
mtexTitle('mis Axis wrt crystal')
figure
plot(aXisKey)
```

All functions that have been used here (e.g. the above example `mis2mean`) can also be programmed by the user and then stored privately or shared with the MTex community.

9.5 ORIENTATION MICROSCOPY IN THE TEM

In contrast to the SEM-based ORM, where only one technique is of universal use, TEM-based ORM can be performed on the basis of a number of different techniques, including the use of Kikuchi patterns, spot patterns of different kind, and reconstructed spot patterns. The two first techniques are classical scanning techniques, where, similar to EBSD-based ORM, a finely focused electron probe is scanned across the sample and orientation and phase are extracted from the resulting diffraction patterns. The microscope is used as a scanning TEM (STEM). The third technique, conical dark-field scanning, works on completely different basis that has been described in some detail in Section 6.2.5. In the last ten years the STEM technique using SCBED patterns (Section 6.2.3), also including precession (Section 6.2.4), has turned out to be the most useful technique, although the Kikuchi diffraction technique has some advantages as well. We will, thus, focus the discussion only on the STEM techniques.

9.5.1 Semi-Automatic TEM ORM

The STEM-based ORM techniques are now fully automated. Nevertheless, we will introduce the section with a kind of an "oldtimer" measurement which was obtained by a "semi-automatic" technique. The diffraction patterns were recorded at manually selected positions, while the orientation determination was performed fully automatically based on either spot patterns or, in some cases, Kikuchi patterns depending on which pattern type was visible. A postprocessing software subsequently assigned the determined orientations to all pixels in the respective subgrain or cell. The example is taken from an investigation on heavily cold-rolled Fe-36% Ni in order to understand the mechanisms of nucleation of cube-oriented grains during recrystallization of this alloy with fcc crystal structure and high stacking fault energy (Zaefferer et al., 2001). In most cases only spot patterns were visible due to the high amount of randomly stored dislocations inside of the subgrains. In areas with cube orientation, (001)[100], however, subgrains with a higher degree of recovery were found and Kikuchi patterns were visible. The areas appear in red color in the orientation map in Figure 9.19. The orientation map shows that the dislocation cell structure appears in cube-oriented grains, whereas crystals with orientations typical for the deformed state show lamellae and microbands.

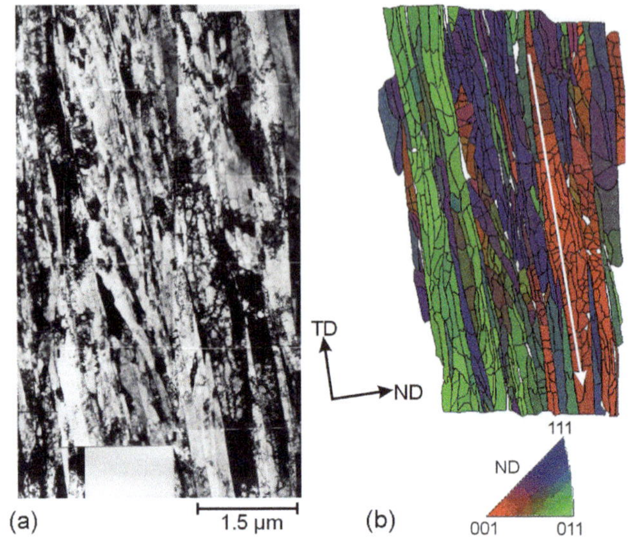

FIGURE 9.19 TEM-based orientation mapping on a 94% cold-rolled Fe36Ni sample, obtained by semiautomatic SCBED-ORM. (a) Bright field image. (b) IPF orientation maps (for ND directions).

9.5.2 STEM Techniques

For the STEM techniques, a focused electron probe is scanned over the sample, the resulting patterns are automatically analyzed and the microstructure is then reconstructed from the measured phase and orientation information, just like in the case of EBSD. The most straightforward technique uses Kikuchi diffraction patterns. As discussed in Section 7.2.3 there have been several attempts to transfer the EBSD technique to the TEM but most of them did not yield a finally running system. As far as the author knows only the efforts in the group of Fundenberger and Morawiec lead to a successful system (Fundenberger et al., 2003). The automatic evaluation is based on application of a dedicated Hough-transform as described in Section 7.2.3. The most important advantages of the transmission Kikuchi patterns (TKP) technique are the high orientation resolution and the rareness of 180°-ambiguities. Reason for the very high angular resolution is the sharpness of the Kikuchi lines. We demonstrate the technique by means of an example provided by E. Bouzy and coworkers with their ORM system. Figure 9.20 shows the TEM bright field image and the respective orientation map of a cellular microstructure of an aluminum sample highly deformed by equal channel angular extrusion (ECAE). The 100 × 100 pixel large map has been measured with a step size of 20 nm in nanoprobe mode with a spot size of 5 nm. Although only boundaries with a minimum misorientation of 0.2° are shown, the angular resolution of the measurement was in fact significantly higher. Selecting a still smaller minimum boundary misorientation would have resulted in the display of the lattice curvature as a consequence of thin-foil bending.

As mentioned in Section 6.2.6, the high angular sensitivity of the Kikuchi technique is at the same time also the reason for the most serious disadvantages of

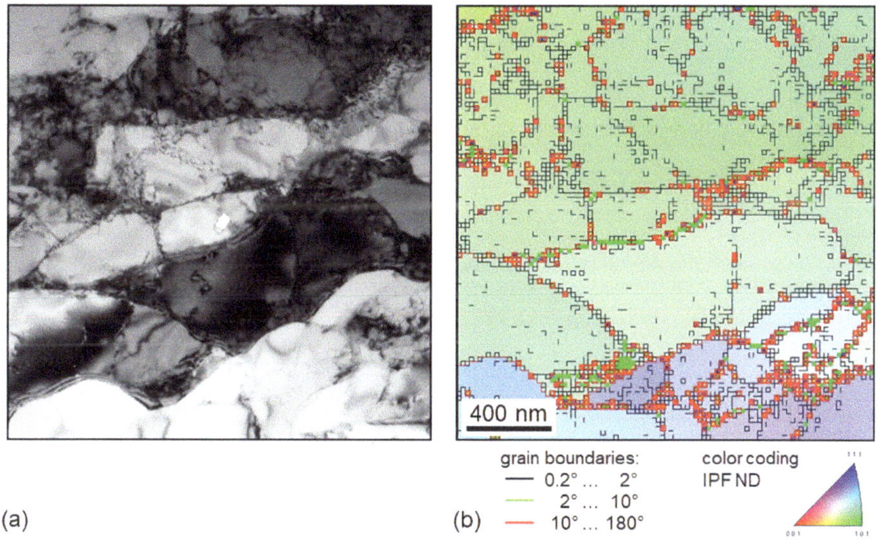

FIGURE 9.20 Results of TEM Kikuchi-diffraction-based orientation microscopy on a highly deformed Al-sample. Mapping parameters: 100 × 100 pixels, step size 20 nm, spot size 5 nm. (a) Bright field image of the microstructure. (b) Orientation map (colored in IPF coloring code) with different boundaries. (Courtesy of E. Bouzy and J.J. Fundenberger.)

the technique, its sensitivity to lattice defects—the patterns quickly blur out with increasing defect density. A second disadvantage of the technique is the difficulty to treat overlaying patterns. Overlaying patterns frequently arise in cases where the grain size is smaller than the specimen thickness (for usual TEM samples in the order of 100 nm) or when grain boundaries are steeply inclined, creating a significantly wide area of pattern overlap. Since the reciprocal vectors belonging to TKP patterns are—in contrast to the EBSD case—narrow-spaced and cover a relatively small solid angle of reciprocal space, and since there are—up to now—no parallel indexing algorithms (e.g. template matching) available for TEM Kikuchi patterns, indexing of these patterns leads to low confidence values.

The second STEM technique is based on the evaluation of spot diffraction patterns. In fact, only SCBED patterns can be used as only those are created by a focused beam. Selected area diffraction patterns (SADP) are created by the position of the field aperture—a scanning technique would therefore have to rely on the mechanical movement of the aperture or the sample. Although the accuracy of orientation determination from TKP is superior to that from spot patterns, the use of spot patterns has significant advantages. First, the sensitivity of spot patterns to lattice defects is significantly reduced compared to TKP. In heavily defected crystals, e.g. heavily deformed metals, spot patterns are still visible while Kikuchi patterns are not (see Sections 6.2.4 and 6.2.6). A second advantage is that spot patterns are visible also on very thin samples when Kikuchi lines are invisible due to insufficient incoherent scattering. The case of thin samples is, however, particularly important for high-resolution ORM because only in this case overlapping crystals can be avoided. Finally, the detection of spots is much simpler and more robust than the line

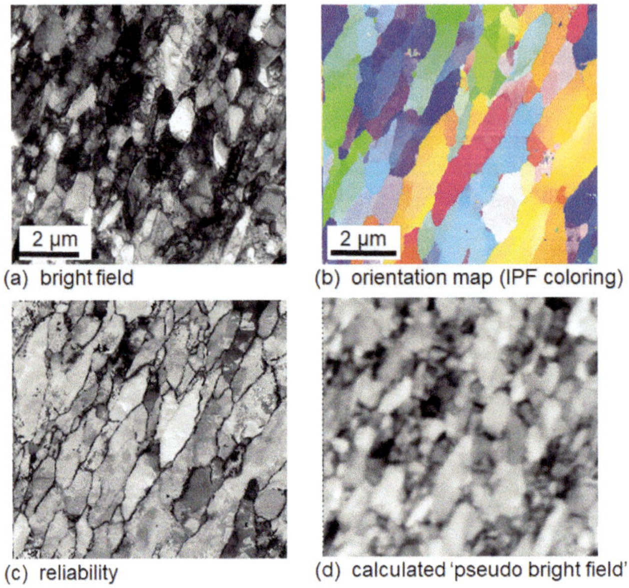

FIGURE 9.21 Results of TEM SCBED-pattern based orientation microscopy using the template matching method developed by Rauch and Dupuy (2005). IF steel severely deformed (ECAE 8 passes route C at 200°C). Mapping parameters: 300 × 255 pixels, step size 28 nm; (a) bright field image, (b) IPF orientation map, (c) reliability map, and (d) calculated bright field map. (Courtesy of E. Rauch.)

detection in TKP and automation can be achieved more easily. Especially the template matching technique—which is a truly parallel indexing technique—is able to allow reliable indexing even in the case of overlapping patterns (Valery et al., 2017). As an illustration of the capabilities of the technique we reproduce in Figure 9.21 an example obtained by Rauch who developed the template matching technique. The example shows an orientation map of an IF steel heavily deformed by eight passes of ECAE. The measurements were carried out with a step size of 28 nm. The microstructure is nicely revealed and the confidence index map shows that the measurements are, in fact, reliable. The smallest visible grains have a diameter of about 100 nm. Nevertheless, a comparison between the real bright field image of the area and a bright field image reconstructed from the scan shows that the TEM image resolution is—of course—significantly better than that of the scan. Other examples on the application of the technique are presented by Shigesato and Rauch (2007) for the study of a heavily deformed ferritic steel and by Cautaerts et al. (2021) for the study of very small precipitations in an austenitic steel.

9.6 WHEN USING TEM- OR SEM-ORM TECHNIQUES?

In the previous sections techniques for TEM and SEM ORM have been presented. In this section, we will discuss when which technique should be applied. Generally, the

TEM techniques are significantly more complicated in terms of sample preparation and use of the technique itself and should therefore be used only in crucial cases. Also, the areas observable by TEM are significantly smaller than those of EBSD. There are, therefore, only few cases of ORM where we think the move to the TEM is really of importance. In these few cases, however, TEM is indispensable.

The most evident case for TEM-based ORM is the investigation of microstructures that are below the resolution limit of EBSD-based ORM. It has been discussed in Section 8.3.5 that the probed volume of TEM ORM is not necessarily much smaller than that of EBSD-based ORM but the probed volumes are differently oriented with respect to the sample surface as illustrated in Figure 8.18. For EBSD the interaction volume spreads parallel to the specimen surface while for TEM it is rather oriented perpendicular to the surface. The thickness of the specimen significantly determines the resolution of TEM ORM. TEM techniques are, therefore, particularly superior to the EBSD on very thin samples, or when samples are prepared such that the main features, e.g. grain boundaries, run parallel to the electron beam. An example for this case is the investigation of heavily rolled or extruded metals where most of the boundaries run parallel to the main deformation direction, forming some kind of columnar structure. We will discuss this particular case in the next section.

9.6.1 Orientation Microscopy on Highly Deformed Metals

During deformation at low homologous temperature the crystal structure of metallic materials is successively subdivided into smaller units, e.g. dislocation cells, subgrains, or grains. The smallest structure sizes occur in materials which recover little after deformation, i.e. metals with a high melting point and a high degree of alloying. A good example is Fe-36%Ni, a ductile fcc alloy, which, after >90% cold rolling develops bands of <50-nm thickness, a width of several micrometer and an extension in rolling direction of several 10 μm. In cube-oriented areas (001)[100], slightly misoriented cells of approximately 100 nm diameter and a length of several 10 micrometer develop. A typical microstructure observed by TEM on a sample cut parallel to the rolling direction, i.e. parallel to the grain or subgrain boundaries has been shown in Figure 9.19a. All boundaries are clearly visible. An orientation map obtained by a semi-automatic STEM technique is shown in Figure 9.19b. A similar area was measured by EBSD-based ORM in a FEGSEM at 15 kV, using a step size of 30 nm; the results are reproduced in Figure 9.22. It is visible in Figure 9.22b that, although the EBSD technique reaches its limits in terms of lateral resolution, most microstructural features are correctly revealed. The positions of the grain boundaries, however, are not as well defined as in the TEM images because for EBSD, because of the strong sample tilt, the beam does not run parallel but almost perpendicular to grain boundary structure, leading to overlap of patterns from neighboring grains. As shown in Section 8.2.9 (Figure 8.13), this leads to orientation averaging and partial disappearance of subgrain boundaries. The overlapping signal from neighboring grains becomes particularly visible in the pattern quality map in Figure 9.22a.

The advantage of TEM for these studies disappears as soon as the crystal orientation is no longer approximately constant along the thickness of the sample. In the present example of cold-rolled Fe36Ni, this occurs in areas of shear bands, observed

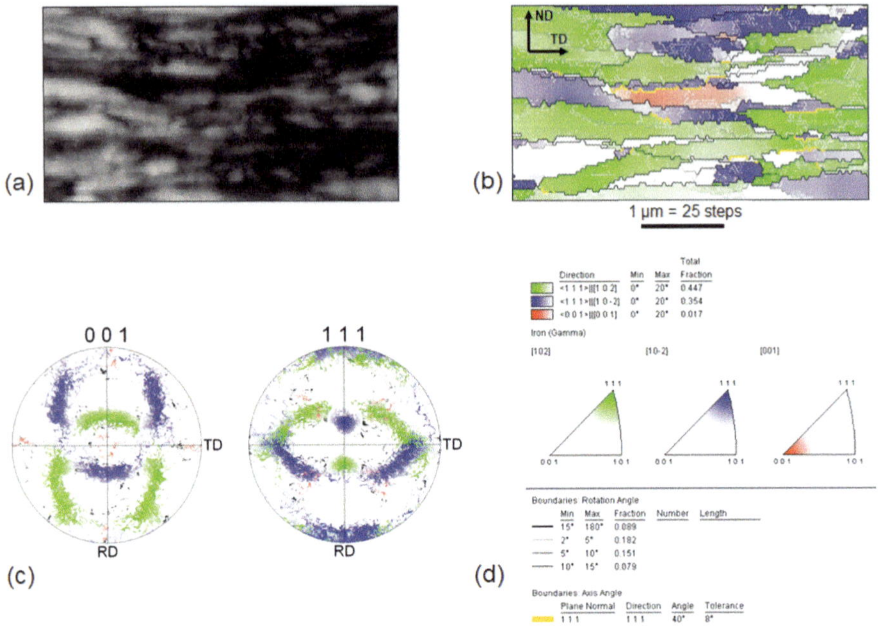

FIGURE 9.22 EBSD-based orientation mapping on a 94% cold-rolled Fe36Ni sample (measurement parameters: JEOL 6500 FEGSEM, 15 kV, step size 30 nm). (a) Pattern quality map (dark corresponds to low quality), (b) orientation map in texture component coloring, indicated in (d). The red area in the center corresponds to a cube-oriented grain. All grain and subgrain boundaries down to 1° are indicated in different gray levels. (c) Normal pole figure from the data displayed in (b) with a coloring according to the data in (d). (d) Inverse pole figures for three different sample directions, [102], [10$\bar{2}$], and [001].

from the transversal direction. An example of a TEM orientation map is given in Figure 9.23a and b: in a certain area inside of the shear band no orientations were measurable, because even the smallest electron probes created only ring diffraction patterns. Here, in fact, the crystals show a continuous rotation along the beam path which leads to the occurrence of ring patterns. The situation is different when using EBSD with its very small penetration depth. An EBSD orientation map from a similar area as that observed by TEM is shown in Figure 9.23c and d.

Finally, we will discuss studies where TEM and SEM techniques were compared with respect to their applicability to medium and highly deformed Al-alloys. The aluminum alloys, due to their low melting point, high stacking fault energy and relatively little alloying content, tend to recover strongly and form relatively large cell structures in the order of 1-μm diameter. Hurley and Humphreys (2003), in their study on an Al–0.13 wt% Mg alloy, deformed up to 90% by cold rolling, found that TEM did not add any information to the knowledge obtained by EBSD ORM. Rather, the much larger areas observable by EBSD-ORM allowed much more statistically significant statements on cell size and misorientation angle distribution. A second study on pure aluminum carried out by Huang and Juul Jensen (2000) came to a similar statement.

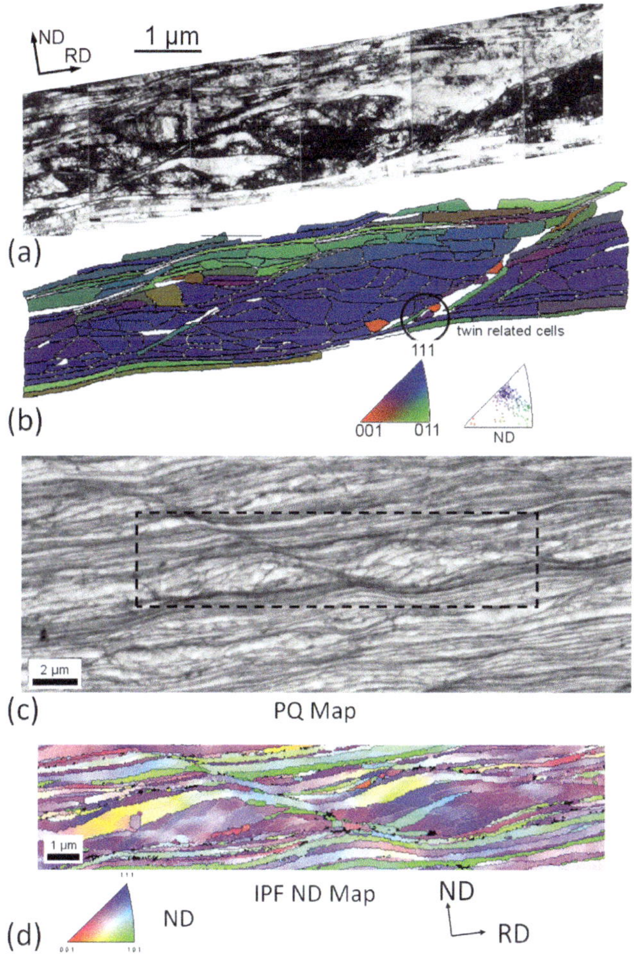

FIGURE 9.23 (a), (b) TEM, and (c), (d) EBSD-based orientation map of shear bands in A 94% cold-rolled Fe-36% Ni sample. The figures show that TEM orientation microscopy may be inferior to EBSD-based microscopy in terms of resolution if several orientations line up along the sample thickness: TEM is not able to resolve all orientations in the shear band while EBSD apparently does.

It is interesting to note that the advent of ECCI in SEM even enables direct observation of dislocation structures in the SEM on bulk samples in a very similar manner as TEM. This is shown in Figure 9.24 which compares ECCI observations of Godec et al. (2020) with TEM results by Wang et al. (2018) on a 316L stainless steel sample, produced by additive manufacturing. The dislocation structure observed with ECCI and those with TEM look very similar (except for the fact that they are inverted in contrast). The TEM allows to go for higher magnification to observe chemical and structural defects, while the SEM enables to study much larger areas (but also still allows to increase magnification).

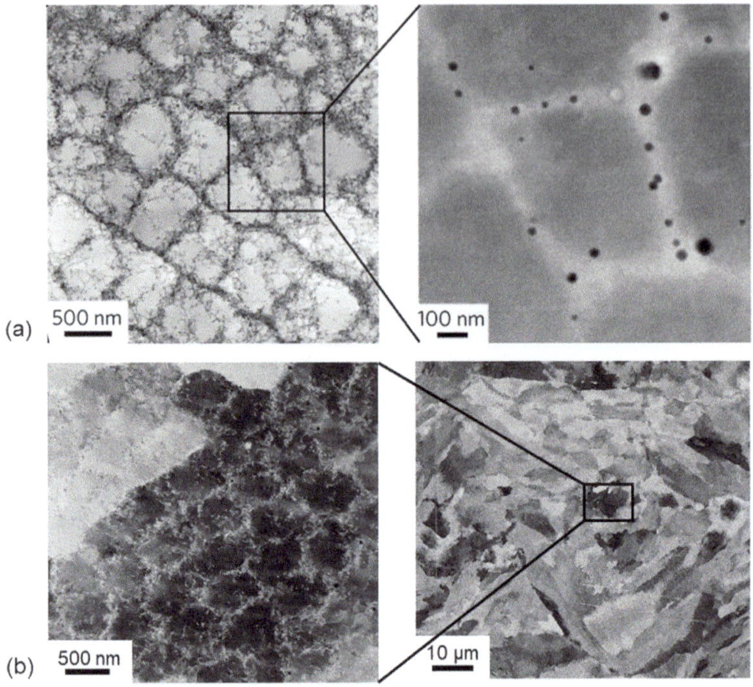

FIGURE 9.24 Comparison of TEM bright field imaging and ECC imaging in SEM. Figures taken from an 316L stainless steel specimen, produced by additive manufacturing. Images on the left side show similar magnifications and similar dislocation structures for both techniques. (a) TEM image extracted from Wang et al. (2018). TEM has the advantage to reach nm resolution also for chemical analysis. (b) ECC imaging taken from Godec et al. (2020). ECCI has the advantage of showing very large specimen areas, all of which can be zoomed in for observation at high magnification.

9.6.2 Other Cases Requiring High Spatial Resolution

A critical case for ORM techniques is that of truly nanocrystalline materials with grain sizes below about 50 nm as it has been sketched in Figure 8.18. Important information that needs to be extracted about the microstructure are the grain size, the grain size distribution, the texture, and the grain boundary character.

The most important difficulty when attempting to measure these materials in TEM is the fact that generally several grains overlap along the beam path, as displayed in Figure 8.18. A typical microstructure example has been shown in Figure 6.6. Consequently, the software for ORM must be able to deal with overlaying diffraction patterns. As shown in Section 7.4.2, a technique which deals with this is template matching on TEM spot patterns (Valery et al., 2017). In this work, stacked nanocrystalline grains were successfully deconvoluted. The most promising approach, however, appears to be the use of TKP in SEM, as it has been introduced in Section 8.2.4. It has been shown first by Suzuki (2013) that the SEM-TKP signal stems from a small volume at the exit surface of the primary beam. This volume, indicated in

Figure 8.18 by gray coloring, may be as small as $10 \times 10 \times 10$ nm³ and, therefore, be even smaller than that of TEM-based ORM. Thus, SEM TKP, for the first time, makes truly nanocrystalline materials accessible to observation with EBSD-based equipment (Trimby, 2012). Figure 9.25 shows an example of a nickel alloy where the smallest visible grain sizes are in the order of 10 nm and even subgrain boundaries are correctly revealed.

9.6.3 Cases Requiring High Angular Resolution or Precision of Structure Analysis

An investigation that has traditionally been studied by TEM orientation measurements is the correct orientation determination for ordered crystal structures that

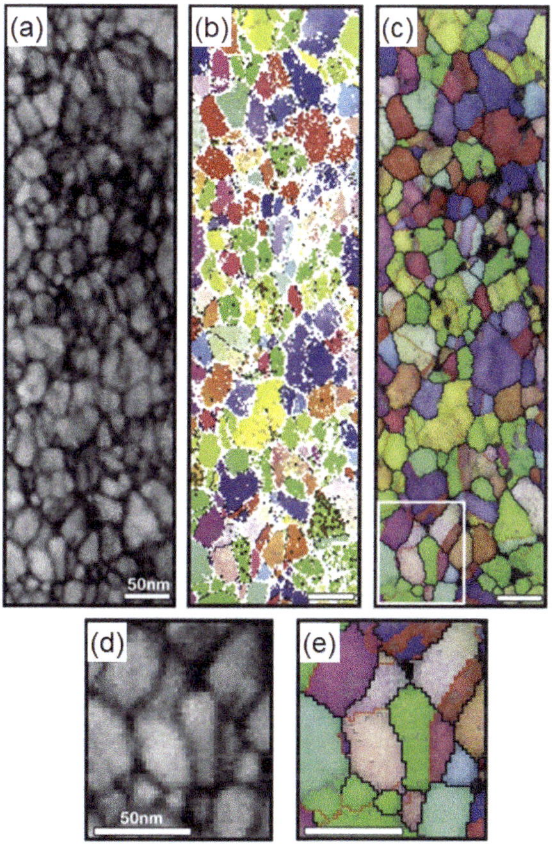

FIGURE 9.25 Results of orientation mapping using SEM-TKD of a nanocrystalline Ni sample. The measurement was performed with 2-nm step size on a Zeiss ultra plus FEGSEM, equipped with an Oxford Instruments Channel 5 EBSD system and a Nordlys-S EBSD detector. (a) Diffraction pattern quality map, (b) raw orientation map, white pixels indicate non-indexed points, and (c) cleaned map of (b). (d) and (e) are from a zoomed-in area of (a) and (c) which is marked by a white rectangle in (c). (Adapted from Trimby, 2012.)

deviate only slightly from cubic symmetry. A prominent example is γ-TiAl which has $L1_0$ structure which has almost fcc structure but shows a tetragonal distortion of c/a = 1.02 due to the stacking of alternating lattice planes of pure Ti and pure Al occupation in c-direction. γ-TiAl is light, strong, and highly high-temperature oxidation- and corrosion resistant (Appel and Wagner, 1998). It is, therefore, a highly interesting candidate, for example, as a replacement of Ni-based superalloys in airplane turbines. The problem, which delays this application, however, is the material's brittleness, which is directly related to its crystal structure (Lasalmonie, 2006). Microstructures of γ-TiAl are often lamellar, with the lamellae inside of one grain being characterized by alternating orientations with exchanged a- and the c-axis, i.e. by a 90°-rotation of the stacking order of the Ti and Al atomic layers. The correct characterization of the lamellar structures, i.e. the correct distinction of the c- and the a-axis, is essential for understanding the deformation mechanisms of the material. TEM can easily determine the correct orientation using superlattice reflections which are, in fact, additional (011) reflections inside of the basic (022) reflections. In contrast, no {110} reflections appear inside the {220} reflections. Figure 9.26 demonstrates the approach: the two patterns taken from two neighboring lamellae show clear superlattice reflections and the correct orientation can therefore be easily determined (Zaefferer 2000). In contrast, variant determination by EBSD was long not

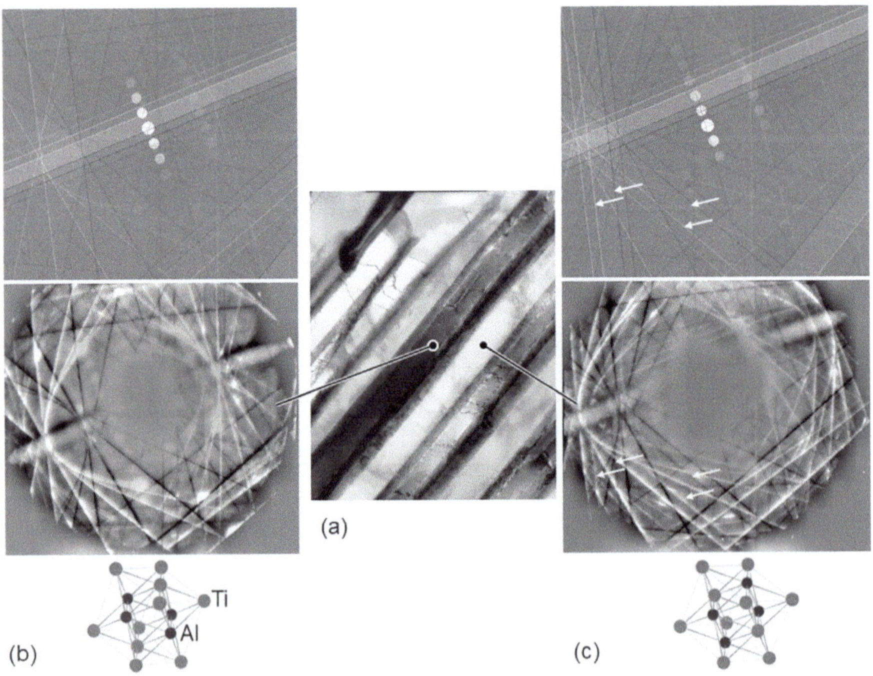

FIGURE 9.26 Determination of tetragonal crystal orientation of γ-TiAl by determination of superlattice reflections in Kikuchi patterns. (a) Lamella microstructure with Lamella of different order variants. (b) and (c) Kikuchi patterns of the two variants. Simulations have been carried out using TOCA, acceleration voltage 200 kV. In (c) the superlattice reflections which serve to distinguish the order variants are marked by white arrows.

possible because, due to the c/a-ratio almost equal to 1, EBSD patterns could only be evaluated for cubic crystal symmetry (Dey et al., 2005). With improved EBSD detector performance and new analysis software the determination of the correct orientation variant has also become possible via EBSD, exploiting the slight tetragonal distortion of the crystal lattice Zambaldi et al. (2009). Since EBSD patterns span a large angular range, the change of the Kikuchi band positions with exchange of c- and a-axis is large enough to be detected by automatic HT-based algorithms. The precise determination of these small differences requires diffraction patterns recorded with a large pixel resolution (e.g. 900 × 900), a high-resolution Hough transform (HT 240 × 0.5°), a very precise pattern center calibration, and an indexing algorithm which checks for the deviation of all experimental and theoretical diffraction vectors. Whether the correct solution is detected in this way can be manually controlled by checking the positions of the superlattice reflections. In fact, also in EBSD patterns superlattice reflections are visible, though they are too weak to be determined with a simple Hough transform. In Figure 9.27 two patterns from neighboring lamellae are shown. Those (022) bands that show (011) superlattice reflections are marked.

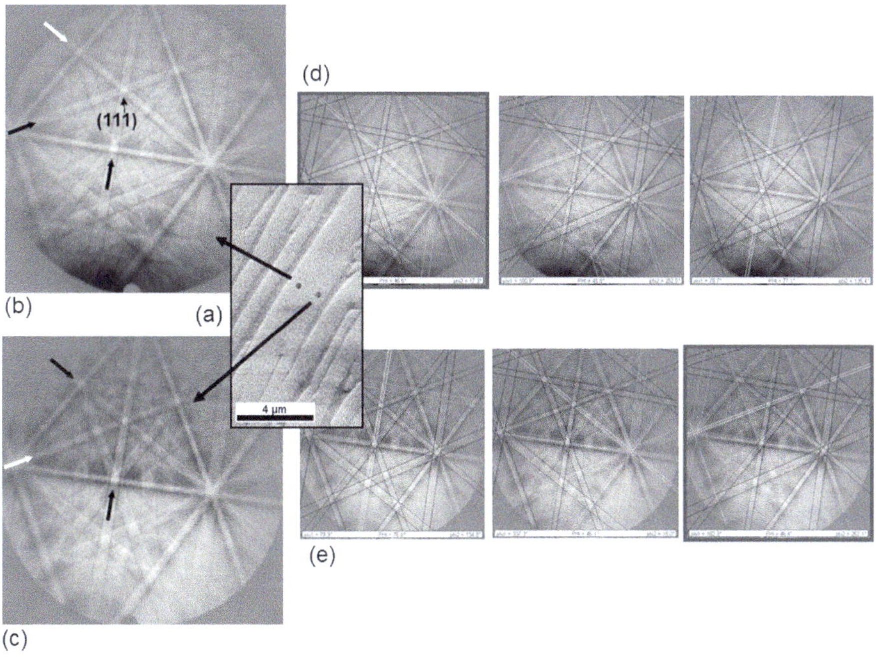

FIGURE 9.27 EBSD-based determination of the correct orientation of the tetragonal unit cell by means of superlattice reflections and by means of distortion measurements. (a) Lamella microstructure of the alloy observed by secondary electrons. Two positions have been measured. (b) and (c) Patterns of positions (1) and (2). The (111) pole where the different {220} and {022} bands cross is marked. The (022) band which contains the superlattice reflections is marked in both patterns by a white arrow. The other bands are marked by black arrows. The three possible indexing solutions for each of the two patterns are shown in (d) and (e). The correct orientation, determined by geometric distortion measurements, is marked by a thick frame. The existence of superlattice reflections is used to proof the correct orientation determination.

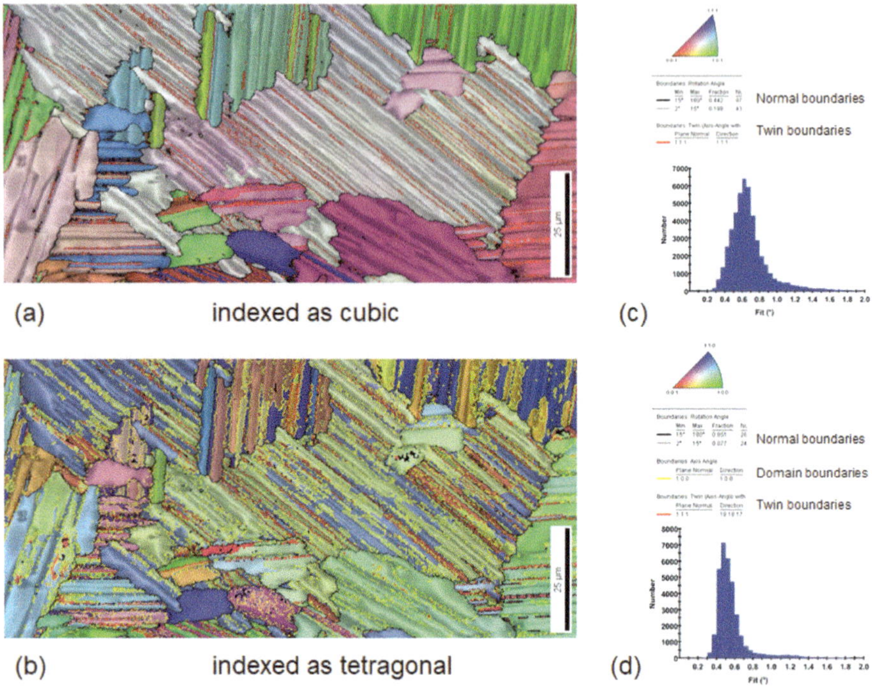

FIGURE 9.28 EBSD-based ORM on a lamellar γ-TiAl microstructure taking into account the tetragonal distortion of the γ-TiAl lattice. (a) Mapping using a cubic lattice description (pseudo-symmetry). (b) Mapping with a tetragonal lattice. (c) and (d) The distribution of the angular fit between measured and calculated diffraction vectors. The tetragonal mapping leads to significantly smaller fit values.

The solution with correct position of the superlattice reflection also corresponds to that with the smallest angular misfit value. Using the abovementioned measurement conditions, it is possible to map the correct variant distribution in a lamellar structure of an as cast, stoichiometric γ-TiAl alloy. Figure 9.28 shows the results: Figure 9.28a shows the cubic indexing, taking into account an fcc lattice. Figure 9.28b shows the correct tetragonal indexing, revealing the distribution of the different variants. A still more accurate and reliable approach to solve the variant problem would be to use the dictionary method (see Section 7.4.3) to fit the experimental patterns to precalculated ones, which incorporate the correct superlattice reflections.

9.7 SUMMATION

EBSD-based ORM clearly is an extremely powerful technique for the rapid, comprehensive, and quantitative characterization of crystalline materials. It covers a large size scale, from areas in the order of several cm^2 for statistical texture determination, here being a very powerful competitor of x-ray diffraction techniques, to the very small scale, covering just few μm^2 at a resolution close to the range of conventional

TEM. With its nowadays speed of data acquisition it becomes almost comparable to conventional SEM-based imaging techniques, with the very important advantage of offering much more and more quantitative information than these. Very important original information are spatially resolved orientation, phase, and crystal quality. The mapping data then allow a large number of partitioning, including neighboring pixels for misorientation calculation, kernels for local averages, grains for all kinds of morphological data, or phases for phase-related properties. Besides this also texture-based, morphology-based, or crystal-quality-based partitions can be built and analyzed. The possibilities with respect to the analysis of deformed structures are very large and ever growing, e.g. by combination with electron channeling imaging or as input for modeling programs (e.g. Section 10.8). The advent of CC EBSD now starts opening the measurement of residual stresses with high local resolution. Dictionary imaging now allows the differentiation of pseudosymmetric structures. Smart probing allows the measurement of macroscopic textures that may be more informative than those measured by x-ray techniques.

The most serious limitation of the technique is its spatial resolution. Besides technical improvements, higher spatial resolution can only be reached by going to thin foil techniques in TEM and SEM. When comparing resolution between TEM and EBSD-based ORM it has, however, often been overlooked that the spatial resolution of an ORM technique is not only given by the size of the beam but, actually, by its total interaction volume. While the lateral beam spread is much smaller in TEM, the depth of probing, determined by the sample thickness, is in most cases larger for the TEM than for the EBSD technique. For the same reason the EBSD-ORM technique has a far higher potential to develop toward still higher spatial and angular resolution (by lower voltage, higher camera resolutions, new algorithms) than the TEM-based techniques (mainly by reduction of sample thickness). In terms of spatial resolution, the use of TKD in SEM appears to give best performance. The enormous potential of the EBSD-ORM technique can, of course, only be fully exploited with the appropriate instrumentation and software, i.e. a thermal field emission gun SEM, a high-resolution highly sensitive EBSD camera, and robust indexing software.

The few cases where we see a true advantage of the TEM techniques over the SEM ones are: (i) the investigation of very heavily deformed materials for those cases where columnar microstructures are formed during deformation (e.g. cold rolling, extrusion) or when very thin samples are available; (ii) the investigation of truly nanocrystalline materials, when techniques are available to deconvolve and display superimposed layers of grains; (iii) the investigation of light elements like Mg, Al, and many minerals in the case that high spatial resolution is required; (iv) the investigation of very small precipitates; (v) the investigation of particular lattice defects in conjunction with their crystallographic and sample orientation. In most of these cases the spot pattern technique combined with template matching and potentially precession is the most powerful technique due to its resolution, robustness, and ability to deal with overlapping patterns. Transmission Kikuchi diffraction on thin foils in the SEM appears to be a serious competitor to the TEM-based techniques, because it reaches the same resolution and is much easier to do than TEM observations. By performing SEM TKP in a focused ion beam instrument the targeted thin foil fabrication can even be directly coupled with the observation process.

10 Evaluation and Representation of Microtexture Data

10.1 INTRODUCTION

A microtexture is, essentially, a population of individual orientations that can be linked to their location in a specimen and are obtained usually by an electron diffraction technique as described in Chapters 6 through 8. These two factors—the discrete nature of the orientation data and their direct link to the microstructure—introduce a much broader and more complex scenario for data evaluation and representation than that for macrotextures, since in the latter case the orientation data evaluation usually involves only direct representation in the form of pole figures and subsequently calculation of an orientation distribution function (ODF) (Chapter 5). The large variety of situations that can be addressed by means of individual orientation measurements and orientation mapping requires many more formats for the representation and interpretation of the measurements.

The nature of microtexture gives rise to two lines of inquiry: statistical distribution of orientation/misorientation data and orientation/misorientation data related to the microstructure. For most investigations both these analyses are required. Chapter 9 describes orientation mapping in detail, whereas this chapter discusses statistical representation of data. There are also many examples in the literature of investigations that involve both orientation mapping and statistical representation of the entailed data. Such investigations are not catalogued here; rather, the principles of data representation are described.

There are two ways in which orientation and misorientation microtexture data can be displayed in a space: as discrete data points or as a continuous density distribution. For simplicity in the case of the latter, we will use the terminology "orientation (or misorientation) distribution function" for *all* continuous distributions, without distinguishing if the function applied involves the series expansion method (as for macrotexture data) or the more simple case of a direct contouring function. This has the advantage of allowing generic terminology to be used for both macrotexture and all continuous microtexture data. Some texts, however, reserve use of the term "(mis)orientation distribution function" [(M)ODF] in microtexture for series expansion methods only and refer to direct contouring as a "(mis)orientation distribution."

This chapter considers data processing for orientations and misorientations separately, since the different physical applications of the measurements in both cases lead to a dichotomy in approach. For example, the angle/axis descriptor (Section 2.7)

DOI: 10.1201/9781003258339-13

has direct relevance to misorientations at interfaces, whereas it is inconvenient for orientation descriptions. Hence, Sections 10.2 through 10.4 and 10.5 through 10.7 deal with orientation and misorientation representations, respectively. For further information, see Gerth and Schwarzer (1993), Engler et al. (1994a), and Randle (2003). The representation methods described could be applied to microtexture data generated from electron backscatter diffraction (EBSD) or transmission electron microscopy (TEM), although the vast majority of investigations are EBSD and only a few are TEM (Paul et al., 2007). Finally, in Section 10.8, we address the utilization of orientation maps as a spatially resolved input for ensuing microstructure modeling activities.

10.1.1 Statistical Distribution of Orientation and Misorientation Data

For this type of microtexture representation, the chief aim is only to represent and quantify the orientation distribution in a region of interest in the microstructure, either by sampling every orientation present or by statistical sampling throughout a large region of the specimen. If only a single orientation is sampled at each point, then there is no access to misorientation or microstructure information—in fact this is equivalent to an "average" texture, that is, the microtexture counterpart to a macrotexture. Some investigations use a combination of macrotexture and microtexture measurements (Rafaja et al., 2008).

Sometimes, since automated EBSD is very rapid and is able to measure directly the "true" grain orientation, the statistical analysis mode is used advantageously in place of a traditional macrotexture technique to generate a measure of the overall texture. However, it is of prime importance to consider both the *size* and the *distribution* of the sample population to obtain a statistically reliable result. The number of grains encountered may be smaller than the number of grains contributing to x-ray or neutron pole figure measurements, depending on the size of the specimen, the grain size, and the mapping speed. Assuming a typical sampling area of 10×10 mm^2 and an average penetration depth of approximately 0.05 mm (Section 4.3), an x-ray macrotexture of a sample with an average grain size of 25 μm represents the orientation information of approximately 300,000 grains. In the case of neutron diffraction, this number is even higher by two orders of magnitude, which is due to the much higher penetration depth in the order of 10 mm (Sections 3.5 and 4.4). In EBSD, despite the large number of orientation data that can nowadays be obtained by fully automated EBSD systems, the number of grains encountered is usually much smaller, which raises the question of the statistical relevance of the data with regard to the description of the orientation distribution of the volume of interest. The success of using EBSD for overall texture estimation, therefore, depends on the ability to assess how many orientation measurements are required and an appropriate sampling schedule. This is discussed in Section 10.3.3.

Turning now to general evaluation of statistical microtexture data in the form of either orientations or misorientations, it is usual to display first the population using one of the methods described in Chapter 2. Secondary processing steps can then be applied to extract quantified information appropriate to the investigation, for example, what proportion of specified orientations within a certain angular tolerance are present, either by number or by projected grain area (Section 10.2.2). Many of these processing steps are available as part of commercial software packages or, alternatively, the user can design customized options. Some of the representation options are those traditionally used for macrotexture, for example, the pole figure/inverse pole figure (Section 2.5) and Euler space (Section 2.6), whereas other options such as Rodrigues space (Section 2.8) and misorientation evaluation (Section 2.7) derive directly from microtexture data. However, the fundamental differences between data collection for microtexture (Chapters 6 through 9) and macrotexture (Chapters 4 and 5) feed through to necessitate different approaches for data handling in both cases.

Although a "raw" microtexture population comprises discrete individual orientations, which are usually displayed as such, when the number of data points is very large (which is often the case for mapping, Chapter 9) or the texture is very sharp, it becomes inconvenient to display the data—in any space—individually because the representation will soon become unacceptably crowded and lucidity will be lost. Instead the data can be shown as density distribution contours (Sections 10.2.2 and 10.3.2).

10.1.2 Orientation and Misorientation Data Related to the Microstructure

The most valuable aspect of microtexture is the direct link that it provides between orientation and microstructure. Consideration of the statistical orientation distribution of the sampled region(s) (or subsets of them), as described in Section 10.1.1, is usually only part of the total microtexture investigation. Almost always an orientation distribution is derived from an orientation map or a selected subset of the map. Several figures in Chapter 9 depict orientation maps accompanied by statistical microtexture output. Sometimes orientation data are accompanied by a manual record of sampling coordinates in the microstructure, and the location of these individual data points is marked on the representation, which is usually a pole figure or inverse pole figure. A similar strategy has traditionally been common practice in TEM. Clearly, this method is only viable for relatively small data subsets. Figures 10.1 and 10.2 depict this strategy, as described in Section 10.2.

Another asset of microtexture is the access it provides to interfacial parameters, orientation relationships, and orientation connectivity in the microstructure. This area gives rise to a further range of data processing and representation options (Section 1.2.1). Even though the representation spaces are the same for orientations and misorientations, they are utilized differently in both cases to highlight the required physical features in the data. Chapter 11 discusses further processing of the crystallography of interfaces and also aspects of microtexture that encompass the way in which the orientation components are distributed in space, that is, orientation connectivity.

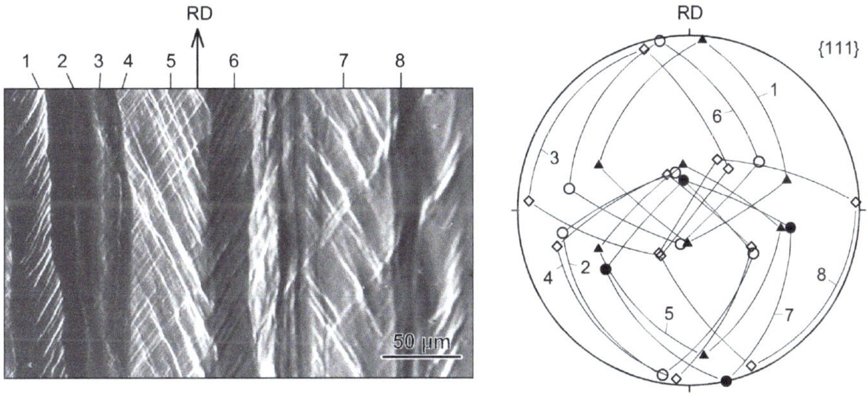

(a) microstructure (b) orientations of a)

FIGURE 10.1 Example of the presentation of microtexture data in a pole figure, showing the influence of the matrix orientation on shear band formation in polycrystalline Al-3%Mg, 90% cold rolled. (a) Microstructure; (b) orientations of (a). Filled symbols—grains with shear bands; open symbols—grains without shear bands.

10.2 REPRESENTATION OF ORIENTATIONS IN A POLE FIGURE OR INVERSE POLE FIGURE

The most obvious method of representing a microtexture is in a pole figure. This is convenient since only a single graphic is required to embody the entire texture information. Local orientation data can be marked in the pole figure (or inverse pole figure) with different symbols or numbers with respect to their location in the microstructure, and the corresponding symbols/numbers are marked on a depiction of the microstructure (Section 10.2.1). This method allows a direct correlation of orientation and site in the microstructure but its application is limited to a quite small number of individual points. For the representation of larger numbers of orientations with statistical relevance in a pole figure, a continuous density distribution is superior (Section 10.2.2), though in this case the direct correlation of orientation and site in the microstructure is lost.

10.2.1 Individual Orientations

An example of pole figure/microstructure correspondence for individual orientations is illustrated in Figure 10.1, which shows a micrograph of several deformed grains in a cold-rolled Al-Mg alloy (Figure 10.1a) and an accompanying pole figure (Figure 10.1b), where the orientations of these grains were determined by EBSD. In Figure 10.1b, the orientations are further distinguished whether or not the corresponding grains developed shear bands during deformation. Orientations of grains that developed shear bands during deformation are marked with filled symbols and those of homogeneously deforming grains with open symbols. From this study it has been concluded that shear band formation is restricted to grains with orientations

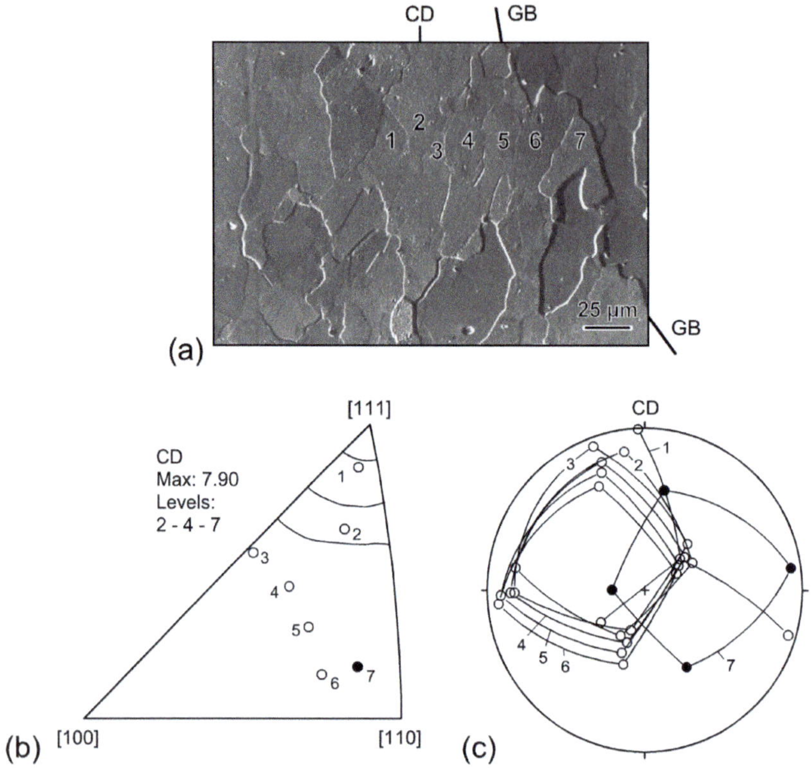

FIGURE 10.2 (a) Microstructure of NiAl deformed at 900°C by 70% in compression, showing a coarse subgrain structure in the vicinity of a grain boundary marked GB. (b) Inverse pole figure of the subgrain orientations shown in (a). The x-ray macrotexture of the sample mainly showed intensities close to [111], as indicated by the iso-intensity lines. EBSD single-grain orientation measurements of subgrains 1–7 close to the GB depict remarkable orientation changes. (c) Pole figure of the subgrain orientations 1–7 showing that the orientations of subgrains 6 and 7 were quite different. (Courtesy of L. Löchte and J. Fischer-Bühner.)

close to the {112}⟨111⟩ orientation, whereas grains with other orientations deform homogeneously (Engler et al., 1992).

Inverse pole figures are often used for axial symmetric samples, where only one of the sample axes is defined (Section 2.5.2). For instance, for compression or tensile samples the orientation changes of the compression/tensile axis are plotted in the crystal coordinate system. According to the crystal symmetry, it is not necessary to show the entire inverse pole figure—one stereographic triangle will suffice. In the case of cubic crystal symmetry, the well-known stereographic triangle [100], [110], and [111] is shown. Figure 10.2a shows the compression axis inverse pole figure of the intermetallic compound NiAl that has been deformed at 900°C. EBSD single-grain orientation measurements close to a grain boundary depict remarkable orientation changes as shown from the position of the subgrain orientations 1–6 in the inverse pole figure, whereas the macrotexture of that sample showed a pronounced ⟨111⟩ texture. In the inverse pole figure (Figure 10.2b) the two subgrains 6 and 7 on different sites of the large-angle grain boundary are very similar, but note that this does not unequivocally prove that these subgrains actually have the same orientation. This is because the inverse pole figure only shows the orientation of *one* reference axis—here the compression axis CD—but rotations about this axis are not considered. In fact, the full grain orientations of subgrains 6 and 7 were quite different as becomes obvious when plotting the same orientations in a pole figure (Figure 10.2c). Hence, for a complete representation of the 3D orientation information, two inverse pole figures showing two of the three different reference axes are required. (The third reference axis is redundant because it is defined by a cross product of the first two.)

10.2.2 Density Distributions

Whether to get more quantitative information on the relative density of microtexture data or to relieve a congestion of individual data points, it is beneficial to represent the density distribution from single-grain orientation data as density levels in the pole figure. Then, however, the direct correlation of orientation and site is lost. Density distributions resulting from a given set of single orientations can readily be determined by subdividing the pole figure into a grid with distinct angular cells, for example, $\alpha \times \beta = 5° \times 5°$. For each orientation g (where g denotes the rotation matrix between the sample and the crystal coordinate system; Section 2.3.2), the pole figure angles α_i and β_i for each pole $(k_ih_il_i)$ (e.g., (111) and $(1\bar{1}1)$) are given by (Section 2.5.1)

$$\begin{pmatrix} \sin\alpha_i \cos\beta_i \\ \sin\alpha_i \sin\beta_i \\ \cos\alpha_i \end{pmatrix} = g^{-1} \cdot \begin{pmatrix} h_i \\ k_i \\ l_i \end{pmatrix} \tag{10.1}$$

In the case of an inverse pole figure, the inverse orientation matrix g^{-1} in Equation 10.1 is replaced by the orientation matrix g, and the vector $(h_ik_il_i)$ is replaced by a vector $(X_SY_SZ_S)$ of the sample coordinate system, for example, X_S = RD = (100), Y_S = TD = (010), and Z_S = ND = (001) (Section 2.5.2). If only one inverse pole figure is presented, then this means that just one axis is considered, that is, $i = 1$. The angles α and β now define the position of the orientation g in the inverse pole figure.

After calculation of the angles α_i and β_i, the value of the corresponding cell (α_i,β_i) is incremented by one. Finally, the data are normalized with the total number of orientations N and the number of poles i. The data can then be plotted in a pole figure or inverse pole figure. As an example, Figure 10.3 displays micro- and macrotexture results of a cold-rolled and partially recrystallized aluminum single crystal. Initially, this single crystal had a (112)⟨111⟩ orientation, which was found to remain stable during deformation by cold rolling. Subsequent recrystallization annealing led to the formation of two symmetrical variants of the orientation {016}⟨461⟩ (Engler et al., 1993a). In Figure 10.3a, some single-grain orientations, as measured by EBSD in the deformed and the recrystallized regions, are plotted in the stereographic projection. Figure 10.3b shows the density distribution calculated from approximately 500

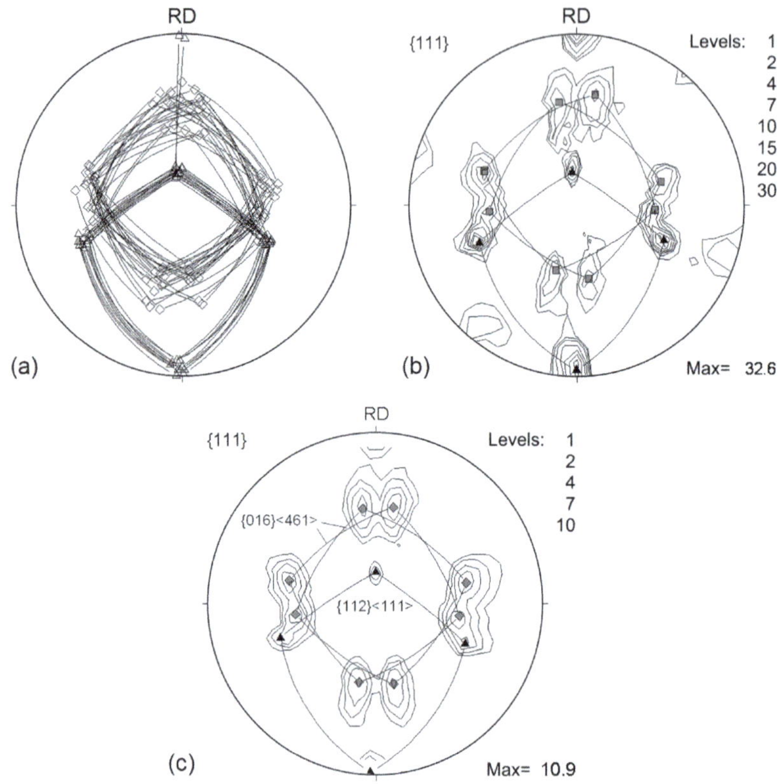

FIGURE 10.3 Comparison of EBSD microtexture and x-ray macrotexture pole figures of a partially recrystallized Al-1.8%Cu single crystal (rolled to 80% reduction, annealed for 300 s at 300°C). (a) Examples of EBSD single-grain orientation measurements in the as-deformed matrix and the new recrystallized grains; (b) continuous density distribution as computed from approximately 500 orientations similar as in (a); (c) x-ray macrotexture of the same sample.

individual orientations as described earlier. Figure 10.3c presents the x-ray macrotexture of the sample.

A comparison of microtexture (Figure 10.3b) and macrotexture (Figure 10.3c) shows very good correspondence in the positions of the main peaks. A quantitative comparison of the orientation density, however, is not possible, as already evident from the different heights of the texture maxima (32.6 in Figure 10.3b compared to 10.9 in Figure 10.3c). This difference in intensity distribution is caused by the different measuring procedures applied: x-ray pole figures are *textures by volume* (Bleck and Bunge, 1981), which means that each orientation contributes to the texture according to its volume. By contrast, during the EBSD single-grain orientation measurements, the orientation of large recrystallized grains was measured only once, yielding a *texture by number*. Therefore, in comparison to the x-ray texture the intensity of these large grains is underestimated. Of course, it is entirely possible to measure textures by volume using single-grain orientation measurements, for example,

0.045	0.122	0.045
0.122	0.332	0.122
0.045	0.122	0.045

(a)

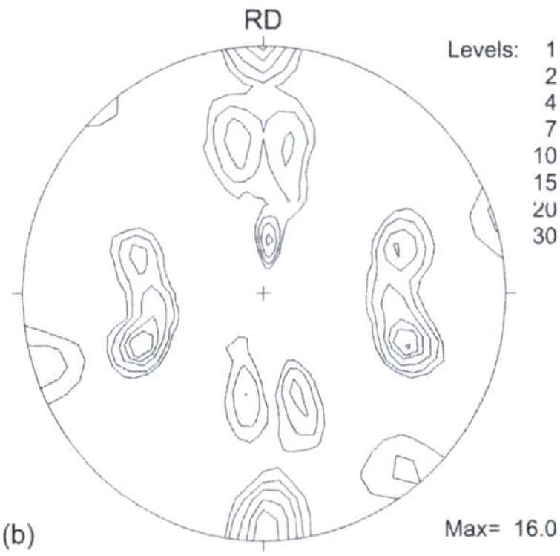

(b)

FIGURE 10.4 Smoothing of pole figure data. (a) Two-dimensional filter with approximate Gauss-shape to smooth pole figures; (b) pole figure as derived from pole figure data as in Figure 10.3c but smoothed with the filter in (a).

by correction of the data with respect to the corresponding grain size. Similarly, orientation mapping as applied during orientation microscopy (Section 9.3) directly yields textures by volume, as in this case the number of measurements per grain is a measure for the volume or, more precisely, the area of the respective grain.

If the number of orientations is smaller or the orientation scatter is larger than that in the example in Figure 10.3, the pole figures may look rather unsystematic or "spiky." In that case, it might be appropriate to represent only the discrete data or to smooth the pole figure data by convolution of the intensity data in the pole figures with a suitable filter. Figure 10.4a shows an appropriate 2D filter with an approximate Gauss-shape. Application of this filter considers the intensity of the current pole figure position α and β with 33%, the intensities of direct neighboring cells with 12%, and that of second-next neighbors with 4.5%. These values do not change the integral intensity (i.e., the normalization) of the pole figure, but in general they reduce the maximum intensity. Figure 10.4b displays the pole figure of Figure 10.3c after smoothing according to this procedure. Another way to derive continuous pole figure or inverse pole figure plots from single-grain orientation data is to calculate an ODF (Section 10.3.2) and then to recalculate the desired (inverse) pole figure (Section 5.3.2). Provided the series expansion method is applied, the pole figure data are automatically smoothed.

10.3 REPRESENTATION OF ORIENTATIONS IN EULER SPACE

Since single-grain orientations already contain the entire orientation information in an unequivocal manner, no further information is yielded if a microtexture is plotted in a 3D space, such as Euler space or Rodrigues space, than if it is plotted in a pole figure. However, representation in a 3D orientation space could be advantageous for complex data sets, and also if direct comparison to x-ray macrotexture data, where the ODF in Euler space is the traditional representation method, is required. This is discussed in Sections 10.3.1 and 10.3.2, and in particular direct and indirect methods to derive a continuous ODF from single-grain orientations are compared.

10.3.1 Individual Orientations

As discussed in Section 2.6, a given orientation can be assigned a point with the coordinates (φ_1, Φ, φ_2) in the 3D Euler angle space. Thus, a set of orientations determined by single-grain orientation measurements can be expressed in terms of the Euler angles and, after appropriate reduction with regard to sample and crystal symmetry, displayed as points or with appropriate symbols in the Euler space. Similarly as discussed for the representation of macrotextures (Section 5.4), microtextures are likewise usually represented as a set of 2D sections along φ_1 or φ_2 through the Euler space. Orientations that do not fall exactly in one of the sections—virtually all orientations—are plotted in the adjacent section. The resulting error can slightly be reduced by projecting the orientations under the condition $\varphi_1 + \varphi_2 =$ constant (which in this strict form only applies for $\Phi = 0$; Section 2.6.1).

Another problem relates to the fact that usually only a subspace of the full Euler space is used for presentation of orientations. For cubic crystals and orthonormal sample symmetry, this is $0° \leq (\varphi_1, \Phi, \varphi_2) \leq 90°$ (1/32 of the full Euler space) that still does not take the threefold axis of cubic crystal symmetry into account (Section 2.6.2). Hence, each individual orientation has to appear three times in this subspace, but the single-grain orientation measurement of course provides only one variant. The two missing variants can readily be generated by successive 120° ⟨111⟩ rotations from the original orientation. Figure 10.5a shows an example of the orientations of

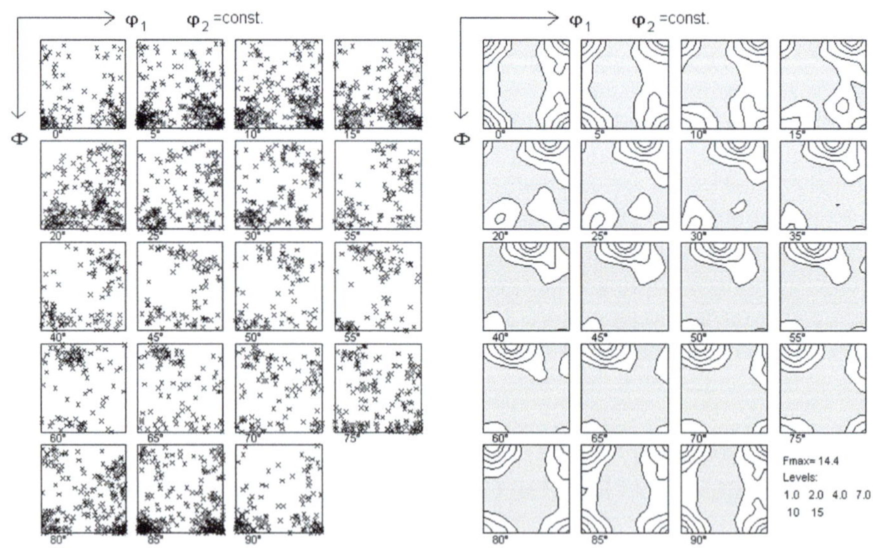

FIGURE 10.5 Presentation in Euler space of microtexture data of a recrystallized aluminum sample. (a) Orientations of 1,000 grains as obtained by EBSD; (b) continuous intensity function (ODF) as derived from x-ray pole figure measurements of the same sample, showing good reproducibility of the main texture features.

1,000 grains in a recrystallized aluminum sample plotted in the Euler angle space. The main textural features strongly resemble the macrotexture of the same sample as obtained by x-ray diffraction (Figure 10.5b), though fine details and smaller intensity variations cannot be resolved.

10.3.2 Continuous Distributions

The easiest way to determine a continuous orientation distribution in Euler space from single-grain orientation data is to apply a *direct method* (cell method) similar to that described earlier for the pole figures. As shown in Section 10.3.1, a point in Euler space is obtained for each orientation. If one now subdivides the Euler angle space into small cells of an appropriate size, the number of points in each cell can be counted and a continuous distribution function is derived. This method was first applied by Perlwitz et al. (1969) to single orientations determined by selected area electron diffraction (SAD) in a TEM.

In the direct methods, the orientations are associated with discrete cells of finite size in the orientation space, so that a direct method provides a discrete ODF. Typically, the Euler angle space is subdivided into cells of size $\Delta\varphi_1 = \Delta\Phi = \Delta\varphi_2 = 5°$ or 10°. For each individual orientation, g, the content of the corresponding cell j with the orientation $g_j - 2.5° < g_i < g_j + 2.5°$ is incremented. Finally, the value of each cell is normalized to the total number of single-grain orientation measurements N and to the cell size Δg_j. If necessary, a filter can be applied to smooth the resulting orientation distribution, as discussed earlier for the pole figures. It is seen that the presentation corresponds to a histogram of orientation densities $\Delta n_j/(N\Delta g_j)$ defined by the number of orientations Δn_j falling into the jth cell and normalized by the number N of measured orientations with regard to the cell size Δg_j.

Although this method appears easy and straightforward, it has several drawbacks. First, the number of measured orientations is often too low for statistical relevance (see Section 10.3.3). As an example, for a cell size of 10° and a statistical reliability of 80%, approximately 25 points per cell must be considered, which for cubic crystals would require approximately 3,000 single-grain orientation measurements (Bunge, 1982). Consequently, in the case of a smaller number of orientations the cell size would become very large, which would result in a falsification of the ODF. Therefore, it may be necessary to use the series expansion method to derive an interpolation function as discussed later in this section.

A second drawback is that in direct methods it is very complicated to take the strong distortion at small angles Φ properly into account, and therefore this is in general not done. This means that an ODF, as derived by means of a direct method from single-grain orientation data, could comprise a strong peak at a given orientation having small Φ, though continuous densities obeying the condition $\varphi_1 + \varphi_2 =$ constant would be more appropriate.

Because of these problems, an *indirect method* might be advantageous (Pospiech and Lücke, 1975; Wagner et al., 1981; Haeßner et al., 1983). As already discussed in more detail in Section 5.3.3, each single grain orientation g_i is associated with a Gauss-type distribution in Euler angle space with the (half) scatter width ψ_0. The distribution of the orientation density S as a function of the angular distance ψ from the exact orientation can be computed according to

$$S(\psi) = S_0 \cdot \exp\left[-\left(\frac{\psi}{\psi_0}\right)^2\right] \tag{10.2}$$

For single-grain orientation measurements, density S_0 is equal to 1. For a texture consisting of N individual orientations, the ODF can be determined by superposition of the corresponding N Gauss peaks. The main advantage of this method is that it provides an easy computation of the C coefficients of the series expansion method by

$$C_l^{mn} = \frac{1}{N} \cdot \frac{\exp\left(-l^2\psi_0^2/4\right) - \exp\left(-(l+1)^2\,\psi_0^2/4\right)}{1-\exp\left(-\psi_0^2/4\right)} \cdot \sum_{i=1}^{N} T_l^{mn}\left(g_i\right) \tag{10.3}$$

With regard to the discussion of the ghost error caused by the lack of odd-order C coefficients in macrotexture experiments (Section 5.3.3), it is noted that Equation 10.3 yields both odd- and even-order C coefficients so that the microtexture approach does not suffer from this error. For a microtexture ODF calculated according to Equation 10.3, each individual orientation i contributes to the final texture with the same weight of $1/N$, that is, it yields a *texture by number.* To determine *textures by volume*, the term $1/N$ has to be replaced by a term that takes the volume of each grain into account. The same would also apply for the scatter widths ψ_0 if it turned out that the ψ_0 is different for different orientations or orientation classes, for example, for the as-deformed matrix and recovered subgrains (Engler et al., 1996b). As an example for calculation of an ODF from single-grain orientation data, Figure 10.6a and b shows the ODFs of a recrystallized aluminum sample, which have been derived from 100 and 1,000 EBSD measurements, respectively (the latter ODF corresponds to the individual data shown in Figure 10.5a). In this example $N = 100$ orientations already yield a fairly good qualitative estimate of the corresponding x-ray texture (Figure 10.5b), whereas $N = 1{,}000$ orientations reproduce the macrotexture with high accuracy.

If the C coefficients are available for microtexture ODFs, then parallel calculations to those derived for macrotexture ODF evaluation can be carried out (Bunge, 1982). For example, plots of individual texture component profiles, volume fractions and scatter widths, subordinate functions such as (inverse) pole figures, and, furthermore, mechanical properties and anisotropic behavior can be predicted (Bunge and Esling, 1982; Bunge, 1988; Wenk and Van Houtte, 2004; Rollett and Gottstein, 2021).

An immediate comparison of ODFs as derived from the direct and indirect methods is difficult for the following reasons. In the case of the (discrete) cell method, the intensities depend on the dimension of the cells and are further locally affected by the distortion of the orientation space. In the continuous representation, the density values strongly depend on the chosen scatter width ψ_0. Moreover, they are locally influenced by the error introduced by the truncation of the series expansion, which is generally the more severe the higher the local concentration of the measured single-grain orientations is.

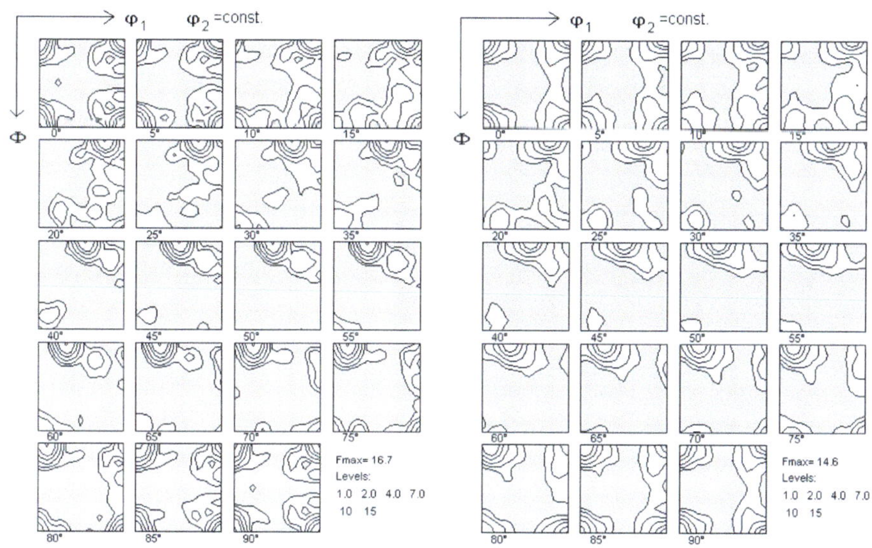

(a) EBSD, N=100 (b) EBSD, N=1000

FIGURE 10.6 ODFs of recrystallized aluminum, as in Figure 10.5, calculated from different numbers N of EBSD single-grain orientation measurements, (a) $N = 100$, (b) $N = 1,000$.

Both the size of the cells in the discrete presentation and the scatter width in the continuous case cannot be chosen arbitrarily, but they should be carefully defined in connection with the resolution of the measurement as well as the computation procedure. In the literature, values ranging from 2° to 10° have been proposed, although larger values should be avoided as this would cause undue smearing of the peaks and errors in the textures (Engler et al., 1994a). As a lower limit for both cell size and scatter width, the absolute accuracy of single-grain orientation measurements, for example, 5°, has been proposed (Engler et al., 1994c). Wagner et al. (1981) have discussed that the scatter width ψ_0 of the series expansion method should depend on the maximum series expansion coefficient l_{max}, which means that for $l_{max} = 22$ and $l_{max} = 34$, ψ_0 should not be smaller than 8° and 5°, respectively. Hutchinson et al. (1999) have pointed out that there is some instrumental broadening of approximately 3° in x-ray diffraction techniques as well. They recommend only small amounts of smoothing of the same order of magnitude (i.e., 2°–4°), if any, to be applied to EBSD data. In general, the degree of smoothing should decrease with increasing number of measurements.

10.3.3 Statistical Relevance of Single-Grain Orientation Measurements

Despite the relatively large number of orientation data, which can nowadays be obtained by the various techniques for single orientation measurements the number of grains encountered is still small compared to the number of grains contributing to x-ray or neutron pole figure measurements. Even in the case of fully automated EBSD measurements, usually less than 10,000 grains are recorded, which raises the question of the statistical relevance of the data with regard to the description of the orientation distribution of the volume of interest.

The number of single-grain orientation measurements, which is necessary to represent adequately the texture, depends on crystal and sample symmetry and, most of all, on the texture sharpness. Although in the ideal case of a single crystal one measurement is sufficient, for weak textures several thousand individual orientations must be measured. As alluded to earlier, Bunge (1982) stated that for determination of an ODF with statistical relevance (i.e., statistical reliability of 80%) approximately 25 points per cell must be considered. For cubic crystal and orthonormal sample symmetry (i.e., ODF size $90° \times 90° \times 90°$) and a cell size of 5°, this would require approximately 10,000 orientation measurements. In a recent experimental study on rolled stainless steel having a medium strength texture, it has also been stated that of the order of 10,000 grain orientations were needed to provide a texture characterization equivalent to that obtained by x-ray diffraction (Wright et al., 2007). In an analytical study by Jura et al. (1996), it was concluded that for weak textures even several tens of thousands of orientations must be considered. Note that in this discussion the number of single-grain orientation measurements N is to be understood as the number of measured *grains*, whereas modern automated EBSD systems usually record many orientations per grain. This will be discussed at the end of this section.

In the early 1990s, several experimental studies were undertaken to estimate the number of single-grain orientation measurements necessary to achieve the statistics of conventional macrotexture measurement techniques using x-ray diffraction. In these studies, the texture of a given sample was determined from 100, 200, 300, …, N individual orientations and compared with the corresponding x-ray macrotextures (Wright and Adams, 1990; Baudin and Penelle, 1993; Engler et al., 1994c). It turned out that for cubic materials, the ODFs derived from only 100 orientations already comprised the main characteristics of the textures, whereas a statistically sound representation of the textures required 500–1,000 orientations.

Because of the strong influence of the texture sharpness on the accuracy of reproducing an ODF from single-grain orientation measurements, it is necessary to develop a quantitative measure for the minimum number of single-grain orientation measurements N_C, which is required to represent adequately the texture of a given sample. Although in all experimental investigations of this problem so far the texture as derived from the single grain orientation data has been compared with x-ray macrotextures, such a criterion must as well hold for the reproduction of a microtexture of a given part of the microstructure.

As such, Pospiech et al. (1994) proposed to analyze the convergence of the statistical parameter ρ, which represents the relative mean square deviation between two textures. To determine the minimum number of single-grain orientation measurements N_C necessary to reproduce appropriately the corresponding texture, ODFs are computed for an increasing number N of single-grain orientation measurements, for example, in successive steps of 100 orientations. A parameter $\rho_{0,N}$ comparing the intensity values f_N of the ODF as computed from N single-grain orientation data with the values f_0 of a reference ODF, for example, the macrotexture, is then calculated according to (see Section 5.3.6)

$$\rho_{0,N} = \frac{\oint (f_0 - f_N)^2 \, dg}{\oint f_0^2 \, dg} \cdot 100\% \tag{10.4}$$

If the two ODFs, f_0 and f_N, are identical, ρ becomes zero. It is noted that the given definition of the parameter ρ is equivalent to the measure introduced by Van Houtte et al. (2004) to assess texture simulations, namely, the texture index $J(\Delta f)$ of the difference ODF Δf between experimental and modeled texture, divided by the texture index J_{exp} of the former (see Section 5.3.6).

Single-grain orientation measurements are used more frequently to determine local microtextures rather than bulk textures, so that no data of a reference texture f_0 may be available. In such cases, a parameter $\rho_{N',N}$ can be calculated from the ODF values f_N and f_N for two consecutive stages of the single-grain orientation measurement procedure, say for $N' = 300$ and $N = 400$. Equation 10.4 is then modified to

$$\rho_{N',N} = \frac{\oint (f_{N'} - f_N)^2 \, dg}{\oint f_N^2 \, dg} \cdot 100\% \tag{10.5}$$

To illustrate the behavior of the statistical parameters ρ with the number of single-grain orientation measurements N, Figure 10.7a displays the evolution of both $\rho_{0,N}$

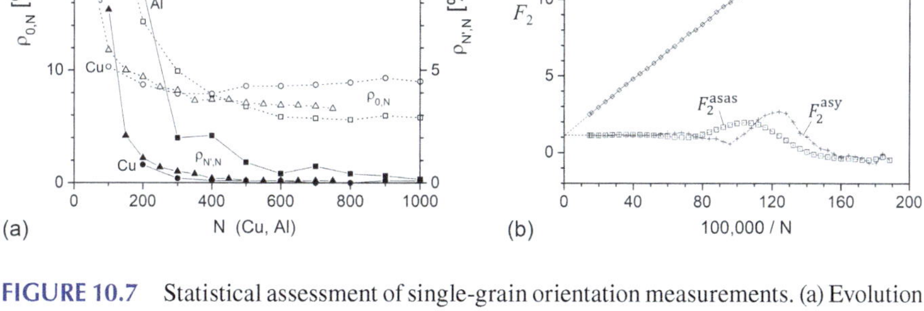

FIGURE 10.7 Statistical assessment of single-grain orientation measurements. (a) Evolution of the statistical parameters $\rho_{0,N}$ and $\rho_{N',N}$ with increasing number of single-grain orientation measurements N for different texture examples and (b) evolution of F_2 with $1/N$ for recrystallized Ni_3Al.

and $\rho_{N',N}$ for three different samples that are differentiated by their texture sharpness. (i) As an example, for a sharp texture, a heavily cold-rolled copper sheet was recrystallization annealed. The resulting texture consisted of a very sharp cube component $\{001\}\langle 100\rangle$ with an intensity as high as $f(g) \infty 120$ plus a minor recrystallization twin component. (ii) Medium texture sharpness with $f(g) \approx 15$ was obtained in a recrystallized aluminum alloy. (iii) A sample of recrystallized Ni_3Al revealed a very weak texture with a maximum intensity of only $f(g) \approx 3.5$, corresponding to a random fraction in excess of 85%.

Analysis of the evolution of the preceding statistical parameters illustrates that, for the Al sample, $\rho_{0,N}$ rapidly decreases with N to values of approximately 6%; convergence is reached at approximately 700 single-grain orientations. For the parameter $\rho_{N',N}$, a similar evolution is observed. Thus, it is concluded that an ODF computed with $N_C = 700$ orientations should be representative of the corresponding macrotexture (Figure 10.5b). The very sharp cube-texture of the Cu sample has already been derived with statistically sufficient accuracy from 100 to 200 orientations. For the weaker twin orientations larger N is required, however, which is reflected in the rather slow convergence of $\rho_{0,N}$ and $\rho_{N',N}$. In contrast, for an adequate reproduction of the weak texture of Ni_3Al, obviously much more orientations are needed ($N_C \approx 4{,}000$).

In an alternative procedure, Matthies and Wagner (1996) developed a parameter $F_2(1/N)$ that approaches linear behavior when the number of orientations N is sufficiently large. Using a discrete structure of the orientation space (e.g., M 5° × 5° × 5° cells at g_i with volume v_i), N measured single orientations can be sorted into these cells. From the resulting "hit-numbers" h_i formally a "hit-ODF" $f_N(g)$ can be constructed as a staircase function:

$$F_m(N) = \int_G \left[f_N(g)\right]^m dg = \sum_{i=1}^{M} \left(\frac{h_i}{v_i N}\right)^m v_i, \quad \text{with} \quad \sum_{i=1}^{M} v_i = 1 \quad \text{and} \quad \sum_{i=1}^{M} h_i = N \tag{10.6}$$

Laws of statistics prescribe a mean linear dependence of the texture index F_2 on $1/N$. Only if this tendency is reached for N greater than the critical number N_C, it makes sense to represent the N orientations in the form of an ODF. From the extrapolation of the asymptote of the least-square fitted $F_2(1/N)$-line to $1/N = 0$ (i.e., $N = \infty$), the unknown "true" F_2 of the sample can be determined, yielding the value $F_2^{asy}(N)$. Similarly as discussed earlier for the parameter ρ, N_C depends on the chosen grid, crystal and sample symmetry, and, most notably, the texture of the sample. Sharp textures with a small isotropic background are equivalent to a lower number of cells that can be hit at all. This means the linear dependency of F_2 on $1/N$ will be reached for lower N than in the case of weak textures. Figure 10.7b shows the $F_2(1/N)$ curve as well as the F_2^{asy} values as derived from the fitted asymptotic $F_2(1/N)$-line for the Ni_3Al specimen described earlier in this section. The fit always considers the last

10 $F_2(1/N)$ points to the right. A second smoothing over the last 10 $F_2^{asy}(1/N)$ values yields $F_2^{asas}(1/N)$. The plot displays a clean $1/N$ tendency with a stable F_2^{asy} of approximately 1.3. Here, N = 100,000/60 ≈ 1,700 grain orientations would suffice for a reasonable ODF.

To summarize the discussion on how many orientation measurements are needed to describe an ODF with statistical relevance, the convergence of the statistical parameters $\rho_{0,N}$ and $\rho_{N',N}$ with increasing number of single-grain orientation measurements N and the linear behavior of the texture index F_2 as a function of the inverse $1/N$ can be analyzed. Among the three quantities described, the parameter $\rho_{0,N}$ comparing the ODF as computed from N single orientations with a reference texture—typically the x-ray macrotexture of the sample—turned out to be most rigid in the sense that it yielded the largest values of N_C. Qualitatively, however, the evolution of $\rho_{N',N}$ and the F_2 versus $1/N$ analysis give very similar results, so that analysis of either parameter yields valuable information on the statistical relevance of single grain orientation data. In conclusion, it is strongly suggested to implement one of these procedures as a mandatory step to assess the statistical significance of microtexture data.

The statistical relevance of microtexture results is also affected by the measurement procedure, which becomes particularly important in the context of automated EBSD measurements. In the preceding discussion, the number of single-grain orientation measurements N is to be understood as the number of measured *grains*, but automated EBSD systems usually record many orientations per grains so as to allow reconstruction of the corresponding microstructure with sufficient accuracy. To illustrate this effect, a recrystallized Al-Mg sample was scanned with a fine step size of 2 μm. Figure 10.8 shows the evolution of the parameter ρ with increasing

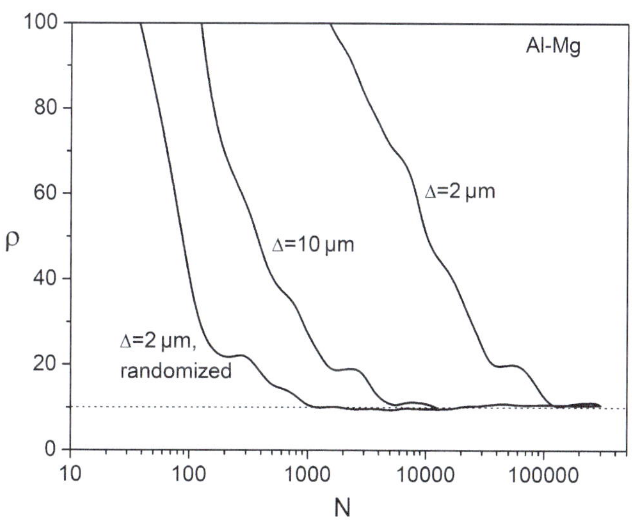

FIGURE 10.8 Evolution of $\rho_{0,}N$ with an increasing number of single-grain orientation measurements N for different scanning procedures in recrystallized Al-Mg.

number of EBSD data points N. Obviously, ρ decreases rather slowly, and it is only at N in excess of 150,000 that the ρ value levels at approximately 10%. Apparently, this value of N_C is larger by several orders of magnitude than the values of the order of 1,000 grains necessary to provide a reasonable macrotexture description reported earlier. The average grain size of the material was approximately 25 μm. Thus, a highly simplified estimate yields that each grain is on average measured 100 times, producing a very large number of redundant data. Despite the quite large number of measurement points (in excess of 300,000), the number of actual *grains* was below 3,000. Obviously, although a 2-μm scan produces a very detailed orientation map, the statistical relevance of the data is not that good, which is reflected in the rather slow convergence of the parameter ρ (Figure 10.8).

To substantiate this point, the sample was scanned again on a 5× coarser grid, that is, with a step size of 10 μm. Figure 10.8 displays that for the 10-μm scan the parameter ρ drops much faster, reaching values of approximately 10% already after 5,000 measurement points. The ratio of 5,000 to 150,000 agrees well with the factor 25 (i.e., 5^2) by which the second map was accelerated. This procedure reduces the number of redundant data, which leads to an increased statistical relevance of the data set. Of course, this procedure is *not* suited for generation of high-quality orientation maps or computation of misorientation parameters, where spatial resolution is important, but it gives a good *statistical* representation of the orientation distribution, that is, the macrotexture (Engler et al., 1999; Schwarzer, 2000; Wright et al., 2007).

A similar effect of breaking up the effect of redundant neighboring data can be achieved by randomizing the arrangement of the EBSD data. For that purpose, the fine EBSD data set (2-μm step width) was reshuffled with a random generator, and the evolution of ρ with N was again computed for these randomized data. Here, ρ converges much faster, and only ~1,000 orientations are required to get a sound description of the corresponding macrotexture (Figure 10.8). This procedure is obviously applicable to all data files and will work as long as enough *grains* (i.e., not measured points) have been recorded in the EBSD scan.

10.4 REPRESENTATION OF ORIENTATIONS IN RODRIGUES SPACE

Use of Rodrigues space, otherwise known as *Rodrigues-Frank space*, wherein a population of R vectors resides (Section 2.8), is a relatively new field compared with the long-established Euler space approach for representation of orientations. The explanation for the late introduction of Rodrigues space is that its application is intimately connected with microtexture, that is, discrete orientation measurements, whereas representation of texture in Euler space relates traditionally to the representation of continuous orientation distributions as calculated from pole figures (Section 5.4). As already stated, microtexture data can be represented in either Euler space or Rodrigues space, whereas macrotexture data tend to be represented only in Euler space because most commercial systems are configured for this approach.

Although Rodrigues space, which is particularly well suited to analysis of misorientations (Section 10.6.3), has not been very widely used to date, it has many advantages over Euler space (Section 2.8.3) and so may well be adopted more in the future (Neumann, 1991; Hughes et al., 2000; Cho, 2007).

Orientations are usually best displayed in the whole of the fundamental zone of Rodrigues space (Section 2.8.2). Since for orientation representation the axes of the zone are coupled to the axes of the specimen, the zone could be divided according to the symmetry of the specimen itself. For example, a rolled specimen has three axes of symmetry, and so in an appropriate reference frame the rolling direction (RD), transverse direction (TD), and direction normal to the rolling plane (ND) should be parallel to the X, Y, and Z axes of the fundamental zone, respectively. For this case, one-eighth of the fundamental zone could be used. The advantage of using this reduced volume is that if the sample population is small, texture trends are more discernable in the smaller volume. However, valuable information can be lost in that the assumed symmetry of the material may not be borne out in practice. For example, a rolling process could bias the textures on one side of the strip or plate more than on the other and this would be evident if the whole fundamental zone were used to represent the data (Section 5.4.2).

Another, and often more important, consideration to be taken into account if orthonormal sample symmetry is enforced by collapsing the fundamental zone into a reduced volume is that related texture variants having the same Miller indices families are not distinguished (Becker, 1991). There are, in the most general case, four different orientations having the same Miller indices families (Section 2.4). As an example, the variants of the *S* orientation are $(123)[63\bar{4}]$, $(\bar{1}\bar{2}3)[634]$, $(1\bar{2}3)[6\bar{3}\bar{4}]$, and $(\bar{1}23)[6\bar{3}4]$ (Table 2.2 and Appendix III), quoted in their lowest-angle forms. These are distinct orientations even though they feature common families of planes that are only symmetrical with regard to orthonormal sample symmetry. The whole of the zone is required to display correctly these distinct variants whose *R* vectors are

$$(0.254,-0.235,-0.291)$$

$$(-0.254,0.235,-0.291)$$

$$(-0.254,-0.235,0.291)$$

$$(0.254,0.235,0.291)$$

Figure 10.9 shows textures from an aluminum–lithium alloy where two different near-*S* orientations were observed in two different parts of the sheet (Randle and Day, 1993). There are 50 grains in each sample population. In the first set (Figure 10.9a), the microtexture comprises 68% near-*S*1 orientations, and in the second set (Figure 10.9b) there is 42% near-*S*2 orientations. Both have {001}⟨100⟩ and {013}⟨100⟩ as secondary components. We can see that the clusters that represent these two textures are occurring in different regions in Rodrigues space. This information would have been lost if the data had been compressed into a single octant.

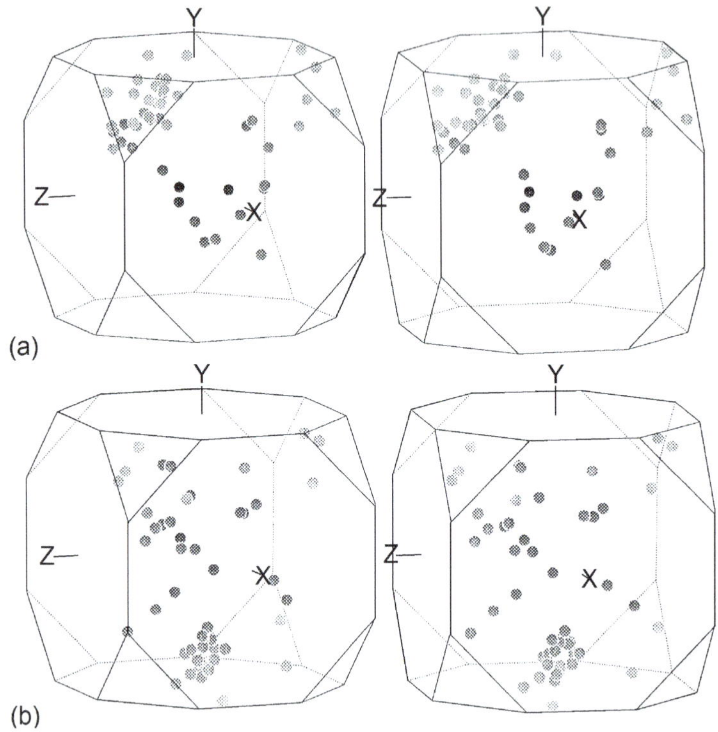

FIGURE 10.9 Textures from an aluminum-lithium alloy displayed in Rodrigues space as stereo pairs. Data sets are of 50 grains having (a) predominantly S1 orientation and (b) predominantly S2 orientation. (Adapted from Randle and Day, 1993.)

Distributions of orientations in Rodrigues space are usually generated by a computer program such that the zone can be rotated on-screen to gain maximum visual appraisal of orientation clusters. However, like any other 3D space, Rodrigues space is problematic to display in the 2D medium of a printed page. Several options have been tried. The first method to be used was to view the whole fundamental zone as a stereo pair (Frank, 1988; Ashbee and Sargent, 1990; Randle and Day, 1993). This is not a very convenient method because most people need a stereo viewer to appreciate the 3D depth perspective (Figure 10.9). Viewing the whole zone in perspective is a feasible option for gaining an appraisal of the data spread especially if either different colors/sizes/shading are attributed to the data points to give pseudo-depth information (Neumann, 1991) or contouring is used (Hughes et al., 2000).

A useful perspective of Rodrigues space is to view all the data by looking directly down one of the reference axes. The Z-axis perspective is then equivalent to a standard {100} pole figure. Figure 10.10a and b shows the Z-axis perspective partners to the views in Figure 10.9a and b. It can be informative to study more than one perspective view. For example, two fiber textures, (103)[*uvw*] and (113)[*uvw*], are

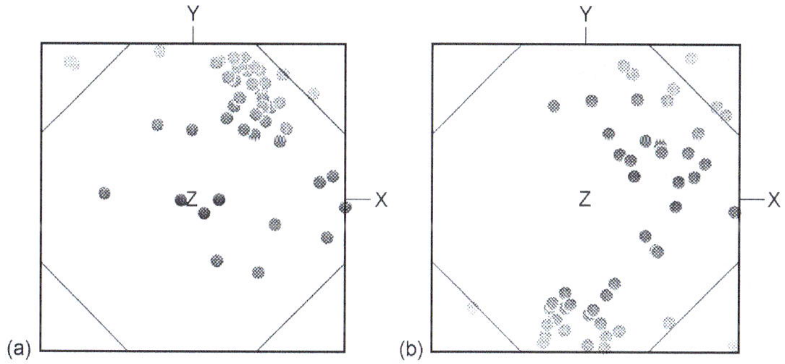

FIGURE 10.10 (a) and (b) two Rodrigues space Z-axis perspective views of the microtextures in Figure 10.9a and b, respectively. (Adapted from Randle and Day, 1993.)

shown by three different perspectives in Rodrigues space in Figure 10.11, and it can be seen that the X-axis rather than the Z-axis perspective is better for appreciating the pattern of how the textures lie on straight lines. The R vectors for these, and other, fiber textures are listed in Table 10.1. Note that one of the orientations, (103)[010], is located outside the fundamental zone to preserve the feature that fiber textures map onto a straight line in Rodrigues space. This point could alternatively be re-expressed in its lowest-angle form (Section 2.3.3) so that it would then plot within the zone.

Rodrigues space can be represented by sections through it, as for Euler space, and the sections reconstructed into a 3D format by eye. Figure 10.12 shows an example of such a representation for cubic symmetry (Becker and Panchanadeeswaran, 1989). The dimensions of Rodrigues space can be represented in absolute R vectors or normalized so that the length of the major axis is unity, and the R vectors scaled accordingly. As for any other representation of microtexture, contouring of the data may be substituted for discrete data points.

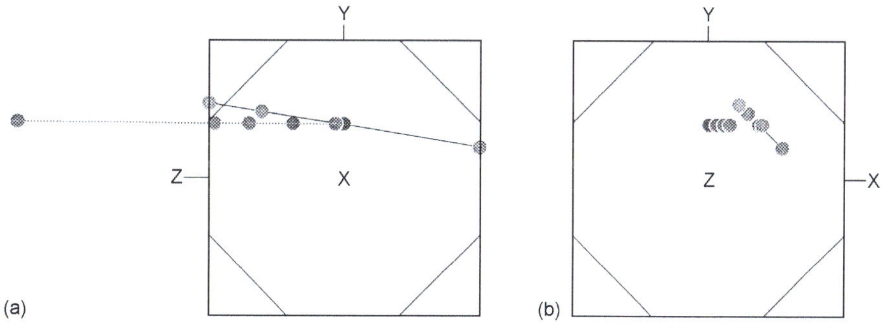

FIGURE 10.11 Fiber textures in Rodrigues space for {103}⟨*hkl*⟩ (dotted line) and {113}⟨*hkl*⟩ (dashed line). Both (a) the X-axis perspective and (b) the Z-axis perspective are shown. (Adapted from Randle and Day, 1993.)

TABLE 10.1
Components of the Rodrigues Vector for Orientations *uvw* along Various Fiber Axes

uvw	R_1	R_2	R_3
001 *Fiber axis*			
100	0	0	0
$3\bar{1}0$	0	0	0.162
$2\bar{1}0$	0	0	0.236
$3\bar{2}0$	0	0	0.303
$1\bar{1}0$	0	0	0.414
$\bar{1}01$ *Fiber axis*			
101	0	0.414	0
212	0.071	0.414	–0.172
111	0.132	0.414	–0.318
121	0.214	0.414	–0.518
131	0.263	0.414	–0.634
$\bar{1}11$ *Fiber axis*			
$1\bar{1}2$	0.214	0.518	0.414
101	0.318	0.414	0.132
312	0.379	0.353	–0.036
211	0.414	0.318	–0.132
110	0.518	0.214	–0.414
103 *Fiber axis*			
010	0.162	0.162	1
$33\bar{1}$	0.065	0.162	0.399
$32\bar{1}$	0.047	0.162	0.290
$31\bar{1}$	0.025	0.162	0.154
301	0	0.162	0
$\bar{1}13$ *Fiber axis*			
$3\bar{3}2$	0.093	0.224	0.414
$2\bar{1}1$	0.119	0.198	0.251
301	0.154	0.162	0.025
110	0.224	0.093	–0.414

Source: Data from Randle and Day (1993).

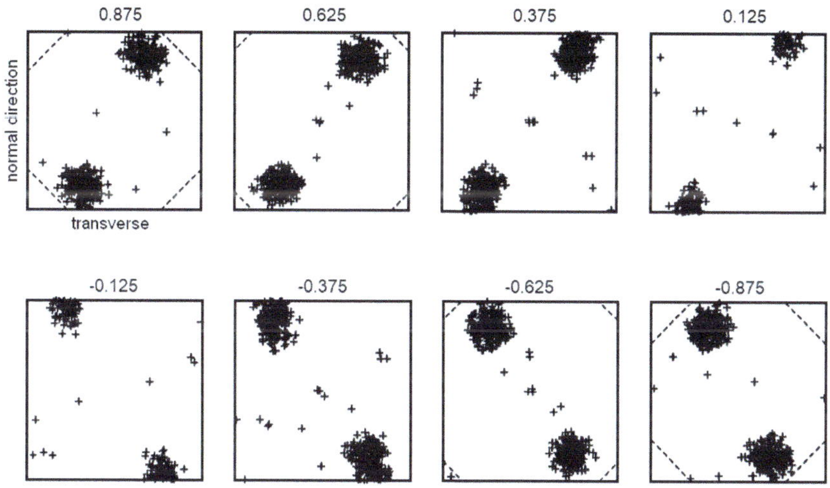

FIGURE 10.12 Example of orientation distributions in Rodrigues space displayed as sections in R_3: copper rolling texture and cubic symmetry. (Data from Becker and Panchanadeeswaran, 1989.)

10.5 GENERAL REPRESENTATION OF MISORIENTATION DATA

The formulation of a misorientation from pairs of orientations is described in Section 2.7. The following categories of misorientation populations can be distinguished:

- Misorientations between nearest-neighbor grains, that is, grain boundary misorientation (also sometimes called *mesotexture*). This is by far the most common case.
- Misorientations between a single, specific grain and other grains, for example, a large grain and all small grains (Shimizu et al., 1990).
- Intragrain misorientations or misorientations sampled from several sites within neighboring grains (Section 10.5.2). It has been shown that even in recrystallized materials small orientation changes, that is, lattice rotation or bending, exist within a grain.
- MODFs and their derivatives (Section 10.7).
- Orientation relationships between different phases (Section 12.2).

Analysis of misorientations at interfaces also gives rise to a further level of processing and crystallographic analysis, for instance, categorizing the statistics according to the coincidence site lattice (CSL) model. These methodologies are discussed in detail in Chapter 11.

10.5.1 Representations Based on the Angle/Axis Descriptor

The angle/axis descriptor is the most popular method for representation of a misorientation. As described in Section 2.3.3, there are a number of crystallographically related, mathematically equivalent solutions for a misorientation—for example, 24 and 12 for cubic and hexagonal crystal symmetry, respectively. To avoid confusion, it has become common practice to select, from all the solutions, the angle/axis pair that has the minimum misorientation angle to represent the misorientation. This solution has been named the *disorientation* to conveniently distinguish it from the remaining ones (Warrington and Boon, 1975). The angle of misorientation and axis of misorientation provide different physical information: a small-angle or large-angle grain boundary is recognized from the angle alone, and certain misorientation axes are considered to be important in processes such as recrystallization. The angle and axis of misorientation can be represented separately, both for convenience on the printed page and to focus on either the angle or axis alone. Alternatively, both parameters can be displayed in cylindrical angle/axis space (Section 10.6.2). A combination of angle and axis is used for representation in Rodrigues space (Section 10.6.3).

Considering first the representation of disorientation axes, these are conveniently represented on the stereographic projection or, more compactly, on a single unit triangle of the stereographic projection since the density of axes becomes multiplied by the number of unit triangles, as for an inverse pole figure. The probability density distribution of disorientation axes [*hkl*] in a unit triangle has been calculated for cubic crystals (Mackenzie, 1958, 1964) and is shown in Figure 10.13a. The plot consists of density contours centered on both [100] and [111]. The density rises as the arc AB of the great circle $2^{1/2} \cdot h = k + l$, where $h \geq k \geq l$, is approached (dashed curve). A more useful form of the probability distribution is shown in Figure 10.13b, which shows the percentages of disorientation axes that are expected to lie in each marked region if the distribution is random. This has come to be known as the "Mackenzie axis distribution" or "Mackenzie triangle" and is used to display disorientation axes and assess the degree of departure from randomness of a sample population.

Frequently disorientation axis distributions are shown in angle ranges. An example is shown in Figure 10.14. Here, the disorientation axis distribution for copper, which has been processed to give a high proportion of CSL boundaries (Section 11.2.3), is shown. There are peaks in the sections, which correspond to 38.9° ⟨110⟩ (Σ9)

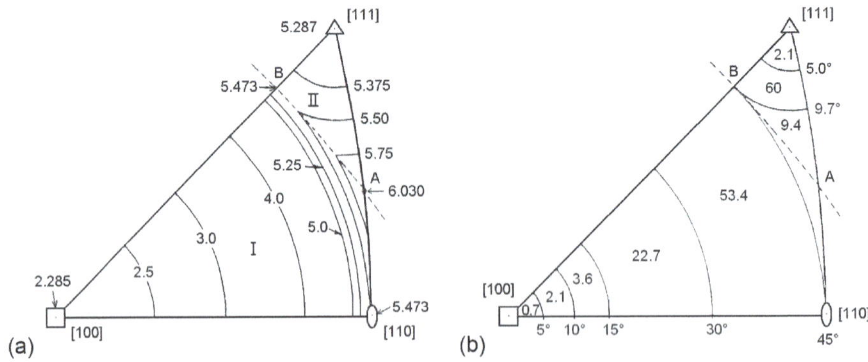

FIGURE 10.13 Distribution of disorientation axes [*hkl*] for cubic crystals in a single unit triangle of the stereo-graphic projection. (a) Probability density plot; (b) percentage of disorientation axes lying in the various regions. (Data from Mackenzie, 1964.)

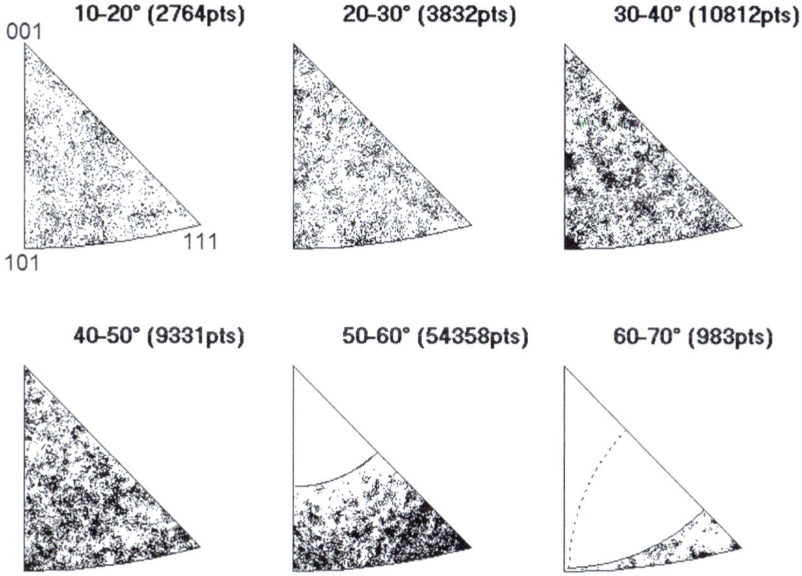

FIGURE 10.14 Distribution of disorientation axes for copper in 10° angle sections.

misorientation and 60° ⟨111⟩ (Σ3) misorientation, as well as a large peak of small-angle boundaries. Another option, although not the one that is used frequently, is to display misorientation axes relative to some external axes, rather than those of the crystal, that is, a major stress axis or a dominant axis associated with the specimen processing, for instance, a rolling direction. Such a distribution is analogous to choosing a pole figure representation rather than an inverse pole figure for orientations (Heidelbach et al., 1996).

Figure 10.15 shows the disorientation angle distribution for copper, which partners the disorientation axis distribution in Figure 10.14. The peaks for small-angle boundaries, Σ3 and Σ9, are evident. The probability distribution for disorientation angles in cubic polycrystals has been derived (Warrington and Boon, 1975), and it has a maximum at 45° and a cutoff at 62.8°. This distribution, often called simply the "Mackenzie distribution," is also shown in Figure 10.15. Similarly to the case for the axis distribution, the Mackenzie angle distribution is used to assess the departure from randomness in a sample population of disorientation angles (Uhbi and Bowen, 1996). The probability distribution for disorientation angles for all point-symmetry groups has also been derived (Morawiec, 1995), and it can be used to assess the departure from randomness, similarly to the case for cubic crystals. For example, Figure 10.16 shows the random distribution for hexagonal crystals, which accompanies an experimental distribution of misorientation angles from commercially pure titanium (Hu and Randle, 2007). The experimental distribution is plotted as "correlated," that is, misorientations as they occur in the microstructure, and "uncorrelated," that is, random pairs selected from the experimental sample population. Section 10.7 addresses normalization of misorientations and correlation functions.

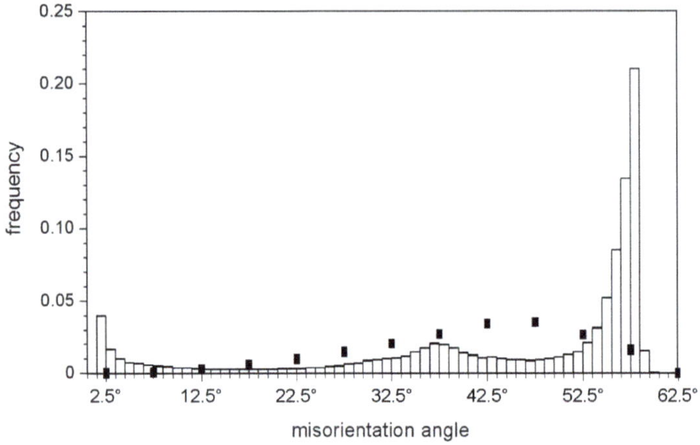

FIGURE 10.15 Distribution of disorientation angles for copper; same data set as in Figure 10.14. The distribution for randomly misoriented crystal pairs that have cubic symmetry (the "Mackenzie distribution") is also included.

It is important to bear in mind that, although it may sometimes be sound to separate the angle and axis of misorientation to concentrate on particular physical aspects of the data, the *total* misorientation embodies three independent variables (Section 2.7). These can be expressed by the same parameters as orientations, and the angle/axis pair is only one descriptor (Chapter 2). Hence, when comparing *amounts* of misorientation it can be misleading to compare angles of misorientation alone, unless the misorientation axis is constant. Furthermore, it may be relevant

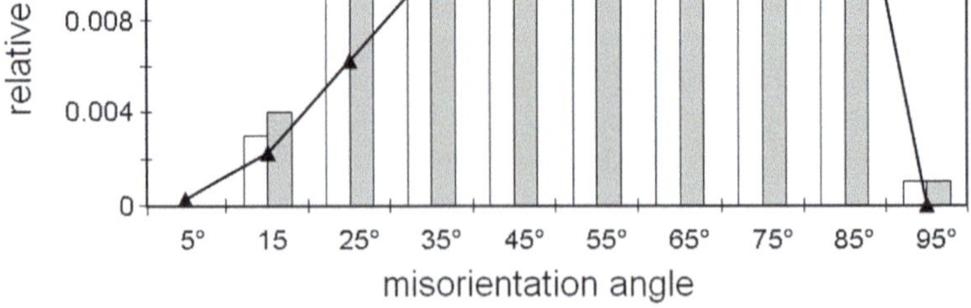

FIGURE 10.16 Disorientation angle distribution histogram for correlated and uncorrelated misorientation pairs in commercially pure titanium. The disorientation angle distribution for randomly misoriented crystal pairs is also included (with triangles).

to assess if the misorientation is in a positive or negative sense, for example, in the evaluation of cumulative misorientations (Driver et al., 1996).

As stated at the beginning of this section, it is a common practice to select the disorientation from all the symmetry-related solutions to describe a misorientation. Although this convention simplifies the data processing, there is a risk that insights into the physics of misorientation-related phenomena may be overlooked if only one angle/axis pair is examined. It is probably true to say that if the misorientation angle alone is of interest in an investigation, it will almost always be the lowest-angle solution that is most meaningful. However, if an investigation concerns the misorientation axis, it may not be that of the disorientation which is of interest (Hutchinson et al., 1996). For example, in bcc metals, it is reported that high-mobility grain boundaries are misoriented on (110). For such investigations, it is clearly better to choose, from the 24 crystallographically related solutions, the one having a misorientation axis closest to (110). The distribution of 10,000 random computer-generated misorientations is displayed as those misorientation angles having axes closest to (110), (111), and (100) in Figure 10.17a, b, and c, respectively, which can be compared with the distribution for the disorientation angles for a random distribution shown in Figure 10.17d. Plots for misorientations within 5° and 10° of the given axis are included. It is clear that even for random misorientations there are peaks in the distributions of misorientation axes, which must be taken into account during data interpretation (Cross and Randle, 2003).

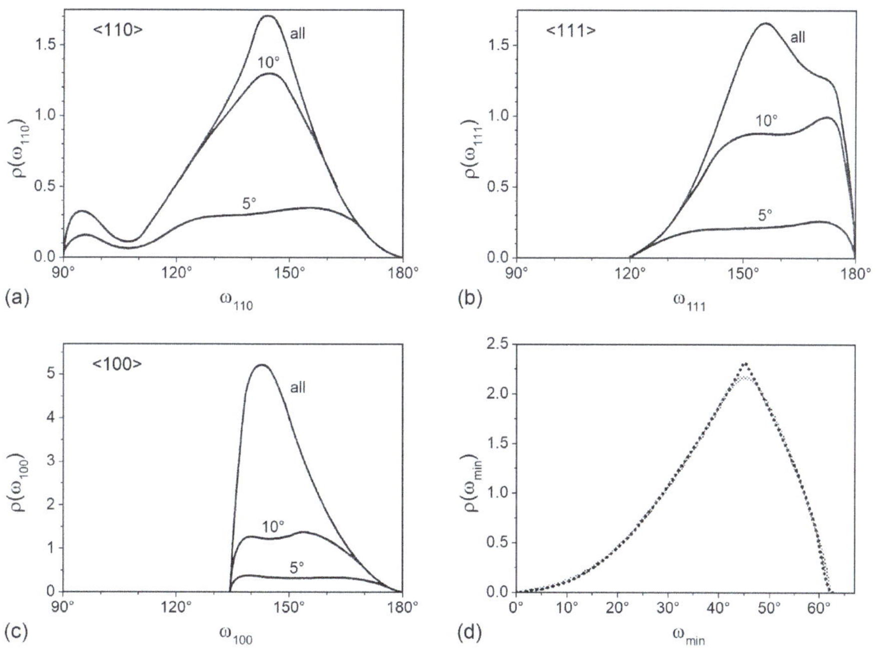

FIGURE 10.17 Distributions of computer-generated misorientation angles for a random distribution plotted for misorientation axes closest to (a) (110), (b) (111), (c) (100), and (d) disorientation axes. (Courtesy of W.B. Hutchinson.)

A second reason for examining misorientations from the standpoint of particular misorientation axes is that where three (or more) grain boundaries adjoin at a grain junction they share a common axis of misorientation (Section 11.5). For example (Randle, 1990, 1993),

$$180°/\langle 210\rangle \leftrightarrow 106.6°/\langle 210\rangle + 73.4°/\langle 210\rangle \qquad (10.7a)$$

refers to the relationship of three grain boundaries at a junction misoriented on (210), according to the "addition rule" (Section 11.5). None of these are in the lowest-angle form and this relationship could not have been recognized from the disorientation since for that case the axes are all different:

$$53.1°/\langle 001\rangle \leftrightarrow 43.3°/\langle 133\rangle + 38.3°/\langle \bar{1}\bar{1}1\rangle \qquad (10.7b)$$

10.5.2 Intragrain Misorientations

Measurements of intragrain misorientations in metals and alloys have revealed the presence of local variations in orientation (Liu and Chakrabarti, 1996). Although this is well known from TEM observations of deformed materials (Hughes et al., 2000), EBSD can be used to study subgrain misorientation and show the presence of local "lattice bending," or intragrain misorientation, even in recrystallized materials (Adams et al., 1993; Thomson and Randle, 1997a). Orientation mapping to reveal intra-grain misorientations in deformed material is discussed in Section 9.4.5. As far as data processing of intragrain misorientations is concerned, it is important to choose a relevant sampling step size to reveal orientation changes. A suitable step size will depend on the scale of the microstructure: typically, a recrystallized microstructure might require a step size of 1 μm, whereas a deformed material might be sampled every 0.01–0.1 μm using EBSD or TEM.

A useful combination of intragrain spatial and numerical information is to show a linear scan of orientation changes across one or more grains. The simplest approach is to use only the misorientation angle. This technique has been used to show, for example, that there is a direct contribution of texture evolution to misorientation accumulation with strain in bcc steel (Kang et al., 2007). Another example is that the technique has been used to show the misorientation profile across banded areas in deformed niobium single crystals (Sandim et al., 2007). This display method has the advantage of both focusing attention on the area of interest and providing concurrently quantification of the orientation change. A disadvantage of the approach is that a compromise must be made to represent compactly the three independent variables that describe the misorientation, in addition to the positional information.

When misorientations between neighboring sampling points are calculated, different values may be obtained depending on the spacing of the sampling points. Although this approach of calculating misorientations between nearest-neighbor points may be valid for some cases, for example, to monitor cumulative misorientations (Hjelen et al., 1991), the ambiguous effect of the sampling step size can be circumvented by plotting an orientation, rather than a misorientation, parameter. Such a parameter can be calculated for each sampling point with reference to a starting point or an average orientation. Alternatively, a compressed orientation parameter can be calculated using either Euler angles or Rodrigues vectors (Randle et al., 1996). The compressed orientation parameter D in Rodrigues space is given by

$$D = 2\tan^{-1}\left[\left(R_1 - R_{1R}\right)^2 + \left(R_2 - R_{2R}\right)^2 + \left(R_3 - R_{3R}\right)^2\right]^{1/2} \tag{10.8}$$

where R_1, R_2, R_3 and R_{1R}, R_{2R}, R_{3R} are the components of the R vector at the measuring point and the reference point (e.g., the grain center), respectively. An inevitable disadvantage of the compressed orientation parameter D is that occasionally different orientations can have the same D value.

10.6 REPRESENTATION OF MISORIENTATIONS IN THREE-DIMENSIONAL SPACES

Misorientations can be displayed in Euler-space (Section 10.6.1) or spaces based on the angle/axis pair, namely, cylindrical angle/axis space (Section 10.6.2) and Rodrigues space (Section 10.6.3). Comparison of the three spaces reveals that representation in Rodrigues space is generally the most advantageous.

10.6.1 Representation of Misorientations in Euler Space

For representation of misorientations in the Euler angle space, from the misorientation matrices M, the Euler angles φ_1, Φ, φ_2 are calculated according to standard procedures (Section 2.7). As has been described in more detail in Section 10.3, for the representation of orientations, the resulting set of Euler angles can be transposed in the Euler space either by direct cell methods or by the series expansion method. As an example, Figure 10.18 shows a misorientation distribution that was derived

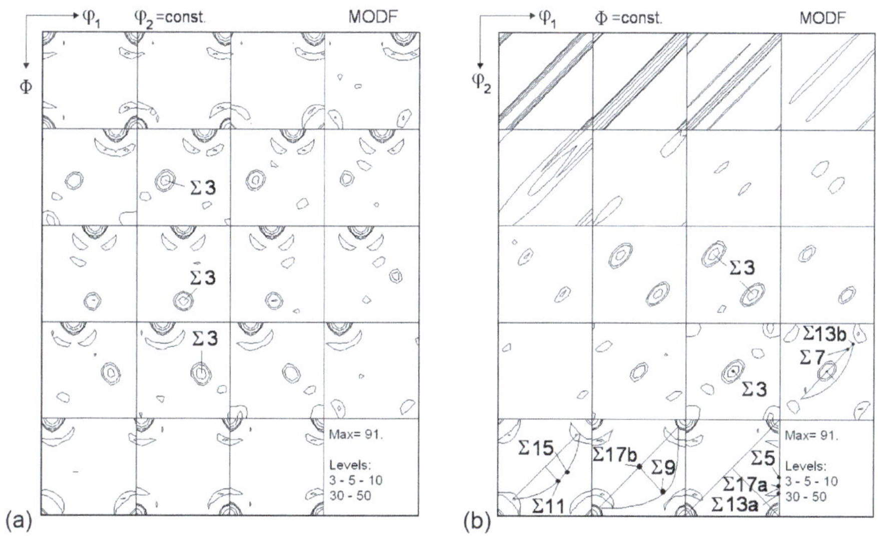

FIGURE 10.18 Representation of the MODF of a cyclically deformed nickel specimen in Euler space. (a) φ_2 = constant; (b) Φ = constant.

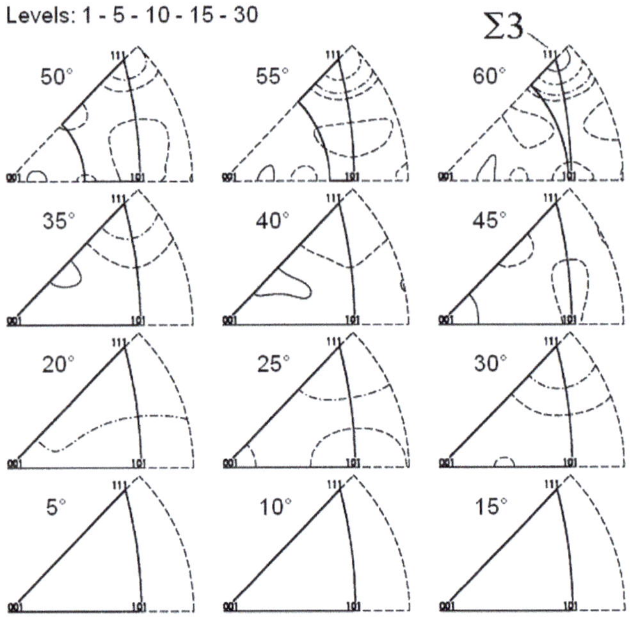

FIGURE 10.19 MODF of a cyclically deformed nickel specimen in cylindrical angle/axis space (see Figure 10.18).

from EBSD single-grain orientation measurements in cyclically deformed nickel. The sample was deformed at 600°C in 300 cycles with a frequency of 13×10^{-3} s^{-1} under a strain amplitude of 0.5%. Under such circumstances the grain boundaries tend to rearrange, and particularly Σ3-twin grain boundaries were observed to remain stable in the microstructure (Brodesser and Gottstein, 1993). Besides the Σ3-peak in the distribution, strong intensities close to the origin of the Euler space, that is, at $\varphi_1 = \Phi = \varphi_2 = 0°$, prevail, representing the small-angle grain boundaries (Figure 10.18). (The same data will also be shown in other spaces in Figures 10.19 and 10.21.)

The main advantages of the representation of misorientation distributions in Euler space are that the computation is relatively easy and the MODF and its derivatives can be obtained, if required (Section 10.7). However, the technique suffers from several disadvantages:

- The Euler angle representation does not yield direct information on the crystallographic orientation relation between the two considered orientations, which means it is difficult to visualize. Thus, evaluation of MODFs in the Euler space needs detailed legends or marking of special misorientation as done in Figure 10.18.
- The orientation space exhibits strong distortions for small angles Φ (Section 5.4.1), as it becomes most evident from representing the MODF in the—highly unusual—Φ sections (Figure 10.18b). To overcome this

shortcoming, Zhao and Adams (1988) proposed to use a subspace with very high Φ angles ($\Phi \geq 70°$), where the distortions are minimal. The borders of this subspace are marked in Figure 10.18b. As the two orientations that define the mutual misorientation can be treated equivalently (Section 2.7.2), this subspace can be further subdivided into two equivalent asymmetric domains, as also indicated in Figure 10.18b. Furthermore, the positions of the special grain boundary orientation relationships with $\Sigma \leq 17$ are marked in Figure 10.18b. It can be seen that each Σ position can be found exactly once in this subspace.

- The surfaces between the various basic domains of the cubic/cubic symmetry are strongly curved (which is induced by the threefold axis), causing difficulties in identification. The $\Sigma3$-twin relation that is located on the interface between adjacent subspaces is visible three times in the standard Euler angle space. Note that all orientation relationships with low Σ lie at the border surfaces of the subspaces rather than in their interior (Figure 10.18b). This is due to the rotation about symmetrical $\langle hk0 \rangle$ or $\langle hll \rangle$ axes for such orientation relationships.

For all these reasons the representation of misorientations in Euler space is very difficult to visualize, which may cause uncertainties in evaluation of misorientations in the Euler space. Rather, misorientations are better represented in angle/axis spaces.

10.6.2 Representation of Misorientations in the Cylindrical Angle/Axis Space

Misorientations can conveniently be represented in an orientation space given by the parameters of the angle/axis pair description. For that purpose, the rotation axis r is expressed in terms of two angles ϑ and ψ (Equation 2.24), such that the misorientation is now given by the three angles ϑ, ψ, and θ. All possible orientations and misorientations are given in the interval ($0° \leq \theta \leq 360°$, $0° \leq \vartheta \leq 90°$, $0° \leq \psi \leq 360°$) and, as shown in Figure 2.17, can be represented in a cylindrically shaped 3D space with the cylinder axis parallel to the θ axis. In this arrangement, the base plane of the cylinder corresponds to a stereographic projection of the rotation axis r, which means that the rotation axes can readily be evaluated.

Figure 10.19 shows the misorientation distribution of a cyclically deformed nickel sample (also shown in Euler angle space, Figure 10.18) in sections parallel to θ ($\Delta\theta = 5°$) through a subsection of the cylindrical orientation space. The basic domain, that is, the smallest subspace in which each possible misorientation would exist exactly once, is marked with thick lines (for cubic/cubic symmetry). The main advantage of this space is given by the fact that for each θ section, the rotation axis r can readily be derived from its position in the stereographic triangle. This is a very familiar way to evaluate misorientation data. Another example of an MODF represented in the cylindrical angle/axis space is given in Section 12.2.

A major disadvantage of the cylindrical angle/axis space is caused by the relatively large distortion of this space. For decreasing θ angles, the resolution for the position of the rotation axis becomes increasingly smaller and is indefinite for $\theta = 0°$ (Pospiech et al., 1986). Since the Rodrigues space described in Section 10.6.3 offers several advantages, the cylindrical angle/axis space will not be discussed here any further.

10.6.3 Representation of Misorientations in Rodrigues Space

When analyzing misorientations, the coordinate frame of Rodrigues space is parallel to the crystal axes. If all the misorientation axes are expressed in the same form—that is, $h \geq k \geq l$—and all are positive, the whole of the fundamental zone does not need to be used to display misorientations. For the cubic system, for instance, the fundamental zone can be divided up along planes of symmetry into 48 identical and equivalent subvolumes (Randle, 1990; Field, 1995). This process is identical to the division of the reference sphere and hence the stereogram into unit triangles (Appendix IV) or division of Euler space into asymmetric domains (Section 10.6.1). The subvolume of the fundamental zone is also sometimes called the *Mackenzie cell* after the Mackenzie triangle that shows the distribution of random misorientations for the cubic system (Section 10.5.1). Table 2.4 includes the size of the subvolume as a fraction of the fundamental zone for each crystal system.

Figure 2.18 shows how the fundamental zone for cubic lattices is divided along the surfaces equivalent to planes perpendicular to the symmetry axes, that is, ⟨100⟩, ⟨111⟩, and ⟨110⟩. The ⟨100⟩, ⟨111⟩, and ⟨110⟩ axes are shown radiating from the origin along edges of the subvolume. All misorientations can be constrained to lie either within the subvolume or on a surface, edge, or vertex. Misorientations that lie on the boundary of the subvolume are "geometrically special," by having a rational, low-index axis of misorientation.

Three types of boundary can be recognized readily in the subvolume of Rodrigues space—small-angle boundaries, boundaries with preferred misorientation axes, and CSL-type boundaries (Section 11.2). Small-angle boundaries are recognized by their proximity to the origin of Rodrigues space, whereas preferred misorientation axes lie along straight lines in the subvolume, analogous to the case of orientation fiber textures. Often, preferred misorientation axes are low-index types, ⟨100⟩, ⟨111⟩, or ⟨110⟩, and so coincide with the edges of the space. It should be noted that distributions of misorientations in Rodrigues space take account only of the lowest-angle solution, and so alternative descriptions, which may show up common misorientation axes, are not recognized. For example, in CSL notation Σ3, Σ5, and Σ15 are all misoriented on a ⟨210⟩ axis (Equation 10.7), yet only Σ15 is in its lowest-angle form.

Rodrigues space is especially apposite for the representation of low-Σ CSL misorientations in cubic systems since the Rodrigues vectors are rational fractions and therefore easy to recognize when plotted (Randle, 1990). The representation of CSLs in Rodrigues space for crystal systems other than cubic has been addressed little as yet and so will not be discussed here. R vectors for CSLs up to $\Sigma = 45$ are listed in Table 10.2 and shown as a Z-axis perspective in the subvolume of the fundamental

TABLE 10.2
Rodrigues Vectors for CSLs up to Σ = 45

Axis/Σ	R_1	R_2	R_3	Axis/Σ	R_1	R_2	R_3
⟨100⟩				⟨311⟩			
5	$^1/_3$	0	0	23	$^3/_9$	$^1/_9$	$^1/_9$
13a	$^1/_5$	0	0	33b	$^3/_{11}$	$^1/_{11}$	$^1/_{11}$
17a	$^1/_4$	0	0				
25a	$^1/_7$	0	0	⟨111⟩			
29a	$^2/_5$	0	0	3	$^1/_3$	$^1/_3$	$^1/_3$
37a	$^1/_6$	0	0	7	$^1/_5$	$^1/_5$	$^1/_5$
41a	$^1/_9$	0	0	13b	$^1/_7$	$^1/_7$	$^1/_7$
				19b	$^1/_4$	$^1/_4$	$^1/_4$
⟨210⟩				21a	$^1/_9$	$^1/_9$	$^1/_9$
15	$^2/_5$	$^1/_5$	0	31a	$^1/_{11}$	$^1/_{11}$	$^1/_{11}$
27b	$^2/_7$	$^1/_7$	0	37c	$^3/_{11}$	$^3/_{11}$	$^3/_{11}$
41b	$^2/_6$	$^1/_6$	0	39a	$^1/_6$	$^1/_6$	$^1/_6$
43b	$^2/_9$	$^1/_9$	0	43a	$^1/_{13}$	$^1/_{13}$	$^1/_{13}$
⟨310⟩				⟨221⟩			
37b	$^3/_8$	$^1/_8$	0	17b	$^2/_5$	$^2/_5$	$^1/_5$
				29b	$^2/_7$	$^2/_7$	$^1/_7$
⟨321⟩				45b	$^2/_9$	$^2/_9$	$^1/_9$
39b	$^3/_8$	$^2/_8$	$^1/_8$	45c	$^2/_6$	$^2/_6$	$^1/_6$
⟨110⟩				⟨331⟩			
9	$^1/_4$	$^1/_4$	0	25b	$^3/_9$	$^3/_9$	$^1/_9$
11	$^1/_3$	$^1/_3$	0	35b	$^3/_{11}$	$^3/_{11}$	$^1/_{11}$
19a	$^1/_6$	$^1/_6$	0				
27a	$^1/_5$	$^1/_5$	0	⟨332⟩			
33a	$^1/_8$	$^1/_8$	0	43c	$^3/_8$	$^3/_8$	$^2/_8$
33c	$^2/_5$	$^2/_5$	0				
41c	$^3/_8$	$^3/_8$	0				
⟨211⟩							
21b	$^2/_6$	$^1/_6$	$^1/_6$				
31b	$^2/_5$	$^1/_5$	$^1/_5$				
35a	$^2/_8$	$^1/_8$	$^1/_8$				

zone in Figure 10.20a. As an example, Figure 10.20b shows a distribution of misorientations from austenitic steel, which again displays a rather high proportion of Σ3 boundaries, in a single subvolume.

The subvolume of the fundamental zone is usually displayed as sections parallel to the XY plane. Either a single subvolume is used that gives a triangular base, or two subvolumes are displayed that give a square base and would allow both right- and

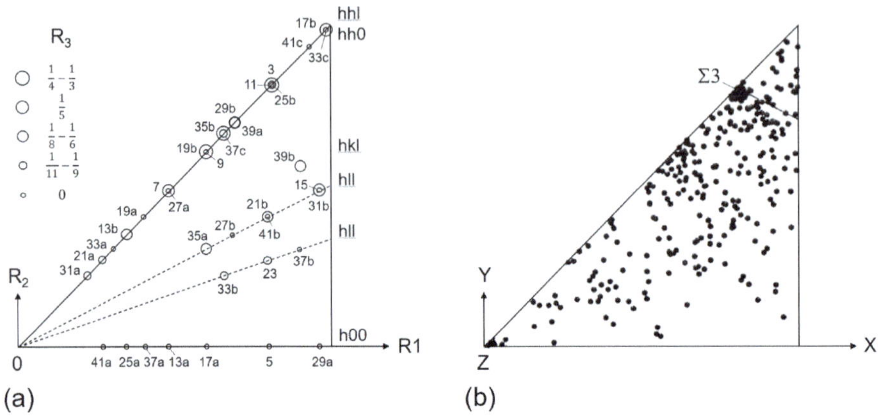

FIGURE 10.20 Representation of misorientations in Rodrigues space. (a) *R* vectors for CSLs having $\Sigma \leq 45$ in a Z-axis perspective of the subvolume. (b) Z-axis perspective of misorientations in an austenitic steel, indicating a high proportion of Σ3s. (Adapted from Randle and Day, 1993.)

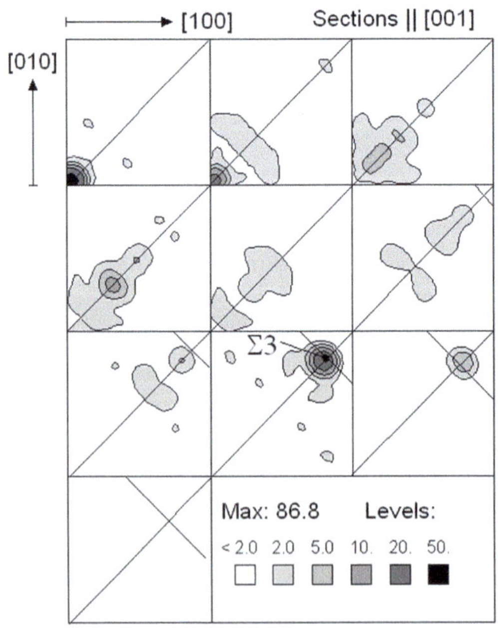

FIGURE 10.21 Distribution of R-vector density in Rodrigues space, displayed in sections through a left- and right-handed subvolume, for cyclically deformed nickel.

left-handed misorientations to be displayed, if required. Figure 10.21 shows a sample population of misorientations from the cyclically deformed nickel sample that has already been shown in the Euler angle space (Figure 10.18) and the cylindrical angle/axis space (Figure 10.19). In Figure 10.21, the MODFs are plotted with contour lines in one-eighth of the fundamental zone by sectioning Rodrigues space perpendicular to the [001] axis in equal distances of tan $\theta/2$. In comparison to constant sections with $\Delta\theta = 5°$ this yields a maximum error of 0.9°, which demonstrates the small distortions of this space. This kind of representation of a cubic space comprising one-eighth of the fundamental zone, which differs from the Mackenzie cell, was chosen because of its similarity to the well-known sections through the Euler space (Engler et al., 1994a).

Rodrigues space has been used to characterize orientation and misorientation distributions of island grains in an iron-silicon alloy (Bennett et al., 2007). It has also proved beneficial for representation of orientation relationships between dissimilar lattices, for example, fcc–bcc or bcc–hcp (e.g. Brückner et al., 2001), which renders it useful for the study of phase transformations or precipitation (see Section 12.2). The formalism of this approach is discussed in detail by He and Jonas (2007). For example, phase transformations in a meteorite have been studied by He et al. (2005). Attempts have been made to superimpose Rodrigues space and real-space coordinates to provide an orientation/microstructure link (Weiland et al., 1991; Hughes et al., 2000), and orientation or misorientation maps can be colored according to Rodrigues vectors (Gerth and Schwarzer, 1993; Quey et al., 2010; Patala and Schuh, 2011).

10.7 NORMALIZATION AND EVALUATION OF THE MISORIENTATION DISTRIBUTION FUNCTION

In analogy to the continuous distribution of the orientations, the ODF, a continuous distribution function of misorientations can be defined, that is, the MODF or, sometimes, MDF (Kallend et al., 1976; Adams, 1986; Pospiech et al., 1986). Thus, the MODF is constituted of the misorientations g_{ij}, between all grains i and all their neighbors j in a given sampling volume. Similarly as described previously for the ODF, continuous misorientation distributions can be derived from the individual misorientations g_{ij}, either by direct methods in an appropriate 3D (mis)orientation space or by the series expansion method in Fourier space. The functions for the series expansion have originally been defined in terms of the Euler angles (Section 5.3.1), whereas comparable functions for the angle/axis and Rodrigues vector parameters have not yet been developed. Therefore, although the angle/axis and Rodrigues vector descriptions of misorientations are generally considered superior to the Euler angle description, the computations are usually performed in Euler space and then the results are transformed into other spaces (Pospiech et al., 1986). Alternatively, the MODF can directly be derived by superposition of the corresponding misorientations in the desired (mis)orientation space (Engler et al., 1994a; Matthies and Vinel, 1994).

Under the assumption of a random arrangement of the orientations in the microstructure, the MODF can directly be derived from the ODF, that is, it can be computed from macrotexture data. The resulting MODF, the so-called texture-derived MODF, statistical MODF, or, following the nomenclature introduced by Pospiech

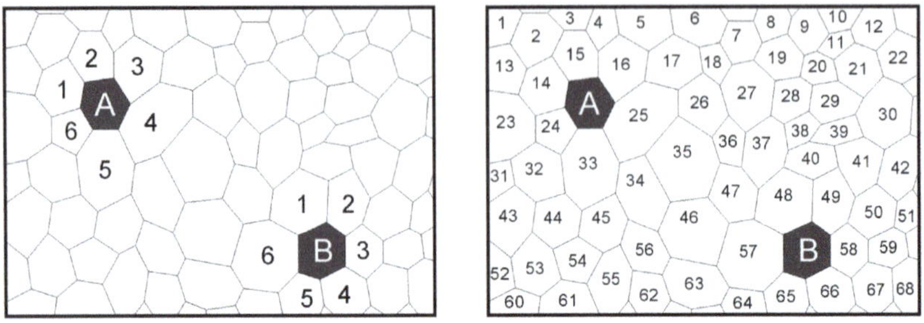

(a) MODF (b) ODDF

FIGURE 10.22 Schematic sketch of the topological features forming the basis of (a) MODF and (b) ODDF.

et al. (1993), orientation difference distribution function (ODDF) can be used to normalize the MODF, but it does not contain any valuable information on the spatial arrangement of the individual orientations. In real microstructures, the orientations are not randomly arranged such that the "real" MODF generally differs from the "statistical" ODDF. Thus, determination of the measured, or physical, MODF requires analysis of the orientations of the corresponding grains in the microstructures together with their spatial arrangements, which can only be accomplished by techniques for microtexture analysis.

The difference between MODF and ODDF is illustrated schematically in Figure 10.22. The MODF is constituted of the misorientations g_{ij} between all grains I and all their *direct* neighbors j. Thus, in the schematic sketch in Figure 10.22a the MODF is made up of the misorientations $g_{\mathrm{A}j}$ between grain A and its neighbors $j = 1, \ldots, 6$ plus the misorientations $g_{\mathrm{B}j}$ between grain B and its neighbors $j = 1, \ldots, 6$, and so on. The ODDF, by contrast, is composed of the mutual misorientations between *all* grains i and j (with $i, j = 1, \ldots, 68$), completely disregarding their actual location in the microstructure (Figure 10.22b).

To distinguish between the features in the MODF that arise from a given local arrangement of the corresponding orientations in the microstructure and the features that are controlled by the macrotexture of the sample, the measured MODF has to be normalized with respect to the macrotexture, which can be accomplished by dividing the MODF by the ODDF (Plege, 1987; Schwarzer and Weiland, 1988; Zhao et al., 1988; Mainprice et al., 1993; Pospiech et al., 1993; Heidelbach et al., 1996). The resulting texture-reduced MODF discloses features arising from a correlation in nearest-neighbor orientation relationships, which gives rise to correlated MODF or orientation correlation function (OCF).

To illustrate the potential applications of MODFs most clearly, in the following paragraphs the various possibilities of normalization of the MODF and the resulting information are discussed using an example with a model distribution of two texture components with a mutual misorientation of 35° ⟨112⟩ (Engler et al., 1994a). The model texture was composed of two components labeled A and B at the Euler angles (75°, 45°, 0°) and (54°, 29°, 78°); both components were associated with a Gauss-type scatter in Euler angle space with volume fraction $M = 50\%$ and a half scatter width

$\psi_0 = 8.5°$. Figure 10.23a shows the corresponding ODF, which was calculated as described in Section 10.3.2. Subsequently, this ODF was discretized, yielding a set consisting of approximately 500 single-grain orientations. With this data set, three different calculations were performed:

1. The ODDF was computed by determination of the orientation difference—that is, the misorientation—between each individual orientation and all other orientations of the entire data set. Thus, the ODDF yields statistical information on the potential misorientations between all orientations in a specimen, without consideration of the actual spatial arrangement of the corresponding grains in the microstructure, in particular of their neighborhood. Thus, the ODDF is only influenced by the macrotexture of the given sample and, as such, it can also be computed directly from macrotexture data (Plege, 1987; Zhao et al., 1988; Pospiech et al., 1993), though it is typically derived from the microtexture data by computing the misorientations between each grain and all other grains irrespective of their local arrangement. The ODDF of the present example—represented in Rodrigues space in Figure 10.23b—comprises two maxima at the position of small-angle grain boundaries (near the origin), indicating A–A and B–B relationships and at the A–B orientation relationship 35° ⟨112⟩.
2. An MODF was computed by artificially "arranging" the grains in such a way that A–B large-angle grain boundaries were preferred over small-angle grain boundaries A–A and B–B. This means that A grains are preferentially surrounded by B grains and vice versa. The corresponding Rodrigues plot reveals a peak at 35° ⟨112⟩ (Figure 10.23c), which is significantly stronger than the corresponding peak in the ODDF (Figure 10.23b). However, there remains a strong peak near the origin, which indicates the presence of still a large number of small-angle grain boundaries. In the case of such a strong macrotexture as in the present example despite the favorization of A–B misorientations, many A–A and B–B grain boundaries will still arise, which is simply due to their statistical frequency. Thus, the MODF yields information about the frequency of orientation relations between neighboring grains in the microstructure. This particularly favors the MODF for problems related to grain boundaries, such as grain boundary diffusion and grain boundary sliding (Pumphrey, 1976). In that context, the special treatment of a material to achieve an optimum grain boundary distribution is referred to as *grain boundary engineering* or *grain boundary design* (Watanabe, 1988, 2011; Palumbo and Aust, 1992; Randle, 2004, 2010). This will be further elaborated on in Section 11.2.6.
3. For an analysis of the orientation correlation, as for example the preferred occurrence of A–B grain boundaries, the MODF is not suited, however, since it is influenced by the texture of the sample. To yield information on orientation correlations, the MODF is divided by the ODDF. This way one obtains a texture-reduced form of the misorientation function, namely, the correlated MODF or OCF. In contrast to the MODF, the OCF yields information about the texture-reduced spatial correlation of misorientations,

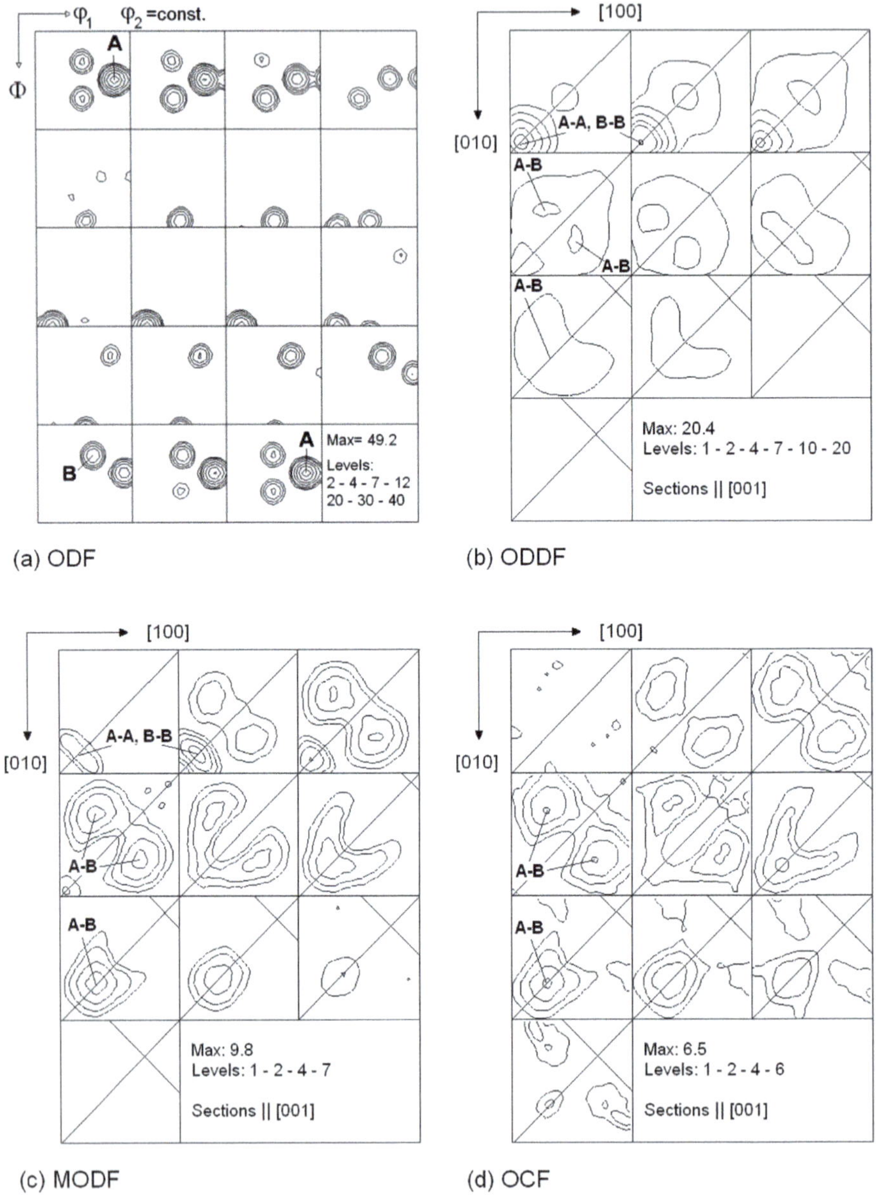

FIGURE 10.23 (a) Model ODF consisting of two components A and B with a 35° (112) orientation relationship, used to illustrate the normalization of MODFs. (b) ODDF, (c) MODF, and (d) OCF of the model texture in (a).

for example, whether deformation or recrystallization leads to a preferred development of distinct neighborhood relations. Occurrence of a given misorientation with $f > 1$ (or $f < 1$) in the OCF indicates its stronger (or weaker) occurrence compared to a statistical arrangement of grains in the microstructure. Hence, the OCF is suitable to allow conclusions to be drawn on the underlying physical mechanisms of microstructure and microtexture development. In the present example, no small-angle grain boundaries arise in the OCF; only the peak at 35° ⟨112⟩ (A–B) remains (Figure 10.23d), which exactly reflects the distribution that is to be expected from the artificial arrangement of the orientations A and B.

The applicability of MODF, ODDF, and OCF is now illustrated by way of an example pertaining to the occurrence of continuous recrystallization in aluminum (Engler and Huh, 1999). High-purity aluminum foil was recrystallization annealed and an EBSD orientation map with size 280 × 220 μm was measured in steps of 1 μm. For each point of the orientation map, the misorientations with regard to the six next neighbors were evaluated in terms of misorientation angle ω and misorientation axis, though the latter will be ignored in the following. Figure 10.24a shows the misorientation distribution as fraction of boundaries in dependence on the misorientation angle ω. The distributions obtained for both the as-deformed state and the recrystallized state were very similar, and in both examples, the majority of boundaries have misorientation angles below 20° and can therewith be classified as small-angle grain boundaries. For an appraisal of the misorientation distribution with regard to the occurrence of continuous recrystallization, these data had to be compared to the average misorientation distribution of the same sample without consideration of the spatial arrangement of the corresponding grains and subgrains in the microstructure, that is, the ODDF. In the present example, the ODDF was derived from the corresponding x-ray macrotexture data and was again expressed in terms of the fraction of misorientation angles ω. Comparison of the resulting curve with the actual misorientation distribution in Figure 10.24a shows that small misorientations, representing small-angle grain boundaries, are much more frequent than they would be in a sample with the same texture but a random spatial arrangement of the orientations. Finally, the OCF was computed, simply by dividing the MODF histogram values by the corresponding values of the ODDF (Figure 10.24a). The resulting OCF, displayed in Figure 10.24b, reveals a very high peak with $f > 10$ at low misorientation angles, whereas all other columns are below 1. This implies that small-angle grain boundaries are strongly preferred over medium- and large-angle boundaries, which proves that recrystallization proceeded by continuous subgrain coarsening rather than the discontinuous growth of some distinct recrystallization nuclei, that is, by continuous recrystallization (Engler and Huh, 1999).

Figure 10.24a shows that the ODDF is quite different from the standard Mackenzie distribution, which is the probability distribution for disorientation angles in randomly oriented cubic polycrystals (Section 10.5.1; Figure 10.15). The difference between ODDF and Mackenzie distribution lies in the occurrence of texture: the Mackenzie distribution is valid for a random orientation distribution, whereas the ODDF is derived for a given texture, though under the assumption of a random

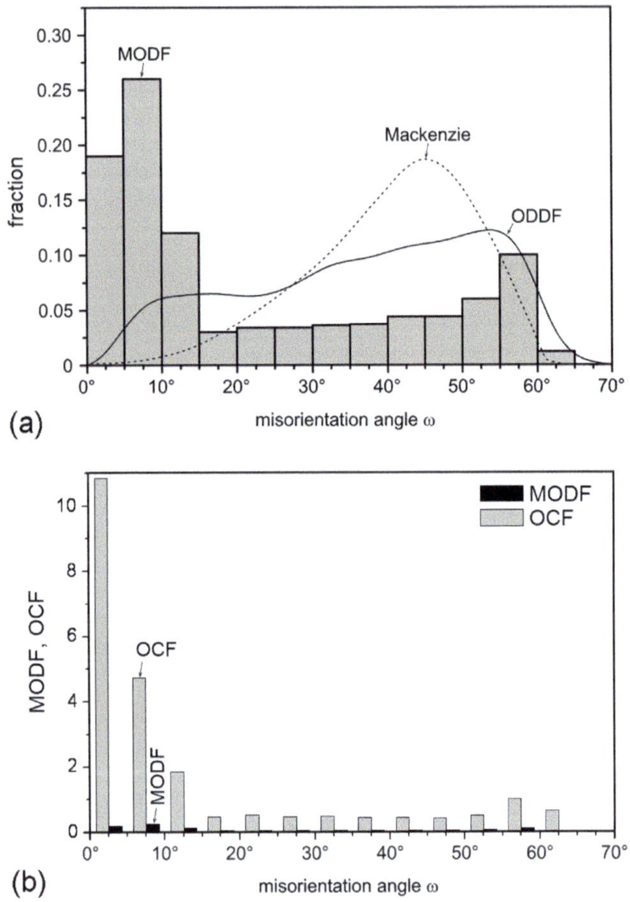

FIGURE 10.24 (a) MODF, ODDF, and Mackenzie distribution and (b) MODF and OCF in terms of fraction of grain boundaries as a function of the misorientation angle ω obtained in high-purity aluminum foil, which had undergone continuous recrystallization.

topological arrangement of the texture constituents. The MODF does incorporate both orientation distribution (i.e., texture) and topological arrangement of the texture constituents, whereas the OCF yields information about the texture-reduced spatial correlation of the texture constituents.

10.8 USE OF ORIENTATION DATA FOR ENSUING MICROSTRUCTURE MODELING

In this section we briefly touch upon methods to utilize orientation maps as an input for spatially resolved microstructure modeling approaches, including models for solidification, phase transformation, deformation, recrystallization, and grain growth. Engler (1998b) used orientation maps obtained by EBSD in cold-rolled aluminum and copper as a starting point for a cellular-automaton-based recrystallization model. Although back then velocity and spatial resolution of EBSD were insufficient to capture sufficiently large maps that still covered all relevant features in the initial as-deformed microstructure, the approach was capable of simulating the evolution of microstructure during recrystallization in a meaningful manner; furthermore, reasonable data of recrystallized grain size and recrystallization texture were obtained. Along similar lines, Chun et al. (2006) and Jedrychowski et al. (2015) adopted a 2D Monte Carlo model combined with a direct mapping technique to simulate recrystallization and grain growth in hexagonal materials. Diehl and Kühbach (2020) coupled 2D EBSD surface maps with a real-time-scaled 3D cellular-automaton with Moore neighborhood rules and software parallelization for analysis of the primary static recrystallization in a low carbon steel. Buffière et al. (2008) demonstrated the contribution of x-ray tomography data to modeling for three different fields of metals research, including deformation of cellular materials, metal solidification, and fatigue crack growth in Al alloys. As will be described in detail in Section 13.2, synchrotron-based x-ray tomography is able to provide the local crystallographic orientation in the bulk of materials, which is a key issue for the modeling of microstructure in metals. Ferrié et al. (2006) applied x-ray microtomography to track the propagation of microcracks by an extended finite element method. Tasan et al. (2014) used 2D EBSD maps of a dual phase steel as an input to calculate potential hot spots for crack initiation using crystal plasticity spectral (CPS) simulations. A comparison of experimental and modeling results showed that the general trends were caught correctly by the CPS simulations but detailed forecasts on the exact location of cracking were not possible. One reason for this was that the microstructure below the surface, which influences the deformation behavior significantly, was not known. As a last example we mention the work of Stechmann et al. (2018) who used molecular statics (MS) modeling to understand the formation mechanisms of textures which were experimentally investigated by 3D EBSD. Although 3D EBSD and MS modeling work on completely different size scales (100-nm resolution for EBSD vs. atomistic resolution for MS), the modeling clearly confirmed the role of defects for the formation of particular texture components.

In the above examples representation of the microstructure in the subsequent modeling was obtained directly by microtexture mapping. However, these approaches are generally very time-consuming and laborious for practical applications. Alternatively, Voronoi tessellation models have been traditionally applied in modeling polycrystalline grain structures for metallurgical applications, as they provide a natural solution to represent grain structures with non-uniform grain shapes (e.g. Lemaitre et al., 1993; Fritzen et al., 2009). Lebensohn et al. (2008) employed a full-field formulation based on fast Fourier transforms (FFTs) for the simulation of the viscoplastic straining of polycrystalline copper. The authors used 2D EBSD maps to construct the initial unit cell via Voronoi tessellation in the third dimension to obtain a full 3D microstructure. Rollett et al. (2007) developed their so-called *Microstructure Builder* to construct statistically representative 3D grain structures by a Voronoi tessellation based on two orthogonal EBSD sections. Therewith, both geometric and crystallographic observations from the polycrystal are mapped to the subsequent computer simulations model. Later, Groeber and Jackson (2014) devised the software package Dream.3D for processing, segmenting, quantifying, representing, and manipulating experimental microstructures. This approach enables transformation of EBSD data into a format which can then be used as an input for advanced, computationally more expensive, microstructure modeling approaches like kinetic Monte Carlo simulation of metal additive manufacturing microstructures (Rodgers et al., 2017), crystal-plasticity based models of deformation (Roters et al., 2019) or phase-field models of recrystallization or grain growth (DeWitt et al., 2020).

10.9 SUMMATION

The aim of this chapter has been to provide a guide to the broad array of choices that exist for evaluation and representation of microtexture data, by giving a concise exposition of the principles and methodology of each. Formalism has been deliberately kept to a minimum, and the reader can obtain fuller information elsewhere (see General Bibliography). There is a host of examples in the literature to illustrate microtexture investigations where the representation methods described in this chapter have been used.

Selection of a method for data representation and analysis will inherently depend on the nature of the inquiry and the user's preference. A key point is that single-grain orientation measurements are *directly 3D* and therefore their representation (and subsequent extraction of statistics) in pole figures, Euler space, or Rodrigues space is equivalent. As far as direct representation of misorientations is concerned, axis/angle representations, including those in Rodrigues space, offer significant advantages over use of Euler space. When analyzing microtexture data, it must always be borne in mind that the quantification can be by *volume* (or area for interfaces) or by *number*.

As a consequence of the speed of EBSD analysis, this technique is now capable of providing an acceptable alternative to macrotexture for the measurement of overall texture, provided that attention is paid to appropriate sampling schemes and population size (see Section 9.3.2). The EBSD-based method is free from the errors introduced in the series expansion method (Sections 5.3.2 and 5.3.3), and the cost of a standard scanning electron microscopy (SEM) plus EBSD system is considerably less than a full x-ray texture goniometer. Furthermore, EBSD gives access to a whole range of local orientation studies, which is usually the main focus.

Display of orientation or misorientation data is only a part of data evaluation; often further processing is required to extract quantified information from the data. To differentiate between the features in the MODF that arise from a certain spatial arrangement of the corresponding orientations in the microstructure and the features that are controlled by the macrotexture of the sample, the MODF has to be normalized with respect to the macrotexture. Finally, microtexture data may serve as a valuable input for ensuing spatially resolved microstructure models. To this end, orientation maps can either be mapped directly onto the simulation grid, or, in more advanced microstructure models, the initial 3D microstructures are constructed from microtexture data with the help of suited statistical reconstruction method.

11 Crystallographic Analysis of Interfaces, Surfaces, and Connectivity

11.1 INTRODUCTION

One of the strongest aspects of orientation microscopy (ORM) is its ability to characterize interfaces between individual crystals or of free surfaces in some crystallographic detail. Such interfaces and surfaces include grain boundaries, phase boundaries, slip traces, cracks, and external surfaces of the specimen, such as facets or fracture surfaces (e.g. Randle, 2000). Grain and phase boundaries as well as slip traces have been traditionally analyzed in transmission electron microscopy (TEM), but with the considerable disadvantage of statistical reliability. Furthermore, analysis of cracks or external specimen surfaces is difficult in TEM because of specimen preparation restrictions. In contrast, scanning electron microscopy (SEM) is ideal for studying the morphology of interfaces and surfaces, and so it is mainly electron backscatter diffraction (EBSD) that has been adapted to crystallographic studies of surfaces.

Section 11.2 provides a short introduction into the atomic structure and properties of grain and phase boundaries, including the concept of small- and large-angle grain boundaries as well as the well-known coincidence site lattice (CSL) model and its offshoots. These models require one or three parameters, namely the misorientation angle, or the full misorientation expressed, for example, in terms by three Euler angles or by misorientation angle and misorientation axis pairs (Chapter 2). However, five independent variables are required to fully specify a plane or surface in the specimen, including, in addition to the misorientation also 2 parameters to specify the position of the grain boundary in sample or crystal space. Section 11.2.7 describes in detail a new means of obtaining distributions of grain boundary (GB) planes in addition to misorientation, called the "five-parameter" analysis.

The various methodologies in connection with such investigations are described in detail in Section 11.3. As a matter of fact, the orientation mapping techniques introduced earlier (SEM, TEM, and also x-ray diffraction (XRD)-based) allow measurement of misorientations across the boundaries and, at least, the position and crystallographic direction of the trace of the interface on the measurement surface. Determination of all five interface parameters mandates use of 3D techniques, like TEM imaging (Zaefferer, 2002) or, more prominently, x-ray tomography (Section 13.2), EBSD-based serial sectioning (Section 11.3.2), or 3D EBSD measurements (Section 11.4), which also allow access to the true boundary planes. Sometimes it is pertinent to examine how the various orientations and misorientations are connected in the microstructure. This is discussed in Section 11.5.

DOI: 10.1201/9781003258339-14

Besides the *crystallographic structure* of boundaries and interfaces, also their *atomic structure* and their *chemical composition* are very important for their properties. The atomic structure of GBs can only be revealed by high-resolution TEM (HRTEM)—and this also only for special cases—while the elemental composition can be tackled either by HRTEM, coupled with energy dispersive x-ray spectroscopy (EDS) or electron energy loss spectroscopy (EELS) or by atom probe tomography (APT). Both, HRTEM and APT, are very laborious and certainly not suited to study and classify a large amount of GBs (i.e. more than a few tens). This is different with the EBSD-based orientation mapping techniques, which may reveal thousands to even hundreds of thousands of GBs. These values can then be analyzed, for example, as GB texture data, using very similar tools as they are used for texture analysis.

Thus, application of EBSD to interfaces has allowed the crystallographic character of GBs to be measured for large sample populations of boundaries. This is illustrated in Figure 11.1 by means of an example of the distribution of macroscopic

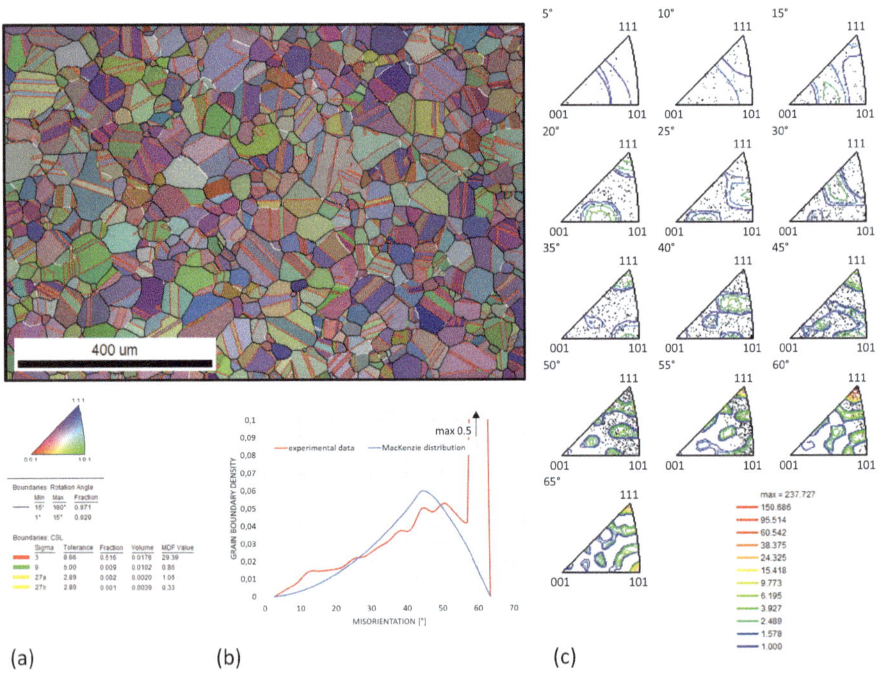

FIGURE 11.1 Grain boundaries and their misorientation data as measured by EBSD on a stainless-steel sample SS316L. (a) Orientation map colored in a (faint) inverse pole figure color code for the normal direction (see legend). Small- and large-angle grain boundaries and some special grain boundaries are marked by colors. (b) Misorientation angle distribution of the map in (a). The experimental data are compared to the theoretical Mackenzie distribution for a random orientation data set. The peak at 60° indicates the presence of $\Sigma 3$ grain boundaries (marked in red in (a)). (c) Axis-angle plot of the grain boundary misorientation distribution. Overlapped are discrete data (black dots) with the misorientation distribution function (MDF) as colored with equal density lines. The MDF is plotted in logarithmic scale. The only strong peak in the distribution corresponds to the $\Sigma 3$ twin boundaries at a misorientation of 60° $\langle 111 \rangle$.

GB parameters in a stainless-steel sample, as measured by EBSD-based ORM. Figure 11.1a shows an orientation map with the GBs being colored by two different principles. First, black and white lines indicate small- and large-angle GBs, respectively. As will be discussed in Section 11.2.2, the misorientation angle is relevant for small-angle GBs, but not of great importance for large-angle GBs, as long as they are not special boundaries. The colored lines then mark the so-called special CSL-type GBs, characterized by their Σ-value, which will be introduced in Section 11.2.3. In the present example first-generation twin boundaries of $\Sigma 3$ type, as well as double and triple twin boundaries of $\Sigma 9$ and $\Sigma 27$ type, are marked. Most of the $\Sigma 3$ boundaries (red lines) appear as straight lines, indicating that they correspond to flat and coherent GBs, with very low energy. A few $\Sigma 3$ boundaries, in contrast, are curved indicating that these are incoherent twin boundaries with high energy. Coherent and incoherent boundaries will be discussed in Sections 11.2.6 and 11.3.3. Figure 11.1b shows the distribution of GB segment lengths over misorientation angles. The values obtained from the map, displayed by a solid line, are compared with the theoretical misorientation value distribution obtained for a completely random distribution of orientations, the so-called Mackenzie distribution (Section 10.5.1), displayed by a dashed line. Differences between the two curves indicate either that the texture is not random or that there is a preference for certain misorientations in the measured volume. For the present example, it is obvious that the curve deviates very strongly from the Mackenzie distribution for angles around 60° for which much more GB segments are present in the material. These GBs mainly belong to the annealing twins, which actually have a 60° misorientation. Also small-angle GBs (up to a misorientation of 15°) appear more frequently than in the random case, which may be due to the overall texture of the material (Section 10.7). Finally, Figure 11.1c displays misorientation axis and misorientation angle for all GB segments in a so-called axis-angle plot (see Section 10.6.2). For every misorientation angle the misorientation axis is plotted in the standard orientation triangle (SOT) of the respective crystal symmetry. In a discrete axis angle plot every GB segment is plotted as one dot in the SOT. If the misorientation distribution function (MDF or MODF) is calculated, the axis-angle plot can be plotted as grain boundary density plot which is displayed in Figure 11.1c by contour lines of equal density simultaneously with the discrete point plot. In this figure two things may be observed: first, most GB axes are randomly distributed. Second, there is one strong peak at 60° misorientation angle and $\langle 111 \rangle$ misorientation axis (written as 60° $\langle 111 \rangle$). This peak corresponds to the $\Sigma 3$ twin boundaries which will be discussed in detail in Sections 11.2.3 and 11.2.6. These boundaries have very special properties and are highly frequent in the stainless steel microstructure that is displayed in Figure 11.1.

11.2 DESCRIPTION OF GRAIN BOUNDARIES

One of the most important defects in polycrystalline materials are the interfaces between the different crystallites, i.e. grain or phase boundaries (Sutton and Balluffi, 1995). These defects significantly influence structural and functional properties and have a very strong influence on thermodynamics and kinetics of materials. Examples for properties that are strongly influenced include, for example, the strength of ductile metals, expressed by the Hall-Petch relation, chemical properties expressed by the susceptibility of certain metals or alloys to GB corrosion, or electronic transport properties expressed by charge carrier recombination processes in solar cell materials. The kinetics of diffusion of impurities and phase transformations may be strongly influenced by the nature of GBs, and the energy stored in GBs is the driving force for grain coarsening processes. Thus, a good understanding of materials properties and processes requires a good understanding of the properties and, therefore, of the atomic structure of grain or phase boundaries.

11.2.1 Characterization of the Grain Boundary Structure

The atomic structure of grain and phase boundaries is influenced by a large number of parameters, some of which are sketched in Figure 11.2. The perhaps most obvious parameters are the crystal orientations under which the two adjacent crystals concur. These relative orientations are expressed by the misorientation between the crystals, which consists of a misorientation angle and a misorientation axis (together three independent parameters; see Section 2.7.2). Next, the position of the interface relative to both lattices is of great importance. This position may be described as the (local) normal vector of the plane given in coordinates of one of the lattices (two independent parameters). Both, misorientation and normal vector are *rotational* parameters, often also called "macroscopic." Besides this, the two adjacent lattices are positioned relative to each other by three atomic coordinates which shift the lattice in *x*, *y*, and *z* direction relative to one of the lattices (three independent, atomic *translational* parameters).

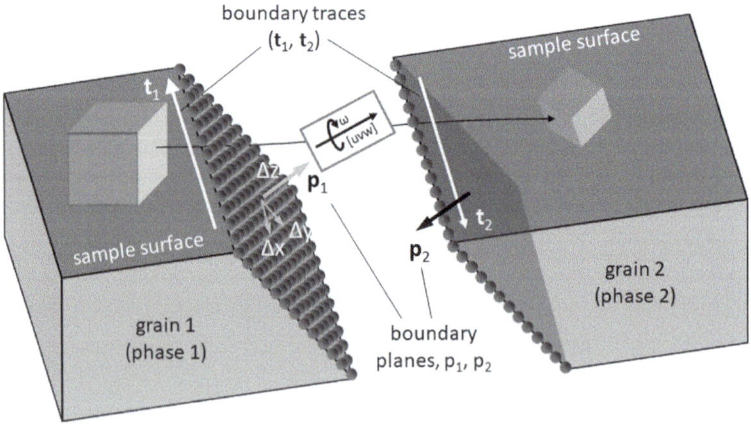

FIGURE 11.2 Important grain boundary parameters. Rotational (or macroscopic) parameters are the misorientation (ω, ⟨uvw⟩) and the grain boundary normal vectors **p**. If grain boundaries are only observed from a free surface, then the trace vectors, **t**, can be measured. Translational (or microscopic) parameters are the shift vectors, measured in fractions of atomic distances in surface plane coordinates, Δ(*x*,*y*,*z*).

In addition to the abovementioned eight crystallographic parameters, also the chemistry of the boundary plays an important role in deciding the structure of the grain or phase boundary which adds an almost infinite number of further parameters. Particularly in materials with (partly) ionic bonding (e.g. in oxides) also the charge of the individual ions may play a very important role to determine GB properties. Finally, on a more macroscopic scale, also morphological parameters, in particular the local curvature of an interface, play a role for the properties of boundaries.

Following Sutton and Balluffi (1995) grain and phase boundaries should be distinguished into (i) special boundaries, (ii) random boundaries, and (iii) vicinal boundaries. Ultimately, this distinction is based on the atomic structure of these boundaries. However, the atomic structure of GBs is difficult to observe and maybe very variable. Therefore, it is more straightforward to use the energy of a GB, as one manifestation of structures, for a more direct classification of boundaries. The energy of an interface may be distinguished as follows:

i. In the "energy landscape" of grain and phase boundaries, *special boundaries* are distinguished by particularly low energies. A low GB energy is indicative of a high degree of atomic order of the boundary, approaching that of the undisturbed lattice. The high degree of order implies a number of special properties, e.g. low degree of segregation, low diffusivity for impurities, "special" electronic properties (i.e. properties that are close to the bulk behavior), and also special mechanical properties. The best example in this respect is the coherent twin boundary, which exists in many fcc and hcp metallic or intermetallic structures. There are probably only few other GB types that are really special (see below).
ii. *Random boundaries* form the majority of GBs in boundary space. Their energy is high and in the energy landscape they form a plateau with little changes. As a consequence also their properties are only little different and it is probably not of great interest to distinguish between any of them.
iii. *Vicinal boundaries* are those boundaries that make the transition between special boundaries and random boundaries. Usually their properties depend strongly on the angular deviation (both in misorientation and in boundary plane) from the special boundary. The transition from special boundaries is accommodated by (GB) dislocations, in the case of misorientation deviation, and/or by boundary faceting in the case of boundary plane deviations. The best-known example of vicinal GBs are small-angle GBs. Their structure is formed by a regular arrangement of lattice dislocations, and their energy is directly related to the density of dislocations building up the boundary.

The most important questions to be answered when distinguishing and evaluating boundaries are thus, which boundaries are "special" (i.e. belong to the first group) and whether it is possible to determine a set of (easily measurable) parameters which allows this distinction in a general way. The text by Wolf (2005) gives detailed insight into these questions. A clear and well-accepted distinction is the one into small- and large-angle GBs (Gleiter, 1983), which is subject of the following section.

11.2.2 Small- and Large-Angle Grain Boundaries

The easiest way to discriminate GBs is simply based on the angle of misorientation between two neighboring grains. In this way, it is possible to separate two physically very different types of boundaries, the *small-angle* GBs, SAGB and the *large-angle* GBs, LAGB. Note that sometimes these GBs are named instead as *low*-angle GBs, LAGB, and *high*-angle GBs, HAGB, which leads to confusing acronyms. SAGB consist of aligned sets of *lattice* dislocations; the spacing of these dislocations decreases with increasing misorientation. Figure 11.3a illustrates such a GB for the case of a regular arrangement of edge dislocations, which builds up a tilt GB (see below for this terminology). The misorientation of this small-angle tilt GB is inversely proportional to the distance between the dislocations and can be calculated according to

$$\theta\,[\mathrm{rad}] = \tan\frac{b}{d} \approx \frac{b}{d} \tag{11.1}$$

where b is the Burgers vector length and d the distance between the dislocations. The GB character changes from SAGB to LAGB when the dislocations get so close together that their dislocation cores overlap and the atoms arrange in new structures. This transition is classically located at about 15° of misorientation for metals. Winning and Rollett (2005), for example, gave a nice example how the thermally activated mobility of GBs changes at these values. Similar transition misorientation angles are claimed for materials with more complicated lattice structures—e.g.

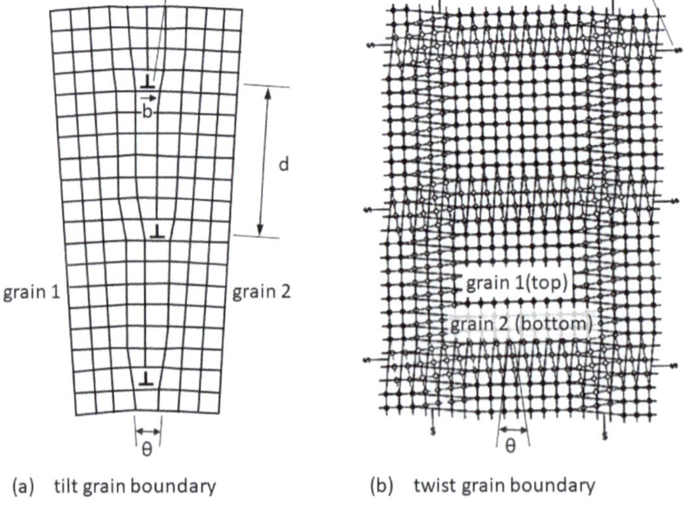

FIGURE 11.3 Dislocation arrangement and misorientations of small-angle grain boundaries. (a) Tilt grain boundary composed of a single set of edge dislocations. (b) Twist grain boundary composed of two sets of crossing screw dislocations. Note that in this case the two grains are on top of each other.

Tochigi et al. (2018) on Al_2O_3 and Ikuhara (2001) on ceramics. The 15° boundary also needs to be considered as a limit of validity of the Read-Shockley equation which calculates the energy, γ_{GB}, of a small-angle GB in dependence of its misorientation, θ:

$$\gamma_{GB} = \gamma_0 \theta [a - \ln \theta] \tag{11.2a}$$

with

$$\gamma_0 = \frac{Gb}{4\pi(1-\nu)} \quad \text{and} \quad a = 1 + \ln\left(\frac{b}{2\pi r_0}\right) \tag{11.2b}$$

G is the shear modulus (which ranges between 40 and 600 GPa for metals and ceramics), ν is the Poisson ratio (0.2–0.45), and r_0 is the radius of the dislocation core ($r_0 \approx 4b \approx 1$ nm).

If not only the angle but also the axis of misorientation and the position of the GB are taken into consideration, subgrain boundaries can be further distinguished into tilt and twist boundaries. For tilt boundaries the rotation axis lies in the boundary plane, in pure twist boundaries it stands perpendicular on the boundary plane. Small-angle tilt and twist boundaries are physically very different, as illustrated in Figure 11.3b. Pure tilt boundaries are constructed out of a regular arrangement of edge dislocations while pure twist boundaries consist of a regular arrangement of crossed screw dislocations. For the same misorientation pure twist GBs require the double amount of dislocations. Though screw dislocations have lower energy than edge dislocations (0.66 times as an ideal value), in total the energy of twist GBs is higher and their mobility lower than that of tilt GBs. Pure tilt and twist GBs can appear in any mix as illustrated in Figure 11.4.

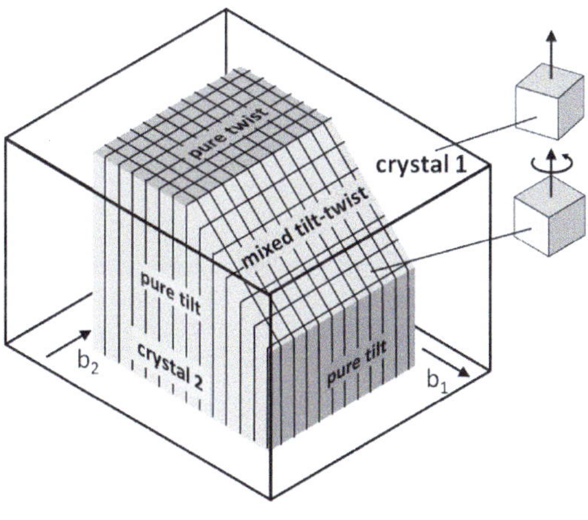

FIGURE 11.4 Two grains with small misorientation and the resulting arrangement of screw and edge dislocations and tilt, twist, and mixed grain boundaries.

In order to determine the type of SAGB, it is important to determine the axis of misorientation and the boundary plane. It has been shown in Section 2.6.2 that, due to crystal symmetry, a misorientation between two crystals of the same phase always has $2n^2$ expressions, where n is the number of symmetry elements of the two crystals. The different misorientations consist of combinations of n misorientation angles with $2n$ misorientation axes. For LAGBs the so-determined different angles and axes are all equally valid and it is only a question of definition to select one of them. This will be discussed a bit further in Section 11.2.3. In contrast, for SAGBs the smallest misorientation angle corresponds directly to the rotation angle caused by the dislocations in the boundary. The correct rotation axis is then the one, which is obtained when selecting the identity matrix, **I**, as symmetry matrix, $\mathbf{S}_i$, for the first crystal orientation matrix:

$$\Delta g_{0j} = \left(g_1\mathbf{I}\right)\left(g_2\mathbf{S}_j\right)^{-1} \tag{11.3}$$

The correct GB plane can only be determined with certainty by a serial sectioning (Section 11.3.2) or by 3D tomography method (Section 11.4). However, the *GB trace* visible on a 2D ORM map can give valuable hints on whether an SAGB has twist, tilt, or mixed character. This trace analysis is displayed in Figure 11.5a, showing two neighboring grains separated by a small-angle GB. In Figure 11.5b, the {111} poles of the two grain orientations are plotted in a common stereographic projection together with the direction of the calculated rotation axis **r**. Next, the trace **t** of the SAGB is plotted into this pole figure as dashed line through the pole figure center. This trace is the common pole of all potential GB plane normals that correspond to this trace. The loci of the potential plane normal (LPN) then all come to lie on a great circle perpendicular to the trace pole as plotted in the figure as solid line. In the present

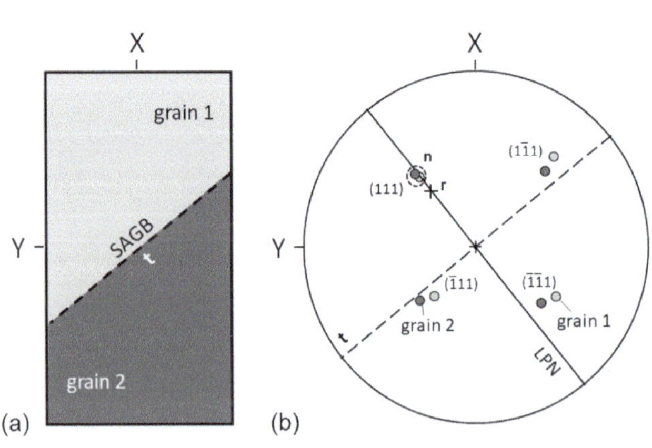

FIGURE 11.5 Trace analysis for determination of grain boundary plane and grain boundary character. (a) Small-angle grain boundary, SAGB, separating two grains labeled 1 and 2, and (b) orientation of the two grains in stereographic projection. The trace of the SAGB is plotted as a dashed line, while the loci of the potential plane normals (LPNs) are plotted at 90° to the grain boundary trace.

example, the LPN intersects the pole of the misorientation axis, **r**, which is close to the normals, **n**, of the (111) planes. Thus, the GB is *probably* an almost pure twist GB on the (111) plane (**r** and **n** almost parallel). Note, however, that a single trace analysis cannot give certainty on the GB plane. For an unambiguous characterization two traces, for example from two perpendicular sides of the sample, are required, which will be described in Section 11.3.2. It shall be emphasized here that other types of GBs cannot be distinguished without further information. This information could be, for example, that pure tilt boundaries are composed of a set of identical edge-type lattice dislocations. Such dislocation arrangements produce well-defined lattice rotations. For example, [110][112]-type edge dislocations (notation: [Burgers vector][line direction]) in the fcc lattice generate rotations about a ⟨112⟩ axis.

While the concept of twist and tilt boundaries is physically meaningful for small-angle and other vicinal GBs connected to special boundaries, it does not appear relevant for general boundaries. This is, first, because there is no mathematically unique solution to the misorientation rotation axis. Morawiec (2009, 2011) showed that every large-angle GB can be described by a number of twist or tilt or mixed boundaries. A second reason is that there is no clear physical concept that would allow a relation of twist-tilt character with GB structure or energy. Wolf (2005) argues that general large-angle GBs actually may consist of a thin (i.e. few nanometer thick) amorphous film which may be responsible for the relatively similar energies of these boundaries. On the other hand, the O-lattice theory (see Section 11.2.5) may actually provide a means to select a physically meaningful misorientation description and, furthermore, assign a potential GB structure also to general LAGB.

11.2.3 Distinction of Grain Boundaries by the CSL Model

One of the earliest and most simple attempts to discriminate between special and non-special boundaries is the coincidence site lattice (CSL) model, which is, in its basic approach, a pure misorientation-based model (Grimmer et al., 1974). It owes its popularity both to convenience, since the analysis is easy to apply, and to the fact that *some* CSLs may indeed be *special* boundaries, that is, they may be associated with special properties (Randle, 1996, 2001). The basic idea is that the lattices of the two abutting crystals are overlapped and it is searched for coinciding lattice points. These lattice points, if they exist, form themselves a new lattice, the CSL. The degree of coincidence of the two lattices is quantified by the ratio, Σ, of the volumes of the coincidence lattice unit cell, V_{CSL} and the base lattice unit cell, V_{BL}:

$$\sum = \frac{V_{CSL}}{V_{BL}} \tag{11.4}$$

The smaller is Σ the higher is the amount of coinciding lattice sites. The most intuitive CSL is Σ = 5 (shortly Σ5) in the cubic lattice, which is illustrated in Figure 11.6. The Σ5 CSL is obtained by selecting a cubic lattice (simple cubic, body centered, or face centered) and rotating it about the [001] axis by exactly 36.87°. As shown in the figure, every fifth lattice point then belongs to both lattices, i.e. the original and the rotated one. A special case is the CSL boundary with Σ = 1 which—strictly speaking—corresponds

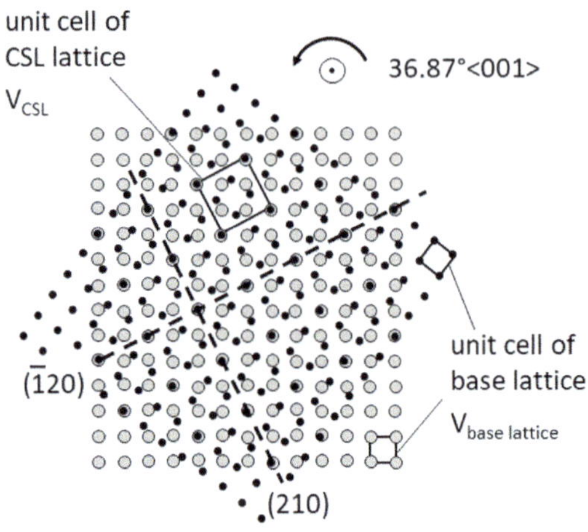

FIGURE 11.6 Construction of a coincidence site lattice, CSL, on example of a Σ5 misorientation. The gray lattice (gray disks indicate lattice positions which may be occupied by atoms) is rotated by 36.87° around the ⟨001⟩ axis (normal of viewing plane) to form the black lattice. Gray and black lattices overlap on every fifth point, leading to a value of Σ = 5. Additionally, two ideal grain boundary traces with particularly high density of coincidence sites are plotted (grain boundary planes stand perpendicular on the viewing plane).

to no misorientation between the lattices. With increasing deviation from the exact Σ1 case one obtains small-angle GBs (see further below for the treatment of deviations from the exact Σ misorientations). The CSL model has universally valid Σ-entries only for cubic crystal symmetries and, with limitation to rotations about the c-axis, for hexagonal, trigonal, and tetragonal crystal symmetries. For all other crystal symmetries and for phase boundaries between different crystal structures the O-lattice theory (see Section 11.2.5) may give valuable classifications. Table 11.1 lists a few regularly discussed CSL misorientations for the cubic symmetry case.

It should be mentioned that if CSL-type GBs meet in a triple junction (for example when forming a twin inside of a twin, i.e. a double twin), then the boundaries are interrelated through a simple multiplication rules. Miyazawa et al. (1983) showed that the following rule applies for the Σ values of the three boundaries (1), (2), and (3) of a triple point:

$$\Sigma_{(3)} = \Sigma_{(1)} \cdot \Sigma_{(2)} / \beta^2 \tag{11.5}$$

where β is a common divisor of $\Sigma_{(1)}$ and $\Sigma_{(2)}$. As an example, if both boundaries (1) and (2) are Σ3 boundaries then the common divisors are 3 and 1, which means that the newly formed boundary can be a Σ9 GB ($9 = 3 \cdot 3/1$) or a small-angle Σ1 GB ($1 = 3 \cdot 3/3^2$). Using and extending this rule, Gertsman (2001) developed a theory of triple line CLSs, which will be further considered in Section 11.5. Based on this theory Jeyaraam et al. (2019) developed a geometrical model, which forecasts the frequency of appearance of $\Sigma 3^n$ GBs (Σ3, Σ9, Σ27, …) in fcc materials that form many Σ3 boundaries.

TABLE 11.1
Properties of Low-Index CSL Misorientations ($\Sigma \leq 27$) for Cubic Symmetry

Σ	ω	Rotation axis [uvw]	$\Delta\omega$	Remarks	Selected References
1	0°	any	15°	Small-angle GBs	
3	60°	111	8.7°	Coherent and incoherent twins in fcc and bcc. Important for many properties and for texture formation.	(1)
3	70.53°	110	8.7°	Alternative but *physically identical* description of the $\Sigma3$ misorientation (as an example). Does not comply with the smallest angle – axis in SOT convention	
5	36.86°	100	6.7°	Forms special boundaries because of low energy anisotropy	(2)
7	38.21°	111	5.7°	Highly mobile in fcc alloys, contributing to the formation of recrystallization textures	(3)
9	39.94°	110	5°	Forms typically as double twin (twin of a twin) in fcc materials. Often thought to form special boundaries Highly mobile in bcc alloys, contributing to the Goss texture formation	(4)
11	50.47°	110	4.5°	O-lattice theory shows that $\Sigma11$ GBs are of low energy, confirmed by measurements	(5)
13a	22.62°	100	4.2°	Not known to create boundaries with special properties	
13b	27.79°	111	4.2°	Physically different boundary from $\Sigma13$a, not just a symmetrically equivalent description	
15…25	var.	var.	var.	Not known to create boundaries with special properties	
27a	31.59°	110	2.9°	Forms as triple twin (twin of twin of twin) in fcc alloys. Relatively frequent in fcc alloys which form annealing twins. Not known to form boundaries with special properties	
27b	35.43°	210	2.9°	See the remark with $\Sigma13$b	

Note: ω: misorientation angle, $\Delta\omega$ maximum allowed deviation from exact Σ according to the Brandon criterion (see below). Selected references: (1) Bozzolo and Bernacki (2020); (2) Kirch et al. (2007); Tsai et al. (2021); (3) Engler et al. (1996c); (4) Li et al. (2009); Homma and Hutchinson (2003); (5) Goukon et al. (2000).

As such, the CSL model makes no statement about GBs, simply because it is only based on misorientations and GB planes do not appear in the model. The sheer existence of a CSL orientation relationship between two crystals, therefore, does not imply any special properties of their boundary. Nevertheless, it is assumed that if a dense CSL exists (indicated by a small Σ value), then one may also position GBs with a high degree of coincidence points into this lattice. Some traces of potential coincidence boundary planes, here the $(\bar{1}20)$ and (210) planes of the gray lattice, are plotted in Figure 11.6 by dashed lines. The high degree of common lattice sites may then imply a high degree of order and, therefore, special properties. We underline here that boundaries with small Σ values *have the potential* to form highly ordered boundaries, though this is not an inevitability.

The best example for a highly ordered CSL GB is the Σ3 twin in fcc materials. It may form either by mechanical twinning, a shear process, or by growth twinning, which is due to the incorporation of stacking faults at the front of growing grains. As displayed in Figure 11.7, the fcc lattice can be described by stacking of densely packed {111} planes. The layers form an A-B-C-A-B-C... stacking sequence. The twin orientation relation is a 60° ⟨111⟩ rotation, as displayed in Figure 11.7 where three layers of each crystal are displayed. It is visible from this graph that the atoms of both crystal lattices coincide in one of the three layers A-B-C. The two crystals therefore form a Σ3 CS lattice. Interestingly, while the two crystals are in a twist rotation relation when looking onto the coincident (111) plane, they have the mirror relationship typical for a twin, when looking perpendicular to the [111] into the [1 $\bar{1}$0] direction. This mirror relationship corresponds to a 70° ⟨110⟩ symmetrical tilt boundary. For the exact twin boundary, it makes no difference whether it is expressed as

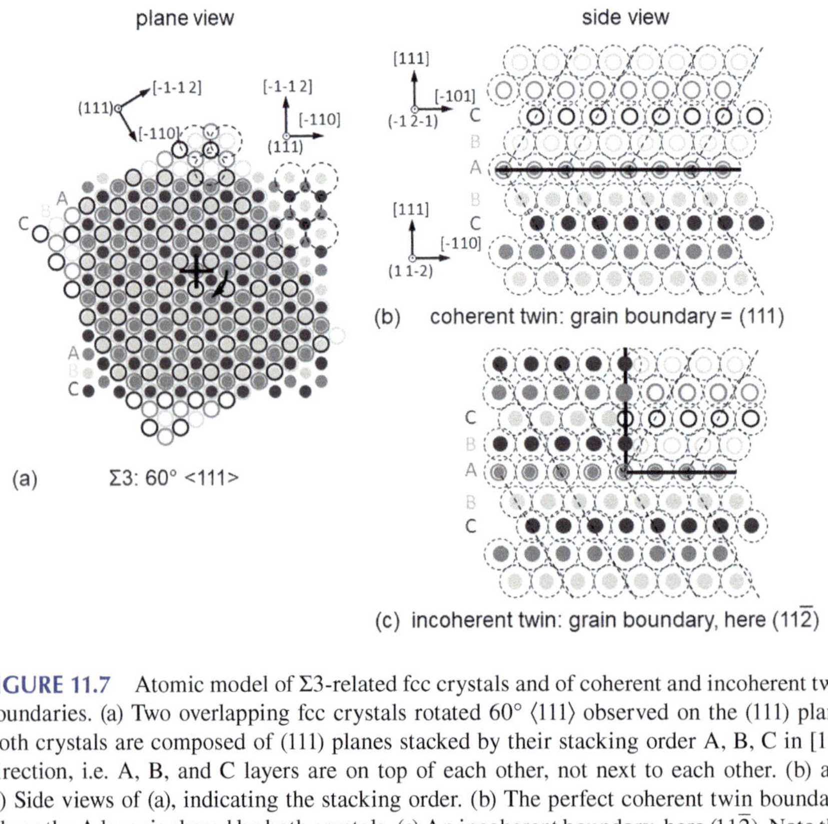

FIGURE 11.7 Atomic model of Σ3-related fcc crystals and of coherent and incoherent twin boundaries. (a) Two overlapping fcc crystals rotated 60° ⟨111⟩ observed on the (111) plane. Both crystals are composed of (111) planes stacked by their stacking order A, B, C in [111] direction, i.e. A, B, and C layers are on top of each other, not next to each other. (b) and (c) Side views of (a), indicating the stacking order. (b) The perfect coherent twin boundary, where the A layer is shared by both crystals. (c) An incoherent boundary, here (11$\bar{2}$). Note that in (b) and (c) atoms are not in the same plane.

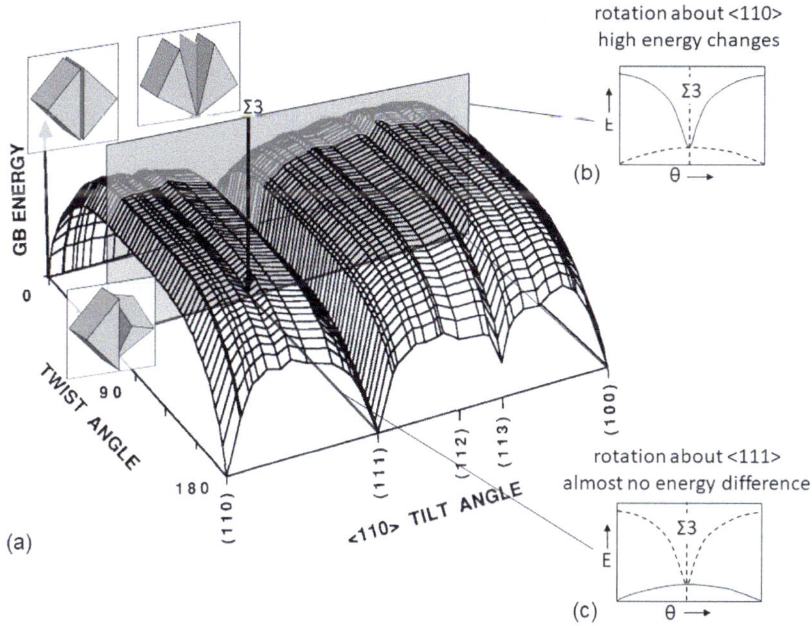

FIGURE 11.8 (a) Grain boundary energy landscape for fcc Cu, adapted from Wolf (1990). The graphic is to be read as indicated by the inscribed cubes: at the origin of the coordinate system (left upper corner) a boundary is created by sectioning a single crystal in two halves about the (110) plane without misorienting them. Along the "twist angle" axis the two crystal halves are misoriented by rotating them about an axis perpendicular to grain boundary plane. Along the "⟨110⟩ tilt angle" axis the two halves are tilted symmetrically to both sides of the grain boundary (plotted as plane in between). The position marked by the two crossing black lines at the bottom of the coordinate system is the coherent Σ3 grain boundary 60° ⟨111⟩, (111). This grain boundary has a particularly low energy. (b) The grain boundary energy changes drastically when tilting the grain boundary about ⟨110⟩ (solid line). (c) The energy changes only moderately when rotating about ⟨111⟩ (and thereby keeping (111) as grain boundary plane) (solid line).

a pure twist or tilt boundary. In contrast, when rotating it away from the exact Σ3 boundary, i.e. creating a vicinal boundary, it makes a great difference whether the rotation occurs about the ⟨111⟩ twist axis or the ⟨110⟩ tilt axis, as shown in the classic work of Wolf (1990) which is reproduced in Figure 11.8. The twist boundary keeps the (111) boundary plane, while the tilt boundary does not. Very similar results on the importance of the twist or tilt character of vicinal CSL GBs for GB properties have been published more recently by Homer et al. (2015).

The graph in Figure 11.8 also illustrates that the vicinal GBs in the proximity of special GBs may show a similar energy increase as small-angle GBs. This indicates that the increasing misorientation is accommodated by dislocations, similar to the case of SAGB. In contrast to these, however, the dislocations building up these deviations are so-called GB dislocations and not lattice dislocations as in the case of SAGB. This observation is expressed, empirically, by the Brandon criterion

(Brandon, 1966), which indicates the angular range, θ_B, in between which a special GB may be considered "special":

$$\theta_B = \theta_0 / \sqrt{\Sigma} \tag{11.6}$$

θ_0 is usually selected to be 15° (resulting in the typical threshold of 15° for SAGB with $\Sigma = 1$). For Σ3 twin boundaries this results in a maximum allowed deviation of 8.7°. Other criteria are also in use, including a narrower $\Sigma^{-5/6}$ dependency (e.g. Palumbo and Aust, 1992), but all of them remain rather simple approximations, as they don't take into account, for example, the axis of rotation around which the deviation occurs.

11.2.4 Refining the CSL Model: The DSC Model

An important consequence of the CSL concept arises when translating the two overlaying rotated lattices with respect to each other: if any of the black lattice sites in Figure 11.9 is translated such that it overlays with any one of the bright gray lattice

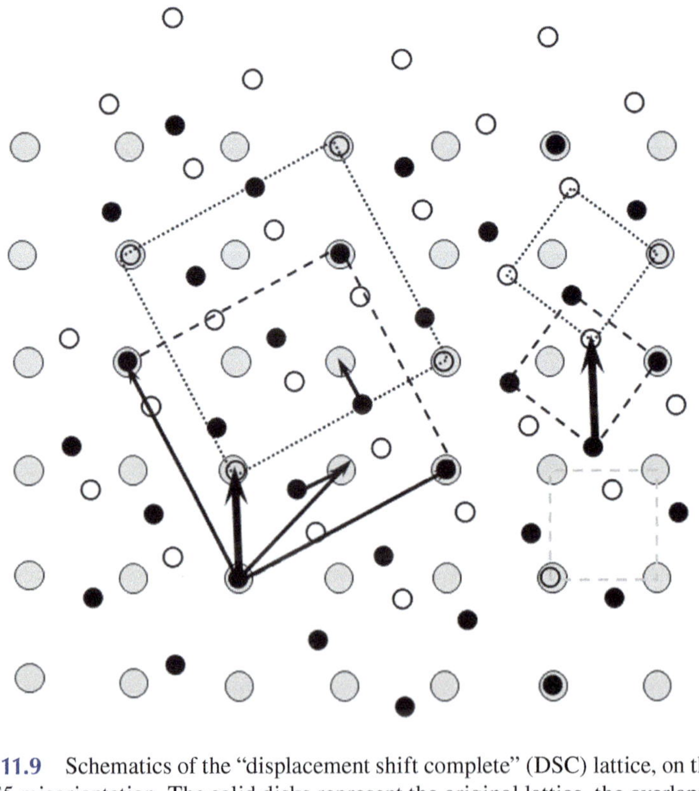

FIGURE 11.9 Schematics of the "displacement shift complete" (DSC) lattice, on the example of a Σ5 misorientation. The solid disks represent the original lattice, the overlapped solid disks the CSL lattice, indicated also by the broken line square. The vectors indicate all potential shift vectors that lead to a new position of the CSL lattice, indicated for one case by open symbols and by the dotted line square. The shortest of the possible shift vectors will act as Burgers vector of grain boundary dislocations in a CSL-type boundary.

sites, for example along one of the vectors indicated in the figure, the coincidence sites move but their relative positions do not change. This means the CS lattice is conserved. The set of vectors created in this way is the basis of the "displacement shift complete," DSC model (Balluffi et al., 1982). Like with lattice translation vectors, the determined DSC translation vectors may serve as Burgers vectors for GB dislocations. Due to reasons of energy minimization, it appears reasonable to assume that only the shortest vectors will be realized as Burgers vectors. If a highly ordered GB, e.g. a low-CSL boundary deviates from its ideal rotation angle, the deviation can be accommodated by GB dislocations, very similar to the accommodation of lattice rotations by lattice dislocations in small-angle GBs. A very educative text explaining the origin and use of the DSC model is found in the hyperscript of Föll (2019). This text also explains that each built-in GB dislocation creates a small step in the boundary, as it is observed, indeed, by TEM. It should be mentioned, once again, that the DSC model, like the CSL model, is purely misorientation based, it does not make a direct statement about the GB. Nevertheless, both models allow forecasting certain ideal GB structures for certain GB planes.

11.2.5 Generalization of CSL Lattices: The O-Lattice Theory

As explained above, the CSL and DSC models allow to identify potentially special GBs. However, because it relies on exact (or nearly exact) lattice site coincidences certain special GBs may be overlooked. The O-lattice model of Bollmann (1970) corrects this shortcoming and extends the concept of coincidences even to boundaries between crystals from different crystal structures or of different lattice constants. The model assumes that coincidence points do not necessarily need to lie on lattice points to guarantee a high order of a resulting interface.

The model is less intuitive than the CSL model, and, in the end, mathematically significantly more complex than the former. The complications of the O-lattice model are very well described by the already mentioned hyperscript of Föll (2019). This may serve as a very valuable introduction to the original reading of Bollmann. Basically the model assumes that there is an affine transformation from one lattice, labeled 1, to another one, labeled 2, which is described by a homogeneous transformation matrix, $\mathbf{A}$ (an affine transformation is a transformation that preserves lines and parallelisms but not necessarily distances and angles). This matrix may be composed of a rotational and a deformational part. Both involved crystal lattices are created from an asymmetric unit by applying translation vectors. The sets of all possible translation vectors for each of the two crystals are named $\mathbf{t}^{(1)}$ and $\mathbf{t}^{(2)}$, respectively. The fundamental equation of the O-lattice theory then claims that all coincidence sites, $\mathbf{x}^{(0)}$, of the two lattices, 1 and 2, are calculated by the following equation:

$$\left(\mathbf{I}-\mathbf{A}^{-1}\right)\cdot\mathbf{x}^{(0)}=\mathbf{t}^{(1)} \tag{11.7a}$$

or, inverted:

$$\mathbf{x}^{(0)}=\left(\mathbf{I}-\mathbf{A}^{-1}\right)^{-1}\cdot\mathbf{t}^{(1)} \tag{11.7b}$$

The vectors $\mathbf{x}^{(0)}$ form the O-lattice. Note that these vectors usually do not coincide with lattice points, but to some position in between the lattice points. Furthermore, it is important to stress that the $\mathbf{x}^{(0)}$ are not all at identical sites within the unit cell, but, depending on the selection of vectors $\mathbf{t}^{(1)}$ one may obtain a lot of different values for $\mathbf{x}^{(0)}$. In some cases, particularly for the case of CSL lattices, the number of $\mathbf{x}^{(0)}$ is small, i.e. the O-lattice is periodic, but in most cases it is aperiodic and there is an infinitive number of $\mathbf{x}^{(0)}$ vectors.

The mathematical and conceptual complication of the O-lattice theory arises with the fact that the matrix **A** is not unambiguous. In fact, even in the simplest case that **A** is just a rotation there are n^2 (n being the number of symmetry elements of the crystals) representations of the rotation matrix. Additionally, the rotation may also be described, if applicable, by a shear transformation. Bollmann argues that the transformation matrix, **A**, which leads to smallest movement of neighboring atoms, is the correct one. This corresponds to the criterion that the determinant, $|\mathbf{I}-\mathbf{A}^{-1}|$, must be minimum. As already mentioned above, for most boundaries, the O-lattice is non-periodic or long-periodic. The number of coincidence points that occur before a coincidence point is repeated at the same place in the unit cell is an important measure that appears in the theory. Only for few, potentially special, boundaries this value becomes very small and one is, for example, the Σ11 GB in cubic lattices. As Föll (2019) points out in his text, the Σ11 GB is indeed of low energy and thus relatively frequent in fcc materials. Finally, it is found that the O-lattice may not always be a point lattice but a lattice of lines or even of planes. These lines are interpreted as dislocation lines and correspond to the GB dislocations forecasted by the DSC model.

Similar to the CSL model, the O-lattice theory, in first place, does not contain information about the boundary plane. However, also similar to the CSL model, it forecasts optimum GB plane positions and, in extension to the CSL model, it also describes potential structures of boundaries and of the dislocation structures therein. Because of its mathematical challenges the O-lattice model is not very popular. Föll, at the end of his highly educational text, is rather pessimistic with respect to practical applications of the model. In contrast, there are a number of recent publications, for example an interesting paper about the origin and prevention of intergranular corrosion in pure iron (Yang et al., 2022) and very useful software developments that make use of the O-lattice theory. We mention here the software tools PTCLab (Gu et al., 2016), which is a computer tool to calculate phase transformation crystallography, and AIMSGB (Cheng et al., 2018), a software to calculate atomic GB structures.

11.2.6 The Importance of the Boundary Plane

The above sections indicated that the boundary plane is of equal or greater importance for the energy and properties of a GB than the misorientation. The role of the misorientation is mainly to provide conditions that allow formation of certain low-energy boundary planes. The probably best example for the importance of GB planes for GB analysis is, again, the Σ3 twin GB in fcc materials (see Bozzolo and Bernacki, 2020; Randle, 2004, for comprehensive overviews on twin boundaries). If the boundary plane follows the plane of coincidence (which is the (111) twist boundary) then a fully coherent boundary is created, where all atoms of the boundary belong to both grains. This boundary is highly ordered and has almost no energy associated with it. Bozzolo and Bernacki (2020) mention values in the order of 5% of the average LAGB energy, which is even lower than the stacking fault energy of these materials (which may be of the order of 15–30%). In contrast, if the boundary plane is not following the coherent plane, incoherent twin GBs of about ten times higher energy are created. The energy of {112} Σ3 may be as high as that of conventional LAGBs. At the same time, coherent {111} Σ3 GBs are very immobile, whereas their incoherent counterpart, e.g. {112} Σ3 shows an extraordinarily high mobility, being even ten times higher than that of random LAGBs. As explained by Bozzolo and Bernacki (2020), this high mobility may be related to the dislocation-based structure of incoherent twin GBs. It is interesting to note that there is at the same time, and in contrast to almost all other GBs in fcc materials, an efficient creation mechanism associated with this GB type, namely growth fault formation during annealing (see Bozzolo and Bernacki, 2020, for an overview on relevant literature) or a shear mechanism for the formation of deformation twins (Christian and Mahjan, 1995; De Cooman et al., 2018).

The combination of very particular properties of coherent twin boundaries in fcc materials, particularly with respect to corrosion, diffusion and mechanical strength, and the availability of special formation mechanisms, is the reason why "grain boundary engineering," GBE has become popular (e.g. Gourgues, 2002; Randle, 2010; Watanabe, 2011). GBE usually makes use of the fact that coherent twin boundaries may have a similar effect with respect to mechanical behavior as random large-angle GBs (rLAGB) (see, e.g., Pande et al., 2004), but show much lower diffusion rates and corrosion activity than the latter. An additional important reason for the beneficial effects of twin GBs with respect to crack growth or GB diffusion is that the coherent twin GBs disrupt the network of rLAGB and, therefore, interrupt diffusion paths or crack growth routes. A striking investigation in this regard is the work of Lehockey et al. (1998) who show that GBE by increasing the amount of coherent twin boundaries leads to much increased fatigue life, much increased creep resistance, higher strength and much longer corrosion resistance in a liquid salt solution at 850°C.

So far, the discussion was about twin boundaries in fcc alloys. But also for other GBs the crystallographic parameters may play an important role for the GB properties. One important example is the strengthening of metallic materials by GBs according to the Hall-Petch relation. There are two competing theories on the origin of the Hall-Petch relation which are critically discussed by Murr (2016): one assumes that GBs are obstacles to dislocation movement, leading to a pile-up of dislocations in front of the boundary. Once the pile-up stresses have become high enough, dislocations can transmit through the GB. The other theory assumes that GBs are acting as sources of dislocations, thus emitting dislocations and, therefore, leading to pile-ups behind the GB. For the first theory, it is possible to estimate the effect of GB plane and misorientation, assuming that the transmission of a dislocation through a GB leads to a residual dislocation in the GB. The situation is traced in Figure 11.10, which was reproduced from Ma et al. (2006). A dislocation arrives at the GB from one crystal on a well-defined slip system A. When transmitting through the GB it changes to a slip system B in the second grain. The difference of the Burgers vectors,

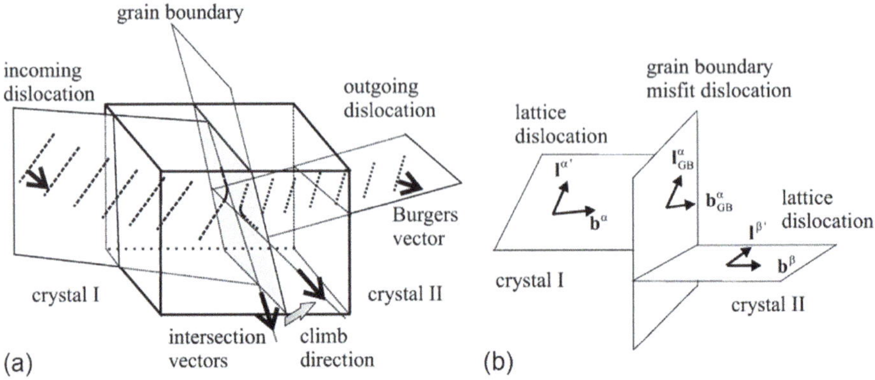

FIGURE 11.10 Schematic drawing of penetration events for mobile dislocations through a grain boundary. l^{α}, l^{β}, l^{α}_{GB} are the line vectors, and b^{α}, b^{β}, b^{α}_{GB} are the Burgers vectors of the dislocations in the moment of the penetration event (superscripts α and β indicate crystals I and II; the index GB indicates the grain boundary dislocation). (a) General geometry of slip systems and grain boundary; (b) vectorial description of the involved slip systems and grain boundary dislocations. (Adapted from Ma et al., 2006.)

$\mathbf{b}_A-\mathbf{b}_B$, remains in the GB, enhances its energy and, thus, leads to the observed increase of stress for dislocation movement. The problem is worked out in detail by De Koning et al. (2002). The paper explains that the slip transmission resistance is a function of only three variables: (i) the ratio of resolved stress on the incoming slip system to that on the outgoing slip system, (ii) the magnitude of any residual Burgers vector content left in the GB, and (iii), the angle between the traces of the incoming and outgoing slip planes in the GB plane. Ma et al. (2006) worked out a unique expression for the energy required for a dislocation to transmit through a boundary. Their calculation is based on the energy required to create a debris boundary dislocation, E^{α}_{GB}, caused by the action of a slip system, α, piling up on one side of the GB and creating dislocations on a slip system β on the other side of the GB. Figure 11.10 reproduces the geometric setup and defines the individual parameters:

$$E^{\alpha}_{GB} = \min_{\beta}\left(c\cdot\left\|\mathbf{b}^{\alpha}\otimes\mathbf{l}^{\alpha'} - \mathbf{b}^{\beta}\otimes\mathbf{l}^{\beta'}\right\|\cdot\frac{1}{2}Gb\right) \tag{11.8}$$

where c is an arbitrary constant, G is the shear modulus, b is the length of the lattice Burgers vector, $\mathbf{b}^{\alpha}$ is the Burgers vector of slip system α, and $\mathbf{l}^{\alpha'}$ the line element of dislocation parallel to the GB (analogous for slip system β). The position of the GB is implicit in the value of $\mathbf{l}^{\alpha'}$ which is the projection of the dislocation line $\mathbf{l}^{\alpha}$ on the GB plane:

$$\mathbf{l}^{\alpha'} = \mathbf{l}^{\alpha}\left(\mathbf{n}^{GB}\times\mathbf{n}^{\alpha}\right) \tag{11.9}$$

where $\mathbf{n}^{GB}$ and $\mathbf{n}^{\alpha}$ are the normal vectors of the GB and of the slip system α, respectively (analogous for slip system β).

$\mathbf{b} \otimes \mathbf{l} = \begin{pmatrix} b_1 l_1 & b_1 l_2 & b_1 l_3 \\ b_2 l_1 & b_2 l_2 & b_2 l_3 \\ b_3 l_1 & b_3 l_2 & b_3 l_3 \end{pmatrix}$ is the so-called dislocation tensor, formed by the dyadic product of **b** and **l**, and $\|\mathbf{t}\| = \sqrt{\Sigma_{i=1}^{3} \Sigma_{j=1}^{3} |t_{i,j}|}$ is the Euclidian norm of the tensor.

With this equation the geometric resistances of different individual GBs to slip transmission can be compared. Of course, details of the atomic order of the GB and or detailed dislocation nucleation mechanisms are not included in this approach. The approach was used, for example, to simulate, using a crystal plasticity finite element framework, the effect of GBs on the evolution of crystal orientations in deformed bicrystals, as investigated experimentally by Zaefferer et al. (2003). A similar approach was taken by Bieler et al. (2009), who developed a semi-empirical so-called "fracture initiation parameter," FIP, for GB fracture in a brittle γ-TiAl intermetallic.

The atomic structure of GBs becomes highly important, when considering, for example, diffusion or segregation problems at GBs. An interesting problem in this regard is intergranular corrosion in stainless steel or stress corrosion cracking in a variety of alloys (e.g. Gertsman and Bruemmer, 2001). In recent studies An et al. (2018) and Tsai et al. (2021) investigated the effect of intergranular corrosion in a 316L stainless steel using 3D EBSD (see Section 11.4.3) and 3D atom probe. The steel sample was sensitized, i.e. annealed at a temperature that leads to formation of $M_{23}C_6$ carbides at the GBs. Subsequently the material was exposed to an aggressive solution of H_2O_2 and HF, which led to the preferred corrosion of certain GBs. The basic mechanism behind this is that the formation of Cr-carbides on the GBs leads to a local depletion of Cr at these GBs. As a consequence, these areas do not form a corrosion-protecting dense Cr-oxide surface film and are thus easily attacked. In the first study by An et al. (2018) it turned out that GBs possessing {111}-type GB planes, irrespective of the particular misorientation, were particularly resistant to corrosion while GBs with {110} GB planes frequently showed strong corrosion. It was concluded that the atomic packing density of these GBs is controlling the diffusion rate of Cr along these boundaries and, therefore, the formation kinetics of Cr-carbides. {111}-type GBs have a very dense structure, while {110}-type GBs have a particularly open atomic packing. Figure 11.11 illustrates the packing density in a stereographic SOT of the GB planes for cubic crystal symmetry.

A follow-up study by Tsai et al. (2021) showed that the relations are not that straightforward, however. For Σ5 GBs, for example, it was found that they corrode very little, irrespective of the GB plane, while Σ7 GBs almost always corrode heavily. The reason appears to be the GB energy anisotropy, which is small for Σ5 GBs and large for Σ7 GBs. As a consequence, Σ7 GBs tend to facet strongly and some of these facets are well suited for nucleation of the cubic $M_{23}C_6$ carbide (with the so-called cube-on-cube orientation relationship, $(001)[100]_{fcc} \parallel (001)[100]_{M23C6}$). Although these studies gave hints on the effect of GB planes on thermodynamic and kinetic properties, science is still far away from understanding these properties in a more general way. One reason for this is the enormously big parameter space for GBs and the difficulty to measure it. As shown by Rohrer et al. (2004), the amount

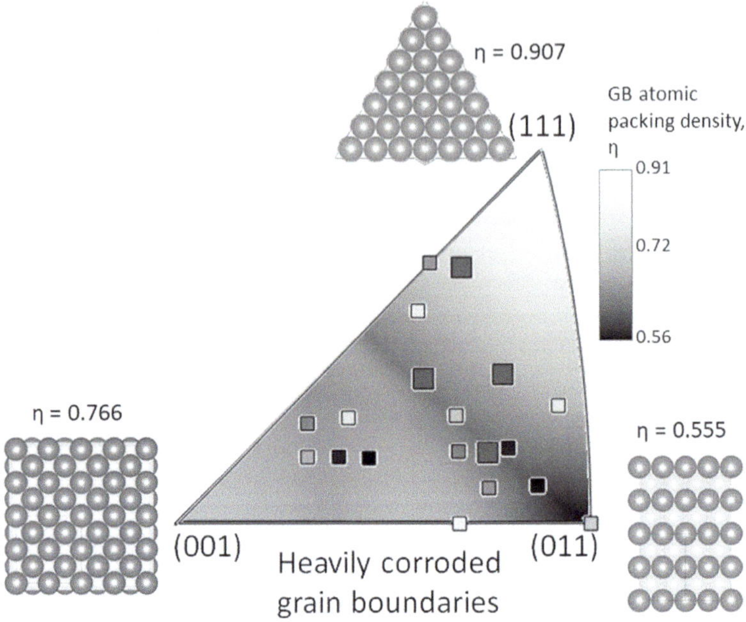

FIGURE 11.11 Grain boundary plane orientations that lead to heavy corrosion on a sensitized stainless steel SS316L. Symbols of similar shape and gray value form the same boundary (left and right side). The standard orientation triangle is shaded according to the GB atomic packing density which ranges between 0.907 for the (111) plane and 0.555 for the (110) plane. Most corroded grain boundaries are in the area of low packing densities. (Data from An et al., 2018.)

of distinguishable GBs for a cubic material ranges in the order of $2 \cdot 10^5$ if all five rotational boundary parameters are measured with a 5° resolution. The amount of measured GBs must be bigger in order to allow a reasonable statistical relevance. The measurement possibilities are subject of Sections 11.3 and 11.4.

11.2.7 Five-Parameter Grain Boundary Description

As described above, the GB plane is more important for determining GB properties than is the misorientation. Nevertheless, the latter is insofar important as it allows the formation of particular boundaries. It appears, therefore, essential to describe GBs by all of its five rotational parameters, namely the misorientation and the GB normal. A comprehensive overview on the five-parameter description has been published by Rohrer (2011). The five-parameter space is usually presented in sections at constant misorientation, i.e. for every selected misorientation, consisting of rotation axis and angle, the GB plane normal distribution is given in a pole figure as sketched in Figure 11.12. The density of GB planes is usually given in multiples of the random distribution (m.r.d.), which needs to be calculated separately for every

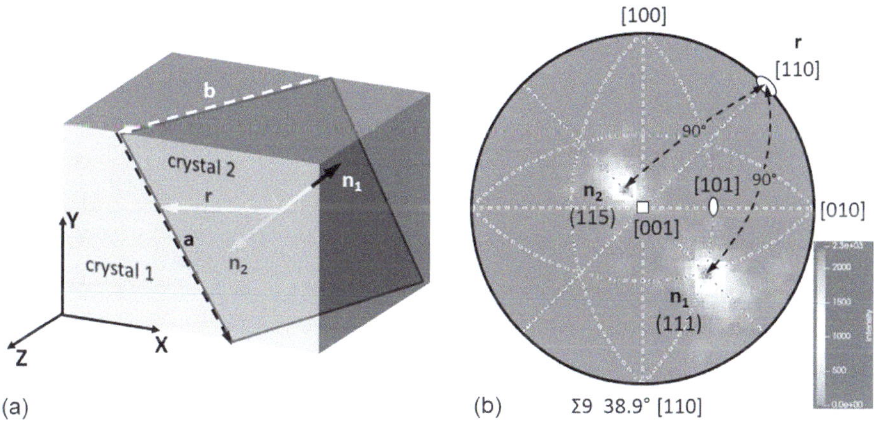

FIGURE 11.12 Principles of five-parameter grain boundary analysis. (a) A boundary between two grains with a given misorientation ω, [uvw]. The grain boundary normals, **n**, with respect to grains 1 and 2, as well as the misorientation axis, **r** (or [uvw]), are displayed. Note that the position of **r** can only be fixed when the procedure described in the text has been performed and a coherent description of GB planes and misorientations has been obtained. In the present example, **r** is in the GB plane, indicating a tilt grain boundary. The two vectors in the GB plane, **a** and **b**, obtained from 3D grain boundary measurement, allow calculation of **n**. (b) Stereographic projection of the GB planes in a misorientation reference coordinate system. The misorientation, here as an example Σ9 38.9° [110], fixes the position of the grain boundary plane, as described in the main text. Here it is found that the misorientation axis is at 90° to both grain boundary normals, i.e. the boundary is a pure tilt boundary. Only after determination of the angle between the misorientation axis and the GB plane, the vector **r** can be drawn in (a).

misorientation. The orientation of the GB normal is obtained from two non-colinear vectors in the boundary, **a** and **b**, as shown in Figure 11.12a. These vectors are measured in sample coordinates and the resulting GB normal, **n**, are also given in sample coordinates. **n** is calculated from the cross product of **a** and **b**:

$$\begin{pmatrix} n_x & n_y & n_z \end{pmatrix} = \begin{bmatrix} a_x \\ a_y \\ a_z \end{bmatrix} \times \begin{bmatrix} b_x \\ b_y \\ b_z \end{bmatrix} \tag{11.10}$$

Note that line vectors are reciprocal vectors while the column vectors are real vectors, which is important in order to keep the correct order for later matrix multiplications. The GB normal is then plotted in a GB pole figure, as shown in Figure 11.12b. It is important to note that the reference system for the pole figure is the crystal coordinate system of one of the two involved crystals. In this pole figure, the misorientation axis needs to be fixed first. In many cases this axis may be selected according to the convention that the misorientation axis should be in the SOT and the rotation angle should be smallest. This is, however, not always the best choice for easy interpretation of the pole figures. For example, the misorientation axis of the Σ9 misorientation would be [101] in the SOT, but it is usually selected as [110], in order to be on the border of the stereographic projection. For a coherent description of GB misorientation and GB plane the correct symmetry operators have to be selected according to the following scheme. In a first step the GB normal, **n**, measured in sample coordinates, is transferred into crystal coordinates of the first crystal with the following operation:

$$\begin{pmatrix} h & k & l \end{pmatrix}_c = \begin{pmatrix} n_x & n_y & n_z \end{pmatrix}_s \mathbf{g}_1^T \tag{11.11}$$

The superscript "T" indicates the transposed matrix. Next, all misorientation variants are calculated:

$$\Delta\mathbf{g}_{ij} = (\mathbf{g}_1 \ \mathbf{S}_i)(\mathbf{g}_2\mathbf{S}_j)^{-1} \quad \text{with} \quad i = 0\ldots n, \quad j = 0\ldots m \tag{11.12}$$

From all these variants the one, $\Delta\mathbf{g}_{ij}$, with the desired misorientation axis and smallest misorientation angle is selected. The so-selected symmetry operator, $\mathbf{S}_i$, is then used to transfer the original GB description into one coherent with the misorientation description:

$$\begin{pmatrix} h & k & l \end{pmatrix}_{c,i} = \begin{pmatrix} h & k & l \end{pmatrix}_{c,0} \mathbf{S}_i \tag{11.13}$$

This GB normal is finally plotted into the GB pole figure. The same procedure can be pursued for the GB described with respect to grain 2. Here, however, the symmetry operator $\mathbf{S}_j$ has to be used.

Homer et al. (2015) discuss the same problem of selecting the correct, misorientation-coherent, GB plane description. They propose to select for every misorientation a special SOT which takes into account the symmetry of the respective misorientation. For example, for Σ3 misorientations the SOT is centered on the [111] rotation axis of this misorientation, while the Σ5 misorientation is centered on [001] and Σ9 on [011]. As the authors argue, their SOT selection facilitates observation of symmetries in GB properties (e.g. energy or mobility). In Figure 11.13 we reproduce some of their results on structure-property relationships—including GB energy, excess volume per unit GB area, and temperature-dependent mobility trend—for several low-Σ CSLs.

The algorithm described above has been used in a large number of papers on GB plane distributions. One of the first materials that was analyzed was a hot-pressed MgO polycrystalline specimen (Saylor et al., 2003). Here GBs were still manually digitized. This paper also discusses the abovementioned symmetry considerations in some detail. In the paper by Rohrer et al. (2006) the stereological method was applied to determine GB plane distributions in GB engineered Brass. This technique and the results from this work will be discussed further in Section 11.3.3. A large number of further references to the five-parameter GB analysis is found in the already mentioned overview by Rohrer in 2011.

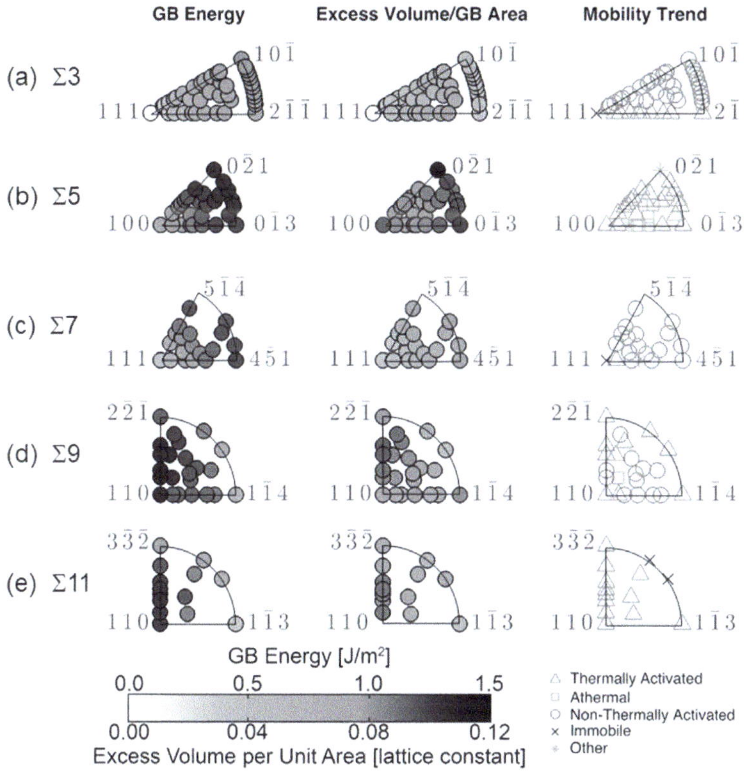

FIGURE 11.13 Various calculated structure-property relationships for low-Σ CSLs displayed in misorientation-adapted SOTs. GB energy, excess volume per unit GB area, and temperature-dependent mobility trend are plotted in the respective SOT for (a) Σ 3, (b) Σ 5, (c) Σ 7, (d) Σ 9, and (e) Σ 11 CSLs, respectively. Note that the SOTs are different for every misorientation. (Adapted from Homer et al., 2015.)

11.3 CRYSTALLOGRAPHIC ANALYSIS OF GRAIN BOUNDARIES AND SURFACES

There are numerous experimental tools to measure interface orientations and their distributions which can be grouped into non-destructive and destructive techniques. For the non-destructive techniques, x-ray based tomography techniques are the most frequently used ones. These techniques will be subject of Section 13.2. The destructive techniques are mostly based on serial sectioning combined with EBSD for orientation measurements; Sections 11.3.2 and 11.4.1 through 11.4.3 will focus on these techniques.

Before continuing the discussion on measurement of GB character, however, we will make a short excursion to the crystallographic characterization of free surfaces as they appear particularly on fracture surfaces or crystalline surface facets of particles or fibers. These measurements are performed by a combination of EBSD and surface inclination measurements using a technique called stereo-photogrammetry.

This technique will be the subject of Section 11.3.1. Next, we will treat the characterization of internal interfaces, in particular GBs, but also, for example, slip planes, for the case that they are visible on two independent surfaces of the sample. The different variations of this approach will be described in Section 11.3.2 on sectioning techniques. Finally, in Section 11.3.3, a technique for the characterization of GBs, called stereological analysis will be introduced. This technique requires only one section of every boundary and uses statistical information obtained from many boundaries to determine the most probable GB character.

11.3.1 Stereo-Photogrammetry for Observation of Free Surfaces

The determination of the crystallographic orientation of free surfaces, e.g. fracture surfaces or crystallographic facets of particles, can be of high interest for the understanding of processes like stress corrosion cracking, hydrogen embrittlement, or crystallographic growth of particles. Such surfaces include intergranular or transgranular microcracks (Liu et al., 1992), fracture surfaces (e.g. fatigue fracture) in an Al-alloy (Slavik and Gangloff, 1996), in a Co-Cr-Mo alloy (Wang et al., 2005) or in an Al-Cu-Li alloy (Ro et al., 2005), cleavage fracture in iron (Ayer et al., 2006) or crystallographic facets of particles (Randle and Hoile, 1998; Kameda et al., 2007).

In order to obtain crystallographic information on free surfaces, it is first necessary to measure the crystal orientation of the underlying crystal. This is in most cases and most conveniently accomplished with EBSD. Next, the surface normal needs to be determined using a photogrammetric technique. Finally, the crystal orientation and the surface normal are combined by straightforward calculations.

Photogrammetry, or stereo-photogrammetry, in the SEM is a well-established technique for determining the coordinates of an exposed facet plane relative to a fixed coordinate system in the microscope (Hilliard, 1972). These data can then be coupled with diffraction data to give the crystallographic orientation of the facet. The photogrammetric procedure involves producing a series of micrographs of the facet using at least two tilt angles (e.g., 0°, 10°, and 20°) and identifying the projected (x,y) coordinates of point features in the plane, relative to the reference axes, at the various tilt angles. If the coordinates of one point at tilt angles θ_1 and θ_2 are (x_1, y_1) and (x_2, y_2), then the true coordinates (x_t, y_t, z_t) of the point in space are given by (Themelis et al., 1990)

$$x_t = \frac{(x_1 \sin\theta_2 - x_2 \sin\theta_1)}{\sin(\theta_2 - \theta_1)} \tag{11.14a}$$

$$y_t = y_1 = y_2 = \frac{(y_1 + y_2)}{2} \tag{11.14b}$$

$$z_t = \frac{(-x_1 \cos\theta_2 + x_2 \cos\theta_1)}{\sin(\theta_2 - \theta_1)} \tag{11.14c}$$

The coordinates of three points on the plane can then be used to compute the plane normal (see Section 11.2.7). This brief description of the photogrammetric procedure is elaborated by Themelis et al. (1990) and Slavik et al. (1993).

The stereographic information on surface inclination is in the next step coupled to orientation information measured by EBSD (Davies and Randle, 2001; Davies et al., 2002). When measuring the crystal orientation care must be taken to ensure that the specimen coordinate system is identical for both photogrammetry and diffraction. In some cases, if the facet surface is not too far off from a 70° tilted surface (EBSD patterns can be seen approximately between 50° and 80° of surface inclination) and if no shadows obscure significant parts of the EBSD pattern it is possible to measure the crystal orientation directly from the surface facet of the observed object. As the geometry of the EBSD pattern mainly depends on the orientation of the crystal and is independent of the surface orientation, it is the then immediately possible to correlate the photogrammetric data with the orientation data. This approach has been demonstrated in great detail by Kameda et al. (2007). In most cases, however, patterns are not immediately visible on the rough or inclined facets. In this case it is either necessary to create, using metallographic techniques, a flat section perpendicular to the facet of interest and measure the orientation of the same grain on this surface. It is, however, more straightforward to use a focused ion beam (FIB) to mill small 70°-inclined, flat segments on the facet surfaces and measure on these surfaces. The geometry that needs to be adapted for the FIB milling and EBSD observations is the same as that required for 3D EBSD on flat surfaces, as it has been described, for example, in Zaefferer et al. (2008). This process, sketched briefly in Figure 11.14, can reveal a number of facet

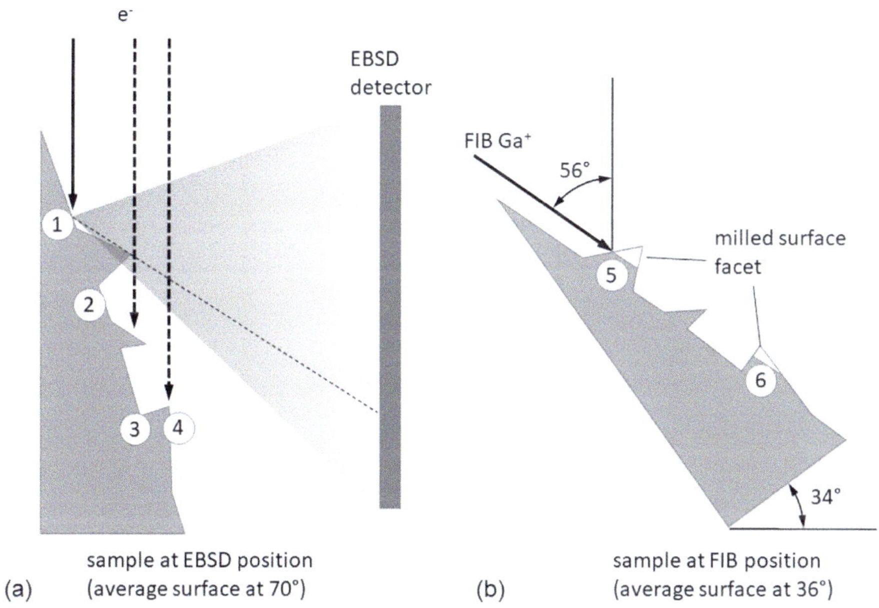

FIGURE 11.14 (a) Observation of a faceted surface by EBSD. Problems are caused at (1): shadowing the detector (along the dotted line to the detector), (2) shadowing of the primary beam, (3) unsuited surface inclination. At (4) EBSD observation is possible directly at the tip of the facets. (b) Same sample tilted to FIB-milling position in a FIB-SEM. The FIB beam is used to cut small facets at (5) and (6) at the correct inclination for EBSD observation.

surfaces in one single measurement approach without taking the sample out of the microscope. A similar direct approach has been proposed by Ro et al. (2005). These authors reported an uncertainty in facet crystallography determination to be 2°–5°. A highly interesting application of the EBSD-coupled stereo-photogrammetry is the correlation of chemical GB segregation with GB plane orientation. This can be achieved by coupling the technique with Auger electron spectroscopy (Pang and Wynblatt, 2005).

11.3.2 Sectioning Technique Principles

The principle of using sectioning techniques to obtain the orientation of a planar surface (e.g., a GB, a slip plane or a free, exposed facet) relies on the trace of the plane being evaluated on at least two sections through the specimen. There are three stages in the analysis (Randle, 1993, 1995b):

- Acquisition of the inclination of the plane with respect to reference axes in the specimen
- Measurement of the orientation of the grain(s) that abut the plane, relative to the *same* reference axes as the inclination measurement
- Calculation of the direction cosines of the plane from the orientation and inclination parameters

The plane inclination is described by the angles α and β, which are shown in Figure 11.15a with respect to the orthogonal reference axes XYZ in the specimen.

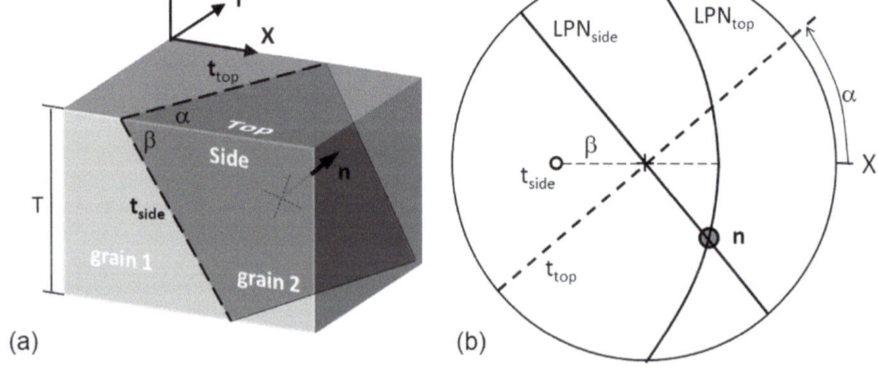

FIGURE 11.15 Trace analysis for determination of grain boundary plane and grain boundary character. (a) Grain boundary separating two grains, observed from two perpendicular surfaces, named "Top" and "Side." XYZ are the specimen axes, α and β are the angles relating the grain boundary trace to the X-axis, and *T* is the experimental section depth. For an exposed rather than an unexposed facet, the right-hand part of the diagram is missing. (b) Stereographic projection showing the relationship among XYZ, α and β from (a). The grain boundary traces $\mathbf{t}_{top}$ and $\mathbf{t}_{side}$ are plotted either as a line (for the top section) or as a point (for the side section). Loci of the potential plane normals (LPNs) are plotted at 90° to the traces. Crossing of the two LPN indicates the grain boundary normal, **n**.

Clearly, the evaluation applies only to surfaces that actually are planar, and not to curved surfaces such as GBs in the course of recrystallization. For convenience, X is horizontal in the microscope and EBSD measurements are made on the XY reference surface. In case of exposed surfaces one of the grains in Figure 11.15a is missing.

Experimentally, the sectioning process is usually carried out by the standard metallographic technique of mounting the specimen in a suitable medium, grinding down the exposed surface of the specimen, followed by appropriate polishing/finishing steps. These procedures can be performed manually or automatically using equipment designed for that purpose. Although sectioning is usually performed in a parallel manner, a series of oblique double sections can also be obtained and combined with EBSD data, which can offer better representation of the bulk microstructure and certain experimental advantages compared to serial sectioning (Homer et al., 2006). A potential alternative to the traditional grinding method for sectioning is ultramicrotomy, that is, sectioning with a diamond knife. This technique is a mainstay of TEM specimen preparation in the life sciences and has been adapted for some materials applications (Malis and Steele, 1990).

Relatively hard materials can be polished on two mutually perpendicular and adjoining surfaces so that the plane trace is revealed on both surfaces, as shown in Figure 11.15a. The GB forms two traces on the XY and the XZ surface, $\mathbf{t}_{XY}$ and $\mathbf{t}_{XZ}$. The angles α and β can easily be measured in the SEM by tilting the specimen about the X-axis so that both angles are visible together. The angles are corrected to account for the microscope tilt angle θ as follows (Randle and Dingley, 1989):

$$\alpha_T = \tan^{-1}\left(\frac{\tan\alpha_M}{\cos\theta}\right), \quad \beta_T = \tan^{-1}\left(\frac{\tan\beta_M}{\cos(90^\circ - \theta)}\right) \tag{11.15}$$

where α_M and β_M are the measured angles after tilting. In Figure 11.15b the GB traces $\mathbf{t}_{XY}$ and $\mathbf{t}_{XZ'}$ are plotted in a pole figure either as a line (for the XY section) or as a point (for the XZ section), respectively. The LPNs are plotted at 90° to the two GB traces (see Figure 11.5). The intersection of the LPNs then indicates the GB normal, $\mathbf{n}$. An example of the application of the two-surface sectioning approach to interface crystallography is a study of fatigue crack initiation in copper (Liu et al., 1992). Fatigue cracks often initiate as a result of impingement of persistent slip bands against GBs, and application of this method allowed the GB crystallography to be determined and related to the slip band interaction.

The above-described technique of measuring planes across two mutually perpendicular interfaces has been named "pseudo 3D EBSD" (see Section 11.4) by Zaefferer and co-workers (Mandal et al., 2014; Afshar and Zaefferer, 2015). Although the technique allows only characterization of those GBs that end at the sharp edge of the sample, the technique has the charm that the so investigated GBs can be investigated further, for example by a corrosion experiment or by APT investigations (Mandal et al. 2014). Wang and Zaefferer (2017) showed by measurements on coherent twin boundaries that GB planes can be determined as accurate as 1°.

The second procedure that can be used to obtain plane inclinations is to section the specimen parallel to the EBSD measurement surface—the XY section in Figure 11.15a (Kral et al., 2000; Spanos et al., 2008; Rowenhorst et al., 2010). If the section depth is known, then the plane inclination can be calculated from accurate measurement of the trace of the plane on the XY surface before and after sectioning. The sectioning depth, where a layer of material is removed by grinding, can be measured by inserting hardness indents into the specimen surface at the start of the procedure. The reduced size of the indent after sectioning, coupled with the knowledge that the angle between opposite faces of the diamond indenter is 136°, allows the depth of material removed to be calculated. The image of the hardness indents also facilitates accurate alignment between serial sections, which is crucial. The thickness of specimen removed, T, is given by

$$T = \frac{(D_b - D_a)}{\tan(136°/2)} \tag{11.16}$$

where D_b and D_a are the average distance from the center to the closest point on the edge of the indent before and after grinding, respectively. If the offset on the specimen surface of a GB before and after grinding is O, then β is given by

$$\beta = \tan^{-1}\left(\frac{O}{T}\right) + 90° \quad \text{or} \quad \beta = 90° - \tan^{-1}\left(\frac{O}{T}\right) \tag{11.17}$$

depending on the sense of the plane inclination in the specimen. The trace angle α is measured directly from the boundary trace on the specimen surface. Once the grain orientation has been measured near to the plane, its crystallographic indices are calculated as follows. Two vectors that lie in the boundary plane are given by A and B, which are obtained from α and β, respectively. The coordinates of A and B are given by

$$A = \begin{pmatrix} \cos\alpha \\ \cos(90° - \alpha) \\ 0 \end{pmatrix}, \quad B = \begin{pmatrix} \cos\beta \\ \cos(90° - \beta) \\ 0 \end{pmatrix} \tag{11.18}$$

The cross product of A and B gives N, the plane normal, which can then be related to the crystal axes of the grains through the orientation matrix. The whole procedure is illustrated on the stereogram in Figure 11.15b.

A valuable application of single section trace analysis is the crystallographic analysis of slip traces (Blochwitz et al., 1996; Raabe et al., 1997) By this method, slip traces in Ni_3Al were found using EBSD to be on {111} (Lin and Pope, 1996). Nandy et al. (2021) applied slip trace analysis in Mg alloys to proof the influence of Ca and Zn on the activation of ⟨c+a⟩ slip. Although the slip trace analysis is not perfectly unique—some slip systems result in similar slip traces—the approach is in most cases significantly better suited for deformation analysis than the tedious slip system analysis in TEM because it allows obtaining much larger and therefore statistically relevant data sets.

With the advent of automated serial sectioning techniques—which will be described in detail in Section 11.4—the above-described manual sectioning appears somewhat outdated and no longer competitive in accuracy and resolution. Nevertheless, the described principles remain true.

11.3.3 Stereological Analysis

Section 11.2 has given good evidence that the boundary plane is of equal or greater importance for the energy and properties of a GB than the misorientation. Given the importance of the boundary plane, it is at first sight surprising that it has been largely omitted from EBSD data. This is because the serial sectioning techniques involved are quite tedious, as described in the previous section. In recent years, interest has been generated in using the information obtained from the GB trace direction on the specimen surface, and dispensing with the serial sectioning step, which has ultimately led to the stereological method for determination of GB crystallography, which is described in the following text.

It has already been mentioned that EBSD, as well as orientation mapping in the TEM, are able to measure up to four out the five parameters of a GB, namely the three misorientation angles and the crystallographic direction of the trace vector, t, of the GB; only the boundary inclination angle is unknown. In some cases, e.g. in the case of coherent twins, the trace alone gives a relatively clear indication on the character (Randle and Davies, 2002), first because the trace of such a coherent twin boundary is almost perfectly straight (it is the intersection of the flat surface with the flat crystal plane), and second it follows accurately the two {111} planes of the abutting crystals. With this knowledge, the correct twin boundary plane can be determined from the trace and its three-dimensional position is uniquely fixed. This situation is sketched in Figure 11.16, where the twin boundary between grains 1 and 2 is a perfect coherent twin, indicated by the fact that the common (111) poles of the two crystals are exactly coinciding with the $\mathrm{LPN}_{1,2}$ (see Figure 11.5) of the GB trace $t_{1,2}$ (Figure 11.16b). The twin GB between grains 1 and 3 shows two facets. The facet $t_{1,3}^{A}$ marked in black is close to a coherent twin boundary but deviates by about 3° (the distance between the common (111) pole (marked by a circle) and the $\mathrm{LPN}_{1,3}^{A}$). The gray facet is an incoherent boundary. It is very probable (but not certain) that it corresponds to the common (012) plane which is close to the gray $\mathrm{LPN}_{1,3}^{B}$ (marked in Figure 11.16c).

An extension of the trace analysis is the so-called *stereological analysis* to extract boundary plane data from a single section. The details of the method—which has been adapted from established techniques to determine habit planes from embedded crystals—are described by Saylor et al. (2004); only a brief summary of the principles is given here. First, one needs to understand that a GB trace is a crystallographic direction within this GB. If a certain type of GB appears frequently in a microstructure then a section through this microstructure will show a number of traces of these GBs. If a sufficiently large number of traces from different boundaries of the same type is measured, then it should be statistically feasible to reconstruct the GB type. For this idea to work, it is important that the GB planes are non-randomly distributed, i.e. that GBs of certain misorientations have a tendency to form particular boundaries.

In practice the process is done as follows (see Figure 11.17 for a graphical illustration): first a sufficiently large orientation map is acquired using EBSD mapping. Saylor et al. (2004) indicate that several tens of thousands GBs are required for the analysis to work reliably. Potentially a subset of GBs with particular misorientations is selected. Figure 11.17a shows an idealized map where four GBs are highlighted.

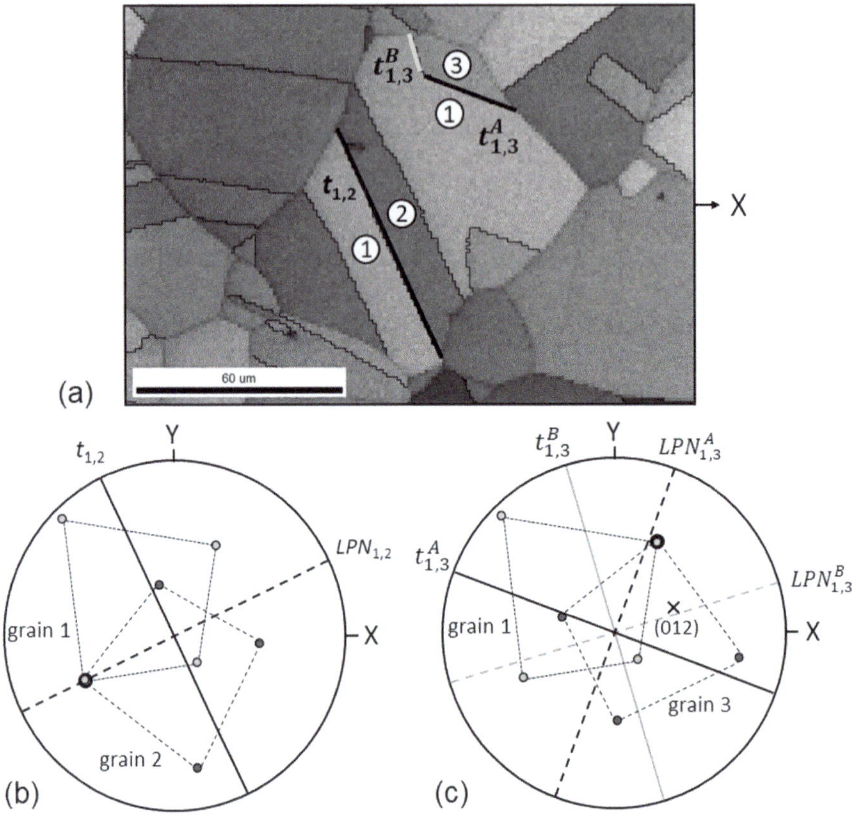

FIGURE 11.16 Discrimination between coherent and incoherent twin planes in fcc materials using stereographic analysis. (a) EBSD map showing a recrystallized structure with several twin boundaries. (b) The twin boundary between grains 1 and 2 is a perfect coherent twin, indicated by the fact that the common (111) poles (displayed as small circles) of the two crystals are exactly coinciding with the LPN of the grain boundary trace. (c) The twin grain boundary between grains 1 and 3 shows two facets. The facet marked in black is close to a coherent twin boundary but deviates by about 3° (the distance between the common (111) pole (marked by a circle) and the LPN). The gray facet is an incoherent boundary. It is very probable (but not certain) that it corresponds to the common (012) plane (marked by a cross) which is on the LPN.

Next, the GBs are identified and vectorized, which means straight boundary segments are approximated by straight lines. The results are trace vectors in sample space, $\left[t_x\ t_y\ 0\right]_s$ (the zero value $t_z = 0$ appears because the vectors are surface vectors). Then the trace vectors are converted into crystal space using the respective crystal orientations, **g**:

$$\begin{bmatrix} t_x \\ t_y \\ t_z \end{bmatrix}_{c,i} = (\mathbf{g} \cdot \mathbf{S}_i) \cdot \begin{bmatrix} t_x \\ t_y \\ 0 \end{bmatrix}_s \tag{11.19}$$

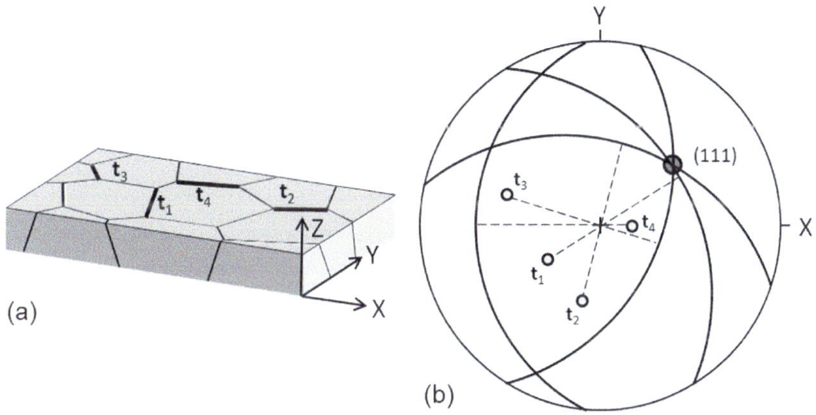

FIGURE 11.17 Principle of stereological analysis of grain boundaries. (a) Sample with 2D orientation map on the surface and unknown grain boundary inclination. Four grain boundary traces, i.e. surface vectors, are selected. (b) Stereographic projection of crystal coordinates. Trace vectors from (a) are plotted in their respective crystal coordinates as disks. Great circles around the individual trace vectors (at 90° from the trace vectors) indicate the potential grain boundary normals. Intersections of great circles indicate probable grain boundary normals. The probability increases with the number of intersecting circles. Here grain boundaries 1, 2, and 3 may be all of (111) type, because they all intersect at (111), while grain boundary 4 remains unknown.

Note that the symmetry of misorientations has to be considered, as already discussed in Section 11.2.7: if only GBs of a particular misorientation are selected (e.g. 60° [111]), then, in order to enforce consistency between misorientation and trace vector, that symmetrically equivalent orientation matrix, $\mathbf{g}\cdot\mathbf{S}_i$, has to be selected, which yields the desired axis-angle pair in the misorientation calculation. If, in contrast, no particular misorientation is prescribed, then all orientation permutations have to be used and all resulting traces need to be considered simultaneously. In the cubic case, for example, 24 symmetrically equivalent trace vectors result from every GB.

In the next step all potential GB plane normals that correspond to the considered trace vector need to be determined. These normals are perpendicular to the trace, i.e. $\mathbf{t}\cdot\mathbf{n}_p = 0$. In a stereographic projection these potential normals lie on great circles around the trace vector. In Figure 11.17b the trace vectors $\mathbf{t}$ of the four marked GBs are plotted as small disks into the stereographic projection of crystal space. The great circles of the potential plane normal vectors are plotted as solid lines. For those GB traces that refer to GBs of the same plane type the great circles intersect in one point, which corresponds to the searched GB plane normal. In the example in Figure 11.17, three great circles, t_1, t_2, and t_3, intersect in one point, corresponding to a (111) plane while a fourth one, t_4, does not, indicating that its GB plane is different from that of the other three and remains unknown. To obtain quantitative results the resulting unit vector space (here plotted into a stereographic projection) is finally binned into small angular sections and its vector density determined for every point.

Note that the stereological approach process is a statistical one and only yields probabilities for the correct boundary traces. Intersections of many great circles potentially correspond to a high probability of frequently occurring GB planes. In fact, the boundary trace generates a set of possible boundary planes. The probability that the set contains the true boundary plane is 1, whereas the probability that other false planes are included in the set is less than 1. Therefore, if a sufficiently large number of observations are made, the true boundary plane will accumulate more than the false planes and form a peak in the distribution. The background (false planes) is then removed, based on the assumption of random sampling. From the remaining peaks, the ratios of the observed line lengths specify the relative areas of each boundary type, that is, the five-parameter GB distribution.

The single-section method for estimating the five-parameter boundary distribution has been tested against both simulated data and known boundary distributions obtained by serial sectioning (Saylor et al., 2004). The comparison was very favorable provided that sufficient data are included in the stereological analysis—at least 50,000 boundary segments for cubic materials, and more for non-cubics. With the advent of rapid EBSD orientation mapping, as described in Section 8.2, such quantities of data, obtained with high spatial resolution, can be collected easily.

As mentioned above, the main advantage of the stereological analysis is its simplicity, requiring only a large 2D EBSD map. Accordingly, it has been applied to a number of materials, both pure metals and alloys and various oxides. Rohrer (2011), in his review on GB character distribution (GBCD) measurements, lists about 20 papers. Here we reproduce an example from a measurement on brass (Rohrer et al., 2006); Figure 11.18 summarizes some of the important results. Each pole figure displays the GB plane distribution (GBPD) in crystal coordinates for one given misorientation. The misorientation axis is displayed on top of every column, the misorientation angle on the right side of every line. For every set of pole figures, the position of the misorientation axis is marked in the legend. Although originally the entire misorientation angle range was inspected, here only a few important angle/axis combinations are shown, for illustration purposes. The respectively closest CSL misorientations are marked below the stereographic projections. From the intensity values shown at the left side of every pole figure in Figure 11.18, it can be concluded that there were few misorientations on [100], but more on [111] and [110]. The distributions also indicate that there is anisotropy in the distribution of boundary planes. The pole figures show, for example, that $\Sigma 9$ GBs (38.9° [110]) are mainly tilt GBs. In contrast, $\Sigma 3$ GBs (60° [111]) and $\Sigma 7$ GBs (38.2° [111]) mainly have twist character. $\Sigma 5$ GBs (36.9° [100]) are exactly at 45° between tilt and twist character. The results, altogether, can be interpreted in the way that GBs that occur frequently have low energies. For brass and many other cases fcc alloys, this seems to be the case for ⟨111⟩-type twist GBs and ⟨110⟩-type tilt GBs. We refrain from citing the many others more recent papers on various materials because it appears that no significantly new results arose from these.

It needs to be emphasized that here EBSD has been used to make a big step forward in GB studies. This is the first time that data on which boundary planes exist in polycrystals are available, which has paved the way for expanding knowledge on GB structure and how it links to properties. Nevertheless, compared to a real 3D reconstruction of the microstructure, the present technique has a number of shortcomings.

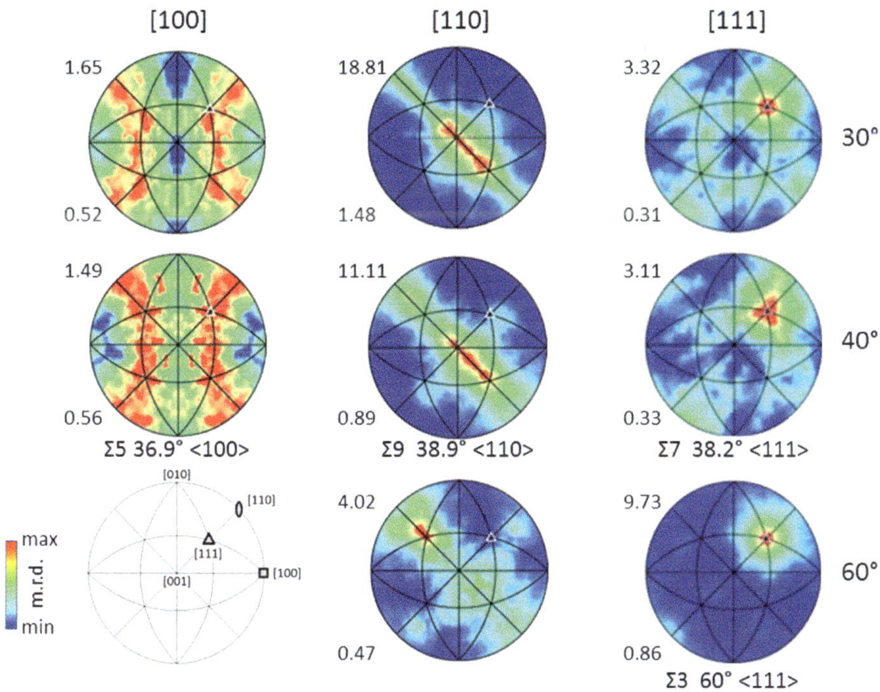

FIGURE 11.18 An example of grain boundary character distribution for Brass. The GBCD is displayed in grain boundary normal pole figures for different misorientation axes and angles. The reference positions of the rotation axes are indicated in the legend. The coherent twins are excluded from the analysis because of their high density. (Adapted from oRohrer et al., 2006.)

The most important one is that the technique is not able to locate and characterize *individual* GBs; it only yields GB plane *distributions*. As a consequence, the GB character can also not be correlated with any other measurement of GB properties, like segregation, corrosion, crack formation, opto-electronic efficiency, and so on. Furthermore, the technique only works for GB planes that appear sufficiently frequently, i.e. that have achieved some sort of energetic equilibrium. GBs with significant curvature cannot be investigated (also because they are not in equilibrium). It is usually assumed that the total number of GBs should be in the order of 50,000. Nevertheless, in such a set there may be, for example, only a few Σ5 GB, which does then not allow determining their character with certainty. For more detailed work on GBs, it is therefore indispensable to perform 3D measurements, which will be treated in the next section.

11.4 THREE-DIMENSIONAL EBSD ORIENTATION MEASUREMENTS

EBSD, by nature, characterizes materials on flat surface sections with a lateral resolution between 50 nm (e.g. for materials with atomic number, $Z > 25$) and 500 nm (for $Z < 14$) (see, for example, the discussion by Tripathi and Zaefferer, 2019) and a depth-probing of a few tens of nanometers. To obtain three-dimensional information on a microstructure one must resort to serial sectioning techniques. In the past, this was achieved by standard metallographical grinding and/or polishing methods, which were described in Section 11.3.2. More recently, several distinct new methods have been developed, which are based on ablation using ion or photon sources, and these techniques are introduced in Section 11.4.1. After each sectioning step, the exposed surface is imaged by SEM using backscatter electron imaging or, to obtain orientation information, EBSD and, optionally, elemental mapping by energy dispersive x-ray spectroscopy (EDS).

Having obtained the 3D EBSD data, reconstruction and visualization in three dimensions is a further step, and there are several commercial or in-house programs available for this 3D rendering (Lewis et al., 2006). Section 11.4.2 describes several steps of data processing necessary to provide reliable 3D orientation data that are required for giving insights into the true microstructure as well as providing input for various computer models. Finally, in Section 11.4.3, we will introduce novel systems for large volume EBSD-based microstructure characterization (Tsai et al., 2022).

11.4.1 3D EBSD Techniques

Generation of 3D microstructure information by means of EBSD mandates the application of serial sectioning techniques. A recent alternative to the traditional mechanical grinding/polishing methods for sectioning is ablation using ion or photon sources. After each sectioning step, the exposed surface is imaged by SEM using backscatter electron imaging or, to obtain orientation information, EBSD and, optionally, elemental mapping by energy dispersive x-ray spectroscopy (EDS). For all ion or photon beam techniques the material removal is performed in gracing incidence, as indicated exemplarily in Figure 11.19a, showing the arrangement of ion beam and electron beam and the shape of the milling volume created by gracing incidence milling starting from a sample edge. Figure 11.19b displays a real sample after cutting of 80 slices of 100 nm. So far, a Ga^+ FI Bmilling has been used most frequently for accurate and plane-parallel removal of material from a surface (Uchic et al., 2006; Xu et al., 2007; Zaefferer et al., 2008; Guyon et al., 2016); other types of radiation will be presented in Section 11.4.3. The Ga^+ ion beam with typical energies between 5 and 30 keV produces some surface damage, depending on the atomic number of the material and the operating conditions. This, in turn, can have a knock-on effect on the clarity and hence solve efficiency of the diffraction patterns. Metals with low atomic numbers are more affected than those with high atomic numbers; non-metallic materials are often completely amorphized and cannot be investigated in this way. Using a 30-kV ion beam and a current of 3 nA a slice from nickel measuring 15 μm × 8 μm × 200 nm would be removed in approximately 25 min. The best spatial resolution of the 3D pixels (voxels) is 50 nm × 50 nm × 50 nm, the angular resolution is 0.5°. However, the total analyzed volume is rather small. The appearance of *curtaining* limits the milling path lengths (X direction in Figure 11.19a) to about 50 μm, where curtaining refers to the occurrence of surface waviness in the direction of the ion beam, reminding of hanging curtains. These artifacts appear due to inhomogeneous milling rates on different crystal orientations, on second-phase particles

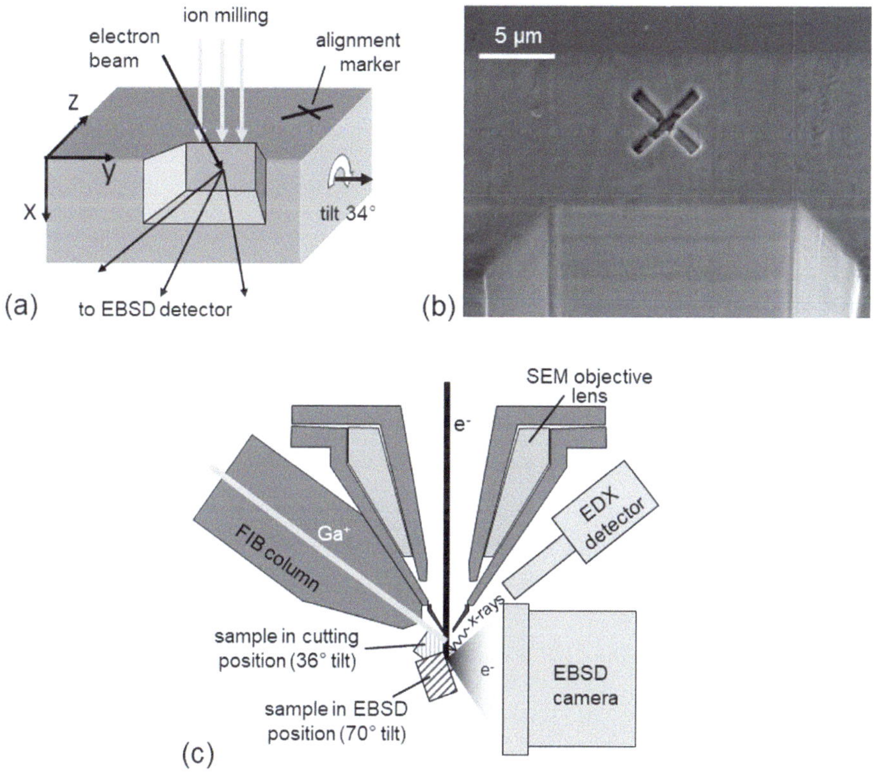

FIGURE 11.19 Geometric setup of 3D EBSD by serial sectioning through Ga+-beam milling in a FIB-SEM. (a) Arrangement of ion beam and electron beam and the shape of the milling volume created by gracing incidence milling starting from a sample edge. (b) Real sample after cutting of 80 slices of 100 nm. Visible is the milling stage and the alignment marker. (c) Schematic cross section through a FIB-SEM with the milling position and the EBSD position marked. (Adapted from Zaefferer et al., 2008.)

with different hardness, or surface roughness. The width (Y in Figure 11.19a) of a milled area cannot be much larger than 50 μm due to time restrictions and the depth of milling (Z in Figure 11.19a) is limited also to about 50 μm because of extensive shadowing.

The combination of FIB-milling and EBSD-based orientation mapping in a dual-beam instrument can be achieved in different geometric setups which are discussed in detail in Zaefferer et al. (2008). Figure 11.19c illustrates the combined geometry of FIB column, SEM column, EBSD and EDS detector, and the specimen stage for the so-called tilt setup. The specimen surface has to be maintained first parallel to the ion beam for milling, and then at 20° (typically) to the electron beam for EBSD. It is therefore necessary to precision rotate and position the specimen between successive operations.

Figure 11.20 shows a typical example of such a FIB-tomography measurement, which is comprised of the 3D grain structure, GB structure, and the *GB plane*

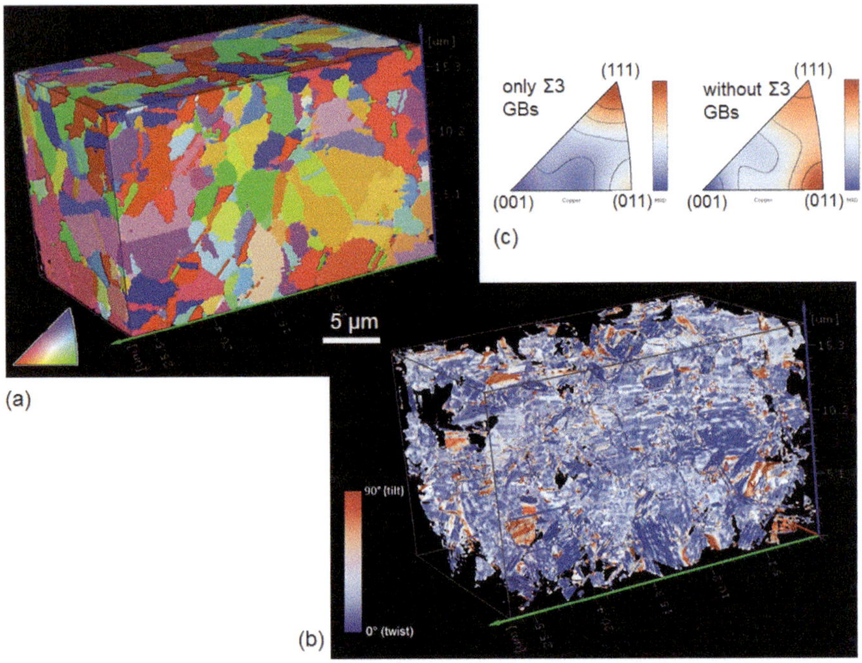

FIGURE 11.20 Results of 3D EBSD by FIB milling on a Cu-3% Zr sample. (a) Inverse pole figure-colored crystal direction map of the sample normal direction (to the top). (b) Twin boundaries colored for their twist and tilt character. Twist boundaries are coherent twin boundaries, tilt boundaries are incoherent facets. (c) Stereographic representation of the grain boundary normal distribution. The left side SOT shows only Σ3 grain boundaries that have mainly (111) boundary planes. The right side shows all boundaries that are not Σ3 twin boundaries. Also these boundaries are not randomly distributed but show a preference for (110) planes. (Adapted from Khorashadizadeh et al., 2011.)

distribution, GBPD, of a fine-grained Cu-Zr alloy (Khorashadizadeh et al., 2011). The total analyzed volume is about 20^3 μm^3 (Figure 11.20a). Figure 11.20b is a display of the Σ3 twin boundary structure found in this material. Σ3 GBs of twist character are colored in blue. These GBs are coherent Σ3 GBs with {111} GB planes. The GBs colored in red are tilt GBs with mainly {110} GB planes. They correspond to incoherent Σ3 GBs. Figure 11.20c shows the GBPD of all other GBs. It is obvious that this distribution is not random but clearly inclined toward a great circle between {111} and {110} planes.

11.4.2 3D EBSD Data Processing

The results displayed in Figure 11.20 were not obtained without extensive further analysis of the acquired raw EBSD data. In a first step, the raw 2D orientation maps need to be aligned for optimum fit onto each other. This alignment may not only require rigid *x-y* shifts or rotations of entire maps but, possible, also non-rigid deformations of the individual maps. This latter may be required because EBSD scanning does not always proceed perfectly rectilinear, and step sizes may vary slightly from point to point because of instabilities of the SEM scanning system. Different approaches for rigid and non-rigid alignment have been proposed, including, for example, triple point alignments (Rohrer et al., 2010) or application of the "smart-align" algorithm (Jones et al., 2015) adapted by Konijnenberg (2024) for 3D EBSD.

Next, the microstructure needs to be segmented into grains. This is usually done, like for 2D EBSD maps, by a 3D flood-fill algorithm. All volume pixels ("voxels") are assigned to grains under certain conditions like maximum misorientation and minimum number of voxels per grain, etc. After this, the recognized GBs are triangulated using, for example, a marching cube (Lee et al., 2014) or marching tetrahedron (Konijnenberg et al., 2012) approach. These algorithms approximate the voxel-based GBs (or other segmented structures) by a set of triangles.

Finally, the triangulated GBs need to be smoothed which, again, can be performed in many different ways. One is the locally constrained vertex model (Konijnenberg et al., 2012), but other more global approaches are also possible.

11.4.3 Large Volume 3D EBSD Techniques

One of the prime interests of microstructure observation systems is to reach a length scale and resolution of investigation, which allow making a connection between *macroscopic properties* and *microscopic features* for the investigated materials. It appears that a volume of 1 mm³ measured at 1-µm step size would be probably the maximum reachable and manageable volume for a large-volume 3D EBSD measurement. For many engineering and geological materials such volume would contain between 100 and 100,000 grains which would yield very good statistics of grains and GBs. Figure 11.21 displays resolution and maximum volume diameter of various 3D techniques, which are able to reveal microstructural information of crystalline materials. The graph, at the same time, compiles typical microstructural features that are investigated with these techniques.

Larger volumes than a few $10 \times 10 \times 10$ µm³ require other types of radiation to replace the Ga^+ beam in FIB tomography, such as a focused Xe^+ plasma beam (Burnett et al., 2016), a broad Ar^+ ion radiation (Gholinia et al., 2020), and femtosecond laser (Echlin et al., 2015). Sometimes even combinations of different radiations are used for coarse and fine milling. A comparison of the different sectioning methods has been presented in the overview article by Echlin et al. (2020). Shortcomings of the mentioned techniques are that the milled surfaces are usually not perfectly flat but show curtaining, that the focused ion radiation may lead to amorphization of non-metallic materials (e.g. intermetallics, oxides) and that generally the specimen surface does not remain flat. To overcome these problems large-area serial mechanical or mechano-chemical polishing combined with multi-modal observations (i.e., EBSD and EDS) in SEM has been introduced, particularly because normal metallographic polishing techniques are well suited to prepare samples of very different

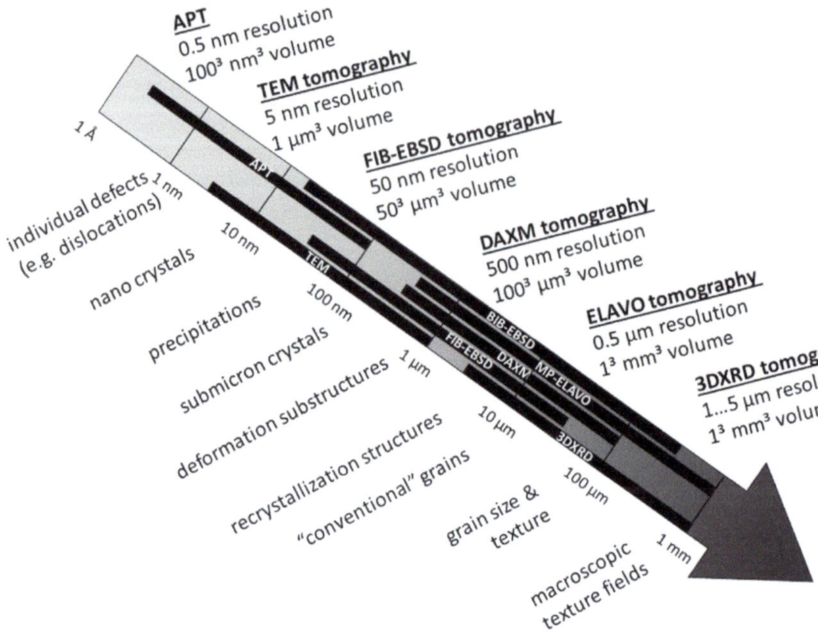

FIGURE 11.21 Overview on resolution, maximum observable volume, and potential applications in materials of different 3D techniques applicable to study crystalline materials. APT: atom probe tomography; TEM: transmission electron microscopy; FIB: focused ion beam; EBSD: electron backscatter diffraction; DAXM: differential aperture x-ray microscopy; ELAVO: EBSD-based large-volume 3D system. This covers MP: mechanical polishing; PFIB: plasma-FIB; LASER FIB; BIB: broad ion beam; 3D XRD: 3-dimensional x-ray diffraction.

materials with very large fields of view up to millimeters or even centimeters in diameter.

There is a long history of mechanical serial sectioning techniques dating back to Hillert's original work on pearlite (Hillert, 1962); further references can be found in Tsai et al. (2022). Rowenhorst et al. (2006) were among the first to combine serial sectioning by polishing with EBSD, investigating the 3D crystallography of coarse auto-tempered martensite. Saylor et al. (2000) just prepared two serial sections to obtain GB information on magnesia. In the beginning (Rowenhorst et al., 2006; Spanos et al., 2008), EBSD scans were still performed manually and not on every section (e.g., on every tenth slice). Next, automation of the laborious serial polishing was achieved, among others, with the Robo-Met.3D (Spowart et al., 2003) and RS^{3D} system (Rowenhorst et al., 2020). The latter work showcases the capability of the RS^{3D} system by analyzing additively manufactured 316L stainless steel. Figure 11.22 reproduces one of the results from this paper.

Recently, the group of Zaefferer devised a fully automated EBSD system for large volume 3D mapping using mechano-chemical polishing for serial sectioning, which

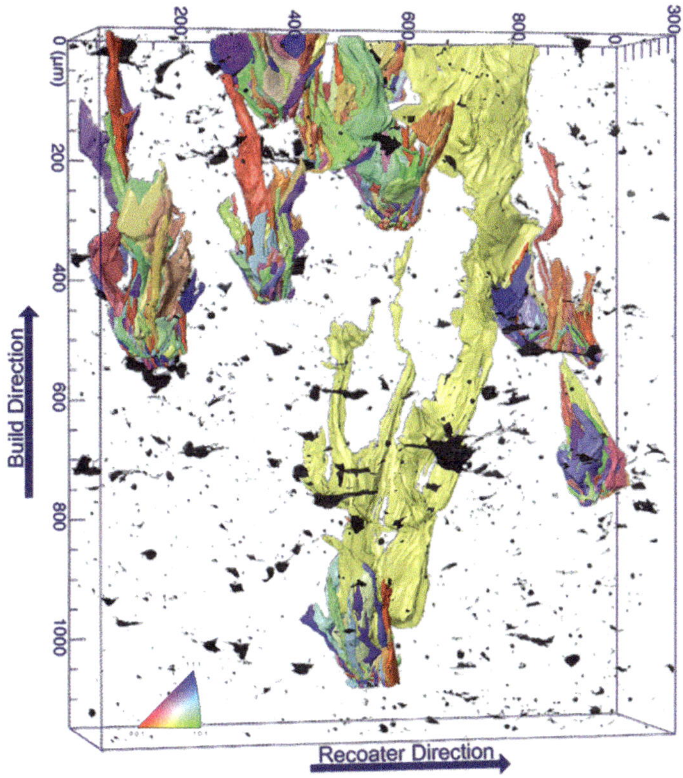

FIGURE 11.22 Reconstruction of the pores (black) in additively manufactured 316L and the grains that grow from the ten largest grain clusters (colored according to the crystallographic direction pointing along the build direction). (Reproduction from Rowenhorst et al., 2020.)

was nicknamed **E**BSD-based **la**rge-**vo**lume **3D** system, ELAVO 3D (Tsai et al., 2022). As an example for the general functioning of such a system the ELAVO 3D setup is displayed in Figure 11.23. It consists of four major components: (i) a polishing automaton which removes layers of approximately constant thickness from the sample surface by mechanical and chemo-mechanical polishing; (ii) a transfer bay with built-in mechanics for automated releasing and fixing the sample in the polishing automaton; (iii) a collaborative robot which transfers the sample between the polishing automaton and the microscope; (iv) a SEM equipped with EBSD, energy dispersive x-ray spectroscopy (EDS), backscattered electron (BSE), and secondary electron (SE) detectors for observation of the serial sections. A master computer receives/sends signals from/to a communication hub and the servers of the SEM and the EBSD/EDS computers, thus controlling the whole process.

A serious challenge that is not addressed in most publications is the design of a proper chemo-mechanical polishing procedure. In Tsai et al. (2022) some of the

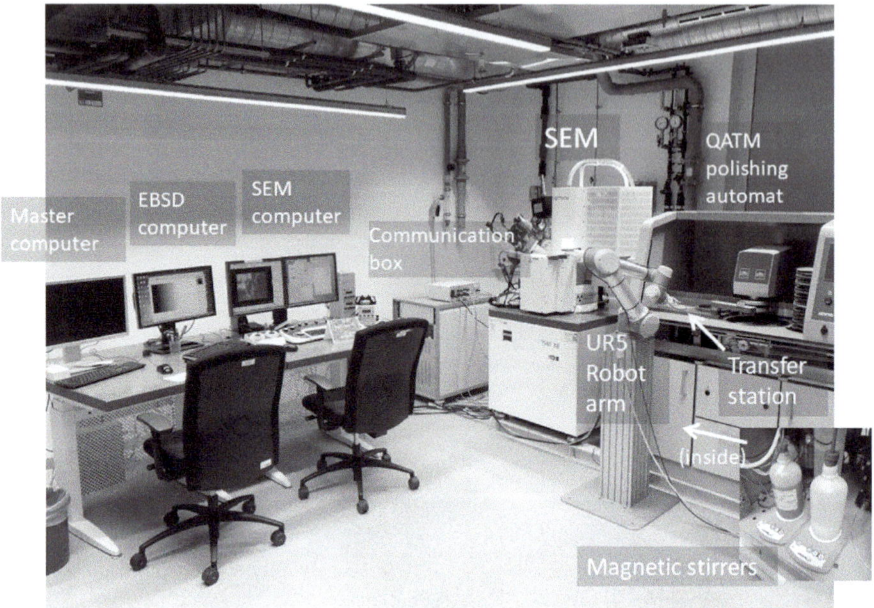

FIGURE 11.23 Overview on the components of the ELAVO 3D system for large volume 3D mapping of crystal orientations using EBSD and mechano-chemical polishing for serial sectioning. All parts are marked.

basic considerations are outlined. It is important to note that polishing with diamond suspension is an abrasive process which does not lead to EBSD-ready surfaces on most materials. It is, rather, required to add a chemical polishing step, for instance application of an oxide particle suspension ("OPS") which is based on a chemical attack of the material's surface by the solvent and a removal of the chemical products by silicon or aluminum oxide particles (Section 8.2.2). The chemical composition of the solvent is of high importance for the success of the polishing process.

If all polishing parameters are correctly selected and controlled then perfectly flat and artifact-free surfaces are achieved for many materials. This is particularly attractive as the polished surface allows to perform further measurements, for example corrosion experiments (An et al., 2018). In this way the five-parameter GB character, obtained through the 3D measurements, can be correlated with the GB corrosion properties for hundreds or thousands of GBs. This approach is illustrated in Figure 11.24: a thin 3D slice of a 316L stainless steel material was characterized by 3D EBSD (~20 serial sections). The resulting grain boundaries are displayed in the left-side inset. Then the material was corroded in a solution of H_2O_2 and HF and the corrosion results were observed by backscatter electron imaging. The right-side inset shows exactly the same area as the left-side one, but now with some grain boundaries strongly corroded. Some of the results have already been discussed above in Section 11.2.6 and displayed in Figure 11.11. Another very interesting result presented by Tsai et al. (2022) is the appearance of Σ9 GBs showing different GB

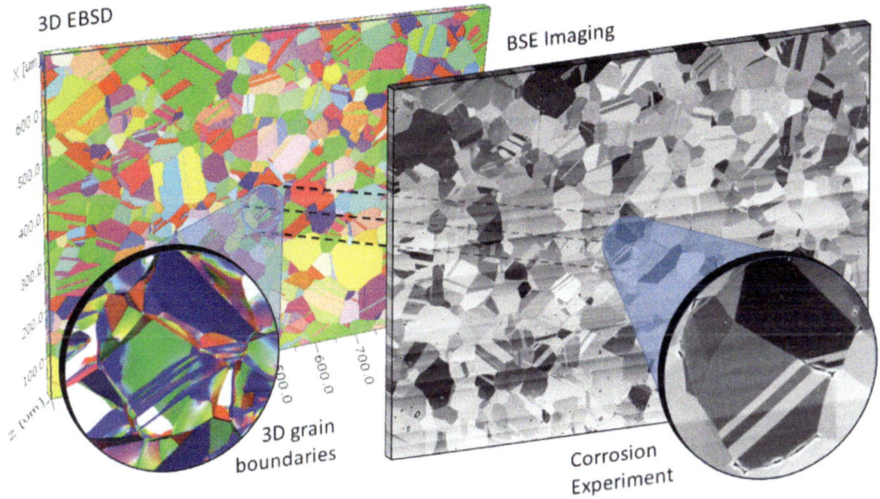

FIGURE 11.24 Example for the scientific use of the large volume 3D EBSD system for research on grain boundary corrosion of a 316L stainless steel. The left side image is a block of a 3D EBSD measured microstructure. The inset shows 3D grain boundaries in the marked volume (3-dimensionality is difficult to recognize in the figure). The right-side figure shows the surface of the 3D EBSD-measured sample after etching. The inset shows grain boundaries that are heavily corroded and those which are not corroded at all in the same area as the one in the inset on the left side. Note that the corrosion experiment was performed after 3D EBSD on the remaining sample.

plane pairings (e.g. {111}||{115}, {110}||{111}, or {114}||{114}) combined with characteristically different morphologies. Here we remind the reader that the stereological method described in Section 11.3.3 would be able to find these pairings as well (provided they appear sufficiently frequently) but would be unable to connect them to morphologies or any other properties.

The chemo-mechanical polishing serial sectioning methods excel by the large and perfectly flat and artifact-free surfaces. Nevertheless, they are not without problems. For example, it is difficult to apply the technique to air or water sensitive materials or to materials that cannot be easily polished mechanically. Also, the process is relatively slow and the removal thickness currently cannot be smaller than about 0.5 μm. An alternative to chemo-mechanical polishing for serial sectioning is the newly developed broad ion beam (BIB) polishing technique (Gholinia et al., 2020) which, at the time of writing this book, seems to be the most promising of all large volume 3D EBSD techniques. It allows artifact-free removal of large volumes, even at very small and well-controllable removal thicknesses. The samples are not exposed to air and the process is very rapid. Whether the technique works well when differently hard phases are involved and how good is the surface flatness after a significant number of layers has been removed remains to be seen, however. Table 11.2, reproduced from Tsai et al. (2022), compiles several key parameters of the most prominent 3D EBSD techniques.

TABLE 11.2
Key Parameters of Different Techniques for Large-Volume 3D EBSD Measurements

Technique	Ga^+-FIB	Xe^+-PFIB	fs-laser FIB	BIB	ELAVO 3D
Removal mechanism	Ga^+ ion sputtering	Xe^+ ion sputtering	fs laser ablation and Ga^+/Xe^+ ion sputtering	Broad Ar^+ ion beam sputtering	Chemo-mechanical polishing
Min. removable slice thickness, z (nm)	20	30	500	20	500
Common size of analyzed volume $x \times y \times z$ (µm³)	50 × 50 × 50	300 × 300 × 300	1,000 × 1,000 × 1,000	1,000 × 1,000 × 300	1,000 × 1,000 × 1,000 (potentially bigger)
Removal rate (µm³/s)	~20 using 65 nA beam	~400 using 65 nA beam	~3,500 (laser + PFIB)	~10^5	>10^4
Estimated removal time per slice (without EBSD)	~15 s for 100 nm	~4 min for 1 µm	~5 min for 1 µm	~10 s for 1 µm	30 min (to be reduced to 22) for 1 µm
Limitations	Very small volume; beam damage; curtaining	Small volume; beam damage; curtaining	Beam damage; curtaining	Curtaining; uneven (concave) area	Corrosion of sensitive samples because of exposure to water and air; inaccuracy of removal rate
Approximate cost in Mio $ (without EBSD)	0.8	1.5	1.8	>0.8	0.5

Source: Reproduced from Tsai et al. (2022).

11.5 ORIENTATION CONNECTIVITY AND SPATIAL DISTRIBUTION

The topological parameters associated with polycrystals—numbers of grains, grain surfaces, edges, and corners—have been studied for many years. Stereological techniques (i.e., the reconstruction of 3D information based on that accrued from 2D sections) have generally been used to investigate the topological connectivity of grains aggregated in a polycrystal (Rhines et al., 1976). The sectioning planes can be random, which gives a statistical result, or parallel to build up an accurate impression of the spatiality of the microstructure. These topological aspects of microstructure have traditionally been studied separately from crystallographic considerations. However, the advent of EBSD allows large regions of microstructure to be accessed in a single section in terms of both the crystallography *and* areal topology. These studies access the way in which grains having various orientations connect together—*orientation connectivity*.

Another aspect of orientation connectivity is the spatial distribution of orientations with respect to some features of the microstructure; for example, how grain textures relate to grain size (Vogel and Klimanek, 1996; Engler, 1998a; Bastos et al., 2006) or the clustering of grains that are separated by small-angle boundaries, where a single cluster is enclosed by large-angle boundaries (West and Adams, 1997). Quantitative cluster structure analysis is carried out using aspects of percolation theory, such as fractal analysis. Examples of specific orientation connectivity topics include the following:

- Grain shape, distribution, and orientation
- Particle distribution with regard to grains of particular orientations
- "Decision trees" for crack paths throughout the microstructure
- Grain boundary connectivity
- Clustering of orientation elements

Any of these examples can be explored on a single section through the microstructure; if sectioning techniques are employed then 3D aspects of orientation connectivity can be addressed, which render the data much more powerful, as described in Section 11.3.3. Other metrics are beginning to be used to assess the connectivity and topology of the GB network. Percolation theory is relevant to the length of connected intergranular pathways or "clusters" through the microstructure. A simple approach is to consider that such pathways consist only of boundaries with random geometries. A "skeletonized" EBSD mapping can be used to show the evolution of boundaries with processing and the data subsequently exported to other applications to quantify the percolation characteristics. This has been carried out, for example, on a GB engineered nickel-base alloy (Schuh et al., 2003).

Aspects of orientation connectivity that relate to *grain junctions*, that is, where more than two grains meet, are often called a "triple line." An "addition rule" applies, which predicts the geometry of the third boundary from the relationship:

$$M_1 M_2 M_3 = I \tag{11.20}$$

where M_{1-3} are the misorientation matrices of three boundaries at a common junction and I is the identity matrix. This rule has the most significant consequences for conjoining CSL boundaries since they share a common misorientation axis; the sum of two of the misorientation angles gives the third multiplied by a common factor, and the product or quotient of two of the Σ values gives the third (Ranganathan, 1966; Gertsman, 2001; Reed and Kumar, 2006). For example,

$$60°/\langle 111\rangle - 21.8°/\langle 111\rangle = 38.2°/\langle 111\rangle$$
$$\text{that is, } \Sigma 3 \cdot \Sigma 21a = \Sigma 7 \tag{11.21a}$$

or more generally

$$\Sigma \mathrm{A} \cdot \Sigma \mathrm{B} = k^2\, \Sigma \mathrm{C} \tag{11.21b}$$

with $\mathrm{A} \cdot \mathrm{B} = k^2\mathrm{C}$ and k being a common factor of A and B. The interest in grain junctions from a practical point of view is that any interfacial propagation phenomenon (diffusion, corrosion, cracking, etc.) is controlled by the juxtaposition of boundaries at grain junctions. The phenomenon will only be likely to propagate along nonspecial boundaries; therefore, it is pertinent to categorize grain junctions in terms of the combinations of special/nonspecial types (Garbacz et al., 1995; Thomson and Randle, 1997b). In recent years, statistics on triple junctions have begun to appear in reports alongside GB proportions (Schuh et al., 2003). For example, Figure 11.25a and b shows the constitution of triple junctions in nickel, in terms of twin-related boundaries (Σ) and random boundaries (R), before and after GBE, respectively. The statistics demonstrate that processing has increased the proportion of triple junctions that contain twin-related CSLs.

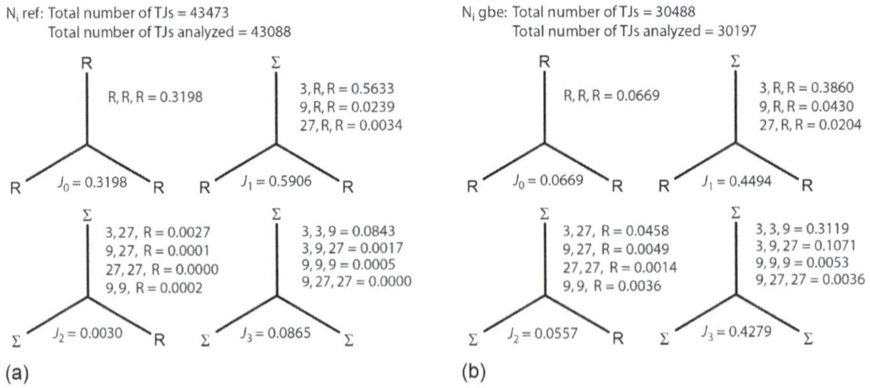

FIGURE 11.25 Constitution of triple junctions in nickel, in terms of twin-related boundaries (Σ) and random boundaries (R), (a) before and (b) after grain boundary engineering. (Courtesy of G. Rohrer.)

Although orientation connectivity aspects of the microstructure can be addressed partially via a single section through the microstructure, far more in-depth data can be obtained from incorporating sectioning techniques to gain a 3D viewpoint. This can be done either statistically or specifically. The *intercrystalline structure distribution function* (ISDF) and related *interface damage function* (IDF) (Adams et al., 1990b) take a statistical approach to couple the spatial distribution of grain boundaries with the MODF (Section 10.7). In one series of experiments, grain misorientations were measured on four section planes, for example, in plate material at 0°, 10°, 30°, and 60° to the plane of the plate. From these data, MODFs for various interface normal inclinations could be plotted in Euler space, which would yield peaks of preferred boundary misorientation. Note that here the attention is focused on *probabilities* rather than on the crystallography of individual boundaries and that the boundary orientation data relate to specimen, rather than crystal, geometry. This methodology has been used to analyze creep damage in copper (Adams et al., 1990b).

11.6 SUMMATION

This chapter has shown how microtexture analysis can be extended to obtain crystallographic characterization of planar structures in the microstructure, namely grain and phase boundaries, cracks and facets, and the whole connectivity of the microstructure. Grain and phase boundaries largely determine the mechanical and functional properties of materials and at the same time play an important role for the formation of textures and microstructures during materials processing. In principle grain and phase boundaries are characterized by five rotational (or macroscopic) parameters and a multitude of morphological and atomistic parameters. Depending on the requested information about a material, it is sufficient to measure and report only one parameter (large- vs. small-angle grain boundaries), three parameters (special vs. non-special boundaries) or five parameters (low vs. high energy boundaries). With increasing request of parameters the measurement and the analysis becomes more demanding. The necessary analysis processes, which have mainly arisen from EBSD, involve nonstandard, multistage methodologies. Particularly the 3D microstructure mapping methods are able to solve long time open problems on the behavior and importance of grain and phase boundaries.

12 Orientation Relationships between Phases and Texture Formation during Phase Transformations

12.1 INTRODUCTION

One of the most important mechanisms for the formation and modification of textures and microstructures is the crystal orientation change occurring during solid-state phase transformations with orientation relationships (ORs) between the involved phases. Probably the technically most important phase transformations are the γ–α transformations in steels (e.g. Pereloma and Edmonds, 2012) and the β–α transformation in titanium alloys (e.g. Banerjee and Williams, 2013). Also in geological materials phase transformations with ORs play an important role for the formation of textures and microstructures (e.g. Habler and Griffith, 2017). The presence of ORs frequently leads to the inheritance of textures from one phase, for example a high-temperature phase, to the transformed phase, e.g. one existing at lower temperatures. During these phase transformations, certain crystallographic selection rules (so called variant selection rules) may be active, leading to very particular textures being formed in the end. In the following section, we will give a brief overview on ORs between different phases, followed by a more detailed discussion of the γ–α transformation in steels and an overview on ORs in other materials. The final section will discuss the principles of texture formation and texture memory by phase transformations.

12.2 ORIENTATION RELATIONSHIPS BETWEEN DIFFERENT PHASES

12.2.1 Introduction to Orientation Relationships between Different Phases

All solid materials undergo phase transformations, at least through different states of aggregation. Many also undergo solid-state phase transformations. The large amount of these transitions and their temperature and composition dependences is reflected in the almost infinite amount of equilibrium and non-equilibrium phase diagrams of materials. When crystalline materials undergo solid-state phase transformations, it is very common that the *parent* phase crystals and the *child* phase crystals observe an OR. ORs form particularly when a phase transformation occurs in a *displacive*

DOI: 10.1201/9781003258339-15

manner, i.e. by a cooperative movement of atoms across the phase boundary. These transformations are usually called *martensitic*, named after the martensite transformation in steels which itself is named after the researcher A. Martens (see, e.g., Christian, 1975; Bhadeshia, 1987; Nishiyama, 2012). Nevertheless, ORs also frequently form during *diffusive* phase transformations where atoms move in an uncorrelated manner from one phase to the other. In this case, it is usually the nucleation process that governs the formation of an OR. ORs are described by defining the *parallel planes* $(hkl)_p \parallel (hkl)_c$ and *parallel directions* $[uvw]_p \parallel [uvw]_c$ in the involved parent and child phases. Note that the plane normals and directions are perpendicular to each other, i.e. $(hkl)_i \cdot [uvw]_i = 0$. An example is the *Kurdjumov–Sachs* (KS) OR frequently occurring in the martensitic γ–α transformation in low-carbon steels (see Section 12.2.2) where the high-temperature fcc phase austenite, γ, transforms into the low-temperature bcc phase ferrite or martensite, α or α'. This OR is defined by the parallel planes $(111)_\gamma \parallel (011)_\alpha$ and parallel directions within these planes $[\bar{1}01]_\gamma \parallel [\bar{1}\bar{1}1]_\alpha$. Because the involved crystal structures both have crystal symmetries, the OR also observes symmetries. That is to say, there are symmetrically equivalent ORs, which create out of one parent crystal a number of child crystals, so called *variants*. The number of variants is calculated from the symmetry elements of the OR. In the case of the KS OR, for example, the fcc phase has four {111} planes, with three ⟨110⟩ directions therein, on each of which two symmetrically equivalent bcc orientations can be connected. This makes $3 \times 4 \times 2 = 24$ variants. The details of this OR will be worked out in more detail in the following section.

The reasons for the formation of ORs are of thermodynamic, kinetic, and mechanical nature (note that the mechanical aspect cannot be easily separated from the thermodynamic one). First, a developing phase boundary should be as low as possible in energy. This may lead to the preference of phase boundary planes of high atomic density and large interplanar distances. A very good example for this is the formation of the KS OR. Here the transformation planes are close to a {111} plane for the fcc parent phase (austenite) and to the {110} plane for the bcc child phase (ferrite or martensite), both these planes are the most densely packed planes in the respective structures. Next, kinetic reasons play a role for martensitic transformations where the atoms must find a way to transform by minimal movement from one to the other crystallographic arrangement. Here again the KS OR displays a good example: the final transformation planes, the so-called *habit planes*, are slightly different from the parallel densely packed $\{111\}_\gamma$ and $\{110\}_\alpha$ planes because the transformation occurs along *invariant lines* or *invariant planes*, which are crystal directions which stay unrotated in both crystal structures. These are directions in which atoms easily shuffle from one crystallographic phase to the other. Finally, the minimization of mechanical stresses and their accommodation by interface and lattice dislocations also plays an important role for OR formation. Again the KS OR is an ideal example: the fcc and the bcc lattice do not fit together without significantly squeezing or stretching one with respect to the other. This leads to significant residual stresses in the transformed lattice and is partly responsible for the high hardness of martensite.

Experimentally, ORs are most conveniently being measured by orientation measurements employing transmission electron microscopy (TEM)-diffraction or electron backscatter diffraction (EBSD). If spatial resolution allows, EBSD is the most convenient technique because of its high degree of automation and large field of view. Gourgues-Lorenzon (2009) presents a very detailed and informative overview on phase transformation studies by EBSD. In this article also 3D EBSD techniques are mentioned in some detail (corresponding to the state of the art in 2009) as a technique that also reveals grain boundary planes (see Sections 11.3 and 11.4). It should be noted, however, that planes between individual variants usually do not correspond to the habit planes of phase transformation, simply because these variants impinge on each other under various constraints (see, e.g., Morito et al., 2006). If parent and child phases are both present, habit planes may be preserved and measurable, however. Sato and Zaefferer (2009) showed that a metastable parent phase may transform into its child phase under the action of the focused ion beam used for 3D sectioning, leading to significant artifacts. The only limit of EBSD for the investigation of ORs is its spatial resolution. In steels and heavier materials lamellae narrower than 50–100 nm cannot be measured reliably; in lighter materials, e.g. in minerals, this limit is even much larger (~500 nm). In these cases, transmission Kikuchi diffraction (TKP) in TEM or scanning electron microscopy (SEM) or spot diffraction in TEM are tools to study the very fine crystal structures caused by phase transformations. The use of SEM-TKP has been discussed by Cayron (2020) to study the 20-30-nm-wide lamella of a TiNi shape memory alloy.

ORs are misorientations between grains of different phases and can, thus, be displayed, similar to other misorientations (Section 10.6), for example in axis-angle plots (see an example in the next section). In these plots well-known ORs can be identified conveniently, but because the plots are reduced to the fundamental zone of crystallographic directions, it is not possible to recognize the symmetry of ORs. This is, in contrast, very well visible when the orientations of variants are displayed in well-selected pole figures (Nolze, 2006; Miyamoto et al., 2009). Brückner et al. (2001), He et al. (2005), and Kestens et al. (2005) showed that ORs can also be described in Rodrigues–Frank (R–F) space. The R–F space is a symmetry-conserving display of (mis)orientations (in contrast to pole figures, which only display directions or planes) and offers, therefore, the chance to conveniently observe symmetries of transformation variants. Nevertheless, the display of the three-dimensional R–F space in printed media is not very convenient.

In the case of two cubic phases—fcc and bcc—the computation and representation of a misorientation distribution function, MODF, proceeds analogous to single-phase materials, whereas in all other conceivable phase combinations, the different crystal symmetries render determination and evaluation of MODFs a complicated task. Field et al. (1996), for example, determined the MODF between ferrite and retained austenite by means of EBSD and presented it in Rodrigues space; surprisingly they observed a Bain-like OR between the two phases. As it will be discussed later, the *Bain* OR is the basic lattice correspondence between the γ and the α phase in steels, but it is probably never observed as the true OR between these phases. It appears, therefore, that either the measurements or the MODFs displayed in the example suffer from too little angular resolution to resolve the subtle changes between the Bain OR and the true ORs grouping around it. Figure 12.1 displays an example of the MODF between bcc tungsten and fcc nickel in a W/Ni composite material. The orientations of both phases were determined by means of automated EBSD-based orientation

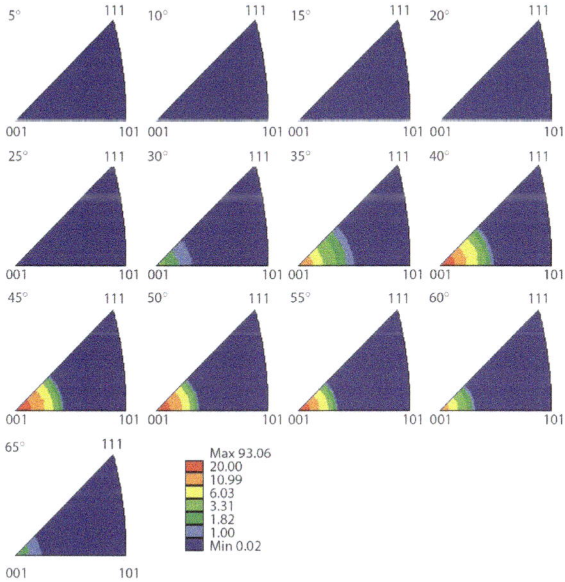

FIGURE 12.1 MODF between two different phases in a nickel/tungsten composite. The MODF was computed from the misorientations between adjacent grains of fcc nickel and bcc tungsten, which were obtained by EBSD (see also Figure 9.7).

microscopy, and the resulting maps—image quality map, orientation map, and phase map—were shown in Figure 9.7. Subsequently, the misorientations between tungsten and nickel across the phase boundaries were determined, expressed in terms of the mutual misorientation axis r and angle θ, and represented in the form of θ sections through the cylindrical angle/axis space (Section 10.6.2). As in the previous example, the resulting MODF, displayed in Figure 12.1, shows a strong maximum close to 45° ⟨100⟩ misorientations, which indicates a Bain, or Pitsch, OR between the two phases (Sinclair et al., 2003).

From a theoretical point of view, ORs, particularly those created by martensitic transformations, are most comprehensively treated by the phenomenological theory of martensite crystallography, PTMC, which was originally developed by Wechsler et al. (1953) and Bowles and Mackenzie (1954).This theory, which will not be worked out here in detail, assumes that the OR between two crystals is formed based on a basic transformation between the two lattices (the so-called Bain strain), a rigid body rotation of one of the lattices, such that an invariant plane is formed along which the atoms shuffle from one structure to the other, and a lattice invariant shear strain, which insures that the external shape of the transforming crystal, constrained by its neighbors, remains the same. The basic equation of this theory is

$$\mathbf{F} = \mathbf{RBL} \tag{12.1}$$

where **F** is the total transformation matrix (from which the OR can be calculated), **R** is the rigid body rotation, **B** the Bain strain (i.e. the lattice correspondence), and

L the lattice invariant shear strain. The task is to find **F** for a parent and a child phase with defined lattice parameters, based on an assumption for the lattice invariant shear accommodation mechanism (by dislocations or twins). As a result, the theory returns the OR, the amount of shear on the lattice invariant shear system, and the habit plane of the transformation. Details of the mathematical treatment can be found, for example, in Kelly (2012). A freely available powerful software that calculates the crystallographic features (ORs, habit planes) for phase transformations between known crystal structures is PTCLab by Gu et al. (2016) which was mentioned already in Section 11.2.5. PTCLab includes the classical PTMC. Recently, Cayron (2020) discussed the pros and cons of the PTMC and proposed modifications to account for unexplained experimental details. One of the most important unexplained detail is that frequently not only one OR is observed experimentally, but a certain spread of ORs, in the case of steels, for example, from the KS OR over the *Greninger–Trojano* (GT) OR to the *Nishiyama–Wassermann* (NW) OR (see the example in the following section). Cayron (2020) proposed the term "natural" OR for that OR, which shows dense parallel planes and dense parallel directions between the two phases. For every displacive phase transformation such a natural OR exists. In the case of steels, it is the KS OR.

One of the critics on the PTMC is that it is phenomenological, i.e. it does not propose a detailed mechanism for the atomic shuffling during the phase transformation (Cayron, 2020). An alternative theory, which calculates dislocation and disclination structures in the habit plane, is the topological model, TM, developed by Pond and co-workers (Pond et al., 2003; Ma and Pond, 2008). Although the TM allows calculating realistic structural features and crystallographic quantities of phase interface planes, the TM appears mathematically more demanding than the PTMC and, to the authors' best knowledge, no general software is available.

12.2.2 Orientation Relationships in Steels

As mentioned above, the phase transformations in steels occurring between the high-temperature fcc austenite phase, γ, and the low-temperature bcc phase ferrite or martensite, α or α', are of highest fundamental and technical importance because almost all steels somehow owe their mechanical properties to it. Its dependence on temperature and alloy composition and the ability to control the kinetics of transformation by the rate of temperature change and interstitial (in particular carbon) and substitutional element composition allows for the extreme diversity of steels. An overview on this enormous field of research and technical applications may be obtained, for example, by the books of Pereloma and Edmonds (2012) and Bhadeshia (2001). Note that the latter book covers most aspects of phase transformations in steels although it is dedicated to bainite, one particular microstructure formed during these transformations. The solid-state phase transformation in steels needs to be distinguished into the displacive or martensitic transformations, leading to the formation of martensite, α' or ε, and the diffusive or ferritic transformations leading to the formation of ferrite, α. During diffusive transformations of carbon-containing steels usually some sort of carbide, e.g. cementite, θ, is involved as well. The bainitic transformation is an intermediate transformation, involving both displacive and diffusive mechanisms.

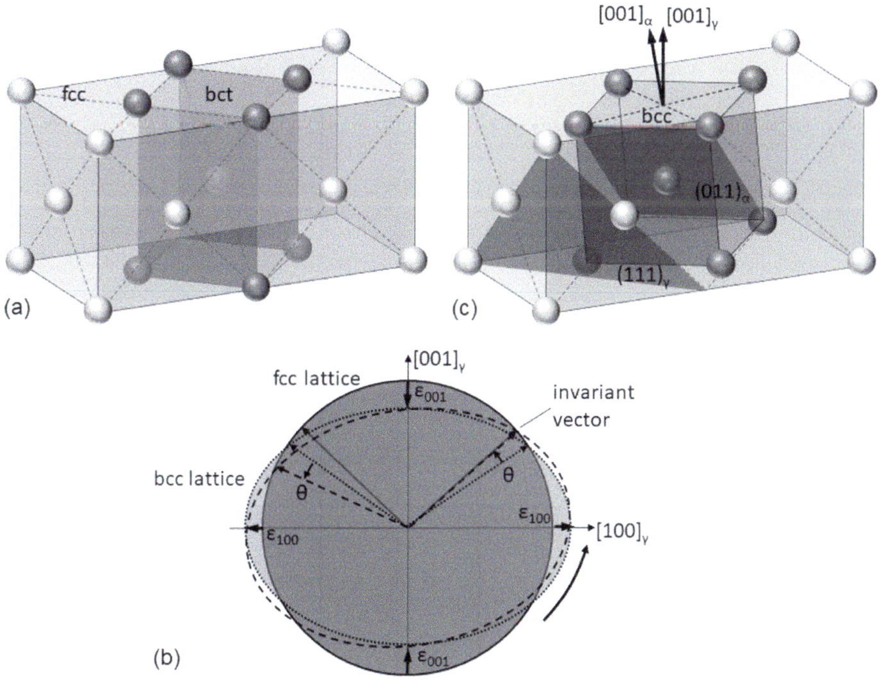

FIGURE 12.2 The Kurdjumov–Sachs orientation relationship. (a) Bain relation between the fcc lattice and the bcc (more precisely bct) lattice. (b) The fcc lattice, displayed as a dark gray sphere with solid line, is deformed by strains ε_{001} (–20%) and ε_{100} (+12%) into the Bain-related bcc ellipsoid displayed in bright gray with dotted line. By this strain two solid line vectors of the fcc lattice are strained into two equally long dotted vectors of the bcc lattice. A rotation (of 9°) of the bcc lattice into the dashed line ellipsoid creates an invariant line vector, displayed by a dashed line. (c) A small rotation of the Bain-related bcc lattice makes (111) planes in fcc and (110) planes in bcc parallel. This leads to a small deviation of the [001] directions.

In the following the crystallographic principles of the orientation relation between austenite and α' martensite are discussed in some detail. The lattice correspondence between α' and γ is given by the Bain relationship, displayed in Figure 12.2a. Here the (tetragonally distorted) bcc lattice (bct) is inscribed into the original fcc lattice. In this Bain relation the c-axis of the bct lattice is parallel to one of the a-axes of the fcc lattice and one {110} plane of bct lattice is parallel to one {100} plane of fcc: $(100)_{fcc} \parallel (110)_{bct}$ and $[001]_{fcc} \parallel [001]_{bct}$. In order to obtain an undistorted bcc lattice the bct lattice needs to contract by about 20% in c-direction and expand by 12% in [110] direction. This is the Bain strain, **B**, mentioned before. Compression and expansion together result in a total volume change of up to 4%. In Figure 12.2b, this distortion is sketched by a circle symbolizing the austenite, deforming into an ellipse, symbolizing the martensite. In order to conserve an invariant line of transformation, the ellipse needs, in the example in Figure 12.2b, to be rotated by 9°. This corresponds to the rigid body rotation, **R**. The natural OR formed in this way is the KS OR mentioned above. The result of the transformation is finally displayed in

TABLE 12.1

Variants of the KS OR, Labeled According to Takayama et al. (2011)

Variant No.	Parallel Planes	Parallel Directions	Close Packed Plane Group	Bain Group	Misorientation with Respect to Variant No. 1: Axis	Misorientation with Respect to Variant No. 1: Angle
V1	$(111)_\gamma \| (011)_\alpha$	$[\bar{1}01]_\gamma \| [\bar{1}\bar{1}1]_\alpha$	CP1	B1	./.	
V2		$[\bar{1}01]_\gamma \| [\bar{1}1\bar{1}]_\alpha$		B2	[0.577 −0.577 0.577]	60.0
V3		$[01\bar{1}]_\gamma \| [\bar{1}\bar{1}1]_\alpha$		B3	[0.000 −0.707 −0.707]	60.0
V4		$[01\bar{1}]_\gamma \| [\bar{1}1\bar{1}]_\alpha$		B1	[0.000 0.707 0.707]	10.5
V5		$[1\bar{1}0]_\gamma \| [\bar{1}\bar{1}1]_\alpha$		B2	[0.000 0.707 0.707]	60.0
V6		$[1\bar{1}0]_\gamma \| [\bar{1}1\bar{1}]_\alpha$		B3	[0.000 −0.707 −0.707]	49.5
V7	$(1\bar{1}1)_\gamma \| (011)_\alpha$	$[10\bar{1}]_\gamma \| [\bar{1}\bar{1}1]_\alpha$	CP2	B2	[−0.577 −0.577 0.577]	49.5
V8		$[10\bar{1}]_\gamma \| [\bar{1}1\bar{1}]_\alpha$		B1	[0.577 −0.577 0.577]	10.5
V9		$[\bar{1}\bar{1}0]_\gamma \| [\bar{1}\bar{1}1]_\alpha$		B3	[−0.186 0.767 0.615]	50.5
V10		$[\bar{1}\bar{1}0]_\gamma \| [\bar{1}1\bar{1}]_\alpha$		B2	[−0.490 −0.463 0.739]	50.5
V11		$[011]_\gamma \| [\bar{1}\bar{1}1]_\alpha$		B1	[0.354 −0.933 −0.065]	14.9
V12		$[011]_\gamma \| [\bar{1}1\bar{1}]_\alpha$		B3	[0.357 −0.714 0.603]	47.1
V13	$(\bar{1}11)_\gamma \| (011)_\alpha$	$[0\bar{1}1]_\gamma \| [\bar{1}\bar{1}1]_\alpha$	CP3	B1	[0.933 0.354 0.065]	14.9
V14		$[0\bar{1}1]_\gamma \| [\bar{1}1\bar{1}]_\alpha$		B3	[−0.739 0.463 −0.490]	50.5
V15		$[\bar{1}0\bar{1}]_\gamma \| [\bar{1}\bar{1}1]_\alpha$		B2	[−0.246 −0.628 −0.738]	57.2
V16		$[\bar{1}0\bar{1}]_\gamma \| [\bar{1}1\bar{1}]_\alpha$		B1	[0.659 0.659 0.363]	20.6
V17		$[110]_\gamma \| [\bar{1}\bar{1}1]_\alpha$		B3	[−0.659 0.363 −0.659]	51.7
V18		$[110]_\gamma \| [\bar{1}1\bar{1}]_\alpha$		B2	[−0.302 −0.626 −0.719]	47.1
V19	$(11\bar{1})_\gamma \| (011)_\alpha$	$[\bar{1}10]_\gamma \| [\bar{1}\bar{1}1]_\alpha$	CP4	B3	[−0.615 0.186 −0.767]	50.5
V20		$[\bar{1}10]_\gamma \| [\bar{1}1\bar{1}]_\alpha$		B2	[−0.357 −0.603 −0.714]	57.2
V21		$[0\bar{1}\bar{1}]_\gamma \| [\bar{1}\bar{1}1]_\alpha$		B1	[0.955 0.000 −0.296]	20.6
V22		$[0\bar{1}\bar{1}]_\gamma \| [\bar{1}1\bar{1}]_\alpha$		B3	[−0.719 0.302 −0.626]	47.1
V23		$[101]_\gamma \| [\bar{1}\bar{1}1]_\alpha$		B2	[−0.738 −0.246 0.628]	57.2
V24		$[101]_\gamma \| [\bar{1}1\bar{1}]_\alpha$		B1	[0.912 0.410 0.000]	21.1

Figure 12.2c. The [001] directions of both phases show a slight misalignment and, instead, one {111} plane of the fcc phase is parallel to one {110} of the bcc phase. The variants formed with the KS OR are displayed in Table 12.1 and Figure 12.3a. In Table 12.1, a classification introduced by Takayama et al. (2012) is adopted: the variants that arise from the same set of closed packed planes are called closed packed plane (CP) groups. Morphologically they often form together and are called packets. The variants that arise from the same Bain transformation (i.e. that have the same c-axis of the bct phase) are called Bain groups or morphological blocks.

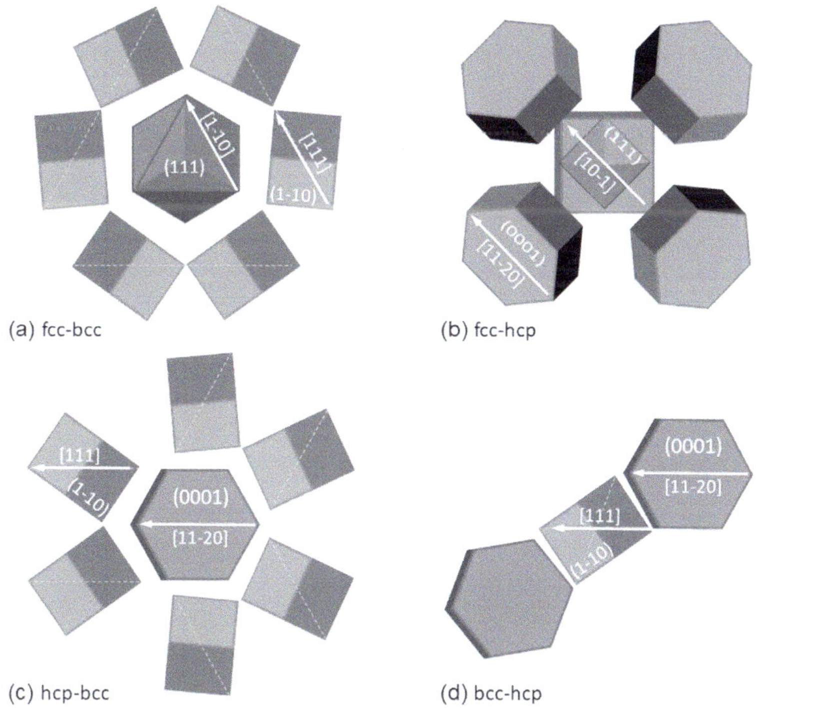

FIGURE 12.3 Graphical description of the most important basic orientation relationships between simple crystal structures. The parent phase is plotted into the center, the child phase outside. (a) Kurdjumov–Sachs OR between fcc and bcc according to $(111)_\gamma \parallel (\bar{1}01)_\alpha$, $[011]_\gamma \parallel [\bar{1}\bar{1}1]$. Only the ORs for one of the four {111} planes are shown. (b) Shoji–Nishiyama OR between fcc and hcp according to $(\bar{1}11)_\gamma \parallel (0001)_\varepsilon$ and $[110]_\gamma \parallel [11\bar{2}0]_\varepsilon$. All four existing variants are displayed. (c) Burgers OR between hcp and bcc according to $(101)_\beta \parallel (0001)_\alpha$ and $[11\bar{1}]_\beta \parallel [11\bar{2}0]_\alpha$, showing all six existing variants. (d) The same Burgers OR as (c), but with the bcc phase as the parent phase (two of six variants; two on every {110} plane).

Depending on the mechanical strength of the related phases, the stacking fault of austenite, and their lattice constants, all of which depend on transformation temperature and composition, the phase transformation creates a number of different ORs and, connected to this, a number of different morphologies. Some well-known α–γ ORs and their related morphologies are compiled in Table 12.2. The most common martensite morphology is *lath martensite* which forms in relatively soft low-carbon steels and steels with negligible carbon content (Morito et al., 2003, 2006). They frequently observe a KS OR. In highly alloyed, high-strength steels, e.g. Fe 28% Ni or high carbon steels, martensite shows plate or lenticular shape (Kitahara et al., 2005). Other forms, for example butterfly or thin plate shapes, appear more rarely (see Maki, 2012). It is important to realize (and will be shown in an example further below) that probably all steels with γ–α transformation usually show a range of ORs appearing simultaneously (Cayron et al., 2010; Cayron, 2020).

TABLE 12.2

Typical Fcc–Bcc Orientation Relations Occurring in Steels and Some of Their Crystallographic Properties, Following Maki (2012) and He et al. (2005)

Name	Orientation Relationship	No. of Variants	Misorientation	Strain Accommodation	Typical Microstructures	M_s and Strength[a]
Bain (B)	$(001)_\gamma \parallel (001)_\alpha$ $[100]_\gamma \parallel [110]_\alpha$	3	45° ⟨1 0 0⟩		Basic fcc–bcc lattice correspondence	
Pitsch (P)	$(100)_\gamma \parallel (110)_\alpha$ $[011]_\gamma \parallel [1\bar{1}1]_\alpha$	12	45.98° ⟨0.08 0.2 0.98⟩		Not important in steels	
Kurdjumov–Sachs (KS)	$(111)_\gamma \parallel (\bar{1}01)_\alpha$ $[011]_\gamma \parallel [\bar{1}\bar{1}1]_\alpha$	24	42.85° ⟨0.97 0.18 0.18⟩	Dislocations	Lath martensite in low-carbon steels	High M_s, low yield strength
Nishiyama–Wasserman (NW)	$(111)_\gamma \parallel (110)_\alpha$ $[11\bar{2}]_\gamma \parallel [1\bar{1}0]_\alpha$	12	45.98° ⟨0.98 0.08 0.2⟩	Dislocations and twins	Plate or lenticular martensite, high alloy steels	Intermediate M_s and yield strength
Greninger–Troiano (GT)	$(111)_\gamma \parallel (110)_\alpha$ $[\overline{17}\ 12\ 5]_\gamma \parallel [17\ \overline{17}\ 7]_\alpha$	24	44.23° ⟨0.97 0.19 0.13⟩	Twins	Thin plate, rarely in steels	Low M_s, high yield strength

[a] M_s: martensite start temperature, the temperature at which the martensitic transformation starts during a quenching experiment.

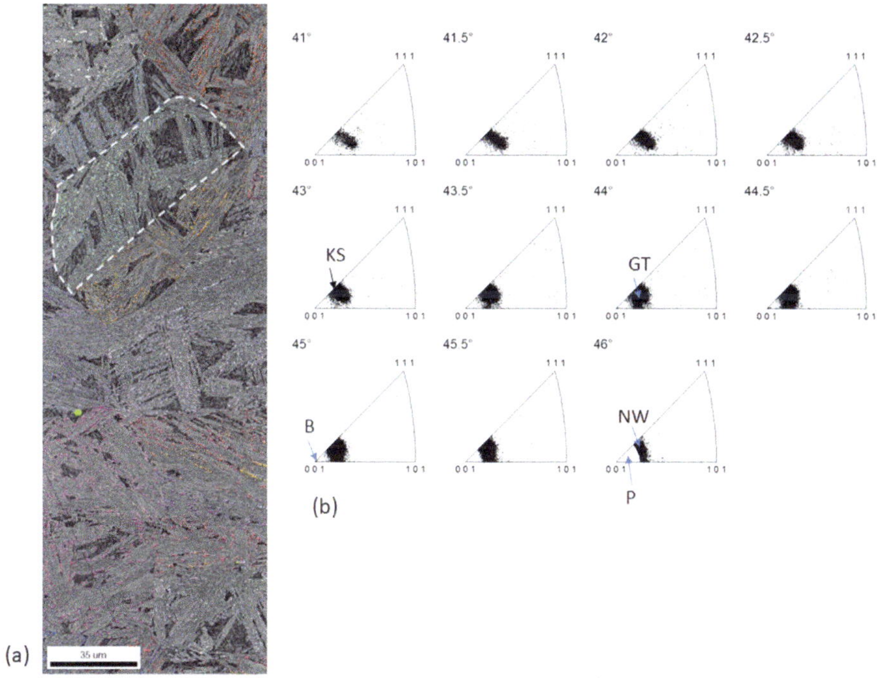

FIGURE 12.4 Microstructure and fcc-bcc orientation relation of a carbide-free bainitic steel. (a) Image quality map of the bcc phase, combined with an IPF map of the fcc phase. Areas with similar coloring have been one austenite grain at high temperature. (b) Orientation relationship between the fcc and bcc phase as axis-angle plot. Only the relevant angular range between 41 and 46 degrees is shown. The positions of classical misorientation relations are indicated: KS: Kurdjumov–Sachs, GT: Greninger–Trojano, B: Bain, and NW: Nishiyama–Wassermann.

An example for the presence and analysis of ORs in a carbide-free bainitic (CFB) steel is displayed in Figure 12.4. This steel microstructure was formed by a quenching and partitioning (Q&P) process with the quenching and holding step conducted at 400°C. The microstructure consists of a bainitic ferrite matrix with thin films of retained austenite between some of the bainite sheaves. Bainitic ferrite is a bcc phase that is formed by a displacive transformation of the iron lattice but whose transformation kinetics is governed by the diffusion of carbon into the residual austenite. This microstructure, as measured by EBSD-based orientation microscopy, is shown in Figure 12.4a. The matrix is displayed by gray values indicating the diffraction pattern quality. The residual austenite is colored by an inverse pole figure color code of the sample normal direction. The plot allows to recognize the original austenite grains as those areas where the thin film lamellae have similar colors. Figure 12.4b shows the axis angle plot, limited to angles between 41° and 46°, of the misorientations between the α and the γ phase. It is obvious that the misorientations cover a continuous fiber across these angles, including all the misorientations compiled in

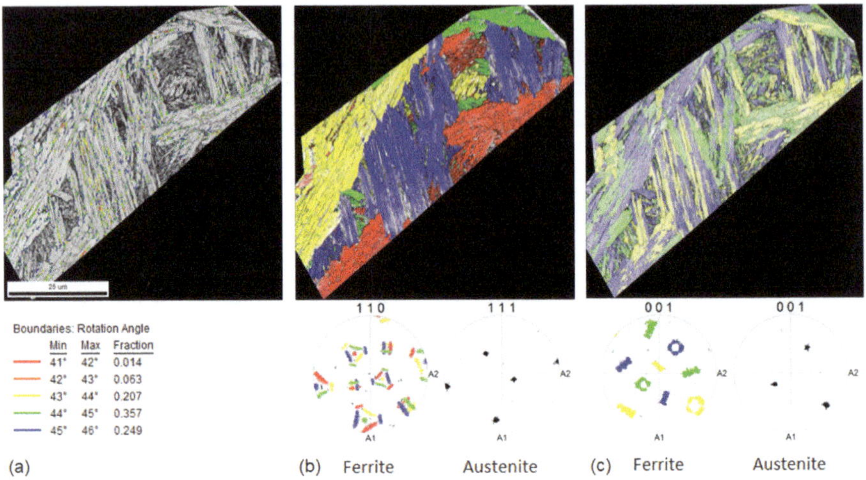

FIGURE 12.5 Microstructure and variant grouping of one grain selected from Figure 12.4. (a) Misorientation coloring of the fcc-bcc phase boundaries. (b) Grouping according to crystals transformed on the same closed-packed plane ($\{111\}_\gamma \parallel \{110\}_\alpha$). Four close packed planes {111} of the fcc lattice require four different colors. (c) Grouping according to the same Bain group ($\langle 001\rangle_\gamma \parallel \langle 001\rangle_\alpha$). Three different ⟨001⟩ directions of the fcc lattice require three different colors.

Table 12.2 (except for the Bain and Pitsch ORs). The positions of these "classical" misorientation are indicated in the figure.

Figure 12.5 displays one single grain out of the large area map in Figure 12.4a. The γ–α phase boundaries are color-coded according to their misorientation angle taking values between 41° and 46°. Clearly even in a single grain a large variety of ORs appears. Cayron (2020) claims that these variations cannot be explained by the PTMC and proposes that they are due to the back stresses caused by the shape change of the transformation and their accommodation by geometrically necessary dislocations. Schumann and Leineweber (2021), on the other hand, show in a detailed study on the reverse α–γ phase transformation between bcc iron and near-fcc Fe_4N that the PTMC is still capable of describing the observed OR variations, provided that the lattice invariant shear strain, **L**, is not assumed to be constant but is allowed to vary by combination of different fractions of individual, coplanar shear systems.

Figure 12.5b and c displays the different groups of variants occurring in the above selected grain. Figure 12.5b shows those variants in similar color that have the same pair of CP $\{111\}_\gamma$ and $\{110\}_\alpha$ in both phases. In the respective pole figures there is a group of α grains for every $\{111\}_\gamma$ plane. Figure 12.5c shows the Bain groups corresponding to a group of α grains for every $\langle 001\rangle_\gamma$ direction. This latter way of coloring allows to figure out the c-axis of the tetragonally distorted martensite: grains of the same color have transformed along the same a-axis and thus have the same c-axis. The small rings in the {001} ferrite pole figure indicate the c-axis directions which are parallel to one a-axis direction in the {001} austenite pole figures.

The existence of ORs and the grouping of variants makes it possible to reconstruct the parent phase microstructure and texture from an orientation map of the child phase (for example in order to determine the difficult-to-measure microstructure and texture of high-temperature austenite). This topic will be discussed in Section 12.3. Before that we will briefly compile ORs that arise in other materials and for other phase pairings.

12.2.3 Orientation Relationships in Other Materials

Besides the fcc–bcc transformation there is one further important martensitic phase transformation in steels which is the one from fcc to the hexagonal closed packed (hcp) ε phase. This phase is of great importance for the transformation induced plasticity (TRIP) effect in high-Mn steels (e.g. Frommeyer et al., 2003). The ε phase transformation occurs similar to the allotropic phase transformation of cobalt (where, however, the high- and low-temperature phases are exchanged, i.e. the high-temperature phase is hcp). The basic OR is the *Shoji–Nishiyama* OR which has four variants, as displayed in Figure 12.3b: $(\bar{1}11)_\gamma \parallel (0001)_\varepsilon$ and $[110]_\gamma \parallel [11\bar{2}0]_\varepsilon$, which corresponds to a misorientation of 56.6° $\langle 0.245\ 0.59\ 0.769 \rangle_\gamma$.

To the knowledge of the present author the detailed mechanism of this transformation is not well established and it appears most probable that it proceeds by some sort of coalescence of stacking faults in the fcc phase (Houska et al., 1960; Tolédano et al., 2001). Also a pole mechanism has been considered (Bollmann, 1961).

Like in steels, the phase transformations occurring in polymorphic titanium and zirconium alloys are intensely exploited for the control of materials properties. Banerjee and Williams (2013) give a comprehensive overview on titanium alloys, including an extensive chapter on the occurring phase transformations. The high-temperature β phase with bcc crystal structure transforms into the low-temperature α phase with hcp structure. The basic OR is the *Burgers* OR with six variants: $(101)_\beta \parallel (0001)_\alpha$ and $[11\bar{1}]_\beta \parallel [11\bar{2}0]_\alpha$, corresponding to a misorientation of 45.25° $\langle 0.046\ 0.11\ 0.993 \rangle_\beta$.

Here again, the most densely packed planes and directions of both phases are parallel to each other as displayed in Figure 12.3b and c. A detailed investigation of the α–β phase transformation in pure titanium has been made by Seward et al. (2004). The Burgers relationship in zirconium alloys has been investigated by Chauvy et al. (2006). The Burgers OR seems to show less variability than the ORs of the martensitic phase transformations in steels, probably because the dislocation shear mechanisms in the relevant phases are less flexible. Nevertheless, in diffusive transformations the deviation can be up to ±5° in order to satisfy a double Burgers OR: in this case the orientation of a newly formed α-phase grain at a β-grain boundary is such that it observes a close-to Burgers OR with both grains (Stanford and Bate, 2004). The occurring deviation may be rationalized by the fact that two of the Burgers OR variants are only 10° $\langle 011 \rangle$ away from each other (Germain et al., 2007) and accommodation of shear occurs by the three $\langle a \rangle_\alpha$-slip systems on the basal plane (or two $\langle 111 \rangle_\beta$-slip systems on the $(011)_\beta$ plane). The Burgers OR is highly relevant for texture formation in titanium alloys as we will briefly discuss in the next section.

TABLE 12.3
Selected Orientation Relationships Occurring in Different Materials Systems (Metals, Ceramics, Minerals)

Material	Phases	Orientation Relation	No. of Variants	Name	Reference
TiAl	α (disordered hcp) or α_2 (ordered hcp), γ tetragonal ($L1_0$) (close to fcc)	$(0002)_\alpha \parallel (111)_\gamma$ and $[11\bar{2}0]_\alpha \parallel [1\bar{1}0]_\gamma$	6 for $\alpha \rightarrow \gamma$ 4 for $\gamma \rightarrow \alpha$	Blackburn	Humbert et al. (2005)
NiTi	B2 (ordered bcc), B19 (orthorhombic) or B19′ (monoclinic)	$(010)_{B19'} \parallel (110)_{B2}$ and $[001]_{B19'} \parallel [110]_{B2}$	12		Ren et al. (2001); Cayron (2020)
CuZnAl	β ($L2_1$) cubic (fcc related), β′ (18R) monoclinic	$[100]_\beta \parallel [100]_{\beta'}$, $[010]_\beta \parallel [01\bar{1}]_{\beta'}$	6		Alkan et al. (2018)
ZrO_2 stabilized by CeO_2	Tetragonal monoclinic	$(001)_{t'} \parallel (001)_m$ and $[100]_{t'} \parallel [100]_m$	4	"cube on cube"	Onda et al. (2010)
Alkali feldspar	Kalsilite (hexagonal), Orthoklas (monoclinic)	$[0001]_{Ks} \parallel [010]_{Or}$ and $[10\bar{1}0]_{Ks} \parallel [100]_{Or}$	6		Ageeva et al. (2012)

There are many other ORs between phases that transform from one to the other and control the texture of the final product. We mention a few further ones in Table 12.3 together with data on their appearance and crystallographic features.

12.3 TEXTURE FORMATION, VARIANT SELECTION, RECONSTRUCTION OF TEXTURE AND MICROSTRUCTURE OF AUSTENITE

It is clear that a phase transformation that occurs with an orientation relation between the involved phases leads to a texture change. In the case of steels, for example, the texture that has been created by hot rolling is partly randomized by the phase transformation, simply because one austenite grain can, in principle, transform into 24 (or 12) different ferrite or martensite variants, the orientations of which are widely distributed in orientation space. Nevertheless, it is frequently observed that the transformation texture is significantly stronger than what would be expected from a non-selective phase transformation from the parent texture. A good example is presented in Tanaka et al. (2017). The reason for this is so-called variant selection, indicating that not all variants are equally possible. The most common reason for variant selection is the so-called double-orientation relation, where a new grain nucleating at the grain boundary between two parent grains observes an OR or a close OR with both grains. This mechanism can lead to very sharp variant selections when a new grain fits to grains of typical texture components of the parent texture, for example by a double KS OR (Tanaka et al., 2017). The double KS OR may also lead to a texture memory effect through cyclic heating and cooling (Tomida et al., 2013) where the texture of the α phase is recovered after $\alpha \rightarrow \gamma \rightarrow \alpha$ phase transformation. Wakita and Suzuki (2017) showed the existence of double KS ORs directly by in-situ heating EBSD mapping. They furthermore show that even the microstructure is largely preserved during the cyclic heating. Variant selection may also be triggered by the presence of certain dislocation structures aligned with the habit plane of variants as shown by Chiba et al. (2013).

For Ti alloys the Burgers OR is important for the generation of the α-phase texture from the hot deformation β-phase texture. A particular example is the occurrence of so-called macro-zones, areas of very similar crystal orientation in forged near-α Ti alloys. These macro-zones are detrimental to the mechanical properties of the material. It was shown by Gey et al. (2012) that these macro-zones are created by the existence of a sharp Burgers OR from the high-temperature β-phase texture. Also here the double Burgers OR leads to a variant selection and texture enforcement. Stanford and Bate (2004) claim that also an incomplete double OR, where the c-axis direction of the α phase is parallel to two $\langle 110 \rangle$ directions of grains at a β-phase grain boundary, leads to a variant selection.

An important problem with the understanding and controlling of the phase transformation texture change is that it is often not easily possible to measure the high-temperature texture and microstructure. A very efficient way out of this problem is the reconstruction of the high-temperature phase texture and microstructure based on the orientation mapping of the low-temperature phase. The existence of an OR between the parent and child phase implies that the child variants have fixed mutual misorientations; an example is given in Table 12.1 for the case of a KS OR. The presence of these misorientations and the orientation of the individual variants can be used to determine the parent phase orientation even if it has completely transformed. If the child orientations (e.g. martensite) have been measured by EBSD orientation microscopy then it is not only possible to calculate the orientations of the parent grains but also to determine their morphology by grouping together neighboring variants with the same parent orientation. Details of the approach will not be discussed here, we rather refer to publications of the groups who developed software for parent phase reconstruction (Cayron, 2007; Miyamoto et al., 2010; Humbert et al., 2011). The basic principle is that every child phase orientation can be traced back to possible parent phase orientations through its orientation relation. The number of possible parent solutions corresponds to the number of variants that can be formed from one parent grain. Of all possible parent grain orientations that one is selected, which leads to a common orientation solution for neighboring child grains. Of course, the grouping has to be done with a certain error allowance. Furthermore, the more variants vote for the same grain the more certain and more accurate the orientations become. The different software tools differ from each other in the way how individual orientations are dealt with (e.g. as average orientation over a whole variant grain or by individual orientations at variant boundaries), whether the software is able to determine or refine the OR by itself, whether parent grains of uniform orientation are created or whether also orientation gradients due to deformation of the parent phase can be recovered (Humbert et al., 2015). As the mentioned software tools are all under development it is not possible to tell which one can do what. The tools have been applied very successfully to determine the parent textures and

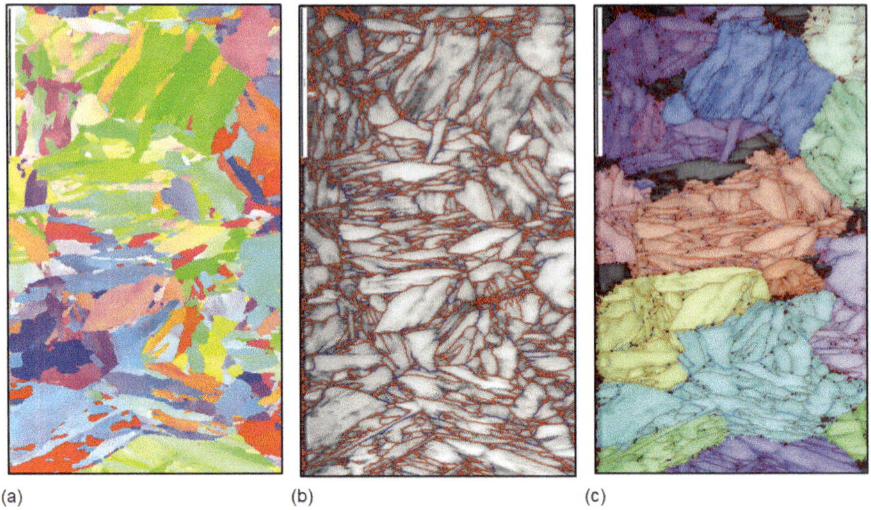

FIGURE 12.6 Austenite grain reconstruction in a fully martensitically transformed low-alloyed TRIP steel. (a) IPF map of the martensite structure. (b) Grain boundary map of the martensite structure. Red boundaries indicate typical variant-variant boundaries according to Table 12.1. Blue are other, non-defined boundaries. (c) Reconstructed austenite grains in IPF coloring. The austenite grains have been reconstructed employing the ARPGE software using the KS orientation relationship for best fitting. Gray areas could not be assigned to any grain.

microstructures of steels (e.g. Cayron et al., 2006). Kubota et al. (2016) determined texture and deformation state of hot-rolled austenite using austenite reconstruction.

To illustrate the functioning of austenite reconstruction we display in Figure 12.6 the reconstruction of the austenite grain structure from a lath martensite microstructure of a low-alloyed TRIP steel. For the reconstruction the software ARPGE of Cayron (2007) was used, employing the KS OR. Figure 12.6a shows the martensite orientation map, colored with an inverse pole figure (IPF) color code of the normal direction of the sheet. In Figure 12.6b, the variant and non-variant grain boundaries are displayed in different colors. Variant grain boundaries are considered those which have one of the misorientations listed in Table 12.1. In Figure 12.6c, finally, the result of austenite reconstruction is shown together with an overlay of the original lath martensite boundaries. Here orientations and grain shapes can be directly observed. A few areas have not been assigned to any of the grains. Possibly they were not reconstructed because they belong to another austenite grain with too few variants being visible in this cross section.

12.4 SUMMATION

Solid-solid phase transformation proceed almost always with crystallographic ORs between the parent phase and the child phase. In the case of diffusive transformations, the ORs may be less pronounced than in the case of displacive or martensitic transformations. The existence of fixed ORs is of great importance for texture formation because it leads to inheritance of textures from one phase to the other, usually together with some characteristic texture weakening. So-called double ORs where a newly formed phase approximately observes ORs with two parent grains lead the phenomenon of variant selection. The crystallographic details of an OR may be used as an indicator of the particular conditions of the phase transformation, as it is the case in steels. Furthermore, the OR may be used to reconstruct the microstructure and texture of the parent phase from an orientation map of the child phase.

13 Synchrotron Radiation, Nondiffraction Techniques, and Comparisons between Methods

13.1 INTRODUCTION

In this chapter, we introduce methods for texture determination that do not fit into the earlier sections on x-ray or neutron macrotexture or electron microscope-based microtexture techniques.

Although x-ray diffraction is usually associated with macrotexture (Section 4.3), such methods can in principle also be used for individual-grain (microtexture) measurements if it can be arranged that the necessary small volumes in a sample are irradiated with a sufficiently fine x-ray beam. X-rays generated in a synchrotron offer this possibility since they are characterized by an intensity several orders of magnitude higher than those generated by conventional x-ray tubes, combined with a high brilliance, that is, low angular divergence. Accordingly, Laue patterns from individual grains or Debye–Scherrer patterns from small polycrystalline volumes can be obtained. The application of synchrotron radiation, based on the Laue and Debye–Scherrer methods (Section 3.4), is one of the latest and most rapidly developing tools for texture and microstructure research, which will be described in Section 13.2.

All techniques for texture analysis mentioned so far are founded on the diffraction of radiation—electrons, x-rays, or neutrons—at the crystal lattice (Chapter 3). Besides those diffraction-based techniques, there are methods (e.g., ultrasonic or magnetic measurements or optical methods) to derive macrotextures or microtextures that are not based on diffraction of radiation. Since such techniques may still be of some interest for special applications, a short discussion is warranted (Section 13.3). Finally, Section 13.4 includes a summary of the various techniques used to determine macrotextures and local orientations, which have been discussed in this book, and a comparison of their main advantages and disadvantages.

13.2 TEXTURE ANALYSIS BY SYNCHROTRON RADIATION

Modern electron synchrotrons with energies in excess of 100 keV yield a white spectrum of x-rays with intensities of many orders of magnitude higher than conventional

DOI: 10.1201/9781003258339-16

x-ray tubes, combined with minimum angular divergence of less than 2 mrad. This unique combination of high photon flux, small beam size, and free choice of wavelength opens a wide range of possibilities for texture analysis. Short-wavelength synchrotron x-rays have similar penetration depths as neutrons, which allow global textures of big centimeter-sized specimens to be measured similarly as in neutron diffraction (Section 4.4). However, because of the local resolution of a high-energy synchrotron beam, measurements can also be performed on a local basis. Especially, the high brilliance of the current third-generation synchrotron radiation sources with x-ray energies in the GeV range enables new applications in x-ray microdiffraction with sub-micrometer beam sizes, where mesoscopic length scales of 0.1–100 μm can be studied. The strongly focused, high-intensity, polychromatic radiation favors synchrotron x-rays for local analysis of individual crystallites by means of the Laue technique (Section 3.4), which is discussed in Section 13.2.1. Alternatively, monochromatic radiation can be produced by using a monochromator crystal, and this radiation can be used to generate Debye–Scherrer patterns from polycrystalline volumes (Section 13.2.2).

In brief, interesting applications of synchrotron radiation may include (e.g. Garbe et al., 1996; Bunge et al., 2003; Brokmeier, 2006; Staron et al., 2017; Zhou et al., 2018)

- Nondestructive local texture measurements
- High-resolution texture analysis
- Investigation of texture gradients
- In-situ texture measurements of texture transformations
- High-resolution, nondestructive orientation and phase mapping, even in 3D, similar to electron backscatter diffraction (EBSD) but with slightly lower resolution (100–500 nm)
- Determination of misorientation and characterization of grain boundaries, including boundary planes
- Internal strain characterization in selected local volumes
- Study of textures of non-centrosymmetric crystals by means of observations near the absorption edge (Section 3.6.1) with high anomalous scattering where Friedel's law does not apply, since it is possible to tune the synchrotron x-ray wavelength in a wide range (Bunge and Esling, 1981; Fuentes, 1989; Hammond, 2009)

The main drawback of the synchrotron techniques is the limited availability of synchrotron sources in large-scale research facilities (there are about 80 synchrotron facilities worldwide). Furthermore, interpretation of the data is often more complicated than for standard x-ray or electron diffraction techniques. Thus, experiments using synchrotron radiation will most likely be limited to special cases that cannot be accomplished by conventional x-ray or electron diffraction techniques.

13.2.1 Individual Orientations from Laue Patterns

The most common technique to determine the orientation of single crystallites by means of x-ray diffraction is the Laue method, which has been introduced in Section 3.4. In this method, the use of white (polychromatic) radiation ensures that for each set of lattice planes, x-rays will exist that fulfill Bragg's condition for diffraction. On intersection with a recording medium, the diffracted beams generate characteristic spot patterns (Figures 3.10 and 3.11) that can be evaluated to derive the crystallographic orientation of the sampled volume. However, the beam diameter of conventional x-ray tubes is approximately 1 mm. As x-ray optics have not been common until recently, x-rays could not be focused easily and small beam diameters of the order of microns—which would be necessary to achieve spatial resolution comparable with electron diffraction techniques—could only be obtained by a collimation of the primary beam. This, however, resulted in a strong reduction in intensity and, consequently, in unacceptably long measuring times. Even under application of rotating anode x-ray tubes, a beam diameter of 10 μm would require illumination times of the order of 10 h, which made large-scale microtexture analysis an unrealistic task. Therefore, the Laue technique was traditionally only applied for analysis of single crystals or very coarse-grained structures with a grain size exceeding 100 μm (Ferran et al., 1971).

To overcome these limitations, Gottstein (1986, 1988) suggested the use of the enormous intensity of synchrotron x-rays for Laue orientation determination. At that time the beam could be collimated to a size of $5 \times 5\ \mu m^2$ (sufficiently fine for many microtexture applications), and illumination times of the order of 1 s could be realized. The sample was mounted in back-reflection geometry on a motorized x/y stage, which allowed scanning of the sample surface in a constant grid with steps of 10 μm. Fairly large numbers of orientations could be determined by recording the diffraction patterns on photographic films or x-ray-sensitive image plates, digitization of the patterns, and a subsequent, automated evaluation similar to that described for Kikuchi patterns (Brodesser et al., 1991). With the advent of charged-coupled device (CCD) cameras, it became possible to record and interpret Laue patterns online (Wenk et al., 1997). For this purpose, the Laue patterns (Figure 13.1a) are converted into gnomonic projection (Appendix IV) where cozonal poles (*hkl*) lie on straight lines (Figure 13.1b). Next the Hough transform is applied, converting lines into points (Section 6.3.3). The computer then detects the points and calculates the interzonal angles that are used to index the patterns, so that finally the orientation can be determined. As an alternative, a template-matching scheme for rapid indexing of Laue diffraction patterns has been developed, which was shown to offer good indexing rates even for low-quality patterns (Gupta and Agnew, 2009).

Thus, the sampling area investigated by this method is given by the selected beam diameter. Meanwhile, the resolution of synchrotron-based Laue diffraction has been improved to almost the lateral size of EBSD, i.e. to a beam size of well below 1 μm, with almost no limitation in the direction along the beam path. Despite the very small beam diameter, patterns can be recorded at about 20 Hz, which is still about one or two orders of magnitude slower than what can be done with EBSD, but with a similar or larger wealth of information, in particular, phase, orientation, and elastic strain tensor of the crystals.

However, the large penetration depth of the high-energy synchrotron x-rays entails that the diffraction patterns contain information stemming from highly elongated volumes of e.g. $1 \times 1 \times 1000\ \mu m^3$ along the beam path. This obviously hampers

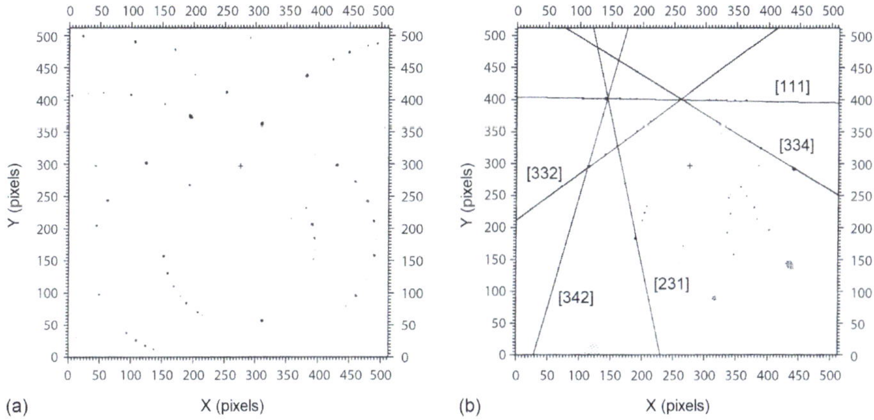

FIGURE 13.1 Laue patterns obtained with polychromatic synchrotron x-radiation. (a) Digitized Laue pattern from an olivine crystal (the small cross represents the pattern center). (b) Pattern from (a) transformed into a gnomonic projection with overlaid zones that were identified by a Hough transform. (Courtesy of H.R. Wenk.)

high-resolution local texture analysis in bulk materials, as long as no other means are taken into account. To overcome this limitation, Wenk et al. (1997) suggested to analyze thin sections of the materials in transmission geometry. This results in a spatial resolution of the order of 10 μm (i.e. the sample thickness) which is sufficient for analysis of many geological materials and many metals in the recrystallized state, but is definitely inferior to electron microscopy-based techniques for orientation determination. However, the signal-to-noise ratio of the synchrotron Laue technique is superior, and furthermore, sample preparation is easy and nonconductive materials can readily be analyzed.

More recently, Larson and coworkers developed a ray-tracing technique to localize the different crystals diffracting onto one detector frame with sub-micrometer resolution (Larson et al., 2002; Yang et al., 2004; Budai et al., 2008). This so-called differential-aperture x-ray microscopy (DAXM) technique was installed at the Advanced Photon Source (APS) at Argonne National Laboratory. In contrast to the Debye–Scherrer techniques described in Section 13.2.2, the use of a Laue-based technique with polychromatic radiation eliminates the need for sample rotation, which facilitates the spatial mapping below 5 μm and provides a truly 3D probe for nondestructively investigating polycrystalline structures on mesoscopic length scales of 0.1–100 μm.

In DAXM, a highly focused, high-energy polychromatic synchrotron x-ray beam with size of $0.5 \times 0.5\ \mu m^2$ is directed onto the sample. The x-ray beam penetrates the sample material and generates Laue diffraction spots over a distance of up to approximately 100 μm. Laue diffraction patterns, emanating from the region of the sample lying in the path of the incident x-ray beam, are detected with a CCD camera located at 90° to the incident beam (Figure 13.2). For very thin samples a single-crystal Laue pattern is generated, as long as the grain size is larger than beam diameter and

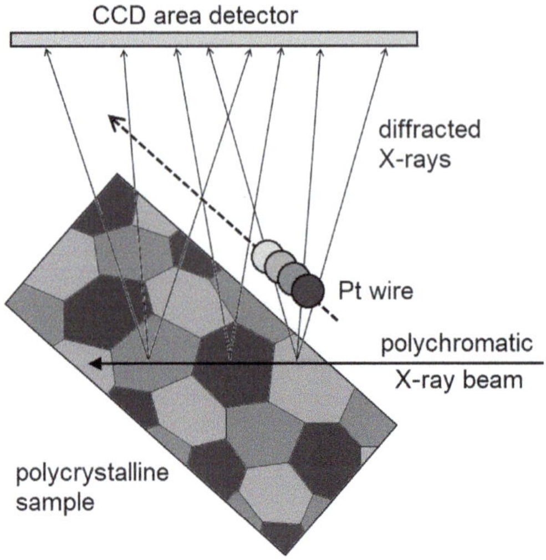

FIGURE 13.2 Schematic of the operation of the differential-aperture x-ray microscope (DAXM).

sample thickness. In this case, the measurement provides the crystal orientation as well as local lattice structure and strain for the given location. Sample translations may be used to generate 2D orientation and strain maps, similarly as in EBSD.

In the case of bulk, polycrystalline samples, Laue diffraction spots will arise from many grains along the beam path. Hence, the diffracted pattern will consist of the superposition of many different Laue patterns of the respective grain orientations. Even if it is possible to index such a pattern consisting of several superimposed Laue peaks, the position of the originating grains in the microstructure is not known. Therefore, in the DAXM technique, a sequence of Laue patterns is recorded as a 50-μm platinum wire is traversed in small steps across the sample surface. The wire acts as a knife-edge profiler, and differences between successive images provide a unique reconstruction of the Laue diffraction patterns for each submicron volume element along the path of the x-ray beam. This allows to establish the location in the sample from which the Laue spots originate. A computer program collates and reconstructs the Laue diffraction images from each microregion of the sample material along the incident beam path. By moving the sample with respect to the incident x-ray beam and, at each position, scanning the Pt wire across the sample, the local crystallographic orientation and residual deviatoric strain tensor can be determined with a spatial resolution of approximately $0.5 \times 0.5 \times 1.0\ \mu m^3$.

13.2.2 Local Textures from Debye–Scherrer Patterns in Polycrystalline Regions

For texture analysis in fine-grained polycrystalline samples by means of the Debye–Scherrer method with synchrotron radiation, a narrow beam of *monochromatic* x-rays is required. To this end, the x-ray beam first passes a cooled copper plate that

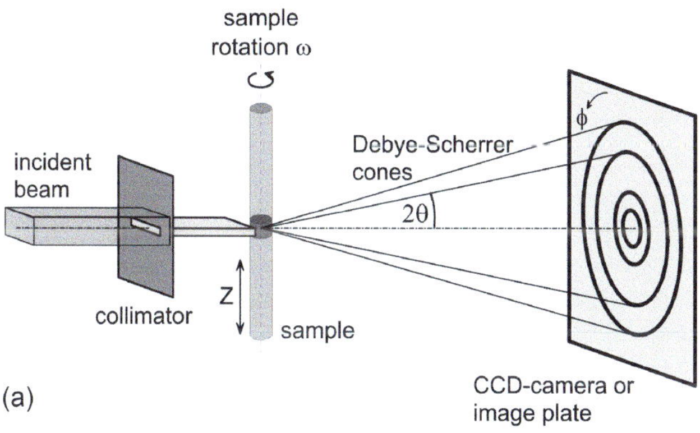

(a)

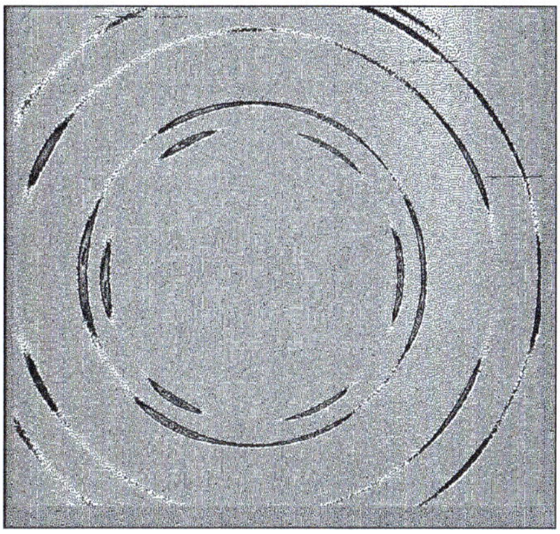

(b)

FIGURE 13.3 Texture analysis by the Debye–Scherrer method with monochromatic synchrotron radiation. (a) Schematic view of the experimental setup. 2θ is the scattering angle, ω describes the sample rotation, and φ denotes the angle along the Debye–Scherrer ring. (b) Example from a cold-rolled aluminum sheet (38% reduction, layer 2850 μm from the surface). (Courtesy of O. Mishin and D. Juul Jensen.)

absorbs radiation with long wavelengths, before a monochromator crystal selects the desired wavelength in the range of 0.01–0.1 nm (Poulsen and Juul Jensen, 1995). Finally, the monochromatic beam is collimated to the desired size and falls onto the sample, where the Debye–Scherrer patterns are generated in transmission geometry (Figure 13.3a). Szpunar and Davies (1984) have demonstrated that it is possible to obtain reliable texture information from the Debye–Scherrer patterns that they recorded on photographic films. Nowadays, the diffraction patterns are recorded either with image plates or with a CCD camera (Section 4.3.4). For online analysis, CCD cameras are superior as they provide digitized intensities with a high spatial resolution and a wide dynamic range within a few seconds (Wenk and Grigull, 2003).

As discussed in Section 3.4, diffraction from a powder sample with random texture gives rise to a set of homogeneous concentric Debye–Scherrer rings that are related to the various reflecting lattice planes (*hkl*) through the Bragg angle θ. If the sample has a pronounced texture, the rings display intensity variations that are characteristic of the texture of the sampled volume (Wever, 1924). As an example, Figure 13.3b illustrates the diffraction pattern recorded with a CCD camera from a cold-rolled aluminum sheet. After appropriate intensity corrections, the texture can be derived from the analysis of the intensity variations along the Debye–Scherrer rings.

If samples with axially symmetric textures, such as wires, are analyzed in transmission with the fiber axis normal to the incident x-ray beam, then a single set of Debye–Scherrer rings contains already the necessary texture information. By this method, texture variations in small rods of Ni-Fe alloys, prepared by electroforming techniques, have been documented and represented in inverse pole figures (Bäckstrøm et al., 1996). For nonaxial textures, one set of Debye–Scherrer rings is not sufficient, but, in analogy to selected area electron diffraction (SAD) pole figures in transmission electron microscopy (TEM) (Section 6.2.2), the sample needs to be rotated by an angle ω (Figure 13.3a) and in each setting a pattern has to be recorded. Geometric expressions for the transformation of the angles φ, θ, and ω into the pole figure coordinates α and β have been given by Poulsen and Juul Jensen (1995) and Heidelbach et al. (1999).

Hard synchrotron x-rays with short wavelengths of approximately 0.01 nm (~120 keV) have similar penetration depths as neutrons, so that global textures of centimeter-sized specimens can be measured (Brokmeier, 2006). Investigations on metals such as iron, titanium, copper, aluminum, and magnesium have been carried out up to 35-mm sample thickness. Because of the high photon flux, the total counting times for texture measurements can be very short, allowing in-situ experiments on a short timescale to measure the texture changes under applied load or at high temperatures (Yi et al., 2006).

Besides, the high brilliance of a synchrotron source makes it ideally suited for measurements on a local scale. With appropriate focusing/collimation of the incident beam, the textures from small mesoscopic volumes can be analyzed. For instance, as sketched in Figure 13.3a, through-thickness texture heterogeneity in a cylindrical sample can be studied by measuring the texture with a finely collimated x-ray beam at several positions along the z axis of the sample (Mishin et al., 2000; Skrotzki et al., 2003).

Because of the extremely small divergence of the incident beam in synchrotron diffraction experiments, such diffraction patterns can be exploited to yield information on microstructure and texture of small volumes (with a diameter of tenths of millimeters), within bulk samples of centimeter size (Black et al., 1991). As illustrated in Figure 13.4a, the radial width of the Debye–Scherrer rings is linked to the sample thickness along the beam path. For a Bragg angle of 5° and a sample thickness of 10 mm, the width of the Debye–Scherrer rings is of the order of 2 mm, and the different parts of the Debye–Scherrer rings disclose the texture information of different layers within the sample volume. Thus, the spatial resolution in terms of the volume analyzed along the beam path is determined by the radial resolution of the Debye–Scherrer rings. Provided a CCD camera with a resolution of 100 μm is used, small volumes with a length of the order of 500 μm along the beam direction can be resolved. By this method, Reeves et al. (1996) were able to differentiate diffraction patterns from three different through-thickness layers in a 15-mm aluminum plate.

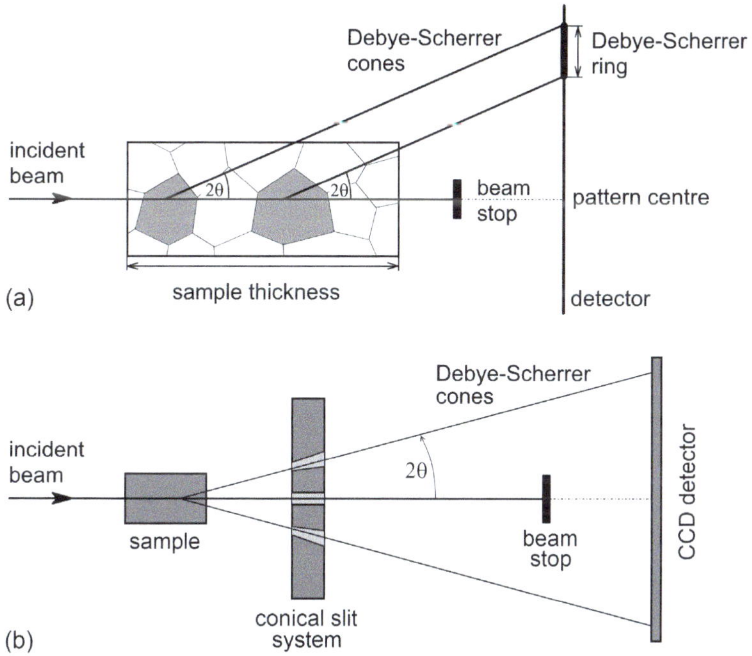

FIGURE 13.4 Analysis of local textures within bulk samples with monochromatic synchrotron radiation. (a) Broadening of the Debye–Scherrer rings diffracted in two grains with identical orientations at different positions along the beam path; (b) use of a conical Debye–Scherrer slit system to define the local volume analyzed.

Garbe et al. (1996) refined this method by introducing a conical Debye–Scherrer slit system in combination with a pinhole to define the local volume (Figure 13.4b). With this setup, pole figures of small volumes with size down to $50 \times 50 \times 600\ \mu m^3$ could be distinguished and the disappearance of the shear texture component at the surface of a 3-mm aluminum sheet could be traced through the depths of 50–850 μm.

Bunge and coworkers have installed a "sweeping" technique at the high-energy (short-wavelength) beamline BW5 at HASYLAB/DESY in Hamburg, Germany (Wcislak et al., 2002; Bunge et al., 2003; Bunge, 2005). This instrument is essentially the advancement of standard diffractometry with an area detector (see Section 4.3.4), where the use of high-brilliance, high-intensity synchrotron radiation allows spatially resolved measurements of textures of materials with very high orientation resolution. The sweeping technique is based on the continuous movement of an area detector during exposure. That is to say, the detector undergoes a vertical translation motion in direction Y with the velocity synchronized with the velocity of sample rotation about angle ω (Figure 13.5a). The sweeping technique yields a continuous 2D image of a pole figure (Figure 13.5b), without the necessity of interpolation as in the discontinuous step-scan technique of conventional pole figure measurements (Section 4.3.3). Similar sweeping modes can be applied to obtain images of other sections through the 6D orientation-location space $\{x,y,z, \varphi_1,\Phi,\varphi_2\}$, that is, two orientation coordinates,

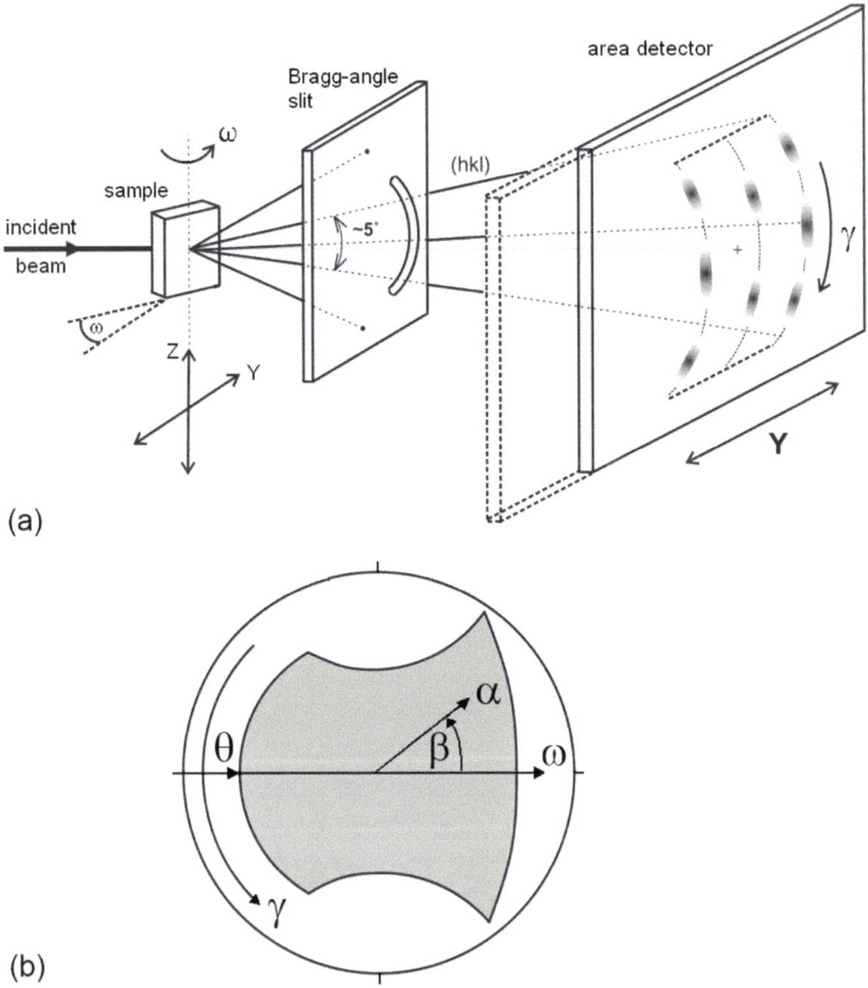

FIGURE 13.5 Texture diffractometer with the sweeping detector and the Bragg-angle slit system. (a) The Debye–Scherrer rings passing through a diffraction angle slit are continuously registered by the sweeping area detector; (b) continuous coverage of a pole figure. (Adapted from Bunge et al., 2003.)

two location coordinates, or a combination thereof. Both the orientational as well as lateral resolution of the instrument is better by at least two orders of magnitude than that of conventional texture goniometers (Wcislak et al., 2002), so that, in many cases, the individual grains of a polycrystalline material can be resolved. Although location resolution is generally limited by the width of the entrance slit to typically 100 μm, finer resolution of the order of 10 μm can be achieved by using other slit systems or a focused incident beam (Figure 13.5a). Then, under certain conditions, the position of individual grains in the polycrystalline structure, and hence the complete 6D "orientation stereology" of grain-resolved microstructures, can be obtained.

One of the latest developments for texture and microstructure research is a technique, referred to as 3D x-ray diffraction (3DXRD), for measuring local crystallographic orientations and thereby generating a truly 3D map of the microstructure (Poulsen et al., 2001; Poulsen, 2004; Juul Jensen et al., 2006; Poulsen, 2012). The 3DXRD microscope has been developed by Risø National Laboratory, Denmark, in cooperation with the European Synchrotron Radiation Facility (ESRF) in Grenoble, France, and is operating at the ESRF materials science beamline (ID11). A similar facility was established at the APS of Argonne National Laboratory (Pantleon et al., 2004; Lienert al., 2007). Based on the use of highly penetrating hard x-rays and a "tomographic" approach to diffraction, the method enables a nondestructive 3D description of the microtexture and microstructure within polycrystals. Hence, the 3DXRD concept can be thought of as a 3D equivalent of orientation microscopy in scanning electron microscopy (SEM) (Juul Jensen et al., 2006).

Figure 13.6 illustrates the principal components of the 3DXRD microscope. The incoming beam is a focused monochromatic x-ray beam tunable to energies in the range 45–100 keV. The focusing is obtained by means of a bent Laue crystal or multilayer mirrors (Lienert et al., 1999; Suter al., 2008). The specimen is mounted on a sample tower that allows translations (x, y, z), rotation (ω), and tilting of the sample with an absolute accuracy of 1 μm. At present, four CCD detectors are available, covering a range of pixel sizes from 1.2 to 225 μm. Detectors with a small pixel size are typically placed close to the sample for optimal spatial resolution, whereas those with a large pixel size are placed far from the sample for strain characterization. For in-situ measurements of the microtexture and microstructure evolution, a cryostat, a tensile machine, and several furnaces are available.

The incoming monochromatic beam penetrates the sample, and all microstructural volume elements within the illuminated gauge volume that fulfill the Bragg condition generate diffracted beams. The incoming beam is confined by focusing, possibly in combination with slits. It determines the gauge volume, which is typically

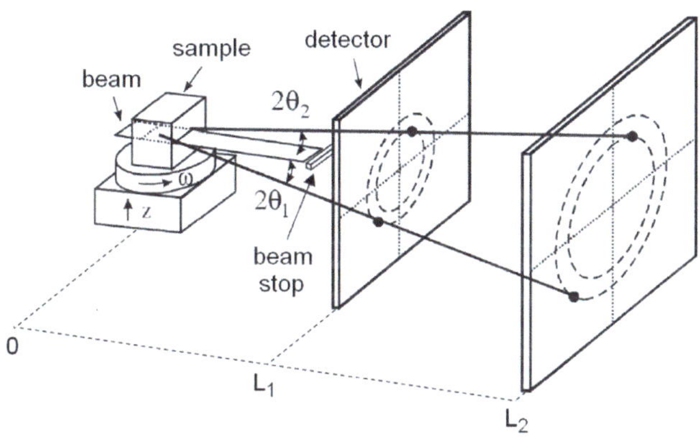

FIGURE 13.6 Sketch of the 3DXRD microscope. The diffraction spots appearing in exposures of the area detector at different sample-detector distances are projected back into the illuminated plane in the sample. (Adapted from Poulsen et al., 2001.)

a plane or a volume within the sample. To probe the complete crystal structure within the illuminated layer or volume, the sample is rotated around the ω axis perpendicular to the incoming beam (Figure 13.6).

A critical issue is the determination of the exact position of the diffracting volume within the gauge volume. The conventional approach for this is to limit the gauge volume to a point by slit(s) and scan the sample with respect to this point. However, this approach is generally too slow for dynamic studies. Hence, in the 3DXRD concept, it is replaced by a tracking method that allows diffraction from large gauge volumes (Lauridsen et al., 2001). Diffraction patterns are recorded at two or more distances between the sample and the detector(s) (distances L_1, L_2, …, etc. in Figure 13.6). The position of the diffracting volume element is then determined by back-extrapolating a straight line through the corresponding diffraction spots at different distances L_i all the way back to the originating volume within the sample.

It should be mentioned that 3DXRD microscopy can be performed in two different modes, a near-field and a far-field mode, depending on the distance of the detector to the sample (Juul Jensen and Poulsen, 2012). In the *near-field mode*, the sample is illuminated with a broad parallel x-ray beam. This leads to the appearance of a projection of the shape of individual crystallites on the detector at the position of the transmitted (forward diffracted) beam and at the position of the Bragg reflected beam. The position of a crystallite in 3D is determined by moving the detector away from the sample and tracing the grain shapes back to the sample. The spatial resolution depends on the pixel resolution of the detector. In the *far-field mode*, the sample is illuminated with a very finely collimated beam, which is scanned across the field of view. Every grain appears only as a small diffraction disk. The 3D-position of a grain is determined in a complicated procedure from the individual Bragg reflections of individual grains appearing on the detector for different rotation angles of the sample. In this case, the spatial resolution is limited by the size of the collimated beam.

The 3D microstructure results can further be refined by a technique referred to as dark-field X-ray microscopy (DFXRM) (Simons et al., 2015). The entire specimen is first mapped on a coarse scale using 3DXRD and, if necessary, classical X-ray tomography. This survey facilitates identification of small-scale elements of interest such as grains or domains. Dark-field microscopy then allows "zooming in" on these elements and recording a magnified high-resolution 3D map.

Ludwig et al. (2008) proposed another techniques employing high-energy X-rays to investigate the internal structure of polycrystalline materials, which is called diffraction contrast tomography (DCT). The principle of DCT is akin to conventional X-ray absorption contrast tomography with which it shares a common experimental setup (the grains are imaged using the occasionally occurring diffraction contribution to the X-ray attenuation coefficient each time a grain fulfils the diffraction condition). The 3D grain shapes are reconstructed from a limited number of projections using an algebraic reconstruction technique.

The major advantage of the synchrotron-based nondestructive methods over the destructive serial section techniques described in Section 11.4 is that the former are able to provide valuable in-situ observations, e.g. during mechanical testing or thermal treatments. Examples for 3D (i.e. x, y, z) and, especially, 4D (i.e. x, y, z, time) microstructure characterization by synchrotron radiation are presented in the review paper by Juul Jensen and Zhang (2020) by taking the example of the nucleation and

TABLE 13.1
Overview of Several Key Parameters for Synchrotron-based Microtexture Techniques (from Juul Jensen and Zhang, 2020)

Method[a]	Spatial Resolution	Angular Resolution	Strain Resolution	Time for Typical 3D Mapping
DFXRM	75–100 nm	0.005°	5×10^{-5}	5–10 h
DAXM	0.2–0.5 μm	0.01°	1×10^{-4}	2–3 days
nf-3DXRD	1 μm	0.1°	5×10^{-4}	1–2 days
DCT	0.5–1 μm	0.1°	–	2–3 h

[a] DFXRM: dark field X-ray microscopy, DAXM: differential aperture X-ray microscopy, nf-3DXRD: near field 3DXRD, and DCT: diffraction contrast tomography.

early growth processes of recrystallization. Another classical application of 4D in-situ X-ray methods is crack growth observation (e.g. Ferrié et al., 2005). The paper by Juul Jensen and Zhang (2020) also provides an overview on the current state-of-the-art of the different synchrotron-based microtexture techniques with respect to in-situ recrystallization experiments, which is reproduced in Table 13.1.

Finally, it shall be noted that the principles of DCT have recently been adapted to allow nondestructive, 3D crystallographic orientation mapping using a powerful laboratory X-ray source, which is known as laboratory X-ray diffraction contrast tomography, LabDCT (McDonald et al., 2015). However, the method still suffers from limited spatial resolution and measurable volume of the present lab-based systems. Hence, grains are mainly reconstructed from their measured centers of gravity. That means the grain boundary positions are most probably incorrect, although grain orientations and grain boundary misorientations may be correct. This has been shown recently by a comparison of LabDCT results with 3D EBSD data on a polycrystalline superalloy (Reschka et al., 2020).

13.3 TEXTURE ANALYSIS BY NONDIFFRACTION TECHNIQUES

A vast majority of techniques that are in use for texture analysis are based on the diffraction of radiation at the crystal lattice planes (Chapter 3). However, besides those diffraction-based techniques, there are a few other methods to derive either macrotextures or microtextures that may be of some interest for special applications. In general, such techniques make use of a crystal property being anisotropic in different crystallographic directions. For instance, measurements of ultrasonic velocity or magnetism yield information on the integral orientation distribution (macrotexture) of the sample analyzed (Section 13.3.1). The optical reflectivity of a crystal depends on the crystallographic nature of its surface, which can be used to derive information on the crystallographic orientation of individual crystals (Section 13.3.2). Alternatively, it has long been known that some etching techniques differently affect grains with different orientations, which can also be used to determine the orientations of individual grains. The following sections introduce some nondiffraction techniques and their main applications.

13.3.1 Ultrasonic Velocity

Methods to derive the crystallographic texture by means of ultrasonic measurements are based on the anisotropy of the ultrasonic velocity, that is, on the fact that ultrasonic waves propagate with different velocities in different crystallographic directions. Both laser ultrasonic (LUS) methods and, more commonly, electromagnetic acoustic transducers (EMAT) can be used to analyze the texture of sheet materials and derive the resulting anisotropic properties (Moreau et al., 2002; Davis, 2009; Bate et al., 2017). Ultrasonic texture measurements involve measuring the angular variations of plate mode velocities. For that purpose, EMAT sends and receives ultrasound in various directions θ in the plane of the sheet (Figure 13.7a). LUS relies on light beams for the transduction of ultrasound. The technique uses one laser with a short pulse (few nanoseconds) for the generation of ultrasound and a longer pulse (duration of tens of microseconds), or a second continuous laser, coupled to an optical interferometer for the detection of ultrasound. In particular, LUS is a practical solution for remote testing at large distances, inspection of fast-moving parts on production lines, and inspection in hostile environments (Hutchinson et al., 2002). Figure 13.7b shows the transit time per unit distance of ultrasonic waves (i.e., the inverse of their velocity or their *slowness*) in different directions of an aluminum hot strip, represented in a polar plot. Obviously, under ±45° and ±135° the waves travel faster, which is attributed to the cube texture of the recrystallized hot strip.

With measurements of the ultrasonic velocity, only the fourth-order C coefficients can be derived, which give a quite poor, low-resolution description of the orientation distribution function (ODF). However, the fourth-order C coefficients already comprise the full elastic and plastic anisotropy of sheet materials (Bunge, 1993); hence, ultrasonic methods are of interest in industrial applications for rapid nondestructive

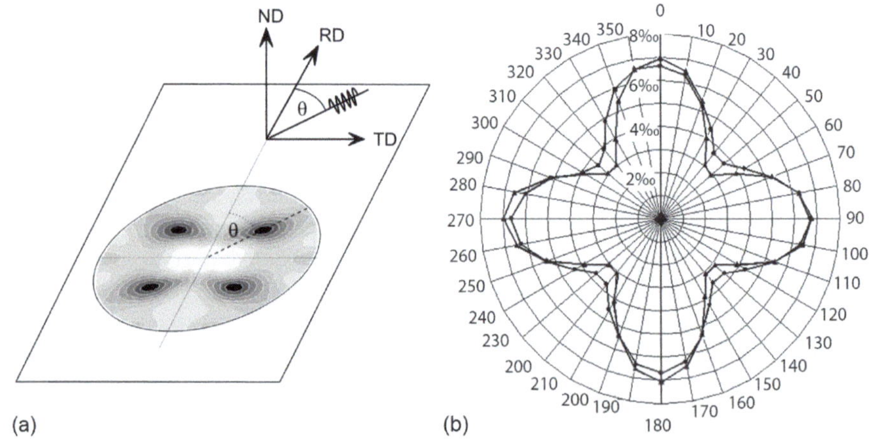

FIGURE 13.7 (a) Anisotropy of the velocity of an ultrasonic wave in different in-plane directions in a textured sheet; (b) directional variation of travel time (slowness) of an SH_0 wave in an aluminum hot strip with a mild cube texture (two different locations; polar plot). (Courtesy of J. Hirsch.)

testing and online determination of sheet quality. For instance, in steel sheets a correlation between ultrasonic velocity and the formability parameters r and Δr has been established (Kawashima et al., 1993; Thompson et al., 1993; Moreau et al., 2002; Davis et al., 2008a). In aluminum, the differences in ultrasonic velocity in the various crystallographic directions are smaller, but, nonetheless, the texture and resultant anisotropy in r value and elastic modulus as well as earing behavior could be determined (Spies and Salama, 1990; Thompson et al., 1993; Schneider and Österlein, 1996; Davis et al., 2008b). Furthermore, it was shown that ultrasonic texture measurements can easily distinguish between rolling and recrystallization textures of aluminum alloys and hence can be used for nondestructive online analysis of the progress of recrystallization (Anderson et al., 1994; Man et al., 2002; Kruger et al., 2003; Lindh-Ulmgren et al., 2004). Similarly, Shinbine et al. (2016) applied in-situ LUS measurements to tackle the hcp to bcc transformation in commercially pure titanium. In conclusion, within some limitations, ultrasonic methods have been developed to yield quantitative texture data and, in particular, information on the elastic and plastic anisotropy of sheet materials.

13.3.2 Optical Methods

When optically anisotropic crystals are studied with polarized light in an optical microscope, the color and the intensity of the reflected or transmitted light are dependent on the crystal orientation. To use this effect for orientation determination, thin sections of the sample are mounted on a microscope with the so-called universal stage (U-stage) that—similar to the texture goniometer—allows rotation of the sample about four or five axes (Figure 13.8). The crystal under investigation is rotated until its optical axis

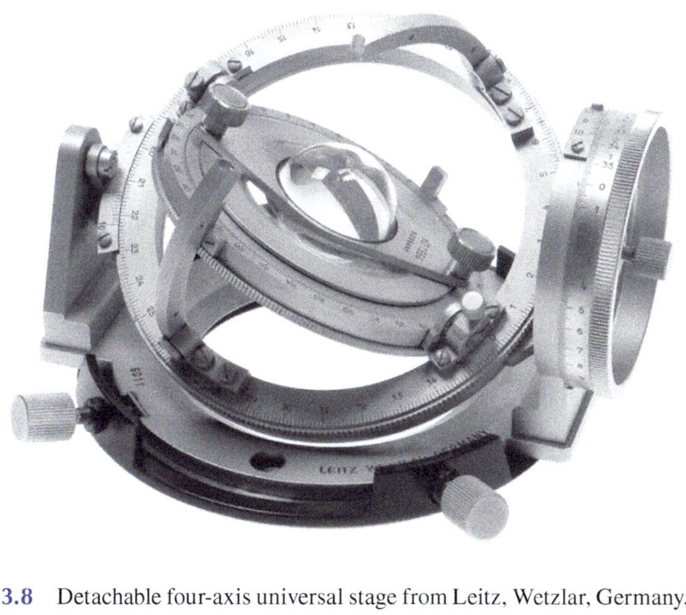

FIGURE 13.8 Detachable four-axis universal stage from Leitz, Wetzlar, Germany, from the 1970s. (Courtesy of C. Gerard.)

is aligned parallel to the microscope axis. The required rotation angles then yield the crystallographic orientation with regard to the external sample frame (see Wenk, 1985; Kile, 2009; Saha, 2021). This method finds some applications in geology where most materials have low crystal symmetry and, hence, are optically anisotropic (Wenk et al., 1986; Siegesmund et al., 1994; Ullemeyer et al., 2000; Ferrière et al., 2009).

Owing to their high symmetry, most metals are optically isotropic, which precludes investigations with the U-stage. In some cases, however, an appropriate chemical etch can introduce an anisotropy of the optical properties, which then can be used to derive the orientation of the individual crystallites (Nauer-Gerhardt and Bunge, 1986). Especially, anodical oxidation of aluminum causes the epitaxial growth of an optically anisotropic oxide layer on the sample surface ("Barker etch," Barker, 1950). For pure aluminum and many aluminum alloys, anodization for 1–2 min at 20 V at room temperature in an aqueous electrolyte containing about 5% tetrafluoroboric acid (HBF_4) usually yields good results. The resultant oxide layers are double-refracting such that different grains appear with different gray shades or, by using a $\lambda/4$ plate, with different colors in a polarization microscope. The different shades or colors produced by the oxide layer are related to the crystallographic orientation of the substrate crystal (Sætre et al., 1986; Kröger et al., 1988). Accordingly, anodical oxidation provides a fast survey on the texture components and, as such, is well suited for a rapid appraisal of the microstructure (Figure 13.9a) and quantitative grain size measurements. Anodical oxidation is also applied to assess the quality of aluminum foil for electrolytic capacitors, where the capacitance of the final product is determined by the fraction of cube-oriented grains. Under polarized light, the cube

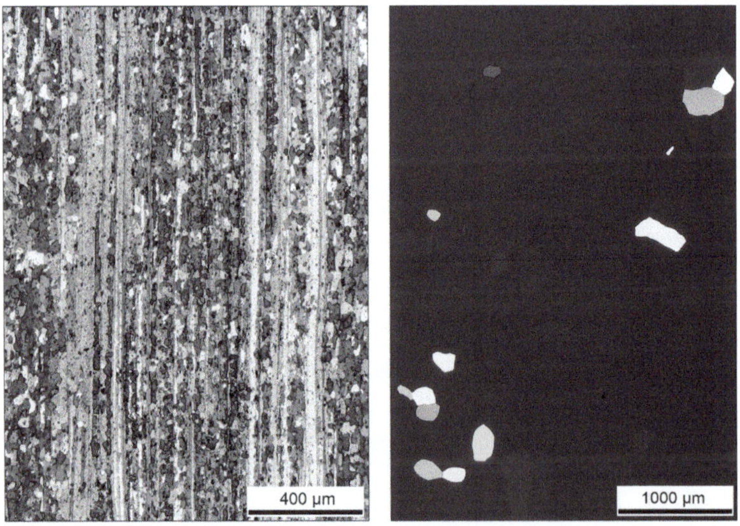

(a) (b)

FIGURE 13.9 Examples of the anodical oxidation of aluminum alloys (Barker etch, investigated with polarized light). (a) Microstructure of recrystallized aluminum, enabling easy analysis of grain size and shape and (b) high-purity aluminum condenser foil, where bright areas indicate non-cube-oriented grains. (Courtesy of W.D. Finkelnburg.)

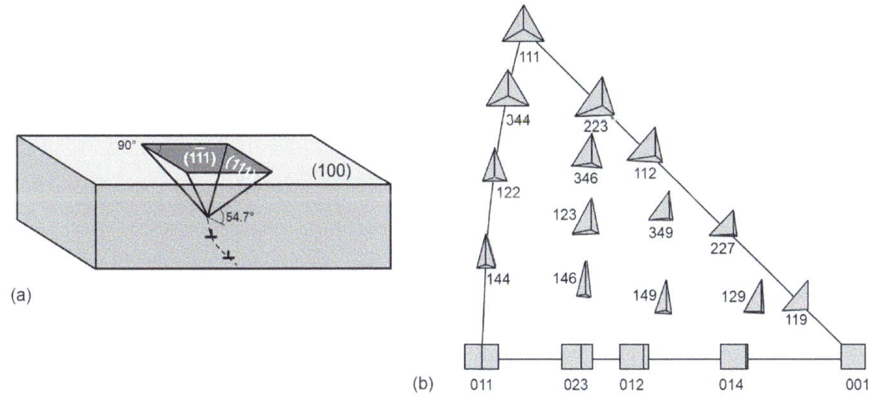

FIGURE 13.10 Sketch of the formation of etch pits. (a) Formation of an etch pit with (111) faces in a (100) oriented crystal at a dislocation line; (b) Representation of etch pit shapes for etching of (100) faces for different surface orientations in the standard triangle. (Adapted from Baudin et al., 1994.)

orientation {001}⟨100⟩ can easily be distinguished from the other recrystallization texture components, mostly the R orientation {124}⟨211⟩ (Ibe et al., 1984; Engler and Hirsch, 2009): because of its symmetry the cube orientation always appears black, independent of sample rotation, whereas grains with other orientations are usually brighter and furthermore change their brightness on sample rotation (Figure 13.9b).

Another optical method to determine individual orientations, which has been known for many decades, is based on the evaluation of so-called *etch pits* (Kostron, 1950; Tucker and Murphy, 1952/53; Nauer-Gerhardt and Bunge, 1986). Here, the etching attack is concentrated at points where dislocations intersect with the sample surface. Since the etching attack is differently rapid on different crystallographic planes the planes with lowest attack rate remain and form characteristically shaped etch pits. In Figure 13.10a this is sketched for the formation of etch pits with (111) faces in the surface of a (100) oriented crystal. (111) faced pits are typically observed on etched diamond-structured crystals (e.g. diamond, silicon, germanium). Depending on the etching conditions etch pits bounded by (111) or (100) planes may form on fcc materials. Figure 13.10b shows the typical shapes that are formed for (100)-plane etch pits developed on different surface normal orientations. The orientation [*uvw*] in the rolling direction can readily be obtained by rotating the ideal etch pits around the normal direction. Figure 13.11 shows an example of the application of etch pits to study the solidification behavior of aluminum alloys. As-cast aluminum rolling slabs can exhibit a grain morphology associated with faceted growth, the so-called "feathery grain" morphology consisting of large, elongated grains of twinned dendrites (Henry et al., 2004). In metallographic micrographs feathery crystals are identified by long straight grain boundaries (marked in Figure 13.11a). The etch pits obtained in two of these grains (Figure 13.11b) reveal that one crystal is in an exact (100) orientation and the neighbor crystal is oriented close to (111), which is indicative of a 60° ⟨111⟩ twin orientation relationship between the two (Figure 13.11c). In the 1990s, etch pits regained some interest via analyzing the evolution of the Goss orientation during

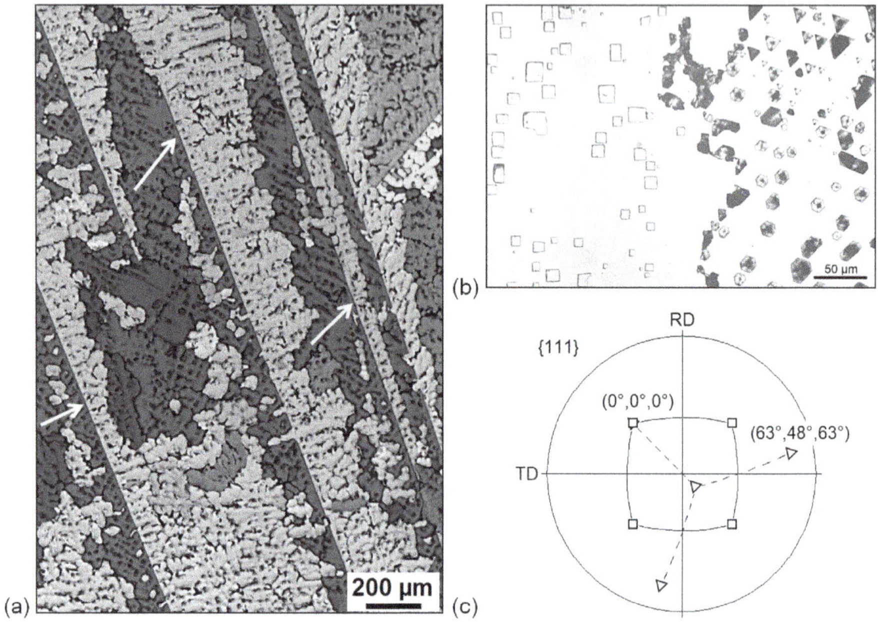

FIGURE 13.11 (a) Microstructure of an as-cast aluminum alloy revealing feathery crystals, (b) (100) and (111) etch pits in two neighboring grains, and (c) ideal orientations revealing a mutual 60° ⟨111⟩ twin orientation relation. (Courtesy of K. Kuhnke.)

secondary recrystallization in iron-silicon transformer steel (Böttcher et al., 1992; Fortunati et al., 1992; Baudin et al., 1994; Lee et al., 1995). However, both the spatial resolution and the accuracy of orientation determination by this technique are limited, so that generally only qualitative or semiquantitative conclusions can be drawn.

For copper and some copper alloys, very satisfactory results can be obtained with another method, the crystallographic "{111}-etching" (Köhlhoff et al., 1988). If copper is etched in hot concentrated HNO_3, only the {111} planes are attacked. The ⟨110⟩ directions along which the corresponding {111} planes intersect then form the valleys and rims of the deep-etched surface. As an example, Figure 13.12a shows a micrograph obtained from a partially recrystallized copper sheet. It is obvious that the etch patterns differ in the various deformed grains, which is a result of their different crystallographic orientations. However, some layers of the sheet reveal a totally different etching pattern. Here, recrystallization has started, leading to the formation of cube-oriented regions (Engler et al., 1993b). Duggan et al. (1993) applied crystallographic etching to analyze the orientation of the substructures surrounding cube-oriented microbands in heavily deformed copper, while Lee and Duggan (1993) established deformation banding as an important deformation mode of fcc metals and alloys. Especially, studying samples etched in such a way in an SEM facilitates analysis of structures down to sizes of the order of 1 µm. Figure 13.12b shows an SEM micrograph obtained at a higher magnification (×5,000), revealing the existence of grain-scale shear bands in Cu-oriented matrix grains (Engler, 2000). Even within

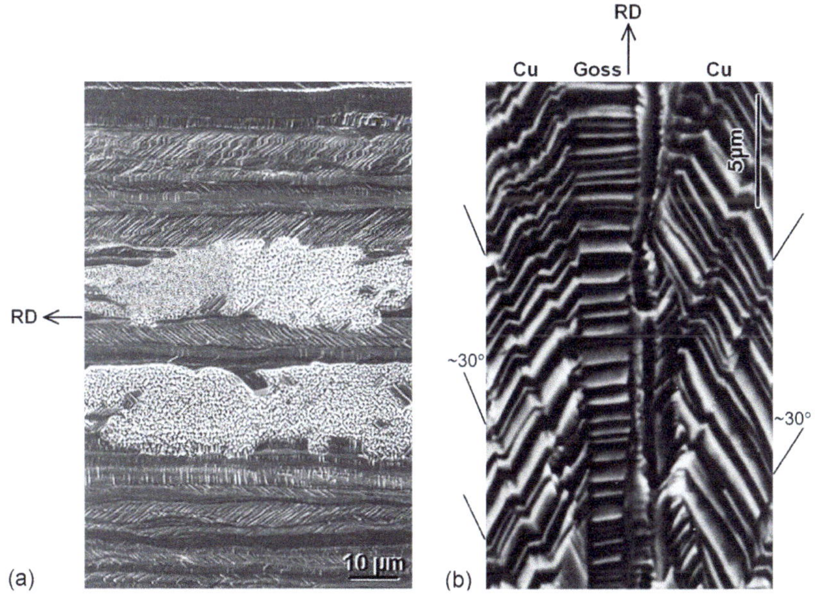

FIGURE 13.12 Examples of the {111}-etching technique to reveal the orientation topography in copper alloys (SEM). (a) Partially recrystallized copper. (b) Shear bands in a 90% cold-rolled copper-manganese alloy. (From Engler, 2000. With permission.)

the shear bands, etch patterns can be resolved. By comparing the etch patterns with an atlas that contains the patterns of all possible orientations, the local orientations can be evaluated to an accuracy of 5°–10° (Wang et al., 1995). A major advantage of this technique is the quasi-continuous imaging of orientations in the microstructure, which enables one to draw conclusions on the orientation topography.

The main drawback of the crystallographic {111}-etching—and in general of all chemical techniques for orientation determination—is their strong material dependence. Suitable etchants to evaluate crystallographic features are only available for a few materials. Methods based on the diffraction of radiation, in contrast, are not only much less material sensitive but in general also much more precise and therefore superior for orientation determination.

13.4 SUMMATION: COMPARISON AND ASSESSMENT OF THE EXPERIMENTAL METHODS FOR TEXTURE ANALYSIS

This book introduces a variety of common techniques to derive the crystallographic texture of crystalline samples. A vast majority of techniques for texture analysis are based on the diffraction of radiation—neutrons, x-rays, or electrons—at the crystal lattice (Chapter 3); these techniques are summarized in Figure 3.16. A few other techniques make use of a crystal property being anisotropic in different crystallographic directions (Section 13.3). In the following text, all the various techniques are summarized and their main advantages and disadvantages compared.

13.4.1 Comparison of Different Techniques and Their Fields of Application

Techniques to obtain the texture of a polycrystalline sample by means of x-ray or neutron diffraction provide the integral texture of the polycrystalline array (Chapter 4). The resulting macrotexture discloses information about the volume fraction associated with each orientation, which can be used to predict texture-related anisotropy of polycrystalline behavior during elastic and plastic straining, and also physical properties such as magnetic permeability and thermal expansion (Chapter 1). Furthermore, analysis of the texture changes during the thermomechanical treatment of materials yields valuable, statistically relevant information about the mechanisms of the underlying metallurgical processes, particularly deformation, recrystallization, or phase transformations (Section 5.5). In geology, macrotexture analysis can provide insight into the geological mechanisms that have led to rock formation in geological periods.

Because of the very limited access to neutron beam lines, macrotexture measurements will, in most cases, be performed by x-ray diffraction. However, the much greater depth of penetration into the sample by neutrons offers some distinct advantages for texture analysis in large grained or irregularly shaped specimens, small volume fractions of a second phase, samples with low crystal symmetry, or multiphase systems (Sections 4.4 and 4.5). In contrast, with a view to determine textures of small volumes, x-rays are superior. In neutron diffraction the entire specimen volume of typically several cubic centimeters contributes to the diffracted intensities, whereas in standard x-ray techniques only volumes of the order of cubic millimeters are sampled, so that the textures of specific layers in a sheet can be discriminated.

Synchrotron sources yield a white spectrum of x-rays with much larger energies and several orders of magnitude higher intensities than conventional x-ray tubes combined with minimum angular divergence. This combination of high energy, high intensity, small beam size, and free choice of wavelength opens a wide range of possibilities for texture analysis (Section 13.2). Experiments that make use of synchrotron diffraction represent a hybrid between the macrotexture techniques covering large bulk samples and the microtexture techniques that record the texture of volumes down to the individual grain level. As in the case of neutron diffraction, the main drawback of synchrotron-based techniques is the rather limited availability of synchrotron sources so that experiments using this radiation are limited to special cases that cannot be accomplished by other techniques but necessitate one of the characteristic advantages of synchrotron radiation. In particular, these are in-situ investigations in 3D and even 4D of mechanical, thermal, or chemical processes that involve changes of crystallographic orientations or phases.

In contrast to the macrotexture methods, microtexture or local texture analysis is performed by sampling the orientations of individual single crystallites—grains and subgrains. The only exception to this is the SAD pole figure technique (Section 6.2.2) that records the texture in a small contiguous polycrystalline volume. The local texture analysis techniques have a number of fundamental advantages over the macroscopic techniques:

- Orientations are measured directly, no pole figure intensity corrections and pole figure inversion techniques are required to obtain full texture information.
- Phases are measured separately, no peak overlap occurs.

- Texture data can be assigned to microstructural features, e.g. deformed vs. recrystallized grains.
- Texture data are measured with spatial assignment, e.g. outer and inner volumes of a sample, without separate measurements or preparation.

Of course, local texture measurements also have disadvantages compared to the macroscopic measurements, most prominently the difficulty to measure a volume large enough to represent the specimen texture and the inability (or difficulty) to measure texture data on very heavily defected or very small (nanocrystalline) crystals. Also, most of the local texture measurement require serious preparation of samples and cannot be applied easily under rough environmental conditions, e.g. at high temperatures, in atmospheres or under severe mechanical loads, where the macrotexture measurements may actually still deliver perfect data.

The main requirements for an experimental technique to determine local, usually individual, orientations are as follows:

- The technique yields the crystallographic orientation and phase unambiguously and with high accuracy.
- The interaction volume of the technique, which represents the spatial resolution of the applied technique, is smaller than the size of the volume with quasi-uniform orientation.
- It is possible to locate and image specific regions of interest in the microstructure of the specimen, from which orientation measurements can be obtained.
- Orientation measurement and evaluation is automated to allow orientation mapping.

The main characteristics that can be addressed by the methods for local texture analysis are as follows:

- The spatial arrangement of the crystallographic orientations and phases in the microstructure, that is, orientation topography
- The intensity distribution of orientations in specific subregions of the microstructure
- Orientation correlations or misorientations, either inside of grains (caused by lattice dislocations) or between neighboring grains of similar or different phase in the microstructure
- Quantification of lattice defects (dislocations, grain boundaries) in correlation with texture and microstructure characteristics.

The most popular techniques for local orientation determination are summarized in Table 13.2. This table, and other work on this subject (Dingley, 1981; Humphreys, 1984; Schwarzer, 1993; Gottstein and Engler, 1993; Zaefferer, 2011), clearly shows that individual orientations are determined most easily and with good accuracy from electron diffraction patterns in SEM or TEM. The mapping techniques—the most well-established method certainly being EBSD in the SEM—all have the great advantage over classical x-ray techniques, that they do not only deliver orientations

TABLE 13.2

Overview of the Most Common Techniques for Determination of Local Orientations

Method	Technique	Spatial Resolution (∅lateral ×, depth)	Angular Accuracy	Application	Limitation
TEM	SCBED (spot diffraction) PED	1 nm × 30⋯200 nm (sample thickness), independent of material	0.2°	Polycrystals of submicron- and nanometer grain size (grains, subgrains, deformed microstructures, inhomogeneities, recrystallization nuclei, …) on materials up to highest defect densities	Measurement only on electron transmissive samples (thin foils), representativeness of selected area
	TKD	10 nm × 100⋯200 nm (sample thickness), independent of material	<0.1°	ORM with high angular and spatial resolution, on material with low to medium defect densities	Sensitive to high lattice defect densities, problems with different sample thickness
	SAD	1 µm × 30⋯200 nm (sample thickness)	5°	Not used for orientation mapping	
SEM	EBSD	<50⋯500 nm × 20⋯100 nm, depending on material (small numbers for	0.5°–0.01°	Polycrystals of sub-micron to sub-millimeter grain size (grains, subgrains, deformation structures, recrystallization structures), bridging scales between sub-micron resolution for microstructure description to millimeter areas for properties, extendable to 3D ORM	Spatial resolution for low-Z materials, relatively high beam current required. Anisotropic interaction volume due to high sample tilt
	SEM-TKP	<10 nm × 30⋯200 nm (sample thickness)	0.1°	Like EBSD but on thin foils, no limitation in terms of material, also on low-Z-material	Thin foil specimens as a disadvantage over bulk samples
	ECP	10 µm	0.5°	Not used for orientation mapping	

(Continued)

TABLE 13.2 (*Continued*)
Overview of the Most Common Techniques for Determination of Local Orientations

Method	Technique	Spatial Resolution (∅lateral ×, depth)	Angular Accuracy	Application	Limitation
	Electron channeling orientation mapping	500 nm × 100 ··· 200 nm, depending on material	0.5°–1°	Large area rapid scanning on materials with low defect densities, no strong sample tilt required	Technique is still new and not well explored, not suited for materials with high defect densities
	Micro-Kossel	10 µm	0.5°–0.0005°	Highest angular resolution elastic strain measurements	Low lateral resolution, no mapping, not good for orientation determination very sensitive to lattice defects
X-ray diffraction	Conventional Laue	100 µm × 20 µm	2°	Single crystals, very coarse grains, sensitive elastic and plastic strain measurements	
	Synchrotron based	0.1–100 µm × 1 ··· 1000 µm (depending on radiation energy and material)	<0.1°	Polycrystals with micron grain sizes (grains, deformation structures), particular for 3D and 4D (in-situ) observations with good spatial and angular resolution	Limited availability, complexity of experiment and analysis
Optical techniques	Selective ({111}) etching	1 µm	5°–10°	Polycrystals (grains), good for a quick overview	Only applicable to special materials and special cases. Not automated, not quantitative
	Etch pits	20–100 µm	>10°	Polycrystals (coarse grains), of historical interest only	

or phase fractions but also the microstructure with unrivaled amount of information. The corresponding evaluation methods are nowadays highly standardized and automated (Chapters 7, 8, and 9), so that local orientation and microtexture determination can be used for routine purposes. Use of other methods, such as selective etching techniques (Section 13.3.2), Laue x-ray diffraction (Section 3.4), and the micro-Kossel technique (Section 6.3.5), is restricted to special applications.

A detailed comparison of SEM- and TEM-based techniques is found already in Section 9.6. We will here just add a rather coarse summary on all microtexture measurement techniques. The most important distinction of the different techniques is made by their spatial resolution and the maximum covered specimen volume. This is displayed in Figure 13.13 for the techniques covering all relevant length scales from the Ångstrom to the millimeter range. Further important distinguishing features are the ease of application and the possibility to perform 3D and/or in-situ investigations.

The currently by far most important technique for orientation and microstructure mapping is EBSD which has developed—within certain limits—into the most quantitative and versatile microtexture and microstructure characterization technique (see below for an overview on applications). The EBSD technique is applied on bulk samples (in contrast to TEM which requires electron transmissive thin foils) with moderate sample preparation efforts (Section 8.2.2). In order to exploit the EBSD technique to the full, it is important to use an SEM with thermal field emission gun which delivers the necessary high beam current with high brilliance and stability (Section 8.2.3). The EBSD technique covers length scales from a few micrometer to

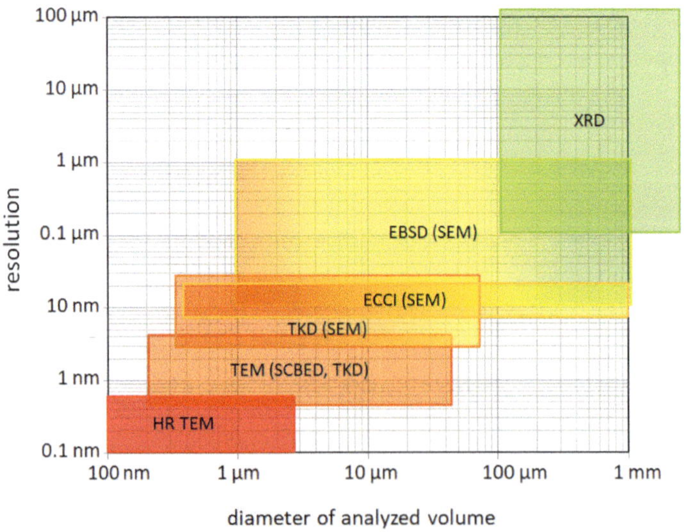

FIGURE 13.13 Spatial resolution and maximum observable volume of different techniques for local orientation measurements. (HR TEM: high-resolution transmission electron diffraction, SCBED: small-angle convergent beam electron diffraction, TKD: transmission Kikuchi diffraction, ECCI: electron channeling contrast imaging, EBSD: electron backscatter diffraction, SEM: scanning electron microscopy, XRD: x-ray diffraction.)

macroscopic samples of millimeter sizes and reaches resolutions down to approximately 50 nm on materials heavier than iron and a few hundred nanometer for lighter materials. The angular resolution of about 0.5° in the standard setup is good enough for many applications, but using cross-correlation pattern analysis (Section 7.4.5) it is even possible to reach resolutions in the order of 0.01° and perform sensitive measurements of the elastic distortion of the crystal lattice ($\Delta\varepsilon/\varepsilon = 10^{-4}$). Due to the very high measurement rates (currently up to several thousand patterns per second) very large samples can be mapped at small step sizes (not smaller, though, than the spatial resolution). On deformed metals the technique delivers data up to a dislocation density of 10^{15} m^{-2}, after that the patterns become too blurred and can usually not be analyzed any longer (Section 9.4.5).

The EBSD technique enables quantification of many texture- and microstructure-relevant features, like grain and phase boundaries, dislocations, and elastic and plastic strain fields. A combination with electron channeling contrast imaging (ECCI) (Section 6.3.4.2) allows even direct observation of lattice defects with a contrast mechanism similar to that used in classical diffraction-contrast based TEM. Furthermore, EBSD can be combined with other characterization methods, e.g. x-ray spectroscopy for elemental composition quantification (Dingley, 2004; Nowell and Wright, 2004), cathodoluminescence for the quantification of opto-electronic properties (Stechmann et al., 2017), or atomic force microscopy (AFM) to correlate crystallography with surface evolution (Chandrasekaran and Nygårds, 2003; Schuh et al., 2003). Even more, EBSD can be extended to a 3D technique by combining it with automated serial sectioning techniques (Section 11.4). Last but not least, EBSD delivers valuable input data for computer modeling of materials processes, e.g. for finite element modeling and other simulations (Section 10.8). A further key feature of microtexture determination is crystallographic analysis of surfaces, including grain or phase boundaries, cracks, slip traces, and external surfaces such as facets or fracture surfaces (Chapter 11). A limited amount of analysis of grain/phase boundaries can be carried out in TEM, but most analysis, because of the 3D nature of the surface/specimen geometry, is the province of SEM and EBSD. Recently crystallographic analysis of inner and outer surfaces and interfaces by EBSD, especially in three dimensions, has become viable and will be a growth area in the future. These and many other applications are augmenting and enriching the use of EBSD as an investigative tool. The most serious shortcoming of the EBSD technique is its limited spatial resolution, particularly for light elements and many minerals. Here the resolution is often not better than half a micrometer or so.

If higher resolution is required one has to resort to transmission techniques where the electron-matter interaction volume is limited by use of a thin sample. A technique which is closely related to EBSD (also known as backscatter Kikuchi diffraction, BKP) is transmission Kikuchi diffraction (TKD) which can be easily applied in every SEM that is equipped with an EBSD system. The basic principles have been laid out in Sections 8.2.4 and 9.6.2. This technique works on thin foils, as used in the TEM, but the acceleration voltage is significantly lower than in the TEM case. As a consequence, the beam spreads more than in TEM techniques, but, at the same time, the diffraction signal only originates in the last 10 or 20 nm of the beam path, which makes SEM-TKP a very high resolution technique. At the same time, it profits from all the advantages of SEM over TEM, like ease of operation, good overview on sample positioning by a large field of view and additional chamber TV cameras, and a large chamber to handle also bulky equipment. Additionally, when performing SEM-TKP in a FIB SEM the sample can be directly target prepared by focused ion beam milling (Schwarz et al., 2018).

Clearly, the TEM techniques should only be used when the spatial resolution of SEM-based techniques is not sufficient or/and when additional information on the crystallographic or compositional nature of lattice defects like dislocations, stacking faults, or small precipitates is required. In TEM the analysis of spot diffraction patterns based on the small-angle convergent beam electron diffraction (SCBED) with or without precession and with template matching as an analysis method has been established as the most robust, most highly automated, and most rapid method for orientation microscopy (Sections 6.2.3, 7.4.2, and 9.5). The technique offers an orientation accuracy of the order of 0.5°. The transmission Kikuchi diffraction (TKD) technique (Sections 6.2.6 and 9.5), which is frequently used for orientation determination, is very much depending on sample thickness and requires relatively high quality crystals, such that it is difficult to automate and its robustness is less than that of SCBED. However, its orientation resolution is at least one order of magnitude better than that of SCBED. One may hear that the spatial resolution of the TEM techniques is in the order of 1 nm, which corresponds to the smallest diameter of the electron beam on the sample. This value, however, does not take into account that diffraction occurs, at least for the case of SCBED, along the whole beam path through the sample. This means that the true spatial resolution of the SCBED technique is limited by the sample thickness. If a 100-nm thick foil contains 30-nm large nano-crystals, one needs to accept that the diffraction patterns of three or more crystals will all overlay in the diffraction pattern. Investigations on the SEM-TKD technique showed that in the case of Kikuchi diffraction the pattern stems only from the last coherently diffracting volume along the beam path which would make the spatial resolution of TKD better than that of SCBED. On the other hand, the formation of TKP requires larger volumes than the formation of SCBED, which may finally counterbalance the resolution of both techniques. To the best knowledge of the authors no systematic comparative investigation on the resolution of the techniques has been performed.

A significant shortcoming of TEM-based techniques is the laborious sample preparation and the related possibility that the microstructure and even crystal structure of the specimen may be changed during sample preparation (Section 8.3.4). Furthermore, the small observable volume makes it impossible to conclude from TEM-based ORM measurements onto the whole specimen. The TEM investigations are, therefore, only suited to perform well-considered target investigations of particular microstructural features.

Besides EBSD and SEM-TKP in SEM and SCBED and TEM-TKP in TEM, we have mentioned a couple of further techniques in the book which are either outdated (ECP in SEM, SAD-based pole figure measurement or dark field mapping in TEM) or are used for very particular applications only. To mention is the Kossel diffraction technique in SEM (Section 6.3.5) which is the technique with the highest sensitivity to lattice parameters. Therefore, it can be used for accurate residual stress measurements ($\Delta\varepsilon/\varepsilon = 10^{-5}$), though with a spatial resolution of merely 5–10 µm and other restrictions. Another strongly growing technique is the orientation measurements with Laue-diffraction by synchrotron radiation (Section 13.2.1). Although this technique has limited accessibility, its high beam energy and intensity allow to perform nondestructive 3D measurements of textures and microstructures, which opens the large field of in-situ observations of materials processes like phase transformation, recrystallization, deformation, or chemical reactions (which are also phase transformations). The spatial resolution of the technique has, in the meantime, reached almost that of EBSD and the angular resolution is even better than this. All these features make synchrotron-based diffraction techniques highly interesting for materials research and push the development of analysis techniques and smaller devices, for example small linear electron accelerators.

It should be emphasized that different techniques should not be considered as competitive but should be applied in a complementary manner. A combination of different techniques for orientation determination with increasing spatial resolution yields much more valuable information on the underlying mechanisms than that accessible by the sole use of the individual techniques.

13.4.2 Overview on the Evolution of EBSD Applications in Literature

The development of fully automated EBSD has led to the implementation of orientation microscopy and orientation mapping (Chapter 9). The ability to automatically produce an orientation map—i.e. a depiction of the microstructure in terms of its orientation and phase constituents—has greatly enhanced the application of EBSD and has become the standard tool for microstructure research. Although the visualization aspects of mapping contribute to its appeal, the opportunity for quantification of the orientation/microstructure relationship on a submicron scale is scientifically a much more powerful feature. Fast mapping of several thousand data points per second on suitable materials has rendered EBSD a very powerful characterization technique indeed.

The trends in EBSD usage have been assessed and quantified by an analysis of the published papers in the scientific literature since 1990 based on output from the scientific platform Web of Science from Clarivate Analytics (www.webofscience.com). Figure 13.14 displays the results of this literature review, showing a more or less continuous increase in the number of EBSD-related papers being published. Depending on the keywords used the absolute number of EBSD-related publications may differ from that quoted on Figure 13.14, but the majority of important journal publications are captured and the growth trend in EBSD-related output is nicely depicted. Up to the year 2022 Web of Science reports altogether almost 20,000 papers with EBSD-related content. It is interesting to see that the slope of the curve of these papers over time is still continuously increasing, and this increase is still mainly carried by investigations on metals and alloys. Overall, metallic materials account for three quarters of the publications which include data obtained by EBSD. A number of the most frequently encountered textures of fcc, bcc, and hexagonal metals was described in Section 5.5. In part the emphasis on metals reflects the relatively simple crystal structures involved and the ease of specimen preparation, in

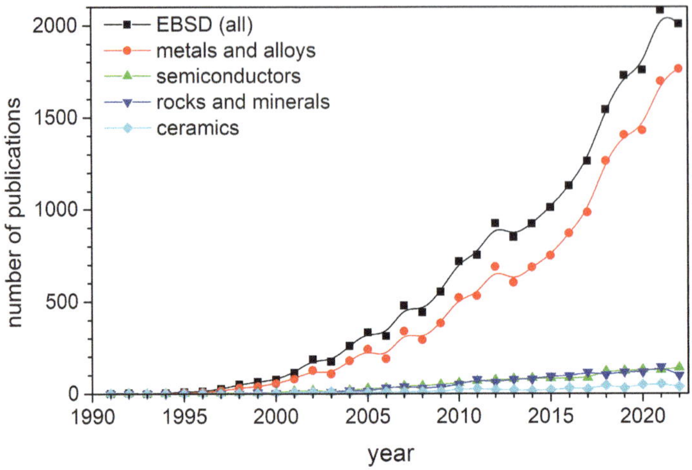

FIGURE 13.14 Number of publications with EBSD-related content per year from 1990 until 2022 for different materials investigations. Literature analysis performed with data from the platform Web of Science.

most cases, compared to the other categories shown in Figure 13.14. The application of EBSD to nominally single-phase metals and alloys is straightforward and data on microtexture and microstructure can be obtained in a routine manner, which is addressed in detail in the present book. Furthermore, study of texture development is highly relevant to control materials properties during metals processing.

Nevertheless, also other materials, like semiconductors (altogether 1,500 papers), geomaterials (1,400), and ceramic materials (500), are increasingly studied by EBSD. This is evidence that EBSD is becoming more widely applicable to the entire range of crystalline materials. It reflects that the described advances of EBSD in resolution, phase identification, and multiphase mapping (Chapter 12) facilitate the analysis of such materials, which are often complex in terms both of the phases present and their morphology.

Compared to EBSD, the application of synchrotron-based orientation measurements or orientation microscopy has yielded much fewer papers. Up to 2022 about 600 papers were published with a slightly increasing trend. Thus, a distinct change from EBSD toward these techniques is not in sight.

Addendum: Spherical Indexing of EBSD Patterns

Only shortly before the printing of the book the author SZ became aware of the enormous power of a new indexing method for EBSD pattern. This technique, named *spherical indexing*, is so powerful that it appears imperative to add at least a short chapter about it, even if it can no longer be properly integrated into the book.

Spherical indexing (SI) is a further development of dictionary indexing (DI) which has been described in some details in Section 7.4.3 and exceeds its performance in terms of orientation accuracy and indexing speed. Already Day, 2008 pointed out that EBSD pattern can be understood as a projection of a Kikuchi sphere on a tangential plane, as it was shown in Figure 7.10. For DI of diffraction patterns a spherical master pattern of the crystal structure under consideration is taken and from this all possible EBSD patterns are created, for a given projection center and in a given angular step size. The resulting large amount of patterns is called a pattern dictionary. Every experimental pattern is then compared to the dictionary patterns until a best match is found. This process is time consuming and can only be performed with the resolution of the dictionary step size. For SI this process, in principle, is inverted: instead of creation of many dictionary patterns from one master pattern, the experimental patterns are distorted such that they fit on the Kikuchi sphere. Then it is searched where the spherical experimental pattern fits best onto the master pattern sphere. Practical implementations of this spherical comparison are known from astronomy. Hielscher et al., 2019 proposed first to use a spherical harmonic transform for this sake. This idea was, parallel to or inspired by Hielscher et al., worked out and implemented as a computer program by Lenthe et al., 2019.

Without going into the mathematical details, which can be found in Lenthe et al (2019), the basic steps are as follows. First a spherical master pattern is calculated using dynamical theory of electron diffraction, as it was discussed in the chapter about dictionary indexing in Section 7.4.3. Next, the intensity distribution of this spherical pattern is fitted by a series of spherical harmonic functions (SHF), a process that is also know from the calculation of orientation distribution functions, ODF, for texture analysis. The desired degree of detail is obtained by the number of increasingly more oscillating spherical harmonics which are added together. The whole process may also be compared to a 3-dimensional Fourier series expansion where every additional sine or cosine term in the series shows higher oscillation frequencies. So far, the authors of the software have proposed a maximum series expansion degree of 255 which is called "bandwidth" in their case. The entry in Wikipedia to spherical harmonic functions (https://en.wikipedia.org/wiki/Spherical_harmonics#) and particular the animated graphics at https://en.wikipedia.org/wiki/Spherical_harmonics#/media/File:Rotating_spherical_harmonics.gif gives an intuitive idea about the nature of SHFs.

Next, the experimental pattern is distorted to a spherical pattern using an inverse gnomonic projection. This pattern, of course, fills only a small part of a complete sphere. The intensities are normalized and the rest of the sphere is filled zero-values. This experimental pattern sphere is now also fitted by a series of SHFs. Finally, the experimental pattern SHF is rotated with respect to the master pattern SHF and both are compared employing a spherical cross correlation (SCC). The best fitting SCC then yields the correct orientation of the experimental pattern.

Similar to the DI approach, but mathematically more elegant and computation time and computer-memory saving, the SI approach is amazingly robust with respect to indexing of low-quality patterns. This is best illustrated in the application examples presented by Lenthe et al., 2019. Orientation maps made from low-quality patterns where the classical Hough transform indexing (HI) virtually yields no correct indexing results are almost fully correctly indexed. Also, as SI and DI, in contrast to HI, are using the intensity distributions of the Kikuchi bands and generally all details in the pattern (determined, however, by the above mentioned bandwidth) SI and DI are able to distinguish closely related phases and pseudo-symmetries significantly more robustly as HI. In contrast to DI, where the orientation resolution is limited to the dictionary step size, SI orientations can be increasingly refined and yield very high orientation resolutions, exceeding that of HI by a factor of 5 to 10. The only serious disadvantage of DI and SI compared to HI is that, at least currently, all patterns need to be stored for post processing (see the remarks in Section 9.3.2). It is, however, possible, that SI, in contrast to DI will become an online indexing routine in the future and will then certainly replace the HI sooner or later.

Appendix I

Miller and Miller–Bravais Indices

Within the crystal coordinate system, both crystallographic directions and lattice planes are commonly described in terms of integer indices, called the *Miller indices*, which are derived as follows.

A crystallographic direction is given by the vector r in the crystal:

$$r = ua + vb + wc \tag{I.1}$$

where a, b, and c are the base vectors of the unit cell (Figure 2.1); thus, the direction is unambiguously described by the values of u, v, and w. It is a common practice to characterize a direction by the triplet of smallest integers having the same ratios, which is denoted by the symbol $[uvw]$. A lattice plane is described according to

$$h\frac{x}{a} + k\frac{y}{b} + l\frac{z}{c} = 1 \tag{I.2}$$

where x, y, and z are the coordinates of any point on that plane. Crystal planes are usually denoted by the reciprocal multiples of the axis intercepts h, k, and l, again reduced to the smallest integers with the same ratios; these are the *Miller indices* (hkl).

Figure I.1a shows the (112) plane in an orthorhombic crystal. The normal to this plane—called its *pole*—is also commonly indexed (112). Note that this normal (112) is not parallel to the direction [112]. Only in cubic crystals are plane normals and directions with same indices parallel, whereas for other crystals this only holds for special cases. Because of the symmetry elements of most crystal structures, a lattice direction or a lattice plane is symmetrically equivalent to some others. For instance, planes like (112), (121), and $(1\bar{2}1)$ all have the same atomic arrangement, lattice spacing, and so on; thus they are summarized by the family {112}—denoted by curly brackets rather than parentheses. Likewise, a generic set of symmetrically equivalent directions is denoted by angular brackets $\langle uvw \rangle$.

Though directions and planes can be described by the Miller indices in *all* crystal structures, for hexagonal and trigonal crystals, it is often more convenient to use a coordinate system with four axes. Commonly, the z axis is chosen along the sixfold or threefold c axis such that the x and y axes come to lie in the basal plane. To account for the symmetry of the basal plane, an additional, redundant axis is introduced in such a way that all the three axes, a_1, a_2, and a_3 lie at 120° to each other in the basal plane (Figure I.1b), leading to the description of directions and planes by a four-index notation, the *Miller–Bravais indices.* Lattice planes can still be described by their reciprocal intercepts on all four axes; they are denoted by the Miller–Bravais indices $(hkil)$ with $h + k + i = 0$. When directions are described by the Miller–Bravais indices

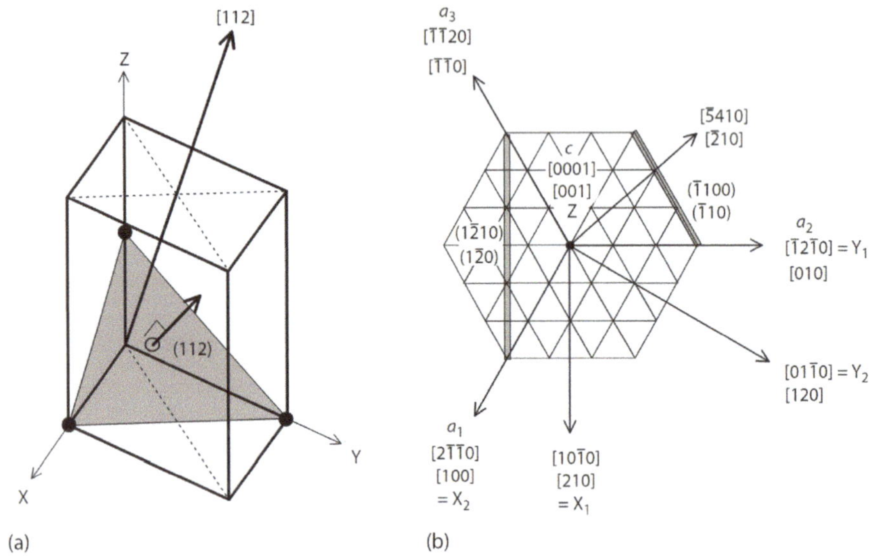

FIGURE I.1 Example of the crystallographic indices for directions and planes. (a) (112) plane and [112] direction in an orthorhombic crystal. Note that the (112) plane normal is not parallel to the [112] direction. (b) Planes and directions in the (0001) basis plane of a hexagonal lattice expressed as both Miller–Bravais and Miller indices.

$[uvtw]$ (with $t = -(u + v)$), the third index has to be considered as an additional vector component (Figure I.1b). Since the third index can readily be derived, it is often replaced by a dot—e.g. $(2\bar{1}.0)$—or sometimes even completely omitted—$(2\bar{1}0)$. However, the resulting three-index set is not to be confused with the Miller indices.

The four-index Miller–Bravais notation has the advantage that symmetrically equivalent directions and planes have similar indices and hence can easily be recognized. As an example, the directions $a_1 = [2\bar{1}\bar{1}0]$, $a_2 = [\bar{1}2\bar{1}0]$, and $a_3 = [\bar{1}\bar{1}20]$ would be indexed in the Miller notation as [100], [010], and $[\bar{1}\bar{1}0]$ so that symmetry is no longer apparent (Figure I.1b). However, some calculations that are important for indexing Kikuchi patterns, including determination of a common zone from two planes, a plane from two directions, and interzonal angles (Section 7.2.1), can be more easily performed by using the three-index Miller indices. The transformations linking the two conventions are (Partridge, 1969):

For planes, that is, $(hkil) \rightarrow (HKL)$:

$$H = h; \quad K = k; \quad L = l \tag{I.3a}$$

$$h = H; \quad k = K; \quad i = -H - K; \quad l = L \tag{I.3b}$$

For directions, that is, $[uvtw] \rightarrow [UVW]$:

$$U = v + 2u; \quad V = u + 2v; \quad W = w \tag{I.4a}$$

$$u = 2U - V; \quad v = 2V - U; \quad t = -U - V; \quad w = 3W \tag{I.4b}$$

Appendix II
Crystallographically Related Operations

As discussed in Section 2.3.3, there are a number of *crystallographically related solutions* for an orientation or misorientation, depending on the crystal system. These crystallographically related solutions are generated by premultiplying the orientation or misorientation matrix M by a symmetry operator T_i:

$$M' = T_i M \tag{II.1}$$

where the value of i depends on the symmetry of the crystal system. The 24 matrices T_i for cubic symmetry are

$$\begin{pmatrix} 1 & 0 & 0 \\ 0 & 1 & 0 \\ 0 & 0 & 1 \end{pmatrix} \quad \begin{pmatrix} 0 & 0 & \bar{1} \\ 0 & \bar{1} & 0 \\ \bar{1} & 0 & 0 \end{pmatrix} \quad \begin{pmatrix} 0 & 0 & \bar{1} \\ 0 & 1 & 0 \\ 1 & 0 & 0 \end{pmatrix} \quad \begin{pmatrix} \bar{1} & 0 & 0 \\ 0 & 1 & 0 \\ 0 & 0 & \bar{1} \end{pmatrix}$$

$$\begin{pmatrix} 0 & 0 & 1 \\ 0 & 1 & 0 \\ \bar{1} & 0 & 0 \end{pmatrix} \quad \begin{pmatrix} 1 & 0 & 0 \\ 0 & 0 & \bar{1} \\ 0 & 1 & 0 \end{pmatrix} \quad \begin{pmatrix} 1 & 0 & 0 \\ 0 & \bar{1} & 0 \\ 0 & 0 & \bar{1} \end{pmatrix} \quad \begin{pmatrix} 1 & 0 & 0 \\ 0 & 0 & 1 \\ 0 & \bar{1} & 0 \end{pmatrix}$$

$$\begin{pmatrix} 0 & \bar{1} & 0 \\ 1 & 0 & 0 \\ 0 & 0 & 1 \end{pmatrix} \quad \begin{pmatrix} \bar{1} & 0 & 0 \\ 0 & \bar{1} & 0 \\ 0 & 0 & 1 \end{pmatrix} \quad \begin{pmatrix} 0 & 1 & 0 \\ \bar{1} & 0 & 0 \\ 0 & 0 & 1 \end{pmatrix} \quad \begin{pmatrix} 0 & 0 & 1 \\ 1 & 0 & 0 \\ 0 & 1 & 0 \end{pmatrix}$$

$$\begin{pmatrix} 0 & 1 & 0 \\ 0 & 0 & 1 \\ 1 & 0 & 0 \end{pmatrix} \quad \begin{pmatrix} 0 & 0 & \bar{1} \\ \bar{1} & 0 & 0 \\ 0 & 1 & 0 \end{pmatrix} \quad \begin{pmatrix} 0 & 1 & 0 \\ 0 & 0 & 1 \\ \bar{1} & 0 & 0 \end{pmatrix} \quad \begin{pmatrix} 0 & 1 & 0 \\ 0 & 0 & \bar{1} \\ \bar{1} & 0 & 0 \end{pmatrix}$$

$$\begin{pmatrix} 0 & 0 & \bar{1} \\ 1 & 0 & 0 \\ 0 & \bar{1} & 0 \end{pmatrix} \quad \begin{pmatrix} 0 & 0 & 1 \\ \bar{1} & 0 & 0 \\ 0 & \bar{1} & 0 \end{pmatrix} \quad \begin{pmatrix} 0 & \bar{1} & 0 \\ 0 & 0 & \bar{1} \\ 1 & 0 & 0 \end{pmatrix} \quad \begin{pmatrix} 0 & 1 & 0 \\ 1 & 0 & 0 \\ 0 & 0 & \bar{1} \end{pmatrix}$$

$$\begin{pmatrix} \bar{1} & 0 & 0 \\ 0 & 0 & 1 \\ 0 & 1 & 0 \end{pmatrix} \quad \begin{pmatrix} 0 & 0 & 1 \\ 0 & \bar{1} & 0 \\ 1 & 0 & 0 \end{pmatrix} \quad \begin{pmatrix} 0 & \bar{1} & 0 \\ \bar{1} & 0 & 0 \\ 0 & 0 & \bar{1} \end{pmatrix} \quad \begin{pmatrix} \bar{1} & 0 & 0 \\ 0 & 0 & \bar{1} \\ 0 & \bar{1} & 0 \end{pmatrix}$$

The 12 symmetry operation matrices for hexagonal symmetry are

$$\begin{pmatrix} 1 & 0 & 0 \\ 0 & 1 & 0 \\ 0 & 0 & 1 \end{pmatrix} \begin{pmatrix} 0.5 & 0.87 & 0 \\ -0.87 & 0.5 & 0 \\ 0 & 0 & 1 \end{pmatrix} \begin{pmatrix} -0.5 & 0.87 & 0 \\ -0.87 & -0.5 & 0 \\ 0 & 0 & 1 \end{pmatrix} \begin{pmatrix} -1 & 0 & 0 \\ 0 & -1 & 0 \\ 0 & 0 & 1 \end{pmatrix}$$

$$\begin{pmatrix} -0.5 & -0.87 & 0 \\ 0.87 & -0.5 & 0 \\ 0 & 0 & 1 \end{pmatrix} \begin{pmatrix} 0.5 & -0.87 & 0 \\ 0.87 & 0.5 & 0 \\ 0 & 0 & 1 \end{pmatrix} \begin{pmatrix} 1 & 0 & 0 \\ 0 & -1 & 0 \\ 0 & 0 & -1 \end{pmatrix} \begin{pmatrix} 0.5 & 0.87 & 0 \\ 0.87 & -0.5 & 0 \\ 0 & 0 & -1 \end{pmatrix}$$

$$\begin{pmatrix} -0.5 & 0.87 & 0 \\ 0.87 & 0.5 & 0 \\ 0 & 0 & -1 \end{pmatrix} \begin{pmatrix} -1 & 0 & 0 \\ 0 & 1 & 0 \\ 0 & 0 & -1 \end{pmatrix} \begin{pmatrix} -0.5 & -0.87 & 0 \\ -0.87 & 0.5 & 0 \\ 0 & 0 & -1 \end{pmatrix} \begin{pmatrix} 0.5 & -0.87 & 0 \\ -0.87 & -0.5 & 0 \\ 0 & 0 & -1 \end{pmatrix}$$

The four symmetry operation matrices for orthorhombic symmetry are

$$\begin{pmatrix} 1 & 0 & 0 \\ 0 & 1 & 0 \\ 0 & 0 & 1 \end{pmatrix} \begin{pmatrix} 1 & 0 & 0 \\ 0 & \bar{1} & 0 \\ 0 & 0 & \bar{1} \end{pmatrix} \begin{pmatrix} \bar{1} & 0 & 0 \\ 0 & 1 & 0 \\ 0 & 0 & \bar{1} \end{pmatrix} \begin{pmatrix} \bar{1} & 0 & 0 \\ 0 & \bar{1} & 0 \\ 0 & 0 & 1 \end{pmatrix}$$

Appendix III

Crystallographically Related Solutions for the Four S-texture Variants

S1	S2	S3	S4
$(123)[63\bar{4}]$[a]	$(123)[\bar{6}\bar{3}4]$	$(213)[36\bar{4}]$	$(213)[\bar{3}\bar{6}4]$
$(32\bar{1})[\bar{4}3\bar{6}]$	$(32\bar{1})[4\bar{3}6]$	$(31\bar{2})[\bar{4}6\bar{3}]$	$(31\bar{2})[4\bar{6}3]$
$(\bar{1}2\bar{3})[\bar{6}34]$	$(\bar{1}2\bar{3})[6\bar{3}\bar{4}]$	$(\bar{2}1\bar{3})[\bar{3}64]$	$(\bar{2}1\bar{3})[3\bar{6}\bar{4}]$
$(\bar{3}21)[436]$	$(\bar{3}21)[\bar{4}\bar{3}\bar{6}]$	$(\bar{3}12)[463]$	$(\bar{3}12)[\bar{4}\bar{6}\bar{3}]$
$(13\bar{2})[6\bar{4}\bar{3}]$	$(13\bar{2})[\bar{6}43]$	$(23\bar{1})[3\bar{4}\bar{6}]$	$(23\bar{1})[\bar{3}46]$
$(1\bar{2}\bar{3})[6\bar{3}4]$	$(1\bar{2}\bar{3})[\bar{6}3\bar{4}]$	$(2\bar{1}\bar{3})[3\bar{6}4]$	$(2\bar{1}\bar{3})[\bar{3}6\bar{4}]$
$(1\bar{3}2)[643]$	$(1\bar{3}2)[\bar{6}\bar{4}\bar{3}]$	$(2\bar{3}1)[346]$	$(2\bar{3}1)[\bar{3}\bar{4}\bar{6}]$
$(2\bar{1}3)[3\bar{6}\bar{4}]$	$(2\bar{1}3)[\bar{3}64]$	$(1\bar{2}3)[6\bar{3}\bar{4}]$[a]	$(1\bar{2}3)[\bar{6}34]$
$(\bar{1}\bar{2}3)[\bar{6}\bar{3}\bar{4}]$	$(\bar{1}\bar{2}3)[634]$[a]	$(\bar{2}\bar{1}3)[\bar{3}\bar{6}\bar{4}]$	$(\bar{2}\bar{1}3)[364]$
$(\bar{2}13)[\bar{3}6\bar{4}]$	$(\bar{2}13)[3\bar{6}4]$	$(\bar{1}23)[\bar{6}3\bar{4}]$	$(\bar{1}23)[6\bar{3}4]$[a]
$(231)[3\bar{4}\bar{6}]$	$(231)[\bar{3}4\bar{6}]$	$(132)[6\bar{4}3]$	$(132)[\bar{6}4\bar{3}]$
$(312)[\bar{4}63]$	$(312)[4\bar{6}\bar{3}]$	$(321)[\bar{4}36]$	$(321)[4\bar{3}\bar{6}]$
$(\bar{2}3\bar{1})[\bar{3}\bar{6}\bar{4}]$	$(\bar{2}3\bar{1})[346]$	$(\bar{1}3\bar{2})[\bar{6}\bar{4}\bar{3}]$	$(\bar{1}3\bar{2})[643]$
$(\bar{3}\bar{1}2)[4\bar{6}3]$	$(\bar{3}\bar{1}2)[\bar{4}6\bar{3}]$	$(\bar{3}\bar{2}1)[4\bar{3}6]$	$(\bar{3}\bar{2}1)[\bar{4}3\bar{6}]$
$(\bar{3}1\bar{2})[46\bar{3}]$	$(\bar{3}1\bar{2})[\bar{4}\bar{6}3]$	$(\bar{3}2\bar{1})[43\bar{6}]$	$(\bar{3}2\bar{1})[\bar{4}\bar{3}6]$
$(2\bar{3}\bar{1})[34\bar{6}]$	$(2\bar{3}\bar{1})[\bar{3}\bar{4}6]$	$(1\bar{3}\bar{2})[64\bar{3}]$	$(1\bar{3}\bar{2})[\bar{6}\bar{4}3]$
$(\bar{2}\bar{3}1)[\bar{3}46]$	$(\bar{2}\bar{3}1)[3\bar{4}\bar{6}]$	$(\bar{1}\bar{3}2)[\bar{6}43]$	$(\bar{1}\bar{3}2)[6\bar{4}\bar{3}]$
$(3\bar{1}\bar{2})[\bar{4}\bar{6}\bar{3}]$	$(3\bar{1}\bar{2})[463]$	$(3\bar{2}\bar{1})[\bar{4}\bar{3}\bar{6}]$	$(3\bar{2}\bar{1})[436]$
$(21\bar{3})[364]$	$(21\bar{3})[\bar{3}\bar{6}\bar{4}]$	$(12\bar{3})[634]$	$(12\bar{3})[\bar{6}\bar{3}\bar{4}]$
$(\bar{1}32)[\bar{6}\bar{4}3]$	$(\bar{1}32)[64\bar{3}]$	$(\bar{2}31)[\bar{3}\bar{4}6]$	$(\bar{2}31)[34\bar{6}]$
$(3\bar{2}1)[\bar{4}\bar{3}6]$	$(3\bar{2}1)[43\bar{6}]$	$(3\bar{1}2)[\bar{4}\bar{6}3]$	$(3\bar{1}2)[46\bar{3}]$
$(\bar{2}\bar{1}\bar{3})[\bar{3}\bar{6}4]$	$(\bar{2}\bar{1}\bar{3})[36\bar{4}]$	$(\bar{1}\bar{2}\bar{3})[\bar{6}\bar{3}4]$	$(\bar{1}\bar{2}\bar{3})[63\bar{4}]$
$(\bar{1}\bar{3}\bar{2})[\bar{6}4\bar{3}]$	$(\bar{1}\bar{3}\bar{2})[6\bar{4}3]$	$(\bar{2}\bar{3}\bar{1})[\bar{3}4\bar{6}]$	$(\bar{2}\bar{3}\bar{1})[3\bar{4}6]$
$(\bar{3}\bar{2}\bar{1})[4\bar{3}\bar{6}]$	$(\bar{3}\bar{2}\bar{1})[\bar{4}36]$	$(\bar{3}\bar{1}\bar{2})[4\bar{6}\bar{3}]$	$(\bar{3}\bar{1}\bar{2})[\bar{4}6\bar{3}]$

[a] The lowest angle solution.

Appendix IV
Spherical Projection and the Stereographic, Equal-Area, and Gnomonic Projections

Projection is used in many branches of crystallography to represent the key features of crystals, namely, the relationships between lattice planes and directions. The subject of this book, texture, is an aspect of crystallography, and to understand such attributes as Kikuchi patterns (Chapter 6) and pole figures (Sections 2.5, 5.2, and 10.2), it is crucial to gain an elementary knowledge of projection geometry.

A crystal can be represented by a set of normals, one from each crystallographic plane. Angular relationships between planes are therefore the same as those between their respective normals. If a sphere having the same origin as the crystal itself is circumscribed on the crystal, the intersections of the plane normals, known as *poles*, with the sphere surface form a *spherical projection*, otherwise known as a *reference sphere* (Figure IV.1). The symmetry of the crystal itself is carried through to the representation on the reference sphere. This means that the reference sphere is divided, by the crystal symmetry planes, into spherical triangles. For example, the cubic, hexagonal, and orthorhombic systems have 48, 24, and 8 crystallographically equivalent *stereographic unit triangles*, respectively. Various other projections are then introduced to project the 3D reference sphere onto a more manageable 2D flat form.

IV.1 THE STEREOGRAPHIC PROJECTION

The *stereographic projection* is a means of representing 3D angular relationships in two dimensions. In the present context, the angular relationships of interest are those between crystallographic plane normals or between crystallographic directions. The main application of the stereographic projection in texture analysis is that pole figures/inverse pole figures, misorientations (Section 10.5.1), and especially one angle/axis space (Sections 2.7.3 and 10.6.2) are usually presented in stereographic projection (although equal-area projections [Section IV.2] are also sometimes used).

The intermediate stage in the construction of a stereographic projection is spherical projection of poles onto the reference sphere as described earlier. Using terminology relating to the earth, the point of projection is the south pole and the plane of projection is the equatorial plane as shown in Figure IV.1. It can be seen that poles in the northern hemisphere fall inside the equator or primitive circle, poles on the primitive circle project onto themselves, and poles in the southern hemisphere fall outside the primitive circle. It is more convenient to project poles in the southern

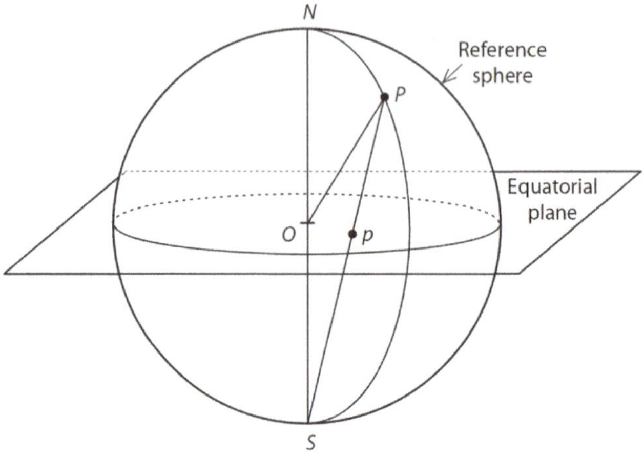

FIGURE IV.1 A spherical projection (reference sphere) with origin, "North Pole," and "South Pole" labeled *O*, *N*, and S, respectively. A crystal is imagined to lie at *O*. The normal of a plane in the crystal intersects the reference sphere at *P*. *P* is projected from the South Pole onto the equatorial plane at *p*; this is the stereographic projection of *P*.

hemisphere from the north pole so that they also fall inside the primitive circle. They are denoted in the stereographic projection by open circles to distinguish them from poles arising from the northern hemisphere, which are denoted by smaller, filled circles. In Figure IV.1, the projection p of the pole P onto the equatorial plane is given by $r\tan\angle OSP$, where r is the radius of the sphere and $\angle OSP = \frac{1}{2}\angle NOP$.

The main attraction for crystallographers of the stereographic projection is that angular relationships in the crystal are preserved in the projection. Consideration of the projection geometry shows that the angle between the center of the projection, *O*, and the primitive circle is 90°. For crystals with orthogonal symmetry, it is conventional to denote the "axes" of the standard stereogram (i.e., the three crystal axes 100, 010, and 001) in the positions shown in Figure IV.2. This figure shows a standard stereographic projection, that is, the equatorial plane, illustrating some of its basic features. p is a pole of a plane, and the trace of the plane itself is a "great circle." A great circle is the projection of any plane that passes through the origin of the reference sphere (whereas a plane that does not pass through the origin projects as a "small circle").

A plane is usually represented by its normal (pole), although sometimes it is more convenient to represent it by the trace of the plane, that is, the great circle or *zone* (Figure IV.2). The angle between two poles/directions is measured along a great circle that passes through both of them. For example, in Figure IV.2, *A* and *B* are directions that lie in the same plane, and the angle between them is measured along the great circle connecting them. Since p is the plane normal, the angle between p and any points on the plane (e.g., *A* and B) is 90°. Angles on the stereogram are measured with the aid of a grid known as a *Wulff net* that is marked in 2° intervals as shown in Figure IV.3a.

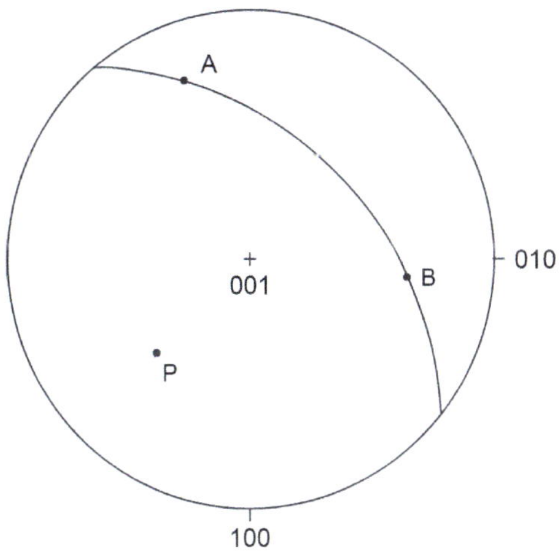

FIGURE IV.2 Stereographic projection showing the projection of the great circle (zone) of which p is the pole. A and B lie on the zone of p, and the angle between them is measured along the great circle that connects them.

The projection of the stereographic unit triangles for the cubic system is shown in Figure IV.4a. There are 24 each of right- and left-handed triangles. A single stereographic unit triangle is usually used for an inverse pole figure (Sections 2.5.2, 5.2.2, and 10.2) and misorientation axis distributions (Section 10.5.1) or interface plane distributions (Sections 11.2.7 and 11.3.3). In general, if the specimen axis that is to be plotted in the inverse pole figure is a mirror (symmetry) plane of the specimen, a

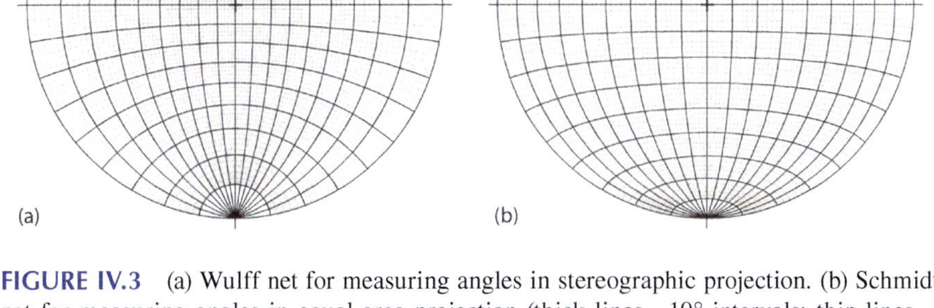

FIGURE IV.3 (a) Wulff net for measuring angles in stereographic projection. (b) Schmidt net for measuring angles in equal-area projection (thick lines—10° intervals; thin lines—2° intervals).

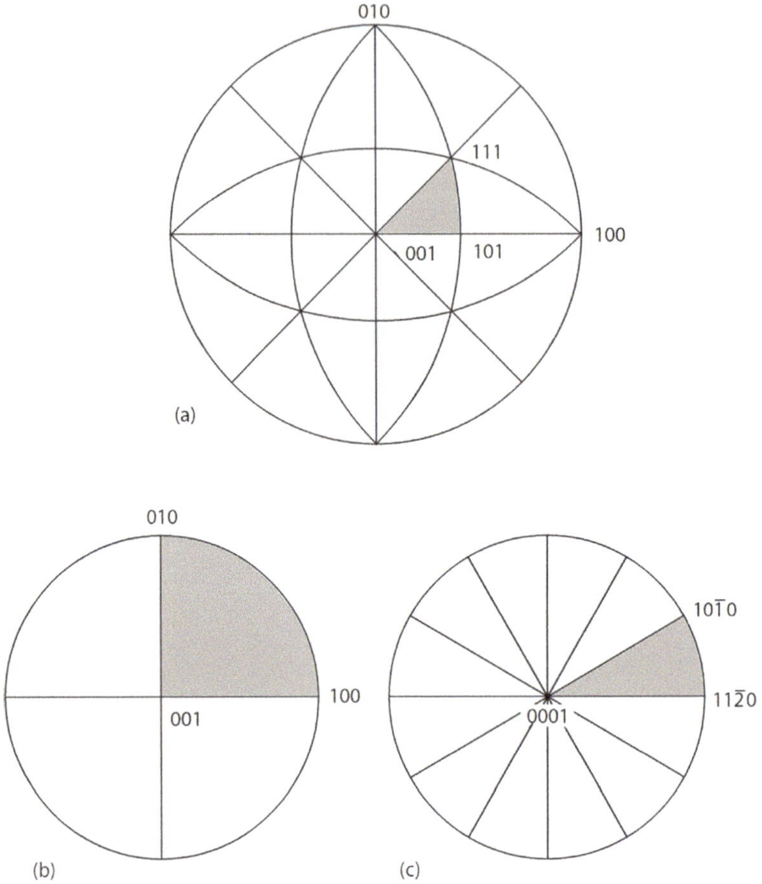

FIGURE IV.4 Unit triangles in stereographic projection for (a) cubic, (b) orthorhombic, and (c) hexagonal crystal symmetry. The apices of one triangle are labeled.

single unit triangle suffices. Otherwise, right- and left-handed triangles should be used. A pole figure is usually plotted in the whole of the stereographic projection, although sometimes "reduced" pole figures in one-quarter of the stereogram are seen. For the sake of completeness, Figure IV.4b and c shows the 4 and 12 stereographic unit triangles of the orthorhombic and hexagonal crystal structures, respectively.

The construction of a stereographic projection and some of its features that are pertinent to texture analysis are amplified and discussed rigorously by Johari and Thomas (1969) and McKie and McKie (1974).

IV.2 THE EQUAL-AREA PROJECTION

In the *equal-area projection*, equal areas on the reference sphere have equal areas in projection. Such a projection is particularly appropriate for the measurement of

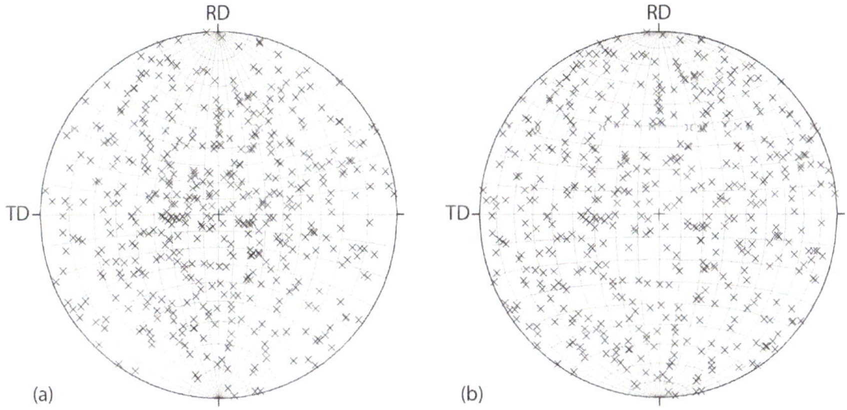

FIGURE IV.5 Projection of a random distribution of 500 poles on the surface of a sphere. (a) Stereographic projection with superimposed Wulff net; (b) equal-area projection with superimposed Schmidt net.

population densities. For example, a random distribution of points (orientations) appears as a uniform density of points on the equal-area projection, whereas this distribution is distorted on the stereographic projection. With reference to Figure IV.1, in an equal-area projection, the projection of P onto the equatorial plane is given by $r\sqrt{2}\sin\angle OSP$. Figure IV.3b shows the *Schmidt net* that is used to determine angular relationships in equal-area projection, and comparison between it and the Wulff net in Figure IV.3a illustrates the equal-area characteristics of the former. Here, the size of the net units represents correctly the fraction of orientation space covered by each.

Figure IV.5 shows the projection of 500 randomly oriented poles in both stereographic and equal-area projections. In the stereographic projection the poles are clearly clustered in the central regions of the pole figure (Figure IV.5a), whereas in the equal-area projection they appear homogeneously distributed (Figure IV.5b). Disadvantages of the equal-area projection are that circles on the sphere do not appear as circles in the projection and angles between great circles are not preserved.

Despite the advantages of equal-area projection for assessing population densities, stereographic projection remains popular in metallurgy and equal-area projection is most widely used in geology (Wenk and Kocks, 1987).

IV.3 THE GNOMONIC PROJECTION

In the *gnomonic projection*, poles are projected from the center of the reference sphere to a tangent plane at the North Pole, as shown in Figure IV.6 (McKie and McKie, 1974). Since every great circle passes through the projection point O, all zones project as straight lines. The projection of P is now given by $r\tan\angle NOP$. This has the disadvantage that as $\angle NOP$ approaches 90°, the distance NP in the projection becomes unmanageably large, until when $\angle NOP = 90°$ P projects at infinity. The

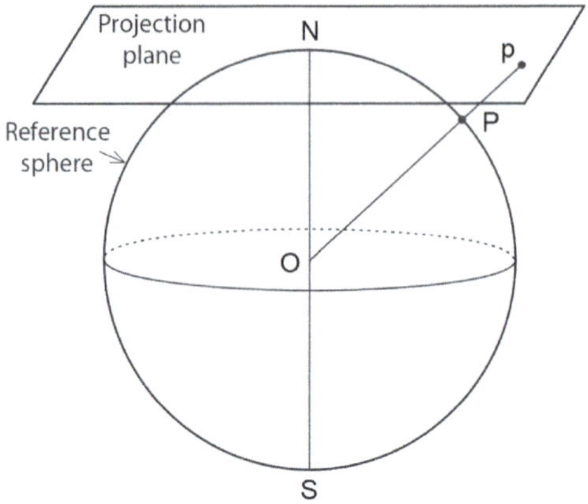

FIGURE IV.6 The relationship between the reference sphere and the projection plane for gnomonic projection. The pole *P* is projected from *O* to a tangent plane at the North Pole.

link between the gnomonic projection and texture analysis is that a Kikuchi diffraction pattern is a gnomonic projection. The distortion at large projection angles is particularly apparent in electron backscatter diffraction (EBSD) patterns, since they have a capture angle of approximately 60° (Figure 7.10).

Appendix V
Indexing a Pole Figure

This appendix describes the methodology for indexing a pole figure, measured by x-ray or neutron diffraction techniques, and evaluating the corresponding orientation. Figure V.1a shows the {111} pole figure of a single crystal with an unknown orientation. To determine this orientation in terms of its Miller indices $(hkl)[uvw]$, the four {111} poles of the orientation have to be indexed first. As described in Section 2.4, in the cubic crystal system there are 48 different possibilities to do so, one of which is applied in Figure V.1b. Remember the crystal orientation $(hkl)[uvw]$ is given by the specimen rolling direction (RD) expressed in crystal coordinates $[uvw]$, and the specimen normal direction (ND) expressed in crystal coordinates (hkl). Hence, to determine the crystal orientation, it is necessary to express the specimen axes RD and ND in crystal coordinates. For that purpose, the angles between the four (indexed) {111} poles and the sample axes RD and ND are determined. As described in Appendix IV, this can be done with the help of a Wulff net (Figure IV.3a) or a Schmidt net if an equal-area pole figure is analyzed (Figure IV.3b). The resulting angles are marked in Figure V.1b with dashed lines for the RD and dotted lines for the ND.

To determine the orientation of the RD in crystal coordinates $[uvw]$, the angles between the {111} poles and the RD are analyzed. For instance, for the (111) pole this reads:

$$\cos \angle\big((111),[uvw]\big) = \frac{\begin{pmatrix} 1 \\ 1 \\ 1 \end{pmatrix} \cdot \begin{pmatrix} u \\ v \\ w \end{pmatrix}}{\sqrt{3}} = \frac{u+v+w}{\sqrt{3}} \tag{V.1}$$

The factor $\sqrt{3}$ is necessary to normalize the vector (111) to unity, whereas the resulting $[uvw]$ will automatically be normalized to unity. This equation is set up three times for three of the four possible {111} poles so as to yield the three unknowns u, v, and w; for example,

$$\begin{aligned} 111 &: u+v+w = \sqrt{3}\cdot\cos 118^\circ \\ 1\bar{1}1 &: u-v+w = \sqrt{3}\cdot\cos 90^\circ \\ \bar{1}\bar{1}1 &: -u-v+w = \sqrt{3}\cdot\cos 19.5^\circ \end{aligned} \tag{V.2}$$

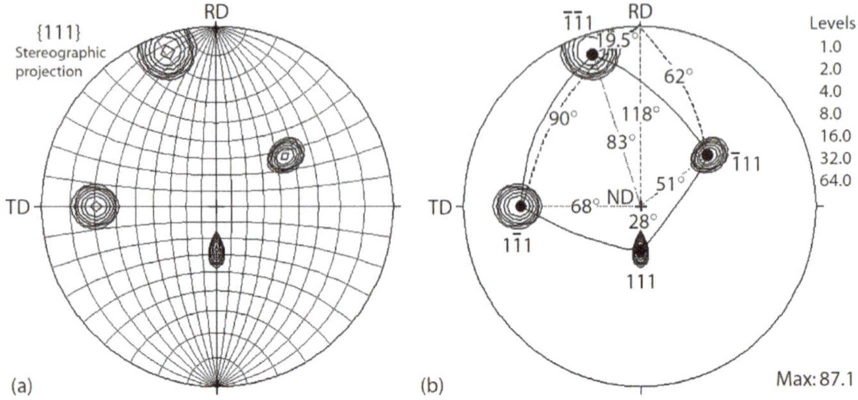

FIGURE V.1 (a) {111} Pole figure of a single crystal, plotted in stereographic projection. (b) Indexed poles and angles between the poles and sample axes RD and ND to determine the single-crystal orientation of (a).

In the present example, the RD is $[uvw] = [-0.816, -0.407, 0.440] \approx [\bar{2}\bar{1}1]$. This procedure is repeated for the ND (hkl):

$$
\begin{aligned}
111 &: h + k + l = \sqrt{3} \cdot \cos 28^\circ \\
1\bar{1}1 &: h - k + l = \sqrt{3} \cdot \cos 68^\circ \\
\bar{1}\bar{1}1 &: -h - k + l = \sqrt{3} \cdot \cos 83^\circ
\end{aligned} \tag{V.3}
$$

These three equations yield (hkl) = (0.219, 0.440, 0.870) ≈ (124).

Thus, the unknown single-crystal orientation displayed in Figure V.1a is $(hkl)[uvw] = (124)[\bar{2}\bar{1}1]$.

Appendix VI
X-ray Counters and Pulse Height Analysis

This appendix describes the different types of counters that may be used for recording the diffracted intensities in x-ray diffraction experiments—Geiger, scintillation, proportional, and semiconductor counters—as well as the use of pulse height analyzers to improve the peak-to-background ratio. For a more thorough treatment, see Cullity (1978) and Barrett and Massalski (1980). All electrical counters suitable for x-ray diffraction experiments depend on the power of the x-rays to ionize atoms, either of a gas (Geiger and proportional counters) or of a solid (scintillation and semiconductor counters). With regard to quantitative texture analysis, mainly three aspects of the different counters are of importance:

- An incoming x-ray quantum results in an electrical pulse of the counter. With increasing number of x-ray quanta, that is, with increasing counting rates, however, the time interval between the pulses decreases and may become so small that successive pulses can no longer be resolved by the counter, which leads to *counting losses* at high counting rates.
- The counter *efficiency* describes the fraction of x-ray quanta that is indeed able to initiate an electrical pulse in the counter. Both absorption of x-rays in the counter window and passing of x-rays through the entire counter without being absorbed by the recording medium decrease the efficiency.
- In most counters, the output voltage is proportional to the energy of the x-ray photon (see below), and the *energy resolution* is a measure for this proportionality between the size of the voltage pulse generated by the counter and the energy of the absorbed x-ray quantum.

VI.1 GEIGER COUNTERS

The *Geiger counter* (also called Geiger–Müller counter) was the first electrical counter used to detect radiation. A Geiger counter consists of a wire electrode in a gas-filled chamber with a voltage at the wire that is high enough (~1,500 V) to attract an electron in the gas so strongly that it would ionize many atoms of the gas along its path (Figure VI.1a). Thus, an ionization effect by a single incident x-ray photon results in a chain of ionizing events that culminate in a strong discharge through the gas. The number of such pulses in a given time interval is a measure of the radiation intensity. The main disadvantage of the Geiger counters for x-ray texture analysis is the long dead time of ~10^{-4} s between initiation of the ionization chain and resetting of the counter, which may lead to losses in the order of 10% at counting rates of only

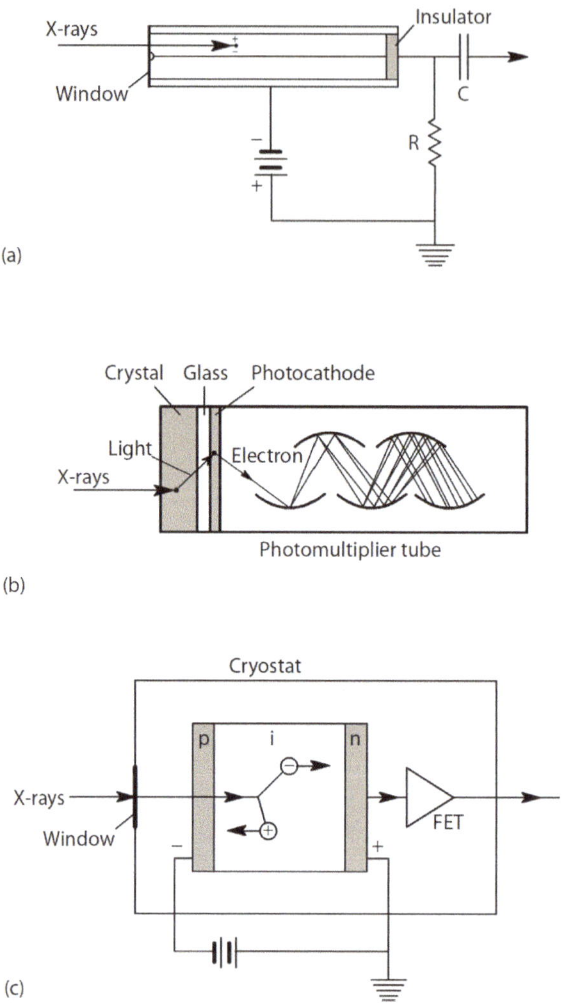

FIGURE VI.1 Schematic representation of x-ray counters: (a) Geiger counter, (b) scintillation counter, and (c) semiconductor detector. (Adapted from Cullity, 1978.)

500 s^{-1}. Another drawback is that the output of Geiger counters does not depend on the incident wavelength, that is, it is not proportional; therefore, Geiger counters are today virtually obsolete in diffractometry.

VI.2 SCINTILLATION COUNTERS

Much more often *scintillation counters* are in use, which consist of a fluorescent crystal mounted at the end of a photomultiplier tube (Figure VI.1b). An x-ray photon absorbed in this crystal—typically a sodium iodide crystal containing ~1% thallium—ejects photoelectrons that energize fluorescence of the thallium ions. The fluorescent light ejects electrons in the photomultiplier tube, which are multiplied and finally transferred to the measuring electronics. Scintillation detectors stand out by a constant, very high efficiency throughout the range of wavelengths used in x-ray diffractometry and by a short dead time of only 10^{-7} s caused by the decay of the fluorescence induced by photon absorption. Thus, x-ray data with rates of 10,000 s^{-1} can be counted with losses below 1%. The output voltage is proportional to the energy of the x-ray photon, so that the reflected peak intensity can be distinguished from background and fluorescent radiation with different wavelengths by setting an appropriate energy window of a subsequent pulse height analyzer (Section VI.5).

VI.3 PROPORTIONAL COUNTERS

In principle, a *proportional counter* resembles a Geiger counter, but it works at a lower voltage of ~1,000 V to ensure proportionality between the energy of the absorbed photon and the amplitude of the output signal. The dead time of proportional counters is comparable to that of scintillation counters, permitting similar counting rates. Their efficiency is lower, however, and depends more strongly on the wavelength of the x-rays. Their amplification is smaller, requiring stronger amplification by the subsequent electronics. The main advantage of proportional counters is that for monochromatic radiation, the distribution of amplitudes of the pulses is narrower than with scintillation counters, thus allowing a better separation of diffracted peak and background intensity.

VI.4 SEMICONDUCTOR DETECTORS

Nowadays, increasingly solid-state *semiconductor detectors* are used, since they produce pulses proportional to the absorbed x-ray energy with better energy resolution than any other counter. Furthermore, they stand out by a high efficiency and by low counting losses. A semiconductor detector basically consists of a silicon single crystal containing small amounts of lithium, and therefore these counters are designated Si(Li). Germanium counters are in use for detection of radiation with shorter wavelengths. The crystal is produced so as to comprise very thin surface layers of *n*-type on one side, meaning it contains free electrons from donor impurities, and *p*-type on the opposite side, meaning it contains free holes from acceptor impurities (Figure VI.1c). The central region is neither *n*- nor *p*-type; that is, it is intrinsic (*i*), which means that it has a very high resistivity because only few electrons are thermally excited from the valence band across the energy gap into the conduction band. However, an incident x-ray quantum can cause excitation and create a multitude (approximately 1,000) of pairs of free electrons (–) in the conduction band and free holes (+) in the valence band. If a high voltage is applied to the opposite faces of the crystal, the electrons and holes will be swept to these faces, creating a small electrical pulse. Subsequently, these very small pulses are amplified by a field-effect transistor (FET). Disadvantages are that the semiconductor counters must be cooled with liquid nitrogen or with a Peltier cooler and they require more complex electronics.

VI.5 PULSE HEIGHT ANALYZERS

In all counters that are commonly used today (scintillation, proportional, and semiconductor), the output voltage depends on the wavelength of the x-ray photon in so far as the size of the resulting pulses is proportional to the energy of the triggering x-ray quanta. This means that if an x-ray quantum of CuKα radiation ($E = h\nu = 9$ keV) generates a pulse of 1 mV, then an MoKα quantum ($h\nu = 20$ keV) would produce a pulse of (20/9) = 2.2 mV. Therefore, electrical units that distinguish between pulses of different height, the so-called *pulse height analyzers*, can distinguish between x-rays of different energy and, therewith, background and fluorescent radiation with different wavelengths can be reduced. Let us assume that x-rays with three different wavelengths $\lambda_A > \lambda_B > \lambda_C$ are detected by a counter, which in turn produces pulses of different heights A < B < C (Figure VI.2). If an electronic circuit that transmits only pulses with height larger than a certain threshold value (given by the voltage V_1) is inserted between counter and rate meter, then pulses with larger wavelengths (pulse A) are not counted. This can be used to reduce background intensities resulting from the continuous spectrum as well as fluorescence with long wavelengths. If the device is additionally able to reject pulses with height larger than a voltage V_2 (pulse C) then diffracted intensities with low wavelengths will be ignored as well (Figure VI.2). Thus, with a proper setting of the *baseline* and the *window* of the pulse height analyzer, the peak-to-background ratio can markedly be improved.

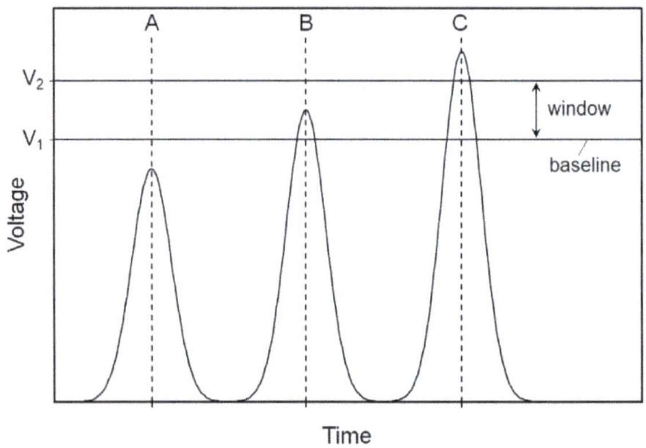

FIGURE VI.2 Pulse height analysis. (Adapted from Cullity, 1978.)

Glossary of Terms

Absolute orientation: *See* crystallographic orientation.

Absorption edge: Drop in the absorption of x-rays of a given wavelength between certain elements, caused by the efficiency of the x-rays in emitting photo-electrons from the electron shells of the sample material.

Angle and axis of rotation: Transformation between the crystal coordinate system and the specimen coordinate system via rotation through an angle about a specific axis.

Angle/axis pair: *See* **angle and axis of rotation**.

Angle and axis of misorientation: Transformation between the crystal coordinate systems of two grains via rotation through an angle about a specific axis.

Angle/axis space: Three-dimensional space wherein angle/axis pairs reside (*see also* **cylindrical angle/axis space**).

Angular-dispersive diffractometry: Conventional technique to separate different reflectors according to their reflection angle for monochromatic radiation.

Area detector: Two-dimensional position sensitive detector.

Asymmetric domain (of Euler or Rodrigues space): Portion of Euler or Rodrigues space defined such that every interior point in it represents a physically distinct orientation or misorientation (*see* **fundamental zone**).

Atomic scattering amplitude: A measure of the intensity scattered by an individual atom in an x-ray or electron diffraction experiment. Mathematically the Fourier transform of the atomic potential.

Automated crystal orientation mapping, ACOM: *See* **orientation mapping**.

Background: Intensities in the diffraction spectra that are caused by incoherent scattering, fluorescence, and reflection of x-rays with other wavelengths which do not contribute to the diffracted information.

Backscatter Kikuchi diffraction, BKD: *See* **electron back-scatter diffraction, EBSD**.

Backscatter Kikuchi pattern, BKP: *See* **electron backscatter diffraction, EBSD**.

Bragg's law: The most basic law of (kinematic) diffraction of radiation (x-rays, neutrons, electrons) on atomic crystal lattices. It claims that for strongly diffracted intensity the incident and emergent angles of radiation on a set of lattice planes must be the same ("reflection") and that constructive interference appears only for angles which fulfil the Bragg equation. Named after the British-Australian physicists William Henry Bragg (1862–1942) and William Lawrence Bragg (1890–1971) (father and son).

Camera constant: Distance between sample and photographic plate L times electron wavelength λ in the transmission electron microscope (TEM); measure for the magnification of diffraction patterns. The distance L is usually not the physical distance, but a virtual distance created by the electron optical system.

Coincidence site lattice, CSL: Lattice formed by notional superposition of lattice sites from two crystals and used in the context of texture analysis to categorize grain boundary geometry.

Continuous scan: Method to scan the pole figure continuously integrating over a range of pole figure angles.

Convergent beam electron diffraction, CBED: Kikuchi-type electron diffraction occurring under irradiation with a convergent rather than parallel electron beam. CBED patterns are mainly used for detailed crystallographic analysis (structure of lattice and defects), not usually for orientation determination. *See also* **small-angle convergent beam electron diffraction, SCBED**.

Crystal coordinate system: Coordinate system chosen as the edges of the crystal unit cell. Two versions exist, a system confining to the symmetry of the lattice and a Cartesian system. The latter is fixed to the former by conventions found in the International Tables of Crystallography.

Crystal matrix: A 3×3 matrix to transform the crystal axes into an orthonormal frame.

Crystallographic orientation: Orientation of the crystal coordinate system of a volume of crystal with respect to the specimen coordinate system. Mathematically described as an (active) rotation that transfers the Cartesian specimen coordinate system into that of the Cartesian crystal coordinate system or as a (passive) coordinate transformation of the crystal coordinate system into the specimen one.

Crystal orientation mapping, COM: *See* **orientation mapping**.

Crystallographically related solutions (of an orientation): *See* **symmetrically equivalent solutions**.

Crystal symmetry: The symmetry of crystallographic planes and directions defined by the space group symmetry of the crystal. For textures, it is only the point or Laue group symmetry which is of importance.

Cylindrical angle/axis space: Three-dimensional orientation or misorientation space wherein the rotation axis is described in polar coordinates on the circular base plane while the rotation angle is displayed along the cylinder axis.

Debye–Scherrer method: Technique for orientation determination in polycrystal volumes which is based on the diffraction of monochromatic x-rays. Named after the Dutch-American physicist Peter J.W. Debye (1884–1966) and the Swiss physicist Paul H. Scherrer (1890–1969).

Defocusing error: Decrease in recorded intensity in pole figure measurements with increasing tilting angle which is caused by broadening of the reflected peak.

Direct methods (of ODF calculation): Methods to calculate the orientation distribution function (ODF) from pole figure data in real space by an iterative fitting procedure.

Disorientation: Of all possible symmetrically equivalent misorientations of a pair of orientations, the one with smallest misorientation angle and axis in the standard orientation triangle.

Dynamical electron diffraction theory: A theory describing the intensity and geometry of electron diffraction, based on the assumption that an electron wave is re-scattered several times in a lattice. Solution of the Schrödinger equation for high energy electrons traveling in the crystal lattice potential.

Electron backscatter diffraction, EBSD: A scanning electron microscopy (SEM)-based technique for obtaining diffraction information from volumes of crystal down to about 50–500-nm diameter, depending on acceleration voltage and average Z-number of the material.

Electron channeling contrast imaging, ECCI: A backscatter electron imaging technique in the scanning electron microscope, used to observe microstructures of bulk samples. Frequently used in conjunction with EBSD.

Electron channeling pattern, ECP: A virtual diffraction pattern obtained by rocking the electron beam across a pivot point on the sample and measuring for every rocking angle the total intensity of backscattered electrons. The intensities are then displayed over polar coordinates.

Energy dispersive diffractometry: Technique in x-ray or electron diffraction to separate different reflectors according to the energy (or wavelength) of the reflected radiation for polychromatic radiation.

Equal-area projection: Projection of a reference sphere onto its equatorial plane such that equal areas on the sphere retain their equality in projection.

Etch pits: Characteristic polygons that form via etching on the surface of a material and whose shape is determined by the crystallographic orientation of the etched grain.

Euler angles: A description of orientation involving sequential rotations of the crystal coordinate system through three angles with respect to the specimen coordinate system. Named after the Swiss mathematician Leonhard Euler (1707–1783).

Euler space: Three-dimensional (usually Cartesian) space wherein Euler angles are represented.

Fundamental zone (of Rodrigues space): Volume of Rodrigues space in which only one of all symmetrically equivalent (mis)orientations of a given symmetry resides. Of all possible symmetrically equivalent (mis)orientations, the one with smallest rotation angle and axis in the standard orientation triangle is selected (*see* **asymmetric domain**).

Geiger counter: Counter to detect x-rays that is based on the potential of x-rays to ionize a gas, today virtually obsolete in diffractometry. Named after the German physicist Hans Geiger (1882–1945).

Ghost error: Error in ODF calculation by the series expansion method caused by the unavailability in the odd-order C-coefficients.

Gnomonic projection: Projection of a sphere on a plane in which the center of the sphere is the projection point and tangential plane is the projection plane. Great circles on the sphere (which correspond to planes through the sphere center) are projected onto straight lines. The center lines of Kikuchi bands correspond to the gnomonic projections of the lattice planes.

Goniometer: Multiple-axes device in a diffractometer to position the sample and the detector with respect to the incident beam.

Harmonic method: *See* **series expansion method** of ODF calculation.

High-resolution transmission electron microscopy, HRTEM: Technique for direct imaging of atoms in a TEM using phase contrast. The technique can be used to determine, with highest spatial resolution, crystal orientations and orientation relationships. Usually a Fourier transform of the lattice is used for this sake.

Hough transform: Transformation of points in an image into sinusoidal curves in an accumulation space. Used for automatic determination of Kikuchi bands in Kikuchi patterns, particularly EBSD patterns. Named after the American physicist Paul V. C. Hough (1925–2021).

Ideal orientation representation: Orientation representation of type (hkl)[uvw], where (hkl) corresponds to the plane normal of a crystal pointing in the direction of the Z-axis of the sample coordinate system and [uvw] corresponds to the crystal direction pointing in the direction of the X-axis, with small integer numbers of the planar (Miller) or directional indices.

Interface-plane scheme: Methodology for characterizing interfaces on the basis of the crystallographic orientation of the planar surface of the interface.

Interplanar angle: Angle between two crystallographic planes.

Interzonal angle: Angle between two crystallographic directions.

Inverse pole figure: Angular distribution of a chosen specimen direction with respect to the crystal coordinate system. Frequently displayed in a stereographic projection.

Kβ-filter: Thin metallic foil positioned between sample and detector in x-ray diffractometry and made of a material whose absorption edge lies just between the Kα and the Kβ peak of the x-ray tube. Used to produce quasi-monochromatic X-radiation.

Kikuchi line: A diffraction features in Kikuchi diffraction patterns. Corresponds to the intersection of the Kikuchi (or Kossel) cone with the detector plane.

Kikuchi diffraction pattern: An electron diffraction pattern that is created from a point source of electrons inside of the crystal. It consists of conical diffraction intensities (Kikuchi cones) around the diffracting lattice planes which forms line or band features on the detector. The point source is created by non-crystallographic diffuse scattering of the primary electron beam, the cones by coherent Bragg scattering. Kikuchi patterns appear in TEM (transmission Kikuchi patterns, TKP) and SEM (backscatter Kikuchi patterns, BKP, also called electron backscatter diffraction (EBSD) patterns). The patterns are well suited for accurate orientation determination with very good spatial resolution. Named after the Japanese physicist Seishi Kikuchi (1902–1972).

Kinematic electron diffraction theory: An electron diffraction theory that mainly describes the geometry of electron diffraction patterns. It is based on the application of Bragg's law.

Kossel or Kikuchi cone: A diffraction double-cone centered on the diffracting lattice plane. The cone mantel includes the Bragg angle with the lattice plane.

Kossel diffraction pattern: A Kikuchi-type diffraction pattern (*see there*) created by x-rays rather than electrons. Used for most accurate orientation and lattice constant measurements at medium spatial resolution. Named after the German physicist Walther Kossel (1888–1956).

Laue method: Technique for orientation determination in single crystal volumes which is based on the diffraction of polychromatic x-rays. Used particularly with synchrotron techniques. Named after the German physicist Max von Laue (1879–1960).

Mackenzie distribution: Distribution of frequency of disorientation angles for a set of random misorientations. The distribution is different for every crystal symmetry and frequently used as a reference for measured misorientation distributions. Named after the Australian physicist James K. Mackenzie (*1920) who worked intensively on martensite.

Macrotexture: An average texture determined from many grains obtained without reference to the location of individual grains within a specimen or the experimental technique used to obtain this information.

Mesotexture: The texture between grains, i.e. the texture associated with grain boundaries.

Metallurgical orientation description: *See* **ideal orientation representation**.

Micro-Kossel diffraction: SEM-based Kossel diffraction technique (*see there*) for obtaining crystallographic information from volumes of crystal down to approximately 10 μm and based on the diffraction of characteristic x-rays generated within the sample under investigation.

Microdiffraction: Generic term for techniques used to generate diffraction patterns in TEM with a spatial resolution beyond the limit of SAD, ~1 μm, by illumination of the specimen with a focused beam. This includes transmission Kikuchi patterns and (S)CBED patterns.

Microstructure: The ensemble of all lattice defects (except for point defects) that disturb the perfect order of a single crystal. This includes dislocations, grain and phase boundaries, precipitations and secondary phases, as well as elastic strain fields.

Microtexture: A sample population of orientation measurements which can be linked individually to their location within a specimen, or the experimental technique used to obtain this information. Often the microtexture is not the same as the macrotexture and not representative for the sample.

Misorientation: The rotation matrix that rotates one crystal orientation into a second one (active description) or, alternatively, that describes one crystal orientation relative to the coordinate system of a second one (passive description). The misorientation is often described as angle and axis of rotation. Misorientation between different phases is often referred to as **orientation relationship** (*see there*).

Misorientation distribution function, MODF or MDF: Continuous density distribution of misorientations. *See also* **orientation correlation function** and **orientation difference distribution function**.

Monochromator: Crystal used in x-ray or neutron diffractometry to produce monochromatic radiation of a certain wavelength in accordance with Bragg's law.

Normalization of pole figures: Normalization of the pole figure intensities recorded in numbers of counts, such that the total intensity over the pole figure is equal to unity. The standard units are commonly referred to as "multiples of a random distribution" (m.r.d.). Normalization is independent of experimental parameters.

O-lattice theory: A theory describing the structure of grain or phase boundaries based on their misorientation and boundary plane. It is the general basis of the CSL lattice model.

Orientation correlation function, OCF: Texture-reduced form of the MODF, derived by division of the MODF by the ODDF. Yields information about the texture-reduced spatial correlation of misorientations.

Orientation difference distribution function, ODDF: Statistical distribution of misorientations in a sample without consideration of the actual location of the orientations, only governed by the texture. Used to normalize the MODF, but does not contain valuable information on the spatial arrangement of the individual orientations.

Orientation distribution function, ODF: A mathematical function that describes the continuous density distribution of orientations, calculated either from discrete orientation data or from pole figures.

Orientation imaging microscopy, OIM™: A term devised by B. Adams et al. for EBSD-based orientation mapping in an SEM. The term is a commercial trademark of Ametek/EDAX but frequently used to describe **orientation mapping** (*see there*).

Orientation mapping: Use of an automated scanning technique in TEM, SEM, or in x-ray/synchrotron diffraction to measure orientations on a two-dimensional or three-dimensional grid. The results can be used to describe textures and microstructures of the measured area. Depiction of the microstructure in terms of its orientation constituents.

Orientation matrix: A mathematical 3×3 rotation matrix that describes the **crystallographic orientation** (*see there*).

Orientation microscopy, ORM: *see* **orientation mapping**.

Orientation relationship: The rotation between two individual crystal orientations, usually from crystals of different phases. Orientation relationships frequently carry information on the mechanism of phase transformation (*see also* **misorientation**).

Orientation topography: Spatial arrangement of the crystallographic orientations in the microstructure.

Orthonormal coordinate system: Right-handed Cartesian coordinate system chosen as three mutually perpendicular unit vectors.

Peak separation: The means to separate intensities stemming from closely spaced reflectors, either experimentally or numerically.

Penetration depth: Depth of the specimen from where the diffracted information arises.

Pole: Normal to a set of crystallographic planes.

Pole figure: Angular distribution of a chosen crystal plane normals (or crystal lattice directions) with respect to the specimen coordinate system. Frequently displayed in a stereographic projection.

Pole figure inversion: Calculation of the ODF from the data of (typically several) pole figures.

Position sensitive detector, PSD: An x-ray detector that that simultaneously records a certain area of a diffractogram. The area can be a linear or an area section.

Proportional counter: Counter to detect x-rays that is based on the potential of x-rays to ionize a gas. Every x-ray quantum entering the detector is recorded by an electrical discharge of a certain electric pulse height.

Pulse-height analyzer: Device that distinguishes between electrical pulses of different size and that is used to distinguish between x-rays of different energy, so to reduce background and fluorescent radiation.

Quaternions: A four-dimensional vector based on an extension of complex numbers, introduced first by O. Rodrigues (*see there*) and later exploited by the Irish mathematician William R. Hamilton (1805–1865). Quaternions are a very convenient way of representation of orientations, because of easy combination of rotations, no need for renormalization and easy averaging of orientations.

Radon transform: Integral transformation using a continuous line integral along analytical curves in a multi-dimensional space. The **Hough transform** (*see there*) is a discrete (i.e. pixel-based) version of the Radon transform. Named after the Austrian mathematician Johann Radon (1887–1956).

Random texture: Sample population of orientations having no preferred crystallographic orientation.

Reflection geometry: Geometrical setup to study diffraction effects on thick bulk samples where the incident radiation is reflected in the surface layers.

Rodrigues or Rodrigues-Frank space: Three-dimensional Cartesian orientation space wherein orientations in form of **Rodrigues vectors** (*see there*) are represented. The Rodrigues space has a number of advantages over Euler space, but cannot be applied to all crystal symmetries. Named after the French mathematician Olinde Rodrigues (1795–1851) and the British physicist Frederick C. Frank (1911–1998).

Rodrigues vector: Combination of the angle and axis of rotation into one three-dimensional vector. The direction of the vector represents the rotation axis, the length the tangent of the half rotation angle. The construct is closely related to **quaternions** (*see there*).

Rotation matrix: *See* **orientation matrix**.

Sample symmetry: A symmetry, visible in orientation space, that is caused by the symmetry of the process that created the sample (e.g. rolling). The sample symmetry is a statistical symmetry, where approximately equal amounts of orientations of symmetrical orientations prevail in a sample.

Scintillation counter: Counter to detect x-rays that is based on the light emission of a scintillating material when hit by x-rays and subsequent light amplification

Selected area electron diffraction, SAD: Characteristic point diffraction pattern in TEM generated by parallel illumination of a thin sample. The area of origin on the sample of the diffraction pattern is selected by an aperture in the first imaging plane. The spatial resolution of the technique is not better than 0.5 μm. The technique is mainly used to select imaging conditions but can also be used to determine crystal orientations.

Semiconductor counter: Counter to detect x-rays that is based on the possibility of x-rays to ionize a semiconductor crystal.

Series expansion method: Method to calculate the ODF from pole figure or discrete orientation data in Fourier space by fitting pole figures and ODF by a series expansion with spherical harmonic functions.

Skeleton line: Representation of the relative intensity maxima in the various ODF-sections in the Euler space as a function of the corresponding Euler angle.

Small-angle convergent beam electron diffraction, SCBED: Electron diffraction occurring under irradiation with a convergent electron beam that covers a small convergence angle. In these patterns no Kikuchi lines but only confined diffraction discs are visible, similar to spot diffraction patterns. The patterns are used (sometimes in combination with beam precession), for robust orientation determination with highest spatial and angular resolution.

Spallation source: Neutron source where the neutrons are produced by spallation, i.e. the interaction of a high energy proton beam with a target material. Interesting for time-of-flight (TOF) applications because of the possibility to pulse the neutron beam.

Specimen coordinate system: Cartesian coordinate system chosen as reference for crystal orientation description. The coordinate system should be selected such as to reflect a major texture-defining process applied to the sample.

Step scan: Method to scan the pole figure in distinct pole figure angles.

Stereographic projection: Projection of a reference sphere onto its equatorial plane such that equal angles between lines on the surface of the sphere retain their equality in the projection. The stereographic projection is the most frequently used basis of pole figures and inverse pole figures.

Structure factor: A measure for the intensity of (electron or x-ray) diffraction obtained from a single unit cell of a crystal. Mathematically it corresponds to the Fourier transform of the lattice potential of a single unit cell.

Symmetrically equivalent solutions (of an orientation): Equivalent representations of an orientation depending on the crystal system symmetry.

Synchrotron radiation: X-radiation generated in an electron synchrotron which is characterized by a high energy, high intensity, and low angular divergence and which can be used for local texture analysis by means of the Laue and Debye–Scherrer method.

Template matching: A technique to determine the orientation of a crystal by matching its diffraction pattern to simulated patterns, precalculated in certain angular steps for all possible crystal orientations.

Texture: Distribution of the crystallographic orientations of a given sample. The texture is usually normalized to a random texture distribution and then quantified as "multiples of a random distribution", m.r.d.

Texture fibers: Intensity distribution along certain predefined paths through the Euler space.

Time-of-flight (TOF) measurements: Method to separate several reflectors by determining the flight time of the neutrons which depends on their energy or wavelength (*see also* **energy dispersive diffractometry**). Requires a pulsed neutron source.

Transmission geometry: Geometrical setup to study diffraction effects on thin samples which are penetrated by the incident radiation.

Transmission Kikuchi diffraction (pattern), TKD: *See* **Kikuchi diffraction (pattern)**.

References

Adams, B.L. (1986), Description of the intercrystalline structure distribution in polycrystalline materials, *Metall. Trans.* **17A**, 2199–2207.

Adams, B.L., Wright, S.I., and Kunze, K. (1993), Orientation imaging: The emergence of a new microscopy, *Metall. Trans.* **24A**, 819–831.

Adams, B.L., Zhao, J., and Grimmer, H. (1990a), Discussion of the representation of intercrystalline misorientation in cubic materials, *Acta Cryst.* **A46**, 620–622.

Adams, B.L., Zhao, J., and O'Hara, D. (1990b), Analysis of interface damage heterogeneity in polycrystalline materials, *Acta Metall. Mater.* **38**, 953–966.

Afshar, M. and Zaefferer, S. (2015), A pseudo-3D approach based on electron backscatter diffraction and backscatter electron imaging to study the character of phase boundaries between Mg and long period stacking ordered phase in a Mg–2Y–Zn alloy, *Mater. Char.* **101**, 130–135.

Ageeva, O.A., Abart, R., Habler, G., Borutzky, B.Y., and Trubkin, N.V. (2012), Oriented feldspar-feldspathoid intergrowths in rocks of the Khibiny massif: Genetic implications, *Mineral. Petrol.* **106**, 1–17.

Agnew, S.R., Tomé, C.N., Brown, D.W., Holden, T.M., and Vogel, S.C. (2003), Study of slip mechanisms in a magnesium alloy by neutron diffraction and modeling, *Scr. Mater.* **48**, 1003–1008.

Alam, M.N., Blackman, M., and Pashley, D.W. (1954), High-angle Kikuchi patterns, *Proc. Roy. Soc. Lond.* **A221**, 224–242.

Alkan, S., Wu, Y., Ojha, A., and Sehitoglu, H. (2018), Transformation stress of shape memory alloy CuZnAl: Non-Schmid behavior, *Acta Mater.* **149**, 220–234.

Allain-Bonasso, N., Wagner, F., Berbenni, S., and Field, D.P. (2012), A study of the heterogeneity of plastic deformation in IF steel by EBSD, *Mater. Sci. Eng.* **A548**, 56–63.

Al-Samman, T. and Gottstein, G. (2008), Dynamic recrystallization during high temperature deformation of magnesium, *Mater. Sci. Eng.* **A490**, 411–420.

Altmann, S.L. (1986), *Rotations, Quaternions and Double Groups*, Clarendon Press, Oxford.

Anderson, A., Thompson, R.B., and Bolingbroke, R. (1994), Ultrasonic characterization of the texture of aluminum alloys, *Mater. Sci. Forum* **157–162**, 295–300.

An, D., Griffiths, T.A., Konijnenberg, P., Mandal, S., Wang, Z., and Zaefferer, S. (2018), Correlating the five parameter grain boundary character distribution and the intergranular corrosion behaviour of a stainless steel using 3D orientation microscopy based on mechanical polishing serial sectioning, *Acta Mater.* **156**, 297–309.

An, D. and Zaefferer, S. (2019), Formation mechanism of dislocation patterns under low cycle fatigue of a high-manganese austenitic TRIP steel with dominating planar slip mode, *Int. J. Plast.* **121**, 244–260.

Ånes, H.W., Hjelen, J., Sørensen, B.E., van Helvoort, A.T.J., and Marthinsen, K. (2020), Processing and indexing of electron backscatter patterns using open-source software, *IOP Conf. Ser.: Mater. Sci. Eng.* **891**, 012002.

Ånes, H.W. (2023), KikuchiPy 0.8.7 documentation: https://kikuchipy.org/en/stable/ or https://doi.org/10.5281/zenodo.3597646.

Antoniadis, A., Berruyer, J., and Filhol, A. (1990), Maximum-likelihood methods in powder diffraction refinements, *Acta Cryst.* **A46**, 692–711.

Appel, F. and Wagner, R. (1998), Microstructure and deformation of two-phase γ-titanium aluminides, *Mater. Sci. Eng.* **R22**, 187–268.

Apps, P.J., Bowen, J.R., and Prangnell, P.B. (2003), The effect of coarse second-phase particles on the rate of grain refinement during severe deformation processing, *Acta Mater.* **51**, 2811–2822.

Ashbee, K.H.G. and Sargent, J.P. (1990), Stereoscopic presentation of Rodrigues vector representation of the full three-dimensional disorientation of iron crystals by rolling, *Metall. Trans.* **21A**, 253–255.

Ayer, R., Mueller, R.R., and Neeraj, T. (2006), Electron backscattered diffraction study of cleavage fracture in pure iron, *Mater. Sci. Eng.* **A417**, 243–248.

Baba-Kishi, K.Z. (1991), A study of the breakdown of Friedel's law in electron backscatter Kikuchi diffraction patterns: Application to zincblende-type structures, *J. Appl. Cryst.* **24**, 38–47.

Baba-Kishi, K.Z. (2002), Electron backscatter Kikuchi diffraction in the scanning electron microscope for crystallographic analysis, *J. Mater. Sci.* **37**, 1715–1746.

Baba-Kishi, K.Z. and Dingley, D.J. (1989), Backscatter Kikuchi diffraction in the SEM for identification of crystallographic point groups, *Scanning* **11**, 305–312.

Bachmann, F., Hielscher, R., and Schaeben, H. (2010), Texture analysis with MTEX— free and open source software toolbox, *Solid State Phenom.* **160**, 63–68.

Bäckstrøm, S.P., Riekel, C., Abel, S., Lehr, H., and Wenk, H.-R. (1996), Microtexture analysis by synchrotron-radiation x-ray diffraction of nickel-iron alloys prepared by microelectroplating, *J. Appl. Cryst.* **29**, 118–124.

Backx, P., Barnett, M.R., and Kestens, L. (2004), Texture changes during uniaxial compression of Mg-3Al-1Zn, *Mater. Sci. Forum* **467–470**, 429–434.

Bacon, G.E. (1975), *Neutron Diffraction* (3rd edition), Clarendon Press, Oxford.

Bacroix, B., Chaveau, T., Gargano, P., and Pochettino, A.A. (1994), Some comments about texture analysis: Comparison between harmonic and vector methods, *Mater. Sci. Forum* **157–162**, 301–308.

Baker, J.L., Jimison, L.H., Mannsfeld, S., Volkman, S., Yin, S., Subramanian, V., Salleo, A., Alivisatos, A.P., and Toney, M.F. (2010), Quantification of thin film crystallographic orientation using x-ray diffraction with an area detector, *Langmuir* **26**, 9146–9151.

Ball, J., Mitteau, R., and Gottstein, G., (1992) Investigation of deformation induced disorder and re-ordering by subsequent annealing in Ni_3Al+B, in: *Ordering and Disordering in Alloys* (Ed. A.R. Yavari), Springer, Berlin, 138–145.

Balluffi, R.W., Brokman, A., and King, A.H. (1982), CSL/DSC lattice model for general crystal-crystal boundaries and their line defects, *Acta Metall.* **30**, 1453–1470.

Banerjee, D. and Williams, J.C. (2013), Perspectives on titanium science and technology, *Acta Mater.* **61**, 844–879.

Barker, L.J. (1950), Revealing the grain structure of common aluminum alloy metallographic specimens, *Trans. Am. Soc. Metals* **42**, 347–356.

Barnard, J.S., Eggeman, A.S., Sharp, J., White, T.A., and Midgley, P.A. (2010), Dislocation electron tomography and precession electron diffraction—minimising the effects of dynamical interactions in real and reciprocal space, *Phil. Mag.* **90**, 4711–4730.

Barnard, J.S., Johnstone, D.N., and Midgley, P.A. (2017), High-resolution scanning precession electron diffraction: Alignment and spatial resolution, *Ultramicroscopy* **174**, 79–88.

Barrett, C.S. and Massalski, T.B. (1980), *Structure of Metals: Crystallographic Methods, Principles and Data*, McGraw-Hill, New York.

Barnett, M.R., Nave, M.D., and Bettles, C.J. (2004), Deformation microstructures and textures of some cold rolled Mg alloys, *Mater. Sci. Eng.* **A386**, 205–211.

Basinger, J., Fullwood, D., Kacher, J., and Adams, B. (2011), Pattern center determination in electron backscatter diffraction microscopy, *Microsc. Microanal.* **17**, 330–340.

Bastos, A., Zaefferer, S., Raabe, D., and Schuh, C. (2006), Characterization of the microstructure and texture of nanostructured electrodeposited NiCo using electron backscatter diffraction (EBSD), *Acta Mater.* **54**, 2451–2462.

Bate, P.S., Knutsen, R.D., Brough, I., and Humphreys, F.J. (2005), The characterization of low-angle boundaries by EBSD, *J. Microsc.* **220**, 36–46.

Bate, P., Lundin, P., Lindh-Ulmgren, E., and Hutchinson, B. (2017), Application of laser-ultrasonics to texture measurements in metal processing, *Acta Mater.* **123**, 329–336.

Bauch, J., Brechbühl, J., Ullrich, H.J., Meinl, G., Lin, H., and Kebede, W. (1999), Innovative analysis of x-ray microdiffraction images on selected applications of the Kossel technique, *Cryst. Res. Technol.* **34**, 81–88.

Baudin, T., Paillard, P., Cruz, F., and Penelle, R. (1994), Microtexture determination in Fe-Si alloy sheets by etch pitting. Comparison with the electron back-scattering pattern technique, *J. Appl. Cryst.* **27**, 924–933.

Baudin, T. and Penelle, R. (1993), Determination of the total texture function from individual orientation measurements by electron backscattering pattern, *Metall. Trans.* **24A**, 2299–2311.

Bauer, G.S. (2010), Overview on spallation target design concepts and related materials issues, *J. Nucl. Mater.* **398**, 19–27.

Becker, R. (1991), Analysis of texture evolution in channel die compression—I. Effects of grain interaction, *Acta Metall. Mater.* **39**, 1211–1230.

Becker, R. and Panchanadeeswaran, S. (1989), Crystal rotations represented as Rodrigues vectors, *Textures Microstruct.* **10**, 167–194.

Bellier, S.P. and Doherty, R.D. (1977), The structure of deformed aluminium and its recrystallization—investigations with transmission Kossel diffraction, *Acta Metall.* **25**, 521–538.

Bennett, T.A., Kalu, P.N., and Rollett, A.D. (2007), On the character of host-island grain interfaces in Fe–1%Si alloy, *Scr. Mater.* **57**, 41–44.

Bessières, J., Heizmann, J., and Eberhardt, J. (1991), XRD quantitative phases analysis of textured materials by using a curved PSD, *Textures Microstruct.* **14–18**, 157–162.

Bevis, M. and Swindells, N. (1967), The determination of the orientation of micro-crystals using a back-reflection Kossel technique and an electron probe microanalyser, *Phys. Stat. Sol.* **20**, 197–212.

Bhadeshia, H.K.D.H. (1987), *Worked Examples in the Geometry of Crystals*, The Institute of Metals, London.

Bhadeshia, H.K.D.H. (2001), *Bainite in Steels: Transformation, Microstructure and Properties*, IOM Communications, London, 237–276.

Bieler, T.R., Eisenlohr, P., Roters, F., Kumar, D., Mason, D.E., Crimp, M.A., and Raabe, D. (2009), The role of heterogeneous deformation on damage nucleation at grain boundaries in single phase metals, *Int. J. Plast.* **25**, 1655–1683.

Birsan, M., Szpunar, J.A., Tun, Z., and Root, J.H. (1996), Magnetic texture determination using nonpolarized neutron diffraction, *Phys. Rev.* **53B**, 6412–6417.

Black, D.R., Burdette, H.E., Kuriyama, M., and Spal, R.D. (1991), Diffraction imaging of polycrystalline materials, *J. Mater. Res.* **6**, 1469–1476.

Bleck, W. and Bunge, H.-J. (1981), The recrystallization of AlMn 1 investigated with pole figures by number, *Acta Metall.* **29**, 1401–1412.

Blochwitz, C., Brechbühl, J., and Tirschler, W. (1996), Analysis of activated slip systems in fatigued nickel polycrystals using the EBSD-technique in the scanning electron microscope, *Mater. Sci. Eng.* **A210**, 42–47.

Boersch, H. (1937), Über Bänder bei Elektronenbeugung, *Z. Techn. Phys.* **18**, 574–578.

Böhlke, T. (2005), Application of the maximum entropy method in texture analysis, *Comp. Mater. Sci.* **32**, 276–283.

Bollmann, W. (1961), On the phase transformation of cobalt, *Acta Metall.* **9**, 972–975.

Bollmann, W. (1970), *Crystal Defects and Crystalline Interfaces*, Springer, Berlin.

Bolmaro, R.E. and Brokmeier, H.-G. (1999), X-ray and neutron diffraction measurement of textures of two-phase materials, in: *Proceedings of ICOTOM 12* (Ed. J.A. Szpunar), NRC Research Press, Ottawa, 211–216.

Bolmaro, R.E., Browning, R.V., Guerra, F.M., and Rollett, A.D. (1994), Texture development in Ag-Ni powder composites, *Mater. Sci. Eng.* **A175**, 113–124.

Booker, G.R., Shaw, A.M.B., Whelan, M.J., and Hirsch, P.B. (1967), Some comments on the interpretation of the 'Kikuchi-like reflection patterns' observed by scanning electron microscopy, *Phil. Mag.* **16**, 1185–1191.

Borbély, A., Maurice, C., and Driver, J.H. (2008), Rotation axis analysis of deformed crystals by X-rays and electrons, *J. Appl. Cryst.* **41**, 747–753.

Böttcher, A., Gerber, T., and Lücke, K. (1992), Determination of microtexture of initial stages of secondary recrystallisation in Fe-3wt-%Si by etch pitting, *Mater. Sci. Technol.* **8**, 16–22.

Bourgeois, D., Moy, J.P., Svensson, S.O., and Kvick, Å. (1994), The point-spread function of X-ray image-intensifiers/CCD-camera and imaging-plate systems in crystallography: Assessment and consequences for the dynamic range, *J. Appl. Cryst.* **27**, 868–877.

Bouscaud, D., Morawiec, A., Pesci, R., Berveiller, S., and Patoor, E. (2014), Strain resolution of scanning electron microscopy based Kossel microdiffraction, *J. Appl. Cryst.* **47**, 1699–1707.

Boutin, F.R. (2010), Industrial development of cubic texture in high-purity aluminium: Discussion of the research and development method, *Adv. Eng. Mater.* **12**, 1060–1067.

Bowen, D.K. and Wormington, M. (1992), Grazing incidence X-ray characterization of materials, *Adv. X-ray Anal.* **36**, 171–184.

Bowles, J.S. and Mackenzie, J.K. (1954), The crystallography of martensitic transformations I, *Acta Metall.* **2**, 129–137.

Bozzolo, N. and Bernacki, M. (2020), Viewpoint on the formation and evolution of annealing twins during thermomechanical processing of fcc metals and alloys, *Metall. Mater. Trans.* **51A**, 2665–2684.

Bozzolo, N., Dewobroto, N., Grosdidier, T., and Wagner, F. (2005), Texture evolution during grain growth in recrystallized commercially pure titanium, *Mater. Sci. Eng.* **A397**, 346–355.

Bozzolo, N., Jacomet, S., and Logé, R.E. (2012), Fast in-situ annealing stage coupled with EBSD: A suitable tool to observe quick recrystallization mechanisms, *Mater. Char.* **70**, 28–32.

Bragg, W.H. and Bragg, W.L. (1913), The reflection of X-rays by crystals, *Proc. Roy. Soc. Lond.* **A88**, 428–438.

Brandon, D.G. (1966), The structure of high-angle grain boundaries, *Acta Metall.* **14**, 1479–1484.

Brechbühl, J., Bauch, J., and Ullrich, H.-J. (1999), Measurements of residual stresses in micron regions by using synchrotron excited Kossel diffraction, *Cryst. Res. Technol.* **34**, 59–70.

Briant, C.L., MacDonald, E., Balliett, R.W., and Luong, T. (2000), Recrystallization textures in tantalum sheet and wire, *Int. J. Refr. Metals Hard Mater.* **18**, 1–8.

Britton, T.B. and Wilkinson, A.J. (2011), Measurement of residual elastic strain and lattice rotations with high resolution electron backscatter diffraction, *Ultramicroscopy* **111**, 1395–1404.

Britton, T.B. and Wilkinson, A.J. (2012), High resolution electron backscatter diffraction measurements of elastic strain variations in the presence of larger lattice rotations, *Ultramicroscopy* **114**, 82–95.

Brochu, M., Yokota, T., and Satoh, S. (1997), Analysis of grain colonies in type 430 ferritic stainless steels by electron back scattering diffraction (EBSD), *ISIJ Int.* **37**, 872–877.

Brockhouse, B.N. (1953), The initial magnetization of nickel under tension, *Can. J. Phys.* **31**, 339–355.

Brodesser, S., Chen, S., and Gottstein, G. (1991), Microstructure and microtexture development during high temperature low cycle fatigue, *Textures Microstruct.* **14–18**, 829–834.

Brodesser, S. and Gottstein, G. (1993), Microstructure and microtexture evolution during high temperature low cycle fatigue of nickel, *Textures Microstruct.* **20**, 179–193.

Brodusch, N., Zaghib, K., and Gauvin, R. (2015), Electron backscatter diffraction applied to lithium sheets prepared by broad ion beam milling, *Microsc. Res. Technique* **78**, 30–39.

Brokmeier, H.-G. (1989), Neutron diffraction texture analysis of multi-phase systems, *Textures Microstruct.* **10**, 325–346.

Brokmeier, H.-G. (1993), Texture analysis of a zinc layer on a steel substrate using neutron diffraction, *Textures Microstruct.* **21**, 71–78.

Brokmeier, H.-G. (1994), Experimental textures of Al-Pb, Al-Cu and Fe-Cu metal-metal composites, *Mater. Sci. Eng.* **A175**, 131–139.

Brokmeier, H.-G. (2006), Global crystallographic textures obtained by neutron and synchrotron radiation, *Physica B* **385–386**, 623–625.

Brokmeier, H.-G. and Ermrich, M. (1994), A new method for texture measurement based on an X-ray imaging plate system, *Mater. Sci. Forum* **157–162**, 119–124.

Brokmeier, H.-G., Gan, W.M., Randau, C., Völler, M., Rebelo-Kornmeier, J., and Hofmann, M. (2011), Texture analysis at neutron diffractometer STRESS-SPEC, *Nucl. Instr. Meth. Phys. Res.* **A 642**, 87–92.

Brokmeier, H.-G. and Ostwaldt, D. (1996), Influence of the sample shape on the global texture determined by neutron diffraction, in: *Proceedings of ICOTOM 11* (Eds. Z. Liang, L. Zou, and Y. Chu), International Academic Publishers, Beijing, 75–80.

Brough, I., Bate, P. S., and Humphreys, F.J. (2006), Optimising the angular resolution of EBSD, *Mater. Sci. Technol.* **22**, 1279–1287.

Brückner, G., Pospiech, J., Seidl, I., and Gottstein, G. (2001), Orientation correlation during diffusional $\alpha \rightarrow \gamma$ phase transformation in a ferritic low carbon steel, *Scr. Mater.* **44**, 2635–2640.

Budai, J.D., Liu, W., Tischler, J.Z., Pan, Z.W., Norton, D.P., Larson, B.C., Yang, W., and Ice, G.E. (2008), Polychromatic X-ray micro- and nanodiffraction for spatially-resolved structural studies, *Thin Solid Films* **516**, 8013–8021.

Buffière, J.Y., Cloetens, P., Ludwig, W., Maire, E., and Salvo, L. (2008), In situ X-ray tomography studies of microstructural evolution combined with 3D modeling, *MRS Bull.* **33**, 611–619.

Bunge, H.-J. (Ed.) (1988), *Directional Properties of Materials*, DGM Inf. Ges., Oberursel, Germany.

Bunge, H.-J. (1965), Zur Darstellung allgemeiner Texturen, *Z. Metallkd.* **56**, 872–874.

Bunge, H.-J. (1982), *Texture Analysis in Materials Science: Mathematical Methods*, Butterworths, London.

Bunge, H.-J. (1985), Textures in multiphase alloys, *Z. Metallkd.* **76**, 92–101.

Bunge, H.-J. (1987), Three-dimensional texture analysis, *Int. Mater. Rev.* **32**, 265–291.

Bunge, H.-J. (1989a), Advantages of neutron diffraction in texture analysis, *Textures Microstruct.* **10**, 265–307.

Bunge, H.-J. (1989b), Texture and magnetic properties, *Textures Microstruct.* **11**, 75–91.

Bunge, H.-J. (1993), On-line determination of texture-dependent materials properties, *J. Nondestructive Eval.* **12**, 3–11.

Bunge, H.-J. (2005), Exploration of the six-dimensional orientation-location space with hard X-rays, *Mater. Sci. Forum* **495–497**, 87–98.

Bunge, H.-J. and Esling, C. (1979), Determination of the odd part of the texture function, *J. Phys. Lett.* **40**, 627–628.

Bunge, H.-J. and Esling, C. (1981), Determination of the odd part of the texture function by anomalous scattering, *J. Appl. Cryst.* **14**, 253–255.

Bunge, H.-J. and Esling, C. (Eds.) (1982), *Quantitative Texture Analysis*, DGM Inf. Ges., Oberursel, Germany.

Bunge, H.-J. and Esling, C. (1985), The harmonic method, in: *Preferred Orientation in Deformed Metals and Rocks: An Introduction to Modern Texture Analysis* (Ed. H.-R. Wenk), Academic Press, Orlando, FL, 109–122.

Bunge, H.-J. and Klein, H. (1996), Determination of quantitative, high-resolution pole figures with the area detector, *Z. Metallkd.* **87**, 465–475.

Bunge, H.-J., Liu, Y.S., and Hanneforth, R. (1987), Anisotropic absorption of x-rays in polyphase materials, *Scripta Metall.* **21**, 1423–1427.

Bunge, H.-J. and Puch, K.H. (1984), Principles of texture goniometer measurements, *Z. Metallkd.* **75**, 124–132.

Bunge, H.-J. and Roberts, W.T. (1969), Orientation distribution, elastic and plastic anisotropy in stabilized steel sheet, *J. Appl. Cryst.* **2**, 116–128.

Bunge, H.-J. and Tobisch, J. (1968), Bestimmung der Walztextur des Kupfers mit Hilfe der Neutronenbeugung, *Z. Metallkd.* **59**, 471–475.

Bunge, H.-J., Wcislak, L., Klein, H., Garbe, U., and Schneider, J.R. (2003), Texture and microstructure imaging in six dimensions with high-energy synchrotron radiation, *J. Appl. Cryst.* **36**, 1240–1255.

Bunge, H.-J., Wenk, H.-R., and Pannetier, J. (1982), Neutron diffraction texture analysis using a 2θ-position sensitive detector, *Textures Microstruct.* **5**, 153–170.

Burnett, T.L., Kelley, R., Winiarski, B., Contreras, L., Daly, M., Gholinia, A., Burke, M.G., and Withers, P.J. (2016), Large volume serial section tomography by Xe Plasma FIB dual beam microscopy, *Ultramicroscopy* **161**, 119–129.

Burton, G.L., Wright, S., Stokes, A., Diercks, D.R., Clarke, A., and Gorman, B.P. (2020), Orientation mapping with Kikuchi patterns generated from a focused STEM probe and indexing with commercially available EDAX software, *Ultramicroscopy* **209**, 112882.

Caleyo, F., Baudin, T., Mathon, M., and Penelle, R. (2001), Comparison of several methods for the reproduction of the orientation distribution function from pole figures in medium to strong textured materials, *Eur. Phys. J. AP* **15**, 85–96.

Callahan, P.G. and De Graef, M. (2013), Dynamical electron backscatter diffraction patterns. Part I: Pattern simulations, *Microsc. Microanal.* **19**, 1255–1265.

Carter, C.B. and Williams, D.B. (Eds.) (2016), *Transmission Electron Microscopy: Diffraction, Imaging, and Spectrometry*, Springer, Berlin.

Cautaerts, N., Rauch, E.F., Jeong, J., Dehm, G., and Liebscher, C.H. (2021), Investigation of the orientation relationship between nano-sized G-phase precipitates and austenite with scanning nano-beam electron diffraction using a pixelated detector, *Scr. Mater.* **201**, 113930.

Cayron, C. (2007), ARPGE: A computer program to automatically reconstruct the parent grains from electron backscatter diffraction data, *J. Appl. Cryst.* **40**, 1183–1188.

Cayron, C. (2020), What EBSD and TKD tell us about the crystallography of the martensitic B2-B19′ transformation in NiTi shape memory alloys, *Crystals* **10**, 562.

Cayron, C., Artaud, B., and Briottet, L. (2006), Reconstruction of parent grains from EBSD data, *Mater. Char.* **57**, 386–401.

Cayron, C., Barcelo, F., and de Carlan, Y. (2010), The mechanisms of the fcc–bcc martensitic transformation revealed by pole figures, *Acta Mater.* **58**, 1395–1402.

Cepeda-Jiménez, C.M., Molina-Aldareguia, J.M., and Pérez-Prado, M.T. (2016), EBSD-assisted slip trace analysis during in situ SEM mechanical testing: Application to unravel grain size effects on plasticity of pure Mg polycrystals, *JOM* **68**, 116–126.

Chandrasekaran, D. and Nygårds, M. (2003), A study of the surface deformation behaviour at grain boundaries in an ultra-low-carbon steel, *Acta Mater.* **51**, 5375–5384.

Chateigner, D., Germi, P., and Pernet, M. (1994), X-ray texture analysis of thin films by the reflection method: Intermediate regime in defocusing corrections, *J. Appl. Cryst.* **27**, 278–282.

Chateigner, D., Lutterotti, L., and Morales, M. (2019), Quantitative texture analysis and combined analysis, in: *International Tables for Crystallography, Vol. H*, Chapter 5.3, Springer, Dordrecht, 555–580.

Chateigner, D., Ouladdiaf, B., and Léon, F. (2010), Magnetic quantitative texture analysis using isotropic thermal neutron beams, *Solid State Phenom.* **160**, 75–82.

Chauvy, C., Barberis, P., and Montheillet, F. (2006), Microstructure transformation during warm working of β-treated lamellar Zircaloy-4 within the upper α-range, *Mater. Sci. Eng.* **A431**, 59–67.

Cheng, J., Luo, J., and Yang, K. (2018), Aimsgb: An algorithm and open-source python library to generate periodic grain boundary structures, *Comp. Mater. Sci.* **155**, 92–103.

Chen, Y.J., Hjelen, J., and Roven, H.J. (2012), Application of EBSD technique to ultrafine grained and nanostructured materials processed by severe plastic deformation: Sample preparation, parameters optimization and analysis, *Trans. Nonferrous Met. Soc. China* **22**, 1801–1809.

Chen, Y.H., Park, S.U., Wei, D., Newstadt, G., Jackson, M.A., Simmons, J.P., De Graef, M., and Hero, A.O. (2015), A dictionary approach to electron backscatter diffraction indexing, *Microsc. Microanal.* **21**, 739–752.

Chiba, T., Miyamoto, G., and Furuhara, T. (2013), Comparison of variant selection between lenticular and lath martensite transformed from deformed austenite, *ISIJ Int.* **53**, 915–919.

Cho, J.-H. (2007), Determination of volume fraction of the texture components in the Rodrigues fundamental region, *Mater. Sci. Eng.* **A465**, 228–237.

Cho, J.-H., Rollett, A.D., and Oh, K.H. (2005), Determination of a mean orientation in electron backscatter diffraction measurements, *Metall. Mater. Trans.* **36A**, 3427–3438.

Chowdhury, S.G., Datta, S., Ravi Kumar, B., De, P.K., and Gosh, R.N. (2007), Randomization of texture during recrystallization of austenite in a cold rolled metastable austenitic stainless steel, *Mater. Sci. Eng.* **A443**, 114–119.

Christian, J.W. (1975), *The Theory of Transformations in Metals and Alloys. I. Equilibrium and General Kinetic Theory*, Elsevier, Amsterdam.

Christian, J.W. and Mahajan, S. (1995), Deformation twinning, *Prog. Mater. Sci.* **39**, 1–157.

Chun, Y.B., Semiatin, S.L., and Hwang, S.K. (2006), Monte Carlo modeling of microstructure evolution during the static recrystallization of cold-rolled, commercial-purity titanium, *Acta Mater.* **54**, 3673–3689.

Coates, D.G. (1967), Kikuchi-like reflection patterns obtained with the scanning electron microscope, *Phil. Mag.* **16**, 1179–1184.

Conrad, H., Brückel, T., Schäfer, W., and Voigt, J. (2008), POWTEX—the high-intensity time-of-flight diffractometer at FRM II for structure analysis of polycrystalline materials, *J. Appl. Cryst.* **41**, 836–845.

Cross, I., and Randle, V. (2003), Lowest angle solution versus low-index axis solution for misorientations, *Scr. Mater.* **48**, 1587–1591.

Cullity, B.D. (1978), *Elements of X-ray Diffraction* (2nd edition), Addison-Wesley, Reading, MA.

Daaland, O. and Nes, E. (1996), Origin of cube texture during hot rolling of commercial Al-Mn-Mg alloys, *Acta Mater.* **44**, 1389–1411.

Dąbrowski, W., Gryboś, P., Hottowy, P., Świentek, K., and Wiącek, P. (2003), Integrated readout of silicon strip detectors for position sensitive measurement of X-rays, *Nucl. Instr. Meth.* **A512**, 213–29.

Dahms, M. (1992a), Introduction of the phone-concept into pole figure inversion using iterative series expansion method, *Textures Microstruct.* **19**, 169–174.

Dahms, M. (1992b), The iterative series-expansion method for quantitative texture analysis. II. Applications, *J. Appl. Cryst.* **25**, 259–267.

Dahms, M., Beaven, P.A., Tobisch, J., Bermig, G., and Helming, K. (1994), Texture analysis of γ/α_2 titanium aluminide—comparison of the component and the positivity method, *Z. Metallkd.* **85**, 536–544.

Dahms, M. and Bunge, H.-J. (1986), Calculation of the normalization factor of incomplete pole figures by cubic extrapolation, *Textures Microstruct.* **6**, 167–179.

Dahms, M. and Bunge, H.-J. (1987), ODF calculation by series expansion from incompletely measured pole figures using the positivity condition. Part I—cubic crystal symmetry, *Textures Microstruct.* **7**, 171–185.

Dahms, M. and Bunge, H.-J. (1988), A positivity method for the determination of complete orientation distribution functions, *Textures Microstruct.* **10**, 21–35.

Dahms, M. and Bunge, H.-J. (1989), The iterative series-expansion method for quantitative texture analysis. I. General outline, *J. Appl. Cryst.* **22**, 439–447.

Därmann-Nowak, C., and Engl, B. (1991), Influence of hot-rolling conditions on texture development in deep-drawing steels, *Steel Res.* **62**, 576–579.

Davies, G.J., Kallend, J.S., and Morris, P.P. (1976), The quantitative prediction of transformation textures, *Acta Mater.* **24**, 159–172.

Davies, P.A., Novikov, M., Randle, V., and Bowen, P. (2002), Application of electron backscatter diffraction (EBSD) to fracture studies of ferritic steels, *J. Microsc.* **205**, 278–284.

Davies, R. and Randle, V. (1996), Secondary processing of electron backscatter data from an aluminum alloy, *Mater. Char.* **37**, 131–141.

Davies, R.K. and Randle, V. (1999), Application of crystal orientation mapping to local orientation perturbations, *Eur. Phys. J. AP* **7**, 25–32.

Davies, P.A., and Randle, V. (2001), Combined application of electron backscatter diffraction and stereo-photogrammetry in fractography studies, *J. Microsc.* **204**, 29–38.

Davis, C.L. (2009), Non-contact non-destructive measurement of texture using an electromagnetic acoustic transducer (EMAT) sensor, in: *Microstructure and Texture in Steels* (Eds. A. Haldar, S. Suwas, and D. Bhattacharjee), Springer, London, 285–305.

Davis, C.L., Mukhopadhyay, P., Strangwood, M., Potter, M., Dixon, S., and Morris, P.F. (2008a), Comparison between elastic modulus and ultrasonic velocity anisotropy with respect to rolling direction in steels, *Ironmaking Steelmaking* **35**, 359–366.

Davis, C.L., Strangwood, M., Potter, M., Dixon, S., and Morris, P.F. (2008b), Prediction of elastic modulus + anisotropy using x-ray and electron backscattered diffraction texture quantification and ultrasonic (electromagnetic acoustic transducer) measurements in aluminum sheets, *Metall. Mater. Trans.* **39A**, 679–687.

Davoudi, K.M. and Vlassak, J.J. (2018), Dislocation evolution during plastic deformation: Equations vs. discrete dislocation dynamics study, *J. Appl. Phys.* **123**, 085302.

Davut, K. and Zaefferer, S. (2010), Statistical reliability of phase fraction determination based on electron backscatter diffraction (EBSD) investigations on the example of an Al-TRIP steel, *Metall. Mater. Trans.* **41A**, 2187–2196.

Davut, K. and Zaefferer, S. (2012), Improving the reliability of EBSD-based texture analysis by a new large area mapping technique, *Mater. Sci. Forum* **702**, 566–569.

Day, A. (2008), Spherical EBSD, *J. Microsc.* **230**, 472–486.

Day, A. (1993), *Developments in the EBSP technique and their application to grain imaging*, PhD Thesis, University of Bristol, UK.

De Cooman, B.C., Estrin, Y., and Kim, S.K. (2018), Twinning-induced plasticity (TWIP) steels, *Acta Mater.* **142**, 283–362.

De Graef, M. (2020), A dictionary indexing approach for EBSD, *IOP Conf. Ser.: Mater. Sci. Eng.* **891**, 012009.

De Koning, M., Miller, R., Bulatov, V.V., and Abraham, F.F. (2002), Modelling grain-boundary resistance in intergranular dislocation slip transmission, *Phil. Mag.* **A82**, 2511–2527.

Deal, A. and Eades, A. (2005), Energy-dependence of an EBSD pattern, *Microsc. Microanal.* **11** (Suppl. 2), 524–525.

Debye, P. and Scherrer, P. (1916), Interferenzen an regellos orientierten Teilchen im Röntgenlicht, *Physikal. Zeitschrift* **17**, 277–283.

Debye, P. and Scherrer, P. (1917), Interferenzen an regellos orientierten Teilchen im Röntgenlicht, *Physikal. Zeitschrift* **18**, 291–301.

Decker, B.F., Asp, E.T., and Harper, D. (1948), Preferred orientation determination using a Geiger counter X-ray diffraction goniometer, *J. Appl. Phys.* **19**, 388–392.

Deininger, C., Necker, G., and Mayer, J. (1994) Determination of structure factors, lattice strains and accelerating voltage by energy filtered convergent beam electron diffraction *Ultramicroscopy* **54**, 15–30.

Del Valle, J.A., Pérez-Prado, M.T., and Ruano, O.A. (2003), Texture evolution during large-strain hot rolling of the Mg AZ61 alloy, *Mater. Sci. Eng.* **A355**, 68–78.

DeWitt, S., Rudraraju, S., Montiel, D., Andrews, W.B., and Thornton, K. (2020), PRISMS-PF: A general framework for phase-field modeling with a matrix-free finite element method, npj Comput. *Mater.* **6**, 29.

Di Gioacchino, F. and Quinta da Fonseca, J. (2013), Plastic strain mapping with sub-micron resolution using digital image correlation, *Exp. Mech.* **53**, 743–754.

Diehl, M., An, D., Shanthraj, P., Zaefferer, S., Roters, F., and Raabe, D. (2017), Crystal plasticity study on stress and strain partitioning in a measured 3D dual phase steel microstructure, *Phys. Mesomech.* **20**, 311–323.

Diehl, M. and Kühbach, M. (2020), Coupled experimental-computational analysis of primary static recrystallization in low carbon steel, *Modell. Simul. Mater. Sci. Eng.* **28**, 014001.

Dillamore, I.L. and Katoh, H. (1974), The mechanisms of recrystallization in cubic metals with particular reference to their orientation-dependence, *Metal Sci.* **8**, 73–83.

Dillamore, I.L. and Roberts, W.T. (1965), Preferred orientation in wrought and annealed metals, *Metall. Rev.* **10**, 271–380.

Dingley, D.J. (1978), Theory and application of Kossel x-ray diffraction in the scanning electron microscope, *Scanning* **1**, 79–99.

Dingley, D.J. (1981), A comparison of diffraction techniques for the SEM, *Scan. Electron Microscopy* No. IV, 273–286.

Dingley, D.J. (1984), Diffraction from sub-micron areas using electron backscattering in a scanning electron microscope, *Scanning Electron Microsc.* No. II, 569–575.

Dingley, D.J. (2000), The development of automated diffraction in scanning and transmission electron microscopy, in: *Electron Backscatter Diffraction in Materials Science* (Eds. A.J. Schwartz, M. Kumar, and B.L. Adams), Kluwer Academic/Plenum Publishers, New York, 1–18.

Dingley, D.J. (2004), Progressive steps in the development of electron backscatter diffraction and orientation imaging microscopy, *J. Microsc.* **213**, 214–224.

Dingley, D.J. (2006), Orientation imaging microscopy for the transmission electron microscope, *Microchim. Acta* **155**, 19–29.

Dingley, D.J. and Field, D.P. (1997), Electron backscatter diffraction and orientation imaging microscopy, *Mater. Sci. Technol.* **13**, 69–78.

Dingley, D.J., Longden, M., Weinbren, J., and Alderman, J. (1987), On-line analysis of electron backscatter diffraction patterns. I. Texture analysis of zone refined polysilicon, *Scanning Microsc.* **1**, 451–456.

Dingley, D.J. and Steeds, J.W. (1974), Application of the Kossel x-ray back reflection technique in the scanning electron microscope, in: *Quantitative Scanning Electron Microscopy* (Eds. D.B. Holt et al.), Academic Press, London, 487–516.

Dingley, D.J., Wilkinson, A.J., Meaden, G., and Karamched, P.S. (2010), Elastic strain tensor measurement using electron backscatter diffraction in the SEM, *J. Electron Microsc.* **59** (Suppl), S155–S163.

Diot, M., Fundenberger, J.J., Philippe, M.J., Wégria, J., and Esling, C. (1998), Texture gradient in rolled zinc sheets, *Scr. Mater.* **39**, 1623–1630.

Doege, E. and Dröder, K. (2001), Sheet metal forming of magnesium wrought alloys – formability and process technology, *J. Mater. Proc. Technol.* **115**, 14–19.

Doherty, R.D. (1997), Recrystallization and texture, *Prog. Mater. Sci.* **42**, 39–58.

Dons, A.L. and Nes, E. (1986), Nucleation of cube texture in aluminium, *Mater. Sci. Technol.* **2**, 8–18.

Dorner, D., Zaefferer, S., Lahn, L., and Raabe, D. (2006), Overview of microstructure and microtexture development in grain-oriented silicon steel, *J. Magn. Magn. Mater.* **304**, 183–186.

Doyle, P.A. and Turner, P.S. (1968), Relativistic Hartree-Fock X-ray and electron scattering factors, *Acta Cryst.* **A24**, 390–397.

Driver, J.H., Theyssier, M.-C., and Maurice, C. (1996), Electron backscattered diffraction microtexture studies on hot deformed aluminium crystals, *Mater. Sci. Technol.* **12**, 851–858.

Duggan, B.J., Lücke, K., Köhlhoff, G.D., and Lee, C.S. (1993), On the origin of cube texture in copper, *Acta Metall. Mater.* **41**, 1921–1927.

Dunst, D. and Mecking, H. (1996), Analysis of experimental and theoretical rolling textures of two-phase titanium alloys, *Z. Metallkd.* **87**, 498–507.

Džubinský, M., Husain, Z., and van Haaften, W.M. (2004), Comparison of recrystallisation kinetics determined by stress relaxation, double hit, optical metallography and EBSD approaches, *Mater. Char.* **52**, 93–102.

Eades, A., Deal, A., Bhattacharyya, A., and Hooghan, T. (2009), Energy filtering in EBSD, in: *Electron Backscatter Diffraction in Materials Science* (2nd edition) (Eds. A.J. Schwartz et al.), Springer, 53–63.

Echlin, M.P., Burnett, T.L., Polonsky, A.T., Pollock, T.M., and Withers, P.J. (2020), Serial sectioning in the SEM for three dimensional materials science, *Curr. Opin. Solid State Mater. Sci.* **24**, 100817.

Echlin, M.P., Straw, M., Randolph, S., Filevich, J., and Pollock, T.M. (2015), The TriBeam system: Femtosecond laser ablation in situ SEM, *Mater. Char.* **100**, 1–12.

Eggeman, A.S., White, T.A., and Midgley, P.A. (2010), Is precession electron diffraction kinematical? Part II: A practical method to determine the optimum precession angle, *Ultramicroscopy* **110**, 771–777.

Eichelkraut, H., Hirsch, J., and Lücke, K. (1984), Rolling and recrystallization textures in copper-germanium alloys, *Z. Metallkd.* **75**, 113–123.

El-Dasher, B.S., Adams, B.L., and Rollett, A.D. (2003), Viewpoint: Experimental recovery of geometrically necessary dislocation density in polycrystals, *Scr. Mater.* **48**, 141–145.

Elf, F., Schäfer, W., Höfler, S., and Will, G. (1990), Texture investigations at low temperatures by neutron diffraction shown on the example of TbAg, *Textures Microstruct.* **13**, 55–58.

Ellis, T., Nanni, L.F., Shrier, A., Weissmann, S., Padawer, G.E., and Hosokawa, N. (1964), Strain and precision lattice parameter measurements by the X-ray divergent beam method, *J. Appl. Phys.* **35**, 3364–3373.

Emren, F., Von Schlippenbach, U., and Lücke, K. (1986), Investigation of the development of the recrystallization textures in deep drawing steels by ODF analysis, *Acta Metall.* **34**, 2105–2117.

Engler, O. (1995), Influence of the initial grain size on the rolling and recrystallization textures in the alloy Al–1.8%Cu, *Textures Microstruct.* **23**, 61–86.

Engler, O. (1996), Nucleation and growth during recrystallisation of aluminium alloys investigated by local texture analysis, *Mater. Sci. Technol.* **12**, 859–872.

Engler, O. (1997), Simulation of the recrystallization textures of Al-alloys on the basis of nucleation and growth probability of the various textures components, *Textures Microstruct.* **28**, 197–209.

Engler, O. (1998a), On the influence of dispersoids on the particle stimulated nucleation of recrystallization in an Al-Fe-Si model alloy, *Mater. Sci. Forum* **273–275**, 483–488.

Engler, O. (1998b), A simulation of recrystallisation based on EBSD orientation microscopy data, in: *Proceedings of 19th Risø International Symposium* (Eds. J.V. Carstensen et al.), Risø Nat. Lab., Roskilde, Denmark, 253–260.

Engler, O. (1999), On the origin of the R-orientation in the recrystallization textures of aluminum alloys, *Metall. Mater. Trans.* **30A**, 1517 1527.

Engler, O. (2000), Deformation and texture of copper-manganese alloys, *Acta Mater.* **48**, 4827–4840.

Engler, O. (2001), An EBSD local texture study on the nucleation of recrystallization at shear bands in the alloy Al–3%Mg, *Scr. Mater.* **44**, 229–236.

Engler, O., Crumbach, M., and Li, S. (2005a), Alloy-dependent rolling texture simulation of aluminium alloys with a grain-interaction (GIA) model, *Acta Mater.* **53**, 2241–2257.

Engler, O., Escher, C., and Gottstein, G. (1993a), Investigation of the nucleation of recrystallization in an Al-Cu single crystal by means of EBSP, *Mater. Sci. Forum* **113**, 127–132.

Engler, O., Escher, C., and Gottstein, G. (1996a), Single grain orientation measurements applied to the formation and growth of recrystallization nuclei, *Textures Microstruct.* **26–27**, 337–359.

Engler, O., Gottstein, G., Pospiech, J., and Jura, J. (1994a), Statistics, evaluation and representation of single grain orientation measurements, *Mater. Sci. Forum* **157–162**, 259–274.

Engler, O., Heckelmann, I., Rickert, T., Hirsch, J., and Lücke, K. (1994b), Effect of pretreatment and texture on recovery and recrystallization in Al–4.5Mg-0.7Mn alloy, *Mater. Sci. Technol.* **10**, 771–781.

Engler, O. and Hirsch, J. (2007), Polycrystal-plasticity simulation of six and eight ears in deep-drawn aluminum cups, *Mater. Sci. Eng.* **A452–453**, 640–651.

Engler, O. and Hirsch, J. (2009), Control of recrystallisation texture and texture-related properties in industrial production of aluminium sheet, *Int. J. Mater. Res.* **100**, 564–575.

Engler, O., Hirsch, J., and Lücke, K. (1989), Texture development in Al–1.8%Cu depending on the precipitation state—I. Rolling textures, *Acta Metall.* **37**, 2743–2753.

Engler, O., Hirsch, J., and Lücke, K. (1995a), Texture development in Al–1.8%Cu depending on the precipitation state—II. Recrystallization textures, *Acta Metall. Mater.* **43**, 121–138.

Engler, O., Hirsch, J., and Lücke, K. (1995b), Texture development in α/β-Brass, Part II, Recrystallization textures, *Z. Metallkd.* **86**, 475–484.

Engler, O. and Huh, M.Y. (1999), Evolution of the cube texture in high purity aluminum capacitor foils by continuous recrystallization and subsequent grain growth, *Mater. Sci. Eng.* **A271**, 371–381.

Engler, O., Huh, M.Y., and Tomé, C.N. (2005b), Crystal-plasticity analysis of ridging in ferritic stainless steel sheets, *Metall. Mater. Trans.* **36A**, 3127–3139.

Engler, O., Jura, J., and Matthies, S. (1999), Influence of number and arrangement of single orientation measurements on their statistical relevance, in: *Proceedings of ICOTOM 12* (Ed. J.A. Szpunar), NRC Research Press, Ottawa, 68–73.

Engler, O. and Juul Jensen, D. (1994), Comparison of texture measurements on two phase α/β-brass obtained by X-ray and neutron diffraction, *Scr. Mater.* **30**, 25–30.

Engler, O. and Knarbakk, K. (2021), Temper rolling to control texture and earing in aluminium alloy AA 5050A, *J. Mater. Proc. Technol.* **288**, 116910.

Engler, O., Löchte, L., and Hirsch, J. (2007), Through-process simulation of texture and properties during the thermomechanical processing of aluminium sheets, *Acta Mater.* **55**, 5449–5463.

Engler, O., Palacios, J., Schäfer, W., Jansen, E., Lücke, K., and Will, G. (1993b), Comparison of results of texture measurements obtained by X-ray and by neutron diffraction, *Textures Microstruct.* **21**, 195–206.

Engler, O., Renoux, C., and Gottstein, G. (1992), Microstructure and microtexture investigations of shear band formation in homogeneous Al-alloys, in: *Proceedings of ICAA3* (Eds. L. Arnberg et al.), NTH, Trondheim, Vol. II, 203–208.

Engler, O., Sachot, E., Ehrström, J.C., Reeves, A., and Shahani, R. (1996b), Recrystallisation and texture in hot deformed aluminium alloy 7010 thick plates, *Mater. Sci. Technol.* **12**, 717–729.

Engler, O., Vatne, H.E., and Nes, E. (1996c), The roles of oriented nucleation and oriented growth on recrystallization textures in commercial purity aluminium, *Mater. Sci. Eng.* **A205**, 187–198.

Engler, O., Yang, P., Pospiech, J., Jura, J., and Gottstein, G. (1994c), Behavior of statistical texture parameters applied to single grain orientation measurements in recrystallized Al-Mn, *Mater. Sci. Forum* **157–162**, 933–938.

English, A.T. and Chin, G. (1965), On the variation of wire texture with stacking fault energy in f.c.c. metals and alloys. *Acta Metall.* **13**, 1013–1016.

Ermrich, M., Hahn, F., and Wölfel, E.R. (1997), Use of imaging plates in x-ray analysis, *Textures Microstruct.* **29**, 89–101.

Eschner, T. and Fundenberger, J.J. (1997), Application of anisotropic texture components, *Textures Microstruct.* **28**, 181–195.

Esling, C., Bechler-Ferry, E., and Bunge, H.-J. (1981), Numerical calculation of the odd part of the texture function, *J. Phys. Lett.* **42**, 141–144.

Euler, L. (1776), Formulae generales pro translatione quacunque corporum rigidorum, *Nov. Comm. Acad. Sci. Imp. Petrop.* **20**, 189–207.

Ewald, P.P. (1921), Die Berechnung optischer und elektrostatischer Gitterpotentiale, *Annalen der Physik* **369**, 253–287.

Feldmann, K. (1989), Texture investigations by neutron time-of-flight diffraction, *Textures Microstruct.* **10**, 309–323.

Feldmann, K., Betzl, M., Andreef, A., Hennig, K., Kleinstück, K., and Matz, W. (1980), Comparison of quantitative texture analysis results from time-of-flight and conventional neutron diffraction, *Texture Cryst. Sol.* **4**, 1–11.

Ferran, G.L., Doherty, R.D., and Cahn, R.W. (1971), The Kossel line determination of the orientation of new grains formed by recrystallization of aluminium, *Acta Metall.* **19**, 1019–1029.

Ferrié, E., Buffière, J.Y., and Ludwig, W. (2005), 3D characterisation of the nucleation of a short fatigue crack at a pore in a cast Al alloy using high resolution synchrotron microtomography, *Int. J. Fatigue* **27**, 1215–1220.

Ferrié, E., Buffière, J.Y., Ludwig, W., Gravouil, A., and Edwards, L. (2006), Fatigue crack propagation: In situ visualization using X-ray microtomography and 3D simulation using the extended finite element method, *Acta Mater.* **54**, 1111–1122.

Ferrière, L., Morrow, J.R., Amgaa, T., and Koeberl, C. (2009), Systematic study of universal-stage measurements of planar deformation features in shocked quartz: Implications for statistical significance and representation of results, *Meteor. Planet. Sci.* **44**, 925–940.

Field, D.P. (1995), On the asymmetric domain of cubic misorientations, *Scr. Metall. Mater.* **32**, 67–70.

Field, D.P. (1997), Recent advances in the application of orientation imaging, *Ultramicroscopy* **67**, 1–9.

Field, D.P., Bradford, L.T., Nowell, M.M., and Lillo, T.M. (2007), The role of annealing twins during recrystallization of Cu, *Acta Mater.* **55**, 4233–4241.

Field, D.P., Magid, K.R., Mastorakos, I.N., Florando, J.N., Lassila, D.H., and Morris Jr, J.W. (2010), Mesoscale strain measurement in deformed crystals: A comparison of X-ray microdiffraction with electron backscatter diffraction, *Phil. Mag.* **90**, 1451–1464.

Field, D.P., Merriman, C.C., Allain-Bonasso, N., and Wagner, F. (2012), Quantification of dislocation structure heterogeneity in deformed polycrystals by EBSD, *Modell. Simul. Mater. Sci. Eng.* **20**, 024007.

Field, D.P., Wright, S.I., and Dingley, D.J. (1996), Multi-phase texture analysis by orientation imaging microscopy, in: *Proceedings of ICOTOM 11* (Eds. Z. Liang, L. Zou, and Y. Chu), International Academic Publishers, Beijing, 94–99.

Fionova, L., Konokenko, O., Matveev, V., Priester, L., Lartigue, S., and Dupau, F. (1994), Heterogeneities of grain boundary arrangement in polycrystals, *Interface Sci.* **1**, 207–211.

Föll, H. (2019), Defects in Crystals; Hyperscript: https://www.tf.uni-kiel.de/matwis/amat/def_en/index.html, University of Kiel, Germany.

Fortunati, S., Abbruzzese, G., and Di Nunzio, P.E. (1992), An etch-pitting technique for statistical analysis of grain size distributions as a function of orientation in Fe-Si alloys, *Mater. Sci. Forum* **94–96**, 431–436.

Frank, F.C. (1988), Orientation mapping, *Metall. Trans.* **19A**, 403–408.

Friedel, G., (1913), Sur les symétries cristallines que peut révéler la diffraction des rayons Röntgen, *Comp. Rend.* **157**, 1533–1536.

Friedrich, W., Knipping, P., and Laue, M. (1912), Interferenzerscheinungen bei Röntgenstrahlen, in: *Sitzungsbericht der mathematisch-physikalischen Klasse der Königlich Bayerischen Akademie der Wissenschaften*, Verlag der Königlich Bayerischen Akademie der Wissenschaften, München, 303–322.

Fritzen, F., Böhlke, T., and Schnack, E. (2009), Periodic three-dimensional mesh generation for crystalline aggregates based on Voronoi tessellations, *Comput. Mech.* **43**, 701–713.

Frommert, M., Zobrist, C., Lahn, L., Böttcher, A., Raabe, D., and Zaefferer, S. (2008), Texture measurement of grain-oriented electrical steels after secondary recrystallization, *J. Magn. Magn. Mater.* **320**, e657–e660.

Frommeyer, G., Brüx, U., and Neumann, P. (2003), Supra-ductile and high-strength manganese-TRIP/TWIP steels for high energy absorption purposes, *ISIJ Int.* **43**, 438–446.

Fuentes, L. (1989), Anomalous scattering and null-domain ghost corrections for fibre textures, *Textures Microstruct.* **10**, 347–360.

Fujimoto, F., Komaki, K., and Takagi, S. (1972), Diffraction Patterns obtained by scanning electron microscope, *Z. f. Naturforschung* **27a**, 441–444.

Fultz, B. and Howe, J.M. (2012), *Transmission Electron Microscopy and Diffractometry of Materials*, Springer Science & Business Media, Berlin.

Fundenberger, J.J., Bouzy, E., Goran, D., Guyon, J., Yuan, H., and Morawiec, A. (2016), Orientation mapping by transmission-SEM with an on-axis detector, *Ultramicroscopy* **161**, 17–22.

Fundenberger, J.J., Morawiec, A., Bouzy, E., and Lecomte, J.S. (2003), Polycrystal orientation maps from TEM, *Ultramicroscopy* **96**, 127–137.

Gale, B. and Griffiths, D. (1960), Influence of instrumental aberrations on the Schultz technique for the measurement of pole figures, *Brit. J. Appl. Phys.* **11**, 96–102.

Gallagher, P.C.J. (1970), The influence of alloying, temperature, and related effects on the stacking fault energy, *Metall. Trans.* **1**, 2429–2461.

Garbacz, A., Ralph, B., and Kurzydlowski, K.J. (1995), On the possible correlation between grain size distribution and distribution of CSL boundaries in polycrystals, *Acta Metall. Mater.* **43**, 1541–1547.

Garbe, S., Poulsen, H.F., and Juul Jensen, D. (1996), Local texture analysis using high energy synchrotron radiation, in: *Proceedings of ICOTOM 11* (Eds. Z. Liang, L. Zou, and Y. Chu), International Academic Publishers, Beijing, 100–105.

Garzón, C.M. and Tschiptschin, A.P. (2006), EBSD texture analysis of a high temperature gas nitrided duplex stainless steel, *Mater. Sci. Eng.* **A441**, 230–238.

Gehrmann, R., Frommert, M., and Gottstein, G. (2005), Texture effects on plastic deformation of magnesium, *Mater. Sci. Eng.* **A395**, 338–349.

Germain, L., Gey, N., and Humbert, M. (2007), Reliability of reconstructed β-orientation maps in titanium alloys, *Ultramicroscopy* **107**, 1129–1135.

Gerth, D. and Schwarzer, R.A. (1993), Graphical representation of grain and hillock orientations in annealed Al–1% Si films, *Textures Microstruct.* **21**, 177–193.

Gertsman, V.Y. (2001), Geometrical theory of triple junctions of CSL boundaries, *Acta Cryst.* **A57**, 369–377.

Gertsman, V.Y. and Bruemmer, S.M. (2001), Study of grain boundary character along intergranular stress corrosion crack paths in austenitic alloys, *Acta Mater.* **49**, 1589–1598.

Gerward, L., Lehn, S., and Christiansen, G. (1976), Quantitative determination of preferred orientation by energy-dispersive x-ray diffraction, *Texture Cryst. Sol.* **2**, 95–111.

Gey, N., Bocher, P., Uta, E., Germain, L., and Humbert, M. (2012), Texture and microtexture variations in a near-α titanium forged disk of bimodal microstructure, *Acta Mater.* **60**, 2647–2655.

Gholinia, A., Curd, M.E., Bousser, E., Taylor, K., Hosman, T., Coyle, S., Shearer, M.H., Hunt, J., and Withers, P.J. (2020), Coupled broad ion beam-scanning electron microscopy (BIB-SEM) for polishing and three dimensional (3D) serial section tomography (SST), *Ultramicroscopy* **214**, 112989.

Ghosh, A., Singh, A., and Gurao, N.P. (2017), Effect of rolling mode and annealing temperature on microstructure and texture of commercially pure-titanium, *Mater. Char.* **125**, 83–93.

Gianola, D.S., Ben Britton, T., and Zaefferer, S. (2019), New techniques for imaging and identifying defects in electron microscopy, *MRS Bull.* **44**, 450–458.

Giessen, B.C. and Gordon, G.E. (1968), X-ray diffraction: New high-speed technique based on X-ray spectrography, *Science* **159**, 973–975.

Gil Sevillano, J., González, D., and Martínez-Esnaola, J.M. (2007), Heterogeneous deformation and internal stresses developed in bcc wires by axisymmetric elongation, *Mater. Sci. Forum* **550**, 75–84.

Gleiter, H. (1983), On the structure of grain boundaries in metals, in: *Atomistics of Fracture* (Eds. R.M. Latanision and J.R. Pickens), Springer, Berlin, 433–464.

Gobernado, P., Petrov, R.H., and Kestens, L.A.I. (2012), Recrystallized $\{3\ 1\ 1\}\langle 1\ 3\ 6\rangle$ orientation in ferrite steels, *Scr. Mater.* **66**, 623–626.

Godec, M., Zaefferer, S., Podgornik, B., Šinko, M., and Tchernychova, E. (2020), Quantitative multiscale correlative microstructure analysis of additive manufacturing of stainless steel 316L processed by selective laser melting, *Mater. Char.* **160**, 110074.

Goehner, R.P. (1975), Orientation of back reflection patterns: A semi-automated approach, *Adv. X-Ray Anal.* **19**, 725–734.

Goehner, R. P., Hatfield, W. T., Rao, P. (1976), An interactive program for the indexing and simulation of electron diffraction patterns, in: *Proceedings of 34th Annual EMSA Meeting* (Ed. G.W. Bailey), San Francisco Press, San Francisco, CA, 542–543.

Goodhew, P.J. (2020), *Specimen Preparation for Transmission Electron Microscopy of Materials*, Garland Science.

Goodhew, P.J., Humphreys, F.J., and Beanland, R. (2001), *Electron Microscopy and Analysis* (3rd edition), Taylor & Francis, London.

Gotthardt, R., Hoschek, G., Reimold, O., and Haeβner, F. (1972), Topographic arrangement of crystallite orientations in rolled copper, *Texture* **1**, 99–109.

Gottstein, G. (1986), Orientation determination in very small volumes using synchrotron radiation, *Scr. Metall.* **20**, 1791–1794.

Gottstein, G. (1988), Automatic microtexture determination with synchrotron radiation, in: *Proceedings of ICOTOM 8* (Eds. J.S. Kallend and G. Gottstein), TMS, Warrendale, PA, 195–202.

Gottstein, G. and Engler, O. (1993), Local texture measurements with the scanning electron microscope, *J. Phys. IV France* **3**, C7-2137–C7-2142.

Goukon, N., Yamada, T., and Kajihara, M. (2000), Boundary energies of $\Sigma 11$ [110] asymmetric tilt boundaries in Cu determined from the shape of boundary silica particles, *Acta Mater.* **48**, 2837–2842.

Gourgues, A.F. (2002), Electron backscatter diffraction and cracking, *Mater. Sci. Technol.* **18**, 119–133.

Gourgues-Lorenzon, A.-F. (2009), Application of electron backscatter diffraction to the study of phase transformations: present and possible future, *J. Microsc.* **233**, 460–473.

Grewen, J. (1973), Textures of hexagonal metals and alloys and their influence on industrial application, in: *Proc. 3ème Colloq. Europ. sur les Textures des Métaux et leurs Application Industrielle, Pont-à-Mousson* (Ed. R. Penelle), Société Française de Métallurgie, Nancy, France, 195–234.

Griffiths, T.A., Habler, G., Rhede, D., Wirth, R., Ram, F., and Abart, R. (2014), Localization of submicron inclusion re-equilibration at healed fractures in host garnet, *Contrib. Mineral. Petrol.* **168**, 1–21.

Grimmer, H., Bollmann, W., and Warrington, D.H. (1974), Coincidence-site lattices and complete pattern-shift in cubic crystals, *Acta Cryst.* **A30**, 197–207.

Groeber, M.A. and Jackson, M.A. (2014), DREAM.3D: A digital representation environment for the analysis of microstructure in 3D, *Integr. Mater. Manuf. Innov.* **3**, 56–72.

Groves, J.R., Yashar, P.C., Arendt, P.N., DePaula, R.F., Peterson, E.J., and Fitzsimmons, M.R. (2001), Ultra-thin bi-axially textured IBAD MgO template layers resolved by grazing incidence X-ray diffraction, *Physica C* **355**, 293–298.

Gu, X.-F., Furuhara, T., and Zhang, W.-Z. (2016), PTCLab: Free and open-source software for calculating phase transformation crystallography, *J. Appl. Cryst.* **49**, 1099–1106.

Guilmeau, E., Chateigner, D., Noudem, J., Funahashi, R., Horii, S., and Ouladdiaf, B. (2005), Rietveld texture analysis of complex oxides: Examples of polyphased Bi2223 superconducting and Co349 thermoelectric textured ceramics characterization using neutron and X-ray diffraction, *J. Appl. Cryst.* **38**, 199–210.

Gupta, V.K. and Agnew, S.R. (2009), Indexation and misorientation analysis of low-quality Laue diffraction patterns, *J. Appl. Cryst.* **42**, 116–124.

Gutierrez-Urrutia, I. (2019), Quantitative analysis of electron channeling contrast of dislocations, *Ultramicroscopy* **206**, 112826.

Guyon, J., Gey, N., Goran, D., Chalal, S., and Pérez-Willard, F. (2016), Advancing FIB assisted 3D EBSD using a static sample setup, *Ultramicroscopy* **161**, 161–167.

Habler, G. and Griffiths, T. (2017), Crystallographic orientation relationships, in: *Mineral Reaction Kinetics: Microstructures, Textures, Chemical and Isotopic Signatures* (Eds. W. Heinrich and R. Abart), *EMU Notes Mineral.* **16**, 541–586.

Haeßner, F., Jakubowski, U., and Wilkens, M. (1966), Topographische Anordnung von Kristallitorientierungen in gewalztem hochreinem Kupfer bestimmt mittels elektronenmikroskopischer Feinbereichsbeugung, *Acta Metall.* **14**, 454–456.

Haeßner, F., Pospiech, J., and Sztwiertnia, K. (1983), Spatial arrangement of orientations in rolled copper, *Mater. Sci. Eng.* **57**, 1–14.

Hahn, T. (Ed.) (2005), *International Tables for Crystallography Vol. A*, Springer, Dordrecht, The Netherlands.

Hamilton, W.R. (1844), On quaternions; or on a new system of imaginaries in algebra, *Phil. Mag.* **25**, 489–495.

Hammond, C. (2009), *The Basics of Crystallography and Diffraction* (3rd edition), Oxford University Press, Oxford.

Han, J.H., Jee, K.K., and Oh, K.H. (2003), Orientation rotation behavior during in situ tensile deformation of polycrystalline 1050 aluminum alloy, *Int. J. Mech. Sci.* **45**, 1613–1623.

Hansen, N. and Juul Jensen, D. (1994), Grain subdivision during deformation of polycrystalline aluminium, *Mater. Sci. Forum* **157–162**, 1211–1218.

Hansen, N., Leffers, T., and Kjems, J.K. (1981), Recrystallization kinetics in copper investigated by in situ texture measurements by neutron diffraction, *Acta Metall.* **29**, 1523–1533.

Hansen, J., Pospiech, J., and Lücke, K. (1978), *Tables for Texture Analysis of Cubic Crystals*, Springer, Berlin.

Hanson, A.J. (2006), *Visualizing Quaternions*, Morgan Kaufmann, San Francisco, CA.

Harland, C.J., Akhter, P., and Venables, J.A. (1981), Accurate microcrystallography at high spatial resolution using electron back-scattering patterns in a field emission gun scanning electron microscope, *J. Phys.* **E14**, 175–182.

Harris, G.B. (1952), Quantitative measurement of preferred orientation in rolled uranium bars, *Phil. Mag.* **43**, 113–123.

Hartig, C., Vogel, S.C., and Mecking, H. (2006), In-situ measurement of texture and elastic strains with HIPPO-CRATES, *Mater. Sci. Eng.* **A437**, 145–150.

Hatherly, M. and Hutchinson, W.B. (1979), *An Introduction to Textures in Metals* (Monograph no. 5), The Institute of Metals, London.

He, B.B. (2003), Introduction to two-dimensional x-ray diffraction, *Powder Diffraction* **18**, 71–85.

He, B.B. (2012), Recent advances in texture measurement using two-dimensional detector, *Mater. Sci. Forum* **702–703**, 507–510.

He, Y., Godet, S., and Jonas, J.J. (2005), Representation of misorientations in Rodrigues-Frank space: Application to the Bain, Kurdjumov-Sachs, Nishiyama-Wassermann and Pitsch orientation relationships in the Gibeon meteorite, *Acta Mater.* **53**, 1179–1190.

Heidelbach, F., Riekel, C., and Wenk, H.-R. (1999), Quantitative texture analysis of small domains with synchrotron radiation X-rays, *J. Appl. Cryst.* **32**, 841–849.

Heidelbach, F., Wenk, H.-R., Chen, S.S., Pospiech, J., and Wright, S.I. (1996), Orientation and misorientation characteristics of annealed, rolled and recrystallized copper, *Mater. Sci. Eng.* **A215**, 39–49.

Heidelbach, F., Wenk, H.-R., Muenchhausen, R.E., Foltyn, S., Nogar, N., and Rollett, A.D. (1992), Textures of laser ablated superconducting thin films of $YBa_2Cu_3O_{7-\delta}$ as a function of deposition temperature, *J. Mater. Res.* **7**, 549–557.

Hielscher, R., Bartel, F., & Britton, T. B. (2019). Gazing at crystal balls: Electron backscatter diffraction pattern analysis and cross correlation on the sphere. *Ultramicroscopy, 207*, 112836.

Heinicke, F., Brokmeier, H.-G., Dahms, M., Bunge, H.-J., Pannetier, J., Ritter, C., and De Wall, H. (1991), Textures of paragneisses from the KTB drilling site, NE Bavaria (FRG), *Textures Microstruct.* **14–18**, 403–408.

Heinz, A. and Neumann, P. (1991), Representation of orientation and disorientation data for cubic, hexagonal, tetragonal and orthorhombic crystals, *Acta Cryst.* **A47**, 780–789.

Heizmann, J.J. and Laruelle, C. (1986), Simultaneous measurement of several X-ray pole figures, *J. Appl. Cryst.* **19**, 467–472.

He, Y. and Jonas, J.J. (2007), Representation of orientation relationships in Rodrigues-Frank space for any two classes of lattice, *J. Appl. Cryst.* **40**, 559–569.

Helming, K. (1996), A nearly equal distant grid of orientations, in: *Proceedings of ICOTOM 11* (Eds. Z. Liang, L. Zou, and Y. Chu), International Academic Publishers, Beijing, 106–111.

Helming, K. (1998), Texture approximations by model components, *Mater. Sci. Forum* **273–275**, 125–132.

Helming, K. and Eschner, T. (1990), A new approach to texture analysis of multiphase materials using a texture component model, *Cryst. Res. Technol.* **25**, K203–K208.

Helming, K. and Prechwinkel, U. (2005), Texture analysis with area detectors, *Solid State Phenom.* **105**, 71–76.

Helming, K. and Schwarzer, R.A. (1994), Texture estimate from minimum ranges of SAD pole figures, *Mater. Sci. Forum* **157–162**, 1219–1224.

Helming, K., Schwarzer, R.A., Rauschenbach, B., Geier, S., Leiss, B., Wenk, H.-R., Ullemeier, K., and Heinitz, J. (1994), Texture estimates by means of components, *Z. Metallkd.* **85**, 545–553.

Helming, K., Voitus, W., and Walther, K. (1992), Progress in texture investigations at the pulsed reactor IBR-2, *Physica B* **180–181**, 1025–1028.

Henry, S., Gruen, G.-U., and Rappaz, M. (2004), Influence of convection on feathery grain formation in aluminum alloys, *Metall. Mater. Trans.* **35A**, 2495–2501.

Hielscher, R. and Schaeben, H. (2008), A novel pole figure inversion method: specification of the MTEX algorithm, *J. Appl. Cryst.* **41**, 1024–1037.

Hillert, M. (1962), The formation of pearlite, in: *Decomposition of Austenite by Diffusional Processes* (Eds. V.F. Zackay and H.I. Aaronson), Interscience, New York, 197–248.

Hilliard, J.E. (1972), Quantitative analysis of scanning electron micrographs, *J. Microsc.* **95**, 45–58.

Hirsch, J., Burmeister, G., Hoenen, L., and Lücke, K. (1986), Autex III—A high speed texture goniometer and data processing system, in: *Experimental Techniques of Texture Analysis* (Ed. H.-J. Bunge), DGM Inf. Ges., Oberursel, Germany, 63–71.

Hirsch, J. and Lücke, K. (1988), Mechanism of deformation and development of rolling textures in polycrystalline f.c.c. metals—I. Description of rolling texture development in homogeneous CuZn alloys, *Acta Metall.* **36**, 2863–2882.

Hirsch, J., Lücke, K. and Hatherly, M. (1988), Mechanism of deformation and development of rolling textures in polycrystalline f.c.c. metals—III. The influence of slip inhomogeneities and twinning, *Acta Metall.* **36**, 2905–2927.

Hjelen, J., Orsund, R., Hoel, E., Runde, P., Furu, T., and Nes, E. (1993), EBSP, progress in technique and applications, *Textures Microstruct.* **20**, 29–40.

Hjelen, J., Ørsund, R., and Nes, E. (1991), On the origin of recrystallization textures in aluminium, *Acta Metall. Mater.* **39**, 1377–1404.

Høier, R., Bentdal, J., Daaland, O., and Nes, E. (1994), A high resolution transmission electron diffraction method for on-line texture analysis, *Mater. Sci. Forum* **157–162**, 143–148.

Hölscher, M., Raabe, D., and Lücke, K. (1991), Rolling and recrystallization textures of bcc steels, *Steel Res.* **62**, 567–575.

Hölscher, M., Raabe, D., and Lücke, K. (1994), Relationship between rolling textures and shear textures in f.c.c. and b.c.c. metals, *Acta Metall. Mater.* **42**, 879–886.

Homer, E.R., Adams, B.L., and Fullwood, D.T. (2006), Recovery of the grain boundary character distribution through oblique double-sectioning, *Scr. Mater.* **54**, 1017–1021.

Homer, E.R., Patala, S., and Priedeman, J.L. (2015), Grain boundary plane orientation fundamental zones and structure-property relationships, *Sci. Reports* **5**, 1–13.

Homma, H. and Hutchinson, B. (2003), Orientation dependence of secondary recrystallisation in silicon-iron, *Acta Mater.* **51**, 3795–3805.

Hosford, W.F. (1964), Microstructural changes during deformation of [011] fiber-textured metals, *Trans. Metall. Soc. AIME* **230**, 12–15.

Hough, P.V.C., Methods and means for recognizing complex patterns, US Patent 3 069.654 December 18 (1962).

Houska, C.R., Averbach, B.L., and Cohen, M. (1960), The cobalt transformation, *Acta Metall.* **8**, 81–87.

Hu, H. (1986), Nucleation of recrystallization for cube texture formation in fcc materials, in: *Proceedings of 7th Risø International Symposium* (Eds. N. Hansen et al.), Risø National Laboratory, Roskilde, Denmark, 75–92.

Huang, X. and Juul Jensen, D. (2000), EBSD contra TEM characterization of a deformed aluminum single crystal, in: *Electron Backscatter Diffraction in Materials Science* (Eds. A.J. Schwartz, M. Kumar, and B.L. Adams), Kluwer Academic/Plenum Publishers, New York, 265–276.

Hughes, D.A., Lebensohn, R.A., Wenk, H.-R., and Kumar, A. (2000), Stacking fault energy and microstructure effects on torsion texture evolution, *Proc. Roy. Soc. Lond.* **A456**, 921–953.

Huh, M.Y. and Engler, O. (2001), Effect of intermediate annealing on texture, formability and ridging of 17%Cr ferritic stainless steel sheet, *Mater. Sci. Eng.* **A308**, 74–87.

Huh, M.Y., Lee, J.H., Park, S.H., Engler, O., and Raabe, D. (2005), Effect of through-thickness macro and micro-texture gradients on ridging of 17%Cr ferritic stainless steel sheet, *Steel Res. Int.* **76**, 797–806.

Huijser-Gerits, E.M.C. and Rieck, G.D. (1974), Defocusing effects in the reflexion technique for the determination of preferred orientation, *J. Appl. Cryst.* **7**, 286–290.

Humbert, M. (1986), Intensity corrections, resolving power and statistical relevance in pole figure measurements, in: *Experimental Techniques of Texture Analysis* (Ed. H.-J. Bunge), DGM Inf. Ges., Oberursel, Germany, 29–50.

Humbert, M. and Bergmann, H.W. (1980), A determination of the normalization factor and its application to the incomplete pole-figure method, *J. Appl. Cryst.* **13**, 500–504.

Humbert, M., Blaineau, P., Germain, L., and Gey, N. (2011), Refinement of orientation relations occurring in phase transformation based on considering only the orientations of the variants, *Scr. Mater.* **64**, 114–117.

Humbert, M., Dey, S.R., Bouzy, E., and Hazotte, A. (2005), Calculation of the parent hcp grain orientation from inherited variants in the hcp to t and in the hcp to cc phase transformations. Application to the α–γ transformation in TiAl-based alloys, *Mater. Sci. Forum* **495**, 1219–1224.

Humbert, M., Germain, L., Gey, N., and Boucard, E. (2015), Evaluation of the orientation relations from misorientation between inherited variants: Application to ausformed martensite, *Acta Mater.* **82**, 137–144.

Humbert, M., Gey, N., Muller, J. and Esling, C. (1996), Determination of a mean orientation from a cloud of orientations. Application to electron back-scattering pattern measurements, *J. Appl. Cryst.* **29**, 662–666.

Humphreys, F.J. (1977), The nucleation of recrystallization at second phase particles in deformed aluminium, *Acta Metall.* **25**, 1323–1344.

Humphreys, F.J. (1983), The determination of crystallographic textures from selected areas of a specimen by electron diffraction, *Textures Microstruct.* **6**, 45–62.

Humphreys, F.J. (1984), The determination of local texture by electron diffraction methods, in: *Proceedings of 5th Risø International Symposium* (Eds. N. Hessel Andersen et al.), Risø National Laboratory, Roskilde, Denmark, 35–50.

Humphreys, F.J. (2001), Review, grain and subgrain characterisation by electron backscatter diffraction, *J. Mater. Sci.* **36**, 3833–3854.

Humphreys, F.J., Bate, P.S., and Hurley, P.J. (2001), Orientation averaging of electron backscattered diffraction data, *J. Microsc.* **201**, 50–58.

Humphreys, F.J., Huang, Y., Brough, I., and Harris, C. (1999), Electron backscatter diffraction of grain and subgrain structures—resolution considerations, *J. Microsc.* **195**, 212–216.

Humphreys, F.J., Rohrer, G.S., and Rollett, A. (2017), *Recrystallization and Related Annealing Phenomena* (3rd edition), Elsevier, Amsterdam.

Hu, Y. and Randle, V. (2007), An electron backscatter diffraction analysis of misorientation distributions in titanium alloys, *Scr. Mater.* **56**, 1051–1054.

Hu, H., Sperry, P.R., and Beck, P.A. (1952), Rolling textures in face-centered cubic metals, *J. Metals* **4**, 76–81.

Hurley, P.J. and Humphreys, F.J. (2003), The application of EBSD to the study of substructural development in a cold rolled single-phase aluminium alloy, *Acta Mater.* **51**, 1087–1102.

Hutchinson, W.B. (1984), Development and control of annealing textures in low-carbon steels, *Int. Met. Rev.* **29**, 25–42.

Hutchinson, W.B. (1989), Recrystallisation textures in iron resulting from nucleation at grain boundaries, *Acta Metall.* **37**, 1047–1056.

Hutchinson, W.B., Duggan, B.J., and Hatherly, M. (1979), Development of deformation texture and microstructure in cold-rolled Cu-30Zn, *Metals Technol.* **6**, 398–403.

Hutchinson, W.B., Lindh, E., and Bate, P. (1999), On the determination of textures from discrete orientation measurements, in: *Proceedings of ICOTOM 12* (Ed. J.A. Szpunar), NRC Research Press, Ottawa, 34–39.

Hutchinson, B., Moss, B., Smith, A., Astill, A., Scruby, C., Engberg, G., and Björklund, J. (2002), Online characterisation of steel structures in hot strip mill using laser ultrasonic measurements, *Ironmaking Steelmaking* **29**, 77–80.

Hutchinson, W.B., Oscarsson, A., and Karlsson, A. (1989), Control of microstructure and earing behaviour in aluminium alloy AA 3004 hot bands, *Mater. Sci. Technol.* **5**, 1118–1127.

Hutchinson, W.B., Ryde, L., Bate, P.S., and Bacroix, B. (1996), On the description of misorientations and interpretation of recrystallisation textures, *Scr. Mater.* **35**, 579–582.

Hünsche, I., Oertel, C.G., Tamm, R., Skrotzki, W., and Knabl, W. (2004), Microstructure and texture development during recrystallization of rolled molybdenum sheets, *Mater. Sci. Forum* **467–470**, 495–500.

Ibe, G., Grzemba, B., Hannibal, W.D., and Scharf, G. (1984), On the origin of cube texture in thin tapes of high-purity aluminium, in: *Proceedings of ICOTOM 7* (Eds. C. M. Brakman, P. Jongenburger, and E.J. Mittemeijer), Netherlands Society for Material Science, Zwijndrecht, 503–508.

Ikuhara, Y. (2001), Grain boundary and interface structures in ceramics, *J. Ceram. Soc. Japan* **109**, S110–S120.

Imhof, J. (1977), Die Bestimmung einer Näherung für die Funktion der Orientierungsverteilung aus einer Polfigur, *Z. Metallkd.* **68**, 38–43.

Imhof, J. (1982), Determination of the orientation distribution function from one pole-figure, *Textures Microstruct.* **5**, 73–86.

Inagaki, H. (1994), Fundamental aspect of texture formation in low carbon steel, *ISIJ Int.* **34**, 313–321.

Inokuti, Y. and Doherty, R.D. (1977), Transmission Kossel study of the structure of cold rolled iron and its nucleation behaviour, *Texture Cryst. Sol.* **2**, 143–168.

Inokuti, Y., Maeda, C., and Ito, Y. (1987), Computer color mapping of configuration of Goss grains after an intermediate annealing in grain oriented silicon steel, *Trans. ISIJ* **27**, 139–144.

Inoue, H., Fukushima, S., and Inakazu, N. (1992), Transformation textures in Ti-15V-3Cr-3Sn-3Al alloy sheets, *Mater. Trans. JIM* **33**, 129–137.

Inoue, H. and Inakazu, N. (1988), Deformation and recrystallization textures in cold rolled titanium, in: *Proceedings of ICOTOM 8* (Eds. J.S. Kallend and G. Gottstein), TMS, Warrendale, PA, 997–1004.

Inoue, H. and Takasugi, T. (2001), Analysis of average orientation distribution in sheet materials with through-thickness texture gradient, *Z. Metallkd.* **92**, 82–88.

Ion, S.E., Humphreys, F.J., and White, S.H. (1982), Dynamic recrystallisation and the development of microstructure during the high temperature deformation of magnesium, *Acta Metall.* **30**, 1909–1919.

Isabell, T.C. and Dravid, V.P. (1997), Resolution and sensitivity of electron backscattered diffraction in a cold field emission gun SEM, *Ultramicroscopy* **67**, 59–68.

Jackson, M.A., Groeber, M.A., Uchic, M.D., Rowenhorst, D.J., and Graef, M.D. (2014), H5ebsd: An archival data format for electron back-scatter diffraction data sets, *Integr. Mater. Manuf. Innov.* **3**, 4.

Jaepel, T. (2015), *Feasibility Study on Local Elastic Strain Measurements with and EBSD Cross-Correlation Method in Elastic-Plastically deforming Materials*, PhD thesis, RWTH Aachen University, Germany.

Jansen, E., Schäfer, W., and Will, G. (1988), Profile fitting and the two-stage method in neutron powder diffractometry for structure and texture analysis, *J. Appl. Cryst.* **21**, 228–239.

Jansen, E., Schäfer, W., and Will, G. (1998), Pole figures from time-of-flight measurements using neutrons of a pulsed spallation source, *Textures Microstruct.* **30**, 145–153.

Jedrychowski, M., Bacroix, B., Salman, O.U., Tarasiuk, J., and Wronski, S. (2015), Investigation of SIBM driven recrystallization in alpha zirconium based on EBSD data and Monte Carlo modeling, *IOP Conf. Ser.: Mater. Sci. Eng.* **89**, 012029.

Jeyaraam, R., Vedantam, S., and Subramanya Sarma, V. (2019), On the significance of misorientation axes of CSL boundaries in triple junctions in cubic materials, *Mater. Char.* **152**, 276–281.

Johari, O. and Thomas, G. (1969), The stereographic projection and its applications, in: *Vol. IIA of Techniques of Metal Research* (Ed. R.F. Bunshah), Interscience, New York.

Jones, J.L., Vogel, S.C., Slamovich, E.B., and Bowman, K.J. (2004), Quantifying texture in ferroelectric bismuth titanate ceramics, *Scr. Mater.* **51**, 1123–1127.

Jones, L., Yang, H., Pennycook, T.J., Marshall, M.S., Van Aert, S., Browning, N.D., Castell, M.R., and Nellist, P.D. (2015), Smart Align—A new tool for robust non-rigid registration of scanning microscope data, *Adv. Struct. Chem. Imag.* **1**, 8.

Joy, D.C. and Newbury, D.E. (1998), Advanced SEM imaging, *AIP Conf. Proc.* **449**, 653–666.

Joy, D.C., Newbury, D.E., and Davidson, D.L. (1982), Electron channeling patterns in the scanning electron microscope, *J. Appl. Phys.* **53**, R81–R122.

Jura, J. and Pospiech, J. (1978), Determination of orientation distribution function from incomplete pole figures—an example of a computer program, *Texture Cryst. Sol.* **3**, 1–25.

Jura, J., Pospiech, J., and Gottstein, G. (1996), Estimation of the minimum number of single grain orientation measurements for ODF determination, *Z. Metallkd.* **87**, 476–480.

Juul Jensen, D. and Kjems, J.K. (1983), Apparatus for dynamical texture measurements by neutron diffraction using a position sensitive-detector, *Textures Microstruct.* **5**, 239–251.

Juul Jensen, D., Lauridsen, E.M., Margulies, L., Poulsen, H.F., Schmidt, S., Sørensen, H.O., and Vaughan, G.B.M. (2006), X-ray microscopy in four dimensions, *Mater. Today* **9**, 18–25.

Juul Jensen, D. and Leffers, T. (1989), Fast texture measurements using a position sensitive detector, *Textures Microstruct.* **10**, 361–373.

Juul Jensen, D. and Poulsen, H.F. (2012), The three dimensional X-ray diffraction technique, *Mater. Char.* **72**, 1–7.

Juul Jensen, D.J. and Zhang, Y.B. (2020), Impact of 3D/4D methods on the understanding of recrystallization, *Curr. Opin. Solid State Mater. Sci.* **24**, 100821.

Kacher, J., Ruggles, T., Key, J., Nowell, M., and Wright, S. (2022), Characterizing defect structures in AM steel using direct electron detection EBSD, *Scr. Mater.* **221**, 114952.

Kallend, J.S. (1998), Determination of the orientation distribution from pole figure data, in: *Texture and Anisotropy, Preferred Orientations in Polycrystals and their Effect on Materials Properties* (Eds. U.F. Kocks, C.N. Tomé, and H.-R. Wenk), Cambridge University Press, Cambridge, 102–124.

Kallend, J.S., Kocks, U.F., Rollett, A.D., and Wenk, H.-R. (1991a), Operational texture analysis, *Mater. Sci. Eng.* **A132**, 1–11.

Kallend, J.S., Morris, P.P., and Davies, G.J. (1976), Texture transformations—the misorientation distribution function, *Acta Metall.* **24**, 361–370.

Kallend, J.S., Schwarz, R.B., and Rollett, A.D. (1991b), Resolution of superimposed diffraction peaks in texture analysis of a $YBa_2Cu_3O_7$ polycrystal, *Textures Microstruct.* **13**, 189–197.

Kamaya, M. (2011), Assessment of local deformation using EBSD: Quantification of accuracy of measurement and definition of local gradient, *Ultramicroscopy* **111**, 1189–1199.

Kamaya, M., Sakakibara, Y., Yoda, R., Suzuki, S., Morita, H., Kobayashi, D., … and Ohtani, T. (2020), A round robin EBSD measurement for quantitative assessment of small plastic strain, *Mater. Char.* **170**, 110662.

Kamaya, M., Wilkinson, A.J., and Titchmarsh, J.M. (2006), Quantification of plastic strain of stainless steel and nickel alloy by electron backscatter diffraction, *Acta Mater.* **54**, 539–548.

Kameda, J., Inoguchi, R., Prior, D.J., and Kogure, T. (2007), Morphological analyses of minute crystals by using stereo-photogrammetric scanning electron microscopy and electron back-scattered diffraction, *J. Microsc.* **228**, 358–365.

Kaneno, Y., Takahashi, A., and Takasugi, T. (2006), Microstructure and texture evolution during cold rolling and annealing of Ni_3Fe alloy, *Mater. Sci. Eng.* **A431**, 328–338.

Kang, J.Y., Bacroix, B., Regle, H., Oh, K.H., and Lee, H.C. (2007), Effect of deformation mode and grain orientation on misorientation development in a body-centered cubic steel, *Acta Mater.* **55**, 4935–4946.

Karlsen, M., Hjelen, J., Grong, Ø., Rørvik, G., Chiron, R., Schubert, U., and Nilsen, E. (2008), SEM/EBSD based in situ studies of deformation induced phase transformations in supermartensitic stainless steels, *Mater. Sci. Technol.* **24**, 64–72.

Kato, N.I., Kohno, Y., and Saka, H. (1999), Side-wall damage in a transmission electron microscopy specimen of crystalline Si prepared by focused ion beam etching, *J. Vacuum Sci. Technol.* **A17**, 1201–1204.

Katrakova, D. and Mücklich, F. (2001), Specimen preparation for electron backscatter diffraction. Part I: Metals, *Pract. Metallogr.* **38**, 547–565.

Katrakova, D. and Mücklich, F. (2002), Specimen preparation for electron backscatter diffraction (EBSD). Part II: Ceramics. *Pract. Metallogr.* **39**, 644–662.

Kawashima, K., Hyoguchi, T., and Akagi, T. (1993), On-line measurement of plastic strain ratio of steel sheet using resonance mode EMAT, *J. Nondestructive Eval.* **12**, 71–77.

Keichel, J., Foct, J., and Gottstein, G. (2003), Deformation and annealing behavior of nitrogen alloyed duplex stainless steels. Part II: Annealing, *ISIJ Int.* **43**, 1788–1794.

Keller, R.R. and Geiss, R.H. (2011), Transmission EBSD from 10 nm domains in a scanning electron microscope, *J. Microsc.* **245**, 245–251.

Kelly, P.M. (2012), Crystallography of martensite transformation in steels, in: *Phase Transformation in Steels, Vol. 2: Diffusionless Transformations, High Strength Steels, Modelling and Advanced Analytical Technique* (Eds. E. Pereloma and D.V. Edmonds), Woodhead Publishing, Cambridge, 3–33.

Kelly, A.M., Bingert, S.R., and Reiswig, R.D. (1996), New metallographic preparation techniques for tantalum and tantalum alloys, *Microstructural Sci.* **23**, 185–195.

Kern, R., Lee, H.P., and Bunge, H.-J. (1986), The rolling texture of iron of different purities, *Steel Res.* **57**, 563–571.

Kestens, L., Verbeken, K., Decocker, R., Petrov, R., Gobernado, P., and Offerman, S.E. (2005), The use of Rodrigues-Frank space for representing discrete misorientation distributions, *Mater. Sci. Forum* **495–497**, 157–166.

Khorashadizadeh, A., Raabe, D., Zaefferer, S., Rohrer, G.S., Rollett, A.D., and Winning, M. (2011), Five-parameter grain boundary analysis by 3D EBSD of an ultra fine grained CuZr alloy processed by equal channel angular pressing, *Adv. Eng. Mater.* **13**, 237–244.

Kikuchi, S. (1928), Diffraction of cathode rays by mica, Parts I – IV, *Proc. Imp. Acad. Japan* **4**, 271–274, 275–278, 354–356, 471–474; summarized in *Japan. J. Phys.* **5**, 83–96.

Kile, D.E. (2009), The universal stage: The past, present, and future of a mineralogical research instrument, *Geochem. News* **140**, 1–7.

Kirch, D.M., Zhao, B., Molodov, D.A., and Gottstein, G. (2007), Faceting of low-angle ⟨100⟩ tilt grain boundaries in aluminum, *Scr. Mater.* **56**, 939–942.

Kitahara, H., Ueji, R., Ueda, M., Tsuji, N., and Minamino, Y. (2005), Crystallographic analysis of plate martensite in Fe–28.5 at.% Ni by FE-SEM/EBSD, *Mater. Char.* **54**, 378–386.

Kleber, W., Bautsch, H.J., and Bohm, J. (2002), *Einführung in die Kristallographie*, Oldenbourg Wissenschaftsverlag, München.

Klein, H., Heubeck, C., and Bunge, H.-J. (1994), Inhomogeneous textures in tantalum sheets, *Mater. Sci. Forum* **157–162**, 1423–1433.

Kleinstück, K. and Tobisch, J. (1968), Die Methode der Texturuntersuchung mittels Neutronenbeugung und ihre besonderen Anwendungsvorteile, *Kristall u. Technik* **3**, 455–466.

Kleinstück, K., Tobisch, J., Betzl, M., Mücklich, A., Schläfer, D., and Schläfer, U. (1976), Texturuntersuchungen von Metallen mittels Neutronenbeugung, *Kristall u. Technik* **11**, 409–429.

Klinkenberg, C., Schmitz, H.P., and Tamler, H. (1999), Texture measurement in steel industry practice, in: *Proceedings of ICOTOM 12* (Ed. J.A. Szpunar), NRC Research Press, Ottawa, 475–480.

Knorr, D.B., Weiland, H., and Szpunar, J.A. (1994), Applying texture analysis to materials engineering problems, *J. Metals* (Sept.), 32–36.

Koblischka-Veneva, A. and Koblischka, M.R. (2008), EBSD analysis of melt-textured YBCO with embedded Ag-2411 nanoparticles, *Mater. Sci. Eng.* **B151**, 65–68.

Kockelmann, W., Chapon, L.C., and Radaelli, P.G. (2006), Neutron texture analysis on GEM at ISIS, *Physica B* **385–386**, 639–643.

Kocks, U.F. (1988), A symmetric set of Euler angles and oblique orientation space sections, in: *Proceedings of ICOTOM 8* (Eds. J.S. Kallend and G. Gottstein), TMS, Warrendale, PA, 31–48.

Kocks, U.F. (1998a), Anisotropy and symmetry, in: *Texture and Anisotropy, Preferred Orientations in Polycrystals and their Effect on Materials Properties* (Eds. U.F. Kocks, C.N. Tomé, and H.-R. Wenk), Cambridge University Press, Cambridge, 10–43.

Kocks, U.F. (1998b), The representation of orientations and textures, in: *Texture and Anisotropy, Preferred Orientations in Polycrystals and their Effect on Materials Properties* (Eds. U.F. Kocks, C.N. Tomé, and H.-R. Wenk), Cambridge University Press, Cambridge, 44–101.

Kocks, U.F. and Mecking, H. (2003), Physics and phenomenology of strain hardening: The FCC case, *Prog. Mater. Sci.* **48**, 171–273.

Kocks, U.F. and Westlake, D. (1967), The importance of twinning for the ductility of CPH polycrystals, *Trans. AIME* **239**, 1107–1109.

Köhlhoff, G.D., Sun, X., and Lücke, K. (1988), The optical determination of crystal orientations, in: *Proceedings of ICOTOM 8* (Eds. J.S. Kallend and G. Gottstein), TMS, Warrendale, PA, 183–188.

Konijnenberg, P. (2024), *A Contribution to Tomography by Analytical Scanning Electron Microscopy*, PhD Thesis, RWTH Aachen University, Germany.

Konijnenberg, P.J., Zaefferer, S., Lee, S.B., Rollett, A.D., Rohrer, G.S., and Raabe, D. (2012), Advanced methods and tools for reconstruction and analysis of grain boundaries from 3D-EBSD data sets, *Mater. Sci. Forum* **702**, 475–478.

Konijnenberg, P.J., Zaefferer, S., and Raabe, D. (2015), Assessment of geometrically necessary dislocation levels derived by 3D EBSD, *Acta Mater.* 99, 402–414.

Kopinek, H.J. (1994), Inspection and control by on-line texture measurement, *Mater. Sci. Forum* **157–162**, 1929–1940.

Kossel, W., Loeck, V., and Voges, H. (1935), Die Richtungsverteilung der in einem Kristall entstandenen charakteristischen Röntgenstrahlung, *Z. Phys.* **94**, 139–144.

Kostron, H. (1950), Über die kristallographische Indizierung von Kornschnittflächen in Metallschliffen, *Z. Metallkd.* **41**, 370–377.

Kral, M.V., Mangan, M.A., Spanos, G., and Rosenberg, R.O. (2000), Three-dimensional analysis of microstructures, *Mater. Char.* **45**, 17–23.

Krämer, S. and Mayer, J. (1999), Using the Hough transform for HOLZ line identification in convergent beam electron diffraction, *J. Microsc.* **194**, 2–11.

Krieger-Lassen, N.C. (1994), *Automated determination of crystal orientations from electron backscattering patterns*, Ph.D. Thesis, Danmarks Tekniske Universitet, Lyngby.

Krieger Lassen, N.C. (1995), Computerized analysis of Kikuchi patterns, in: *Proceedings of 16th Risø Internatioal Symposium* (Eds. N. Hansen et al.), Risø National Laboratory, Roskilde, Denmark, 405–411.

Krieger Lassen, N.C. (1996), Automatic localisation of electron backscattering pattern bands from Hough transform, *Mater. Sci. Technol.* **12**, 837–843.

Krieger Lassen, N.C. (1998), Automatic high-precision measurements of the location and width of Kikuchi bands in electron backscatter diffraction patterns, *J. Microsc.* **190**, 375–391.

Krieger Lassen, N.C. and Bilde-Sørensen, J.B. (1993), Calibration of an electron back-scattering pattern set-up, *J. Microsc.* **170**, 125–129.

Krieger Lassen, N.C., Juul Jensen, D., and Conradsen, K. (1992), Image processing procedures for analysis of electron back scattering patterns, *Scan. Microsc.* **6**, 115–121.

Krieger Lassen, N.C., Juul Jensen, D., and Conradsen, K. (1994a), Automated recognition of deformed and recrystallized regions in partly recrystallized samples using electron back scattering patterns, *Mater. Sci. Forum* **157–162**, 149–158.

Krieger Lassen, N.C., Juul Jensen, D., and Conradsen, K. (1994b), On the statistical analysis of orientation data, *Acta Cryst.* **A50**, 741–748.

Kröger, S., Dahlem-Klein, E., Weiland, H., and Bunge, H.-J. (1988), Texture determination in aluminium alloys using colour metallography, *Textures Microstruct.* **10**, 41–48.

Kröner, E. (1959), Allgemeine Kontinuumstheorie der Versetzungen und Eigenspannungen, *Arch. Ration. Mech. Anal.* **4**, 273–334.

Kruger, S.E., Moreau, A., Militzer, M., and Biggs, T. (2003), In-situ, laser-ultrasonic monitoring of the recrystallization of aluminum alloys, *Mater. Sci. Forum* **426–432**, 483–488.

Kubota, M., Ushioda, K., Miyamoto, G., and Furuhara, T. (2016), Analysis of recrystallization behavior of hot-deformed austenite reconstructed from electron backscattering diffraction orientation maps of lath martensite, *Scr. Mater.* **112**, 92–95.

Lafond, C., Douillard, T., Cazottes, S., De Graef, M., Steyer, P., and Langlois, C. (2020), Towards large scale orientation mapping using the eCHORD method, *Ultramicroscopy* **208**, 112854.

Lafond, C., Douillard, T., Cazottes, S., Steyer, P., and Langlois, C. (2018), Electron CHanneling ORientation Determination (eCHORD): An original approach to crystalline orientation mapping, *Ultramicroscopy* **186**, 146–149.

Lafond, C., Douillard, T., Saad, H., Deville, S., Meille, S., Steyer, P., Cazottes, S., and Langlois, C. (2021), eCHORD orientation mapping of bio-inspired alumina down to 1 kV, *Materialia* **20**, 101207.

Lange F.F. (1967), Mathematical characterization of a general bicrystal, *Acta Metall.* **15**, 311–318.

Larsen, A.W. and Juul Jensen, D. (2003), Automatic determination of recrystallization parameters in metals by EBSP line scans, *Mater. Char.* **51**, 271–282.

Larson, B.C., Yang, W.G., Ice, G.E., Budai, J.D., and Tischler, J.Z. (2002), Three-dimensional x-ray structural microscopy with submicrometre resolution, *Nature* **415**, 887–890.

Lasalmonie, A. (2006), Intermetallics: Why is it so difficult to introduce them in gas turbine engines? *Intermetallics* **14**, 1123–1129.

Laue, M. (1912), Eine quantitative Prüfung der Theorie für die Interferenz-Erscheinungen bei Röntgenstrahlen, in: *Sitzungsbericht der mathematisch-physikalischen Klasse der Königlich Bayerischen Akademie der Wissenschaften*, Verlag der Königlich Bayerischen Akademie der Wissenschaften, München, 363–373.

Lauridsen, E.M., Schmidt, S., Suter, R.M., and Poulsen, H.F. (2001), Tracking: A method for structural characterization of grains in powders or polycrystals, *J. Appl. Cryst.* **34**, 744–750.

Lazik, S. and Esling, C. (1993), Textures in multiphase materials—a review, *Textures Microstruct.* **22**, 1–15.

Lebensohn, R.A., Brenner, R., Castelnau, O., and Rollett, A.D. (2008), Orientation image-based micromechanical modelling of subgrain texture evolution in polycrystalline copper, *Acta Mater.* **56**, 3914–3926.

Lee, K.T., de Wit, G., Morawiec, A., and Szpunar, J.A. (1995), The application of the etch-pit method to quantitative texture analysis, *J. Mater. Sci.* **30**, 1327–1332.

Lee, C.S. and Duggan, B.J. (1993), Deformation banding and copper-type rolling textures, *Acta Metall. Mater.* **41**, 2691–2699.

Lee, H.P., Esling, C., and Bunge, H.-J. (1988), Development of the rolling texture in titanium, *Textures Microstruct.* **7**, 317–337.

Lee, S.B., Rohrer, G.S., and Rollett, A.D. (2014), Three-dimensional digital approximations of grain boundary networks in polycrystals, *Modell. Simul. Mater. Sci. Eng.* **22**, 025017.

Leffers, T. and Bilde-Sørensen, W.B. (1990), Intra- and intergranular heterogeneities in the plastic deformation of brass during rolling, *Acta Metall. Mater.* **38**, 1917–1926.

Leffers, T. and Juul Jensen, D. (1986), Evaluation of the effect of initial texture on the development of deformation texture, *Textures Microstruct.* **6**, 231–254.

Leffers, T. and Ray, R.K. (2008), The brass-type texture and its deviation from the copper-type texture, *Prog. Mater. Sci.* **54**, 351–396.

Lehockey, E.M., Palumbo, G., and Lin, P. (1998), Improving the weldability and service performance of nickel-and iron-based superalloys by grain boundary engineering, *Metall. Mater. Trans.* **29A**, 3069–3079.

Lemaitre, J., Gervois, A., Troadec, J.P., Rivier, N., Ammi, M., Oger, L., and Bideau, D. (1993), Arrangement of cells in Voronoi tesselations of monosize packing of discs, *Phil. Mag.* **B67**, 347–362.

Lens, A., Maurice, C., and Driver, J.H. (2005), Grain boundary mobilities during recrystallization of Al-Mn alloys as measured by in situ annealing experiments, *Mater. Sci. Eng.* **A403**, 144–153.

Lenthe, W. C., Singh, S., & De Graef, M. (2019). A spherical harmonic transform approach to the indexing of electron back-scattered diffraction patterns. *Ultramicroscopy, 207*, 112841.

Lewis, A.C., Bingert, J.F., Rowenhorst, D.J., Gupta, A., Geltmacher, A.B., and Spanos, G. (2006), Two- and three-dimensional microstructural characterization of a super-austenitic stainless steel, *Mater. Sci. Eng.* **A418**, 11–18.

Li, J., Dillon, S.J., and Rohrer, G.S. (2009), Relative grain boundary area and energy distributions in nickel, *Acta Mater.* **57**, 4304–4311.

Liebmann, B., Lücke, K., and Masing, G. (1956), Orientation dependency of the rate of growth during primary recrystallization of A1 single crystals, *Z. Metallkd.* **47**, 57–63.

Lienert, U., Almer, J., Jakobsen, B., Pantleon, W., Poulsen, H.F., Hennessy, D., Xiao, C., and Suter, R.M. (2007), 3-dimensional characterization of polycrystalline bulk materials using high-energy synchrotron radiation, *Mater. Sci. Forum* **539–543**, 2353–2358.

Lienert, U., Poulsen, H.F., Honkimäki, V., Schulze, C., and Hignette, O. (1999), A focusing multilayer analyser for local diffraction studies, *J. Synchr. Rad.* **6**, 979–684.

Lin, H. and Pope, D.P. (1996), Formation of slip traces during grain growth of thin metal sheets, *Mater. Sci. Forum* **204–206**, 399–404.

Lindh-Ulmgren, E., Ericsson, M., Artymowicz, D., and Hutchinson, B. (2004), Laser-ultrasonics as a technique to study recrystallisation and grain growth, *Mater. Sci. Forum* **467–470**, 1353–1362.

Lischewski, I., Kirch, D.M., Ziemons, A., and Gottstein, G. (2008), Investigation of the α–γ–α phase transformation in steel: High-temperature in situ EBSD measurements, *Texture, Stress and Microstructure*, 294508.

Lisowski, P.W. and Schoenberg, K.F. (2006), The Los Alamos neutron science center, *Nucl. Instr. Meth.* **A562**, 910–914.

Liu, W., Bayerlein, M., Mughrabi, H., Day, A., and Quested, P.N. (1992), Crystallographic features of intergranular crack initiation in fatigued copper polycrystals, *Acta Metall. Mater.* **40**, 1763–1771.

Liu, J. and Chakrabarti, D.J. (1996), Grain structure and microtexture evolution during superplastic forming of a high strength Al-Zn-Mg-Cu alloy, *Acta Mater.* **44**, 4647–4661.

Lobanov, M.L., Danilov, S.V., Pastukhov, V.I., Averin, S.A., Khrunyk, Y.Y., and Popov, A.A. (2016), The crystallographic relationship of molybdenum textures after hot rolling and recrystallization, *Mater. Des.* **109**, 251–255.

Lonardelli, I., Wenk, H.-R., Lutterotti, L., and Goodwin, M. (2005), Texture analysis from synchrotron diffraction images with the Rietveld method: Dinosaur tendon and salmon scale, *J. Synchr. Rad.* **12**, 354–360.

Lücke, K. (1974), The orientation dependence of grain boundary motion and the formation of recrystallization textures, *Canad. Met. Quart.* **13**, 261–274.

Lücke, K., Därmann, C., and Hirsch, J. (1985), Rolling textures in FCC and BCC metals investigated by means of ODF's, *Trans. Ind. Inst. Met.* **38**, 496–509.

Lücke, K. and Engler, O. (1990), Effects of particles on the development of microstructure and texture during rolling and recrystallization in fcc alloys, *Mater. Sci. Technol.* **6**, 1113–1130.

Lücke, K., Pospiech, J., Jura, J., and Hirsch, J. (1986), On the presentation of orientation distribution-functions by model functions, *Z. Metallkd.* **77**, 312–321.

Lücke, K., Pospiech, J., Virnich, K.H., and Jura, J. (1981), On the problem of the reproduction of the true orientation distribution from pole figures, *Acta Metall.* **29**, 167–185.

Ludwig, W., Schmidt, S., Lauridsen, E.M., and Poulsen, H.F. (2008), X-ray diffraction contrast tomography: A novel technique for three-dimensional grain mapping of polycrystals. I. Direct beam case, *J. Appl. Cryst.* **41**, 302–309.

Lutterotti, L., Matthies, S., Wenk, H.-R., Schultz, A.J., and Richardson Jr. J.W. (1997), Combined texture and structure analysis of deformed limestone from time-of-flight neutron diffraction spectra, *J. Appl. Phys.* **81**, 594–600.

Luzin, V. and Brokmeier, H.-G. (2002), Attenuation corrections in neutron texture experiment, *Mater. Sci. Forum* **408–412**, 191–196.

Mackenzie, J.K. (1958), Second paper on statistics associated with the random disorientation of cubes, *Biometrika* **45**, 229–240.

Mackenzie, J.K. (1964), The distribution of rotation axes in a random aggregate of cubic crystals, *Acta Metall.* **12**, 223–225.

Mainprice, D., Bachmann, F., Hielscher, R., and Schaeben, H. (2015), Descriptive tools for the analysis of texture projects with large datasets using MTEX: Strength, symmetry and components, *Geol. Soc. Spec. Publ.* **409**, 251–271

Mainprice, D., Lloyd, G.E., and Casey, M. (1993), Individual orientation measurements in quartz polycrystals: Advantages and limitations for texture and petrophysical property determinations, *J. Struct. Geol.* **15**, 1169–1187.

Maki, T. (2012), Morphology and substructure of martensite in steels, in: *Phase Transformations in Steels, Vol. 2: Diffusionless Transformations, High Strength Steels, Modelling and Advanced Analytical Technique* (Eds. E. Pereloma and D.V. Edmonds), Woodhead Publishing, Cambridge, 34–58.

Makineni, S.K., Kumar, A., Lenz, M., Kontis, P., Meiners, T., Zenk, C., Zaefferer, S., Eggeler, G., Neumeier, S., Spiecker, E., Raabe, D., and Gault, B. (2018), On the diffusive phase transformation mechanism assisted by extended dislocations during creep of a single crystal CoNi-based superalloy, *Acta Mater.* **155**, 362–371.

Malis, T.F. and Steele, D. (1990), Ultramicrotomy for materials science, in: *MRS Symposium Proceedings 199, Workshop on Specimen Preparation for TEM Materials* (Ed. R. Anderson), MRS, Pittsburgh PA, 3–43.

Mamtani, M.A., Chakraborty, R., Biswas, S., Suryawanshi, A., Goswami, S., and Bhatt, S. (2020), SEM-EBSD analysis of broad ion beam polished rock thin sections—the MFAL protocol, *J. Geol. Soc. India* **95**, 337–342.

Man, C.-S. (2022), Crystallographic texture and group representations, *J. Elast.* **149**, 3–445.

Mandal, S., Pradeep, K.G., Zaefferer, S., and Raabe, D. (2014), A novel approach to measure grain boundary segregation in bulk polycrystalline materials in dependence of the boundaries' five rotational degrees of freedom, *Scr. Mater.* **81**, 16–19.

Man, C.-S., Fan, X., Morris, J.C., and Kawashima, K. (2002), Monitoring of primary recrystallization in aluminum alloy by resonance EMAT spectroscopy, *Mater. Sci. Forum* **408–412**, 875–880.

Mansfield, J. (1984), *Convergent Beam Electron Diffraction of Alloy Phases*, Adam Hilger Ltd., Bristol and Boston

Mao, W., Hirsch, J., and Lücke, K. (1988), Influence of the cube starting texture on rolling and recrystallization texture development, in: *Proceedings of ICOTOM 8* (Eds. J.S. Kallend and G. Gottstein), TMS, Warrendale, PA, 613–618.

Ma, X. and Pond, R.C. (2008), Defect modelling of martensitic interfaces in plate martensite, *Mater. Sci. Eng.* **A481**, 404–408.

Ma, A., Roters, F., and Raabe, D. (2006), On the consideration of interactions between dislocations and grain boundaries in crystal plasticity finite element modeling—Theory, experiments, and simulations, *Acta Mater.* **54**, 2181–2194.

Marthinsen, K. and Hoier, R. (1988), On the breakdown of Friedel's law in electron backscattering channeling patterns, *Acta Cryst.* **A44**, 700–707.

Masiello, F., Cembali, G., Chumakov, A.I., Connell, S.H., Ferrero, C., Härtwig, J., Sergeev, I., and Van Vaerenbergh, P. (2014), Rocking curve measurements revisited, *J. Appl. Cryst.* **47**, 1304–1314.

Mason, T.A. and Adams, B.L. (1994), The application of orientation imaging microscopy, *JOM* **46**, 43–45.

Matthies, S. (1979), On the reproducibility of the orientation distribution function of texture samples from pole figures (Ghost phenomena), *Phys. Stat. Sol. (b)* **92**, K135–K138.

Matthies, S. (1982), Form effects in the description of the orientation distribution function (ODF) of texturized materials by model components, *Phys. Stat. Sol. (b)* **112**, 705–716.

Matthies, S. (2002), 20 years WIMV, history, experience and contemporary developments, *Mater. Sci. Forum* **408–412**, 95–100.

Matthies, S., Helming, K., and Kunze, K. (1990), On the representation of orientation distributions in texture analysis by σ-sections, *Phys. Stat. Sol. (b)* **157**, 71–83, 489–507.

Matthies, S., Lutterotti, L., and Wenk, H.-R. (1997), Advances in texture analysis from diffraction spectra, *J. Appl. Cryst.* **30**, 31–42.

Matthies, S., Pehl, J., Wenk, H.-R., Lutterotti, L., and Vogel, S.C. (2005), Quantitative texture analysis with the HIPPO neutron TOF diffractometer, *J. Appl. Cryst.* **38**, 462–475.

Matthies, S. and Vinel, G.W. (1982), On the reproduction of the orientation distribution function of texturized samples from reduced pole figures using the conception of a conditional ghost correction, *Phys. Stat. Sol. (b)* **112**, K111–K114; K115–K120.

Matthies, S. and Vinel, G.W. (1994), On some methodical developments concerning calculations performed directly in the orientation space, *Mater. Sci. Forum* **157–162**, 1641–1646.

Matthies, S., Vinel, G.W., and Helming, K. (1987), *Standard Distributions in Texture Analysis*, Akademie-Verlag, Berlin.

Matthies, S. and Wagner, F. (1996), On a 1/n law in texture related single orientation analysis, *Phys. Stat. Sol. (b)* **196**, K11–K15

Matthies, S. and Wenk, H.-R. (1985), ODF reproduction with conditional ghost correction, in: *Preferred Orientation in Deformed Metals and Rocks: An Introduction to Modern Texture Analysis* (Ed. H.-R. Wenk), Academic Press, Orlando, FL, 139–147.

Matthies, S. and Wenk, H.-R. (1992), Optimization of texture measurements by pole figure coverage with hexagonal grids, *Phys. Stat. Sol. (a)* **133**, 253–257.

Matthies, S., Wenk, H.-R., and Vinel, G.W. (1988), Some basic concepts of texture analysis and comparison of three methods to calculate orientation distributions from pole figures, *J. Appl. Cryst.* **21**, 285–304.

Maurice, C. and Driver, J.H. (1997), Hot rolling textures of f.c.c. metals. Part 1. Experimental results on Al single and polycrystals, *Acta Mater.* 45, 4627–4638.

Maurice, C. and Fortunier, R. (2008), A 3D Hough transform for indexing EBSD and Kossel patterns, *J. Microsc.* 230, 520–529.

Maurice, C., Driver, J.H., and Fortunier, R. (2012), On solving the orientation gradient dependency of high angular resolution EBSD, *Ultramicroscopy* **113**, 171–181.

Maurice, C., Quey, R., Fortunier, R., and Driver, J.H. (2013), High angular resolution EBSD and its materials applications, in: *Microstructural Design of Advanced Engineering Materials* (Ed. D.A. Molodov), Wiley-VCH, 339–365.

McDonald, S., Reischig, P., Holzner, C., Lauridsen, E.M., Withers, P.J., Merkle, A.P., and Feser, M. (2015), Non-destructive mapping of grain orientations in 3D by laboratory X-ray microscopy, *Sci. Rep.* **5**, 14665.

McIntyre, G.J. (2015), Area detectors in single-crystal neutron diffraction, *Eur. Phys. J. AP* **48**, 504002.

McKie, C. and McKie, D. (1974), *Crystalline Solids*, Nelson, London.

Michael, J.R. and Goehner, R.P. (1994), Advances in backscattered-electron Kikuchi patterns for crystallographic phase identification, in: *Proceedings of Microscopy Society of America 52nd Annual Meeting* (Eds. G.W. Bailey and A.J. Garratt-Read), San Francisco Press, San Francisco, CA, 596–597.

Michael, J.R. and Kotula, P.G. (2008), Low Energy Ga+ and Ar+ ion milling for improved EBSD sample preparation, *Microsc. Microanal.* **14** (Suppl. 2), 976–977.

Mingard, K.P., Stewart, M., Gee, M.G., Vespucci, S., and Trager-Cowan, C. (2018), Practical application of direct electron detectors to EBSD mapping in 2D and 3D, *Ultramicroscopy* **184**, 242–251.

Mirpuri, K., Wendrock, H., Menzel, S., Wetzig, K., and Szpunar, J. (2006), Texture evolution in copper film at high temperature studied in situ by electron back-scatter diffraction, *Thin Solid Films* **496**, 703–717.

Mishin, O.V., Lauridsen, E.M., Krieger Lassen, N.C., Brückner, G., Tschentscher, T., Bay, B., Juul Jensen, D., and Poulsen, H.F. (2000), Application of high-energy synchrotron radiation for texture studies, *J. Appl. Cryst.* **33**, 364–371.

Miyamoto, G., Iwata, N., Takayama, N., and Furuhara, T. (2010), Mapping the parent austenite orientation reconstructed from the orientation of martensite by EBSD and its application to ausformed martensite, *Acta Mater.* **58**, 6393–6403.

Miyamoto, G., Takayama, N., and Furuhara, T. (2009), Accurate measurement of the orientation relationship of lath martensite and bainite by electron backscatter diffraction analysis, *Scr. Mater.* **60**, 1113–1116.

Miyazawa, K., Ishida, Y., and Mori, M. (1983), *Seisan-Kenkyu* **35**, 470–473 (in Japanese).

Molodov, D.A., Czubayko, U., Gottstein, G., and Shvindlerman, L.S. (1995), Mobility of <111> tilt grain boundaries in the vicinity of the special misorientation $\Sigma = 7$ in bicrystals of pure aluminium, *Scr. Metall. Mater.* **32**, 529–534.

Montesin, T. and Heizmann, J.J. (1992), Evolution of crystallographic texture in thin wires, *J. Appl. Cryst.* **25**, 665–673.

Morawiec, A. (1995), Misorientation angle distribution of randomly oriented symmetric objects, *J. Appl. Cryst.* **28**, 289–293.

Morawiec, A. (2004), *Orientations and Rotations: Computations in Crystallographic Textures*, Springer, Berlin, Germany.

Morawiec, A. (2009), Tilt and twist grain boundaries, *J. Appl. Cryst.* **42**, 308–311.

Morawiec, A. (2011), Low-Σ twist and tilt grain boundaries in cubic materials, *J. Appl. Cryst.* **44**, 1152–1156.

Morawiec, A. (2016), A program for refinement of lattice parameters and strain determination using Kossel diffraction patterns, *J. Appl. Cryst.* **49**, 322–329.

Morawiec, A. and Bieda, M. (2005), On algorithms for indexing of K-line diffraction patterns, *Arch. Metall. Mater.* **50**, 47–56.

Morawiec, A., Bouzy, E., Paul, H., and Fundenberger, J.J. (2014), Orientation precision of TEM-based orientation mapping techniques, *Ultramicroscopy* **136**, 107–118.

Morawiec, A. and Field, D.P. (1996), Rodrigues parameterization for orientation and misorientation distributions, *Phil. Mag.* **A73**, 1113–1130.

Morawiec, A. and Pospiech, J. (1989), Some information on quaternions useful in texture calculations, *Texture Microstruct.* **10**, 211–216.

Moreau, A., Lévesque, D., Lord, M., Dubois, M., Monchalin, J.-P., Padioleau, C., and Bussière, J.F. (2002), On-line measurement of texture, thickness and plastic strain ratio using laser-ultrasound resonance spectroscopy, *Ultrasonics* **40**, 1047–1056.

Morito, S., Huang, X., Furuhara, T., Maki, T., and Hansen, N. (2006), The morphology and crystallography of lath martensite in alloy steels, *Acta Mater.* **54**, 5323–5331.

Morito, S., Tanaka, H., Konishi, R., Furuhara, T., and Maki, A.T. (2003), The morphology and crystallography of lath martensite in Fe-C alloys, *Acta Mater.* **51**, 1789–1799.

Morris, P.R. (1959), Reducing the effects of nonuniform pole distribution in inverse pole figure studies, *J. Appl. Phys.* **30**, 595–596.

Morris, P.R. (1975), Crystallite orientation analysis from incomplete pole figures, *Adv. X-ray Anal.* **18**, 514–534.

Morris, P.R. and Heckler, A.J. (1968), Crystallite orientation analysis for rolled cubic materials, *Adv. X-Ray Anal.* **11**, 454–472.

Morris, P.R. and Hook, R.E. (1992), Comparison of incomplete pole-figure methods for surfaces perpendicular to rolling, transverse and normal directions, *Textures Microstruct.* **19**, 75–80.

Müller, S., Müller, K., Rosumek, M., and Reimers, W. (2006), Microstructure development of differently extruded Mg alloys Part II, *Aluminium* **82**, 438–441.

Murr, L.E., (2016), Dislocation ledge sources: Dispelling the myth of Frank-Read source importance, *Metall. Mater. Trans.* **47A**, 5811–5826.

Mykura, H. (1980), Checklist of cubic coincidence site lattice relations, in: *Grain Boundary Structure and Kinetics*, ASM, Metals Park, Novelty, OH, 445–455.

Nakamichi, H., Humphreys, F.J., and Brough, I. (2008), Recrystallization phenomena in an IF steel observed by in situ EBSD experiments, *J. Microsc.* **230**, 464–471.

Nandy, S., Tsai, S.P., Stephenson, L., Raabe, D., and Zaefferer, S. (2021), The role of Ca, Al and Zn on room temperature ductility and grain boundary cohesion of magnesium, *J. Magnes. Alloy.* **9**, 1521–1536.

Nauer-Gerhardt, C.U. and Bunge, H.-J. (1986), Orientation determination by optical methods, in: *Experimental Techniques of Texture Analysis* (Ed. H.-J. Bunge), DGM Inf. Ges., Oberursel, Germany, 125–142.

Neumann, P. (1991), Graphical representations of orientations and ODFs by Rodrigues vectors, *Steel Res.* **62**, 560–566.

Nielsen, S.F., Wolf, A., Poulsen, H.F., Ohler, M., Lienert, U., and Owen, R.A. (2000), A conical slit for three-dimensional XRD mapping, *J. Synchr. Rad.* **7**, 103–109.

Nishikawa, S. and Kikuchi, S. (1928), The diffraction of cathode rays by calcite, *Proc. Imp. Acad. Japan* **4**, 475–477 and *Nature* **121**, 1019–1020.

Nishiyama, Z. (2012), *Martensitic Transformation*, Elsevier, Amsterdam.

Nolze, G. (2006), Improved determination of fcc/bcc orientation relationships by use of high-indexed pole figures, *Cryst. Res. Technol.* **41**, 72–77.

Nolze, G. (2007), Image distortions in SEM and their influences on EBSD measurements, *Ultramicroscopy* **107**, 172–183.

Nolze, G., Hielscher, R., and Winkelmann, A. (2017), Electron backscatter diffraction beyond the mainstream, *Cryst. Res. Technol.* **52**, 1600252.

Nolze, G. and Winkelmann, A. (2012), Dynamical simulation of electron backscatter diffraction (EBSD) patterns of imperfect crystals, *Microsc. Microanal.* **18** (Suppl. 2), 1246–1247.

Nolze, G., Winkelmann, A., and Boyle, A.P. (2016), Pattern matching approach to pseudosymmetry problems in electron backscatter diffraction, *Ultramicroscopy* **160**, 146–154.

Norton, J.T. (1948), A technique for quantitative determination of texture of sheet metals, *J. Appl. Phys.* **19**, 1176–1178.

Nowakowski, P., Schlenker, J., Ray, M.L., and Fischione, P. (2016), Sample preparation using broad argon ion beam milling for electron backscatter diffraction (EBSD) analysis, *Microsc. Microanal.* **22** (Suppl. 3), 12–13.

Nowell, M.M. (2022), Cutting-edge EBSD detector technology, Microsc. *Anal.* **63**, 14

Nowell, M.M., Witt, R.A., and True, B. (2005), EBSD sample preparation: Techniques, tips, and tricks, *Microsc. Today* (July), 44–48.

Nowell, M.M. and Wright, S.I. (2004), Phase differentiation via combined EBSD and XEDS, *J. Microsc.* **213**, 296–305.

Nye, J.F. (1953), Some geometrical relations in dislocated crystals, *Acta Metall.* **1**, 153–162.

Obasi, G.C., Moat, R.J., Leo Prakash, D.G., Kockelmann, W., Quinta da Fonseca, J., and Preuss M. (2012), In situ neutron diffraction study of texture evolution and variant selection during the $\alpha \rightarrow \beta \rightarrow \alpha$ phase transformation in Ti-6Al-4V, *Acta Mater.* **60**, 7169–7182.

Obst, B., Nast, R., Schlachter, S., de Boer, B., Holzapfel, B., Nemetschek, R., and Prusseit, W. (2003), Application of electron backscatter diffraction in the SEM to textural problems of coated high-temperature superconductors, *Z. Metallkd.* **94**, 580–586.

Ogata, Y., Tsuda, K., and Tanaka, M. (2008), Determination of the electrostatic potential and electron density of silicon using convergent-beam electron diffraction, *Acta Cryst.* **A64**, 587–597.

Ojima, M., Adachi, Y., Suzuki, S., and Tomota, Y. (2011), Stress partitioning behavior in an fcc alloy evaluated by the in situ/ex situ EBSD-Wilkinson method, *Acta Mater.* **59**, 4177–4185.

Omoto, K., Tsuda, K., and Tanaka, M. (2002), Simulation of Kikuchi patterns due to thermal diffuse scattering on MgO crystals, *J. Electron. Mic.* 51, 67–78.

Onda, T., Iwagaki, S., Morito, S., and Hayakawa, M. (2010), Shape-strain analysis of martensite in ceria-doped zirconia, *Mater. Trans.* **51**, 899–905.

Ortiz, M. and Hermida, J.D. (1981), A method for obtaining the complete pole figure with a single sample using the Schulz technique, *Texture Cryst. Sol.* **4**, 159–169.

Own, C.S.Y. (2005), *System Design and Verification of the Precession Electron Diffraction Technique*, Dissertation, Northwestern University, Evanston, IL.

Palumbo, G. and Aust, K.T. (1992), Special properties of Σ grain boundaries, in: *Materials Interfaces: Atomic Level Structure and Properties* (Eds. D. Wolf and S. Yip), Chapman & Hall, London, 190–211.

Pande, C.S., Rath, B.B., and Imam, M.A. (2004), Effect of annealing twins on Hall-Petch relation in polycrystalline materials, *Mater. Sci. Eng.* **A367**, 171–175.

Pang, E.L., Larsen, P.M., and Schuh, C.A. (2020), Global optimization for accurate determination of EBSD pattern centers, *Ultramicroscopy* **209**, 112876.

Pang, Y. and Wynblatt, P. (2005), Correlation between grain-boundary segregation and grain-boundary plane orientation in Nb-doped TiO_2, *J. Am. Ceram. Soc.* **88**, 2286–2291.

Pantleon, W., Poulsen, H.F., Almer, J., and Lienert, U. (2004), In situ X-ray peak shape analysis of embedded individual grains during plastic deformation of metals, *Mater. Sci. Eng.* **A387–389**, 339–342.

Park, N.J. and Bunge, H.-J. (1994), Texture transformation due to martensitic phase transformation in CuZnAl shape memory alloys, *Mater. Sci. Forum* **157–162**, 563–570.

Park, S.H., Kim, K.Y., Lee, Y.D., and Park, C.G. (2002), Evolution of microstructure and texture associated with ridging in ferritic stainless steels, *ISIJ Int.* **42**, 100–105.

Partridge, P.G. (1969), The crystallography and deformation modes of hexagonal close-packed metals, *Metall. Rev.* **118**, 169–194.

Patala, S. and Schuh, C.A. (2011), A continuous and one-to-one coloring scheme for grain boundary misorientations, *Acta Mater.* **59**, 554–562.

Paul, H., Driver, J.H., Maurice, C., and Piątkowski, A. (2007), The role of shear banding on deformation texture in low stacking fault energy metals as characterized on model Ag crystals, *Acta Mater.* **55**, 575–588.

Pawlik, K. (1986), Determination of the orientation distribution function from pole figures in arbitrarily defined cells, *Phys. Stat. Sol (b)* **134**, 477–483.

Pawlik, K. (1994), The ODF calculation from pole figures for different types of crystal and sample symmetries, *Mater. Sci. Forum* **157–162**, 401–406.

Pawlik, K., Pospiech, J., and Lücke, K. (1991), The ODF approximation from pole figures with the aid of the ADC method, *Textures Microstruct.* **14–18**,) 25–30.

Pereloma, E. and Edmonds, D.V. (Eds.) (2012), *Phase Transformation in Steels, Vol. 2: Diffusionless Transformations, High Strength Steels, Modelling and Advanced Analytical Technique*, Woodhead Publishing, Cambridge.

Perlwitz, H.J., Lücke, K., and Pitsch, W. (1969), Determination of the orientation distribution of the crystallites in rolled copper and brass by electron microscopy, *Acta Metall.* **17**, 1183–1195.

Peters, E.T. and Ogilvie, R.E. (1965), X-ray orientation and diffractions studies by Kossel lines, *Trans. AIME* **233**, 89–95.

Petrov, R., Kestens, L., Wasilkowska, A., and Houbaert, Y. (2007), Microstructure and texture of a lightly deformed TRIP assisted steel characterized by means of the EBSD technique, *Mater. Sci. Eng.* **A447**, 285–297.

Philippe, M.J. (1994), Texture formation in hexagonal materials, *Mater. Sci. Forum* **157–162**, 1337–1350.

Philippe, M.J., Esling, C., and Hocheid, B. (1988), Role of twinning in texture development and in plastic deformation of hexagonal materials, *Textures Microstruct.* **7**, 265–301.

Philippe, M.J., Wagner, F., Mellab, F.E., Esling, C., and Wegria, J. (1994), Modelling of texture evolution for materials of hexagonal symmetry—I. Application to zinc alloys, *Acta Metall Mater.* **42**, 239–250.

Phillips, F.C. (1971), *An Introduction to Crystallography* (4th edition), Longman Group, London.

Pitsch, W., Perlwitz, H.J., and Lücke, K. (1964), Elektronenmikroskopische Bestimmung der Orientierungsverteilung der Kristallite in vielkristallinen Metallen, *Phys. Stat. Sol.* **7**, K105–K108.

Plege, P. (1987), On orientation correlations and their relationship to texture, in: *Theoretical Methods of Texture Analysis* (Ed. H.-J. Bunge), DGM Inf. Ges., Oberursel, Germany, 393–403.

Pond, R.C., Celotto, S., and Hirth, J.P. (2003), A comparison of the phenomenological theory of martensitic transformations with a model based on interfacial defects, *Acta Mater.* **51**, 5385–5398.

Pospiech, J. (1972), Die Parameter der Drehung und die Orientierungsverteilungsfunktion (OVF), *Kristall und Technik* **7**, 1057–1072.

Pospiech, J. (1982), Symmetry analysis in the space of Euler angles, in: *Quantitative Texture Analysis* (Eds. H.-J. Bunge and C. Esling), DGM Inf. Ges., Oberursel, Germany, 293–317.

Pospiech, J. and Jura, J. (1974), Determination of the orientation distribution function from incomplete pole figures, *Z. Metallkd.* **65**, 324–330.

Pospiech, J. and Jura, J. (1975), Fourier coefficients of the generalized spherical function and an exemplary computer program, *Kristall Technik* **10**, 783–787.

Pospiech, J., Jura, J., and Gottstein, G. (1994), Statistical analysis of single grain orientation data generated from model textures, *Mater. Sci. Forum* **157–162**, 407–412.

Pospiech, J. and Lücke, K. (1975), The rolling textures of copper and α-brasses discussed in terms of the orientation distribution function, *Acta Metall.* **23**, 997–1007.

Pospiech, J. and Lücke, K. (1979), Comparison between the orientation distribution functions of the recrystallization texture and the 40° <111>-transformed rolling texture of Cu 20% Zn, *Z. Metallkd.* **70**, 567–572.

Pospiech, J., Lücke, K., and Sztwiertnia, K. (1993), Orientation distribution and orientation correlation functions for description of microstructures, *Acta Metall. Mater.* **41**, 305–321.

Pospiech, J., Sztwiertnia, K., and Haeßner, F. (1986), The misorientation distribution function, *Textures Microstruct.* **6**, 201–215.

Pouchou, J.L. (Ed.) (2004), *L'Analyse EBSD: Principes et Applications*, EDP Sciences, Les Ulis, France.

Poulsen, H.F. (2004), *Three-dimensional X-ray Diffraction Microscopy: Mapping Polycrystals and Their Dynamics*, Springer Science & Business Media, Berlin.

Poulsen, H.F. (2012), An introduction to three-dimensional X-ray diffraction microscopy, *J. Appl. Cryst.* **45**, 1084–1097.

Poulsen, H.F. and Juul Jensen, D. (1995), Synchrotron radiation diffraction; A novel tool for recrystallization studies in bulk μm³ sized local areas, in: *Proceedings of 16th Risø International Symposium* (Eds. N. Hansen et al.), Risø National Laboratory, Roskilde, Denmark, 503–508.

Poulsen, H.F., Nielsen, S.F., Lauridsen, E.M., Schmidt, S., Suter, R.M., Lienert, U., Margulies, L., Lorentzen, T., and Juul Jensen, D. (2001), Three-dimensional maps of grain boundaries and the stress state of individual grains in polycrystals and powders, *J. Appl. Cryst.* **34**, 751–756.

Prince, E., Ed. (2004), *International Tables for Crystallography Vol. C*, Springer, Dordrecht, The Netherlands.

Prior, D.J. Boyle, A.P., Brenker, F., Cheadle, M.C., Day, A., Lopez, G., Peruzzo, L., Potts, G.J., Reddy, S., Spiess, R., Timms, N.E., Trimby, P., Wheeler, J., and Zetterström, L. (1999), The application of electron backscatter diffraction and orientation contrast imaging in the SEM to textural problems in rocks, *Amer. Mineral.* **84**, 1741–1759.

Pumphrey, P.H. (1976), Special high angle grain boundaries, in: *Grain Boundary Structure and Properties* (Eds. G.A. Chadwick and D.A. Smith), Academic Press, London, 139–200.

Quey, R., Piot, D., and Driver, J.H. (2010), Microtexture tracking in hot-deformed polycrystalline aluminium: Experimental results, *Acta Mater.* **58**, 1629–1642.

Raabe, D. (1992), *Texturen kubisch-raumzentrierter Übergangsmetalle*, PhD dissertation, RWTH Aachen, Germany.

Raabe, D. (1995), Investigation of the orientation dependence of recovery in low-carbon steel by use of single orientation determination, *Steel Res.* **66**, 222–229.

Raabe, D. (1996), On the influence of the chromium content on the evolution of rolling textures in ferritic stainless steels, *J. Mater. Sci.* **31**, 3839–3845.

Raabe, D., Keichel, J., and Gottstein, G. (1997), Investigation of crystallographic slip in polycrystalline Fe_3Al using slip trace measurement and microtexture determination, *Acta Mater.* **45**, 2839–2849.

Raabe, D. and Lücke, K. (1992), Selective particle drag during primary recrystallization of Fe-Cr alloys, *Scr. Metall.* **26**, 19–24.

Raabe, D. and Lücke, K. (1994), Rolling textures of niobium and molybdenum, *Z. Metallkd.* **85**, 302–307.

Raabe, D., Schlenkert, G., Weisshaupt, H., and Lücke, K. (1994), Texture and microstructure of rolled and annealed tantalum, *Mater. Sci. Technol.* **10**, 299–305.

Radon, J. (1917), Über die Bestimmung von Funktionen durch ihre Integralwerte längs gewisser Mannigfaltigkeiten, *Ber. Verh. Sächs. Akad. Wiss. Leibzig Math.-Phys. Klasse* **69**, 262–277; English translation by P.C. Parks (1986), On the determination of functions from their integral values along certain manifolds, *IEEE Trans. Med. Imaging* **5**, 170–176.

Rafaja, D., Dopita, M., Masimov, M., Klemm, V., Wendt, N., and Lengauer, W. (2008), Analysis of local composition gradients in the hard-phase grains of cermets using a combination of X-ray diffraction and electron microscopy, *Int. J. Ref. Met. Hard Mater.* **26**, 263–275.

Rajmohan, N., Hayakawa, Y., Szpunar, J.A., and Root, J.H. (1997), Neutron diffraction method for stored energy measurement in interstitial free steel, *Acta Mater.* **45**, 2485–2494.

Ram, F. and De Graef, M. (2018), Energy dependence of the spatial distribution of inelastically scattered electrons in backscatter electron diffraction, *Phys. Rev.* **97B**, 134104.

Ram, F., Zaefferer, S., Jäpel, T., and Raabe, D. (2015), Error analysis of the crystal orientations and disorientations obtained by the classical electron backscatter diffraction technique, *J. Appl. Cryst.* **48**, 797–813.

Randle, V. (1990), Representation of grain misorientations (mesotexture) in Rodrigues-Frank space, *Proc. Roy. Soc. Lond.* **A431**, 61–69.

Randle, V. (1993), *Grain Boundary Geometry in Polycrystals*, IOP Publ., Bristol.

Randle, V. (1995a), The influence of grain junctions and boundaries on superplastic deformation, *Acta Metall. Mater.* **43**, 1741–1749.

Randle, V. (1995b), Crystallographic characterization of planes in the scanning electron microscope, *Mater. Char.* **34**, 29–34.

Randle, V. (1996), *The Role of the Coincidence Site Lattice in Grain Boundary Engineering*, Institute of Materials, London.

Randle, V. (2000), Analysis of facets and other surfaces using electron backscatter diffraction, in: *Electron Backscatter Diffraction in Materials Science* (Eds. A.J. Schwartz, M. Kumar, and B.L. Adams), Kluwer Academic/Plenum Publishers, New York, 291–298.

Randle, V. (2001), The coincidence site lattice and the 'sigma enigma', *Mater. Char.* **47**, 411–416.

Randle, V. (2003), *Microtexture Determination and its Applications* (2nd edition), Institute of Materials, London.

Randle, V. (2004), Twinning-related grain boundary engineering, *Acta Mater.* **52**, 4067–4081.

Randle, V. (2007), Recent developments in electron backscatter diffraction, *Adv. Imag. Electron Phys.* **151**, 363–416.

Randle, V. (2009), Applications of electron backscatter diffraction to materials science: Status in 2009, *J. Mater. Sci.* **44**, 4211–4218.

Randle, V. (2010), Grain boundary engineering: An overview after 25 years, *Mater. Sci. Technol.* **26**, 253–261.

Randle, V. and Davies, H. (2002), A comparison between three-dimensional and two-dimensional grain boundary plane analysis, *Ultramicroscopy* **90**, 153–162.

Randle, V. and Day, A. (1993), Use of Rodrigues-Frank space for representation of microtexture and grain boundary parameters, *Mater. Sci. Technol.* **9**, 1069–1078.

Randle, V. and Dingley, D.J. (1989), Measurement of boundary plane inclination in a scanning electron microscope, *Scr. Metall.* **23**, 1565–1570.

Randle, V. and Hoile, C. (1998), Crystallographic analysis of facets using electron backscatter diffraction, *Mater. Sci. Forum* **273–275**, 183–190.

Randle, V., Juul Jensen, D., and Hansen, N. (1996), The deformation behaviour of grain boundary regions in polycrystalline aluminium, *Phil. Mag.* **A73**, 265–282.

Randle, V. and Ralph, B. (1988), Local texture changes associated with grain growth, *Proc. Roy. Soc. Lond.* **A415**, 239–256.

Ranganathan, S. (1966), On the geometry of coincidence-site lattices, *Acta Cryst.* **21**, 197–199.

Rauch, E.F. and Dupuy, L. (2005), Rapid spot diffraction pattern identification through template matching, *Arch. Metall. Mater.* **50**, 87–99.

Rauch, E.F., Harrison, P., Zhou, X., Herbig, M., Ludwig, W., and Véron, M. (2021), New features in crystal orientation and phase mapping for transmission electron microscopy, *Symmetry* **13**, 1675.

Rauch, E.F. and Véron, M. (2014), Automated crystal orientation and phase mapping in TEM, *Mater. Char.* **98**, 1–9.

Ray, R.K., Jonas, J.J., and Hook, R.E. (1994), Cold rolling and annealing textures in low carbon and extra low carbon steels, *Int. Mater. Rev.* **39**, 129–172.

Reed, B.W. and Kumar, M. (2006), Mathematical methods for analyzing highly-twinned grain boundary networks, *Scr. Mater.* **54**, 1029–1033.

Reeves, A., Le Gouic, F., Pannetier, J., Anne, M., Bordet, P., and Hodeau, J.L. (1996), Diffraction imaging of the grain structure and texture of aluminium alloys using high energy synchrotron x rays at the E.S.R.F., *Mater. Sci. Forum* **217–222**, 595–600.

Reimer, L. (2013a), *Scanning Electron Microscopy: Physics of Image Formation and Microanalysis* (2nd edition), Springer, Berlin.

Reimer, L. (2013b), *Transmission Electron Microscopy: Physics of Image Formation and Microanalysis* (4th edition), Springer, Berlin.

Ren, S.X., Kenik, E.A., Alexander, K.B., and Goyal, A. (1998), Exploring spatial resolution in electron back-scattered diffraction experiments via Monte-Carlo simulations, *Microsc. Microanal.* **4**, 13–22.

Ren, X., Miura, N., Zhang, J., Otsuka, K., Tanaka, K., Koiwa, M., Suzuki, T., Chumlyakov, Y.I., and Asai, M. (2001), A comparative study of elastic constants of Ti–Ni-based alloys prior to martensitic transformation, *Mater. Sci. Eng.* **A312**, 196–206.

Reschka, S., Gerstein, G., Herbst, S., Munk, L., Wriggers, P., and Maier, H.J. (2020), 3D orientation data—A comparison of diffraction contrast tomography and serial sectioning electron backscatter diffraction for the nickel-base superalloy IN738LC, *Mater. Lett.* **262**, 127177.

Rhines, F.N., Craig, K.R., and Rousse, D.A. (1976), Measurement of average grain volume and certain topological parameters by serial section analysis, *Metall. Trans.* **7A**, 1729–1734.

Rhoades, B.L. (1976), *XIDENT – A Computer Technique for the Direct Indexing of Electron Diffraction Spot Patterns*, Research Report 70/76, Univ. Canterbury, Christchurch, New Zealand.

Ridha, A.A. and Hutchinson, W.B. (1982), Recrystallisation mechanisms and the origin of cube texture in copper, *Acta Metall.* **30**, 1929–1939.

Riecke, W.D. (1962), Feinstrahl-Elektronenbeugung mit dreistufigem Kondensor und langbrennweitiger letzter Kondensorstufe, *Optik* **19**, 81–116.

Rietveld, H.M. (1969), A profile refinement method for nuclear and magnetic structures, *J. Appl. Cryst.* **2**, 65–71.

Ro, Y.J., Agnew, S.R., and Gangloff, R.P. (2005), Uncertainty in the determination of fatigue crack facet crystallography, *Scr. Mater.* **52**, 531–536.

Rodgers, T.M., Madison, J.D., and Tikare, V. (2017), Simulation of metal additive manufacturing microstructures using kinetic Monte Carlo, *Comp. Mater. Sci.* **135**, 78–89.

Rodrigues, O. (1840), Des lois géométriques qui régissent les déplacements d'un système solide dans l'espace, et de la variation des coordonnées provenant de ces déplacements considérés indépendants des causes qui peuvent les produire, *J. des Mathématiques Pures et Appliquées* **5**, 380–440.

Rodriguez-Navarro, A.B. (2006), XRD2DScan: New software for polycrystalline materials characterization using two-dimensional X-ray diffraction, *J. Appl. Cryst.* **39**, 905–909.

Rodriguez-Navarro, A.B. (2007), Registering pole figures using an X-ray single-crystal diffractometer equipped with an area detector, *J. Appl. Cryst.* **40**, 631–634.

Roe, R.J. (1965), Description of crystallite orientation in polycrystalline materials. III. General solution to pole figure inversion, *J. Appl. Phys.* **36**, 2024–2031.

Rohrer, G.S. (2011), Measuring and interpreting the structure of grain-boundary networks, *J. Amer. Ceramic Soc.* **94**, 633–646.

Rohrer, G.S., Li, J., Lee, S., Rollett, A.D., Groeber, M., and Uchic, M.D. (2010), Deriving grain boundary character distributions and relative grain boundary energies from three-dimensional EBSD data, *Mater. Sci. Technol.* **26**, 661–669.

Rohrer, G.S., Randle, V., Kim, C., and Hu, Y. (2006), Changes in the five-parameter grain boundary character distribution in α-brass brought about by iterative thermomechanical processing, *Acta Mater.* **54**, 4489–4502.

Rohrer, G.S., Saylor, D.M., El-Dasher, B.S., Adams, B.L., Rollett, A.D., and Wynblatt, P. (2004), The distribution of internal interfaces in polycrystals, *Z. Metallkd.* **95**, 197–214.

Rollett, A.D. and Gottstein, G. (2021), A brief overview of texture and anisotropy, *IOP Conf. Ser.: Mater. Sci. Eng.* **1121**, 012001.

Rollett, A.D., Lee, S.-B., Campman, R., and Rohrer, G.S. (2007), Three-dimensional characterization of microstructure by electron back-scatter diffraction, *Annu. Rev. Mater. Res.* **37**, 627–658.

Rollett, A.D. and Wright, S.I. (1998), Typical textures in metals, in: *Texture and Anisotropy, Preferred Orientations in Polycrystals and their Effect on Materials Properties* (Eds. U.F. Kocks, C.N. Tomé, and H.-R. Wenk), Cambridge University Press, Cambridge, 178–239.

Röntgen, W.C. (1896), Über eine neue Art von Strahlen, *Sitzungsbericht der Würzburger Physikalisch-Medicinischen Gesellschaft*, 11–17; English translation "On a new kind of rays", *Science* **3**, 227–231.

Roşca, D., Morawiec, A., and De Graef, M. (2014), A new method of constructing a grid in the space of 3D rotations and its applications to texture analysis, *Modell. Simul. Mater. Sci. Eng.* **22**, 075013.

Rossouw, C.J., Miller, P.R., Josefsson, T.W., and Allen, L.J. (1994), Zone-axis back-scattered electron contrast for fast electrons, *Phil. Mag.* **A70**, 985–998.

Roters, F., Diehl, M., Shanthraj, P., Eisenlohr, P., Reuber, C., Wong, S.L., Maiti, T., Ebrahimi, A., Hochrainer, T., Fabritius, H.-O., …, and Raabe, D. (2019), DAMASK—the Düsseldorf advanced material simulation kit for modeling multi-physics crystal plasticity, thermal, and damage phenomena from the single crystal up to the component scale, *Comp. Mater. Sci.* **158**, 420–478.

Rowenhorst, D.J., Gupta, A., Feng, C.R., and Spanos, G. (2006), 3D crystallographic and morphological analysis of coarse martensite: Combining EBSD and serial sectioning, *Scr. Mater.* **55**, 11–16.

Rowenhorst, D.J., Lewis, A.C., and Spanos, G. (2010), Three-dimensional analysis of grain topology and interface curvature in a β-titanium alloy, *Acta Mater.* **58**, 5511–5519.

Rowenhorst, D.J., Nguyen, L., Murphy-Leonard, A.D., and Fonda, R.W. (2020), Characterization of microstructure in additively manufactured 316L using automated serial sectioning, *Curr. Opin. Solid State Mater. Sci.* **24**, 100819.

Ruer, D. and Baro, R. (1977), A new method for the determination of the texture of materials of cubic structure from incomplete reflection pole figures, *Adv. X-ray Anal.* **20**, 187–200.

Ruer, D. and Vadon, A. (1982), The vector method, in: *Quantitative Texture Analysis* (Eds. H.J. Bunge, and C. Esling), DGM Inf. Ges., Oberursel, Germany, 343–381.

Sætre, T.O., Solberg, J.K., and Ryum, M. (1986), Variation in polarized light intensity with grain-orientation in anodized aluminum, *Metallography* **19**, 345–357.

Saha, D. (2021), Universal stage measurements in petrofabric analysis revisited, *J. Earth Syst. Sci.* **130**, 116.

Saleh, A.A., Vu, V.Q., and Gazder, A.A. (2016), Correcting intensity loss errors in the absence of texture-free reference samples during pole figure measurement, *Mater. Char.* **118**, 425–430.

Samajdar, I., Verlinden, B., Van Houtte, P., and Vanderschueren, D. (1997), γ-Fibre recrystallization texture in IF-steel: An investigation on the recrystallization mechanisms, *Mater. Sci. Eng.* **A238**, 343–350.

Sandim, H.R.Z., Bernardi, H.H., Verlinden, B., and Raabe, D. (2007), Equal channel angular extrusion of niobium single crystals, *Mater. Sci. Eng.* **A467**, 44–52.

Santoro, A. and Mighell, A.D. (1973), Coincidence-site lattices, *Acta Cryst.* **A29**, 169–175.

Sato, H. and Zaefferer, S. (2009), A study on the formation mechanisms of butterfly-type martensite in Fe–30% Ni alloy using EBSD-based orientation microscopy, *Acta Mater.* **57**, 1931–1937.

Savoie, J. and Lücke, K. (1991), Influence of the grain size and of the rolling degree on the recrystallization texture of low alloyed brass, *Textures Microstruct.* **14–18**, 897–902.

Saylor, D.M., El-Dasher, B.S., Adams, B.L., and Rohrer, G.S. (2004), Measuring the five-parameter grain-boundary distribution from observations of planar sections, *Metall. Mater. Trans.* **35A**, 1981–1989.

Saylor, D.M., Morawiec, A., Adams, B.L., and Rohrer, G.S. (2000), Misorientation dependence of the grain boundary energy in magnesia, *Interface Sci.* **8**, 131–140.

Saylor, D.M., Morawiec, A., and Rohrer, G.S. (2003), Distribution of grain boundaries in magnesia as a function of five macroscopic parameters, *Acta Mater.* **51**, 3663–3674.

Schaeben, H. (1988), Entropy optimization in texture goniometry. I. Methodology, *Phys. Stat. Sol. (b)* **148**, 63–72.

Schaeben, H. (1991), Entropy optimization in quantitative texture analysis: II. Application to pole-to-orientation density inversion, *J. Appl. Phys.* **69**, 1320–1329.

Schaeben, H., Hielscher, R., Fundenberger, J.-J., Potts, D., and Prestin, J. (2007), Orientation density function-controlled pole probability density function measurements: Automated adaptive control of texture goniometers, *J. Appl. Cryst.* **40**, 570–579.

Schaeben, H. and Van den Boogaart, K.G. (2003), Spherical harmonics in texture analysis, *Tectonophys.* **370**, 253–268.

Schaeben, H., Wenk, H.-R., and Vadon, A. (1985), Vector method, in: *Preferred Orientation in Deformed Metals and Rocks: An Introduction to Modern Texture Analysis* (Ed. H.-R. Wenk), Academic Press, Orlando, FL, 123–127.

Schäfer, B. and Schwarzer, R.A. (1998), SAD pole figures in transmission and reflection using a high-grade CCD camera as an area detector, *Mater. Sci. Forum* **273–275**, 223–228.

Schäffer, A.K., Jäpel, T., Zaefferer, S., Abart, R., and Rhede, D. (2014), Lattice strain across Na–K interdiffusion fronts in alkali feldspar: An electron back-scatter diffraction study, *Phys. Chem. Mineral.* **41**, 795–804.

Schemmann, L. (2014), *The Inheritance of Different Microstructures Found after Hot Rolling on the Properties of a Completely Annealed Dual Phase Steel*, PhD dissertation, RWTH Aachen University, Germany.

Scheriau, S. and Pippan, R. (2008), Influence of grain size on orientation changes during plastic deformation, *Mater. Sci. Eng.* **A493**, 48–52.

Schmidt, N.H., Bilde-Sørensen, J.B., and Juul Jensen, D. (1991), Band positions used for on-line crystallographic orientation determination from electron back scattering patterns, *Scan. Microsc.* **5**, 637–643.

Schmidt, U. and Lücke, K. (1979), Recrystallization textures of silver, copper and α-brasses with different zinc-contents as a function of the rolling temperature, *Texture Cryst. Sol.* **3**, 85–112.

Schmidt, N.H. and Olesen, N.O. (1989), Computer-aided determination of crystal-lattice orientation from electron-channeling patterns in the SEM, *Can. Mineral.* **27**, 15–22.

Schneider, E. and Österlein, L. (1996), Ultrasonic characterization of texture in aluminum rolled products, *Mater. Sci. Forum* **210–213**, 405–410.

Schuh, C.A., Kumar, M., and King, W.E. (2003), Analysis of grain boundary networks and their evolution during grain boundary engineering, *Acta Mater.* **51**, 687–700.

Schulz, L.G. (1949a), A direct method of determining preferred orientation of a flat reflection sample using a Geiger counter x-ray spectrometer, *J. Appl. Phys.* **20**, 1030–1033.

Schulz, L.G. (1949b), Determination of preferred orientation in flat transmission samples using a Geiger counter x-ray spectrometer, *J. Appl. Phys.* **20**, 1033–1036.

Schumann, H. and Leineweber, A. (2021), Crystallography of γ′-Fe4N formation on bulk polycrystalline α-Fe substrates, *Materialia* **17**, 101119.

Schwartz, A.J., Kumar, M., Adams, B.L., and Field, D.P. (Eds.) (2009), *Electron Backscatter Diffraction in Materials Science* (2nd edition), Springer, Berlin.

Schwarz, T, Stechmann, G, Gault, B, Cojocaru-Mirédin, O, Wuerz, R, and Raabe, D. (2018), Correlative transmission Kikuchi diffraction and atom probe tomography study of Cu(In,Ga)Se2 grain boundaries, *Prog. Photovolt. Res. Appl.* **26**, 196–204.

Schwarzer, R.A. (1983), Intensitätskorrektur für die Messung von Polfiguren im TEM, *Beitr. elektronenmikr. Direktabb. Oberfl. (BEDO)* **16**, 131–134.

Schwarzer, R.A. (1988), Quantitative TEM pole-figures of deformed titanium, *Inst. Phys. Conf. Ser.* **93**, Vol. 2, 23–24

Schwarzer, R.A. (1993), The determination of local texture by electron diffraction—A tutorial review, *Textures Microstruct.* **20**, 7–27.

Schwarzer, R.A. (1997a), Advances in crystal orientation mapping with the SEM and TEM, *Ultramicroscopy* **67**, 19–24.

Schwarzer, R.A. (1997b), Automated crystal lattice orientation mapping using a computer-controlled SEM, *Micron* **28**, 249–265.

Schwarzer, R.A. (2000), Measurement of macrotexture by automated crystal orientation mapping: an alternative to X-ray diffraction, *Mater. Sci. Technol.* **16**, 1384–1388.

Schwarzer, R.A. (2006), Automated crystal orientation measurement by backscatter Kikuchi diffraction, *Z. Kristallogr. Suppl.* **23**, 163–168.

Schwarzer, R.A., Field, D.P., Adams, B.L., Kumar, M., and Schwartz, A.J. (2009), Present state of electron backscatter diffraction and prospective developments, in: *Electron Backscatter Diffraction in Materials Science* (2nd edition) (Eds. A.J. Schwartz et al.), Springer, Berlin, 1–20.

Schwarzer, R.A. and Hjelen, J. (2010), Orientation microscopy with fast EBSD, *Mater. Sci. Technol.* **26**, 646–649.

Schwarzer, R.A., Krieger Lassen, N.C., and Schäfer, B. (1996a), Electron texture goniometer consisting of a TEM with side entry goniometer and a CCD camera as an area detector, in: *Proceedings of ICOTOM 11* (Eds. Z. Liang, L. Zou, and Y. Chu), International Academic Publishers, Beijing, 170–175.

Schwarzer, R.A., Singh, A.K., and Sukkau, J. (2000), Discrimination and mapping of phase distributions by automated crystal orientation measurement, *Mater. Sci. Technol.* **16**, 1389–1392.

Schwarzer, R.A., Springer, F., and Zaefferer, S. (1996b), Crystal orientation mapping by digital beam scan and automated interpretation of backscatter Kikuchi patterns in the SEM, in: *Proceedings of ICOTOM 11* (Eds. Z. Liang, L. Zou, and Y. Chu), International Academic Publishers, Beijing, 43–52.

Schwarzer, R.A. and Sukkau, J. (1998), Automated crystal orientation mapping (ACOM) with a computer-controlled TEM by interpreting transmission Kikuchi patterns, *Mater. Sci. Forum* **273–275**, 215–222.

Schwarzer, R.A. and Weiland, H. (1984), On-line computerized evaluation of Kikuchi patterns for the determination of preferred orientations and orientation correlations, in: *Proceedings of ICOTOM 7* (Eds. C.M. Brakman, P. Jongenburger, and E.J. Mittemeijer), Netherlands Society for Material Science, Zwijndrecht, 839–843.

Schwarzer, R.A. and Weiland, H. (1986), Electron diffraction pole figure measurements, in: *Experimental Techniques of Texture Analysis* (Ed. H.-J. Bunge), DGM Inf. Ges., Oberursel, Germany, 287–313.

Schwarzer, R.A. and Weiland, H. (1988), Texture analysis by the measurement of individual grain orientations—Electron microscopical methods and application on dual-phase steel, *Textures Microstruct.* **8&9**, 551–577.

Schwarzer, R.A. and Zaefferer, S. (1994a), An inexpensive CCD camera system for the recording and on-line interpretation of TEM Kikuchi patterns, *Mater. Sci. Forum* **157–162**, 189–194.

Schwarzer, R.A. and Zaefferer, S. (1994b), On-line interpretation of SAD channeling patterns, *Mater. Sci. Forum* **157–162**, 195–200.

Schwarzer, R.A., Zaefferer, S., and Kunze, K. (1995), The characterization of microtexture by orientation mapping, *Adv. X-Ray Anal.* **38**, 547–550.

Sears, V.F. (1992), Neutron scattering lengths and cross sections, *Neutron News* **3**, 26–37.

Seidel, L., Hölscher, M., and Lücke, K. (1989), Rolling and recrystallization textures in iron–3% silicon, *Textures Microstruct.* **11**, 171–185.

Seward, G.G.E., Celotto, S., Prior, D.J., Wheeler, J., and Pond, R.C. (2004), In situ SEM-EBSD observations of the hcp to bcc phase transformation in commercially pure titanium, *Acta Mater.* **52**, 821–832.

Shigesato, G. and Rauch, E.F. (2007), Dislocation structure misorientations measured with an automated electron diffraction pattern indexing tool, *Mater. Sci. Eng.* **A462**, 402–406.

Shimizu, R., Harase, J., and Dingley, D.J. (1990), Prediction of secondary recrystallization texture in Fe–3% Si by three-dimensional texture analysis, *Acta Metall. Mater.* **38**, 973–978.

Shimizu, M., Katayama, M., and Kitagawa, H. (1986), Inverse pole figure measurements in steels by energy-dispersive x-ray diffraction, in: *Experimental Techniques of Texture Analysis* (Ed. H.-J. Bunge), DGM Inf. Ges., Oberursel, Germany, 107–114.

Shinbine, A., Garcin, T., and Sinclair, C. (2016), In-situ laser ultrasonic measurement of the hcp to bcc transformation in commercially pure titanium, *Mater. Char.* **117**, 57–64.

Shoemake, K. (1992), Uniform random rotations, in: *Graphics Gems III* (Ed. D. Kirk), Academic Press, Boston, MA, 124–132.

Siegesmund, S., Helming, K., and Kruse, R. (1994), Complete texture analysis of a deformed amphibolite: comparison between neutron diffraction and U-stage data, *J. Struct. Geol.* **16**, 131–142.

Simkin, B.A., Ng, B.C., Bieler, T.R., Crimp, M.A., and Mason, D.E. (2003), Orientation determination and defect analysis in the near-cubic intermetallic γ-TiAl using SACP, ECCI, and EBSD, *Intermetallics* **11**, 215–223.

Simons, H., King, A., Ludwig, W., Detlefs, C., Pantleon, W., Schmidt, S., Stöhr, F., Snigireva, I., Snigirev, A., and Poulsen, H.F. (2015), Dark-field X-ray microscopy for multiscale structural characterization, *Nature Commun.* **6**, 6098.

Sinclair, C.W., Embury, J.D., Weatherly, G.C., Conlon, K.T., and Engler, O. (2003), Heterogeneous deformation of a two phase nickel-tungsten alloy, *Mater. Sci. Technol.* **19**, 1321–1329.

Sinclair, C.W., Mithieux, J.D., Schmitt, J.H., and Bréchet, Y. (2005), Recrystallization of stabilized ferritic stainless steel sheet, *Metall. Mater. Trans.* **36A**, 3205–3215.

Singh, S. and De Graef, M. (2016), Orientation sampling for dictionary-based diffraction pattern indexing methods, *Modell. Simul. Mater. Sci. Eng.* **24**, 085013.

Singh, S., Guo, Y., Winiarski, B., Burnett, T.L., Withers, P.J., and De Graef, M. (2018), High resolution low kV EBSD of heavily deformed and nanocrystalline aluminium by dictionary-based indexing, *Sci. Reports* **8**, 10991.

Singh, A.K. and Schwarzer, R.A. (2000), Texture and anisotropy of mechanical properties in titanium and its alloys, *Z. Metallkd.* **91**, 702–716.

Skrotzki, W., Klöden, B., Tamm, R., Oertel, C.-G., Garbe, U., and Rybacki, E. (2003), Torsion texture measurements with high-energy synchrotron radiation on NiAl, *Textures Microstruct.* **35**, 163–173.

Slavik, D.C. and Gangloff, R.P. (1996), Environment and microstructure effects on fatigue crack facet orientation in an Al-Li-Cu-Zr alloy, *Acta Mater.* **44**, 3515–3534.

Slavik, D.C., Wert, J.A., and Gangloff, R.P. (1993), Determining fracture facet crystallography using electron backscatter patterns and quantitative tilt fractography, *J. Mater. Res.* **8**, 2482–2491.

Smith, K.L. and Ortega, R.B. (1993), Use of a two-dimensional, position sensitive detector for collecting pole figures, *Adv. X-ray Anal.* **36**, 641–647.

Sneddon, G.C., Trimby, P.W., and Cairney, J.M. (2016), Transmission Kikuchi diffraction in a scanning electron microscope: A review, *Mater. Sci. Eng.* **R110**, 1–12.

Sous, S.A., Hildmann, B.O., and Kaysser, W.A. (1997), Characterization of dislocation densities in germanium and silicon single crystals by high resolution x-ray diffraction, *Phys. Stat. Sol. (a)* **159**, 343–353.

Spanos, G., Rowenhorst, D.J., Lewis, A.C., and Geltmacher, A.B. (2008), Combining serial sectioning, EBSD analysis, and image-based finite element modeling, *MRS Bull.* **33**, 597–602.

Spies, M. and Salama, K. (1990), Relationship between elastic anisotropy and texture in metal-matrix composites, *Ultrasonics* **28**, 370–374.

Spowart, J.E., Mullens, H.E., and Puchala, B.T. (2003), Collecting and analyzing microstructures in three dimensions: A fully automated approach. *JOM* **55**, 35–37.

Stanford, N. and Bate, P.S. (2004), Crystallographic variant selection in Ti–6Al–4V, *Acta Mater.* **52**, 5215–5224.

Staron, P., Schreyer, A., Clemens, H., and Mayer, S. (Eds.) (2017), *Neutrons and Synchrotron Radiation in Engineering Materials Science, from Fundamentals to Applications* (2nd edition), Wiley-VCH, Weinheim.

Stechmann, G., Zaefferer, S., and Raabe, D. (2018), Molecular statics simulation of CdTe grain boundary structures and energetics using a bond-order potential, *Modell. Simul. Mater. Sci. Eng.* **26**, 045009.

Stechmann, G., Zaefferer, S., Schwarz, T., Konijnenberg, P., Raabe, D., Gretener, C., Kranz, L., Perrenoud, J., Buecheler, S., and Nath Tiwari, A. (2017), A correlative investigation of grain boundary crystallography and electronic properties in CdTe thin film solar cells, *Sol. Energy Mater. Sol. Cells* **166**, 108–120.

Steeds, J.W. and Morniroli, J.P. (1992), Selected area electron diffraction (SAED) and convergent beam electron diffraction (CBED), *Rev. Mineral. Geochem.* **27**, 37–84.

Stout, M.G. and O'Rourke, J.A. (1989), Experimental deformation textures of OFE copper and 70:30 brass from wire drawing, compression, and torsion, *Metall. Trans.* **20A**, 125–131.

Straumal, B.B., Protasova, S.G., Mazilkin, A.A., Tietze, T., Goering, E., Schütz, G., Straumal, P.B., and Baretzky, B. (2013), Ferromagnetic behaviour of Fe-doped ZnO nanograined films, *Beilstein J. Nanotechnol.* **4**, 361–369.

Sukhopar, O. and Gottstein, G. (2016), In-situ annealing and computation study of cube texture development in a commercial aluminum alloy, *Int. J. Mater. Res.* **107**, 979–987.

Suter, R.M., Hefferan, C.M., Li, S.F., Hennessy, D., Xiao, C., Lienert, U., and Tieman, B. (2008), Probing microstructure dynamics with x-ray diffraction microscopy, *J. Eng. Mater. Technol.* **130**, 021007.

Sutton, A.P. and Balluffi, R.W. (1995), *Interfaces in Crystalline Materials*, Clarendon Press, Oxford.

Suzuki, S. (2013), Features of transmission EBSD and its application, *JOM* **65**, 1254–1263.

Szpunar, J. (1984), Texture studies using neutron diffraction, *J. Mater. Sci.* **19**, 3467–3476.

Szpunar, J. (1990), Energy dispersive diffractometry for quantitative texture studies, *Textures Microstruct.* **12**, 243–247.

Szpunar, J.A., Ahlroos, S., and Tavernier, P. (1993), Method of measurement and analysis of texture in thin films, *J. Mater. Sci.* **28**, 2366–2376.

Szpunar, J.A. and Davies, S.T. (1984), Application of synchrotron radiation to texture experiments, in: *Proceedings of ICOTOM 7* (Eds. C.M. Brakman, P. Jongenburger, and E.J. Mittemeijer), Netherlands Society for Material Science, Zwijndrecht, 845–849.

Szpunar, J. and Gerward, L. (1980), Energy-dispersive X-ray diffraction studies of the texture in cold-rolled alpha-beta brass, *J. Mater. Sci.* **15**, 469–476.

Takata, N., Ikeda, K., Yoshida, F., Nakashima, H., and Abe, H. (2004), Influence of purity on the formation of cube texture in aluminum foils for electrolytic capacitors, *Mater. Trans. JIM* **45**, 1687–1692.

Takayama, N., Miyamoto, G., and Furuhara, T. (2012), Effects of transformation temperature on variant pairing of bainitic ferrite in low carbon steel, *Acta Mater.* **60**, 2387–2396.

Tanaka, Y., Inamura, T., Hosoda, H., Suwa, Y., and Tomida, T. (2017), Temperature dependency of diffusional transformation texture development in steel sheet, *Mater. Trans.* **58**, 554–560.

Tanaka, M., Sekii, H., and Nagasawa, T. (1983), Space-group determination by dynamic extinction in convergent-beam electron diffraction, *Acta Cryst.* **A39**, 825–837.

Tanaka, M., Terauchi, M., and Kaneyama, T. (1988), *Convergent-Beam Electron Diffraction II*, JEOL Ltd., Tokyo.

Tanaka, M., Terauchi, M., and Tsuda, K. (1994), *Convergent-Beam Electron Diffraction III*, JEOL-Maruzen, Tokyo.

Tarkowski, T., Bonarski, J., and Dabrowski, W. (2005), Application of the Si-strip detector in X-ray crystallographic texture measurements, *Nucl. Instr. Meth.* **A551**, 178–182.

Tasan, C.C., Hoefnagels, J.P., Diehl, M., Yan, D., Roters, F., and Raabe, D. (2014), Strain localization and damage in dual phase steels investigated by coupled in-situ deformation experiments and crystal plasticity simulations, *Int. J. Plast.* **63**, 198–210.

Tenckhoff, E. (1970), Defocusing for the Schulz technique of determining preferred orientation, *J. Appl. Phys.* **41**, 3944–3948.

Tenckhoff, E. (1988), *Deformation Mechanisms, Texture, and Anisotropy in Zirconium and Zircaloy*, ASTM, Philadelphia, PA.

Themelis, G., Chikwembani, S., and Weertman, J. (1990), Determination of the orientation of Cu-Bi grain boundary facets using a photogrammetric technique, *Mater. Char.* **24**, 27–40.

Thomas, G. and Goringe, M.J. (1979), *Transmission Electron Microscopy of Materials*, Wiley, New York.

Thomas, I., Zaefferer, S, Friedel, F., and Raabe, D. (2003), High resolution EBSD investigation of deformed and partially recrystallized IF steel, *Adv. Eng. Mat.* **5**, 566–570.

Thompson, R.B., Papadakis, E.P., Bluhm, D.D., Alers, G.A., Forouraghi, K., Skank, H.D., and Wormley, S.J. (1993), Measurement of texture and formability parameters with a fully automated, ultrasonic instrument, *J. Nondestructive Eval.* **12**, 45–62.

Thompson-Russell, K.C. and Edington, J.W. (1977), *Electron Microscope Specimen Preparation Techniques in Materials Science*, Macmillan Education, London.

Thomsen, K., Schmidt, N.H., Bewick, A., Larsen, K., and Goulden, J. (2013), Improving the accuracy of orientation measurements using EBSD, *Microsc. Microanal.* 19 (Suppl. 2), 724–725.

Thomson, C.B. and Randle, V. (1997a), "Fine tuning" at $\Sigma3^n$ boundaries in nickel, *Acta Mater.* **45**, 4909–4916.

Thomson, C.B. and Randle, V. (1997b), The effects of strain annealing on grain boundaries and secure triple junctions in nickel 200, *J. Mater. Sci.* **32**, 1909–1914.

Tixier, R. and Wache, C. (1970), Kossel patterns, *J. Appl. Cryst.* **3**, 466–485.

Tizliouine, A., Bessières, J., Heizmann, J.J., and Bobo, J.F. (1996), Thin-film study using low-incidence and Bragg-Brentano texture goniometry. Application to mono- and bilayers of Al and Al/Fe, *J. Appl. Cryst.* **29**, 531–539.

Tobisch, J. and Bunge, H.-J. (1972), The spherical sample method in neutron diffraction texture determination, *Texture* **1**, 125–127.

Tochigi, E., Nakamura, A., Shibata, N., and Ikuhara, Y. (2018), Dislocation structures in low-angle grain boundaries of α-Al_2O_3, *Crystals* **8**, 133.

Tolédano, P., Krexner, G., Prem, M., Weber, H.P., and Dmitriev, V.P. (2001), Theory of the martensitic transformation in cobalt, *Phys. Review* **64B**, 144104.

Tomé, C.N., Lebensohn, R.A., and Kocks, U.F. (1991), A model for texture development dominated by deformation twinning: Application to zirconium alloys, *Acta Metall. Mater.* **39**, 2667–2680.

Tomida, T., Wakita, M., Yasuyama, M., Sugaya, S., Tomota, Y., and Vogel, S.C. (2013), Memory effects of transformation textures in steel and its prediction by the double Kurdjumov–Sachs relation, *Acta Mater.* **61**, 2828–2839.

Trimby, P.W. (2012), Orientation mapping of nanostructured materials using transmission Kikuchi diffraction in the scanning electron microscope, *Ultramicroscopy* **120**, 16–24.

Tripathi, A. and Zaefferer, S. (2019), On the resolution of EBSD across atomic density and accelerating voltage with a particular focus on the light metal magnesium, *Ultramicroscopy* **207**, 112828.

Tsai, S.P., Konijnenberg, P.J., Gonzalez, I., Hartke, S., Griffiths, T.A., Herbig, M., Kawano-Miyata, K., Taniyama, A., Sano, N., and Zaefferer, S. (2022), Development of a new, fully automated system for electron backscatter diffraction (EBSD)-based large volume three-dimensional microstructure mapping using serial sectioning by mechanical polishing, and its application to the analysis of special boundaries in 316L stainless steel, *Rev. Sci. Instrum.* **93**, 093707.

Tsai, S.P., Makineni, S.K., Gault, B., Kawano-Miyata, K., Taniyama, A., and Zaefferer, S. (2021), Precipitation formation on $\Sigma5$ and $\Sigma7$ grain boundaries in 316L stainless steel and their roles on intergranular corrosion, *Acta Mater.* **210**, 116822.

Tse, Y.Y., Liu, G.L., and Duggan, B.J. (2000), Deformation banding and nucleation of recrystallisation in IF steel, *Scr. Mater.* **42**, 25–30.

Tucker, G.E.G. and Murphy, P.C. (1952/53), A method of determining orientations in aluminium single crystals and polycrystalline aggregates, *J. Inst. Met.* **8**, 235–244.

Ubhi, H.S., Parsons, J., Othen, N., Campbell, S., Poole, R., and Gholinia, A. (2014), In-situ EBSD phase transformation and recrystallisation, *J. Phys.: Conf. Ser.* **522**, 012011.

Uchic, M.D., Groeber, M.A., Dimiduk, D.M., and Simmons, J.P. (2006), 3D microstructural characterization of nickel superalloys via serial-sectioning using a dual beam FIB-SEM, *Scr. Mater.* **55**, 23–28.

Uhbi, H.S. and Bowen, A. (1996), Analysis and representation of electron backscattered diffraction texture data—examples from heat treated Al-Li alloy sheet, *Mater. Sci. Technol.* **12**, 880–886.

Ullemeyer, K., Braun, G., Dahms, M., Kruhl, J.H., Olesen, N.Ø., and Siegesmund, S. (2000), Texture analysis of a muscovite-bearing quartzite: A comparison of some currently used techniques, *J. Struct. Geol.* **22**, 1541–1557.

Ullemeyer, K., Spalthoff, P., Heinitz, J., Isakov, N.N., Nikitin, A.N., and Weber, K. (1998), The SKAT texture diffractometer at the pulsed reactor IBR-2 at Dubna: Experimental layout and first measurements, *Nucl. Instr. Meth.* **A412**, 80–88.

Ullrich, H.-J., Friedel, F., Bauch, J., Meinl, G., Geise, G., Lengauer, W., and Ettmayer, P. (1992), Phase identification in Ti/TiN diffusion couples with the Kossel technique, *Mikrochim. Acta* **109**, 233–242.

Ullrich, H.-J., Schlaubitz, M., Friedel, F., Spann, T., Bauch, J., Wroblewski, T., Garbe, S., Gaul, G., Knöchel, A., Lechtenberg, F., Rossmanith, E., Kumpat, G., and Ulrich, G. (1994), Excitation of Kossel patterns by synchrotron radiation, *Nuclear Instrum. Methods Phys. Res.* **A349**, 269–273.

Ullrich, H.-J. and Schulze, G.E.R. (1972), Röntgenographische Mikrobeugungsuntersuchungen an kristallinen Festkörpern mittels Gitterquellen-interferenzen (Kossel-Linien) und Weitwinkel-Interferenzen (Pseudo-Kossel-Linien), *Kristall Technik* **7**, 207–220.

Vadon, A. and Heizmann, J.J. (1991), A new program to calculate the texture vector for the vector method, *Textures Microstruct.* **14–18**, 37–44.

Valery, A., Rauch, E.F., Clément, L., and Lorut, F. (2017), Retrieving overlapping crystals information from TEM nano-beam electron diffraction patterns, *J. Microsc.* **268**, 208–218.

Van Acker, K., De Buyser, L., Celis, J.P., and Van Houtte, P. (1994), Characterization of thin nickel electrocoatings by the low-incident-beam-angle diffraction method, *J. Appl. Cryst.* **27**, 56–66.

Van Houtte, P. (1980), A method for orientation distribution function analysis from incomplete pole figures normalized by an iterative method, *Mater. Sci. Eng.* **43**, 7–11.

Van Houtte, P. (1983), The use of a quadratic form for the determination of non-negative texture functions, *Textures Microstruct.* **6**, 1–19.

Van Houtte, P. (1984), A new method for the determination of texture functions from incomplete figures, comparison with older methods, *Textures Microstruct.* **6**, 137–162.

Van Houtte, P. (1987), On the representation of texture functions of cubic metals in Euler space, *Textures Microstruct.* **7**, 187–205.

Van Houtte, P. (1991), A method for the generation of various ghost correction algorithms—the example of the positive method and the exponential method, *Textures Microstruct.* **13**, 199–212fv.

Van Houtte, P., Li, S., and Engler, O. (2004), Modelling deformation texture of aluminium alloys using grain interaction models, *Aluminium* **80**, 702–706.

Vander Voort, G. F. (2011). Metallographic specimen preparation for Electron backscattered diffraction part I. *Praktische Metallographie*, 48(9), 454–473.

Vander Voort, G. F. (2011). Metallographic specimen preparation for electron backscattered diffraction part II. *Practical Metallography*, 48(10), 527–543.

Vandermeer, R.A. and Ogle, J.C. (1969), Texture inhomogeneities in cold-rolled niobium, *Trans. Metall. Soc. AIME* **245**, 1511–1518.

Vandermeer, R.A. and Snyder, W.B. (1979), Recovery and recrystallization in rolled tantalum single crystals, *Metall. Trans.* **10A**, 1031–1044.

Vasin, R.N. (2022), Quantitative texture analysis: Implementation on the SKAT neutron diffractometer and application in materials and earth sciences, *Cryst. Reports* **67**, 64–80.

Vatne, H.E., Furu, T., Ørsund, R., and Nes, E. (1996), Modelling recrystallization after hot deformation of aluminium, *Acta Mater.* **44**, 4463–4473.

Vaudin, M.D., Gerbig, Y.B., Stranick, S.J. and Cook, R.F. (2008), Comparison of nanoscale measurements of strain and stress using electron back scattered diffraction and confocal Raman microscopy, *Appl. Phys. Lett.* **93**, 193116.

Venables, J.A. and Bin-Jaya, R. (1977), Accurate microcrystallography using electron backscattering patterns, *Phil. Mag.* **35**, 1317–1332.

Venables, J.A. and Harland, C.J. (1973), Electron back-scattering patterns—A new technique for obtaining crystallographic information in the scanning electron microscope, *Phil. Mag.* **27**, 1193–1200.

Viladot, D., Véron, M., Gemmi, M., Peiró, F., Portillo, J., Estradé, S., Mendoza, J., Llorca-Isern, N., and Nicolopoulos, S. (2013), Orientation and phase mapping in the transmission electron microscope using precession-assisted diffraction spot recognition: State-of-the-art results, *J. Microsc.* **252**, 23–34.

Vincent, R. and Midgley, P.A. (1994), Double conical beam-rocking system for measurement of integrated electron diffraction intensities, *Ultramicroscopy* **53**, 271–282.

Vogel, S.C., Hartig, C., Lutterotti, L., Von Dreele, R.B., Wenk, H.-R., and Williams, D.J. (2004), Texture measurements using the new neutron diffractometer HIPPO and their analysis using the Rietveld method, *Powder Diffr.* **19**, 65–68.

Vogel, S. and Klimanek, P. (1996), Computer-aided investigation of isothermal grain growth in copper and brass, *Mater. Sci. Forum* **204–206**, 449–454.

Von Dreele, R.B. (1997), Quantitative texture analysis by Rietveld refinement, *J. Appl. Cryst.* **30**, 517–525.

von Laue, M. (1948). Materiewellen und ihre Interferenzen. Akademische Verlagsges., Leipzig

Von Meibom, R. and Rupp, E. (1933), Elektronenbeugung in weiten Winkeln, *Z. Phys.* **82**, 690–696.

Von Schlippenbach, U., Emren, F., and Lücke, K. (1986), Investigation of the development of the cold rolling texture in deep drawing steels by ODF-analysis, *Acta Metall.* **34**, 1289–1301.

Vos, M. and Winkelmann, A. (2019), Effects of multiple elastic and inelastic scattering on energy-resolved contrast in Kikuchi diffraction, *New J. Phys.* **21**, 123018.

Wagner, F., Bechler-Ferry, E., Esling, C., and Baro, R. (1982), A generalized library program for texture calculations with even and odd coefficients, *J. Appl. Cryst.* **15**, 122–125.

Wagner, F., Bozzolo, N., Van Landuyt, O., and Grosdidier, T. (2002), Evolution of recrystallisation texture and microstructure in low alloyed titanium sheets, *Acta Mater.* **50**, 1245–1259.

Wagner, F. and Dahms M. (1991), Determination of complete O.D.F.s with an optimized positivity method, in: *Advances and Applications of Quantitative Texture Analysis* (Eds. H.J. Bunge and C. Esling), DGM, Oberursel, Germany, 101–108.

Wagner, P., Engler, O., and Lücke, K. (1995), Formation of Cu-type shear bands and their influence on deformation and texture of rolled f.c.c. {112}<111> single crystals, *Acta Metall. Mater.* **43**, 3799–3812.

Wagner, F., Humbert, M., Muller, J., and Esling, C. (1990), Optimization of the positivity method in quantitative texture analysis, *Europhys. Lett.* **11**, 479–483.

Wagner, F., Wenk, H.-R., Esling, C., and Bunge, H.-J. (1981), Importance of odd coefficients in texture calculations for trigonal–triclinic symmetries, *Phys. Stat. Sol. (a)* **67**, 269–285.

Wakita, M. and Suzuki, S. (2017), In-situ observation of microstructure change in steel by EBSD, *Nippon Steel & Sumitomo Metal Technical Report* **114**, 32–37.

Walde, T. and Riedel, H. (2007), Modeling texture evolution during hot rolling of magnesium alloy AZ31, *Mater. Sci. Eng.* **A443**, 277–284.

Walther, K., Heinitz, J., Ullemeyer, K., Betzl, M., and Wenk, H.-R. (1995), Time-of-flight texture analysis of limestone standard: Dubna results, *J. Appl. Cryst.* **28**, 503–507.

Wang, Z.L. (1992), Dynamical simulations of energy-filtered inelastic electron diffraction patterns, *Acta Cryst.* **A48**, 674–688.

Wang, Z.L. (2003), Thermal diffuse scattering in sub-angstrom quantitative electron microscopy—phenomenon effects and approaches, *Micron* **34**, 141–155.

Wang, Z.L. (2013), *Elastic and Inelastic Scattering in Electron Diffraction and Imaging*, Springer, Berlin.

Wang, S.C., Browne, M., Ubhi, H.S., and Starink, M.J. (2005), Determination of the fatigue fracture planes of Co–Cr–Mo biomedical alloys using electron backscatter diffraction, *J. Microsc.* **217**, 118–121.

Wang, Y.N., Chang, C.I., Lee, C.J., Lin, H.K., and Huang, J.C. (2006), Texture and weak grain size dependence in friction stir processed Mg-Al-Zn alloy, *Scr. Mater.* **55**, 637–640.

Wang, F., Echlin, M., Shin, J., Bammes, B., De Graef, M., Pollock, T., and Gianola, D. (2020), Opportunities for electron backscattered diffraction enabled by direct electron detection, *Microsc. Microanal.* **26** (Suppl. 2), 1164–1165.

Wang, Y.N. and Huang, J.C. (2003), Texture analysis in hexagonal materials, *Mater. Chem. Phys.* **81**, 11–26.

Wang, W.H., Sun, X., Köhlhoff, G.D., and Lücke, K. (1995), Orientation determination by continuous etching patterns in copper and copper alloys, *Textures Microstruct.* **24**, 199–219.

Wang, Y. M., Voisin, T., McKeown, J. T., Ye, J., Calta, N. P., Li, Z., … and Zhu, T. (2018), Additively manufactured hierarchical stainless steels with high strength and ductility, *Nature Mater.* **17**, 63–71.

Wang, D., Wang, D., Wu, B., Wang, F., and Liang, Z.D. (2014), An evolutionary algorithm for the texture analysis of cubic system materials derived by the maximum entropy principle, *Entropy* **16**, 6477–6496.

Wang, F., Xu, J.Z., and Liang, Z.D. (1992), Determination of the complete ODF of cubic system materials by the maximum entropy method, *Textures Microstruct.* **19**, 55–58.

Wang, Z. and Zaefferer, S. (2017), On the accuracy of grain boundary character determination by pseudo-3D EBSD, *Mater. Char.* **130**, 33–38.

Wang, Y.D., Zuo, L., Liang, Z.D., Laruelle, C., Vadon, A., and Heizmann, J.J. (1997), Quantitative texture analysis from x-ray diffraction spectra, *J. Appl. Cryst.* **30**, 433–438.

Warrington, D.H. and Boon, M. (1975), Ordered structures in random grain boundaries; some geometrical probabilities, *Acta Metall.* **23**, 599–607.

Wassermann, G. and Grewen, J. (1962), *Texturen Metallischer Werkstoffe*, Springer, Berlin.

Watanabe, T. (1988), The potential for grain boundary design in materials development, *Mater. Forum* **11**, 284–303.

Watanabe, T. (2011), Grain boundary engineering: Historical perspective and future prospects, *J. Mater. Sci.* **46**, 4095–4115.

Wcislak, L. and Bunge, H.-J. (1991), Simultaneous measurement of pole figures by x-ray diffraction using a 2θ-position sensitive detector, *Textures Microstruct.* **14–18**, 257–271.

Wcislak, L. and Bunge, H.-J. (1996), *Texture Analysis with a Position Sensitive Detector*, Cuvillier Verlag, Göttingen, Germany.

Wcislak, L., Bunge, H.-J., and Nauer-Gerhardt, C.U. (1993), X-ray diffraction texture analysis with a position sensitive detector, *Z. Metallkd.* **84**, 479–493.

Wcislak, L., Klein, H., Bunge, H.-J., Garbe, U., Tschentscher, T., and Schneider, J.R. (2002), Texture analysis with high-energy synchrotron radiation, *J. Appl. Cryst.* **35**, 82–95.

Wechsler, M.S., Liebermann, D.S., and Read, T.A. (1953), On the theory of the formation of martensite, *Trans. AIME* **197**, 1503–1515.

Weiland, H. and Field, D.P. (1994), Automated indexing of Kikuchi patterns for orientation determination, in: *Proceedings of the 52nd Annual Meeting of the Microscopy Society of America* (Eds. G.W. Bailey and A.J. Garratt-Reed), San Francisco Press, San Francisco, CA, 900–901.

Weiland, H., Liu, J., and Nes, E. (1991), Substructure and local orientation evolution of aluminum single crystals during deformation, *Textures Microstruct.* **14–18**, 109–114.

Weiland, H. and Panchanadeeswaran, S. (1993), Local pole figures in the TEM—Measurement and ODF determination, *Textures Microstruct.* **20**, 67–86.

Weiss, L.E. and Wenk, H.-R. (1985), Symmetry of pole figures and textures, in: *Preferred Orientation in Deformed Metals and Rocks: An Introduction to Modern Texture Analysis* (Ed. H.-R. Wenk), Academic Press, Orlando, FL, 49–72.

Welch, P.I. (1986), Neutron diffraction texture analysis, in: *Experimental Techniques of Texture Analysis* (Ed. H.-J. Bunge), DGM Inf. Ges., Oberursel, Germany, 183–207.

Wenk, H.-R. (1985), Measurement of pole figures, in: *Preferred Orientation in Deformed Metals and Rocks: An Introduction to Modern Texture Analysis* (Ed. H.-R. Wenk), Academic Press, Orlando, FL, 11–47.

Wenk, H.-R. (1992), Advantages of monochromatic X-rays for texture determination of superconducting thin films, *J. Appl. Cryst.* **25**, 524–530.

Wenk, H.-R. (1994), Texture analysis with TOF neutrons, in: *Time-of-Flight Diffraction at Pulsed Neutron Sources* (Eds. J.D. Jorgensen and A.J. Schultz), *Trans. Am. Cryst. Assoc.* **29**, 95–108.

Wenk, H.-R. (1998), Pole figure measurements with diffraction techniques, in: *Texture and Anisotropy, Preferred Orientations in Polycrystals and their Effect on Materials Properties* (Eds. U.F. Kocks, C.N. Tomé, and H.-R. Wenk), Cambridge University Press, Cambridge, 126–177.

Wenk, H.-R. (2006), Neutron diffraction texture analysis, *Rev. Mineral. Geochem.* **63**, 399–426.

Wenk, H.-R., Bunge, H.-J., Jansen, E., and Pannetier, J. (1986), Preferred orientation of plagioclase—neutron diffraction and U-stage data, *Tectonophys.* **126**, 271–284.

Wenk, H.-R., Bunge, H.-J., Kallend, J.S., Lücke, K., Matthies, S., and Pospiech, J. (1988), Orientation distribution: Representation and determination, in: *Proceedings of ICOTOM 8* (Eds. J.S. Kallend and G. Gottstein), TMS, Warrendale, PA, 17–30.

Wenk, H.-R., Chateigner, D., Pernet, M., Bingert, J.F., Hellstrom, E., and Ouladdiaf, B. (1996), Texture analysis of Bi–2212 and 2223 tapes and wires by neutron diffraction, *Physica C* **272**, 1–12.

Wenk, H.-R., Cont, L., Xie, Y., Lutterotti, L., Ratschbacher, L., and Richardson, J. (2001), Rietveld texture analysis of Dabie Shan eclogite from TOF neutron diffraction spectra, *J. Appl. Cryst.* **34**, 442–453.

Wenk, H.-R. and Grigull, S. (2003), Synchrotron texture analysis with area detectors, *J. Appl. Cryst.* **36**, 1040–1049.

Wenk, H.-R., Heidelbach, F., Chateigner, D., and Zontone, F. (1997), Laue orientation imaging, *J. Synchr. Rad.* **4**, 95–101.

Wenk, H.-R. and Kocks, U.F. (1987), The representation of orientation distributions, *Metall. Trans.* **18A**, 1083–1092.

Wenk, H.-R., Larson, A.C., Vergamini, P.J., and Schultz, A.J. (1991), TOF of pulsed neutrons and 2d detectors for texture analysis of deformed polycrystals, *J. Appl. Phys.* **70**, 2035–2040.

Wenk, H.-R., Lutterotti, L., and Vogel, S. (2003), Texture analysis with the new HIPPO TOF diffractometer, *Nucl. Instr. Meth.* **A515**, 575–588.

Wenk, H.-R., Matthies, S., Donovan, J., and Chateigner, D. (1998), BEARTEX, a Windows-based program system for quantitative texture analysis, *J. Appl. Cryst.* **31**, 262–269.

Wenk, H.-R., Matthies, S., and Lutterotti, L. (1994), Texture analysis from diffraction spectra, *Mater. Sci. Forum* **157–162**, 473–479.

Wenk, H.-R. and Pannetier, J. (1990), Texture development in deformed granodiorites from the Santa Rosa mylonite zone, southern California, *J. Struct. Geol.* **12**, 177–184.

Wenk, H.-R. and Phillips, D.S. (1992), Highly textured Bi–2223 aggregates produced by cold pressing of powders, *Physica C* **200**, 105–112.

Wenk, H.-R. and Van Houtte, P. (2004), Texture and anisotropy, *Rep. Prog. Phys.* **67**, 1367–1428.

West, D.A. and Adams, B.L. (1997), Analysis of orientation clustering in a directionally solidified nickel-based ingot, *Metall. Mater. Trans.* **28A**, 229–236.

Wever, F. (1924), Über die Walzstruktur kubisch kristallisierender Metalle, *Z. f. Physik* **28**, 69–90.

Wheeler, J., Mariani, E., Piazolo, S., Prior, D.J., Trimby, P., and Drury, M.R. (2009), The weighted Burgers vector: A new quantity for constraining dislocation densities and types using electron backscatter diffraction on 2D sections through crystalline materials, *J. Microsc.* **233**, 482–494.

Wilkinson, A.J. (1997), Methods for determining elastic strains from electron backscatter diffraction and electron channelling patterns, *Mater. Sci. Technol.* **13**, 79–84.

Wilkinson, A.J. (2000), Advances in SEM-based diffraction studies of defects and strains in semiconductors, *J. Electron Microsc.* **49**, 299–310.

Wilkinson, A.J. (2001), A new method for determining small misorientations from electron back scatter diffraction patterns, *Scr. Mater.* **44**, 2379–2385.

Wilkinson, A.J., and Britton, T.B. (2012), Strains, planes, and EBSD in materials science, *Mater. Today* **15**, 366–376.

Wilkinson, A.J., Britton, T.B., Jiang. J., and Karamched, P.S. (2014), A review of advances and challenges in EBSD strain mapping, *IOP Conf. Ser.: Mater. Sci. Eng.* **55**, 012020.

Wilkinson, A.J. and Hirsch, P.B. (1997), Electron diffraction based techniques in scanning electron microscopy of bulk materials, *Micron* **28**, 279–308.

Wilkinson, A.J., Meaden, G., and Dingley, D.J. (2006), High-resolution elastic strain measurement from electron backscatter diffraction patterns: New levels of sensitivity, *Ultramicroscopy* **106**, 307–313.

Wilkinson, A.J., Moldovan, G., Britton, T.B., Bewick, A., Clough, R., and Kirkland, A.I. (2013), Direct detection of electron backscatter diffraction patterns, *Phys. Rev. Let.* **111**, 065506.

Will, G., Schäfer, W., and Merz, P. (1989), Texture analysis by neutron diffraction using a linear position sensitive detector, *Textures Microstruct.* **10**, 375–387.

Williams, R.O. (1968), Analytical methods for representing complex textures by biaxial pole figures, *J. Appl. Phys.* **39**, 4329–4335.

Williams, D.B. and Carter, C.B. (2009), *Transmission Electron Microscopy: A Textbook for Materials Science* (2nd edition), Springer, Berlin.

Winiarski, B., Gholinia, A., Mingard, K., Gee, M., Thompson, G., and Withers, P.J. (2021), Correction of artefacts associated with large area EBSD, *Ultramicroscopy* **226**, 113315.

Winkelmann, A. (2009a), Dynamical simulation of electron backscatter diffraction patterns, in: *Electron Backscatter Diffraction in Materials Science* (2nd edition) (Eds. A.J. Schwartz et al.), Springer, Berlin, 21–33.

Winkelmann, A. (2009b), Principles of depth-resolved Kikuchi pattern simulation for electron backscatter diffraction, *J. Microsc.* **239**, 32–45.

Winkelmann, A., Britton, T.B., and Nolze, G. (2019), Constraints on the effective electron energy spectrum in backscatter Kikuchi diffraction, *Phys. Review* **99B**, 064115.

Winkelmann, A. and Nolze, G. (2015a), Chirality determination of quartz crystals using Electron Backscatter Diffraction, *Ultramicroscopy* **149**, 58–63.

Winkelmann, A. and Nolze, G. (2015b), Point-group sensitive orientation mapping of non-centrosymmetric crystals, *Appl. Phys. Lett.* **106**, 072101.

Winkelmann, A., Trager-Cowan, C., Sweeney, F., Day, A.P., and Parbrook, P. (2007), Many-beam dynamical simulation of electron backscatter diffraction patterns, *Ultramicroscopy* **107**, 414–421.

Winning, M. and Rollett, A.D. (2005), Transition between low and high angle grain boundaries, *Acta Mater.* **53**, 2901–2907.

Wolf, D. (1990), Correlation between structure, energy, and ideal cleavage fracture for symmetrical grain boundaries in fcc metals, *J. Mater. Res.* **5**, 1708–1730.

Wolf, D. (2005), Structure and energy of grain boundaries, in: *Handbook of Materials Modeling* (Ed. S. Yip), Springer, Berlin, 1953–1983.

Wölfel, E.R. (1983), A novel curved position-sensitive proportional counter for X-ray diffractometry, *J. Appl. Cryst.* **16**, 341–348.

Wöllmer, S., Zaefferer, S., Göken, M., Mack, T., and Glatzel, U. (2003), Characterization of phases of aluminized nickel base superalloys, *Surf. Coat. Technol.* **167**, 83–96.

Wright, S.I. (1992), *Individual Lattice Orientation Measurements: Development and Application of a Fully Automatic Technique*, PhD Dissertation, Yale University, New Haven, CT.

Wright, S.I. and Adams, B.L. (1990), An evaluation of the single orientation method for texture determination in materials of moderate texture strength, *Textures Microstruct.* **12**, 65–76.

Wright, S.I. and Adams, B.L. (1992), Automatic analysis of electron backscatter diffraction patterns, *Metall. Trans.* **23A**, 759–767.

Wright, S.I. and Dingley, D. (1998), Orientation imaging in the transmission electron microscope, *Mater. Sci. Forum* **273–275**, 209–214.

Wright, S.I., Gray, G.T., and Rollett, A.D. (1994), Textural and microstructural gradient effects on the mechanical behavior of a tantalum plate, *Metall. Mater. Trans.* **25A**, 1025–1031.

Wright, S. and Nowell, M. (2008), High-speed EBSD, *Adv. Mater. Process.* **166**, 29–31.

Wright, S.I., Nowell, M.M., and Bingert, J.F. (2007), A comparison of textures measured using x-ray and electron backscatter diffraction, *Metall. Mater. Trans.* **38A**, 1845–1855.

Wright, S.I., Nowell, M.M., de Kloe, R., Camus, P., and Rampton, T. (2015a), Electron imaging with an EBSD detector, *Ultramicroscopy* **148**, 132–145.

Wright, S.I., Nowell, M.M., and Field, D.P. (2011), A review of strain analysis using electron backscatter diffraction, *Microsc. Microanal.* **17**, 316–329.

Wright, S.I., Nowell, M.M., Lindeman, S.P., Camus, P.P., De Graef, M., and Jackson, M.A. (2015b), Introduction and comparison of new EBSD post-processing methodologies, *Ultramicroscopy* **159**, 81–94.

Wright, S.I., Suzuki, S., and Nowell, M.M. (2016), In situ EBSD observations of the evolution in crystallographic orientation with deformation, *JOM* **68**, 2730–2736.

Wu, G. and Zaefferer, S. (2009), Advances in TEM orientation microscopy by combination of dark-field conical scanning and improved image matching, *Ultramicroscopy* **109**, 1317–1325.

Wynick, G.L. and Boehlert, C.J. (2005), Use of electropolishing for enhanced metallic specimen preparation for electron backscatter diffraction analysis, *Mater. Char.* **55**, 190–202.

Xie, Y., Wenk, H.-R., and Matthies, S. (2003), Plagioclase preferred orientation by TOF neutron diffraction and SEM-EBSD, *Tectonophys.* **370**, 269–286.

Xu, W., Ferry, M., Mateescu, N., Cairney, J.M., and Humphreys, F.J. (2007), Techniques for generating 3-D EBSD microstructures by FIB tomography, *Mater. Char.* **58**, 961–967.

Yang, W., Larson, B.C., Tischler, J.Z., Ice, G.E., Budai, J.D., and Liu, W. (2004), Differential-aperture X-ray structural microscopy: A submicron-resolution three-dimensional probe of local microstructure and strain, *Micron* **35**, 431–439.

Yang, X., Wang, W., and Gu, X. (2022), The near singular boundaries in BCC iron, *Phil. Mag.* **102**, 440–466.

Yang, P., Zhao, Z., Ren, X., and Huang, S. (2005), Microstructure, textures and deformation behaviors of fine-grained magnesium alloy AZ31, *J. Mater. Sci. Technol.* **21**, 331–337.

Yi, S.-B., Davies, C.H.J., Brokmeier, H.-G., Bolmaro, R.E., Kainer, K.U., and Homeyer, J. (2006), Deformation and texture evolution in AZ31 magnesium alloy during uniaxial loading, *Acta Mater.* **54**, 549–562.

Young, C.T. and Lytton, J.L. (1972), Computer generation and identification of Kikuchi projections, *J. Appl. Phys.* **43**, 1408–1417.

Young, C.T., Steele, J.H., and Lytton, J.L. (1973), Characterization of bicrystals using Kikuchi patterns, *Metall. Trans.* **4**, 2081–2089.

Zaefferer, S. (2000), New developments of computer-aided crystallographic analysis in transmission electron microscopy, *J. Appl. Cryst.* **33**, 10–25.

Zaefferer, S. (2002), Computer-aided crystallographic analysis in the TEM, *Adv. Imaging Electron Phys.* **125**, 355–415.

Zaefferer, S. (2003), A study of active deformation systems in titanium alloys: dependence on alloy composition and correlation with deformation texture, *Mater. Sci. Eng.* **A344**, 20–30.

Zaefferer, S. (2007), On the formation mechanisms, spatial resolution and intensity of backscatter Kikuchi patterns, *Ultramicroscopy* **107**, 254–266.

Zaefferer, S. (2011), A critical review of orientation microscopy in SEM and TEM, *Crystal Res. Technol.* **46**, 607–628.

Zaefferer, S., Baudin, T., and Penelle, R. (2001), A study on the formation mechanisms of the cube recrystallization texture in cold rolled Fe–36%Ni alloys, *Acta Mater.* **49**, 1105–1122.

Zaefferer, S. and Elhami, N.-N. (2014), Theory and application of electron channelling contrast imaging under controlled diffraction conditions, *Acta Mater.* **75**, 20–50.

Zaefferer, S. and Habler, G. (2017), Scanning electron microscopy and electron backscatter diffraction, in: *Mineral Reaction Kinetics: Microstructures, Textures, Chemical and Isotopic Signatures* (Eds. W. Heinrich and R. Abart), *EMU Notes Mineral.* **16**, 37–95.

Zaefferer, S., Kuo, J.-C., Zhao, Z., Winning, M., and Raabe, D. (2003), On the influence of the grain boundary misorientation on the plastic deformation of aluminum bicrystals, *Acta Mater.* **51**, 4719–4735.

Zaefferer, S. and Schwarzer, R.A. (1994), Automated measurement of single grain orientations in the TEM, *Z. Metallkd.* **85**, 585–591.

Zaefferer, S., Wright, S.I., and Raabe, D. (2008), Three-dimensional orientation microscopy in a focused ion beam-scanning electron microscope: A new dimension of microstructure characterization, *Metall. Mater. Trans.* **39A**, 374–389.

Zaid Ahmed, M. and Bhattacharjee, P.P. (2014), Evolution of microstructure and texture during isothermal annealing of a heavily warm-rolled duplex steel, *ISIJ Int.* **54**, 2844–2853.

Zambaldi, C., Zaefferer, S., and Wright, S. (2009), Characterization of order domains in γ-TiAl by orientation microscopy based on electron backscatter diffraction, *J. Appl. Cryst.* **42**, 1092–1101.

Zanini, L., Andersen, K.H., Batkov, K., Klinkby, E.B., Mezei, F., Schönfeldt, T., and Takibayev, A. (2019), Design of the cold and thermal neutron moderators for the European Spallation Source, *Nucl. Instr. Meth.* **A925**, 33–52.

Zhang, S., Godfrey, A., Zhang, C., Liu, W., and Juul Jensen, D. (2020a), Surface patterning for combined digital image correlation and electron backscatter diffraction in-situ deformation experiments, *Mater. Char.* **164**, 110332.

Zhang, X., Gong, Z., Huang, J., and Yu, B. (2020b), Texture evolution of tungsten materials during recrystallization, *Mater. Res. Express* **7**, 056513.

Zhao, J. and Adams, B.L., Definition of an asymmetric domain for intercrystalline misorientation in cubic materials in the space of Euler angles, *Acta Cryst.* **A44** (1988) 326–336.

Zhao, J., Adams, B.L., and Morris, P.R. (1988), A Comparison of measured and texture-estimated misorientation distributions in type 304 stainless steel tubing, *Textures Microstruct.* **8&9**, 493–508.

Zhao, Y., Zhang, J., Xu, H., Lokshin, K.A., He, D., Qian, J., Pantea, C., Daemen, L.L., Vogel, S.C., Ding, Y., and Xu, J. (2010), High-pressure neutron diffraction studies at LANSCE, *Appl. Phys.* **A99**, 585–599.

Zhou, G., Kou, J., Li, Y., Zhu, W., Chen, K., and Tamura, N. (2018), Quantitative scanning Laue diffraction microscopy: Application to the study of 3D printed nickel-based superalloys, *Quantum Beam Sci.* **2**, 13.

Zhu, K.Y., Bacroix, B., Chaveau, T., Chaubet, D., and Castelnau, O. (2009), Texture evolution and associated nucleation and growth mechanisms during annealing of a Zr alloy, *Metall. Mater. Trans.* **40A**, 2423–2434.

Zhu, Y., Zhang, H., Wang, H., and Suenaga, M. (1991), Grain boundary in textured $YBa_2Cu_3O_{7-\delta}$ superconductor, *J. Mater. Res.* **12**, 2507–2518.

Ziegler, J.F., Ziegler, M.D., and Biersack, J. P. (2010), SRIM—The stopping and range of ions in matter, *Nucl. Instr. Meth. Phys. Res.* **B 268**, 1818–1823.

Zink, U., Brokmeier, H.-G., and Bunge, H.-J. (1994), Determination of magnetic textures in ferromagnetics by means of neutron diffraction, *Mater. Sci. Forum* **157–162**, 251–256.

General Bibliography

SERIES, INTERNATIONAL CONFERENCE ON TEXTURE OF MATERIALS, ICOTOM

Textures in Research and Practice, Clausthal, Eds. J. Grewen and G. Wassermann, Springer, Berlin (1969).

Quantitative Analysis of Texture, Cracow, Eds. J. Karp et al., The Society of Polish Metallurgical Engineers, Krakow, Poland (1971).

Proc. 3ème Colloque Européen sur les Textures des Métaux et leurs Applications Industrielles, Pont-à-Mousson, Ed. R. Penelle, Société Française de Métallurgie, Nancy, France (1973).

Texture and the Properties of Materials, Proceedings of the 4th International Conference on Texture, Cambridge, Eds. G.J. Davies et al., The Metals Society, London (1976).

Proceedings of the 5th International Conference on Textures of Materials, Aachen (ICOTOM 5), Eds. G. Gottstein and K. Lücke, Springer, Berlin (1978).

Proceedings of the 6th International Conference on Textures of Materials, Tokyo (ICOTOM 6), Ed. S. Nagashima, The Iron and Steel Institute of Japan, Tokyo (1981).

Proceedings of the 7th International Conference on Textures of Materials, Noordwijkerhout (ICOTOM 7), Eds. C.M. Brakman, P. Jongenburger, and E.J. Mittemeijer, Netherlands Society for Materials Science, Zwijndrecht (1984).

Proceedings of the 8th International Conference on Textures of Materials, Santa Fe, NM (ICOTOM 8), Eds. J.S. Kallend and G. Gottstein, TMS, Warrendale, PA (1988).

Proceedings of the 9th International Conference on Textures of Materials, Avignon (ICOTOM 9), Ed. H.-J. Bunge, Special Issue, *Textures Microstruct.* **14–18** (1991).

Proceedings of the 10th International Conference on Textures of Materials, Clausthal (ICOTOM 10), Ed. H.-J. Bunge, *Mater. Sci. Forum* **157–162** (1994).

Proceedings of the 11th International Conference on Textures of Materials, Xi'an (ICOTOM 11), Eds. Z. Liang, L. Zou, and Y. Chu, International Academic Publishers, Beijing (1996).

Proceedings of the 12th International Conference on Textures of Materials, Montreal (ICOTOM 12), Ed. J.A. Szpunar, NRC Research Press, Canada (1999).

Proceedings of the 13th International Conference on Textures of Materials, Seoul (ICOTOM 13), Ed. D.N. Lee, *Mater. Sci. Forum* **408–412** (2002).

Proceedings of the 14th International Conference on Textures of Materials, Leuven (ICOTOM 14), Eds. P. Van Houtte and L. Kestens, *Mater. Sci. Forum* **495–497** (2005).

Proceedings of the 15th International Conference on Textures of Materials, Pittsburgh, PA, 2008 (ICOTOM 15), Ed. A.D. Rollett, *Ceramic Trans.* **200–201** (2009).

Proceedings of the 16th International Conference on Textures of Materials, Mumbai, 2011 (ICOTOM 16), Eds. A. Tewari et al., *Mater. Sci. Forum* **702–703** (2012).

Proceedings of the 17th International Conference on Textures of Materials, Dresden, 2014 (ICOTOM 17), Eds. W. Skrotzki and C.-G. Oertel, *IOP Conf. Series: Mater. Sci. Eng.* **82** (2015).

Proceedings of the 18th International Conference on Textures of Materials, St. George, UT, 2017 (ICOTOM 18), Eds. S.I. Wright, D.T. Fullwood, and M.M. Nowell, *IOP Conf. Series: Mater. Sci. Eng.* **375** (2018).

Proceedings of the 19th International Conference on Textures of Materials, Sakai (ICOTOM 19), Eds. H. Inoue, Y. Takayama, and N. Yoshinaga, *IOP Conf. Series: Mater. Sci. Eng.* **1121** (2021).

CONFERENCES AND SYMPOSIA

Recrystallization, Grain Growth and Textures, Ed. H. Margolin, ASM, Metals Park, OH (1965).

Proceedings of the Symposium *Textures in Non-Ferrous Metals and Alloys, Detroit*, Eds. H.D. Merchant and J.G. Morris, TMS, Warrendale, PA (1985).

Proceedings of the Symposia *Textures in Non-Metallic Materials* and *Microstructure and Texture Evolution During Annealing of Deformed Materials,* Indianapolis, Eds. D.B. Knorr, J. Hirsch, and H.-J. Bunge, Special Issue, *Textures Microstruct.* **13** (1991).

Proceedings of the Symposium *Microscale Textures of Materials*, Cincinnati, Eds. B.L. Adams and H. Weiland, Special Issue, *Textures Microstruct.* **20** (1993).

Proceedings of the Symposium *Interfacial Engineering for Optimized Properties,* MRS Proc. No. 458, Eds. C.L. Briant, C.B. Carter, and E.L. Hall, The Materials Research Society, Pittsburgh, PA (1996).

Proceedings of a One-Day Conference *Textures on a Microscale*, Eds. H.S. Uhbi and A.W. Bowen, Special Issue, *Mater. Sci. Technol.* **12** (1996).

Proceedings of the International Conference on Texture and Anisotropy of Polycrystals, Clausthal (ITAP 1), Ed. R.A. Schwarzer, *Mater. Sci. Forum* **273–275** (1998).

Proceedings of the Symposium *Interfacial Engineering for Optimized Properties II,* MRS Proc. No. 586, Eds. C.B. Carter et al., Cambridge University Press, Cambridge (2000).

Proceedings of the Symposium *Interfacial Engineering for Optimized Properties III,* MRS Proc. No. 819, Eds. C.A. Schuh et al., Cambridge University Press, Cambridge (2004).

Proceedings of the International Conference on Texture and Anisotropy of Polycrystals II, Metz (ITAP 2), Eds. C. Esling et al., *Solid State Phenom.* **105** (2005).

Proceedings of the International Conference on Microstructure and Texture in Steels and Other Materials, Jamshedpur, Eds. A. Haldar, S. Suwas, and D. Bhattacharjee, Springer, London (2009).

Proceedings of the International Conference on Texture and Anisotropy of Polycrystals III, Göttingen (ITAP 3), Eds. H. Klein and R.A. Schwarzer, *Solid State Phenom.* **160** (2010).

TEXTBOOKS, MONOGRAPHS

Altmann, S.L., *Rotations, Quaternions and Double Groups*, Clarendon Press, Oxford (1986).

Bacon, G.E., *Neutron Diffraction* (3rd edition), Clarendon Press, Oxford (1975).

Barrett, C.S. and Massalski T.B., *Structure of Metals: Crystallographic Methods, Principles and Data*, McGraw-Hill, New York (1980).

Bunge, H.-J., *Texture Analysis in Materials Science: Mathematical Methods*, Butterworths, London (1982).

Bunge, H.-J. (Ed.), *Experimental Techniques of Texture Analysis*, DGM Inf. Ges., Oberursel, Germany (1986).

Bunge, H.-J. (Ed.), *Theoretical Methods of Texture Analysis*, DGM Inf. Ges., Oberursel, Germany (1987).

Bunge, H.-J. (Ed.), *Directional Properties of Materials*, DGM Inf. Ges., Oberursel, Germany (1988).

Bunge, H.-J. (Ed.), *Advances and Applications of Quantitative Texture Analysis*, DGM Inf. Ges., Oberursel, Germany (1989).

Bunge, H.-J. and Esling, C. (Eds.), *Quantitative Texture Analysis*, DGM Inf. Ges., Oberursel, Germany (1982).

Bunge, H.-J., Siegesmund, S., Skrotzki, W., and Weber, K. (Eds.), *Textures in Geological Materials*, DGM Inf. Ges., Oberursel, Germany (1994).

Carter, C.B. and Williams, D.B. (Eds.), *Transmission Electron Microscopy: Diffraction, Imaging, and Spectrometry*, Springer, Berlin (2016).

Cullity, B.D., *Elements of X-ray Diffraction* (2nd edition), Addison-Wesley, Reading, MA (1978).
De Graef, M., *Introduction to Conventional Transmission Electron Microscopy*, Cambridge University Press, Cambridge (2003).
Dingley, D.J., Baba-Kishi, K., and Randle, V., *Atlas of Backscatter Kikuchi Diffraction Patterns*, IOP Publ., Bristol (1994).
Edington, J.W., *Electron Diffraction in the Electron Microscope*, Macmillan, London (1975).
Edington, J.W., *Practical Electron Microscopy in Materials Science*, Van Nostrand Reinhold, New York (1976).
Goldstein, J.I., Newbury, D.E., Michael, J.R., Ritchie, N.W.M., Scott, J.H.J, and Joy, D.C., *Scanning Electron Microscopy and X-ray Microanalysis* (4th edition), Springer, Berlin (2018).
Goodhew, P.J., Humphreys F.J., and Beanland R., *Electron Microscopy and Analysis* (3rd edition), Taylor & Francis, London (2001).
Haldar, A., Suwas, S., and Bhattacharjee, D., *Microstructure and Texture in Steels and Other Materials*, Springer, Berlin (2009).
Hammond, C., *The Basics of Crystallography and Diffraction* (4th edition), Oxford University Press, Oxford (2015).
Hansen, J., Pospiech, J., and Lücke, K., *Tables for Texture Analysis of Cubic Crystals*, Springer, Berlin (1978).
Hatherly, M. and Hutchinson, W.B., *An Introduction to Textures in Metals*, Monograph no. 5, The Institute of Metals, London (1979).
Hirsch, P.B., Howie, A., Nicholson, R.B., Pashley, D.W., and Whelan, M.J., *Electron Microscopy of Thin Crystals*, Krieger Publishing, Malabar, FL (1977).
Johari, O. and Thomas, G., *The Stereographic Projection and Its Applications*, Vol. IIA of *Techniques of Metal Research*, Ed. R.F. Bunshah, Interscience, New York (1969).
Kocks, U.F., Tomé, C.N., and Wenk, H.-R. (Eds.), *Texture and Anisotropy, Preferred Orientations in Polycrystals and Their Effect on Materials Properties*, Cambridge University Press, Cambridge (1998).
Loretto, M.H., *Electron Beam Analysis of Materials* (2nd edition), Chapman & Hall, London (1994).
McKie, C. and McKie, D., *Crystalline Solids*, Nelson, London (1974).
Mittemeijer, E.J. and Welzel, U. (Eds.), *Modern Diffraction Methods*, Wiley-VCH, Weinheim (2012).
Phillips, F.C., *An Introduction to Crystallography* (4th edition), Longman Group, London (1971).
Randle, V., *Grain Boundary Geometry in Polycrystals*, IOP Publ., Bristol (1993).
Randle, V., *The Role of the Coincidence Site Lattice in Grain Boundary Engineering*, Institute of Materials, London (1996).
Randle, V., *Microtexture Determination and Its Applications* (2nd edition), Institute of Materials, London (2003).
Randle, V. (Ed.), *The Measurement of Grain Boundary Geometry*, CRC Press, Boca Raton, FL (1993).
Reimer, L., *Scanning Electron Microscopy: Physics of Image Formation and Microanalysis* (2nd edition), Springer, Berlin (2013).
Reimer, L., *Transmission Electron Microscopy: Physics of Image Formation and Microanalysis* (4th edition), Springer, Berlin (2013).
Schumann, H., *Kristallgeometrie*, VEB-Verlag, Leipzig (1979).
Schwartz, A.J., Kumar, M., Adams, B.L., and Field, D.P. (Eds.), *Electron Backscatter Diffraction in Materials Science* (2nd edition), Springer, Berlin (2009).
Smallman, R.E. and Bishop, R.J., *Modern Physical Metallurgy and Materials Engineering* (6th edition), Butterworth-Heinemann, Oxford (1999).

Snyder, R.L., Fiala, J., and Bunge, H.-J. (Eds.), *Defect and Microstructure Analysis by Diffraction*, Inter. Union Cryst., Oxford University Press, New York, NY (1999).

Spieß, L., Teichert, G., Behnken, H., and Schwarzer, R.A., *Moderne Röntgenbeugung—Röntgendiffraktometrie für Materialwissenschaftler*, Reihe Teubner-Lehrbücher, B.G. Teubner-Verlag, Stuttgart-Leipzig-Wiesbaden, Germany (2005).

Sutton, A.P. and Balluffi, R.W., *Interfaces in Crystalline Materials*, Clarendon Press, Oxford (1995).

Suwas, S. and Ray, R.K., *Crystallographic Texture of Materials*, Springer, Berlin (2014).

Von Heimendahl, M., *Electron Microscopy of Materials: An Introduction*, Academic Press, San Diego, CA (1980).

Wassermann, G. and Grewen, J., *Texturen metallischer Werkstoffe*, Springer, Berlin (1962).

Wenk, H.-R. (Ed.), *Electron Microscopy in Mineralogy*, Springer, Berlin (1976).

Wenk, H.-R. (Ed.), *Preferred Orientation in Deformed Metals and Rocks: An Introduction to Modern Texture Analysis*, Academic Press, Orlando, FL (1985).

Wenk, H.-R. (Ed.), *Neutron Scattering in Earth Sciences*, De Gruyter, Berlin (2006).

Williams, D.B. and Carter, C.B., *Transmission Electron Microscopy: A Textbook for Materials Science* (2nd edition), Plenum Press, New York (2009).

Index

K

L

M

P

Q

R